Fundamentals of Structural Dynamics

Fundamentals of Structural Dynamics

Second Edition

Roy R. Craig, Jr.
Andrew J. Kurdila

WILEY
JOHN WILEY & SONS, INC.

Copyright © 2006 by John Wiley & Sons, Inc. All rights reserved.

Published by John Wiley & Sons, Inc., Hoboken, New Jersey
Published simultaneously in Canada.

No part of this publication may be reproduced, stored in a retrieval system, or transmitted in any form or by any means, electronic, mechanical, photocopying, recording, scanning, or otherwise, except as permitted under Section 107 or 108 of the 1976 United States Copyright Act, without either the prior written permission of the Publisher, or authorization through payment of the appropriate per-copy fee to the Copyright Clearance Center, Inc., 222 Rosewood Drive, Danvers, MA 01923, 978-750-8400, fax 978-646-8600, or on the web at www.copyright.com. Requests to the Publisher for permission should be addressed to the Permissions Department, John Wiley & Sons, Inc., 111 River Street, Hoboken, NJ 07030, (201) 748-6011, fax (201) 748-6008.

Limit of Liability/Disclaimer of Warranty: While the publisher and author have used their best efforts in preparing this book, they make no representations or warranties with respect to the accuracy or completeness of the contents of this book and specifically disclaim any implied warranties of merchantability or fitness for a particular purpose. No warranty may be created or extended by sales representatives or written sales materials. The advice and strategies contained herein may not be suitable for your situation. You should consult with a professional where appropriate. Neither the publisher nor author shall be liable for any loss of profit or any other commercial damages, including but not limited to special, incidental, consequential, or other damages.

For general information on our other products and services please contact our Customer Care Department within the U.S. at 877-762-2974, outside the U.S. at 317-572-3993 or fax 317-572-4002.

Wiley also publishes its books in a variety of electronic formats. Some content that appears in print, however, may not be available in electronic format.

Library of Congress Cataloging-in-Publication Data:

Craig, Roy R., 1934–
   Fundamentals of structural dynamics / by Roy R. Craig, Jr. and Andrew J. Kurdila.—2nd ed.
     p. cm.
   Rev. ed. of: Structural dynamics, c1981.
   Includes bibliographical references and index.
   ISBN 13: 978-0-471-43044-5
   ISBN 10: 0-471-43044-7 (cloth)
   1. Structural dynamics—Data processing. 2. Structural dynamics—Mathematical models.
  I. Kurdila, Andrew. II. Craig, Roy R., 1934-. Structural dynamics. III. Title.

TA654.C72 2006
624.1′71—dc22

2005043679

Printed in the United States of America.

10 9 8 7 6 5

The first author dedicates his work on this edition to his grandchildren: *Talia*, *Kyle*, and *Hart Barron*, and *Alex*, *Brandon*, and *Chase Lemens*. The second author dedicates his work to his wife, *Jeannie*, and to his children: *Patrick*, *Hannah*, and *Justin*.

Contents

Preface to Structural Dynamics—An Introduction to Computer Methods   xi
Preface to Fundamentals of Structural Dynamics   xiii
About the Authors   xv

1 The Science and Art of Structural Dynamics   1

1.1 Introduction to Structural Dynamics   1
1.2 Modeling of Structural Components and Systems   2
1.3 Prototype Spring–Mass Model   7
1.4 Vibration Testing of Structures   12
1.5 Scope of the Book   12
1.6 Computer Simulations; Supplementary Material on the Website   15
References   16
Problems   16

Part I   Single-Degree-of-Freedom Systems   19

2 Mathematical Models of SDOF Systems   21

2.1 Brief Review of the Dynamics of Particles and Rigid Bodies   21
2.2 Elements of Lumped-Parameter Models   24
2.3 Application of Newton's Laws to Lumped-Parameter Models   27
2.4 Application of the Principle of Virtual Displacements to Lumped-Parameter Models   34
2.5 Application of the Principle of Virtual Displacements to Continuous Models: Assumed-Modes Method   41
References   50
Problems   51

3 Free Vibration of SDOF Systems   56

3.1 Free Vibration of Undamped SDOF Systems   58
3.2 Free Vibration of Viscous-Damped SDOF Systems   61
3.3 Stability of Motion   66
3.4 Free Vibration of an SDOF System with Coulomb Damping   70
3.5 Experimental Determination of the Natural Frequency and Damping Factor of an SDOF System   72
References   77
Problems   78

4 Response of SDOF Systems to Harmonic Excitation   81

4.1 Response of Undamped SDOF Systems to Harmonic Excitation   82
4.2 Response of Viscous-Damped SDOF Systems to Harmonic Excitation: Frequency-Response Functions   87
4.3 Complex Frequency Response   93
4.4 Vibration Isolation: Force Transmissibility and Base Motion   96
4.5 Vibration Measuring Instruments: Accelerometers and Vibrometers   101
4.6 Use of Frequency-Response Data to Determine the Natural Frequency and Damping Factor of a Lightly Damped SDOF System   104
4.7 Equivalent Viscous Damping   107
4.8 Structural Damping   111
References   112
Problems   113

5 Response of SDOF Systems to Nonperiodic Excitation   117

5.1 Response of a Viscous-Damped SDOF System to an Ideal Step Input   117
5.2 Response of Undamped SDOF Systems to Rectangular Pulse and Ramp Loadings   119
5.3 Response of Undamped SDOF Systems to a Short-Duration Impulse: Unit Impulse Response   123
5.4 Response of SDOF Systems to General Dynamic Excitation: Convolution Integral Method   125
5.5 Response Spectra   128

5.6	System Response by the Laplace Transform Method: System Transfer Function 136		References 240	
	References 142		Problems 241	
	Problems 143			

6	Numerical Evaluation of the Dynamic Response of SDOF Systems 147

9	Vibration of Undamped 2-DOF Systems 248

6.1	Integration of Second-Order Ordinary Differential Equations 148
6.2	Integration of First-Order Ordinary Differential Equations 159
6.3	Nonlinear SDOF Systems 171
	References 181
	Problems 182

9.1	Free Vibration of 2-DOF Systems: Natural Frequencies and Mode Shapes 249
9.2	Beat Phenomenon 254
9.3	Additional Examples of Modes and Frequencies of 2-DOF Systems: Assumed-Modes Models 258
9.4	Free Vibration of Systems with Rigid-Body Modes 266
9.5	Introduction to Mode Superposition: Frequency Response of an Undamped 2-DOF System 268
9.6	Undamped Vibration Absorber 272
	Reference 275
	Problems 275

7	Response of SDOF Systems to Periodic Excitation: Frequency-Domain Analysis 184

7.1	Response to Periodic Excitation: Real Fourier Series 184
7.2	Response to Periodic Excitation: Complex Fourier Series 189
7.3	Response to Nonperiodic Excitation: Fourier Integral 195
7.4	Relationship Between Complex Frequency Response and Unit Impulse Response 199
7.5	Discrete Fourier Transform and Fast Fourier Transform 200
	References 205
	Problems 205

10	Vibration Properties of MDOF Systems: Modes, Frequencies, and Damping 281

10.1	Some Properties of Natural Frequencies and Natural Modes of Undamped MDOF Systems 282
10.2	Model Reduction: Rayleigh, Rayleigh–Ritz, and Assumed-Modes Methods 298
10.3	Uncoupled Damping in MDOF Systems 302
10.4	Structures with Arbitrary Viscous Damping: Complex Modes 307
10.5	Natural Frequencies and Mode Shapes of Damped Structures with Rigid-Body Modes 316
	References 322
	Problems 322

Part II	Multiple-Degree-of-Freedom Systems— Basic Topics 209

8	Mathematical Models of MDOF Systems 211

8.1	Application of Newton's Laws to Lumped-Parameter Models 212
8.2	Introduction to Analytical Dynamics: Hamilton's Principle and Lagrange's Equations 218
8.3	Application of Lagrange's Equations to Lumped-Parameter Models 223
8.4	Application of Lagrange's Equations to Continuous Models: Assumed-Modes Method 228
8.5	Constrained Coordinates and Lagrange Multipliers 238

11	Dynamic Response of MDOF Systems: Mode-Superposition Method 325

11.1	Mode-Superposition Method: Principal Coordinates 325
11.2	Mode-Superposition Solutions for MDOF Systems with Modal Damping: Frequency-Response Analysis 330
11.3	Mode-Displacement Solution for the Response of MDOF Systems 342

11.4	Mode-Acceleration Solution for the Response of Undamped MDOF Systems 349
11.5	Dynamic Stresses by Mode Superposition 351
11.6	Mode Superposition for Undamped Systems with Rigid-Body Modes 353
	References 359
	Problems 360

Part III Continuous Systems 365

12 Mathematical Models of Continuous Systems 367

12.1	Applications of Newton's Laws: Axial Deformation and Torsion 367
12.2	Application of Newton's Laws: Transverse Vibration of Linearly Elastic Beams (Bernoulli–Euler Beam Theory) 374
12.3	Application of Hamilton's Principle: Torsion of a Rod with Circular Cross Section 379
12.4	Application of the Extended Hamilton's Principle: Beam Flexure Including Shear Deformation and Rotatory Inertia (Timoshenko Beam Theory) 382
	References 385
	Problems 385

13 Free Vibration of Continuous Systems 388

13.1	Free Axial and Torsional Vibration 388
13.2	Free Transverse Vibration of Bernoulli–Euler Beams 392
13.3	Rayleigh's Method for Approximating the Fundamental Frequency of a Continuous System 398
13.4	Free Transverse Vibration of Beams Including Shear Deformation and Rotatory Inertia 400
13.5	Some Properties of Natural Modes of Continuous Systems 401
13.6	Free Vibration of Thin Flat Plates 405
	References 409
	Problems 409

Part IV Computational Methods in Structural Dynamics 415

14 Introduction to Finite Element Modeling of Structures 417

14.1	Introduction to the Finite Element Method 418
14.2	Element Stiffness and Mass Matrices and Element Force Vector 419
14.3	Transformation of Element Matrices 430
14.4	Assembly of System Matrices: Direct Stiffness Method 438
14.5	Boundary Conditions 445
14.6	Constraints: Reduction of Degrees of Freedom 447
14.7	Systems with Rigid-Body Modes 451
14.8	Finite Element Solutions for Natural Frequencies and Mode Shapes 453
	References 462
	Problems 463

15 Numerical Evaluation of Modes and Frequencies of MDOF Systems 469

15.1	Introduction to Methods for Solving Algebraic Eigenproblems 469
15.2	Vector Iteration Methods 471
15.3	Subspace Iteration 480
15.4	QR Method for Symmetric Eigenproblems 483
15.5	Lanczos Eigensolver 489
15.6	Numerical Case Study 496
	References 498
	Problems 498

16 Direct Integration Methods for Dynamic Response of MDOF Systems 500

16.1	Damping in MDOF Systems 501
16.2	Numerical Integration: Mathematical Framework 504
16.3	Integration of Second-Order MDOF Systems 510
16.4	Single-Step Methods and Spectral Stability 516
16.5	Numerical Case Study 525
	References 527
	Problems 528

17 Component-Mode Synthesis 531

17.1 Introduction to Component-Mode Synthesis 532
17.2 Component Modes: Normal, Constraint, and Rigid-Body Modes 534
17.3 Component Modes: Attachment and Inertia-Relief Attachment Modes 539
17.4 Flexibility Matrices and Residual Flexibility 544
17.5 Substructure Coupling Procedures 549
17.6 Component-Mode Synthesis Methods: Fixed-Interface Methods 557
17.7 Component-Mode Synthesis Methods: Free-Interface Methods 559
17.8 Brief Introduction to Multilevel Substructuring 564
References 571
Problems 572

Part V Advanced Topics in Structural Dynamics 577

18 Introduction to Experimental Modal Analysis 579

18.1 Introduction 580
18.2 Frequency-Response Function Representations 584
18.3 Vibration Test Hardware 590
18.4 Fourier Transforms, Digital Signal Processing, and Estimation of FRFs 594
18.5 Modal Parameter Estimation 604
18.6 Mode Shape Estimation and Model Verification 612
References 615
Problems 616

19 Introduction to Active Structures 617

19.1 Introduction to Piezoelectric Materials 617
19.2 Constitutive Laws of Linear Piezoelectricity 620
19.3 Application of Newton's Laws to Piezostructural Systems 624
19.4 Application of Extended Hamilton's Principle to Piezoelectricity 627
19.5 Active Truss Models 630
19.6 Active Beam Models 637
19.7 Active Composite Laminates 641
References 646
Problems 647

20 Introduction to Earthquake Response of Structures 650

20.1 Introduction 650
20.2 Response of a SDOF System to Earthquake Excitation: Response Spectra 652
20.3 Response of MDOF Systems to Earthquake Excitation 660
20.4 Further Considerations 664
References 665
Problems 666

A Units 667

B Complex Numbers 671

C Elements of Laplace Transforms 674

D Fundamentals of Linear Algebra 682

E Introduction to the Use of MATLAB 697

Index 715

Preface to Structural Dynamics—An Introduction to Computer Methods*

The topic of structural dynamics has undergone profound changes over the past two decades. The reason is the availability of digital computers to carry out numerical aspects of structural dynamics problem solving. Recently, the extensive use of the fast Fourier transform has brought about even more extensive changes in structural dynamics analysis, and has begun to make feasible the correlation of analysis with structural dynamics testing. Although this book contains much of the material that characterizes standard textbooks on mechanical vibrations, or structural dynamics, its goal is to present the background needed by an engineer who will be using structural dynamics computer programs or doing structural dynamics testing, or who will be taking advanced courses in finite element analysis or structural dynamics.

Although the applications of structural dynamics in aerospace engineering, civil engineering, engineering mechanics, and mechanical engineering are different, the principles and solution techniques are basically the same. Therefore, this book places emphasis on these principles and solution techniques, and illustrates them with numerous examples and homework exercises from the various engineering disciplines.

Special features of this book include: an emphasis on mathematical modeling of structures and experimental verification of mathematical models; an extensive introduction to numerical techniques for computing natural frequencies and mode shapes and for computing transient response; a systematic introduction to the use of finite elements in structural dynamics analysis; an application of complex frequency-response representations for the response of single- and multiple-degree-of-freedom systems; a thorough exposition of both the mode-displacement and mode-acceleration versions of mode superposition for computing dynamic response; an introduction to practical methods of component-mode synthesis for dynamic analysis; and the introduction of an instructional matrix algebra and finite element computer code, *ISMIS* (Interactive Structures and Matrix Interpretive System), for solving structural dynamics problems.

Although the emphasis of this book is on linear problems in structural dynamics, techniques for solving a limited class of nonlinear structural dynamics problems are also introduced. On the other hand, the topic of random vibrations is not discussed, since a thorough treatment of the subject is definitely beyond the scope of the book, and a cursory introduction would merely dilute the emphasis on numerical techniques for structural dynamics analysis. However, instructors wishing to supplement the text

---

*Copyright 1981, John Wiley & Sons, Inc.

with material on random vibrations will find the information on complex frequency response to be valuable as background for the study of random vibrations.

A primary aim of the book is to give students a thorough introduction to the numerical techniques underlying finite element computer codes. This is done primarily through "hand" solutions and the coding of several subroutines in FORTRAN (or BASIC). Use of the *ISMIS* computer program extends the problem-solving capability of the student while avoiding the "black box" nature of production-type finite element codes. Although the *ISMIS* computer program is employed in Chapters 14 and 17, its use is by no means mandatory. The FORTRAN source code and a complete User's Manual for *ISMIS* are available for a very nominal fee and can be obtained by contacting the author directly at The University of Texas at Austin (Austin, TX 78712).

Computer graphics is beginning to play an important role in structural dynamics, for example, in computer simulations of vehicle collisions and animated displays of structural mode shapes. One feature of this book is that all figures that portray functional representations are direct computer-generated plots.

The text of this book has been used for a one-semester senior-level course in structural dynamics and a one-semester graduate-level course in computational methods in structural dynamics. The undergraduate course typically covers the following material: Chapters 1 through 6, Sections 9.1, 9.2, 10.1, 10.2, 11.1 through 11.4, and Chapter 12. The graduate course reviews the topics above (i.e., it assumes that students have had a prior course in mechanical vibrations or structural dynamics) and then covers the remaining topics in the book as time permits. Both undergraduate and graduate courses make use of the *ISMIS* computer program, while the graduate course also includes several FORTRAN coding exercises.

Portions of this text have been used in a self-paced undergraduate course in structural dynamics. This led to the statements of objectives at the beginning of each chapter and to the extensive use of example problems. Thus, the text should be especially valuable to engineers pursuing a study of structural dynamics on a self-study basis.

I express appreciation to my students who used the notes that led to the present text. Special thanks are due to Arne Berg, Mike Himes, and Rick McKenzie, who generated most of the computer plots, and to Butch Miller and Rodney Rocha, who served as proctors for the self-paced classes. Much of the content and "flavor" of the book is a result of my industrial experience at the Boeing Company's Commercial Airplane Division, at Lockheed Palo Alto Research Laboratory, and at NASA Johnson Space Center. I am indebted to the colleagues with whom I worked at these places.

I am grateful to Dr. Pol D. Spanos for reading Chapter 20 and making helpful comments. Dean Richard Gallagher reviewed the manuscript and offered many suggestions for changes, which have been incorporated into the text. This valuable service is greatly appreciated.

This book might never have been completed had it not been for the patience and accuracy of its typist, Mrs. Bettye Lofton, and to her I am most deeply indebted.

Finally, many of the hours spent in the writing of this book were hours that would otherwise have been spent with Jane, Carole, and Karen, my family. My gratitude for their sacrifices cannot be measured.

ROY R. CRAIG, JR., AUSTIN, TX

Preface to Fundamentals of Structural Dynamics

Although there has been a title change to *Fundamentals of Structural Dynamics*, this book is essentially the 2nd edition of *Structural Dynamics—An Introduction to Computer Methods*, published in 1981 by the senior author. As a textbook and as a resource book for practicing engineers, that edition had a phenomenal run of a quarter century. Although this edition retains the emphasis placed in the first edition on the topics of mathematical modeling, computer solution of structural dynamics problems, and the relationship of finite element analysis and experimental structural dynamics, it takes full advantage of the current state of the art in each of those topics. For example, whereas the first edition employed *ISMIS*, a FORTRAN-based introductory matrix algebra and finite element computer code, the present edition employs a MATLAB-based version of *ISMIS* and provides many additional structural dynamics solutions directly in MATLAB.

The new features of this edition are:

1. A coauthor, Dr. Andrew Kurdila, who has been responsible for Chapters 6, 15, 16, and 19 and Appendices D and E in this edition.
2. A greater emphasis on computer solutions, especially MATLAB-based plots; numerical algorithms in Chapters 6, 15, and 16; and digital signal-processing techniques in Chapter 18.
3. A new section (Section 5.6) on system response by the Laplace transform method, and a new appendix, Appendix C, on Laplace transforms.
4. An introduction, in Sections 10.4 and 10.5, to state-space solutions for complex modes of damped systems.
5. Greatly expanded chapters on eigensolvers (Chapter 15), numerical algorithms for calculating dynamic response (Chapters 6 and 16), and component-mode synthesis (Chapter 17).
6. New chapters on experimental modal analysis (Chapter 18) and on smart structures (Chapter 19).
7. A revised grouping of topics that places vibration of continuous systems after basic multiple-DOF topics, but before the major sections on computational methods and the advanced-topics chapters.
8. Many new or revised homework problems, including many to be solved on the computer.
9. A supplement that contains many sample MATLAB .m-files, the MATLAB-based *ISMIS* matrix structural analysis computer program, notes for an extensive short course on finite element analysis and experimental modal analysis, and other

study aids. This supplement, referred to throughout the book as the "book's website," is available online from the Wiley Web site www.wiley.com/college/craig.

The senior author would like to acknowledge the outstanding wealth of knowledge that has been shared with him by authors of papers presented at the many International Modal Analysis Conferences (IMACs) that he has attended over the past quarter century. Special appreciation is due to Prof. David L. Brown and his colleagues from the University of Cincinnati; to numerous engineers from Structural Dynamics Research Corporation, ATA Engineering, Inc., and Leuven Measurement Systems; and to the late Dominick J. (Dick) DeMichele, the founder of IMAC. Professor Eric Becker, a colleague of the senior author at The University of Texas at Austin, was responsible for many features of the original *ISMIS* (Interactive Structures and Matrix Interpretive System) FORTRAN code.

The authors would like to express appreciation to the following persons for their major contributions to this edition:

- Prof. Peter Avitabile: for permission to include on the book's website his extensive short course notes on finite element analysis and experimental modal analysis.
- Mr. Charlie Pickrel: for providing Boeing GVT photos (Fig. 1.9*a,b*) and his journal article on experimental modal analysis, the latter for inclusion on the book's website.
- Dr. Matthew F. Kaplan: for permission to use substantial text and figures from his Ph.D. dissertation as the basis for the new Section 17.8 on multilevel substructuring.
- Dr. Eric Blades: for conversion of the *ISMIS* FORTRAN code to form the MATLAB toolchest that is included on the book's website.
- Mr. Sean Regisford: for assistance with the finite element case studies in Sections 15.6 and 16.5.
- Mr. Garrett Moran: for assistance with solutions to new homework problems and for generating MATLAB plots duplicating the figures in the original *Structural Dynamics* book.

Their respective chapters of this edition were typeset in LaTeX by the authors.

ROY R. CRAIG, JR., AUSTIN, TX

ANDREW J. KURDILA, BLACKSBURG, VA

About the Authors

Roy R. Craig, Jr. is the John J. McKetta Energy Professor Emeritus in Engineering in the Department of Aerospace Engineering and Engineering Mechanics at The University of Texas at Austin. He received his B.S. degree in civil engineering from the University of Oklahoma, and M.S. and Ph.D. degrees in theoretical and applied mechanics from the University of Illinois at Urbana–Champaign. Dr. Craig's research and publications have been principally in the areas of structural dynamics analysis and testing, structural optimization, control of flexible structures, and the use of computers in engineering education. He is the developer of the *Craig–Bampton Method* of component-mode synthesis, which has been used extensively throughout the world for analyzing the dynamic response of complex structures, and he is the author of many technical papers and reports and of one other textbook, *Mechanics of Materials*. His industrial experience has been with the U.S. Naval Civil Engineering Laboratory, the Boeing Company, Lockheed Palo Alto Research Laboratory, Exxon Production Research Corporation, NASA, and IBM.

Dr. Craig has received numerous teaching awards and faculty leadership awards, including the General Dynamics Teaching Excellence Award in the College of Engineering, the John Leland Atwood Award presented jointly by the Aerospace Division of the American Society for Engineering Education and by the American Institute of Aeronautics and Astronautics "for sustained outstanding leadership and contributions in structural dynamics and experimental methods," and the D. J. DeMichele Award of the Society for Experimental Mechanics "for exemplary service and support in promoting the science and educational aspects of modal analysis technology." He is a member of the Society for Experimental Mechanics and a Fellow of the American Institute of Aeronautics and Astronautics.

Andrew J. Kurdila is the W. Martin Johnson Professor of Mechanical Engineering at the Virginia Polytechnic Institute and State University. He received his B.S. degree in applied mechanics in 1983 from the University of Cincinnati in the Department of Aerospace Engineering and Applied Mechanics. He subsequently entered The University of Texas at Austin and was awarded the M.S. degree in engineering mechanics the following year. He entered the Department of Engineering Science and Mechanics at the Georgia Institute of Technology as a Presidential Fellow and earned his Ph.D. in 1989.

Dr. Kurdila joined the faculty of the Aerospace Engineering Department at Texas A&M University in 1990 as an assistant professor. He was tenured and promoted to associate professor in 1993. He joined the faculty of the University of Florida in 1997 and was promoted to full professor in 1998. In 2005 he joined the faculty of the Virginia Polytechnic Institute and State University. He was recognized as a Select Faculty Fellow at Texas A&M University in 1994 and as a Faculty Fellow in 1996 and was awarded the Raymond L. Bisplinghoff Award at the University of Florida in 1999 for Excellence in Teaching.

Dr. Kurdila is the author of over 50 archival journal publications, 100 conference presentations and publications, four book chapters, two edited volumes, and two books. He has served as an associate editor of the *Journal of Vibration and Control* and of the *Journal of Guidance, Control and Dynamics*. He was named an Associate Fellow of the AIAA in 2001. His current research is in the areas of dynamical systems theory, control theory, and computational mechanics. His research has been funded by the Army Research Office, the Office of Naval Research, the Air Force Office of Scientific Research, the Air Force Research Laboratory, the National Science Foundation, the Department of Energy, the Army Research and Development Command, and the State of Texas.

1

The Science and Art of Structural Dynamics

What do a sport-utility vehicle traveling off-road, an airplane flying near a thunderstorm, an offshore oil platform in rough seas, and an office tower during an earthquake all have in common? One answer is that all of these are structures that are subjected to *dynamic loading*, that is, to time-varying loading. The emphasis placed on the safety, performance, and reliability of mechanical and civil structures such as these has led to the need for extensive analysis and testing to determine their response to dynamic loading. The structural dynamics techniques that are discussed in this book have even been employed to study the dynamics of snow skis and violins.

Although the topic of this book, as indicated by its title, is *structural dynamics*, some books with the word *vibrations* in their title discuss essentially the same subject matter. Powerful computer programs are invariably used to implement the modeling, analysis, and testing tasks that are discussed in this book, whether the application is one in aerospace engineering, civil engineering, mechanical engineering, electrical engineering, or even in sports or music.

1.1 INTRODUCTION TO STRUCTURAL DYNAMICS

This introductory chapter is entitled "The Science and Art of Structural Dynamics" to emphasize at the outset that by studying the principles and mathematical formulas discussed in this book you will begin to understand the *science* of structural dynamics analysis. However, structural dynamicists must also master the *art* of creating mathematical models of structures, and in many cases they must also perform dynamic tests. The cover photo depicts an automobile that is undergoing such dynamic testing. *Modeling*, *analysis*, and *testing* tasks all demand that skill and judgment be exercised in order that useful results will be obtained; and all three of these tasks are discussed in this book.

A *dynamic load* is one whose magnitude, direction, or point of application varies with time. The resulting time-varying displacements and stresses constitute the *dynamic response*. If the loading is a known function of time, the loading is said to be *prescribed loading*, and the analysis of a given structural system to a prescribed loading is called a *deterministic analysis*. If the time history of the loading is not known completely but only in a statistical sense, the loading is said to be *random*. In this book we treat only prescribed dynamic loading.

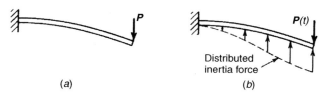

Figure 1.1 Cantilever beam under (a) static loading and (b) dynamic loading.

A structural dynamics problem differs from the corresponding static problem in two important respects. The first has been noted above: namely, the time-varying nature of the excitation. Of equal importance in a structural dynamics problem, however, is the role played by *acceleration*. Figure 1.1a shows a cantilever beam under static loading. The deflection and internal stresses depend directly on the static load P. On the other hand, Fig. 1.1b shows a similar cantilever beam subjected to a time-varying load $P(t)$. The acceleration of the beam gives rise to a distributed *inertia force*, as indicated in the figure. If the inertia force contributes significantly to the deflection of the structure and the internal stresses in the structure, a dynamical investigation is required.

Figure 1.2 shows the typical steps in a complete dynamical investigation. The three major steps, which are outlined by dashed-line boxes, are: *design*, *analysis*, and *testing*. The engineer is generally required to perform only one, or possibly two, of these steps. For example, a civil engineer might be asked to perform a dynamical analysis of an existing building and to confirm the analysis by performing specific dynamic testing of the building. The results of the analysis and testing might lead to criteria for retrofitting the building with additional bracing or damping to ensure safety against failure due to specified earthquake excitation.[1.1,1.2] Automotive engineers perform extensive analysis and vibration testing to determine the dynamical behavior of new car designs.[1.3,1.4] Results of this analysis and testing frequently lead to design changes that will improve the ride quality, economy, or safety of the vehicle.

In Section 1.2 we introduce the topic of mathematical models. In Section 1.3 we introduce the *prototype single-degree-of-freedom model* and indicate how to analyze the dynamic response of this model when it is subjected to certain simple inputs. Finally, in Section 1.4 we indicate some of the types of vibration tests that are performed on structures.

1.2 MODELING OF STRUCTURAL COMPONENTS AND SYSTEMS

Perhaps the most demanding step in any dynamical analysis is the creation of a *mathematical model* of the structure. This process is illustrated by steps 2a and 2b of Fig. 1.2. In step 2a you must contrive an idealized model of the structural system to be studied, a model essentially like the real system (which may already exist or may merely be in the design stages) but easier to analyze mathematically. This *analytical model* consists of:

1. A list of the simplifying assumptions made in reducing the real system to the analytical model
2. Drawings that depict the analytical model (e.g., see Fig. 1.3)
3. A list of the design parameters (i.e., sizes, materials, etc.)

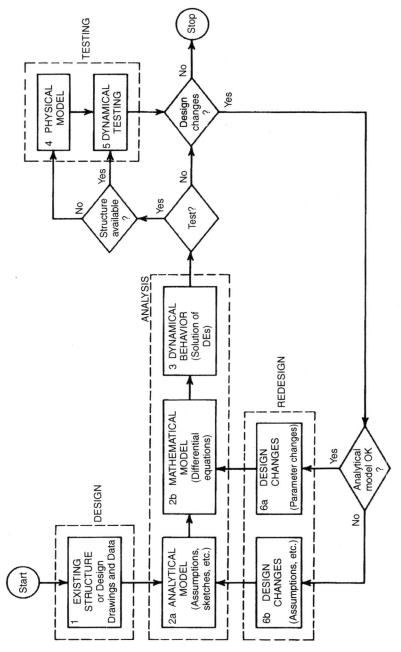

Figure 1.2 Steps in a dynamical investigation.

4 The Science and Art of Structural Dynamics

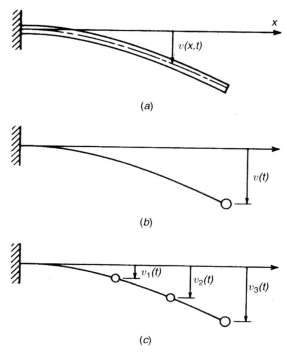

Figure 1.3 Analytical models of a cantilever beam: (*a*) distributed-mass cantilever beam, a continuous model (or distributed-parameter model); (*b*) one-degree-of-freedom model, a discrete-parameter model; (*c*) three-degree-of-freedom model, a more refined discrete-parameter model.

Analytical models fall into two basic categories: *continuous models* and *discrete-parameter models*. Figure 1.3*a* shows a continuous model of a cantilever beam. The number of displacement quantities that must be considered to represent the effects of all significant inertia forces is called the *number of degrees of freedom* (DOF) of the system. Thus, a continuous model represents an infinite-DOF system. Techniques for creating continuous models are discussed in Chapter 12. However, Fig. 1.3*b* and *c* depict finite-DOF systems. The discrete-parameter models shown here are called *lumped-mass models* because the mass of the system is assumed to be represented by a small number of point masses, or particles. Techniques for creating discrete-parameter models are discussed in Chapters 2, 8, and 14.

To create a useful analytical model, you must have clearly in mind the intended use of the analytical model, that is, the types of behavior of the real system that the model is supposed to represent faithfully. The complexity of the analytical model is determined (1) by the types and detail of behavior that it must represent, (2) by the computational analysis capability available (hardware and software), and (3) by the time and expense allowable. For example, Fig. 1.4 shows four different analytical models used in the 1960s to study the dynamical behavior of the *Apollo Saturn V* space vehicle, the vehicle that was used in landing astronauts on the surface of the moon. The 30-DOF beam-rod model was used for preliminary studies and to determine full-scale testing requirements. The 300-DOF model on the right, on the other hand, was required to give

1.2 Modeling of Structural Components and Systems   5

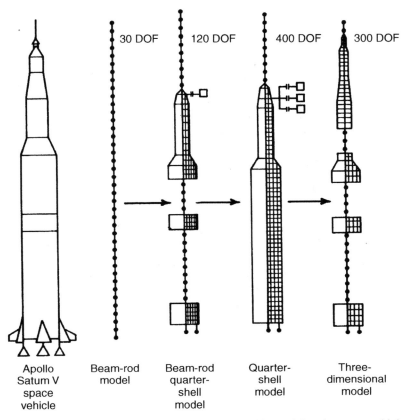

Figure 1.4 Analytical models of varying complexity used in studying the space vehicle dynamics of the *Apollo Saturn V*. (From C. E. Green et al., *Dynamic Testing for Shuttle Design Verification*, NASA, Washington, DC, 1972.)

a more accurate description of motion at the flight sensor locations. All of these *Saturn V* analytical models are extremely small compared with the multimillion-DOF models that can be analyzed now (see Section 17.8). However, supported by extensive dynamical testing, these analytical models were sufficient to ensure successful accomplishment of *Apollo V*'s moon-landing mission. Simplicity of the analytical model is very desirable as long as the model is adequate to represent the necessary behavior.

Once you have created an analytical model of the structure you wish to study, you can apply physical laws (e.g., Newton's Laws, stress–strain relationships) to obtain the differential equation(s) of motion that describe, in mathematical language, the analytical model. A continuous model leads to partial differential equations, whereas a discrete-parameter model leads to ordinary differential equations. The set of differential equations of motion so derived is called a *mathematical model* of the structure. To obtain a mathematical model, you will use methods studied in *dynamics* (e.g., Newton's Laws, Lagrange's Equations) and in *mechanics of deformable solids* (e.g., strain–displacement relations, stress–strain relations) and will combine these to obtain differential equations describing the dynamical behavior of a deformable structure.

6 The Science and Art of Structural Dynamics

In practice you will find that the entire process of creating first an analytical model and then a mathematical model may be referred to simply as *mathematical modeling*. In using a finite element computer program such as ABAQUS[1.5], ANSYS[1.6], MSC-Nastran[1.7], OpenFEM[1.8], SAP2000[1.9], or another computer program to carry out a structural dynamics analysis, your major modeling task will be to simplify the system and provide input data on dimensions, material properties, loads, and so on. This is

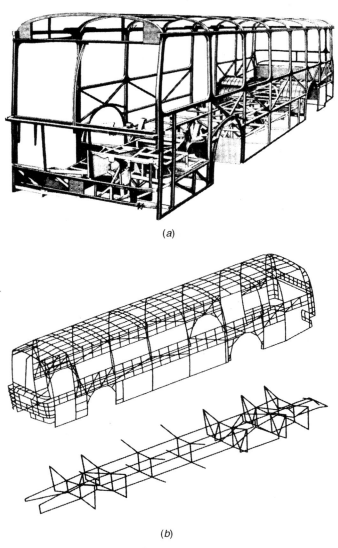

(a)

(b)

Figure 1.5 (*a*) Actual bus body and frame structure; (*b*) finite element models of the body and frame. (From D. Radaj et al., *Finite Element Analysis: An Automobile Engineer's Tool*, Society of Automotive Engineers, 1974. Used with permission of the Society of Automotive Engineers, Inc. Copyright © 1974 SAE.)

where the "art" of structural dynamics comes into play. On the other hand, actual creation and solution of the differential equations is done by the computer program. Figure 1.5 shows a picture of an actual bus body and a computer-generated plot of the idealized structure, that is, analytical model, which was input to a computer. Computer graphics software (e.g., MSC-Patran[1.7]) has become an invaluable tool for use in creating mathematical models of structures and in displaying the results of the analyses that are performed by computers.

Once a mathematical model has been formulated, the next step in a dynamical analysis is to solve the differential equation(s) to obtain the dynamical response that is predicted. (*Note:* The terms *dynamical response* and *vibration* are used interchangeably.) The two types of dynamical behavior that are of primary importance in structural applications are *free vibration* and *forced vibration* (or *forced response*), the former being the motion resulting from specified initial conditions, and the latter being the motion resulting directly from specified inputs to the system from external sources. Thus, you solve the differential equations of motion subject to specified initial conditions and to specified inputs from external sources, and you obtain the resulting time histories of the motion of the structure and stresses within the structure. This constitutes the behavior predicted for the (real) structure, or the *response*.

The analysis phase of a dynamical investigation consists of the three steps just described: defining the *analytical model*, deriving the corresponding *mathematical model*, and solving for the *dynamical response*. This book deals primarily with the second and third steps in the analysis phase of a structural dynamics investigation. Section 1.3 illustrates these steps for the simplest analytical model, a lumped-mass single-DOF model. Section 1.4 provides a brief discussion of dynamical testing.

1.3 PROTOTYPE SPRING–MASS MODEL

Before proceeding with the details of how to model complex structures and analyze their dynamical behavior, let us consider the simplest structure undergoing the simplest forms of vibration. The structure must have an *elastic component*, which can store and release potential energy; and it must have *mass*, which can store and release kinetic energy. The simplest model, therefore, is the *spring–mass oscillator*, shown in Fig. 1.6a.

1.3.1 Simplifying Assumptions: Analytical Model

The simplifying assumptions that define this *prototype analytical model* are:

1. The mass is a point mass that is confined to move along one horizontal direction on a frictionless plane. The displacement of the mass in the x direction from the position where the spring is undeformed is designated by the displacement variable $u(t)$.
2. The mass is connected to a fixed base by an idealized massless, linear spring. The fixed base serves as an inertial reference frame. Figure 1.6b shows the linear relationship between the *elongation* (u positive) and *contraction* (u negative) of the spring and the force $f_S(t)$ that the spring exerts on the mass. When the spring is in tension, f_S is positive; when the spring is in compression, f_S is negative.
3. A specified external force $p(t)$ acts on the mass, as shown in Fig. 1.6a.

8 The Science and Art of Structural Dynamics

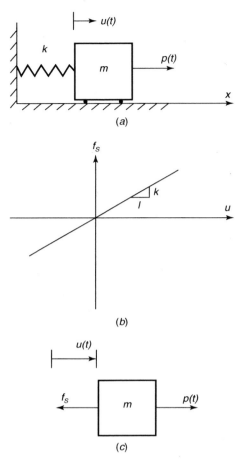

Figure 1.6 (a) Spring–mass oscillator; (b) force–elongation behavior of a linear spring; (c) free-body diagram of the spring–mass oscillator.

Since it takes only one variable [e.g., $u(t)$] to specify the instantaneous position of the mass, this is called a *single-degree-of-freedom* (SDOF) *system*.

1.3.2 Mathematical Model: Equation of Motion

Newton's Second Law   To obtain a mathematical model describing the behavior of the spring–mass oscillator, we start by drawing a *free-body diagram* of the mass (Fig. 1.6c) and applying *Newton's Second Law*,

$$\sum F_x = ma_x \tag{1.1}$$

where m is the mass and a_x is the acceleration of the mass, taken as positive in the $+x$ direction. Acceleration a_x is given by the second derivative of the displacement, that is, $a_x = \ddot{u}(t)$; similarly, the velocity is given by $\dot{u}(t)$. By assuming that the mass is displaced u to the right of the position where the spring force is zero, we can say that the spring will be in tension, so the spring force will act to the left on the mass, as shown on the free-body diagram. Thus, Eq. 1.1 becomes

$$-f_S + p(t) = m\ddot{u} \tag{1.2}$$

Force–Displacement Relationship As indicated in Fig. 1.6b, there is assumed to be a linear relationship between the force in the spring and its elongation u, so

$$f_S = ku \tag{1.3}$$

where k is the *stiffness* of the spring.

Equation of Motion Finally, by combining Eqs. 1.2 and 1.3 and rearranging to place all u-terms on the left, we obtain the *equation of motion* for the prototype undamped SDOF model:

$$\boxed{m\ddot{u} + ku = p(t)} \tag{1.4}$$

This equation of motion is a linear second-order ordinary differential equation. It is the *mathematical model* of this simple SDOF system.

Having Eq. 1.4, the equation of motion that governs the motion of the SDOF spring–mass oscillator in Fig. 1.6a, we now examine the dynamic response of this prototype system. The response of the system is determined by its *initial conditions*, that is, by the values of its displacement and velocity at time $t = 0$:

$$u(0) = u_0 = \text{initial displacement}, \qquad \dot{u}(0) = v_0 = \text{initial velocity} \tag{1.5}$$

and by $p(t)$, the *external force* acting on the system. Here we consider two simple examples of vibration of the spring–mass oscillator; a more general discussion of SDOF systems follows in Chapters 3 through 7.

1.3.3 Free Vibration Example

The spring–mass oscillator is said to undergo *free vibration* if $p(t) \equiv 0$, but the mass has nonzero initial displacement u_0 and/or nonzero initial velocity v_0. Therefore, the equation of motion for free vibration is the homogeneous second-order differential equation

$$m\ddot{u} + ku = 0 \tag{1.6}$$

The general solution of this well-known simple differential equation is

$$u = A_1 \cos \omega_n t + A_2 \sin \omega_n t \tag{1.7}$$

where ω_n is the *undamped circular natural frequency*, defined by

$$\boxed{\omega_n = \sqrt{\frac{k}{m}}} \tag{1.8}$$

The units of ω_n are radians per second (rad/s).

The constants A_1 and A_2 in Eq. 1.7 are chosen so that the two initial conditions, Eqs. 1.5, will be satisfied. Thus, *free vibration of an undamped spring–mass oscillator* is characterized by the time-dependent displacement

$$u(t) = u_0 \cos \omega_n t + \frac{v_0}{\omega_n} \sin \omega_n t \qquad (1.9)$$

It is easy to show that this solution satisfies the differential equation, Eq. 1.6, and the two initial conditions, Eqs. 1.5.

Figure 1.7 depicts the response of a spring–mass oscillator released from rest from an initial displacement of u_0. Thus, the motion depicted in Fig. 1.7 is given by

$$u(t) = u_0 \cos \omega_n t = u_0 \cos \frac{2\pi t}{T_n} \qquad (1.10)$$

From Eq. 1.10 and Fig. 1.7, free vibration of an undamped SDOF system consists of harmonic (sinusoidal) motion that repeats itself with a *period* (in seconds) given by

$$T_n = \frac{2\pi}{\omega_n} \qquad (1.11)$$

as illustrated in Fig. 1.7. The *amplitude* of the vibration is defined as the maximum displacement that is experienced by the mass. For the free vibration depicted in Fig. 1.7, the amplitude is equal to the initial displacement u_0.

Free vibration is discussed further in Chapter 3.

1.3.4 Forced Response Example

The spring–mass oscillator is said to undergo *forced vibration* if $p(t) \neq 0$ in Eq. 1.4. Solution of the differential equation of motion for this case, Eq. 1.4, requires both a *complementary solution* u_c and a *particular solution* u_p. Thus,

$$u(t) = u_c(t) + u_p(t) \qquad (1.12)$$

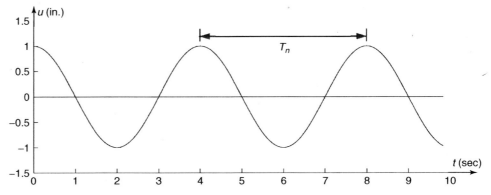

Figure 1.7 Free vibration of a spring-mass oscillator with $u_0 = 1.0$ in., $v_0 = 0$, and $T_n = 4.0$ sec.

As a simple illustration of forced vibration we consider *ramp response*, the response of the spring–mass oscillator to the linearly varying ramp excitation force given by

$$p(t) = p_0 \frac{t}{t_0}, \qquad t > 0 \tag{1.13}$$

and illustrated in Fig. 1.8a. (The time t_0 is the time at which the force reaches the value p_0.) The particular solution, like the excitation, varies linearly with time. The complementary solution has the same form as given in Eq. 1.7, so the total response has the form

$$u(t) = A_1 \cos \omega_n t + A_2 \sin \omega_n t + \frac{p_0}{k} \frac{t}{t_0} \tag{1.14}$$

where the constants A_1 and A_2 must be selected so that the initial conditions $u(0)$ and $\dot{u}(0)$ will be satisfied.

Figure 1.8b depicts the response of a spring–mass oscillator that is initially at rest at the origin, so the initial conditions are $u(0) = \dot{u}(0) = 0$. The corresponding *ramp response* is thus given by

$$\boxed{u(t) = \frac{p_0}{k} \left(\frac{t}{t_0} - \frac{1}{\omega_n t_0} \sin \omega_n t \right)} \tag{1.15}$$

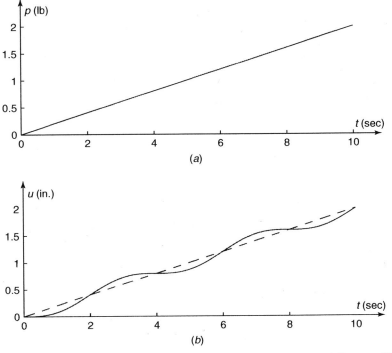

Figure 1.8 (a) Ramp excitation $p(t) = p_0(t/t_0)$ for $t > 0$, with $p_0 = 2$ lb, $t_0 = 10$ sec; (b) response of a spring–mass oscillator to ramp excitation. For (b), $k = 1$ lb/in. and $T_n = 4$ sec.

Clearly evident in this example of forced response are two components: (1) a linearly time-varying displacement (dashed curve), which is due directly to the linearly time-varying ramp excitation, and (2) an induced oscillatory motion at the undamped natural frequency ω_n. Of course, this ramp response is only valid as long as the spring remains within its linearly elastic range.

1.3.5 Conclusions

In this section we have taken a preliminary look at several characteristics that are typical of the response of structures to nonzero initial conditions and/or to time-varying excitation. We have especially noted the oscillatory nature of the response. In Chapters 3 through 7, we consider many additional examples of free and forced vibration of SDOF systems, including systems with damping.

1.4 VIBRATION TESTING OF STRUCTURES

A primary purpose of dynamical testing is to confirm a mathematical model and, in many instances, to obtain important information on loads, on damping, and on other quantities that may be required in the dynamical analysis. In some instances these tests are conducted on reduced-scale *physical models*: for example, wind tunnel tests of airplane models. In other cases, when a full-scale structure (e.g., an automobile) is available, the tests may be conducted on it.

Aerospace vehicles (i.e., airplanes, spacecraft, etc.) must be subjected to extensive static and dynamic testing on the ground prior to actual flight of the vehicle. Figure 1.9*a* shows a *ground vibration test* in progress on a Boeing 767 airplane. Note the electrodynamic shaker in place under each wingtip and the special soft support under the nose landing gear.

Dynamical testing of physical models may be employed for determining qualitatively and quantitatively the dynamical behavior characteristics of a particular class of structures. For example, Fig. 1.9*b* shows an aeroelastic model of a Boeing 777 airplane undergoing ground vibration testing in preparation for testing in a wind tunnel to aid in predicting the dynamics of the full-scale airplane in flight. Note the soft bungee-cord distributed support of the model and the two electrodynamic shakers that are attached by stingers to the engine nacelles. Figure 1.10 shows a fluid-filled cylindrical tank structure in place on a shake table in a university laboratory. The shake table is used to simulate earthquake excitation at the base of the tank structure.

Chapter 18 provides an introduction to *Experimental Modal Analysis*, a very important structural dynamics test procedure that is used extensively in the automotive and aerospace industries and is also used to test buildings, bridges, and other civil structures.

1.5 SCOPE OF THE BOOK

Part I, encompassing Chapters 2 through 7, treats single-degree-of-freedom (SDOF) systems. In Chapter 2 procedures are described for developing SDOF mathematical models; both Newton's Laws and the Principle of Virtual Displacements are employed. The free vibration of undamped and damped systems is the topic of Chapter 3, and

(a)

(b)

Figure 1.9 (a) Ground vibration testing of the Boeing 767 airplane; (b) aeroelastic wind-tunnel model of the Boeing 777 airplane. (Courtesy of Boeing Commercial Airplane Company.)

in Chapter 4 you will learn about the response of SDOF systems to harmonic (i.e., sinusoidal) excitation. This is, perhaps, the most important topic in the entire book, because it describes the fundamental characteristics of the dynamic response of flexible structures. Chapters 5 and 6 continue the discussion of dynamic response of SDOF systems. In Chapter 5 we treat closed-form methods for evaluating the response of SDOF systems to nonperiodic inputs, and Chapter 6 deals with numerical techniques. Part I concludes with Chapter 7, where the response of SDOF systems to periodic excitation is considered. Frequency-domain techniques, which are currently used widely

Figure 1.10 Fluid-filled tank subjected to simulated earthquake excitation. (Courtesy of R. W. Clough.)

in dynamic testing of structures, are introduced as a natural extension of the discussion of periodic excitation.

Analysis of the dynamical behavior of structures of even moderate complexity requires the use of multiple-degree-of-freedom (MDOF) models. Part II, which consists of Chapters 8 through 11, discusses procedures for obtaining MDOF mathematical models and for analyzing their free- and forced-vibration response. Mathematical modeling is treated in Chapter 8. While Newton's Laws can be used to derive mathematical models of some MDOF systems, the primary procedure employed for deriving such models is the use of Lagrange's Equations. Continuous systems are approximated by MDOF models through the use of the Assumed-Modes Method. Chapter 9, which treats vibration of undamped two-degree-of-freedom (2-DOF) systems, introduces many of the concepts that are important to a study of the response of MDOF systems: for example, natural frequencies, normal modes, and mode superposition. In Chapter 10 we discuss many of the mathematical properties related to modes and frequencies of both undamped and damped MDOF systems and introduce several popular schemes for representing damping in structures. In addition, the subject of complex modes is introduced in Chapter 10. Finally, in Chapter 11 we discuss mode superposition, the most widely used procedure for determining the response of MDOF systems.

Part III treats structures that are modeled as continuous systems. Although important topics such as the determination of partial differential equation mathematical models

(Chapter 12) and determination of modes and frequencies (Chapter 13) are discussed, the primary purpose of Part III is to provide several "exact solutions" that can be used for evaluating the accuracy of the approximate MDOF models analyzed in Parts II and IV.

Part IV presents computational techniques for handling structural dynamics applications in engineering by the methods that are widely used for analyzing and testing the behavior of complex structures (e.g., airplanes, automobiles, high-rise buildings). In Chapter 14, a continuation of the mathematical modeling topics in Chapter 8, you are introduced to the important Finite Element Method (FEM) for creating MDOF mathematical models. Chapter 14 also includes several FEM examples involving mode shapes and natural frequencies, extending the discussion of those topics in Chapters 9 and 10. Prior study of the Finite Element Method, or of Matrix Structural Analysis, is not a prerequisite for Chapter 14. Chapters 15 through 17 treat advanced numerical methods for solving structural dynamics problems: eigensolvers for determining modes and frequencies of complex structures (Chapter 15), direct integration methods for computing dynamic response (Chapter 16), and substructuring (Chapter 17).

Finally, in Part V, Chapters 18 through 20, we introduce advanced structural dynamics applications: experimental modal analysis, or modal testing (Chapter 18); structures containing piezoelectric members, or "active structures" (Chapter 19); and earthquake response of structures (Chapter 20).

A reader who merely wishes to attain an introductory level of understanding of current structural dynamics analysis techniques need only study Chapters 2 through 11 and Chapter 14.

1.6 COMPUTER SIMULATIONS; SUPPLEMENTARY MATERIAL ON THE WEBSITE

The objective of this book is to introduce you to computer methods in structural dynamics. Therefore, most chapters contain computer plots of one or more of the following: mode shapes, response time histories, frequency-response functions, and so on. Some of these are plots of closed-form mathematical expressions; others are generated by algorithms that are presented in the text. Also, a number of homework problems ask for similar computer-generated results.

To facilitate your understanding of the procedures discussed in the book, a number of supplementary materials are made available to you on the book's website www.wiley.com/college/craig. Included are the following:

- MATLAB[1] .m-files that were used to create the plots included in this book, and MATLAB .m-files for many of the numerical analysis algorithms presented in Chapters 6, 7, 15, and 16.
- Line drawings and plots in the form of .eps files.
- *ISMIS*, a MATLAB-based matrix structural analysis computer program. Included are the MATLAB .m-files, a .pdf file entitled *Special ISMIS Operations for Structural Analysis*, and two *ISMIS* examples. This software supplements Chapter 14.

---

[1]MATLAB is the most widely used computer software for solving dynamics problems, including structural dynamics problems of the size and scope of those in this book[1.10]. If you are inexperienced in the use of MATLAB, you should read Appendix E, "Introduction to the Use of MATLAB," in its entirety.

- *TUTORIAL NOTES: Structural Dynamics and Experimental Modal Analysis* by Peter Avitabile. These short course notes supplement Chapter 18.
- "Airplane Ground Vibration Testing—Nominal Modal Model Correlation," by Charles Pickrel. This technical article supplements Chapter 18.
- Other selected items.

REFERENCES

[1.1] T. T. Soong and G. F. Dargush, *Passive Energy Dissipation Systems in Structural Engineering*, Wiley, New York, 1997.

[1.2] G. C. Hart and K. Wong, *Structural Dynamics for Structural Engineers*, Wiley, New York, 2000.

[1.3] LMS International, http://www.lmsintl.com.

[1.4] MTS, http://www.mts.com.

[1.5] ABAQUS, Inc., http://www.abaqus.com.

[1.6] ANSYS, http://www.ansys.com.

[1.7] MSC Software, http://www.mscsoftware.com.

[1.8] OpenFEM, http://www-rocq.inria.fr/OpenFEM.

[1.9] SAP2000, http://www.csiberkeley.com.

[1.10] MATLAB, http://www.mathworks.com.

PROBLEMS

Problem Set 1.3[2]

1.1 (**a**) What is the natural frequency in hertz of the spring–mass system in Fig. 1.6a if $k = 40$ N/m and the mass is $m = 2.0$ kg? (**b**) What is the natural frequency if $k = 100$ lb/in. and the mass weighs $W = 50$ lb? Recall that $m = W/g$, where the value of g must be given in the proper units.

> Use Newton's Laws to determine the equations of motion of the SDOF spring–mass systems in Problems 1.2 and 1.3. Show all necessary free-body diagrams and deformation diagrams.

1.2 As shown in Fig. P1.2, a mass m is connected to rigid walls on both sides by identical massless, linear springs, each having stiffness k. (**a**) Following the steps in Section 1.3.2, determine the equation of motion of mass m. (**b**) Determine expressions for the undamped circular natural frequency ω_n and the period T_n of this spring–mass system. (**c**) If $k = 40$ lb/in. and the mass weighs $W = 20$ lb, what is the resulting natural frequency of this system in hertz? Recall that $m = W/g$, where the value of g must be given in the proper units.

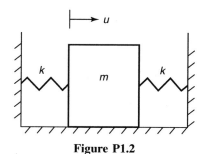

Figure P1.2

1.3 As shown in Fig. P1.3, a mass m is suspended from a rigid ceiling by a massless linear spring of stiffness k. (**a**) Following the steps in Section 1.3.2, determine the equation of motion of mass m. Measure the displacement u of the mass vertically downward from the position where the spring is unstretched. Write the equation of

---
[2]Problem Set headings refer to the text section to which the problem set pertains.

motion in the form given by Eq. 1.4. **(b)** Determine expressions for the undamped circular natural frequency ω_n and the period T_n of this spring–mass system. **(c)** If $k = 1.2$ N/m and the mass is $m = 0.8$ kg, what is the resulting natural frequency of this system in hertz?

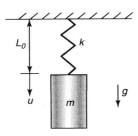

Figure P1.3

1.4 (a) Determine an expression for the *free vibration* of the undamped SDOF spring–mass system in Fig. 1.6a if $T_n = 4$ sec, $u_0 = 0$, and $v_0 = 10$ in./sec. **(b)** What is the maximum displacement of the mass as it vibrates?

1.5 Determine an expression for the *free vibration* of the undamped SDOF spring–mass system in part **(c)** of Problem 1.2 if $u_0 = 2$ in. and $v_0 = 0$.

> For problems whose number is preceded by a **C**, you are to write a computer program and use it to produce the plot(s) requested. *Note*: MATLAB .m-files for many of the plots in this book may be found on the book's website.

C1.6 (a) Determine an expression for the *free vibration* of the undamped SDOF system in Fig. 1.6a if $T_n = 2$ sec, $u_0 = 0$, and $v_0 = 10$ in./sec. **(b)** Using your answer to part (a), write a computer program (e.g., a MATLAB .m-file), and generate a plot similar to Fig. 1.7.

PART I

Single-Degree-of-Freedom Systems

2

Mathematical Models of SDOF Systems

In Chapter 1 you were introduced to the concept of mathematical modeling and to vibration of a simple single-degree-of-freedom (SDOF) spring–mass model. To create an analytical model of a structure, we make simplifying assumptions (e.g., the assumptions that defined the spring–mass model in Section 1.3). Then, to derive equations of motion for this model, we must apply *three fundamental types of equations*:

1. *Newton's Laws* (or equivalent energy principles) must be satisfied.
2. The *force–deformation behavior* of the elastic element(s) and the *force–velocity behavior* of the damping element(s) must be characterized.
3. The *kinematics of deformation* must be incorporated.

Section 2.1 provides a brief review of elementary dynamics needed for item 1, and Section 2.2 summarizes the behavior of some common lumped-parameter elements referred to in item 2. The remaining sections of Chapter 2 cover specific procedures for deriving SDOF mathematical models of various structures, that is, models that are described by a single second-order ordinary differential equation. Both Newton's Laws and the Principle of Virtual Displacements are employed to derive the equation of motion of various lumped-parameter systems. Then the Principle of Virtual Displacements is extended to approximate a continuous system by an SDOF generalized-parameter model.

Upon completion of this chapter you should be able to:

- Use Newton's Laws to derive the equation of motion of a single particle or rigid body having one degree of freedom.
- Use the Principle of Virtual Displacements, in conjunction with d'Alembert forces, to derive the equation of motion of a particle, a rigid body, or an assemblage of bodies having a single degree of freedom.
- Use the Principle of Virtual Displacements, together with an assumed mode, to derive the equation of motion for a generalized-parameter SDOF model of a structure.

2.1 BRIEF REVIEW OF THE DYNAMICS OF PARTICLES AND RIGID BODIES

There are two ways of formulating mathematical models governing the dynamics of particles and rigid bodies. One, *Vectorial Mechanics* or *Newtonian Mechanics*, uses

22 Mathematical Models of SDOF Systems

Newton's Laws and deals with force and acceleration, both vector quantities. The other, *Analytical Mechanics* or *Lagrangian Mechanics*, is based on Work–Energy Methods (the Principle of Virtual Displacements, Lagrange's Equations, and Hamilton's Principle) and deals with work and energy quantities. Both approaches are employed in this book (in Chapters 2, 8, and 12) to derive the equations of motion of vibrating structures. In this section we provide a brief review of Newtonian Mechanics; Lagrangian Mechanics concepts are introduced in Sections 2.4, 8.2, and 12.3.

2.1.1 Particle Dynamics

Figure 1.6a illustrates how a spring–mass oscillator can be modeled as a mass particle attached to a massless spring. As in Fig. 1.6c, a *free-body diagram* is drawn of the particle, and all forces that act on the particle are shown. In writing the equation of motion of a mass particle, *Newton's Second Law*,

$$\sum \boldsymbol{F} = m\boldsymbol{a} \qquad (2.1)$$

is used, where m is the mass of the particle and $\boldsymbol{a}$ is its acceleration relative to an inertial reference frame.[1] The units of mass are (lb-sec^2/in.) or (N·s^2/m). For motion of a particle in a plane, the component form of Eq. 2.1 is

$$\boxed{\begin{aligned}\sum F_x &= ma_x \\ \sum F_y &= ma_y\end{aligned}} \qquad (2.2)$$

For structural dynamics problems it is frequently useful to introduce the *d'Alembert force*, or *inertia force*

$$\boldsymbol{f}_I = -m\boldsymbol{a} \qquad (2.3)$$

Then Eq. 2.1 can be written as an equation of *dynamic equilibrium*,

$$\sum \boldsymbol{F}^* \equiv \boldsymbol{f}_I + \sum \boldsymbol{F} = \boldsymbol{0} \qquad (2.4)$$

with the asterisk indicating that the resultant inertia force has been added to the resultant of the other forces acting on the particle. Figure 2.1 shows the inertia forces for a particle and for two rigid bodies in general plane motion.

The *kinetic energy* of a particle is given by

$$\boxed{T = \tfrac{1}{2}mv^2 = \tfrac{1}{2}m(v_x^2 + v_y^2)} \qquad (2.5)$$

where v is the magnitude of the velocity of the particle.

---

[1] An *inertial reference frame* is an imaginary set of rectangular axes assumed to have no translation or rotation in space.

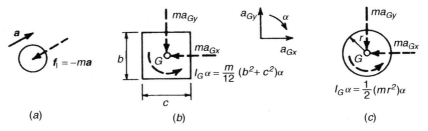

Figure 2.1 Inertia forces acting on (*a*) a particle, (*b*) a rectangular block in plane motion, and (*c*) a circular disk in plane motion.

2.1.2 Rigid-Body Dynamics: Plane Motion

Newton's Laws for a particle may be extended to give the following equations for a rigid body in plane motion. The force equation is

$$\sum \mathbf{F} = m\mathbf{a}_G \tag{2.6}$$

where $\mathbf{a}_G$ is the acceleration vector of the mass center of the body. The component form of Eq. 2.6 is

$$\boxed{\begin{aligned}\sum F_x &= ma_{Gx} \\ \sum F_y &= ma_{Gy}\end{aligned}} \tag{2.7}$$

One of the following three moment equations is also needed:

$$\boxed{\sum M_G = I_G \alpha} \tag{2.8a}$$

where moments are summed about the mass center G, I_G is the mass moment of inertia about G, and α is the angular acceleration of the body[2]; or

$$\sum M_O = I_O \alpha \tag{2.8b}$$

where O is a fixed axis about which the body is rotating; or

$$\sum M_C = I_C \alpha \tag{2.8c}$$

where C is a point whose acceleration vector passes through G. Equation 2.8a is a general equation and should be used in most cases. Equation 2.8b is valid only for fixed-axis rotation, and Eq. 2.8c is of limited applicability (e.g., when a uniform disk is rolling without slipping).

---

[2]The Greek letter α is used later as a phase angle. However, this apparent duplication should not cause confusion.

When inertia force components ma_{Gx} and ma_{Gy} and inertia couple $I_G\alpha$ have been added to the free-body diagram, as in Fig. 2.1b and c, Eqs. 2.7 and 2.8 can be written as equations of dynamic equilibrium

$$\sum F_x^* = 0$$
$$\sum F_y^* = 0 \tag{2.9}$$

and

$$\sum M_z^* = 0 \tag{2.10}$$

respectively. In Eq. 2.10, the z-moment summation may be taken about any point.

For rigid bodies in plane motion, the kinetic energy is given by

$$T = \tfrac{1}{2} m \,(v_{Gx}^2 + v_{Gy}^2) + \tfrac{1}{2} I_G \omega^2 \tag{2.11}$$

where v_G is the magnitude of the linear velocity of the mass center of the rigid body and ω is the magnitude of the angular velocity of the rigid body.

In Section 2.3 the equations of dynamics that have been reviewed in this section are applied to derive mathematical models for several lumped-parameter models.

2.2 ELEMENTS OF LUMPED-PARAMETER MODELS

The physical components that constitute a lumped-parameter model of a structure are those that relate force to displacement, velocity, and acceleration, respectively. In the preceding section the inertial properties of particles and rigid bodies, which relate force to acceleration, were reviewed. In this section several common elastic and damping elements are presented.

2.2.1 Elastic Elements

Spring One type of element that relates force to deformation is called a *spring*. Figure 2.2a shows the symbol that is used to represent an idealized massless spring, and Fig. 2.2b is a plot of spring force versus elongation. The spring force always acts along the line joining the two ends of the spring. For most structural materials, for small values of the elongation $e \equiv u_2 - u_1$, there is a linear relationship between force and elongation (contraction). This relationship is given by

$$f_s = ke \tag{2.12}$$

where k is called the *spring constant*. The units of k are pounds per inch (lb/in.) or newtons per meter (N/m).[3] Nonlinear behavior is treated briefly in Section 6.3.

---

[3] The units of various quantities will be given in the fundamental English and SI units. Problems will, however, also be solved using related units: for example, kip ($\equiv$ 1000 lb), foot, kilonewton, millimeter, and so on.

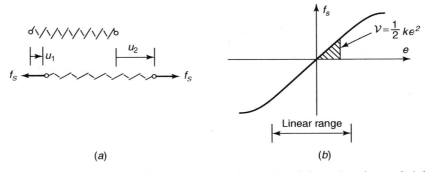

Figure 2.2 Force–deformation behavior of a spring: (*a*) undeformed spring and deformed spring; (*b*) force-deformation diagram.

Within its elastic range a spring serves as an energy storage device. The energy stored in a spring when it is elongated (contracted) an amount e within its linearly elastic range is called *strain energy* and is given by

$$\mathcal{V} = \tfrac{1}{2} k e^2 \qquad (2.13)$$

Within the linear range of the f_s versus e curve, the strain energy is indicated by the area under the curve, as in Fig. 2.2*b*.

Axial–Deformation Member Figure 2.3*a* shows a uniform linearly elastic member with an axial load P that causes an elongation e. The relationship between the axial force P and the elongation e is given by[2.1]

$$P = \frac{EA}{L} e \qquad (2.14a)$$

where E is the modulus of elasticity of the material from which the member is made, A is the cross-sectional area of the member, and L is the length of the member. Therefore, this type of member acts like a spring with spring constant

$$k = \frac{EA}{L} \qquad (2.14b)$$

Torsion Rod Figure 2.3*b* shows a uniform linearly elastic torsion member with circular cross section subjected to a torque T that causes a relative angle of twist ϕ between ends of the rod. The relationship between the torque T and the twist angle ϕ is given by[2.1]

$$T = \frac{GJ}{L} \phi \qquad (2.15a)$$

where G is the shear modulus of elasticity of the material from which the member is made, J is the polar moment of inertia of the cross section of the member,[4] and L

---

[4]If the cross section is not circular, J stands for a torsion constant that is not the polar moment of inertia (see Ref. [2.1]).

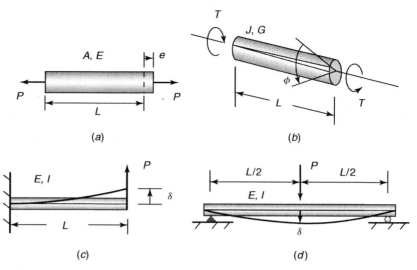

Figure 2.3 Force–deformation behavior of elastic elements: (*a*) axial-deformation bar; (*b*) torsion rod; (*c*) cantilever beam; (*d*) simply supported beam.

is the total length of the member. Therefore, this type of member acts like a torsional spring with torsional-spring constant

$$k_\phi = \frac{GJ}{L} \tag{2.15b}$$

Cantilever Beam Figure 2.3*c* shows a uniform cantilever beam with a transverse tip load P that causes a tip deflection δ. The relationship between the force P and the deflection δ is given by[2.1]

$$P = \frac{3EI}{L^3}\delta \tag{2.16a}$$

where E is the modulus of elasticity of the material from which the member is made, I is the moment of inertia of the cross section of the member, and L is the length of the member. Therefore, this type of member acts like a spring with spring constant

$$k = \frac{3EI}{L^3} \tag{2.16b}$$

Simply Supported Beam Figure 2.3*d* shows a uniform simply supported beam with a transverse midspan load P that causes a deflection δ at the point of application of the load. The relationship between the force P and the deflection δ is given by[2.1]

$$P = \frac{48EI}{L^3}\delta \tag{2.17a}$$

where E is the modulus of elasticity of the material from which the member is made, I is the moment of inertia of the cross section of the member, and L is the length of

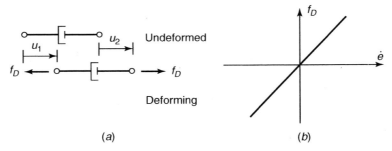

Figure 2.4 (*a*) Undeformed and deforming linear viscous dashpots; (*b*) plot of force versus elongation rate.

the member. Therefore, this type of member acts like a spring with spring constant

$$k = \frac{48EI}{L^3} \qquad (2.17b)$$

2.2.2 Viscous-Damping Element

Just as a spring-type member serves as an energy storage device, there are means by which energy is dissipated from a vibrating structure. These are called *damping mechanisms*, or simply *dampers*. Although research studies have proposed numerous ways to describe material damping mathematically, the exact nature of damping in a structure is usually impossible to determine. There are many references that discuss the general topic of damping in vibrating structures (e.g., Refs. [2.2] to [2.4]).

The simplest analytical model of damping employed in structural dynamics analyses is the *linear viscous dashpot* model, which is illustrated in Fig. 2.4. The damping force f_D is given by

$$f_D = c\dot{e} = c(\dot{u}_2 - \dot{u}_1) \qquad (2.18)$$

and is thus a linear function of the relative velocity between the two ends of the dashpot. The constant c is called the *coefficient of viscous damping*, and its units are pounds per inch per second (lb-sec/in.) or newtons per meter per second (N·s/m).

In Section 2.3 the constitutive equations that have been reviewed in this section (i.e., force–deformation equations of elastic elements and the force–velocity equation of a linear viscous dashpot) are applied to derive mathematical models for several lumped-parameter models.

2.3 APPLICATION OF NEWTON'S LAWS TO LUMPED-PARAMETER MODELS

In this section the equation of motion of several lumped-parameter models (i.e., particle models and rigid-body models) are derived by using Newton's Laws or, equivalently, the d'Alembert Force Method. Utilizing the principles of dynamics from Section 2.1

and the constitutive equations from Section 2.2, these derivations will serve as a review of your previous studies of dynamics and will introduce you to procedures that may be employed to determine the *mathematical model of an SDOF system*.

***

Example 2.1 Use Newton's Laws to derive the equation of motion of the simple mass–spring–dashpot system shown in Fig. 1. Assume only vertical motion, and assume that the spring is linear with spring constant k. Neglect air resistance, the mass of the spring, and any internal damping in the spring. The force labeled $p(t)$ is the force applied to the mass by an external source.

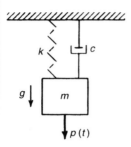

Figure 1 Mass–spring–dashpot SDOF system.

SOLUTION 1. Establish a reference frame and a displacement coordinate (Fig. 2).

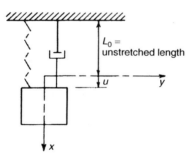

Figure 2 Mass–spring–dashpot SDOF system with reference frame.

Choose the x axis along the line of motion, and choose the origin (i.e., $x = 0$) at the location where the spring is unstretched. Let u be the displacement in the x direction.

2. Draw a free-body diagram of the particle in an arbitrary displaced position (Fig. 3).

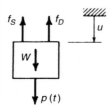

Figure 3 Free-body diagram of a mass–spring–dashpot SDOF system.

3. Apply Newton's Second Law,

$$+\downarrow \sum F_x = m\ddot{u} \quad (1)$$

(*Note*: + sense downward since u is positive downward.) From the free-body diagram, determine the forces for the left-hand side of Eq. 1.

$$p(t) - f_S - f_D + W = m\ddot{u} \quad (2)$$

4. Relate the forces to the motion variables:

$$f_S = ke = ku \quad (3)$$

$$f_D = c\dot{e} = c\dot{u} \quad (4)$$

(Note that $u = 0$ corresponds to the unstretched spring.)

5. Combine and simplify, arranging unknowns on the left-hand side of the equation:

$$m\ddot{u} + c\dot{u} + ku = W + p(t) \quad (5)$$

Note that this is a second-order linear nonhomogeneous ordinary differential equation with constant coefficients.

Equation 5 can be simplified by the following considerations. The *static displacement* of the weight W on the linear spring is denoted by u_{st} and is given by

$$u_{st} = \frac{W}{k} \quad (6)$$

Let the *displacement measured relative to the static equilibrium position* be denoted by u_r. That is,

$$u \equiv u_r + u_{st} \quad (7)$$

Since u_{st} is a constant, *the differential equation of motion of the SDOF system* (Eq. 5) can be rewritten as

$$m\ddot{u}_r + c\dot{u}_r + ku_r = p(t) \quad \textbf{Ans. (8)}$$

Thus, the weight force can be eliminated from the differential equation of motion of an SDOF system if the *displacement of the mass is measured relative to its static equilibrium position*.

Equation 8 in Example 2.1 is an example of the second-order differential equation that defines the *prototype SDOF system*: namely,

$$\boxed{m\ddot{u} + c\dot{u} + ku = p(t)} \quad (2.19)$$

This equation may be considered to be the *fundamental equation in structural dynamics and linear vibration theory*. In the remainder of this chapter you will encounter several additional examples of SDOF systems described by various forms of this equation. Section 1.3 introduced you to solutions of this equation without the damping term. From Chapter 3 onward you will be spending a great deal of time determining solutions of Eq. 2.19 and applying them to structural dynamics problems for both SDOF and MDOF systems.

In Example 2.1, Newton's Second Law was used directly, so no inertia forces were shown on the free-body diagram. Example 2.2 demonstrates the use of inertia forces and illustrates the important feature of support excitation, or *base motion*, such as that a building structure would experience during an earthquake.

Example 2.2

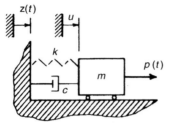

Figure 1 Mass–spring–dashpot SDOF system with a moving base.

Use the d'Alembert Force Method to determine the equation of motion of the mass m in Fig. 1. Assume that the damping forces in the system can be represented by a linear viscous dashpot, as shown, and assume that the support motion, $z(t)$, is known. When $u = z = 0$, the spring is unstretched.

SOLUTION   1. Draw a free-body diagram of the mass; include the inertia force along with the real forces (Fig. 2).

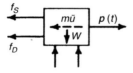

Figure 2 Free-body diagram, including the d'Alembert force.

2. Write the dynamic equilibrium equation,

$$\overset{+}{\rightarrow} \sum F_x^* = 0 \tag{1}$$

From the free-body diagram,

$$p(t) - f_S - f_D - m\ddot{u} = 0 \tag{2}$$

3. Relate forces to motion variables, and simplify the resulting equation.

$$m\ddot{u} + c(\dot{u} - \dot{z}) + k(u - z) = p(t) \tag{3}$$

Note that the damping force and spring force are related to the motion of the mass relative to the moving support. Equation 3 can be written with all known quantities on the right-hand side. Thus,

$$m\ddot{u} + c\dot{u} + ku = c\dot{z} + kz + p(t) \qquad \textbf{Ans.} \quad (4)$$

Equation 4 is the *equation of motion for the actual displacement of the mass relative to an inertial reference frame*, that is, for $u(t)$. This form of the base-motion equation of motion is useful in studies of the ride quality of vehicles or the motion of buildings due to earthquake excitation.

It is frequently desirable to formulate the base-motion equation of motion in terms of the displacement of the mass <u>relative to the moving base</u>, given by

$$w \equiv u - z \qquad (5)$$

since the forces applied to the mass are directly related to this relative displacement. By subtracting $m\ddot{z}$ from both sides of Eq. 3 and making use of Eq. 5, we obtain the following equation:

$$m\ddot{w} + c\dot{w} + kw = p(t) - m\ddot{z} \qquad \textbf{Ans.} \quad (6)$$

Thus, base motion has the effect of adding a *reversed inertia force* $p_{\text{eff}}(t) \equiv -m\ddot{z}$ to the other applied forces. Since relative motion may be more important than absolute motion, and since the base acceleration, $\ddot{z}$, is much easier to measure than the base velocity or base displacement, there are many applications where Eq. 6 is more useful than Eq. 4.

Notice that Eqs. 4 and 6 have the same form of left-hand side, which is the same as the left-hand side of Eq. 2.19.

Next, let us apply Newton's Laws for rigid bodies in plane motion, which are summarized in Eqs. 2.6–2.8, to determine mathematical models for SDOF systems that involve rigid bodies in plane motion.

Example 2.3 An airplane engine and the pylon that attaches it to the wing are modeled, for a particular study of lateral motion, as a rigid body attached to a rigid weightless beam, which is restrained elastically as shown in Fig. 1. Derive the equation of motion for small-angle motion. Neglect damping. The rotational spring shown exerts a restoring moment on the beam that is proportional to the angle the beam makes with the vertical. The engine has a mass moment of inertia I_G about an axis through its mass center G, which is located as shown.

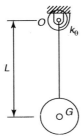

Figure 1 Spring–mass pendulum system.

SOLUTION   1. Draw a free-body diagram of the pendulum system in an arbitrary displaced position, and identify the displacement coordinate (Fig. 2).

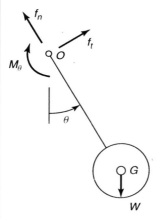

Figure 2 Free-body diagram of a spring–mass pendulum system.

2. Write the equation of motion. The force equation, Eq. 2.6, will not be needed for this problem since the pin forces f_n and f_t are not required. Since this is a fixed-axis rotation problem, Eq. 2.8b will be used for the required moment equation:

$$\circlearrowleft_+ \sum M_O = I_O \ddot{\theta} \qquad (+ \text{ sense in direction of } + \theta) \qquad (1)$$

From the free-body diagram in Fig. 2,

$$-M_\theta - WL \sin \theta = I_O \ddot{\theta} \qquad (2)$$

3. Write the force–displacement (i.e., moment–rotation) equation.

$$M_\theta = k_\theta \theta \qquad (3)$$

4. Combine and simplify the equations. From the parallel-axis theorem,

$$I_O = I_G + mL^2 \qquad (4)$$

From Eqs. 1, 2, and 3, the final equation of motion is obtained.

$$(I_G + mL^2)\ddot{\theta} + k_\theta \theta + WL \sin \theta = 0 \qquad (5)$$

Note that because of the $\sin \theta$ term, this is a <u>nonlinear</u> ordinary differential equation. However, since the problem statement requests the equation of motion for <u>small-angle motion</u>, and since $\sin \theta \approx \theta$ when θ is small, for small-angle motion Eq. 5 becomes the linear differential equation

$$(I_G + mL^2)\ddot{\theta} + (k_\theta + WL)\theta = 0 \qquad \textbf{Ans.} \quad (6)$$

2.3 Application of Newton's Laws to Lumped-Parameter Models

As a final example of using Newton's Laws directly (or using inertia forces in a dynamic equilibrium equation) to derive the equation of motion of an SDOF system, consider Example 2.4.

Example 2.4 A uniform disk of radius R is attached at the midpoint of a circular shaft of radius r and length $2L$, as shown in Fig. 1. Neglecting the thickness of the disk and the weight of the shaft, and neglecting damping, derive the equation of motion for torsional oscillation of the disk. The shaft is made of material with a shear modulus of elasticity G.

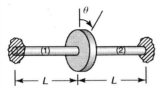

Figure 1 Disk–shaft SDOF system.

SOLUTION 1. Draw a free-body diagram of the disk in an arbitrary displaced position and identify the displacement coordinate (Fig. 2).

Figure 2 Free-body diagram of the disk.

2. Write the equation of motion. From Eq. 2.8b,

$$\circlearrowleft + \sum M_O = I_O \ddot{\theta} \tag{1}$$

From the free-body diagram,

$$-M_1 - M_2 = I_O \ddot{\theta} \tag{2}$$

and from Fig. 2.1c,

$$I_O = \tfrac{1}{2} m R^2 \tag{3}$$

3. Write the force–displacement equation (i.e., the torque–twist equation) for the two segments of the shaft. From mechanics of materials (Eq. 2.15a), the relationship between the torque (or twisting moment) M_i that each segment of the shaft exerts on the disk, and the angle of rotation θ at the disk end of the shaft is

$$M_i = \frac{G_i J_i}{L_i} \theta \tag{4}$$

where J_i is the polar moment inertia of the shaft's circular cross section, given by

$$J_i = \frac{\pi r_i^4}{2} \tag{5}$$

Since the shaft segments are identical,

$$M_1 = M_2 = \frac{GJ}{L}\theta \tag{6}$$

4. Combine and simplify.

$$I_o \ddot{\theta} + 2\frac{GJ}{L}\theta = 0 \tag{7}$$

or

$$\frac{mR^2}{2}\ddot{\theta} + 2\frac{\pi r^4}{2}\frac{G}{L}\theta = 0 \qquad \text{Ans.} \tag{8}$$

Notice that this has the form

$$m_\theta \ddot{\theta} + k_\theta \theta = 0$$

which is similar to the form obtained in previous examples.

2.4 APPLICATION OF THE PRINCIPLE OF VIRTUAL DISPLACEMENTS TO LUMPED-PARAMETER MODELS

In Section 2.3 we briefly reviewed elementary dynamics and saw that Newton's Laws can be used directly or with the introduction of inertia forces to obtain the *equation of motion of an SDOF system*. In this section you will be introduced to the *Principle of Virtual Displacements*. Although it can be applied to problems such as those in Section 2.3, it is particularly useful when applied to connected rigid bodies or when used to approximate continuous bodies by finite-DOF models (Section 2.5). More powerful methods of analytical mechanics, Hamilton's Principle and Lagrange's Equations, are introduced in Section 8.2.

Several important definitions are needed at the outset; these definitions are illustrated by examples that follow.

- A *displacement coordinate* is a quantity used in specifying a change of configuration of a system.
- A *constraint* is a kinematical restriction on the possible configurations that a system may assume.
- A *virtual displacement* is an <u>arbitrary</u>, <u>infinitesimal</u>, <u>imaginary</u> change of configuration of a system, consistent with all displacement constraints on the system. Virtual displacements are not true displacements but infinitesimally small variations in the system coordinates that are imagined to occur instantaneously. Hence, virtual displacements are not functions of time.

2.4 Application of the Principle of Virtual Displacements to Lumped-Parameter Models

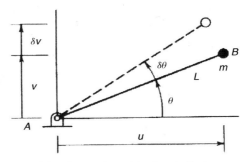

Figure 2.5 Constraint with virtual displacement.

To illustrate the meaning of the term *virtual displacement*, Fig. 2.5 shows a rigid rod of length L that can rotate about a fixed pin at end A and has a point mass m attached at end B. Since Cartesian coordinates $u(t)$ and $v(t)$ can be used to locate the position of the point mass m, they can be used as its displacement coordinates. However, mass m is constrained to move in a circle by the *equation of constraint*

$$u^2 + v^2 = L^2 \tag{2.20}$$

Because of this constraint, displacements $u(t)$ and $v(t)$ are not independent. In fact, they can both be related to the single displacement coordinate $\theta(t)$ by

$$u(t) = L\cos\theta(t), \qquad v(t) = L\sin\theta(t) \tag{2.21}$$

A small change in configuration is also shown in Fig. 2.5. The (infinitesimally small) angle $\delta\theta$ represents a virtual angular displacement of the system. The corresponding virtual displacements δu and δv can be obtained graphically or by using Eqs. 2.21. For example,

$$\begin{aligned} v + \delta v &= L\sin(\theta + \delta\theta) \\ &= L(\sin\theta \cos\delta\theta + \sin\delta\theta \cos\theta) \end{aligned} \tag{2.22}$$

Since $\delta\theta$ is infinitesimal, $\cos\delta\theta \approx 1$ and $\sin\delta\theta \approx \delta\theta$. Thus,

$$v + \delta v = L\sin\theta + (L\cos\theta)\,\delta\theta$$

or

$$\delta v = (L\cos\theta)\,\delta\theta \tag{2.23a}$$

Similarly,

$$\delta u = -(L\sin\theta)\,\delta\theta \tag{2.23b}$$

Note that δu and δv behave like differentials. From the form of Eqs. 2.23 it would appear that the virtual displacements δu and δv are functions of time t, since the displacement coordinate θ is a function of time. However, that is not the case; the angle θ in Eqs. 2.23 simply identifies the reference position from which the virtual angular displacement $\delta\theta$ is measured. It does not matter what time(s) the rod AB might actually occupy the position at the given angle θ. That is, $\delta\theta \equiv \delta[\theta(t)]$ is not a function of time.

We continue with the definitions:

- A *set of generalized coordinates* is a set of linearly independent displacement coordinates that (1) are consistent with the constraints, and (2) are sufficient in number to describe an arbitrary configuration of the system. For example, θ is a generalized coordinate for the SDOF system in Fig. 2.5. The symbols $(q_1, q_2, \ldots, q_N)$ are frequently employed as the labels for the generalized coordinates of an N-DOF system.
- The *virtual work* δW is the work of the forces acting on a system as the system undergoes a virtual displacement. The virtual work can be written as the following sum:

$$\delta W = \sum_{i=1}^{N} Q_i \, \delta q_i \tag{2.24}$$

That is, corresponding to each generalized coordinate q_i there is a contribution δW_i to the total virtual work δW.

- The *generalized force* Q_i is the quantity that multiplies the virtual displacement δq_i in forming the virtual work term δW_i. That is, Q_i is the virtual work done when $\delta q_i = 1$ and $\delta q_j = 0$ for $j \neq i$. The concept of a generalized force is very important in structural dynamics, so you should become very familiar with this way of identifying generalized forces by noting the virtual work they do.

Example 2.5 Determine the virtual work done by the distributed force acting on the rigid beam shown in Fig. 1. Assume that the force remains perpendicular to the beam.

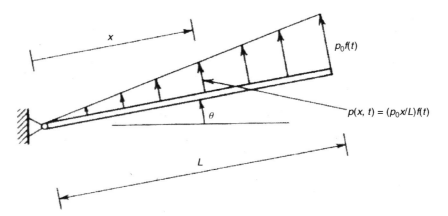

Figure 1 Rigid beam with a distributed load.

SOLUTION From the basic definition of work (force times distance moved),

$$\delta W = \int_0^L p(x, t) \, \delta v(x, t) \, dx \tag{1}$$

where $\delta v(x, t)$ is the virtual displacement of the bar due to a virtual angular displacement $\delta\theta$. Since virtual displacements are infinitesimal, δv can be approximated by

$$\delta v(x, t) = x\, \delta\theta \tag{2}$$

Note that $\delta v(x, t) \equiv \delta[v(x, t)]$ is a function of x but not of time t. Then

$$\delta W = \int_0^L \left[\frac{p_0 x}{L} f(t)\right] (x\, \delta\theta)\, dx \tag{3}$$

or

$$\delta W = \frac{p_0 f(t) L^2}{3} \delta\theta = \frac{p_0 L f}{2} \frac{2L\, \delta\theta}{3} \qquad \textbf{Ans.} \tag{4}$$

Two things should be noted about this answer: (1) the generalized force Q_θ is given by $Q_\theta = p_0 f L^2/3$, and (2) the result is the same as would be obtained by placing the statically equivalent load $p_0 f L/2$ at its centroid ($2L/3$).

Again, continuing with the definitions:

- The *Principle of Virtual Displacements* as applied to dynamics problems can now be stated as follows: *For any arbitrary virtual displacement of a system, the combined virtual work of real forces and inertia forces must vanish.* That is,

$$\boxed{\delta W^* \equiv \delta W_{\substack{\text{real}\\ \text{forces}}} + \delta W_{\substack{\text{inertia}\\ \text{forces}}} = 0} \tag{2.25}$$

Consider the following examples. Example 2.6 could easily be treated by using Newton's Laws, as in Section 2.3. However, Example 2.7 shows you the advantage of the virtual work formulation.

Example 2.6 Use the Principle of Virtual Displacements to derive the equation of motion of the idealized SDOF system shown in Fig. 1. Neglect gravity, and assume small-angle rotation of the beam.

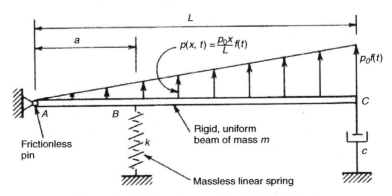

Figure 1 Uniform beam with a distributed load.

SOLUTION 1. Sketch the body in an arbitrary displaced configuration and also with an additional virtual displacement (Fig. 2), and write the necessary kinematic equations.

$$v(x, t) = x \tan \theta(t) \tag{1}$$

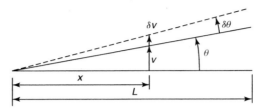

Figure 2 Deflection diagram of the beam. (*Note:* Displacement of the beam is exaggerated.)

For small θ,

$$v(x, t) = x\theta(t), \qquad \delta v(x, t) = x \, \delta\theta \tag{2}$$

2. On a sketch of the body show all of the forces that can do work, including inertia forces (Fig. 3).

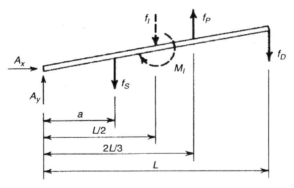

Figure 3 Force diagram for the beam.

3. Write the virtual work equation.

$$\delta W^* = 0 \tag{3}$$

From Eq. 2b and the figures above,

$$\delta W^* = -f_S(a \, \delta\theta) - f_D(L \, \delta\theta) + f_P \frac{2L \, \delta\theta}{3}$$

$$- f_I \frac{L \, \delta\theta}{2} - M_I(\delta\theta) = 0 \tag{4}$$

4. Relate forces to motion variables.

$$f_S = ka\theta, \qquad f_D = cL\dot\theta, \qquad f_P = \frac{p_0 L}{2} f(t)$$

$$f_I = \frac{mL}{2}\ddot\theta, \qquad M_I = \frac{mL^2}{12}\ddot\theta \tag{5}$$

2.4 Application of the Principle of Virtual Displacements to Lumped-Parameter Models

5. Combine Eqs. 4 and 5 and simplify.

$$\left[\frac{mL^2}{3}\ddot{\theta} + (cL^2)\dot{\theta} + (ka^2)\theta - \frac{p_0 L^2}{3}f(t)\right]\delta\theta = 0 \qquad (6)$$

Since $\delta\theta \neq 0$, the expression in brackets must be equal to zero, giving the following equation of motion:

$$\frac{mL^2}{3}\ddot{\theta} + (cL^2)\dot{\theta} + (ka^2)\theta = \frac{p_0 L^2}{3}f(t) \qquad \textbf{Ans.} \quad (7)$$

Note that the equation of motion above has the same form as that of Eq. 2.19, the equation of motion of the prototype damped SDOF system.

Notice in Example 2.6 that the solution procedure for applying the Principle of Virtual Displacements is slightly different from the solution procedure for applying Newton's Laws. Since the key equation is the virtual work equation, $\delta W^* = 0$, the needed quantities are (1) *forces that do work*, including inertia forces, and (2) *kinematics of displacement*, including virtual displacements. This is the reason that a sketch showing forces that do work and a sketch showing deformation replace the free-body diagram, which is so essential when Newton's Laws are being used directly.

Of course, in Example 2.6 it would have been easier just to draw a free-body diagram and use Newton's Laws. When there are connected rigid bodies, however, several free-body diagrams and many equations would be required in a solution using Newton's Laws. The virtual work approach greatly simplifies the solution, as you will see in Example 2.7.

Example 2.7 An instrument package of mass M is attached to the wall of a moving vehicle by two identical thin rigid beams of mass m, as shown in Fig. 1. Use the Principle of Virtual Displacements to derive the equation of motion for vertical motion of the instrument package. Neglect gravity and damping forces, and assume small-angle motions of the support beams.

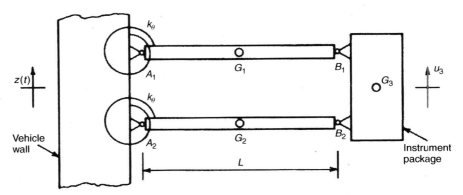

Figure 1 Rigid body supported by two rigid beams.

SOLUTION 1. Draw a deformation diagram; that is, sketch the system in an arbitrary displaced configuration and also with an additional virtual displacement (Fig. 2).

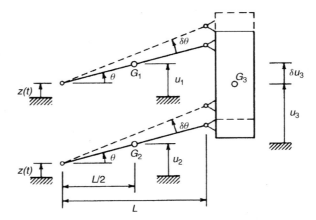

Figure 2 Deformation diagram.

2. Write the necessary kinematical equations for real displacements and for virtual displacements. For small θ and infinitesimally small $\delta\theta$,

$$u_1 = u_2 = z + \frac{L}{2}\theta, \qquad \delta u_1 = \delta u_2 = \frac{L}{2}\delta\theta \qquad (1)$$
$$u_3 = z + L\theta, \qquad \delta u_3 = L\,\delta\theta$$

[Notice that since $z(t)$ is a specified function of time, no virtual displacement δz appears above. Also notice that the motions of G_1, G_2, and G_3 can all be expressed in terms of the known displacement $z(t)$ and the single unknown displacement θ. Therefore, this is an SDOF system even though it involves motion of several rigid bodies.]

3. On a sketch of the displaced system, show all of the forces that do work, including all inertia forces (Fig. 3).

4. Write the virtual work equation, Eq. 2.25, and use Fig. 3 to write down all of the relevant spring–force terms and inertia–force terms in this equation.

$$\delta W^* = 0 \qquad (2)$$

$$-(2k_\theta \theta)\,\delta\theta - (m\ddot{u}_1)\,\delta u_1 - (I_{G1}\ddot{\theta})\,\delta\theta - (m\ddot{u}_2)\,\delta u_2 - (I_{G2}\ddot{\theta})\,\delta\theta - (M\ddot{u}_3)\,\delta u_3 = 0 \qquad (3)$$

5. Combine Eqs. 1 and 3 and simplify.

$$(2k_\theta \theta)\,\delta\theta + 2m[\ddot{z} + (L/2)\ddot{\theta}](L/2)\,\delta\theta + 2\frac{mL^2}{12}\ddot{\theta}\,\delta\theta + M(\ddot{z} + L\ddot{\theta})(L\,\delta\theta) = 0 \qquad (4)$$

or

$$\left[\left(M + \frac{2m}{3}\right)L^2\ddot{\theta} + 2k_\theta \theta + (m+M)L\ddot{z}\right]\delta\theta = 0 \qquad (5)$$

2.5 Application of the Principle of Virtual Displacements to Continuous Models: Assumed-Modes Method

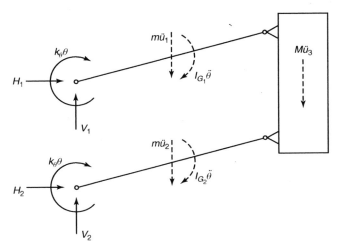

Figure 3 Force diagram.

Finally, since $\delta\theta \neq 0$, Eq. 5 requires that the expression in brackets vanish. This gives the final equation of motion of the system.

$$\left(M + \frac{2m}{3}\right) L^2 \ddot{\theta} + 2k_\theta \theta = -(m + M) L \ddot{z} \qquad \textbf{Ans.} \quad (6)$$

From Examples 2.6 and 2.7 you can see that applying the Principle of Virtual Displacements to an SDOF system will lead to a single virtual work equation of the form

$$[F(\ddot{q}, \dot{q}, q\,;\, t)]\,\delta q = 0 \qquad (2.26)$$

Since δq is an arbitrary infinitesimally small displacement that is not zero, the expression in brackets must be zero. This expression, then, is the *differential equation of motion* for the SDOF system. As you have seen in Examples 2.6 and 2.7, it has the prototype-SDOF form given in Eq. 2.19.

2.5 APPLICATION OF THE PRINCIPLE OF VIRTUAL DISPLACEMENTS TO CONTINUOUS MODELS: ASSUMED-MODES METHOD

In Section 2.4 you observed that using the Principle of Virtual Displacements greatly simplifies the task of deriving the equation of motion of connected rigid bodies. But the idea of *rigid body*, in itself, is an idealization of a system. For example, if the "rigid" beam in Example 2.6 were to be excited by a time-dependent forcing function $f(t)$, the beam would not actually remain rigid but would deform to a greater or lesser extent. That is, the deflection curve would assume a time-varying curved shape. Hence, to assume that it is a rigid body would generally not be a suitable analytical model. The cantilever beam of Fig. 1.3a would not be free to deform at all if we were to model it as a rigid body!

Fortunately, the Principle of Virtual Displacements can be extended to produce a generalized discrete-parameter model, or simply *generalized-parameter model*, of a continuous system in a manner that approximates the deformation of the system. The procedure is referred to as the *Assumed-Modes Method*. It is employed here to create an SDOF generalized-parameter model and is extended to MDOF systems in Chapter 8.

A few definitions are needed at the outset.

- A *continuous system* is a system whose deformation is described by one or more functions of one, two, or three spatial variables and time. For example, the deformation of the cantilever beam in Fig. 1.3a is specified in terms of the deflection curve $v(x, t)$ of the neutral axis.
- A *geometric boundary condition* is a specified kinematical constraint placed on displacement and/or slope on portions of the boundary of a body.
- A *virtual displacement* of a continuous system is an infinitesimal, imaginary change in the displacement function(s) consistent with all geometric boundary conditions.

Figure 2.6 illustrates the foregoing definitions. The deformation of the propped cantilever beam in this figure is given by the *deflection curve* $v(x, t)$. The geometric boundary conditions are the cantilevered (i.e., fixed) end at $x = 0$ and the prop that prevents vertical motion of the beam at $x = L$. These geometric boundary conditions are given by the following three equations:

$$v(0, t) = 0, \quad v'(0, t) \equiv \frac{\partial v(0, t)}{\partial t} = 0, \quad v(L, t) = 0 \quad \text{for all } t \quad (2.27)$$

The dashed curve shows a possible virtual displacement, $\delta v(x, t)$, of the beam. The only condition on $\delta v(x, t)$ is that it satisfy the same geometric boundary conditions as $v(x, t)$; that is,

$$\delta v(0, t) = \delta v'(0, t) = \delta v(L, t) = 0 \quad (2.28)$$

Note that $\delta v(x, t) \equiv \delta[v(x, t)]$ is <u>not a function of time</u>, as is $v(x, t)$. However, the notation $\delta v(x, t)$ means an arbitrary small change of configuration relative to the configuration of the beam at time t.

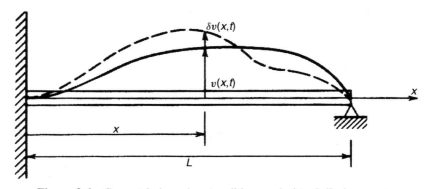

Figure 2.6 Geometric boundary conditions and virtual displacements.

2.5 Application of the Principle of Virtual Displacements to Continuous Models: Assumed-Modes Method

We continue with the definitions:

- An *admissible function* is a function that satisfies the geometric boundary conditions of the system under consideration and possesses derivatives of order at least equal to that appearing in the strain energy expression for the system (e.g., see Eqs. 2.33).
- An *assumed mode* (also called a *shape function*) $\psi(x)$ is an admissible function that is selected by the user for the purpose of approximating the deformation of a continuous system.

To create a generalized-parameter SDOF model of a continuous system, a single assumed mode is used. For example, the deflection curve of a beam may be approximated by

$$v(x,t) = \psi(x) q_v(t) \tag{2.29}$$

Any admissible function may be employed as the shape function $\psi(x)$, but a shape that can be expected to be similar to the shape of the deforming structure should be chosen. The time-dependent function $q_v(t)$ in Eq. 2.29 is called the *generalized displacement coordinate* for this SDOF model; the subscript v identifies it as a generalized coordinate related to the physical displacement $v(x,t)$. It will be determined as the solution to an ordinary differential equation, just as in the case of lumped-parameter models.

The Principle of Virtual Displacements, as given by Eq. 2.25, may now be employed to create a SDOF generalized-parameter model based on the assumed displacement of Eq. 2.29. For continuous systems, however, it is convenient to introduce *potential energy*, that is, the work done by conservative forces. Let

$$\delta \mathcal{W}_{\substack{\text{real} \\ \text{forces}}} = \delta \mathcal{W}_{\text{cons}} + \delta \mathcal{W}_{\text{nc}} \tag{2.30}$$

where $\delta \mathcal{W}_{\text{cons}}$ is the *virtual work of conservative forces* and $\delta \mathcal{W}_{\text{nc}}$ is the *virtual work of nonconservative forces*. From the definition of potential energy,

$$\delta \mathcal{W}_{\text{cons}} = -\delta \mathcal{V} \tag{2.31}$$

where $\delta \mathcal{V}$ is the change in potential energy as the conservative forces acting on the system move through a virtual change of configuration, for example, through $\delta v(x,t)$ for a beam. Then the Principle of Virtual Displacements, Eq. 2.25, becomes

$$\delta \mathcal{W}^* = \delta \mathcal{W}_{\text{nc}} - \delta \mathcal{V} + \delta \mathcal{W}_{\text{inertia}} = 0 \tag{2.32}$$

The *strain energy in a bar undergoing axial deformation* $u(x,t)$ is given by

$$\mathcal{V}_{\text{axial}} = \tfrac{1}{2} \int_0^L AE(u')^2 \, dx \tag{2.33a}$$

and the *strain energy in a beam undergoing transverse deflection* $v(x,t)$ is given by[2.1]

$$\mathcal{V}_{\text{bending}} = \tfrac{1}{2} \int_0^L EI(v'')^2 \, dx \tag{2.33b}$$

Consequently, for virtual displacements $\delta u(x, t)$ and $\delta v(x, t)$, the corresponding changes in the strain energy, $\delta \mathcal{V}$, are given by

$$\delta \mathcal{V}_{\text{axial}} = \int_0^L (AEu')\, \delta u'\, dx \qquad (2.34a)$$

and

$$\delta \mathcal{V}_{\text{bending}} = \int_0^L (EIv'')\, \delta v''\, dx \qquad (2.34b)$$

respectively.

Example 2.8 In Fig. 1, a time-dependent axial force $P(t)$ is applied to the end of a linearly elastic uniform rod. Choose a simple function for the shape function $\psi(x)$, and use the Assumed-Modes Method to derive the equation of motion of the resulting SDOF model of the rod.

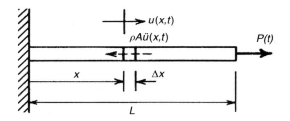

Figure 1 Uniform rod undergoing axial deformation.

SOLUTION   1. Choose a shape function $\psi(x)$ consistent with the geometric boundary condition. The only geometric boundary condition here is

$$u(0, t) = 0 \qquad (1)$$

Thus, the shape function $\psi(x)$ must satisfy

$$\psi(0) = 0 \qquad (2)$$

The simplest function satisfying this condition is $\psi(x) = x$. However, it is convenient to normalize $\psi(x)$ so that it has a value of 1 at some convenient place, say at $x = L$. Therefore, choose

$$\psi(x) = \frac{x}{L} \qquad \textbf{Ans.} \quad (3)$$

Thus, the axial displacement $u(x, t)$ has the assumed form

$$u(x, t) = \frac{x}{L} q_u(t) \qquad (4)$$

2.5 Application of the Principle of Virtual Displacements to Continuous Models: Assumed-Modes Method

where $q_u(t)$, the generalized displacement, is actually the time-dependent tip displacement.

2. Set up the virtual work equation, Eq. 2.32.

$$\delta W^* = \delta W_{nc} - \delta V + \delta W_{\text{inertia}} = 0 \quad (5)$$

The virtual work of nonconservative forces is just the virtual work done by the end force $P(t)$, or

$$\delta W_{nc} = P(t)\,\delta u(L,t) = P(t)\,\delta q_u \quad (6)$$

From Eq. 2.33a, the virtual change in strain energy is

$$\delta V_{\text{axial}} = \int_0^L (AEu')\,\delta u'\,dx$$

$$= \int_0^L AE\,\frac{q_u}{L}\,\frac{\delta q_u}{L}\,dx \quad (7)$$

$$= \frac{AE}{L}q_u\,\delta q_u$$

The inertia force per unit length $[-\rho A\ddot{u}(x,t)]$ is distributed along the rod. Hence, the virtual work of inertia forces is

$$\delta W_{\text{inertia}} = \int_0^L [-\rho A\ddot{u}(x,t)]\,\delta u(x,t)\,dx$$

$$= -\frac{\rho A\,\ddot{q}_u\,\delta q_u}{L^2}\int_0^L x^2\,dx \quad (8)$$

$$= -\frac{\rho AL}{3}\ddot{q}_u\,\delta q_u$$

3. Combine and simplify.

$$\left[P(t) - \frac{AE}{L}q_u - \frac{\rho AL}{3}\ddot{q}_u\right]\delta q_u = 0 \quad (9)$$

Equation 9 has the form of Eq. 2.26. The condition that δq_u be arbitrary leads to the following ordinary differential equation as the equation of motion for the assumed-mode SDOF model:

$$\frac{\rho AL}{3}\ddot{q}_u + \frac{AE}{L}q_u = P(t) \qquad \textbf{Ans.} \quad (10)$$

Note that this equation has the familiar form

$$m\ddot{u} + ku = p(t)$$

seen in previous examples involving lumped-parameter models.

It is very important that you remember that Eq. 10 is based on Eq. 4, which is only an SDOF approximation to the real time-dependent axial motion $u(x,t)$ that the bar would actually experience as a result of the applied force $P(t)$.

Example 2.8 shows how a particular choice for the shape function $\psi(x)$ leads to an ordinary differential equation representing an SDOF mathematical model of a continuous system. Next, a very important general example of the Assumed-Modes Method is presented. You should study this example carefully.

Example 2.9 For the system shown in Fig. 1, determine expressions for m, c, k, k_G, and $Q_v(t)$ in the following generalized-parameter equation of motion:

$$m\ddot{q}_v + c\dot{q}_v + (k - k_G)q_v = Q_v(t)$$

For the transverse displacement $v(x, t)$, use the general assumed-modes form

$$v(x, t) = \psi(x)q_v(t)$$

Assume that $N(x)$, the compressive axial force at cross section x, may vary along the length of the beam, but that it remains horizontal. (For example, if the beam were a vertical column, the axial force due to the weight of the column above a particular section would vary with the position of that section from the top of the column.)

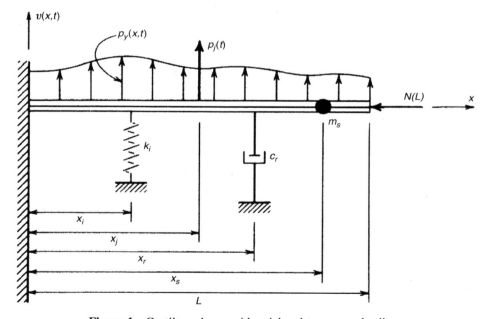

Figure 1 Cantilever beam with axial and transverse loading.

SOLUTION 1. Start with the virtual work equation, Eq. 2.32.

$$\delta W^* = \delta W_{\text{nc}} - \delta V + \delta W_{\text{inertia}} = 0 \tag{1}$$

2. Sketch the beam in its deflected position and with an additional virtual deflection (Fig. 2).

2.5 Application of the Principle of Virtual Displacements to Continuous Models: Assumed-Modes Method

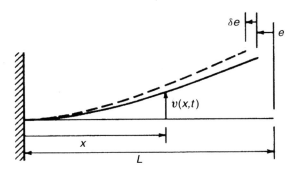

Figure 2 Axial and transverse displacement of a cantilever beam.

3. From the virtual work of inertia forces, determine the *generalized mass m*.

$$\delta W_{\text{inertia}} = \int_0^L (-\rho A \ddot{v}) \, \delta v(x, t) \, dx - m_s \ddot{v}(x_s, t) \, \delta v(x_s, t) \tag{2}$$

Substituting $\ddot{v}$ and δv into the equation above, we obtain

$$\delta W_{\text{inertia}} = -\left[\int_0^L \rho A \psi^2(x) \, dx + m_s \psi^2(x_s) \right] \ddot{q}_v \, \delta q_v \tag{3}$$

In anticipation of the final equation to be derived, let

$$\delta W_{\text{inertia}} = -m \ddot{q}_v \, \delta q_v \tag{4}$$

Then the generalized mass associated with $\psi(x)$ is

$$m = \int_0^L \rho A \psi^2(x) \, dx + m_s \psi^2(x_s) \quad \textbf{Ans.} \tag{5}$$

4. From the (nonconservative) virtual work done by the viscous dashpot, determine the *generalized viscous-damping coefficient c*.

$$\delta W_{\text{damping}} = -c_r \dot{v}(x_r, t) \, \delta v(x_r, t) \tag{6}$$

Let

$$\delta W_{\text{damping}} = -c \dot{q}_v \, \delta q_v \tag{7}$$

Then the generalized viscous-damping coefficient is given by

$$c = c_r \psi^2(x_r) \quad \textbf{Ans.} \tag{8}$$

5. From the strain energy of the linear spring and the strain energy of the elastic beam, determine the *generalized stiffness coefficient k*.
From Eq. 2.13, the equation for strain energy of a linear spring,

$$\delta V_{\text{spring}} = k_i e_i \, \delta e_i = k_i v(x_i, t) \, \delta v(x_i, t) \tag{9}$$

So
$$\delta V_{\text{spring}} = k_i \psi^2(x_i) q_v \, \delta q_v \tag{10}$$

From Eq. 2.34b, the equation for the variation in strain energy stored in a beam,

$$\delta V_{\text{bending}} = \int_0^L (EIv'') \, \delta v'' \, dx \tag{11}$$

Thus,

$$\delta V_{\text{bending}} = \left[\int_0^L EI(\psi'')^2 \, dx \right] q_v \, \delta q_v \tag{12}$$

Let
$$\delta V_{\text{bending}} + \delta V_{\text{spring}} = k q_v \, \delta q_v \tag{13}$$

Then the *generalized stiffness coefficient* k is given by

$$k = \int_0^L EI(\psi'')^2 \, dx + k_i \psi^2(x_i) \qquad \textbf{Ans.} \tag{14}$$

6. From the virtual work done by the compressive axial force, $N(x)$, determine the *generalized geometric stiffness coefficient* k_G.

$$\delta W_N = k_G q_v \, \delta q_v \tag{15}$$

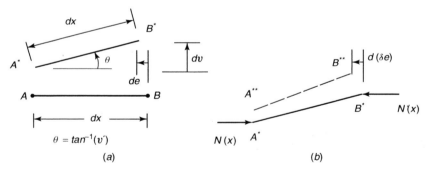

Figure 3 Axial and transverse displacement of a beam element dx: (a) actual displacements; (b) virtual displacements added. Left ends A, A^*, and A^{**} are shown in vertical alignment to facilitate visualization of the shortening effects.

To determine the virtual work of the axial force $N(x)$ due to the shortening of the beam that accompanies transverse virtual deflection $\delta v(x, t)$, consider the element dx shown in Fig. 3. Figure 3a shows the horizontal shortening $de(x)$ that occurs due to the transverse displacement $v(x, t)$. Figure 3b shows that the axial force $N(x)$ acting on the element of length dx will do virtual work

$$d(\delta W_N) = N(x) \, d(\delta e) \tag{16}$$

which can be summed up over the length of the beam to obtain δW_N.

2.5 Application of the Principle of Virtual Displacements to Continuous Models: Assumed-Modes Method

It is permissible to exchange the operations $d(\cdot)$ and $\delta(\cdot)$. That is,

$$d(\delta e) = \delta(de)$$

so we need an expression for de. Therefore, assume that the element AB of length dx remains dx in length but rotates to the position A^*B^* due to the transverse deflection, $v(x,t)$. From Fig. 3a,

$$\overline{AB} = \overline{A^*B^*} \cos(\tan^{-1} v') + de \tag{17}$$

For $v' \ll 1$,

$$\cos(\tan^{-1} v') \approx \cos v' \approx 1 - \tfrac{1}{2}(v')^2 \tag{18}$$

Therefore, from Eqs. 17 and 18,

$$dx = dx[1 - \tfrac{1}{2}(v')^2] + de$$

or

$$de = \tfrac{1}{2}(v')^2 \, dx \tag{19}$$

Hence,

$$d(\delta e) = \delta(de) = v' \, \delta v' \, dx \tag{20}$$

so the total virtual work done by the axial force is given by

$$\delta W_N = \int_0^L N(x) v' \, \delta v' \, dx \tag{21}$$

Substituting the assumed mode into this equation, and referring to Eq. 15, we get the following expression for the *generalized geometric stiffness coefficient*:

$$k_G = \int_0^L N(x)[\psi'(x)]^2 \, dx \qquad \textbf{Ans.} \tag{22}$$

7. Finally, from the virtual work done by the distributed transverse force $p_y(x,t)$ and the concentrated transverse force $P_j(t)$, determine the *generalized external force* $Q_v(t)$.

$$\delta W_P = \int_0^L p_y(x,t) \, \delta v(x,t) \, dx + P_j \, \delta v(x_j, t) \tag{23}$$

Substituting δv into the equation above, we get

$$\delta W_P = Q_v(t) \, \delta q_v \tag{24}$$

where the generalized external force is given by

$$Q_v(t) = \int_0^L p_y(x,t) \psi(x) \, dx + P_j \psi(x_j) \qquad \textbf{Ans.} \tag{25}$$

8. Combine the expressions above for $\delta \mathcal{V}$ and $\delta \mathcal{W}$ with Eq. 1, and simplify.

$$[-c\dot{q}_v + Q_v(t) + k_G q_v - k q_v - m \ddot{q}_v] \, \delta q_v = 0 \tag{26}$$

Since $\delta q_v \neq 0$, we can write the equation of motion of this SDOF assumed-mode model of the beam as

$$m\ddot{q}_v + c\dot{q}_v + (k - k_G)q_v = Q_v(t) \qquad \textbf{Ans. (27)}$$

as required, where the coefficients and generalized force are as defined above.

Notice that with the exception of the k_G term, the differential equation determined above by the Assumed-Modes Method is identical in form to Eq. 2.19. The latter equation was derived for a simple lumped-parameter spring–mass–dashpot system. The Assumed-Modes Method leads to a *generalized-parameter model* whose displacement function is the *generalized (displacement) coordinate*, $q(t)$. Correspondingly, m is called the generalized mass coefficient, or simply *generalized mass*, c is the *generalized viscous damping*, k is the *generalized stiffness*, k_G is the *generalized geometric stiffness*, and $Q_v(t)$ is the *generalized force*. In more general form, these would be given by the following expressions:

$$m = \int_0^L \rho A \psi^2(x)\, dx + \sum_s m_s \psi^2(x_s) \qquad (2.35)$$

$$c = \int_0^L c(x) \psi^2(x)\, dx + \sum_r c_r \psi^2(x_r) \qquad (2.36)$$

$$k = \int_0^L EI(\psi'')^2\, dx + \int_0^L k(x)\psi^2(x)\, dx + \sum_i k_i \psi^2(x_i) \qquad (2.37)$$

$$k_G = \int_0^L N(x)[\psi'(x)]^2\, dx \qquad (2.38)$$

$$Q_v(t) = \int_0^L p_y(x,t)\psi(x)\, dx + \sum_j P_j \psi(x_j) \qquad (2.39)$$

where the summation signs $\sum$ indicates that discrete elements or forces may be present at several locations along the beam, where distributed damping $c(x)$ and distributed elastic foundation $k(x)$ are permitted, and where the internal (compressive) axial load $N(x)$ may vary with position. The same symbols m, c, and k have been used here for generalized parameters as were introduced earlier for lumped-parameter models, since the symbols serve the same function in the final SDOF differential equation of motion.

REFERENCES

[2.1] R. R. Craig, Jr., *Mechanics of Materials*, 2nd ed., Wiley, New York, 2000.

[2.2] C. W. Bert, "Material Damping: An Introductory Review of Mathematical Models, Measures and Experimental Techniques," *Journal of Sound and Vibration*, Vol. 29, 1973, pp. 129–153.

[2.3] A. D. Nashif, D. I. G. Jones, and J. P. Henderson, *Vibration Damping*, Wiley, New York, 1985.

[2.4] C. T. Sun and Y. Lu, *Vibration Damping of Structural Elements*, Prentice Hall, Englewood Cliffs, NJ, 1995.

PROBLEMS

Problem Set 2.3

> Use *Newton's Laws* to determine the equations of motion of the SDOF systems in *Problems 2.1 through 2.6*. Show all necessary free-body diagrams and deformation diagrams.

2.1 The building frame structure in Fig. P2.1 consists of a rigid horizontal member BC of mass m supported by two vertical columns each of whose bending stiffness is EI. Neglect the effect of the weight of BC on the stiffness of the vertical members. (**a**) Determine the equation of motion of mass m. (**b**) How would the equation of motion differ if the left-hand column had a stiffness $2EI$ with no other changes to the structure?

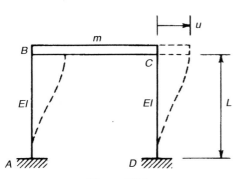

Figure P2.1

2.2 A heavy disk of radius R and mass m is attached at the $\frac{1}{3}$-point of a uniform shaft of length L, as shown in Fig. P2.2. The shaft is clamped to a rigid base at A and at C. The radius of the shaft is r, and it is made of material whose shear modulus of elasticity is G. Letting the rotation angle θ be the displacement coordinate, determine the equation of motion of the disk. Neglect the thickness of the disk.

2.3 The rigid beam BC in Fig. P2.3 is hinged at end C and is excited by the force in spring AB, where the motion of A is specified to be $z(t)$. (**a**) Determine the equation of motion of the beam in terms of the vertical motion $v(t)$ at end B. Assume small motions. (**b**) Write the equation of motion in terms of the force f_1 in spring AB. [*Hint:* You can write displacement $v(t)$ in terms of $z(t)$ and the spring force f_1, and then substitute this into the equation of motion determined in part (a).]

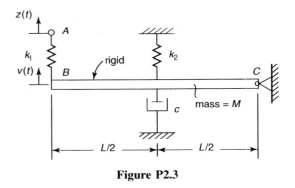

Figure P2.3

2.4 An automobile is modeled as a lumped mass m, with input from the road via a suspension system that consists of a linear spring and a linear viscous dashpot, as shown in Fig. P2.4. Assume that the automobile travels

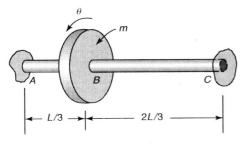

Figure P2.2

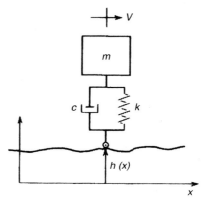

Figure P2.4

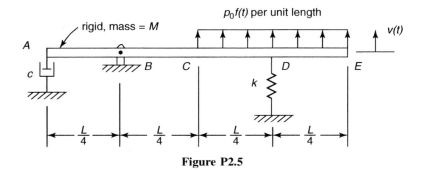

Figure P2.5

at constant speed V over a road whose roughness is a known function $h(x)$ of position along the roadway. (a) Determine the equation of motion for vertical motion of mass m. You may leave $h(x)$ and $dh(x)/dx$ in your answer, since it would be possible to eliminate the position coordinate x by letting time t_0 be the time that the car passes the origin $x = 0$. (b) Discuss some of the possible limitations of this model.

2.5 For the system shown in Fig. P2.5, determine the equation of motion in the form

$$m_v \ddot{v} + c_v \dot{v} + k_v v = p_v(t)$$

where $v(t)$ is the vertical displacement of end E. Assume small rotation of the thin, rigid, uniform beam AE, whose total mass is M.

2.6 The inertia properties of a control tab of an airplane elevator may be determined by weighing the control tab, determining its center of gravity, and performing a free vibration test from which the moment of inertia I_A can be determined. Figure P2.6 shows the setup for the vibration test. Using Newton's Laws, determine the equation of motion for small rotations of the control tab about the hinge A.

Problem Set 2.4

Use the *Principle of Virtual Displacements* to determine the equations of motion of the SDOF systems in *Problems 2.7 through 2.12*. Show all necessary deformation diagrams.

2.7 Derive the equation of motion for the system in Fig. P2.3. Let $v(t)$, the vertical displacement at B, be the displacement coordinate.

2.8 Derive the equation of motion of the system in Fig. P2.5. Let $v(t)$, the vertical displacement at end E, be the displacement coordinate.

2.9 Derive the equation of motion of the system shown in Fig. P2.9. Let the angular displacement of the rigid beam, θ (counterclockwise positive), be the displacement coordinate. Assume small-angle rotation. The triangularly distributed transverse load varies in amplitude with time, but compressive force N remains constant in magnitude and remains horizontal.

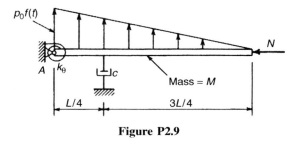

Figure P2.9

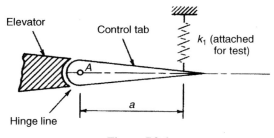

Figure P2.6

2.10 An airplane landing gear system is modeled as a lumped mass m attached to an airplane (mass $M \gg m$)

by a spring and a dashpot, as illustrated in Fig. P2.10. The tire forces transmitted to the mass m are modeled as resulting from the motion of the bottom of the spring k_2. Assume that the airplane travels at a constant speed V over a sinusoidally rough runway. Determine the equation of motion of the landing gear mass m. Since $M \gg m$, assume horizontal motion of the heavy airplane mass.

2.12 A uniform disk m_1 is connected to a rigid, uniform beam AD of mass m_2 by a rigid, massless rod CE, as shown in Fig. P2.12. The disk rolls without slipping on the horizontal surface, and the beam rotates about a frictionless pin at A. Determine the equation of motion for this system. Use the displacement u of point E as the displacement coordinate. When $u = 0$, both springs are taut but are unstretched.

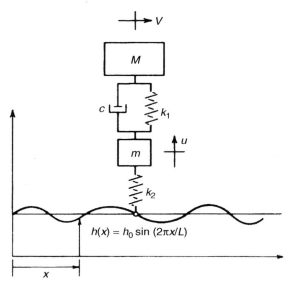

Figure P2.10

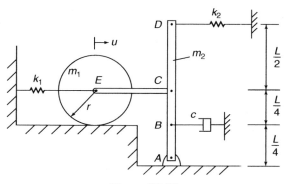

Figure P2.12

Problem Set 2.5

Use the *Assumed-Modes Method* to determine the equations of motion of the continuous systems in Problems. 2.13 through 2.20.

2.11 Inertia properties of masses are frequently measured by performing a pendulum test. In Fig. P2.11, a uniform bar of weight W is suspended by two identical weightless, inextensible, vertical wires of length L. The bar performs small rotational oscillations about the vertical axis. Determine the equation of motion. Let the angle of rotation of the bar about the vertical axis be the displacement coordinate.

2.13 A uniform bar of mass density ρ per unit volume and length L has a tip mass M, as shown in Fig. P2.13. With $\psi(x) = x/L$ as the shape function, derive the SDOF equation of motion for free axial vibration of this system. $AE = $ constant. Express your answer in terms of L, A, E, ρ, and M.

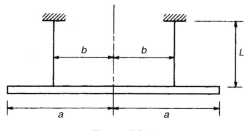

Figure P2.11

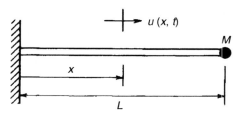

Figure P2.13

2.14 Repeat Problem 2.13 using $\psi(x) = \sin(\pi x/2L)$ as the shape function.

2.15 The tapered flat bar in Fig. P2.15 has a uniform thickness h and a mass density ρ per unit volume. With $\psi(x) = \sin(\pi x/2L)$ as the shape function, derive the SDOF equation of motion for free axial vibration of the rod. Express your answer in terms of a, L, E, and ρ.

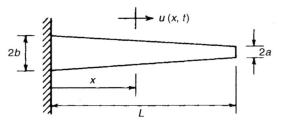

Figure P2.15

2.16 The tapered rod in Fig. P2.16 has a circular cross section and has a mass density ρ per unit volume. That is, the rod has the shape of a conical frustum. With $\psi(x) = 1 - x/L$ as the shape function, derive the SDOF equation of motion for free axial vibration of the rod. Express your answer in terms of a, L, E, and ρ.

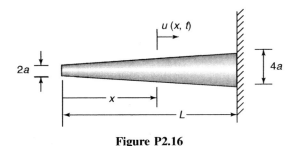

Figure P2.16

2.17 A uniform circular rod of diameter d and length L is attached to rigid supports at both ends, as shown in Fig. P2.17. The rod has a mass density ρ per unit volume, and its modulus of elasticity is E. With $\psi(x) = (x/L)(1 - x/L)$ as the shape function, derive the SDOF equation of motion for free axial vibration of the rod.

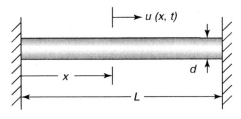

Figure P2.17

2.18 A tapered flat bar with thickness t and length L is attached to rigid supports at both ends, as shown in Fig. P2.18. The bar has a mass density ρ per unit volume, and its modulus of elasticity is E. Determine appropriate shape functions to be used in representing the axial motion of this bar by SDOF models: (**a**) if $\psi(x)$ is a simple polynomial function; and (**b**) if $\psi(x)$ is sinusoidal.

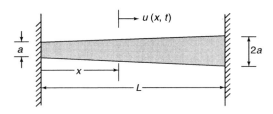

Figure P2.18

2.19 A uniform cantilever beam has a constant horizontal tip force N and a triangularly distributed time-varying transverse load $p(x, t) = (x/L)[p_0 f(t)]$, as shown in Fig. P2.19. The beam has a mass density ρ per unit volume, and its modulus of elasticity is E. Using a simple polynomial function for the shape function $\psi(x)$, derive the corresponding assumed-modes SDOF equation of motion for transverse vibration of the beam.

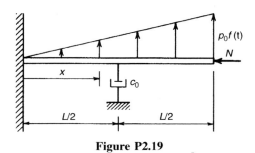

Figure P2.19

2.20 As illustrated in Fig. P2.20a, an observation tower in an amusement park can be modeled as a rigid uniform cylinder of mass M_C and radius R atop a slender uniform column. The total mass of the column is M_T, and its flexural stiffness is EI. Assume that $H \ll R$ and $H \ll L$; that is, the cylinder can be considered to be a "thin disk" whose center of gravity is at the top of the column. However, translational and rotational inertia effects of this disk must be included in your expression for kinetic energy. Derive an equation of motion for lateral vibration of an SDOF model of the system. For the shape function $\psi(x)$ use the static deflection curve of a uniform beam with concentrated transverse tip force, as illustrated in Fig. P2.20b. Include the geometric stiffness contributions from both the tip mass M_C and the distributed column mass M_T.

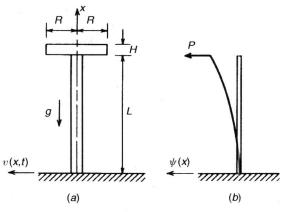

Figure P2.20

3

Free Vibration of SDOF Systems

In Chapter 2 we presented several techniques for deriving the equation of motion of an SDOF system. You observed that in all cases the equation of motion for a linear SDOF system was a *second-order ordinary differential equation* of the form

$$m\ddot{u} + c\dot{u} + ku = p(t) \tag{3.1}$$

where m, u, and so on, might refer to generalized quantities. Equation 3.1 is the equation of motion for the system of Fig. 3.1, which can be considered to be the *prototype SDOF system*. In Section 1.3 we gave you an introduction to vibration by discussing the undamped version of this prototype SDOF system. We now begin our study of the *response* of such an SDOF system to *excitation* $p(t)$ and to *initial conditions*, the velocity and displacement at time $t = 0$, given by

$$u(0) = u_0, \qquad \dot{u}(0) = v_0 \tag{3.2}$$

where u_0 and v_0 are the specified *initial displacement* and *initial velocity*, respectively.

It is convenient to divide Eq. 3.1 by m and to rewrite the *prototype SDOF equation of motion* in the form

$$\boxed{\ddot{u} + 2\zeta\omega_n\dot{u} + \omega_n^2 u = \omega_n^2 \frac{p(t)}{k}} \tag{3.3}$$

where ω_n is defined by

$$\boxed{\omega_n^2 = \frac{k}{m}} \tag{3.4a}$$

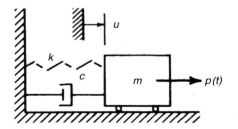

Figure 3.1 Prototype SDOF system.

and ζ is defined by

$$\zeta = \frac{c}{c_{cr}} \qquad (3.4b)$$

where

$$c_{cr} = 2m\omega_n = 2\sqrt{km} \qquad (3.4c)$$

Equations 3.4 define the <u>two key system parameters</u> that characterize the vibration of linear SDOF systems. The parameter ω_n is called the *undamped circular natural frequency*; its units are radians per second (rad/s). The parameter ζ is a dimensionless quantity called the *viscous damping factor*, which is defined in terms of c_{cr}, the *critical damping coefficient*.

Equation 3.3 is a linear ordinary differential equation with constant coefficients. Its solution, the *total response*, consists of the linear sum of two distinct parts, a *forced motion* $u_p(t)$, which is related directly to $p(t)$, and a *natural motion* $u_c(t)$, which makes it possible to satisfy arbitrary initial conditions. Thus,

$$u(t) = u_p(t) + u_c(t) \qquad (3.5)$$

In mathematical terminology, the general solution of the differential equation consists of a *particular solution* $u_p(t)$ and a *complementary solution* $u_c(t)$.

In this chapter we consider only *free vibration*, that is, the natural motion $u_c(t)$ that is the solution of Eq. 3.3 when $p(t) \equiv 0$. Then the equation of motion for free vibration of a viscous-damped SDOF system is

$$\ddot{u} + 2\zeta\omega_n\dot{u} + \omega_n^2 u = 0 \qquad (3.6)$$

The general technique for solving Eq. 3.6 is to assume a solution of the form[1]

$$u(t) = \overline{C} e^{\bar{s}t} \qquad (3.7)$$

When Eq. 3.7 is substituted into Eq. 3.6, we obtain

$$(\bar{s}^2 + 2\zeta\omega_n\bar{s} + \omega_n^2)\overline{C}e^{\bar{s}t} = 0 \qquad (3.8)$$

For Eq. 3.8 to be valid for all values of t, we must set

$$\bar{s}^2 + 2\zeta\omega_n\bar{s} + \omega_n^2 = 0 \qquad (3.9)$$

Equation 3.9 is called the *characteristic equation*.

In Section 1.3 we have already considered a particular case of *undamped free vibration*, the solution of Eq. 3.6 when $\zeta = 0$, $u(0) = u_0$, and $\dot{u}(0) = 0$. Here we expand on that by considering the response of an undamped SDOF system when $\dot{u}(0) = v_0 \neq 0$. Then, in Section 3.2 we consider the solution when $\zeta \neq 0$, that is, when viscous damping is present.

---

[1] As will soon be seen, although $u(t)$ is a real quantity, it is possible for C and s to be complex numbers. Here, complex numbers will be designated by an overbar (¯); later in the book this notation will be dropped. In Appendix B we summarize the basics of the algebra of complex numbers used in this book.

58 Free Vibration of SDOF Systems

Upon completion of this chapter you should be able to:

- Determine, for an undamped SDOF system, the following: the undamped natural frequency and period and the motion resulting from specified initial conditions.
- Determine, for a viscously damped SDOF system, the following: the damping factor, the damped natural frequency, the time constant, and the motion resulting from specified initial conditions.
- List three classes of stability of motion and describe the type(s) of motion experienced by SDOF systems in each class.
- Describe how to determine experimentally, for a simple lightly damped SDOF system, approximate values of the natural frequency and the damping factor.

3.1 FREE VIBRATION OF UNDAMPED SDOF SYSTEMS

The equation of motion for an undamped SDOF system is

$$\ddot{u} + \omega_n^2 u = 0 \tag{3.10}$$

and the corresponding *characteristic equation* is

$$\bar{s}^2 + \omega_n^2 = 0 \tag{3.11}$$

The roots of Eq. 3.11 are[2]

$$\bar{s}_{1,2} = \pm i\omega_n, \qquad \text{where } i = \sqrt{-1} \tag{3.12}$$

When these roots are substituted into Eq. 3.7, we get the *general solution*

$$u = \overline{C}_1 e^{i\omega_n t} + \overline{C}_2 e^{-i\omega_n t} \tag{3.13}$$

By introducing Euler's equation,

$$\boxed{e^{\pm i\theta} = \cos\theta \pm i\,\sin\theta} \tag{3.14}$$

we can rewrite Eq. 3.13 in terms of trigonometric functions as

$$\boxed{u = A_1 \cos\omega_n t + A_2 \sin\omega_n t} \tag{3.15}$$

were A_1 and A_2 are real constants to be determined from the initial conditions, Eqs. 3.2. Equations 3.2 and 3.15 lead to

$$u(0) = u_0 = A_1, \qquad \dot{u}(0) = v_0 = A_2\,\omega_n \tag{3.16}$$

Thus,

$$\boxed{u(t) = u_0 \cos\omega_n t + \frac{v_0}{\omega_n} \sin\omega_n t} \tag{3.17}$$

---

[2] The roots are also labeled $\bar{\lambda}_{1,2}$ in some texts, and the notation $j = \sqrt{-1}$ is sometimes used.

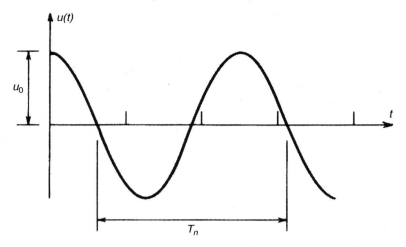

Figure 3.2 Free vibration of an undamped SDOF system with $\dot{u}(0) = 0$.

is the *free vibration response of an undamped SDOF system*. Whereas Eq. 3.17 holds only for free vibration, Eq. 3.15 is a more general solution that will be used to determine the complementary solution when the excitation $p(t)$ is not zero.

Consider first the case of a system that is displaced from its equilibrium position by an amount u_0 and released. (This is the case that was discussed in Section 1.3.) Then $v_0 = 0$, so

$$u(t) = u_0 \cos \omega_n t \qquad (3.18)$$

which is plotted in Fig. 3.2. From Eq. 3.18 it can be seen that the resulting motion is *simple harmonic motion* with an *amplitude* of u_0, an *undamped natural frequency* in cycles per second of

$$\boxed{f_n = \frac{\omega_n}{2\pi}} \quad \text{(Hz)} \qquad (3.19)$$

and an *undamped natural period* in seconds per cycle of

$$\boxed{T_n = \frac{1}{f_n} = \frac{2\pi}{\omega_n}} \quad \text{(s)} \qquad (3.20)$$

The symbol Hz stands for hertz, with 1 Hz ≡ 1 cycle/s.

Figure 3.3 shows a plot of Eq. 3.17 when both u_0 and v_0 are nonzero. This is still simple harmonic motion with period T_n. The displacement $u(t)$ can be expressed by Eq. 3.17 or by the following equation:

$$\boxed{u(t) = U \cos(\omega_n t - \alpha) = U \cos \omega_n \left(t - \frac{\alpha}{\omega_n} \right)} \qquad (3.21)$$

where the *amplitude* is U and where the *phase angle* α determines the amount by which $u(t)$ *lags* the function $\cos \omega_n t$. To relate these quantities to u_0 and v_0, it is convenient to

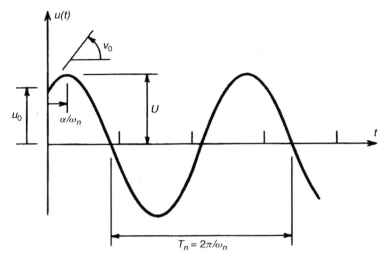

Figure 3.3 General free vibration of an undamped SDOF system.

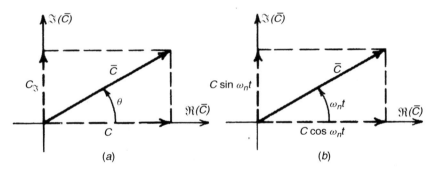

Figure 3.4 Vectors in the complex (Argand) plane: (*a*) complex number representation; (*b*) rotating vector representation of simple harmonic motion (SHM).

employ the rotating-vector representation, or complex-plane representation, of simple harmonic motion. Figure 3.4*a* shows how a complex number $\overline{C}$ can be represented in the complex plane by its polar form,

$$\overline{C} = Ce^{i\theta} \qquad (3.22a)$$

or by its rectangular form (see Appendix B)

$$\overline{C} = \Re(\overline{C}) + i\,\Im(\overline{C}) \equiv C_\Re + i\,C_\Im \qquad (3.22b)$$

where $\Re(\cdot)$ stands for "the real part of" and $\Im(\cdot)$ for "the imaginary part of." The two forms are related by Euler's formula, Eq. 3.14. Thus,

$$C_\Re = C\cos\theta, \quad C_\Im = C\sin\theta \qquad (3.23)$$

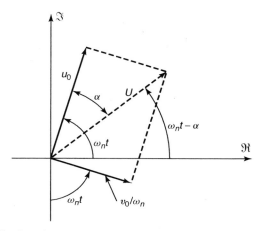

Figure 3.5 Rotating vector representation of undamped free vibration.

Figure 3.4b shows that when the angle θ is taken to be $\omega_n t$, then $\overline{C}$ becomes a *rotating vector*,

$$\overline{C} = Ce^{i\omega_n t} = C\cos\omega_n t + iC\sin\omega_n t \tag{3.24}$$

The horizontal projection, or real component, of $\overline{C}$ represents the harmonic motion $C\cos\omega_n t$, and the vertical projection, or imaginary component, represents the harmonic motion $C\sin\omega_n t$.

Figure 3.5 shows how two rotating vectors can be arranged so that the sum of their projections onto the real axis corresponds to the expression for $u(t)$ on the right-hand side of Eq. 3.17, and so that the projection of their resultant corresponds to the expression for $u(t)$ on the right-hand side of Eq. 3.21. Thus, from Fig. 3.5 it can be seen that the amplitude U and phase angle α in Eq. 3.21 are given by

$$\boxed{U = \sqrt{u_0^2 + \left(\frac{v_0}{\omega_n}\right)^2}} \tag{3.25a}$$

and

$$\boxed{\tan\alpha = \frac{v_0/\omega_n}{u_0}} \tag{3.25b}$$

respectively.

3.2 FREE VIBRATION OF VISCOUS-DAMPED SDOF SYSTEMS

From Eq. 3.21 and Fig. 3.3 you can see that once an undamped SDOF system is set into motion with an initial displacement u_0 and/or an initial velocity v_0, that motion will continue (theoretically) indefinitely. In actuality, all systems have some damping that dissipates energy and causes the motion to die out eventually.

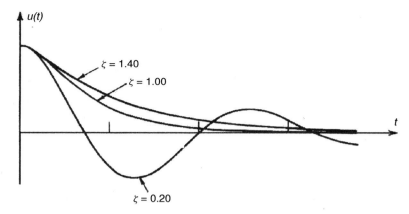

Figure 3.6 Response of a viscous-damped SDOF system with various levels of damping.

Linear viscous damping is the easiest form of damping to handle mathematically, so consider now the free vibration of a SDOF system with linear viscous damping. Its equation of motion is Eq. 3.6, repeated here for convenience.

$$\ddot{u} + 2\zeta\omega_n\dot{u} + \omega_n^2 u = 0 \qquad (3.6)$$

Assuming a solution of the form (Eq. 3.7)

$$u(t) = \overline{C}e^{\overline{s}t} \qquad (3.7)$$

we obtain the characteristic equation (Eq. 3.9)

$$\overline{s}^2 + 2\zeta\omega_n\overline{s} + \omega_n^2 = 0 \qquad (3.9)$$

whose roots $\overline{s}_1$ and $\overline{s}_2$ are given by

$$\left.\begin{array}{c}\overline{s}_1\\ \overline{s}_2\end{array}\right\} = -\zeta\omega_n \pm \omega_n\sqrt{\zeta^2 - 1} \qquad (3.26)$$

The magnitude of the damping factor ζ can be used to distinguish three cases: *underdamped* $(0 < \zeta < 1)$, *critically damped* $(\zeta = 1)$, and *overdamped* $(\zeta > 1)$. Figure 3.6 illustrates these three cases. For the underdamped case the motion is oscillatory with a decaying amplitude. For the overdamped case there is no oscillation, and the amplitude slowly decays. For the critically damped system there is no oscillation, and the amplitude decays more rapidly than in either the underdamped or overdamped cases. Since the underdamped case is the most important case for structural dynamics applications, it is treated first.

3.2.1 Underdamped Case ($\zeta < 1$)

The underdamped case is the most important of the three cases. For $\zeta < 1$ it is convenient to write Eq. 3.26 in the form

$$\left.\begin{array}{c}\overline{s}_1\\ \overline{s}_2\end{array}\right\} = -\zeta\omega_n \pm i\omega_d \qquad (3.27)$$

3.2 Free Vibration of Viscous-Damped SDOF Systems

where ω_d is the *damped circular natural frequency*, given by

$$\omega_d = \omega_n \sqrt{1 - \zeta^2} \quad \text{(rad/s)} \tag{3.28a}$$

with corresponding *damped period*, T_d, given by

$$T_d = \frac{2\pi}{\omega_d} \quad \text{(s)} \tag{3.28b}$$

With the aid of Euler's formula, the general solution $u(t)$ can be written in the form

$$u(t) = e^{-\zeta \omega_n t}(A_1 \cos \omega_d t + A_2 \sin \omega_d t) \tag{3.29}$$

Again, the initial conditions u_0 and v_0 are used to evaluate A_1 and A_2, with the result that *free vibration of an underdamped viscous-damped SDOF system* is given by

$$u(t) = e^{-\zeta \omega_n t}\left(u_0 \cos \omega_d t + \frac{v_0 + \zeta \omega_n u_0}{\omega_d} \sin \omega_d t\right) \tag{3.30}$$

Equation 3.30 can be written in the form

$$u(t) = U e^{-\zeta \omega_n t} \cos(\omega_d t - \alpha) \tag{3.31}$$

and the rotating-vector technique can be employed to show that the amplitude U and phase angle α are given by

$$U = \sqrt{u_0^2 + \left(\frac{v_0 + \zeta \omega_n u_0}{\omega_d}\right)^2} \tag{3.32a}$$

and

$$\tan \alpha = \frac{v_0 + \zeta \omega_n u_0}{\omega_d u_0} \tag{3.32b}$$

respectively.

Figure 3.7 shows a comparison of the responses of SDOF systems having different levels of subcritical damping. In each case, since $u_0 = 0$, the response is given by

$$u(t) = \frac{v_0}{\omega_d} e^{-\zeta \omega_n t} \sin \omega_d t \tag{3.33}$$

Although the value of ζ has an effect on the frequency ω_d, the most pronounced effect of the damping is on the rate at which the motion dies out, that is, on the $e^{-\zeta \omega_n t}$ term. This effect is considered further in Section 3.5, which treats the measurement of damping.

Example 3.1 An SDOF system has an undamped natural frequency of 5 rad/sec and a damping factor of 20%. It is given the initial conditions $u_0 = 0$ and $v_0 = 20$ in./sec. Determine the damped natural frequency and the expression for the motion of the system for $t > 0$.

SOLUTION

$$\omega_n = 5 \text{ rad/sec}, \quad \zeta = 0.2, \quad u_0 = 0, \quad v_0 = 20 \text{ in./sec}$$

From Eq. 3.28a,

$$\omega_d = \omega_n \sqrt{1 - \zeta^2} = 5 \sqrt{1 - (0.2)^2} = 4.8990 \text{ rad/sec} \tag{1}$$

$$f_d = \frac{\omega_d}{2\pi} = 0.7797 \text{ Hz} \tag{2}$$

$$f_d = 0.78 \text{ Hz} \quad\quad\quad \textbf{Ans.} \tag{3}$$

From Eq. 3.33,

$$u(t) = \frac{v_0}{\omega_d} e^{-\zeta \omega_n t} \sin \omega_d t \tag{4}$$

Thus,

$$u(t) = \frac{20}{4.90} e^{-(0.2)(5.0)t} \sin(4.90t) \tag{5}$$

or

$$u(t) = 4.08 e^{-t} \sin(4.90t) \quad \text{in.} \quad\quad \textbf{Ans.} \tag{6}$$

The $\zeta = 0.2$ curve of Fig. 3.7 is a plot of this result.

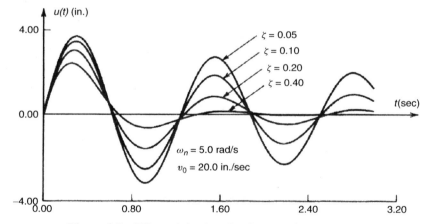

Figure 3.7 Effect of the damping level on free vibration.

Unless significant amounts of damping are intentionally incorporated into a structural system, the damping will fall into the underdamped category, generally in the range 0.5% to 5%. However, the critical damping case and the overdamped case will now be treated briefly for completeness.

3.2.2 Critically Damped Case ($\zeta = 1$)

When $\zeta = 1$, Eq. 3.9 gives only one solution,

$$\bar{s} = -\omega_n \tag{3.34}$$

Then the response takes the form

$$u(t) = (C_1 + C_2 t)e^{-\omega_n t} \tag{3.35}$$

rather than the form given in Eq. 3.7. When the initial conditions are taken into account, the *response of a critically damped SDOF system* is found to be

$$u(t) = [u_0 + (v_0 + \omega_n u_0)t]e^{-\omega_n t} \tag{3.36}$$

Examples of this type of nonoscillatory response are seen in Figs. 3.6 and 3.8.

3.2.3 Overdamped Case ($\zeta > 1$)

When $\zeta > 1$, Eq. 3.26 gives two distinct negative real roots. Let

$$\omega^* = \omega_n \sqrt{\zeta^2 - 1} \tag{3.37}$$

Then the response of an overdamped system can be written in the form

$$u(t) = e^{-\zeta \omega_n t}(C_1 \cosh \omega^* t + C_2 \sinh \omega^* t) \tag{3.38}$$

where C_1 and C_2 depend on the initial conditions. Finally,

$$u(t) = e^{-\zeta \omega_n t} \left[u_0 \cosh \omega^* t + \frac{v_0 + \zeta \omega_n u_0}{\omega^*} \sinh \omega^* t \right] \tag{3.39}$$

Figure 3.8 shows the effect of damping level on the response, indicating that the initial overshoot is greater for the smaller damping levels, but the final decay is more rapid for damping levels approaching $\zeta = 1$.

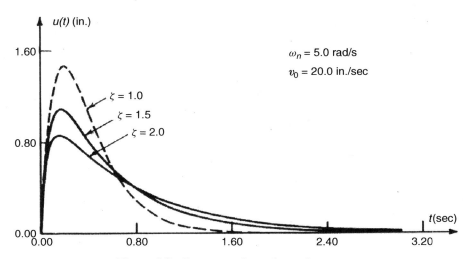

Figure 3.8 Response of overdamped systems.

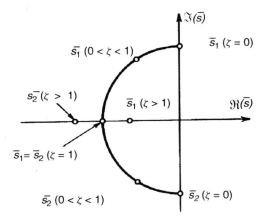

Figure 3.9 $\bar{s}$-plane plot of the roots of the characteristic equation.

From the previous discussion you can see that the form, or character, of the response depends solely on the roots $\bar{s}_i$ of the characteristic equation. Furthermore, these roots depend solely on the two system parameters ω_n and ζ, not on the initial conditions. The dependence of the roots on the damping factor ζ is summarized in the $\bar{s}$-plane plot of Fig. 3.9. By comparing this figure with Fig. 3.6, you can see how the position of the roots in the $\bar{s}$ plane is related to the character of the free-vibration response.

3.3 STABILITY OF MOTION

We have been studying solutions of the equation of motion for linear SDOF systems described by Eq. 3.1, repeated here:

$$m\ddot{u} + c\dot{u} + ku = p(t) \tag{3.1}$$

where the mass m and stiffness k are both positive, and the damping coefficient c satisfies $c \geq 0$. That is, either the system is undamped (Sections 1.3 and 3.1) or has positive viscous damping (Section 3.2). You should be aware that there are also SDOF systems whose equation of motion can be written in the form

$$\ddot{u} + a\dot{u} + bu = 0 \tag{3.40}$$

where the coefficients a and b are not necessarily positive. Since this is a linear differential equation with constant coefficients, we can again assume a solution of the form (Eq. 3.7)

$$u = \overline{C}e^{\bar{s}t} \tag{3.7}$$

Substituting this into Eq. 3.40, we get the *characteristic equation*

$$\bar{s}^2 + a\bar{s} + b = 0 \tag{3.41}$$

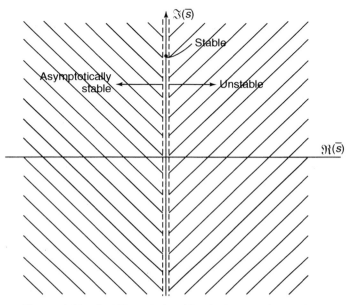

Figure 3.10 Stability relationships in the complex $\bar{s}$ plane.

The two roots of this equation are

$$\left.\begin{array}{c}\bar{s}_1\\ \bar{s}_2\end{array}\right\} = -\frac{a}{2} \pm \sqrt{\left(\frac{a}{2}\right)^2 - b} \qquad (3.42)$$

The general solution, then, has the form

$$u = \overline{C}_1 e^{\bar{s}_1 t} + \overline{C}_2 e^{\bar{s}_2 t} \qquad (3.43)$$

The motion of a system governed by Eq. 3.40 is classified according to the following stability categories: (1) *asymptotically stable*, (2) *stable*, or (3) *unstable* (Fig. 3.10). The character of the motion depends on the roots $\bar{s}_1$ and $\bar{s}_2$, which can be real, purely imaginary, or complex. Let the roots have the general form

$$\bar{s} = \Re(\bar{s}) + i\,\Im(\bar{s}) \equiv \alpha + i\beta \qquad (3.44)$$

Since $u(t)$ is real, roots that are pure imaginary or complex must occur in complex-conjugate pairs.

1. *Asymptotically stable motion.* If both roots of the characteristic equation lie in the left half-plane (i.e., $\alpha_1 < 0$ and $\alpha_2 < 0$), the motion of the system is said to be *asymptotically stable*. That is, with time the motion will die out. Included are the behavior of underdamped systems, critically damped systems, and overdamped systems, as discussed in Section 3.2. Response of an underdamped system is illustrated in Fig. 3.11a.

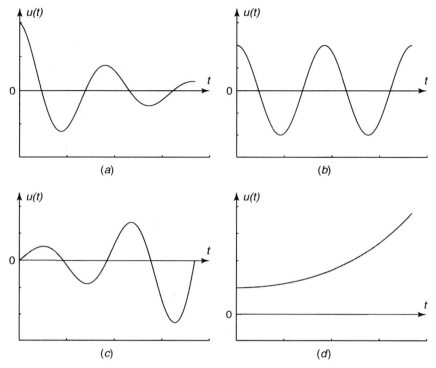

Figure 3.11 Response of four SDOF systems: (*a*) underdamped (decaying) oscillation (asymptotically stable); (*b*) undamped (harmonic) oscillation (stable); (*c*) flutter (diverging, unstable); (*d*) nonoscillatory divergence (unstable).

2. *Stable motion.* If the two roots of the characteristic equation are pure imaginary complex conjugates (i.e., $\alpha_1 = \alpha_2 = 0$), the motion is said to be *stable*. The simple harmonic motion of an undamped SDOF system is illustrated in Fig. 3.11*b*.
3. *Unstable motion.* If either of the two roots of the characteristic equation has a positive real part (i.e., $\alpha_1 > 0$, or $\alpha_2 > 0$, or both) the motion is said to be *unstable*. There are two types of unstable motion.
   (a) *Flutter.* If the two roots are complex conjugates that lie in the right half-plane, the motion will be a diverging oscillation, as illustrated in Fig. 3.11*c*. Flutter avoidance is an essential design consideration in the design of all airplanes.
   (b) *Divergence.* If both roots lie on the real axis and at least one of them has a positive real part, nonoscillatory divergent motion will occur. Divergence is illustrated in Fig. 3.11*d* and in the following example, which illustrates the free-vibration solution for an SDOF system with $a = 0$, $b \neq 0$.

Example 3.1 Figure 1 shows an inverted simple pendulum that consists of a mass m at the upper end of a rigid, massless rod whose lower end is connected to a pin support at A. The mass m is also supported laterally by two linear springs of spring constant k. (a) Determine the linearized equation of motion of this system for small-angle oscillation, that is, for $\theta \ll 1$. (b) Solve for the free vibration of the pendulum with initial conditions $\theta(0) = \theta_0$ and $\dot{\theta}(0) = 0$.

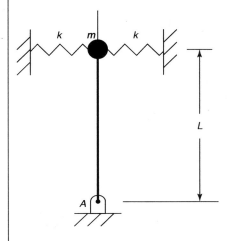

Figure 1 Inverted-pendulum SDOF system.

SOLUTION (a) Derive the equation of motion for small-angle oscillations. First we draw a free-body diagram of the pendulum in a slightly displaced configuration (Fig. 2). Next we apply the equation of motion for fixed-axis rotation about the pin at A (Eq. 2.8b).

$$\circlearrowleft +\ \sum M_A = I_A \ddot{\theta} \qquad (1)$$

$$mg(L \sin\theta) - 2f_S(L \cos\theta) = mL^2 \ddot{\theta} \qquad (2)$$

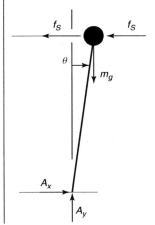

Figure 2 Free-body diagram.

The spring forces are given by

$$f_S = k(L \sin \theta) \tag{3}$$

For small values of θ, we can use the approximations $\sin \theta \approx \theta$ and $\cos \theta \approx 1$. Then Eq. 3 can be substituted into Eq. 2 and the resulting equation linearized to give

$$\ddot{\theta} + \left(\frac{2k}{m} - \frac{g}{L}\right)\theta = 0 \qquad \text{Ans. (a)} \tag{4}$$

which has the form

$$\ddot{\theta} + b\theta = 0 \tag{5}$$

(b) Obtain the free-vibration solution of Eq. 4. As before, we can assume the solution of this equation to be of the form

$$u = \overline{C} e^{\overline{s}t} \tag{6}$$

Substituting this into Eq. 4, we get

$$\overline{s}^2 + \left(\frac{2k}{m} - \frac{g}{L}\right) = 0 \tag{7}$$

Clearly, the solutions that satisfy Eq. 7 will depend on the sign of the term in parentheses, that is, on the sign of the *effective stiffness*

$$b = \frac{2k}{m} - \frac{g}{L} \tag{8}$$

If b is positive, the solution of Eq. 4 will be oscillatory at a natural frequency of $\sqrt{2k/m - g/L}$ rad/s. However, if $2k/m - g/L < 0$, the solution of Eq. 4 will have the form

$$\theta = C_1 e^{\sqrt{(g/L - 2k/m)}\,t} + C_2 e^{-\sqrt{(g/L - 2k/m)}\,t} \tag{9}$$

Finally, the solution that corresponds to the initial conditions $\theta(0) = \theta_0$ and $\dot{\theta}(0) = 0$ is

$$\theta = \frac{\theta_0}{2}\left[e^{\sqrt{(g/L - 2k/m)}\,t} + e^{-\sqrt{(g/L - 2k/m)}\,t}\right] \qquad \text{Ans. (b)} \tag{10}$$

Clearly, the second term in Eq. 10 dies out with time, but the first term grows with time in a nonoscillatory fashion. This type of behavior, called *divergence*, is illustrated in Fig. 3.11*d*.

3.4 FREE VIBRATION OF AN SDOF SYSTEM WITH COULOMB DAMPING

Since viscous damping leads to a linear differential equation of motion, Eq. 3.1, which is relatively easily solved for free or forced response, this model of damping is the one used most frequently in analytical studies. However, the actual damping in a structure

3.4 Free Vibration of an SDOF System with Coulomb Damping

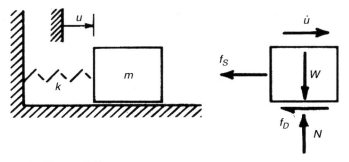

Figure 3.12 SDOF system with Coulomb friction.

may result from looseness of joints, dry friction between components, material damping, and many other complex causes, any of which would lead to nonlinear behavior of the structure. These forms of damping occur naturally, but some are also employed in the design of passive energy dissipation systems.[3.1–3.3] It is not possible to treat all of these forms of damping, but it is instructive to compare and contrast viscous damping with *Coulomb damping*, or dry-friction damping.

Figure 3.12 shows a mass sliding on a rough surface, which produces a force of sliding friction

$$f_D = \mu_k N = \mu_k mg \tag{3.45}$$

where μ_k is the *coefficient of kinetic friction*, or coefficient of sliding friction. The friction force always opposes the motion; that is, its sense is opposite to that of the velocity $\dot{u}$. Using Newton's Second Law, we get

$$-f_S - f_D = m\ddot{u} \tag{3.46}$$

But the spring force is given by

$$f_S = ku \tag{3.47a}$$

and the damping force by

$$f_D = \mu_k mg \,\text{sgn}(\dot{u}) \tag{3.47b}$$

where $\text{sgn}(\dot{u})$ means "the sign of the velocity." Then

$$\begin{aligned} m\ddot{u} + ku &= -\mu_k mg, & \dot{u} &> 0 \\ m\ddot{u} + ku &= +\mu_k mg, & \dot{u} &< 0 \end{aligned} \tag{3.48}$$

Let

$$u_D \equiv |f_D|\frac{1}{k} = \frac{\mu_k g}{\omega_n^2} \tag{3.49}$$

Equations 3.48 and 3.49 may be combined to give

$$\begin{aligned} \ddot{u} + \omega_n^2 u &= -\omega_n^2 u_D, & \dot{u} &> 0 \\ \ddot{u} + \omega_n^2 u &= +\omega_n^2 u_D, & \dot{u} &< 0 \end{aligned} \tag{3.50}$$

72 Free Vibration of SDOF Systems

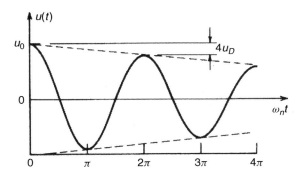

Figure 3.13 Free vibration response of a system with Coulomb friction.

The resulting motion is plotted in Fig. 3.13. Note in the figure that the Coulomb-damped system behaves like an undamped SDOF system whose equilibrium position is shifted at the end of each half-cycle. A distinguishing feature of the response, as shown in Fig. 3.13, is that the <u>amplitude decays linearly with time</u>, not exponentially as in the case of viscous damping.

3.5 EXPERIMENTAL DETERMINATION OF THE NATURAL FREQUENCY AND DAMPING FACTOR OF AN SDOF SYSTEM

It is frequently necessary to determine the dynamical properties (natural frequency, damping factor, etc.) of a given system by experimental methods. Chapter 18 provides an introduction to advanced methods that are used for determining the dynamical properties of complex structures such as automobiles and airplanes. In this section, however, simpler methods that are applicable to SDOF systems with assumed viscous-type damping are presented.

It may be possible to measure the spring constant k and the mass m of a simple SDOF system, but damping seldom arises in a manner that allows the damping coefficient c to be measured directly. The damping of a real system usually results from looseness of joints, internal damping in the material, and so on, not from a viscous dashpot. However, as long as the amplitude of vibration decays exponentially (or approximately so), as shown in Fig. 3.7, it may be assumed that viscous damping can be used in a mathematical model of the system. A damping factor ζ is usually measured, and if desired, an effective value of c can be computed from Eqs. 3.4b and c.

3.5.1 Experimental Determination of the Undamped Natural Frequency

The undamped natural frequency of a simple SDOF system may be determined from a static-displacement measurement or free-vibration experiment. Example 3.3 illustrates the static-displacement method; Example 3.4 illustrates the free-vibration method.

3.5 Experimental Determination of the Natural Frequency and Damping Factor of an SDOF System

Undamped Natural Frequency: Static-Displacement Method

Example 3.2 Determine the natural frequency of the simple spring–mass system of Fig. 1 by measuring the static deflection.

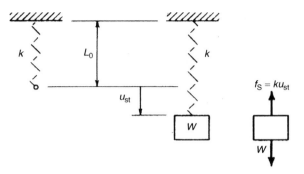

Figure 1 Spring–mass SDOF system.

SOLUTION   From Eq. 3.4a,

$$\omega_n^2 = \frac{k}{m} \qquad (1)$$

The equilibrium of the mass as it hangs on the spring is expressed by

$$+\downarrow \sum F = 0 \qquad (2)$$

or

$$W - f_S = 0 \qquad (3)$$

From the force–elongation equation for the spring,

$$f_S = k u_{st} \qquad (4)$$

Combining Eqs. 3 and 4, we obtain

$$k u_{st} = W = mg \qquad (5)$$

Thus, from Eqs. 1 and 5, we obtain the following expression for the undamped natural frequency:

$$\omega_n^2 = \frac{g}{u_{st}} \qquad \textbf{Ans.} \quad (6)$$

Undamped Natural Frequency: Free-Vibration Method   If the damping in the system is small ($\zeta < 0.2$), Eq. 3.28a shows that ω_d is approximately equal to ω_n. Example 3.4 shows how a free-vibration experiment could be used to determine the natural frequency of a lightly damped SDOF system.

Example 3.3 The natural frequency of a cantilever beam with a lumped mass at its tip (Fig. 1) is to be determined dynamically. The mass is deflected by an amount $A = 1$ in. and released. The ensuing motion, shown in Fig. 2, indicates that the damping in the system is very small. Compute the natural frequency in radians per second and in hertz. What is the period?

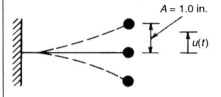

Figure 1 Cantilever-beam SDOF system.

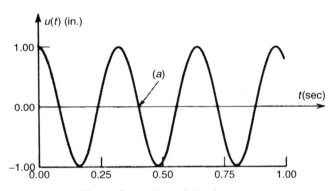

Figure 2 Motion of the tip mass.

SOLUTION At point (*a*) the mass has executed $1\frac{1}{4}$ cycles in approximately 0.4 sec. Therefore, the undamped natural frequency in hertz is

$$f_n = \frac{1.25 \text{ cycles}}{0.4 \text{ sec}} = 3.125 \approx 3.1 \text{ Hz} \qquad \textbf{Ans.}$$

and in rad/sec it is

$$\omega_n = 2\pi f_n = 2\pi(3.125) = 19.64 \approx 20 \text{ rad/sec} \qquad \textbf{Ans.}$$

The undamped natural period is

$$T_n = \frac{1}{f_n} = \frac{1}{3.125} = 0.320 \approx 0.32 \text{ sec} \qquad \textbf{Ans.}$$

Note: Estimates are rounded off to two decimal places.

3.5.2 Experimental Determination of the Damping Factor

There are two methods for determining the damping factor ζ from the decay record of free vibration of an SDOF system: the *logarithmic decrement method* and the *half-amplitude method*. Both are based on Eq. 3.31, repeated here:

$$u(t) = U e^{-\zeta \omega_n t} \cos(\omega_d t - \alpha) \qquad (3.31)$$

3.5 Experimental Determination of the Natural Frequency and Damping Factor of an SDOF System

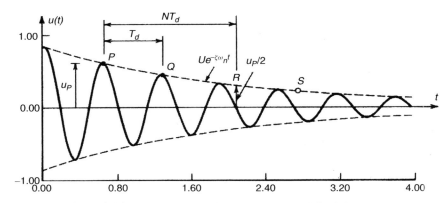

Figure 3.14 Decay record of a viscous-damped SDOF system.

Damping Factor: Logarithmic Decrement Method In the logarithmic decrement method, the amplitude of motion u_P at the beginning of a cycle and the amplitude u_Q at the end of the cycle are measured (Fig. 3.14). At the end of a period (i.e., one cycle), the value of $\cos(\omega_d t - \alpha)$ returns to the value it had at the beginning of the cycle. Hence, from Eq. 3.31 is obtained the expression

$$\frac{u_P}{u_Q} = e^{\zeta \omega_n T_d} \qquad (3.51)$$

The *logarithmic decrement* δ is defined by

$$\delta = \ln \frac{u_P}{u_Q} = \zeta \omega_n T_d \qquad (3.52)$$

where T_d is the *damped natural period*, given by

$$T_d = \frac{2\pi}{\omega_d} = \frac{2\pi}{\omega_n \sqrt{1-\zeta^2}} \qquad (3.53)$$

Thus, from Eqs. 3.52 and 3.53, we obtain

$$\delta = \zeta \omega_n T_d = \frac{2\pi \zeta}{\sqrt{1-\zeta^2}} \qquad (3.54)$$

For small damping ($\zeta < 0.2$) the approximation

$$\delta \approx 2\pi \zeta \qquad (3.55)$$

is acceptable, enabling the damping factor to be obtained from the equation

$$\boxed{\zeta = \frac{1}{2\pi} \ln \frac{u_P}{u_Q}} \qquad (3.56)$$

Damping Factor: Half-Amplitude Method A similar procedure leads to the *half-amplitude method*, which results in a much simpler calculation for the damping factor. The half-amplitude method is based on the amplitude of the envelope curve,

$$\hat{u}(t) = U e^{-\zeta \omega_n t} \qquad (3.57)$$

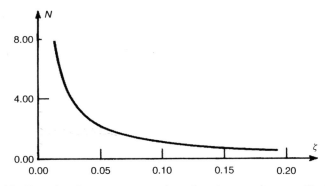

Figure 3.15 Damping factor versus number of cycles to reduce amplitude by 50%.

at two points P and R, where

$$\hat{u}_R = \frac{\hat{u}_P}{2} \tag{3.58}$$

As shown on Fig. 3.14, half-amplitude point R is N damped periods after reference point P, where N is not necessarily an integer. Then, from Eq. 3.57,

$$\frac{\hat{u}_P}{\hat{u}_R} = e^{\zeta \omega_n N T_d} = 2 \tag{3.59}$$

From Eqs. 3.53 and 3.59,

$$\frac{2\pi N \zeta}{\sqrt{1-\zeta^2}} = \ln(2) \tag{3.60}$$

Figure 3.15 shows the relationship between ζ and N, based on Eq. 3.60. However, for small values of damping ($\zeta^2 \ll 1$), Eq. 3.60 gives

$$2\pi N \zeta \approx \ln(2) \tag{3.61}$$

So for a lightly damped SDOF system, the measured damping factor ζ can be closely approximated by

$$\boxed{\zeta \approx \frac{0.11}{N}} \tag{3.62}$$

Equation 3.62 provides a very convenient means of estimating the damping in a system that is lightly damped ($\zeta < 0.1$, i.e., $N > 1$).

Example 3.4 Use the *half-amplitude method* to estimate the damping of the system whose motion is recorded in Fig. 3.14.

SOLUTION   1. Sketch the envelope curve (already on the figure).
 2. Pick point P at one peak and measure u_P.

$$u_P = 0.50 \text{ in.}$$

3. Locate point R, where the amplitude of the envelope curve is

$$\hat{u}_R = \frac{u_P}{2} = 0.25 \text{ in.}$$

4. Estimate the number of cycles between P and R.

$$N = 2.25 \text{ cycles}$$

5. Use Eq. 3.62 to estimate ζ.

$$\zeta = \frac{0.11}{2.25} = 0.049 \qquad \textbf{Ans.}$$

Time Constant The level of damping in a system is also reflected in a quantity called the *time constant* τ. It is defined as the time required for the amplitude to be reduced by a factor of $1/e$. In a manner similar to the way in which the half-amplitude formula was derived, an expression for the time constant can be obtained. Use the envelope curve of Fig. 3.14 again, and let S be the point such that

$$\frac{u_P}{\hat{u}_S} = \frac{u_P}{u_P(1/e)} = e \tag{3.63}$$

Thus,

$$\frac{u_P}{\hat{u}_S} = \frac{U \exp(-\zeta \omega_n t_P)}{U \exp[-\zeta \omega_n (t_P + \tau)]} = e \tag{3.64}$$

or

$$e^{\zeta \omega_n \tau} = e \tag{3.65}$$

By taking the logarithm of both sides, we get

$$\tau = \frac{1}{\zeta \omega_n} \tag{3.66}$$

Therefore, the time constant τ is given by

$$\boxed{\tau = \frac{T_n}{2\pi \zeta}} \tag{3.67}$$

Recall that $1/e = 1/2.718 = 0.368$. Hence, the time constant τ is the time required for the amplitude of motion (i.e., of the envelope curve) to be reduced by 63%.

REFERENCES

[3.1] T. T. Soong and G. F. Dargush, *Passive Energy Dissipation Systems in Structural Engineering*, Wiley, New York, 1997.

[3.2] A. D. Nashif, D. I. G. Jones, and J. P. Henderson, *Vibration Damping*, Wiley, New York, 1985.

[3.3] C. T. Sun and Y. Lu, *Vibration Damping of Structural Elements*, Prentice Hall, Englewood Cliffs, NJ, 1995.

78 Free Vibration of SDOF Systems

PROBLEMS

> For problems whose number is preceded by a **C**, you are to write a computer program and use it to produce the plot(s) requested. *Note*: MATLAB .m-files for many of the plots in this book may be found on the book's website.

Problem Set 3.1

3.1 The structure in Fig. P2.1 has the following properties: $m = 0.15$ kip-sec^2/ft, $L = 12$ ft, $EI = 1800$ kip-ft^2. Determine its undamped natural period, T_n.

3.2 The control surface of Fig. P2.6 has the following properties: $k_1 = 600$ N/m and $a = 0.10$ m. When the surface is displaced slightly and released from rest, it is observed to have very low damping and to have an "undamped" natural frequency of 5.0 Hz. Determine the mass moment of inertia I_A.

3.3 The slender column in Fig. P3.3 has a total mass M, and its bending stiffness is EI. A lumped mass μM is located at the top of the column. Determine an expression for the approximate change in ω_n^2 due to the inertia and geometric stiffness effects of the tip mass. Use the assumed-modes expressions of Eqs. 2.35 through 2.38 (see also Example 2.9), and use $\psi(x) = (x/L)^2$ as the shape function. Include the geometric-stiffness effect of the lumped mass at the top, but neglect the geometric-stiffness effect of the distributed column weight.

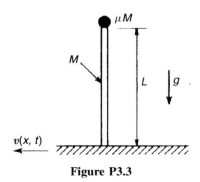

Figure P3.3

3.4 In Fig. P3.4, a 20-kg mass m_1 hangs from a spring whose spring constant is $k = 15$ kN/m. A second mass $m_2 = 10$ kg drops from a height $h = 0.1$ m and sticks to mass m_1. Subsequently, the two vibrate up and down together. (a) Determine an expression for the motion $u(t)$ of the two masses after the moment of impact. Measure the displacement $u(t)$ positive downward from the rest position of mass m_1. (b) Determine the natural frequency and the maximum displacement of the two-mass system.

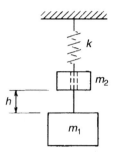

Figure P3.4

3.5 As shown in Fig. P3.5, mass m_1 is attached directly to a linear spring with spring constant k, and it slides on a frictionless horizontal surface. A second mass, m_2, is attached to mass m_1 by a weightless inextensible cable that passes over a pulley that has a radius r and mass moment of inertia I_3 about its axis. (a) Use the *Principle of Virtual Displacements* to determine an equation of motion for this system. Measure the displacement $u(t)$ from the equilibrium position of mass m_2. (b) Determine an expression for the natural frequency of this system. Express your answer in terms of m_1, m_2, I_3, k, and r.

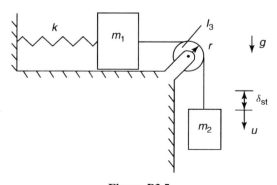

Figure P3.5

Problem Set 3.2

3.6 The system in Fig. P2.3 has the following properties: $M = 0.01$ kip-sec^2/ft, $k_1 = 15.0$ kips/ft, $k_2 = 10.0$ kips/ft, and $L = 10.0$ ft. The damping coefficient c is such that $\zeta = 0.10$. (a) Determine the numerical value of c. (b) Determine an expression for $v(t)$ if A is held fixed [i.e., $z(t) = 0$] and beam end B is displaced vertically upward 0.10 ft and released from rest at $t = 0$.

C 3.7 The system in Fig. P2.3 has the following properties: $M = 0.01$ kip-sec^2/ft, $k_1 = 15.0$ kips/ft, $k_2 = 10.0$ kips/ft, and $L = 10.0$ ft. The damping coefficient c is such that $\zeta = 0.10$. (a) Determine an expression for $v(t)$ if A is held fixed [i.e., $z(t) = 0$] and beam end B is displaced vertically upward 0.10 ft and released from rest at $t = 0$. (b) Plot at least three cycles of the tip motion $v(t)$ resulting from the "pluck test" described in part (a). (c) How many complete cycles of vibration will have to occur before the maximum force in spring 1 stays below 0.2 kips?

C 3.8 A machine that weighs 1800 lb is mounted on springs whose total stiffness is 4200 lb/ft, and the total damping is 50 lb-sec/ft (Fig. P3.8). Determine expressions for the motion $u(t)$ for the following two initial conditions, and plot at least four cycles of each function $u(t)$. (a) $u_a(0) = 0.5$ in., $\dot{u}_a(0) = 0$; and (b) $u_b(0) = 0$, $\dot{u}_b(0) = 4$ in./sec.

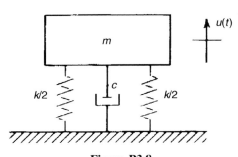

Figure P3.8

3.9 A wind turbine is modeled as a concentrated mass (the turbine) atop a weightless tower of height L (Fig. P3.9). To determine the dynamic properties of the system, a large crane is brought alongside the tower and a lateral force $P = 200$ lb is exerted along the turbine axis as shown. This causes a horizontal displacement of the turbine of 1.0 in. The cable attaching the turbine to the crane is severed instantaneously, and the resulting free vibration of the turbine is recorded. At the end of two complete cycles, the time is 1.25 sec and the vibration amplitude is 0.64 in. From the data above, determine (a) the undamped natural frequency ω_n (rad/sec); (b) the effective stiffness k (lb/in.); (c) the effective mass m (lb-sec^2/in.); and (d) the effective damping factor ζ.

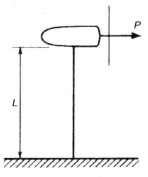

Figure P3.9

3.10 Determine an expression for the critical damping coefficient, $(c_v)_{cr}$, for the system in Fig. P2.5.

Problem Set 3.3

C 3.11 The spring–mass oscillator in Fig. P3.11 has a velocity-feedback force generator that exerts a force f_v on the mass that is proportional to the velocity of the mass. The sign of the force can be either positive or negative. For a particular setup of this SDOF system, the spring, mass, and feedback force parameters lead to the following differential equation of motion:

$$\ddot{u} - 2\dot{u} + 5u = 0$$

where u is the displacement of the mass in inches. (a) If the initial conditions are

$$u(0) = 0, \qquad \dot{u}(0) = 0.10 \text{ in./sec}$$

determine the motion function $u(t)$. (b) Using your answer to part (a), write a computer program (e.g., a MATLAB .m-file), and obtain a plot similar to the appropriate one of the four plots in Fig. 3.11.

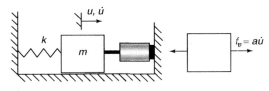

Figure P3.11

Problem Set 3.4

3.12 A SDOF system with Coulomb friction is shown in Fig. 3.12. (a) Starting with Eqs. 3.50, determine expressions for $u(t)$ for the two cases $\dot{u} < 0$ and $\dot{u} > 0$. (b) If $u(0) = u_0$ and $\dot{u}(0) = 0$, what is the expression for $u(t)$: (1) for $0 \leq t \leq \pi/\omega_n$, (2) for $\pi/\omega_n \leq t \leq 2\pi/\omega_n$? Assume that $u_D < u_0$. (c) Determine an expression for the amount of amplitude decay per cycle.

3.13 As shown in Fig. P3.5, mass m_1 is attached directly to a linear spring with spring constant k. This mass slides on a horizontal surface with coefficient of sliding friction μ_k. A second mass, m_2, is attached to mass m_1 by a weightless, inextensible cable that passes over a massless pulley; that is, $m_3 = 0$. Use free-body diagrams and Newton's Laws to determine the two equations of motion for this system, like Eqs. 3.48.

$$m_u\ddot{u} + k_u u = -\mu_k m_1 g, \quad \dot{u} > 0$$
$$m_u\ddot{u} + k_u u = +\mu_k m_1 g, \quad \dot{u} < 0$$

Measure the displacement $u(t)$ from the equilibrium position of mass m_2. Express your answer in terms of m_1, m_2, k, and μ_k.

Problem Set 3.5

3.14 For the time history plotted in Fig. P3.14: (a) estimate the damped natural frequency in hertz; and (b) use the half-amplitude method to estimate the damping factor ζ.

3.15 Figure P3.15 is a simulation of a virtual oscilloscope trace of the displacement of an SDOF system. (a) Assuming viscous damping, estimate the damped natural frequency in hertz; and (b) use the half-amplitude method to estimate the damping factor ζ.

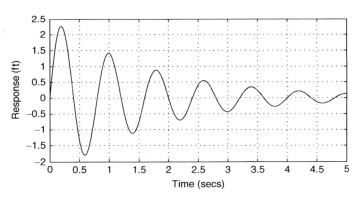

Figure P3.14

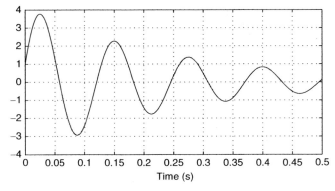

Figure P3.15

4
Response of SDOF Systems to Harmonic Excitation

In Chapter 2 several methods were presented for deriving the differential equation of motion of a SDOF system. In Chapter 3 you began to consider the solution of this differential equation by studying the free vibration response of a SDOF system. In this chapter you will begin your study of forced motion by studying the response of undamped and viscous-damped SDOF systems to harmonic (i.e., sinusoidal) excitation. This topic is <u>extremely important</u>, not only because many SDOF structures are subjected to harmonic excitation, but also because (1) the results of this chapter can be extended to treat multiple-DOF (MDOF) structures and structures subjected to more complex types of excitation, and (2) the results of this chapter form the basis for *Experimental Structural Dynamics* (Chapter 18).

Upon completion of this chapter you should be able to:

- Determine the steady-state response and total response of an undamped SDOF system subjected to simple harmonic excitation.
- Determine the steady-state response and the total response of a viscous-damped SDOF system subjected to simple harmonic excitation.
- Describe the effect of damping factor and frequency ratio on the amplitude and phase of the steady-state response of a viscous-damped SDOF system subjected to a harmonic excitation force.
- Sketch, for various frequency ratios, force vector polygons that show the phase and amplitude relationships of the steady-state forces acting on an SDOF system.
- Use the complex frequency response method to obtain expressions for the amplitude and phase of the steady-state response of a viscous-damped SDOF system.
- Set up the equation of motion for a base excitation problem and obtain the steady-state solutions (frequency-response functions) for absolute motion and motion of the mass relative to the base.
- Obtain numerical values for the magnification factor and phase angle of a system with given base excitation.
- Discuss the principle of operation of seismic transducers: namely, vibrometers and accelerometers.
- Describe how frequency-response data can be used to determine the undamped natural frequency and damping factor of a viscous-damped SDOF system.
- Calculate the work done per cycle of harmonic motion by a specified dissipative force.

- Calculate the equivalent viscous damping coefficient for a specified dissipative force.
- Discuss how structural damping differs from viscous damping, and list the principal features of vector response plots for systems with structural damping.

4.1 RESPONSE OF UNDAMPED SDOF SYSTEMS TO HARMONIC EXCITATION

As noted in Section 1.3 and again in Chapter 3, the total response of a linear system consists of the superposition of a forced motion and a natural motion. In the case of simple harmonic excitation, forced motion is referred to as the *steady-state response*. In this section we treat the response of undamped SDOF systems to harmonic excitation, and in Sections 4.2 through 4.8, the response of damped SDOF systems.

Consider the undamped SDOF system shown in Fig. 4.1. It is assumed that the system is linear and that the excitation force amplitude p_0 and the forcing frequency Ω (in rad/s) are constants.[1] The equation of motion is

$$m\ddot{u} + ku = p_0 \cos \Omega t \tag{4.1}$$

(We could equally well have taken the excitation to be $p_0 \sin \Omega t$.) From the fact that only even-order derivatives appear on the left-hand side of Eq. 4.1, it is seen that the forced motion, or *steady-state response*, will have the form

$$u_p = U \cos \Omega t \tag{4.2}$$

To determine the (signed) *amplitude U* of the steady-state response, Eq. 4.2 is substituted into Eq. 4.1, giving[2]

$$\boxed{U = \frac{p_0}{k - m\Omega^2}} \tag{4.3}$$

provided that $k - m\Omega^2 \neq 0$. Let

$$U_0 \equiv \frac{p_0}{k} \tag{4.4}$$

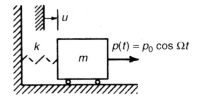

Figure 4.1 Harmonic excitation of an undamped SDOF system.

---

[1] In this chapter, capital omega (Ω) is used as the symbol for the forcing frequency in radians per second to distinguish it clearly from the natural frequency ω_n. In some other places in the book, lowercase omega (ω) is used as the symbol for the forcing frequency.

[2] In this section U can be positive or negative, depending on the frequency ratio Ω/ω_n. In Section 4.2 and later sections, U will stand for the magnitude of the displacement. See footnote 3 in Section 4.2.

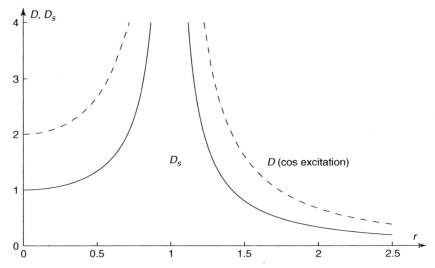

Figure 4.2 Dynamic magnification factors for an undamped SDOF system with $p(t) = p_0 \cos \Omega t$: steady-state magnification factor D_s (solid curve), and total dynamic magnification factor D (dashed curve).

which is the *static displacement*, that is, the displacement that the mass would undergo if a force of magnitude p_0 were to be applied "statically." Then Eq. 4.3 may be written

$$\frac{U}{U_0} = \frac{1}{1-r^2}, \qquad r \neq 1 \tag{4.5}$$

where the ratio of the forcing frequency Ω to the undamped natural frequency ω_n is called the *frequency ratio*

$$\boxed{r = \frac{\Omega}{\omega_n}} \tag{4.6}$$

and

$$\boxed{H(r) \equiv \frac{U}{U_0}} \tag{4.7}$$

is called the nondimensionalized *frequency-response function*. The frequency-response function, $H(r)$, gives the magnitude and sign of the steady-state motion as a function of the frequency ratio r. The magnitude

$$D_s(r) \equiv |H(r)| \tag{4.8}$$

is called the *steady-state magnification factor*, or gain. It is plotted in Fig. 4.2.

Equations 4.2 through 4.5 may be combined to give the *steady-state response*

$$u_p = \frac{U_0}{1-r^2} \cos \Omega t, \qquad r \neq 1 \tag{4.9}$$

If $r < 1$, the response is in phase with the excitation, since $1 - r^2$ is positive. If $r > 1$, the response is 180° out of phase with the excitation; that is, u_p can be written

$$u_p = \frac{U_0}{r^2 - 1}(-\cos \Omega t) \tag{4.10}$$

The general form of the natural motion of an undamped SDOF system is given by Eq. 3.15. (Note that Eq. 3.17 is valid only for free vibration!) Hence, Eqs. 3.15 and 4.9 may be combined to give the *total response* due to excitation $p(t) = p_0 \cos \Omega t$ as

$$\boxed{u(t) = \frac{U_0}{1 - r^2} \cos \Omega t + A_1 \cos \omega_n t + A_2 \sin \omega_n t} \tag{4.11}$$

with the constants A_1 and A_2 to be determined by the initial conditions.

Example 4.1 illustrates the use of Eq. 4.11 with $\cos \Omega t$ excitation. Homework Problem 4.4 examines the sine excitation case, and shows that $r = 0.5$ is a special case.

Example 4.1 The system shown in Fig. 4.1 has a spring stiffness $k = 40$ lb/in., and the mass weighs 38.6 lb. If the system is initially at rest, that is, $u(0) = \dot{u}(0) = 0$, when an excitation $p(t) = 10\cos(10t)$ lb begins, determine an expression for the resulting motion. Sketch the resulting motion.

SOLUTION From Eq. 4.11 the total response is given by

$$u = \frac{U_0}{1 - r^2} \cos \Omega t + A_1 \cos \omega_n t + A_2 \sin \omega_n t \tag{1}$$

This may be differentiated to give the velocity

$$\dot{u} = \frac{-U_0 \Omega}{1 - r^2} \sin \Omega t - A_1 \omega_n \sin \omega_n t + A_2 \omega_n \cos \omega_n t \tag{2}$$

From Eq. 3.4a, the undamped natural frequency is

$$\omega_n = \left(\frac{k}{m}\right)^{1/2} = \left(\frac{kg}{W}\right)^{1/2} = \left(\frac{40(386)}{38.6}\right)^{1/2} = 20 \text{ rad/sec} \tag{3}$$

From Eq. 4.4, the static displacement is

$$U_0 = \frac{p_0}{k} = \frac{10 \text{ lb}}{40 \text{ lb/in.}} = 0.25 \text{ in.} \tag{4}$$

From Eqs. 4.6 and 4.9,

$$r = \frac{\Omega}{\omega_n} = \frac{10 \text{ rad/sec}}{20 \text{ rad/sec}} = 0.5 \tag{5}$$

and

$$U = \frac{U_0}{1 - r^2} = \frac{0.25 \text{ in.}}{1 - (0.5)^2} = \frac{0.25 \text{ in.}}{0.75} = \frac{1}{3} \text{ in.} \tag{6}$$

Use the initial conditions to evaluate A_1 and A_2.

$$u(0) = 0 = U + A_1, \qquad \dot{u}(0) = 0 = A_2 \omega_n \tag{7}$$

Therefore,
$$A_1 = -U = -\tfrac{1}{3} \text{ in.}, \qquad A_2 = 0 \tag{8}$$

Finally, the total response to the given sinusoidal excitation is

$$\boxed{u(t) = \tfrac{1}{3}[\cos(10t) - \cos(20t)]} \qquad \textbf{Ans.} \quad (9)$$

This equation for the total response is plotted in Fig. 1, together with its two constituents.

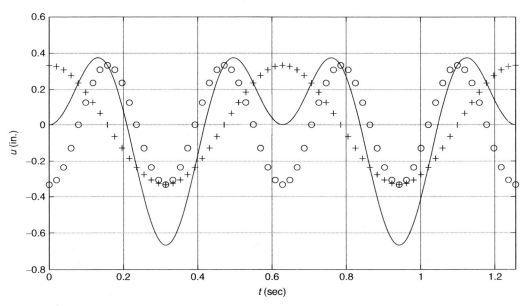

Figure 1 Forced response $u_p(t)$ (pluses), free response $u_c(t)$ (circles), and total response $u(t)$ (solid curve) of a SDOF system subjected to sinusoidal excitation.

From the responses plotted in Example 4.1, observe the following:

1. The steady-state response has the same frequency as the excitation and is in phase with the excitation since $r < 1$.
2. The total response is not simple harmonic motion. However, since $\omega_n = 2\Omega$, the total response is *periodic* with a period of $\pi/5$ sec.
3. The magnitude of the maximum total response $(\max|u(t)| = \tfrac{2}{3}$ in. at $t = \pi/10$ sec) is greater than the maximum steady-state response $[\max|u_p(t)| = \tfrac{1}{3}$ in. at $t = 0]$.

The *total dynamic magnification factor* is defined as

$$\boxed{D = \frac{\max_t |u(t)|}{U_0}} \tag{4.12}$$

In Example 4.1, where $r = 0.5$, it has the value 2/3 in./1/4 in. $= \frac{8}{3}$. Figure 4.2 shows a plot of the total dynamic magnification factor D and the steady-state magnification factor D_s as functions of the frequency ratio r for an undamped system that is initially at rest and is subjected to cosine excitation $p_0 \cos \Omega t$. Although the steady-state frequency response is the same for either cosine or sine excitation, the total response curves for cosine and sine excitation are different.

Resonance Equations 4.9 and 4.11 are not valid at $r = 1$. The condition $r = 1$, or $\Omega = \omega_n$, is called *resonance*, and it is obvious from Fig. 4.2 that at excitation frequencies near resonance the response becomes very large. The importance attached to the study of the response of structures to harmonic excitation stems in large part from the necessity to avoid the resonance condition, when large-amplitude motion can occur.

If the excitation is $p_0 \cos \omega_n t$, that is, if the forcing frequency Ω in Eq. 4.1 is equal to the undamped natural frequency ω_n (i.e., when $r = 1$), it is necessary to replace Eq. 4.2 by the assumed particular solution

$$u_p = Ct \sin \omega_n t \tag{4.13}$$

Then, by substituting Eq. 4.13 into 4.1, we obtain

$$C = \frac{p_0}{2m\omega_n} \tag{4.14}$$

so the particular solution for cosine-type excitation at resonance is

$$\boxed{u_p(t) = \frac{U_0}{2}(\omega_n t) \sin \omega_n t} \tag{4.15}$$

This response is plotted in Fig. 4.3. Note that although Fig. 4.2 indicates an infinite amplitude at $r = 1$, Eq. 4.15 and Fig. 4.3 show that the amplitude of the particular solution builds as a linear function of time.

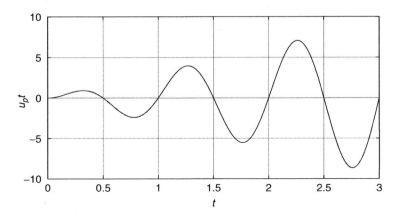

Figure 4.3 Response $u_p(t)$ at resonance: $p(t) = p_0 \cos \omega_n t$.

4.2 RESPONSE OF VISCOUS-DAMPED SDOF SYSTEMS TO HARMONIC EXCITATION: FREQUENCY-RESPONSE FUNCTIONS

The classical analytical model of a linear SDOF system is the spring–mass–dashpot model of Fig. 3.1. When this system is subjected to harmonic excitation $p_0 \cos \Omega t$, its equation of motion is

$$m\ddot{u} + c\dot{u} + ku = p_0 \cos \Omega t \qquad (4.16)$$

Due to the presence of the damping term in Eq. 4.16, the *steady-state response* will not be in phase (or 180° out of phase) with the excitation but will be given by

$$u_p = U \cos(\Omega t - \alpha) \qquad (4.17)$$

where U is the *steady-state amplitude* and α is the *phase (lag) angle* of the steady-state response relative to the excitation.[3]

The determination of amplitude U and phase angle α is facilitated by the use of rotating vectors. The steady-state velocity and acceleration, which are needed in Eq. 4.16, are given by

$$\begin{aligned} \dot{u}_p &= -\Omega U \sin(\Omega t - \alpha) \\ \ddot{u}_p &= -\Omega^2 U \cos(\Omega t - \alpha) \end{aligned} \qquad (4.18)$$

Figure 4.4 shows the relationship of rotating vectors such that their projections onto the real (horizontal) axis are $p_0 \cos \Omega t$, and the displacement, velocity, and acceleration expressions as indicated in Eqs. 4.17 and 4.18.

When Eqs. 4.17 and 4.18 are substituted into Eq. 4.16, the result is

$$-m\Omega^2 U \cos(\Omega t - \alpha) - c\Omega U \sin(\Omega t - \alpha) + kU \cos(\Omega t - \alpha) = p_0 \cos \Omega t \qquad (4.19)$$

This equation is represented conveniently by a *force vector polygon*, since each term in Eq. 4.19 represents a force acting on the mass of Fig. 3.1. Figure 4.5 shows a force vector polygon for the case $m\Omega^2 U < kU$, that is, $\Omega < \omega_n$. The projections of the

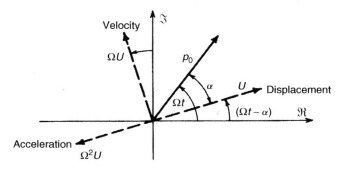

Figure 4.4 Rotating vectors representing p, u, $\dot{u}$, and $\ddot{u}$.

---

[3] In this and later sections, U is the (positive) magnitude of the steady-state response.

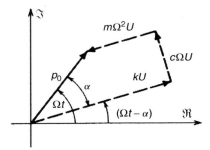

Figure 4.5 Force vector polygon.

dashed-line vectors in Fig. 4.5 onto the horizontal (real) axis are the terms on the left-hand side of Eq. 4.19; the projection of the solid vector onto the real axis gives the right-hand side of Eq. 4.19. From Fig. 4.5 it is easily seen that

$$(kU - m\Omega^2 U)^2 + (c\Omega U)^2 = p_0^2 \tag{4.20a}$$

and

$$\tan \alpha = \frac{c\Omega}{k - m\Omega^2} \tag{4.20b}$$

These can be written as

$$D_s(r) \equiv \frac{U(r)}{U_0} = \frac{1}{[(1-r^2)^2 + (2\zeta r)^2]^{1/2}} \tag{4.21a}$$

and

$$\tan \alpha(r) = \frac{2\zeta r}{1 - r^2} \tag{4.21b}$$

where ζ, r, and U_0 are as defined previously. The *steady-state magnification factor* D_s and the *phase angle* α are plotted versus the nondimensionalized forcing frequency $r = \Omega/\omega_n$ in Fig. 4.6a and 4.6b, respectively. The plots in Figs. 4.6 are called linear plots since both horizontal and vertical axes have linear scales.

The combination of amplitude versus frequency and phase versus frequency information is called the *frequency response* of the system, and figures such as Fig. 4.6 are called *frequency-response plots*. Analyzing the frequency response of systems is one of the most important topics in structural dynamics, as you will see in the remainder of this chapter and in several later chapters. From Eqs. 4.17 and 4.21 and Fig. 4.6, the following significant features of steady-state response of a viscously damped SDOF system can be observed:

1. The steady-state motion described by Eq. 4.17 is sinusoidal and is of the same frequency as the excitation.
2. The amplitude of the steady-state response is a function of both the amplitude and frequency of the excitation as well as of the natural frequency and damping factor of the system. The steady-state magnification factor can be considerably greater than unity or can be less than unity.

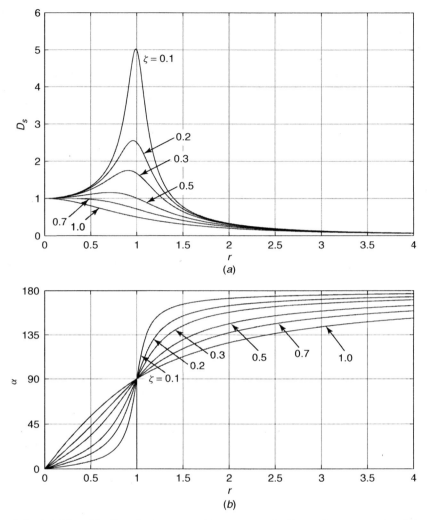

Figure 4.6 (*a*) Magnification factor versus frequency ratio for various amounts of damping (linear plot); (*b*) phase angle versus frequency ratio for various amounts of damping (linear plot).

3. The excitation, $p = p_0 \cos \Omega t$, and the steady-state response, $u_p = U \cos(\Omega t - \alpha)$, are not in phase; that is, they do not attain their maximum values at the same time. The steady-state response <u>lags</u> the excitation by the phase angle α. This corresponds to a time lag of α/Ω.

4. At resonance ($r = 1$) the amplitude is limited only by the damping force, and

$$(D_s)_{r=1} = \frac{1}{2\zeta} \qquad (4.22)$$

Also, at resonance $\alpha = 90°$; that is, the response lags the excitation by $90°$.

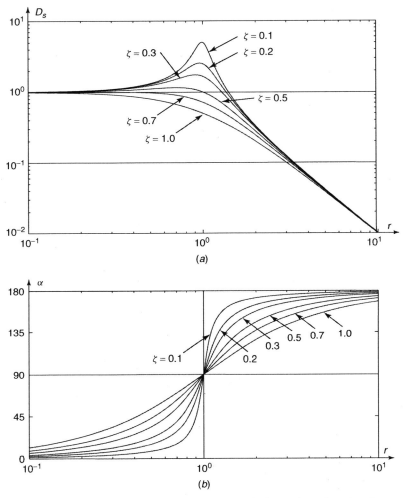

Figure 4.7 (*a*) Magnification factor versus frequency ratio for various damping factors (log-log plot); (*b*) phase angle versus frequency ratio for various damping factors (logarithmic frequency scale).

Since the dynamic magnification factor may be quite large near resonance, and since excitation covering a broad range of frequencies may be of interest, the curves of Fig. 4.6 are frequently plotted to logarithmic scales, as shown in Fig. 4.7. This form of frequency-response plot is often referred to as a *Bode plot*.

The *total response* is given by $u(t) = u_p(t) + u_c(t)$. From Eqs. 4.17, 4.21, and 3.29 it can be written

$$u(t) = \frac{U_0}{[(1-r^2)^2 + (2\zeta r)^2]^{1/2}} \cos(\Omega t - \alpha) + e^{-\zeta \omega_n t}(A_1 \cos \omega_d t + A_2 \sin \omega_d t)$$

(4.23)

where α is given by Eq. 4.21b, and A_1 and A_2 are constants to be determined from the initial conditions. Since the natural motion in Eq. 4.23 dies out with time, it is referred to as the *starting transient*.

Example 4.2 If damping equivalent to $\zeta = 0.2$ is added to the system in Example 4.1, and the same excitation and initial conditions prevail, determine an expression for the resulting motion. Sketch the motion.

SOLUTION   The total response function is given by Eq. 4.23,

$$u(t) = U \cos(\Omega t - \alpha) + e^{-\zeta \omega_n t}(A_1 \cos \omega_d t + A_2 \sin \omega_d t) \tag{1}$$

where the first term is the steady-state response, and the $e^{-\zeta \omega_n t}$ term is the starting-transient term.

The amplitude of the steady-state response is

$$U = \frac{U_0}{[(1-r^2)^2 + (2\zeta r)^2]^{1/2}} \tag{2a}$$

and the phase lag angle is given by

$$\tan \alpha = \frac{2\zeta r}{1 - r^2} \tag{2b}$$

Numerical values for ω_n, U_0, and r may be found in Example 4.1.

$$\left. \begin{aligned} \Omega &= 10 \text{ rad/sec} \\ \omega_n &= \sqrt{\frac{k}{m}} = 20 \text{ rad/sec} \\ U_0 &= \frac{p_0}{k} = 0.25 \text{ in.} \\ r &= \frac{\Omega}{\omega_n} = 0.5 \\ \zeta \omega_n &= 0.2(20) = 4 \text{ rad/sec} \end{aligned} \right\} \tag{3}$$

Therefore, from Eqs. 2a and 3,

$$\begin{aligned} U &= \frac{0.25}{\{[1-(0.5)^2]^2 + [2(0.2)(0.5)]^2\}^{1/2}} \\ &= \frac{0.25}{[(0.75)^2 + (0.2)^2]^{1/2}} = \frac{0.25}{0.776} = 0.322 \text{ in.} \end{aligned} \tag{4}$$

From Eqs. 2b and 3,

$$\tan \alpha = \frac{2(0.2)(0.5)}{1 - (0.5)^2} = \frac{0.2}{0.75} = 0.26\overline{6} \tag{5}$$

so the phase lag angle is

$$\alpha = 0.261 \text{ rad} = 14.93° \tag{6}$$

Now we use the initial conditions to determine the constants A_1 and A_2. From Eq. 3.28a, the damped natural frequency is

$$\omega_d = \omega_n\sqrt{1-\zeta^2} = 20\sqrt{1-(0.2)^2} = 19.60 \text{ rad/sec} \tag{7}$$

Equation 1 is differentiated with respect to time to give

$$\begin{aligned}\dot{u} &= -\Omega U \sin(\Omega t - \alpha) \\ &+ e^{-\zeta\omega_n t}[(A_2\omega_d - A_1\zeta\omega_n)\cos\omega_d t - (A_1\omega_d + A_2\zeta\omega_n)\sin\omega_d t]\end{aligned} \tag{8}$$

Evaluate Eqs. 1 and 8 at $t = 0$ and set equal to the initial conditions.

$$u(0) = 0 = 0.322\cos(-0.261) + A_1 \tag{9}$$

so

$$A_1 = -0.322\cos(-0.261) = -0.311 \text{ in.} \tag{10}$$

$$\dot{u}(0) = 0 = -0.322(10)\sin(-0.261) + [A_2(19.60) - (-0.311)(0.2)(20)] \tag{11}$$

or

$$A_2 = -0.1059 \text{ in.} \tag{12}$$

Therefore, Eqs. 1, 4, 6, 10, and 12 may be combined to give

$$\begin{aligned}u(t) &= 0.322\cos(10t - 0.261) \\ &- e^{-4t}[0.311\cos(19.60t) + 0.1059\sin(19.60t)] \text{ in.}\end{aligned} \quad \text{Ans.} \tag{13}$$

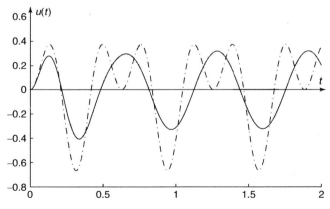

Figure 1 Total responses of undamped and viscous-damped SDOF systems to sinusoidal excitation: Example 4.1, $\zeta = 0$ (dash-dotted curve); Example 4.2, $\zeta = 0.2$ (solid curve).

The response is plotted in Fig. 1. To see the significant effect of damping, compare the solid curve (damped system of Example 4.2) with the dash-dotted curve (undamped system of Example 4.1). Note that for the damped system, the starting transient dies out very quickly, and the steady-state response dominates.

4.3 COMPLEX FREQUENCY RESPONSE

In Section 4.2 the amplitude U and phase angle α of the steady-state response to excitation $p_0 \cos \Omega t$ were determined by projecting rotating vectors onto the real axis to obtain the terms in Eq. 4.19. The use of vectors in the complex plane is pursued further in this section because it greatly simplifies many frequency-response calculations and provides a useful way of examining experimental results, as will be seen later, particularly in Chapter 18.[4]

Consider again the steady-state response of the prototype spring–mass–dashpot SDOF system in Fig. 3.1. Let the subscript $\Re$ (for projection onto the real axis) be used to designate the steady-state (i.e., forced) motion due to $\cos \Omega t$ excitation. Thus, the equation of motion for a viscous-damped SDOF system with $\cos \Omega t$ excitation is

$$m\ddot{u}_\Re + c\dot{u}_\Re + ku_\Re = p_0 \cos \Omega t \tag{4.24}$$

and the corresponding steady-state solution is

$$u_\Re = U \cos(\Omega t - \alpha) \tag{4.25}$$

Similarly, for $\sin \Omega t$ excitation the equation of motion and corresponding steady-state solution are

$$m\ddot{u}_\Im + c\dot{u}_\Im + ku_\Im = p_0 \sin \Omega t \tag{4.26}$$

and

$$u_\Im = U \sin(\Omega t - \alpha) \tag{4.27}$$

where the subscript $\Im$ stands for projection onto the imaginary axis. Now, if Eq. 4.26 is multiplied by $i = \sqrt{-1}$ and added to Eq. 4.24, and Euler's formula, Eq. 3.14, is used, there results

$$m\ddot{\bar{u}} + c\dot{\bar{u}} + k\bar{u} = \bar{p} = p_0 e^{i\Omega t} \tag{4.28}$$

where the overbar denotes a vector in the complex plane. Equation 4.28 is called the *complex equation of motion*, and the vector

$$\bar{u} = u_\Re + iu_\Im \tag{4.29}$$

is called the *complex response*. It is understood that the actual steady-state motion will be given by either the real part of $\bar{u}$ or its imaginary part, depending on whether the excitation is of $\cos \Omega t$ or $\sin \Omega t$ type.

The steady-state solution of Eq. 4.28 may be assumed to have the form

$$\bar{u}(t) = \bar{U} e^{i\Omega t} \tag{4.30}$$

where $\bar{U}$ is the *complex amplitude*, which may also be written

$$\bar{U} = U e^{-i\alpha} \tag{4.31}$$

---

[4]Topics in the algebra of complex numbers that are used in this chapter are reviewed in Appendix B.

where U and α are the same response amplitude and phase (lag) angle introduced in Eq. 4.17. By substituting Eq. 4.30 into Eq. 4.28, we obtain directly the complex fraction

$$\overline{U} = \frac{p_0}{(k - m\Omega^2) + ic\Omega} \tag{4.32}$$

which can be written in the nondimensionalized form

$$\boxed{\overline{H}_0(\Omega) \equiv \frac{\overline{U}}{U_0} = \frac{1}{(1 - r^2) + i(2\zeta r)}} \tag{4.33}$$

where $\overline{H}_0(\Omega)$ is called the (nondimensionalized) *complex frequency response*.[5]

From Eqs. 4.31 and 4.33 it is clear that to determine the amplitude and phase of the steady-state response we need only to find the amplitude and phase of the complex expression on the right-hand side of Eq. 4.33. Let us first summarize a couple of results from the theory of complex numbers (see Appendix B):

- *Rectangular and polar representation.* If a complex number (vector) $\overline{A}$ is represented in rectangular form by

$$\overline{A} = A_\Re + iA_\Im \tag{4.34a}$$

and in polar form by

$$\overline{A} = Ae^{i\alpha} \tag{4.34b}$$

then the amplitude of $\overline{A}$ is given by

$$A \equiv |\overline{A}| = \sqrt{A_\Re^2 + A_\Im^2} \tag{4.34c}$$

and the phase angle by

$$\tan\alpha = \frac{A_\Im}{A_\Re} \tag{4.34d}$$

- *Quotient of two complex numbers.* If $\overline{A}$ and $\overline{B}$ are two complex numbers,

$$\frac{\overline{B}}{\overline{A}} = \frac{Be^{i\beta}}{Ae^{i\alpha}} = \frac{B}{A}e^{i(\beta-\alpha)} \quad \text{and} \quad \left|\frac{\overline{B}}{\overline{A}}\right| = \frac{B}{A} \tag{4.35}$$

By using Eqs. 4.34 to express the denominator of Eq. 4.33 in polar form and Eq. 4.35 to obtain the amplitude and phase of the quotient in Eq. 4.33, we obtain

$$\boxed{H_0(r) = |\overline{H}_0(r)| = \frac{U(r)}{U_0} = \frac{1}{[(1-r^2)^2 + (2\zeta r)^2]^{1/2}}} \tag{4.36a}$$

and

$$\boxed{\tan\alpha(r) = \frac{2\zeta r}{1 - r^2}} \tag{4.36b}$$

---

[5] In later chapters the dimensional form of the frequency-response function, $\overline{H}_{u/p}$, is employed.

4.3 Complex Frequency Response

which are the same results as those obtained in Eqs. 4.21. Thus, the complex frequency response $\overline{H}_0(r)$ contains both the magnitude and phase of the steady-state response, and Eqs. 4.34 and 4.35 can be used to extract this magnitude and phase information quite easily.

In summary, the four steps employed in using complex vectors to determine the steady-state response are:

1. Write the differential equation in terms of complex excitation and complex response, Eq. 4.28.
2. Assume a solution with complex amplitude $\overline{U}$, as in Eq. 4.30.
3. Substitute the assumed response into the differential equation to get an expression for the complex frequency response $\overline{H}_0(r)$.
4. Use Eqs. 4.34 and 4.35 to obtain the amplitude and phase of the complex frequency response.

The force vector polygon employed in Section 4.2 can now be related directly to the complex differential equation, Eq. 4.28. By differentiating Eq. 4.30, we obtain

$$\dot{\overline{u}} = i\Omega\overline{U}e^{i\Omega t} = i\Omega\overline{u}$$
$$\ddot{\overline{u}} = -\Omega^2\overline{U}e^{i\Omega t} = -\Omega^2\overline{u}$$
(4.37)

Figures 4.4 and 4.5 can thus be relabeled in terms of complex vectors as shown in Fig. 4.8a, and the force vector polygon in Fig. 4.8b represents Eq. 4.28 directly. The resulting frequency-response magnitude and phase, given by Eqs. 4.21a and b, were plotted in Fig. 4.6a and b, respectively.

Equation 4.33 can be converted to the rectangular (vector) form of Eq. 4.34a by multiplying the numerator and denominator of Eq. 4.33 by the complex conjugate of the denominator. Then

$$\overline{H}_0(r) = \frac{1}{(1-r^2) + i(2\zeta r)} \frac{(1-r^2) - i(2\zeta r)}{(1-r^2) - i(2\zeta r)}$$

or

$$\overline{H}_0(r) = \Re(\overline{H}_0) + i\Im(\overline{H}_0) = \frac{1-r^2}{(1-r^2)^2 + (2\zeta r)^2} + i\left[\frac{-2\zeta r}{(1-r^2)^2 + (2\zeta r)^2}\right] \quad (4.38)$$

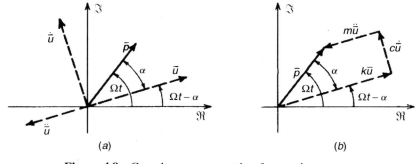

Figure 4.8 Complex vector notation for rotating vectors.

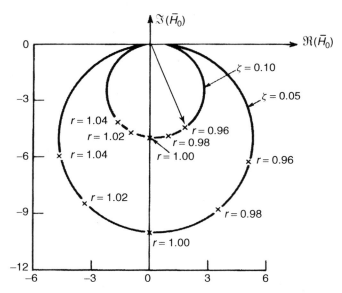

Figure 4.9 Vector-response plots for the steady-state vibration of two viscous-damped systems.

The magnitude and phase information can be combined in a single plot by plotting the vector components of Eq. 4.38 on the complex plane. The resulting plot is called a *vector-response plot* or *Nyquist plot*. Figure 4.9 shows vector-response plots for damping factors $\zeta = 0.05$ and $\zeta = 0.10$. One vector, for $\zeta = 0.10$ and $r = 0.96$, is shown. Note that in these plots the frequency ratio r is a parameter, so frequency ratio values must be marked at points along the plotted curves.

The vector-response diagram is very useful in examining experimental results in structural dynamics, as will be seen later, especially in Chapter 18. It can also be helpful to plot $\Re(\overline{H}_0)$ versus frequency and $\Im(\overline{H}_0)$ versus frequency, that is, the real part of $\overline{H}_0$ versus frequency and the imaginary part of $\overline{H}_0$ versus frequency.

4.4 VIBRATION ISOLATION: FORCE TRANSMISSIBILITY AND BASE MOTION

Having considered the basic frequency-response behavior of an SDOF system, we now consider two important topics related to frequency response: (1) the force that is transmitted through the spring and dashpot of an SDOF system to the supporting fixed base, and (2) the motion imparted to a mass when the base to which its spring and dashpot supports are attached is moving. Figure 4.10 illustrates these situations. The former situation arises when, for example, a machine is attached to a floor structure by shock-isolation mounts, which may be modeled as a combination of a spring and a viscous dashpot. There are numerous applications of the moving-base problem: earthquake motion of a building, motion of a car over a rough road, and so on. It is convenient, although not essential, to use the complex frequency response technique to solve these two vibration-isolation problems.

4.4 Vibration Isolation: Force Transmissibility and Base Motion

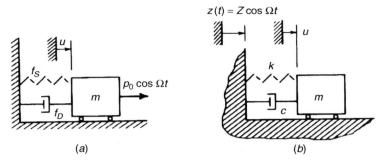

Figure 4.10 Vibration-isolation situations: (*a*) Force transmitted to stationary base, and (*b*) excitation by a moving base.

4.4.1 Force Transmissibility

From Fig. 4.8*b* the force transmitted to the base can be written in vector form as

$$\overline{f}_{tr} = \overline{f}_S + \overline{f}_D = k\overline{u} + c\dot{\overline{u}} \tag{4.39}$$

Incorporating Eqs. 4.30 and 4.37*a*, we get

$$\overline{f}_{tr} = (k + ic\Omega)\overline{U}e^{i\Omega t} \tag{4.40}$$

The expression for $\overline{U}$ in Eq. 4.33 can be inserted into Eq. 4.40 to give

$$\overline{f}_{tr} = \frac{(k + ic\Omega)U_0}{(1 - r^2) + i(2\zeta r)} e^{i\Omega t} \tag{4.41}$$

or

$$\overline{f}_{tr} = \frac{1 + i(2\zeta r)}{(1 - r^2) + i(2\zeta r)} kU_0 e^{i\Omega t} \tag{4.42}$$

Since $|e^{i\Omega t}| = 1$, and since from Eq. 4.35, $|\overline{B}/\overline{A}| = |\overline{B}|/|\overline{A}|$, the magnitude of the force transmitted is found to be

$$\boxed{|\overline{f}_{tr}| = \frac{[1 + (2\zeta r)^2]^{1/2}}{[(1 - r^2)^2 + (2\zeta r)^2]^{1/2}} p_0} \tag{4.43}$$

The *transmissibility* is defined as the ratio of the magnitude of the dynamic force $\overline{f}_{tr}$ transmitted to the force $p_0 = kU_0$ that would be transmitted to the base if the force p_0 were applied statically. Thus,

$$\text{TR} \equiv \frac{|\overline{f}_{tr}|}{p_0} = H_0[1 + (2\zeta r)^2]^{1/2} \tag{4.44}$$

Figure 4.11 shows the value of the transmissibility TR as a function of frequency ratio r. Two important conclusions can be drawn from the curves of Fig. 4.11: (1) the force transmitted to the base dynamically is less than the static force only if $r > \sqrt{2}$,

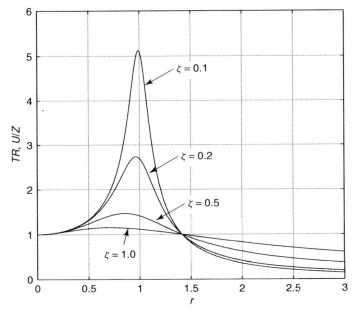

Figure 4.11 Transmissibility and absolute response to base excitation.

and (2) <u>decreasing</u> the damping decreases the transmitted force, provided that $r > \sqrt{2}$. Further details on the design of vibration isolation systems may be found in Refs. [4.1] to [4.3].

4.4.2 Base Motion

The equations of motion for the absolute motion, u, and the relative motion, $w = u - z$, were derived in Example 2.2. We consider the case when $p(t) = 0$ but the base undergoes simple harmonic motion $z = Z \cos \Omega t$. Based on Eq. 4.28, the complex equations of motion for base excitation can be written

$$m\ddot{\bar{u}} + c\dot{\bar{u}} + k\bar{u} = k\bar{z} + c\dot{\bar{z}} = (k + ic\Omega)Ze^{i\Omega t} \tag{4.45}$$

for absolute motion u, and

$$m\ddot{\bar{w}} + c\dot{\bar{w}} + k\bar{w} = -m\ddot{\bar{z}} = m\Omega^2 Z e^{i\Omega t} \tag{4.46}$$

for relative motion w.

As in Eq. 4.30, let us assume complex steady-state responses of the form

$$\bar{u} = \overline{U}e^{i\Omega t}, \qquad \bar{w} = \overline{W}e^{i\Omega t} \tag{4.47}$$

Then we get the complex frequency-response functions

$$\boxed{\overline{H}_{u/z} \equiv \frac{\overline{U}}{Z} = \frac{k + ic\Omega}{(k - m\Omega^2) + ic\Omega} = \frac{1 + i(2\zeta r)}{(1 - r^2) + i(2\zeta r)}} \tag{4.48}$$

and

$$\overline{H}_{w/z} \equiv \frac{\overline{W}}{\overline{Z}} = \frac{m\Omega^2}{(k - m\Omega^2) + ic\Omega} = \frac{r^2}{(1 - r^2) + i(2\zeta r)} \quad (4.49)$$

From the complex frequency-response functions above we can determine the magnitude of the absolute and relative responses as functions of frequency:

$$\frac{U}{Z} = \frac{|\overline{U}|}{Z} = \frac{[1 + (2\zeta r)^2]^{1/2}}{[(1 - r^2)^2 + (2\zeta r)^2]^{1/2}} = H_0[1 + (2\zeta r)^2]^{1/2} \quad (4.50)$$

and

$$\frac{W}{Z} = \frac{|\overline{W}|}{Z} = \frac{r^2}{[(1 - r^2)^2 + (2\zeta r)^2]^{1/2}} = r^2 H_0 \quad (4.51)$$

Since the expression for U/Z given by Eq. 4.50 is the same as the expression for the transmissibility TR, the plot of the absolute response is given by Fig. 4.11. The relative response is shown in Fig. 4.12.

The following important conclusions can be drawn from the relative motion frequency-response function shown in Fig. 4.12:

1. When $\Omega \ll \omega_n$ (i.e., $r \ll 1$), there is little relative motion between the mass and the base; that is, the mass moves with the base.
2. When $\Omega \approx \omega_n$, the usual resonance phenomenon is observed. That is, for small base motion there is a large amplitude of relative motion with only the damping force limiting the amplitude.

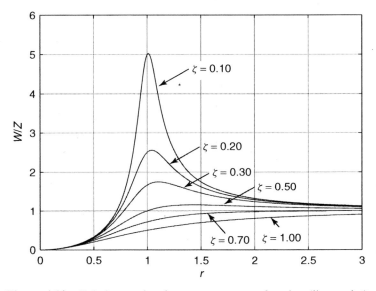

Figure 4.12 Relative motion frequency-response function (linear plot).

3. When $\Omega \gg \omega_n$, the inertia of the mass keeps it from moving much, so that the relative motion consists primarily of the base moving relative to the mass.

The analysis of base motion above will be very important to the study of motion-measuring devices in Section 4.5.

Example 4.3 A vehicle is a complex system with many degrees of freedom. However, the following SDOF analytical model may be employed in a "first-approximation" study of the ride quality of a vehicle. The steady-state magnification factor for the vehicle's absolute motion is to be determined for the vehicle when fully loaded and when empty if the vehicle is traveling at 100 km/h over a road whose surface has a sinusoidally varying roughness with a spatial "period" (see Fig. 1) of 4 m. The mass of the vehicle is 1200 kg when fully loaded and 400 kg when empty, and the effective spring constant is 400 kN/m. The damping factor is $\zeta_f = 0.4$ when the vehicle is fully loaded.

Make the following assumptions:

- As the vehicle moves forward at constant speed, only the vertical motion, $u(t)$, needs to be considered.
- The tires are infinitely stiff; that is, $z(t)$ represents the motion of the axle of the vehicle.
- The tires remain in contact with the road.

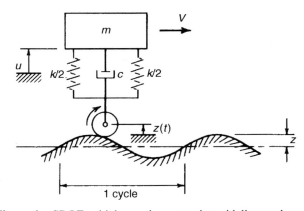

Figure 1 SDOF vehicle moving over sinusoidally rough road.

SOLUTION   1. Compute the excitation frequency. The speed of the vehicle and the period of the bumpiness can be combined to give

$$\Omega = \frac{(100{,}000 \text{ m/h})(2\pi \text{ rad/cycle})}{(4 \text{ m/cycle})(3600 \text{ s/h})} = 43.63 \text{ rad/s} \tag{1}$$

2. Compute the damping factor for the empty vehicle. From Eqs. 3.4,

$$c = 2\zeta\sqrt{km} \tag{2}$$

Table 1 Magnification-Factor Calculations for Two Automobiles

Formula	Empty	Full
$\omega_n = \sqrt{k/m}$	$[400(10^3)/400]^{1/2} = 31.62$ rad/s	$[400(10^3)/1200]^{1/2} = 18.26$ rad/s
$r = \Omega/\omega_n$	$43.63/31.62 = 1.380$	$43.63/18.26 = 2.390$
$2\zeta r$	$2(0.693)(1.380) = 1.912$	$2(0.4)(2.390) = 1.912$
$(1-r^2)^2$	$[1-(1.380)^2]^2 = 0.8170$	$[1-(2.390)^2]^2 = 22.20$
$[1+(2\zeta r)^2]^{1/2}$	$[1+(1.912)^2]^{1/2} = 2.158$	$[1+(1.912)^2]^{1/2} = 2.158$
$[(1-r^2)^2 + (2\zeta r)^2]^{1/2}$	$[0.8170+(1.912)^2]^{1/2} = 2.115$	$[22.20+(1.912)^2]^{1/2} = 5.085$
U/Z	$2.158/2.115 = 1.020$	$2.158/5.085 = 0.4243$

Since c and k do not change, but m does,

$$c = 2\zeta_f\sqrt{km_f} = 2\zeta_e\sqrt{km_e} \tag{3}$$

where the subscripts designate the full and empty vehicle configurations. Thus,

$$\zeta_e = \zeta_f \left(\frac{m_f}{m_e}\right)^{1/2} = 0.4\left(\frac{1200}{400}\right)^{1/2} = 0.693 \tag{4}$$

Note that the damping factor is less for the full vehicle than for the empty vehicle.

3. Compute the magnification factors. From Eqs. 4.50 and 4.36a,

$$\frac{U}{Z} = \frac{[1+(2\zeta r)^2]^{1/2}}{[(1-r^2)^2 + (2\zeta r)^2]^{1/2}} \tag{5}$$

The calculations are presented in Table 1. Note that both the damping factor and the natural frequency change when the vehicle weight changes, and that both of these factors enter into the calculation of the magnification factor.

4.5 VIBRATION MEASURING INSTRUMENTS: ACCELEROMETERS AND VIBROMETERS

In Chapter 1 the importance of vibration testing was discussed. Figure 4.13 shows the major stages of a vibration measurement system. The system consists of a motion detector-transducer, an intermediate signal modification system (e.g., amplifier), and a display system (e.g., oscilloscope). The motion quantity to be measured may be displacement, velocity, or acceleration. The purpose of the detector is to detect the desired motion quantity; the *transducer* produces an output that is proportional to the input motion but of different form. The most widely used motion transducer is the accelerometer, a device that senses acceleration and produces an output voltage or an output charge that is proportional to the input acceleration. Other motion transducers are the vibrometer for measuring displacement, the electrodynamic velocity transducer, the eddy-current displacement sensor, and the optical interferometer.[4.4,4.5]

102   Response of SDOF Systems to Harmonic Excitation

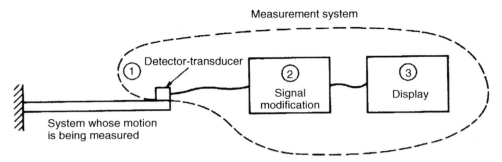

Figure 4.13   Stages in a motion measurement system.

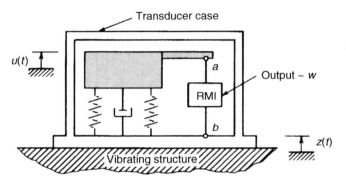

Figure 4.14   Seismic transducer.

Vibrometers and accelerometers are called *seismic transducers*, transducers that employ a spring–mass system to measure motion. The behavior of seismic transducers can be analyzed using the results of Section 4.4. Figure 4.14 shows a schematic diagram of a seismic transducer, whose principal ingredient is an SDOF vibratory system. A mass is attached by a spring to the base of the transducer. Viscous damping is sometimes, but not always, provided.

The base of the transducer is firmly attached to the specimen whose motion is to be measured. A relative motion instrument (RMI) measures, as a function of time, the relative motion between the moving mass (point a) and the base (point b). The RMI usually provides an electrical signal as its output, the signal being proportional to the relative displacement between points a and b. Thus, it is necessary to consider how the relative motion

$$w = u - z \tag{4.52}$$

is related to the base motion $z(t)$. The response to harmonic base motion $z = Z \cos \Omega t$ is considered.

Vibrometer   A *vibrometer* is a seismic instrument whose output is to be proportional to the displacement of the base; that is, $w(t)$ is to be proportional to $z(t)$. From Eq. 4.51, the steady-state relative displacement amplitude is related to the input displacement amplitude by

$$\frac{W}{Z} = \frac{r^2}{[(1-r^2)^2 + (2\zeta r)^2]^{1/2}} \tag{4.53}$$

Figure 4.12 is a plot of this frequency-response function. From this figure it can be seen that $W/Z \to 1$ as r becomes very large. To be useful in measuring the displacement $z(t)$, then, a vibrometer must have a low natural frequency relative to the excitation frequency. Since a system with a low natural frequency has a (relatively) large mass and soft spring, and consequently has a large static displacement, vibrometers are seldom used other than for some seismological measurements.

Accelerometer An *accelerometer* is a seismic instrument whose output is proportional to the base acceleration, $\ddot{z}(t)$. Consider first the steady-state output, $w_p(t)$, due to harmonic base motion. The base acceleration is

$$a_z(t) = \ddot{z} = -\Omega^2 Z \cos \Omega t \tag{4.54}$$

We need to determine how the output $w_p(t)$ can be employed to represent the time history of the acceleration $a_z(t)$, whose amplitude is $A = \Omega^2 Z$. Equation 4.53 gives the amplitude of w_p, which we can write in the nondimensionalized *acceleration frequency-response function* form

$$\boxed{\frac{\omega_n^2 W}{A} = \frac{1}{[(1-r^2)^2 + (2\zeta r)^2]^{1/2}} = H_0} \tag{4.55}$$

where H_0 is the nondimensionalized displacement frequency-response function plotted in Fig. 4.6a. From Eq. 4.49 it can be seen that $w_p(t)$ has a phase lag relative to $z(t)$ given by

$$\tan \alpha_w = \frac{2\zeta r}{1-r^2} \tag{4.56}$$

Thus,

$$w_p(t) = \frac{1}{\omega_n^2} A H_0 \cos \Omega \left(t - \frac{\alpha_w}{\Omega} \right) \tag{4.57}$$

For the accelerometer to be useful over a range of frequency without having to calibrate it at each frequency of interest [in effect, determining $H_0(\Omega)$], it is desirable to select an operating frequency range within which $H_0(\Omega)$ is approximately constant. From Fig. 4.6a it can be seen that this condition holds for $r \ll 1$, where $H_0(\Omega) \approx 1$.

Consider a more general acceleration input of the form

$$a_z(t) = A_1 \cos \Omega_1 t + A_2 \cos \Omega_2 t \tag{4.58}$$

The desired form of the steady-state output is

$$w_p(t) = C \left[A_1 \cos \Omega_1 (t-\tau) + A_2 \cos \Omega_2 (t-\tau) \right] \tag{4.59}$$

that is, the output may be shifted along the time axis by an amount τ and may be scaled by a constant C, but the components should not be shifted in time by different amounts or scaled by different factors. If the former occurs, the transducer is said to introduce *phase distortion*; if the latter occurs, it is said to produce *amplitude distortion*.

Most modern accelerometers have very low damping. Hence, they produce little phase distortion. To minimize amplitude distortion, the natural frequency of the accelerometer must be much greater than the highest input frequency, for example, $\omega_n > 10\,\Omega_{\max}$. By introducing damping of $\zeta \approx 0.7$, the frequency range of an accelerometer can be extended to about $0.6\,\omega_n$. References [4.6] and [4.7] may be consulted for further discussion of accelerometer performance characteristics and the factors to be considered in selecting an accelerometer.

4.6 USE OF FREQUENCY-RESPONSE DATA TO DETERMINE THE NATURAL FREQUENCY AND DAMPING FACTOR OF A LIGHTLY DAMPED SDOF SYSTEM

In Section 3.5, procedures for using free vibration response to determine the natural frequency and damping factor of a lightly damped SDOF system were discussed. Forced vibration using harmonic or nonharmonic excitation may also be employed to determine these system parameters experimentally. Forced-vibration techniques that may be employed to determine system parameters for MDOF systems are discussed in Chapter 18 and in Refs. [4.8] and [4.9]. Here we consider the viscous-damped SDOF system shown in Fig. 3.1.

It is assumed that frequency-response information has been determined experimentally. This may be plotted in nondimensionalized magnitude/phase form as in Fig. 4.6 or 4.7, or in vector response plot form as in Fig. 4.9. However, the static displacement $U_0 = p_0/k$ is seldom available, and the undamped natural frequency ω_n is not known until the experimental data are analyzed. Therefore, experimental data are usually plotted in dimensional form of magnitude U/p_0 versus the forcing frequency f and phase angle α versus f. Simulated experimental magnitude data and phase data taken at a uniform frequency spacing of Δf hertz are shown in Fig. 4.15a and b, respectively. From this type of frequency-response data we wish to determine the undamped natural frequency ω_n, the damping factor ζ, and possibly the stiffness k or mass m.

From Eq. 4.21a (or 4.36a), we can obtain the following *frequency-response function* in dimensional form:

$$\boxed{\,|\overline{H}_{u/p}| \equiv \frac{U}{p_0} = \frac{1/k}{[(1-r^2)^2 + (2\zeta r)^2]^{1/2}}\,} \qquad (4.60a)$$

and from Eq. 4.21b (or 4.36b),

$$\boxed{\,\tan \alpha = \frac{2\zeta r}{1 - r^2}\,} \qquad (4.60b)$$

where, in terms of the forcing frequency f,

$$\boxed{\,r = \frac{\Omega}{\omega_n} = \frac{2\pi f}{\omega_n}\,} \qquad (4.60c)$$

4.6 Frequency-Response Data to Determine the Natural Frequency and Damping Factor

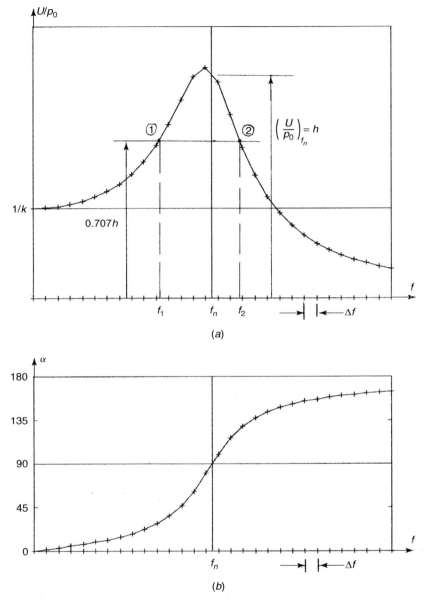

Figure 4.15 (a) FRF magnitude U/p_0 versus frequency f for an SDOF system (linear plot); (b) phase lag angle versus frequency (linear plot).

Undamped Natural Frequency As shown in Fig. 4.6a, when the damping is small, a sharp resonance peak occurs, with the maximum amplitude occurring at an excitation frequency just slightly below the undamped natural frequency. From Fig. 4.6b it is seen that the phase lag of the steady-state response is 90° when the excitation frequency is equal to the natural frequency of the system. From Fig. 4.9 it can be seen that in the vicinity of $r = 1$ (i.e., $\Omega = \omega_n$) the increment, Δs, of arc length along the frequency response curve corresponding to a given frequency increment (e.g., $\Delta r = 0.01$ or $\Delta \Omega = 0.01\omega_n$) is a maximum; that is, $\Delta s/\Delta\Omega = $ max when $\Omega = \omega_n$. Thus, the undamped natural frequency of an SDOF system may be determined from the frequency-response data by noting the excitation frequency at which any of the following occurs:

- The response lags the input by 90° [this also implies that $\Re(\overline{U}) = 0.0$]. In Figs. 4.15a and b this frequency is indicated by a vertical line at the frequency corresponding to 90° phase lag.
- The response magnitude is at its maximum value. (Depending on the level of damping, maximum response occurs at a frequency slightly lower than the natural frequency, as it does in Fig. 4.15a.)
- The spacing on the vector-response plot, $\Delta s/\Delta f$, is a maximum.

Damping Factor From Eq. 4.60a, the amplitude of U/p_0 at resonance is given by

$$\left(\frac{U}{p_0}\right)_{r=1} = \frac{1/k}{2\zeta} \tag{4.61}$$

However, the use of this equation to determine ζ would require determination of k, which is rarely available. Consequently, this equation cannot be employed for determining the damping factor ζ. Another method, called the *half-power method*, will be described. A portion of a frequency-response curve is shown in Fig. 4.15a. The frequencies above and below resonance at which the response amplitude is $\sqrt{2}/2$ times the resonant response amplitude are referred to as the *half-power points*. Let their frequencies be called f_1 and f_2, with corresponding frequency ratios r_1, and r_2. These frequencies may be obtained by letting $(U/p_0)_i = (\sqrt{2}/2)(U/p_0)_{r=1}$ and using Eq. 4.60a. Upon squaring Eq. 4.60a, we get

$$\left(\frac{U}{p_0}\right)^2 = \frac{(1/k)^2}{(1-r^2)^2 + (2\zeta r)^2} \tag{4.62}$$

Letting $(U/p_0)_i = (\sqrt{2}/2)(U/p_0)_{r=1}$, and using Eq. 4.61, we obtain

$$\frac{1}{2}\left(\frac{1/k}{2\zeta}\right)^2 = \frac{(1/k)^2}{(1-r_i^2)^2 + (2\zeta r_i)^2} \tag{4.63}$$

where $r_i = r_1$ or r_2. Equation 4.63 may be rewritten

$$r_i^4 - 2(1 - 2\zeta^2)r_i^2 + (1 - 8\zeta^2) = 0 \tag{4.64}$$

whose roots are given by

$$r_i^2 = (1 - 2\zeta^2) \pm 2\zeta\sqrt{1 + \zeta^2} \tag{4.65}$$

Assuming that $\zeta \ll 1$ and neglecting higher-order terms in ζ, we arrive at the result

$$r_i^2 = 1 \pm 2\zeta \tag{4.66}$$

Using the binomial expansion, we get

$$r_2 = (1 + 2\zeta)^{1/2} = 1 + \tfrac{1}{2}(2\zeta) + \cdots$$
$$r_1 = (1 - 2\zeta)^{1/2} = 1 - \tfrac{1}{2}(2\zeta) + \cdots \quad (4.67)$$

or since $\zeta \ll 1$,

$$r_2 - r_1 = 2\zeta \quad (4.68)$$

The *half-power method* for measuring the damping factor is based on Eq. 4.68. The procedure is as follows:

1. Determine the undamped natural frequency, f_n hertz, by one of the methods described above and then determine the resonant response amplitude, $(U/p_0)_{r=1}$.
2. Note the points on the response amplitude curve where the amplitude is $(\sqrt{2}/2)(U/p_0)_{r=1}$. Call the corresponding frequencies (in hertz) f_1 and f_2.
3. Then, from Eq. 4.68,

$$\boxed{\zeta = \frac{1}{2}\frac{f_2 - f_1}{f_n}} \quad (4.69)$$

The accuracy with which ζ is determined using Eq. 4.69 depends on the frequency resolution in the original frequency-response data.

A related procedure using spacing of points along the vector response plot may also be used in determining damping.[4.10,4.11] The procedure is sometimes referred to as the *Kennedy–Pancu Method* or *Circle-Fit Method*. This method is described in Section 18.5.2.

By the methods indicated above, with good-quality FRF data it is feasible to obtain the natural frequency of a SDOF system to about ±5% and the damping factor to about ±20%.

Modal Stiffness and Modal Mass   As indicated in Fig. 4.15a and given by Eq. 4.60a, it is theoretically possible to obtain the stiffness k from the measured value of U/p_0 at $r = 0$, that is, from a static-deflection test. However, the transducers that are used to acquire displacement and force frequency-response data do not permit data to be acquired below a forcing frequency of about 10 Hz, so k and m are seldom determined from frequency-response data by simple procedures such as those described in this section.

4.7 EQUIVALENT VISCOUS DAMPING

Damping is present in all oscillatory systems. The primary effect of the damping is to remove energy from the system. This loss of energy from the damped system results in the decay of amplitude of free vibration, as shown in Chapter 3. In steady-state forced vibration the loss of energy is balanced by the energy that is supplied by the excitation. There are many different mechanisms that can cause damping in a system: internal friction, fluid resistance, sliding friction, and so on. *Linear viscous damping* provides the simplest mathematical model of damping: namely, a force that is directly

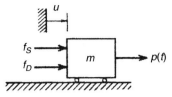

Figure 4.16 Damping and elastic forces acting on a mass.

proportional to the velocity. Even when the true damping in a system is far more complex than linear viscous damping, it may be possible to retain the simplicity of the linear viscous damping model by introducing an "equivalent viscous damping." In this section we indicate how such equivalent viscous damping may be determined.

As indicated above, one of the principal effects of damping in a system is to remove energy from the system. Hence, the concept of *equivalent viscous damping* is based on the equivalence of energy removed by a viscous damping mechanism and the actual energy removed by the given nonviscous damping mechanism.

Figure 4.16 shows two forces acting on a free body of a mass, the elastic spring force f_S, which is associated with the potential energy of the system, and the damping force f_D, which is associated with energy dissipation. The work W_S done by the spring force and the work W_D done by the damping force are given by

$$W_S = \int_{u_i}^{u_f} f_S \, du = \int_{t_i}^{t_f} f_S \dot{u} \, dt \tag{4.70}$$

and

$$W_D = \int_{t_i}^{t_f} f_D \dot{u} \, dt \tag{4.71}$$

respectively.

Energy dissipation is usually calculated for one cycle of harmonic motion. Consider the energy dissipation in an SDOF system with a viscous dashpot. Then the damping force is

$$f_D = -c\dot{u} \tag{4.72}$$

For steady-state motion the displacement and velocity are given by

$$\begin{aligned} u &= U \cos(\Omega t - \alpha) \\ \dot{u} &= -\Omega U \sin(\Omega t - \alpha) \end{aligned} \tag{4.73}$$

The work done per cycle by the viscous-damping force $f_D = -c\dot{u}$ is thus

$$\begin{aligned} W_D &= \int_0^{2\pi/\Omega} (-c\dot{u})\dot{u} \, dt \\ &= -c\Omega^2 U^2 \int_0^{2\pi/\Omega} \sin^2(\Omega t - \alpha) \, dt \end{aligned}$$

or

$$W_D = -\pi c \Omega U^2 \tag{4.74}$$

4.7 Equivalent Viscous Damping

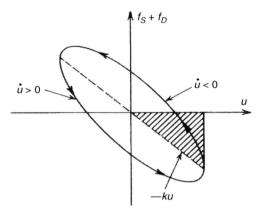

Figure 4.17 Force versus displacement for a system with a linear spring and a viscous dashpot.

The energy dissipated per cycle by viscous damping is $\pi c \Omega U^2$. Thus, the energy loss per cycle due to linear viscous damping is proportional to the damping coefficient c, the excitation frequency Ω, and the square of the amplitude U.

On Fig. 4.17 is plotted the total force $f_S + f_D$ acting on the SDOF system in Fig. 4.16, where $f_S = -ku$, $f_D = -c\dot{u}$, and the system is executing steady-state harmonic motion. The lower portion of the curve is for motion with $\dot{u} > 0$ (i.e., when the mass is moving to the right). The upper segment of the curve is for the portion of the cycle when $\dot{u} < 0$ (i.e., when the mass is moving to the left). The triangular area marked by diagonal crosshatching is the peak potential energy stored elastically; the area inside the elliptical curve is the energy dissipated per cycle by the damping force, that is, $|\mathcal{W}_D|$. The loop, which is frequently referred to as a *hysteresis loop*, is characteristic of dissipative forces.

If the damping in a system is not of the linear viscous-damping type, an *equivalent viscous-damping coefficient* may be defined by

$$c_{eq} = -\frac{\mathcal{W}_D}{\pi \Omega U^2} \tag{4.75}$$

where $\mathcal{W}_D$ is the energy dissipated by the nonviscous-damping mechanism. Example 4.4 illustrates the calculation of c_{eq}.

Example 4.4 Bodies moving with moderate speed in a fluid (Fig. 1) experience a resisting force that is proportional to the square of the speed. Determine the equivalent viscous-damping coefficient for this type of damping.

SOLUTION Let the damping force be expressed by the equation

$$f_D = \pm a\dot{u}^2 \tag{1}$$

where the negative sign is used when $\dot{u}$ is positive, and the positive sign when $\dot{u}$ is negative. Assume harmonic motion with time measured from the position of extreme

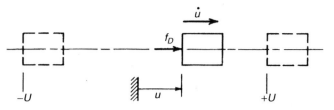

Figure 1 Damping force acting on a moving mass.

negative displacement. Then

$$u = -U\cos\Omega t, \qquad \dot{u} = \Omega U \sin\Omega t \tag{2}$$

The energy dissipated per cycle is

$$W_D = -2\int_{-U}^{U} a\dot{u}^2\, du = -2\int_{0}^{\pi/\Omega} a\dot{u}^3\, dt \tag{3}$$
$$= -2a\Omega^3 U^3 \int_{0}^{\pi/\Omega} \sin^3\Omega t\, dt$$

Then

$$W_D = -\tfrac{8}{3}a\Omega^2 U^3 \tag{4}$$

From Eq. 4.75,

$$c_{eq} = -\frac{W_D}{\pi\Omega U^2} = \frac{\tfrac{8}{3}a\Omega^2 U^3}{\pi\Omega U^2}$$

or

$$c_{eq} = \frac{8}{3\pi}a\Omega U \qquad\qquad \textbf{Ans.} \tag{5}$$

It is convenient to define an equivalent damping factor ζ_{eq} by using Eq. 4.75 and Eqs. 3.4b and c. Then

$$\zeta_{eq} \equiv \frac{c_{eq}}{c_{cr}} = -\frac{W_D/\pi\Omega U^2}{2k/\omega_n} \tag{4.76}$$

But k can be determined from the elastic energy, shown crosshatched in Fig. 4.17.

$$W_S = \tfrac{1}{2}k U^2 \tag{4.77}$$

Thus,

$$\boxed{\zeta_{eq} = -\frac{W_D}{W_S}\frac{\omega_n}{4\pi\Omega}} \tag{4.78}$$

where Ω is the forcing frequency at which W_D is obtained.

4.8 STRUCTURAL DAMPING

The complex-frequency-response notation introduced in Section 4.3 is particularly well suited to the introduction of a type of damping frequently employed in structural dynamics analysis: for example, in aircraft vibration and flutter studies. This type of damping, called *structural damping*, is proportional to displacement but 180° out of phase with the velocity of a harmonically oscillating system. Theodorsen and Garrick, in an early study of flutter, used this form of damping.[4.12] The usual way in which it is introduced into an SDOF system's equation of motion is to write

$$\ddot{\bar{u}} + k(1 + i\gamma)\bar{u} = p_0 e^{i\Omega t} \quad (4.79)$$

where γ is the *structural damping factor*.[6] The quantity $k(1 + i\gamma)$ is called the *complex stiffness*. Assuming a solution of the form

$$\bar{u} = \overline{U} e^{i\Omega t} \quad (4.80)$$

and substituting this solution into Eq. 4.79, we obtain the following expression for the complex amplitude $\overline{U}$:

$$\overline{U} = \frac{p_0}{(k - m\Omega^2) + ik\gamma} \quad (4.81)$$

In nondimensionalized form, the *complex-frequency-response function* for an SDOF system with structural damping is

$$\boxed{\overline{H}_0 \equiv \frac{\overline{U}}{U_0} = \frac{1}{(1 - r^2) + i\gamma}} \quad (4.82)$$

By comparing the denominators of Eqs. 4.33 and 4.82, we see that the factor γ in the latter corresponds to the factor $2\zeta r$ in the former. Since, when damping factors are small (as is generally the case in structures), damping is effective primarily at frequencies in the vicinity of resonance, it can be seen that under harmonic excitation conditions, structural damping is effectively equivalent to viscous damping with

$$\zeta_{eq} = \frac{\gamma}{2} \quad (4.83)$$

From Eq. 4.82 the amplitude and phase angle of the frequency-response function for an SDOF system with structural damping may be found to be

$$H_0 \equiv \frac{U}{U_0} = \frac{1}{[(1 - r^2)^2 + \gamma^2]^{1/2}} \quad (4.84)$$

and

$$\tan \alpha = \frac{\gamma}{1 - r^2} \quad (4.85)$$

where, again, $\overline{U} = U e^{-i\alpha}$. An excellent study of harmonic excitation of systems having structural damping is that of Bishop and Gladwell.[4.11]

---

[6]In some of the literature the symbol g is used for the structural damping factor. However, because of possible confusion with the gravitational constant, the symbol γ is used here.

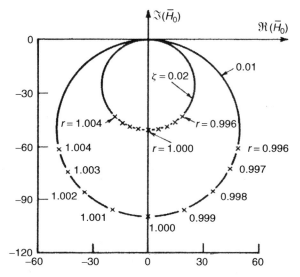

Figure 4.18 Vector-response plots of the steady-state response of structurally damped SDOF systems.

Figure 4.18 is a vector-response plot of Eq. 4.82. The following facts may be noted:

- At $r = 1$ (resonance),

$$\frac{\overline{U}}{U_0} = \frac{1}{i\gamma} = -i\frac{1}{\gamma}$$

Therefore,

$$\frac{U}{U_0} = \frac{1}{\gamma} \quad \text{and} \quad \alpha = 90° \quad \text{at } r = 1$$

- The vector-response plot for structural damping is a circle. The diameter of the circle is determined by the damping factor.
- The spacing between points of equal frequency difference is greatest in the vicinity of $r = 1$. That is, for a given $\Delta r = r_2 - r_1$, the spacing on the circle is greatest near $r = 1$.

These are some of the factors that are considered further in Chapter 18, *Introduction to Experimental Modal Analysis*.

REFERENCES

[4.1] A. D. Nashif, D. I. G. Jones, and J. P. Henderson, *Vibration Damping*, Wiley, New York, 1985.

[4.2] T. T. Soong and G. F. Dargush, *Passive Energy Dissipation Systems in Structural Engineering*, Wiley, New York, 1997.

[4.3] D. J. Inman, *Engineering Vibration*, 2nd ed., Prentice Hall, Upper Saddle River, NJ, 2001, pp. 359–419.

[4.4] A. D. Khazan, *Transducers and Their Elements*, Prentice Hall, Englewood Cliffs, NJ, 1994.

[4.5] R. Pallás-Areny and J. G. Webster, *Sensors and Signal Conditioning*, Wiley, New York, 1991.

[4.6] J. S. Wilson, "Performance Characteristics and the Selection of Accelerometers," *Sound and Vibration*, Vol. 12, 1978, pp. 24–29.

[4.7] K. G. McConnell, *Vibration Testing: Theory and Practice*, Wiley, New York, 1995.

[4.8] D. J. Ewins, *Modal Testing: Theory, Practice and Application*, 2nd ed., Research Studies Press, Baldock, Hertfordshire, England, 2000.

[4.9] N. M. M. Maia and J. M. M. Silva, eds., *Theoretical and Experimental Modal Analysis*, Wiley, New York, 1997.

[4.10] C. C. Kennedy and C. D. P. Pancu, "Use of Vectors in Vibration Measurement and Analysis," *Journal of Aeronautical Science*, Vol. 14, 1947, pp. 603–625.

[4.11] R. E. D. Bishop and G. M. L. Gladwell, "An Investigation into the Theory of Resonance Testing," *Philosophical Transactions*, Vol. 255, 1963, pp. 241–280.

[4.12] T. Theodorsen and I. E. Garrick, *Mechanism of Flutter: A Theoretical and Experimental Investigation of the Flutter Problem*, NACA Report 685, National Advisory Committee on Aeronautics, Washington, DC, 1940.

PROBLEMS

For problems whose number is preceded by a **C**, you are to write a computer program and use it to produce the plot(s) requested. *Note:* MATLAB .m-files for many of the plots in this book may be found on the book's website.

Problem Set 4.1

4.1 What frequency range $\Omega_l/\omega_n \leq \Omega/\omega_n \leq \Omega_u/\omega_n$ must be avoided if the steady-state response of an undamped system must be less than, or equal to, two times the static deflection due to p_0? Indicate how you arrived at your answer.

4.2 The undamped SDOF system shown in Fig. 4.1 has the following properties: The spring constant is $k = 50$ lb/in. and the mass m weighs $W = 8$ lb. The system is at rest when an excitation $p(t) = 20\sin(40t)$ lb begins. (**a**) Determine an expression for the steady-state response. (**b**) Determine an expression for the total response.

C 4.3 The undamped SDOF system shown in Fig. 4.1 has a spring stiffness $k = 40$ lb/in., and the mass weighs 38.6 lb. The system is initially at rest [i.e., $u(0) = 0$, $\dot{u}(0) = 0$], when an excitation $p(t) = 10\sin(\Omega t)$ lb begins. (**a**) Determine an expression for the steady-state response. (**b**) Determine an expression for the total response. (**c**) For the excitation frequency $\Omega = 10$ rad/sec, create plots of these expressions (i.e., the steady-state response and the total response). (See the two curves in Fig. 1 of Example 4.1.) Plot the curves for $0 \leq t \leq 2$ sec. (**d**) What is the ratio of the maximum total response to the maximum steady-state response for the excitation frequency in part (c)?

C 4.4 The undamped SDOF system of Fig. 4.1 is at rest at $t = 0$. Subsequently, it is subjected to harmonic excitation $p(t) = p_0 \sin \Omega t$. [The plots in Fig. 4.2 are for $\cos \Omega t$ excitation, that is, for $p(t) = p_0 \cos \Omega t$.] (**a**) Create similar "steady-state response" and "total response" plots for $\sin \Omega t$ excitation. Your plot for steady-state response should be identical to the plot in Fig. 4.2. Is that also true for your plot of total response? (**b**) Show that the total response values for $r = 0.5$ and $r = 2.0$ are special cases.

4.5 Harmonic excitation is to be used to determine the natural frequency and the mass of a nominally "undamped" SDOF system similar to the one shown in Fig. 4.1. At a frequency $\Omega = 6$ rad/s, the resonance condition is achieved; that is, the response tends to increase without bound. This frequency is therefore taken to be the undamped natural frequency; that is, $\omega_n = 6$ rad/s. Next, a mass of $\Delta m = 1$ kg is attached to mass m and the resonance test is repeated. This time resonance occurs at $\Omega = 5.86$ rad/s. **(a)** Determine the value of the mass m in kilograms. **(b)** Determine the value of the spring constant k in N/m.

4.6 The undamped SDOF system of Fig. 4.1 is at rest at $t = 0$. **(a)** Starting with Eq. 4.11, determine an expression for the response of this undamped SDOF system when, beginning at $t = 0$, it is subjected to harmonic excitation $p_0 \cos \Omega t$. **(b)** Let $\Omega = \omega_n + \Delta\omega$, where $\Delta\omega \ll \omega_n$. Show that the response can be written in the form

$$u(t) = A \sin\left(\frac{\Delta\omega\, t}{2}\right) \sin\left(\frac{2\omega_n + \Delta\omega}{2}\right) t$$

which appears to be an oscillation $A \sin[(2\omega_n + \Delta\omega)/2]t$ with slowly varying amplitude. (This is called a *beat phenomenon*.)

4.7 A control tab of an airplane elevator is hinged about axis A in the elevator. Although Problem 2.6 suggested a free vibration test for determining the mass moment of inertia I_A, the dynamic behavior of the tab also depends on the elastic stiffness of the control linkage, which is modeled in Fig. P4.7 as a torsional spring. Since k_θ cannot be measured statically, a resonance test is to be used. The elevator is held fixed, the tab is supported by spring k_1, and harmonic excitation is applied through spring k_2 as shown. **(a)** Show that the undamped natural frequency ω_n for the elevator and its control linkage alone is given by

$$\omega_n = \sqrt{\frac{k_\theta}{I_A}}$$

(b) If the resonance frequency obtained in the test using the setup in Fig. P4.7 is ω_r, determine an expression for k_θ in terms of ω_r, I_A, and other parameters shown in the sketch.

Problem Set 4.2

4.8 Figure 4.5 is a force vector polygon that corresponds to harmonic excitation with $\Omega < \omega_n$. Sketch similar force vector polygons for the following excitation frequencies: **(a)** the resonance case $\Omega = \omega_n$, and **(b)** a typical case for which $\Omega > \omega_n$. Discuss how the phase angle for this case differs from the phase angle in the $\Omega < \omega_n$ case shown in Fig. 4.5.

Problem Set 4.3

4.9 Imbalance in rotating machines is a common source of vibration excitation. Such a situation is illustrated schematically in Fig. P4.9. Let $M - m$ be the mass of the machine alone, and let m be the mass of the rotating imbalance, which rotates at speed Ω rad/s at eccentricity e. **(a)** Derive the equation of motion for vertical motion, $u(t)$, of the machine. **(b)** Use the complex-frequency-response method to derive an expression for the amplitude of the steady-state response of this system, expressed in the nondimensional form MU/me, and show that Fig. 4.12 is the frequency-response plot for this situation.

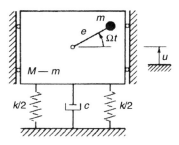

Figure P4.9

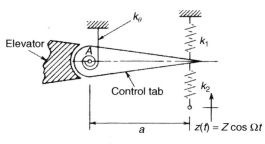

Figure P4.7

Problem Set 4.4

4.10 The SDOF system in Fig. P4.10 is subjected to harmonic excitation $z(t) = Z \cos \Omega t$ applied at point P. Express your answers to the following in terms of the givens: m, c, k, Z, and Ω. **(a)** Derive the equation of motion of the system with the absolute displacement $u(t)$ as the unknown. **(b)** Derive the equation of motion of the system with the relative displacement $w(t) = z - u$ as the unknown. **(c)** Determine expressions for ω_n and ζ for this system. **(d)** Determine expressions for the following complex-frequency-response functions: $\overline{U}/Z$ and $\overline{W}/Z$.

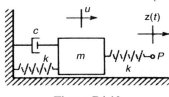

Figure P4.10

4.11 For the system shown in Fig. P2.3, use the complex-frequency-response method to determine expressions for the following: (a) $\overline{V}/Z$, where the excitation is $z(t) = Z\cos\Omega t$; and (b) amplitude V and phase angle α, where the steady-state response is $v(t) = V\cos(\Omega t - \alpha)$.

4.12 For the system shown in Fig. P2.3, use the complex-frequency-response method to obtain an expression for the maximum steady-state force in spring AB due to harmonic excitation $z(t) = Z\cos\Omega t$.

C 4.13 Imbalance in rotating machines is a common source of vibration excitation. Such a situation is illustrated schematically in Fig. P4.9. Let $M - m$ be the mass of the machine alone, and let m be the mass of the rotating unbalance, which rotates at speed Ω rad/s at eccentricity e. (a) Derive the equation of motion for vertical motion, $u(t)$, of the machine. (b) Derive an expression for the magnitude of the steady-state force that is transmitted from the mass to the base through the springs, expressed in the nondimensional form $|(\overline{f}_s)_{tr}|/ke$ (c) For a system with damping factor $\zeta = 0.1$, create a plot of the frequency response obtained in part (b), similar to the curves plotted in Fig. 4.11, except plotted to $r = 20$.

4.14 As illustrated in Fig. P4.14, a vibration-isolation block is to be installed in a laboratory so that the vibration from adjacent factory operations will not disturb certain experiments. If the isolation block weighs 1000 lb and the surrounding floor and foundation vibrate at 30 Hz with an amplitude of 0.01 in., determine the stiffness k of the isolation system, in lb/in., such that the isolation block will have a steady-state amplitude of only 0.002 in.

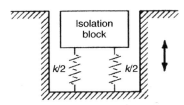

Figure P4.14

4.15 Rotating imbalance, as illustrated in Fig. P4.15, leads to harmonic forces. The nonrotating mass is $M - m$, and counterrotating masses ($m/2$) at eccentricity e rotate at an angular rate of Ω rad/s. (a) Determine the equation of motion for the system. (b) Use the complex-frequency-response method to determine an expression for $M\overline{u}/me$, where $\overline{u}$ is the steady-state response $\overline{u}(t) = \overline{U}e^{i\Omega t}$. (c) Use the complex-frequency-response method to obtain the total (dashpot + spring) force transmitted to the base in the nondimensional form $\overline{f}_{tr}/me\Omega^2$.

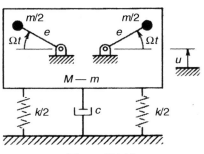

Figure P4.15

4.16 For the vehicle and "sinusoidal" road shown in Fig. 1 of Example 4.3, the following parameters are given ($f \to$ full; $e \to$ empty):

$$W_f = 3860 \text{ lb}, \; W_e = 2680 \text{ lb},$$
$$k = 2000 \text{ lb/in.}, \; \zeta_f = 0.2,$$
$$L = 10 \text{ ft/cycle}$$

Determine the speed of the vehicle in miles per hour that would produce a resonance condition (a) if the vehicle were empty, and (b) if the vehicle were fully loaded.

Problem Set 4.5

4.17 At a point on a vibrating structure the motion is given by

$$z = Z_1 \sin(\Omega_1 t) + Z_2 \sin(\Omega_2 t)$$
$$= 0.05 \sin(100\pi t) + 0.02 \sin(200\pi t)$$

(a) Determine an expression for the vibration record $w_p(t)$ that would be obtained with an accelerometer having a damping factor $\zeta = 0.70$ and a resonant frequency of 20 kHz. (b) Is any significant amplitude or phase distortion produced by the accelerometer?

Problem Set 4.6

4.18 (a) Using the data on the outer "circle" of Fig. 4.9, sketch a curve of $D_s \equiv |\bar{H}_0|$ versus the frequency ratio r for $0.96 \leq r \leq 1.04$. (b) Using the half-power method, verify that $\zeta = 0.05$, as shown on the curve. (If you need to do so, you may extrapolate the curve below $r = 0.96$ and above $r = 1.04$ and then estimate values for r_2 and r_1.)

4.19 Experimental vibration data might be obtained in the form of amplitude $U(f)$ versus frequency in hertz ($f = \Omega/2\pi$) (Fig. P4.19). If a system has 1% damping, estimate the frequency resolution, Δf, with which data should be taken in order to obtain a reasonably accurate estimate of the damping factor ζ for a SDOF system. Explain how you arrived at your answer.

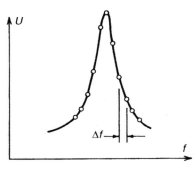

Figure P4.19

Problem Set 4.7

4.20 A force $p_0 \cos \Omega t$ acts on a mass that slides on a surface having a coefficient of kinetic friction of μ_k (Fig. P4.20; see Section 3.4). (a) Determine the equivalent viscous-damping coefficient. (b) Determine the response amplitude, U, of the system with equivalent viscous damping.

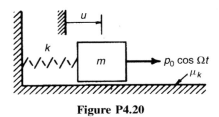

Figure P4.20

Problem Set 4.8

4.21 For harmonic motion, the structural damping force can be represented by the complex form shown in Eq. 4.79, that is, by the equation

$$\bar{f}_d = -i\gamma k \bar{u}$$

or by the real form

$$f_d = -\gamma k \dot{u} \frac{|u|}{|\dot{u}|} = -\gamma k |u| \, \text{sgn}(\dot{u})$$

(a) Sketch f_d versus u for one cycle of harmonic motion, where

$$u = -U \cos(\Omega t - \alpha)$$

(b) Determine an expression for the work done per cycle.
(c) Determine an expression for c_{eq}. (d) By substituting c_{eq} into Eq. 4.20a, determine an expression for U/U_0. Compare this with Eq. 4.84 and discuss any differences that you observe.

4.22 Starting with Eq. 4.82, show that the vector-response plot for structural damping is a circle with center on the imaginary axis and passing through the origin, as shown in Fig. 4.18 or P4.22.

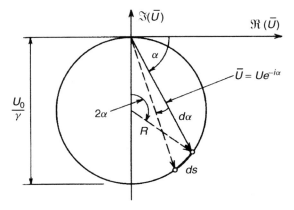

Figure P4.22

4.23 (a) Determine the rate of change of arc length along the circle with respect to frequency change by determining an expression for $ds/d(r^2)$. Use Fig. P4.22. (b) Illustrate how the damping level γ affects the derivative above by evaluating $ds/d(r^2)$ at $r = 1$.

5

Response of SDOF Systems to Nonperiodic Excitation

In Chapter 4 you studied the response of SDOF systems to sinusoidal excitation and learned that the steady-state response depends primarily on the damping factor of the system and the excitation frequency ratio. In this chapter we first consider the response of SDOF systems to several simple forms of excitation in order to gain some insight into what system parameters and excitation parameters govern that response. Then we consider two analytical procedures for determining the response of SDOF systems to more general excitation time histories: the *Convolution Integral Method* and the *Laplace Transform Method*. In Chapter 6 we discuss numerical procedures for determining the response of SDOF systems to general excitation histories, and in Chapter 7 treat both analytical and numerical procedures for determining the response of SDOF systems to periodic excitation.

Upon completion of this chapter you should be able to:

- Apply the classical differential equations method to determine the response of an SDOF system to an impulse loading at $t = 0$, a step loading applied at $t = 0$, or a ramp loading.
- Describe the difference in the response of a system loaded "rapidly" as contrasted to the response of the same system when loaded "slowly."
- Explain why it is that the maximum dynamic response to a pulse may occur after the excitation has ceased to act on the system.
- Use the convolution integral to determine the response of an SDOF system to simple loading histories.
- Use the Laplace transform to determine the response of an SDOF system to simple loading histories.

5.1 RESPONSE OF A VISCOUS-DAMPED SDOF SYSTEM TO AN IDEAL STEP INPUT

Let the prototype viscous-damped SDOF system shown in Fig. 3.1 be subjected to an ideal step input as shown in Fig. 5.1. The equation of motion is given by Eq. 3.1, which is repeated here:

$$m\ddot{u} + c\dot{u} + ku = p_0, \qquad t \geq 0 \tag{5.1}$$

Response of SDOF Systems to Nonperiodic Excitation

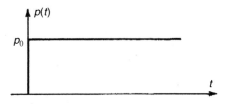

Figure 5.1 Ideal step input.

Let the system be at rest at $t = 0$, that is,

$$u(0) = \dot{u}(0) = 0 \tag{5.2}$$

The solution of Eq. 5.1 consists of a *particular solution* which, from Eq. 5.1, can be seen to be

$$u_p = \frac{p_0}{k} \tag{5.3}$$

and a *complementary solution* given (for $\zeta < 1$) by Eq. 3.29. Then

$$u = \frac{p_0}{k} + e^{-\zeta \omega_n t}(A_1 \cos \omega_d t + A_2 \sin \omega_d t) \tag{5.4}$$

Using the initial conditions to evaluate A_1 and A_2, we obtain the following *step response of an underdamped SDOF system*:

$$\boxed{u(t) = \frac{p_0}{k}\left[1 - e^{-\zeta \omega_n t}\left(\cos \omega_d t + \frac{\zeta \omega_n}{\omega_d} \sin \omega_d t\right)\right]} \tag{5.5}$$

A useful way of examining dynamic response is to consider the *response ratio*, or *dynamic load factor*, defined by[1]

$$R(t) = \frac{ku(t)}{p_{\max}} \tag{5.6}$$

Thus, the response ratio is the ratio of dynamic response to static deformation. For the ideal step input, $R(t)$ is given by

$$R(t) = 1 - e^{-\zeta \omega_n t}\left(\cos \omega_d t + \frac{\zeta \omega_n}{\omega_d} \sin \omega_d t\right) \tag{5.7}$$

A typical response-ratio plot is shown in Fig. 5.2. On the response-ratio plot, $R(t) = 1$ corresponds to the static displacement position. Because the load was applied instantaneously, there is an *overshoot*, and the system settles to the static value of 1 after a number of cycles of damped oscillation. The damping level determines the amount of the overshoot and the rate of decay of the oscillation about the static equilibrium position.

For an undamped system, Eq. 5.5 becomes

$$\boxed{u(t) = \frac{p_0}{k}(1 - \cos \omega_n t)} \tag{5.8}$$

---

[1] $(\cdot)_{\max}$ will be used as an abbreviation for *maximum absolute value*, that is, $\max_t |(\cdot)|$.

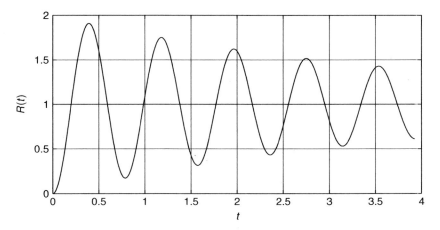

Figure 5.2 Response-ratio plot for a step input.

and $R_{\max} = 2$. Thus, when a load is applied instantaneously to an undamped system, a maximum displacement of twice the static displacement is attained. This is the reason that a safety factor of 2 is frequently applied to the design of structures that will be subjected to rapidly applied loads.

In the next section we consider the effect of terminating the load after a duration of t_d, or of linearly increasing the load over a rise time of t_r.

5.2 RESPONSE OF UNDAMPED SDOF SYSTEMS TO RECTANGULAR PULSE AND RAMP LOADINGS

In Section 5.1 the response of a damped SDOF system to an ideal step input was considered. It was noted that the maximum overshoot occurs after one-half cycle of the resulting oscillation, and that the presence of damping reduces the overshoot. Now we consider two questions: "What effect does the duration of the loading have on the response of an undamped system?" and "What effect does the rise time of the loading have on the response?"

5.2.1 Rectangular Pulse Loading

Figure 5.3 shows a rectangular pulse input and the response ratio for an undamped system for two representative cases:

$$\text{(a)} \quad t_d = \frac{5}{4} T_n > \frac{T_n}{2}; \qquad \text{(b)} \quad t_d = \frac{1}{8} T_n < \frac{T_n}{2}$$

where t_d is the *duration* of the rectangular pulse. The circles on the two curves in Fig. 5.3b indicate the respective t_d points on the curves.

From Fig. 5.3 it is clear that when $t_d \geq T_n/2$, the maximum occurs during the *forced-vibration era*, that is, prior to t_d. On the other hand, if $t_d < T_n/2$, the maximum occurs in the *residual-vibration era*, that is, after t_d. We can determine expressions for the maximum response that occurs in each case.

120 Response of SDOF Systems to Nonperiodic Excitation

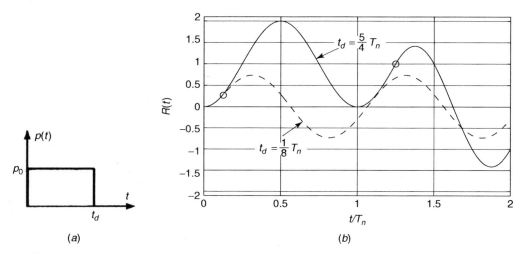

Figure 5.3 Response to a rectangular pulse input: (a) rectangular pulse; (b) response ratios.

Case 1: Forced-Vibration Era $(0 \leq t \leq t_d)$ For this case, $R(t)$ is the same as for an ideal step: namely,

$$R_1(t) = 1 - \cos \omega_n t \qquad (0 \leq t \leq t_d) \qquad (5.9)$$

and the maximum value is

$$(R_1)_{max} = R_1\left(\frac{T_n}{2}\right) = 2 \qquad (5.10)$$

Case 2: Residual-Vibration Era $(t_d \leq t)$ Figure 5.3b shows $R(t)$ for a pulse of duration $t_d = T_n/8$. Note that R_{max} occurs during the residual-vibration era. Since the response for $t > t_d$ is free vibration with "initial" conditions $R_1(t_d)$ and $\dot{R}_1(t_d)$, Eq. 3.17 can be used in the form

$$R_2(t) = R_1(t_d) \cos \omega_n (t - t_d) + \frac{\dot{R}_1(t_d)}{\omega_n} \sin \omega_n (t - t_d) \qquad (5.11)$$

for $t \geq t_d$, where $R_1(t_d)$ and $\dot{R}_1(t_d)$ are obtained from Eq. 5.9. Equation 3.25a can be used to determine the amplitude of this response.

$$(R_2)_{max} = \left\{[R_1(t_d)]^2 + \left[\frac{\dot{R}_1(t_d)}{\omega_n}\right]^2\right\}^{1/2} \qquad (5.12)$$

which can be simplified to

$$(R_2)_{max} = 2 \sin \frac{\pi t_d}{T_n} \qquad (5.13)$$

Figure 5.4a shows the maximum response ratio as a function of pulse duration. From this figure, or from Eq. 5.13, it can be seen that any pulse of duration longer than $T_n/6$ will cause a displacement larger than the static displacement, p_0/k, and for

5.2 Response of Undamped SDOF Systems to Rectangular Pulse and Ramp Loadings

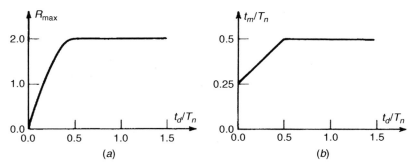

Figure 5.4 Maximum response ratio for rectangular pulse excitation: (*a*) maximum response amplitude; (*b*) time at which maximum response occurs.

any pulse longer than $T_n/2$, the maximum displacement will be twice the static value. Figure 5.4*b* shows the time at which the maximum response occurs. Expressions plotted in Fig. 5.4*b* are to be derived as an exercise (Problem 5.5).

5.2.2 Ramp Loading

Having considered the effect of the duration of a load on the maximum response, let us now examine the effect of rise time. In Section 1.3.4 we examined the response of an undamped SDOF system to an excitation that increased linearly with time. Now we examine what happens if the excitation ever levels off.

Consider a *ramp input* with *rise time* t_r, as shown in Fig. 5.5, applied to an undamped SDOF system that is at rest prior to application of the load. The equation of motion and initial conditions are

$$m\ddot{u} + ku = \begin{cases} \dfrac{t}{t_r} p_0, & 0 \leq t \leq t_r \\ p_0, & t_r \leq t \end{cases} \quad (5.14)$$

$$u(0) = \dot{u}(0) = 0 \quad (5.15)$$

For $0 \leq t \leq t_r$, the particular solution is seen to be

$$u_p = \frac{t}{t_r} \frac{p_0}{k} \quad (5.16)$$

Then the total solution has the form

$$u = \frac{t}{t_r} \frac{p_0}{k} + A_1 \cos \omega_n t + A_2 \sin \omega_n t \quad (5.17)$$

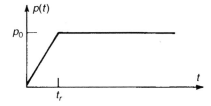

Figure 5.5 Ramp input function.

Using the initial conditions of Eq. 5.15, we get the following total response for $0 \leq t \leq t_r$:

$$u(t) = \frac{p_0}{k}\left(\frac{t}{t_r} - \frac{1}{\omega_n t_r}\sin\omega_n t\right) \qquad (5.18)$$

For $t \geq t_r$, the solution of Eq. 5.14b can be shown to be

$$u(t) = \frac{p_0}{k}\left\{1 + \frac{1}{\omega_n t_r}[\sin\omega_n(t - t_r) - \sin\omega_n t]\right\} \qquad (5.19)$$

Figure 5.6a shows the response to an input with $t_r > T_n$ and the response to an input with $t_r < T_n$. Figure 5.6b summarizes the effect of rise time on maximum response. From Fig. 5.6 it can be seen that the maximum response value of $R_{\max} = 2$ occurs for an ideal step input (i.e., for $t_r = 0$). For ramps with $t_r \gg T_n$ there will be little overshoot and the system will just undergo small oscillations about the pseudostatic

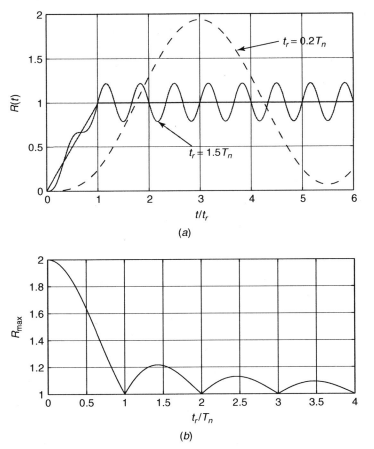

Figure 5.6 (a) Response of an undamped SDOF system to ramp inputs; (b) maximum response to ramp inputs.

deflection curve:

$$u_{\text{pseudostatic}} = u_p = \frac{t}{t_r}\frac{p_0}{k}, \qquad 0 \le t \le t_r \tag{5.20}$$

Thus, a load can be considered to be "applied slowly," and dynamic effects can generally be ignored, if the rise time is longer than about $3T_n$.

Figures such as 5.4a and 5.6b, called *response spectra*, are discussed in greater detail in Section 5.5.

5.3 RESPONSE OF UNDAMPED SDOF SYSTEMS TO A SHORT-DURATION IMPULSE: UNIT IMPULSE RESPONSE

A very important special form of excitation is the short-duration impulse (Fig. 5.7b), since it is also frequently used in determining the response of the system to more general forms of excitation, as you will see in the next section. Consider an undamped SDOF system subjected to a force of duration $t_d \ll T_n$ having an impulse

$$I \triangleq \int_0^{t_d} p(t)\, dt \tag{5.21}$$

Let the system be at rest for $t \le 0$, that is, prior to application of the excitation. The equation of motion and initial conditions are

$$m\ddot{u} + ku = \begin{cases} p(t), & 0 < t \le t_d \\ 0, & t_d < t \end{cases} \tag{5.22}$$

$$u(0) = \dot{u}(0) = 0 \tag{5.23}$$

By integrating Eq. 5.22a with respect to time and incorporating the initial conditions, we get

$$m\dot{u}(t_d) + k u_{\text{avg}} t_d = I \tag{5.24}$$

where u_{avg} is the (small) average displacement in the time interval $0 < t \le t_d$. For $t_d \to 0$, that is, $t_d \ll T_n$, the second term in Eq. 5.24 can be ignored, leaving

$$m\dot{u}(0^+) = I \tag{5.25}$$

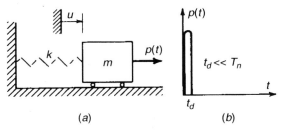

Figure 5.7 Undamped SDOF system subjected to a short-duration impulse.

Thus, an impulse consisting of a large force acting for a very short time has the effect of giving the mass an initial velocity of

$$\dot{u}(0^+) = \frac{I}{m} \qquad (5.26a)$$

but leaving it with an initial displacement of

$$u(0^+) = 0 \qquad (5.26b)$$

Equations 5.26 can be used as "initial" conditions for the free vibration problem of Eq. 5.22b. Using Eq. 3.17, we get the *impulse response*

$$u(t) = \frac{I}{m\omega_n} \sin \omega_n t \qquad (5.27)$$

The *unit impulse response function* for an undamped SDOF system is obtained from Eq. 5.27 by letting $I = 1$. By convention, the unit impulse response function is frequently called $h(t)$. Thus,

$$\boxed{h(t) = \frac{1}{m\omega_n} \sin \omega_n t} \qquad (5.28)$$

For a viscous-damped SDOF system with $\zeta < 1$, the impulse response function can be shown to be

$$u(t) = \frac{I}{m\omega_d} e^{-\zeta \omega_n t} \sin \omega_d t \qquad (5.29)$$

The corresponding unit impulse response function is therefore

$$\boxed{h(t) = \frac{1}{m\omega_d} e^{-\zeta \omega_n t} \sin \omega_d t} \qquad (5.30)$$

Example 5.1 Assume that the impulse $I = \int p(t)\, dt$ is due to a constant force $p(t) = p_0$ applied over a time interval $0 < t \leq t_d$ to an undamped SDOF system that is at rest at $t = 0$. Show that for $t_d \ll T_n$, Eq. 5.11 reduces to Eq. 5.27.

SOLUTION   1. Determine $u(t_d)$ and $\dot{u}(t_d)$ from Eq. 5.8.

$$u(t_d) = \frac{p_0}{k}(1 - \cos \omega_n t_d) \qquad (1)$$

$$\dot{u}(t_d) = \frac{\omega_n p_0}{k} \sin \omega_n t_d \qquad (2)$$

Since $\omega_n T_n = 2\pi$ and $t_d \ll T_n$, $\omega_n t_d \ll 2\pi$. Therefore,

$$1 - \cos \omega_n t_d \approx \tfrac{1}{2}(\omega_n t_d)^2$$

$$\sin \omega_n t_d \approx \omega_n t_d$$

Thus, since $I = p_0 t_d$,

$$u(t_d) \approx \frac{1}{2} \frac{p_0}{k} (\omega_n t_d)^2 = \frac{I}{2} \frac{t_d}{m}$$

$$\dot{u}(t_d) \approx \frac{\omega_n p_0}{k} \omega_n t_d = \frac{I}{m}$$

(3)

2. Evaluate $u(t)$ from Eq. 5.11, letting $t_d \to 0$.

$$u(t) = \lim_{t_d \to 0} \left[\frac{I t_d}{2m} \cos \omega_n (t - t_d) + \frac{I}{m \omega_n} \sin \omega_n (t - t_d) \right]$$

(4)

Thus, for $t_d \to 0$,

$$u(t) = \frac{I}{m \omega_n} \sin \omega_n t \qquad \textbf{Ans.}$$

(5)

as found in Eq. 5.27.

In Section 5.4 you will learn how to determine the response of an SDOF system to general excitation. You will then be able to show that the form of $p(t)$ is not very important as long as the duration is very short, that is, $t_d \ll T_n$. Consequently, Eq. 5.27 for undamped SDOF systems, or Eq. 5.29 for viscous-damped systems, gives a very close approximation to the actual response determined using $p(t)$.

In summary, you have learned that an excitation can be considered to be an impulse if $t_d \ll T_n$. On the other hand, a load can be considered to be applied "statically" if its rise time, t_r, is much greater than T_n. Finally, dynamic overshoot must be considered for loads that are applied rapidly.

5.4 RESPONSE OF SDOF SYSTEMS TO GENERAL DYNAMIC EXCITATION: CONVOLUTION INTEGRAL METHOD

The *Convolution Integral Method* (or Duhamel Integral Method) for determining the response of an SDOF system to general dynamic excitation can be developed from the impulse response function derived in Section 5.3. In some cases, an analytical expression for the response can be obtained in this manner. Alternatively, the Laplace Transform Method, discussed in Section 5.6, is useful for obtaining the response of SDOF systems to relatively simple forms of excitation. For more complex forms of excitation, numerical evaluation of the response is necessary, as described in Chapter 6.

The convolution integral is based on the Principle of Superposition, which is only valid for linear systems. Figure 5.8 shows an undamped SDOF system that is initially at rest and is then subjected to an input $p(t)$, as shown. The response of the system to an impulse $dI = p(\tau) d\tau$ at time τ is called $du(t)$ and is given by

$$du(t) = \frac{dI}{m \omega_n} \sin \omega_n (t - \tau) \tag{5.31}$$

The total response of the undamped SDOF system at time t will be the sum of the responses due to all incremental impulses at times τ prior to time t. Therefore,

$$\boxed{u(t) = \int_0^t p(\tau) h(t - \tau) \, d\tau} \tag{5.32}$$

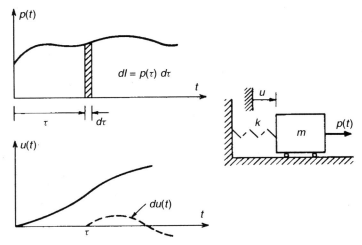

Figure 5.8 Incremental response of an undamped SDOF system.

where $h(t - \tau)$, obtained from Eq. 5.28, is the *unit impulse response function* for an undamped system, given by

$$h(t - \tau) = \frac{1}{m \omega_n} \sin \omega_n (t - \tau) \tag{5.33}$$

Equation 5.32 is equally valid for viscous-damped SDOF systems if Eq. 5.30 is used to obtain $h(t - \tau)$. Thus, for viscous-damped systems the unit impulse response function is given by

$$h(t - \tau) = \frac{1}{m \omega_d} e^{-\zeta \omega_n (t-\tau)} \sin \omega_d (t - \tau) \tag{5.34}$$

Equation 5.32 is referred to as the *Duhamel integral expression* for the response of undamped or damped SDOF systems, with $h(t - \tau)$ given by Eq. 5.33 or 5.34, respectively. Equation 5.32 is frequently referred to as a *convolution integral*, a more general form of which is

$$x(t) = \int_{-\infty}^{\infty} f_1(\tau) f_2(t - \tau) \, d\tau \tag{5.35}$$

Equation 5.32, with Eq. 5.33 or 5.34, may thus be used to determine the response of an SDOF system to general dynamic excitation if the system is initially at rest. If the system has nonzero initial conditions, the response to the initial conditions is determined: for undamped systems from Eq. 3.17 or, for viscous-damped systems with $\zeta < 1$, from Eq. 3.30. Thus, for an undamped system,

$$\boxed{\begin{aligned} u(t) &= \frac{1}{m \omega_n} \int_0^t p(\tau) \sin \omega_n (t - \tau) \, d\tau \\ &\quad + u_0 \cos \omega_n t + \frac{v_0}{\omega_n} \sin \omega_n t \end{aligned}} \tag{5.36}$$

5.4 Response of SDOF Systems to General Dynamic Excitation: Convolution Integral Method

and for an underdamped system,

$$u(t) = \frac{1}{m\omega_d} \int_0^t p(\tau) e^{-\zeta\omega_n(t-\tau)} \sin \omega_d(t-\tau) \, d\tau$$
$$+ e^{-\zeta\omega_n t} \left(u_0 \cos \omega_d t + \frac{v_0 + \zeta\omega_n u_0}{\omega_d} \sin \omega_d t \right) \quad (5.37)$$

It is convenient to use the following trigonometric identity when evaluating the Duhamel integrals above:

$$\sin \omega_n(t-\tau) = \sin \omega_n t \cos \omega_n \tau - \cos \omega_n t \sin \omega_n \tau \quad (5.38)$$

Example 5.2 (a) Use the Duhamel integral to determine the response of an undamped SDOF system to a "blast" loading specified by the triangular pulse shown in Fig. 1. (b) Obtain expressions that are valid for $t \leq t_d$ and for $t \geq t_d$. The system is initially at rest.

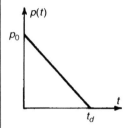

Figure 1 Triangular pulse excitation force.

SOLUTION Use Eq. 5.36 with $u_0 = v_0 = 0$, and with excitation given by

$$p_1(t) = p_0 \left(1 - \frac{t}{t_d}\right), \qquad 0 \leq t \leq t_d \quad (1)$$

$$p_2(t) = 0, \qquad t_d \leq t \quad (2)$$

(a) For $0 \leq t \leq t_d$, the Duhamel integral expression for the response is

$$u_1(t) = \frac{1}{m\omega_n} \int_0^t p_0 \left(1 - \frac{\tau}{t_d}\right) \sin \omega_n(t-\tau) \, d\tau$$
$$= \frac{p_0}{k} \left[\sin \omega_n t \int_0^t \left(1 - \frac{\tau}{t_d}\right) \cos \omega_n \tau \, d(\omega_n \tau) \right.$$
$$\left. - \cos \omega_n t \int_0^t \left(1 - \frac{\tau}{t_d}\right) \sin \omega_n \tau \, d(\omega_n \tau) \right] \quad (3)$$

Using integration by parts, we get

$$\int \tau \cos \omega_n \tau \, d(\omega_n \tau) = \tau \sin \omega_n \tau - \frac{1}{\omega_n} \int \sin \omega_n \tau \, d(\omega_n \tau)$$
$$= \tau \sin \omega_n \tau + \frac{1}{\omega_n} \cos \omega_n \tau \quad (4)$$

Similarly,
$$\int \tau \sin\omega_n\tau \, d(\omega_n\tau) = -\tau \cos\omega_n\tau + \frac{1}{\omega_n}\sin\omega_n\tau \tag{5}$$

Hence, for $0 \leq t \leq t_d$, the displacement is given by
$$u_1(t) = \frac{p_0}{k}\left\{\sin\omega_n t\left[\sin\omega_n t - \frac{t}{t_d}\sin\omega_n t + \frac{1}{\omega_n t_d}(1 - \cos\omega_n t)\right]\right.$$
$$\left. - \cos\omega_n t\left(1 - \cos\omega_n t + \frac{t}{t_d}\cos\omega_n t - \frac{1}{\omega_n t_d}\sin\omega_n t\right)\right\} \tag{6}$$

Simplifying this expression, we get the following expression for the response ratio for $0 \leq t \leq t_d$:
$$R_1(t) \equiv \frac{ku_1(t)}{p_0} = 1 - \cos\omega_n t - \frac{t}{t_d} + \frac{1}{\omega_n t_d}\sin\omega_n t \qquad \text{Ans. (a)} \tag{7}$$

(b) For $t_d < t$, the Duhamel integral expression for the response is
$$u_2(t) = \frac{1}{m\omega_n}\int_0^{t_d} p_0\left(1 - \frac{\tau}{t_d}\right)\sin\omega_n(t - \tau)\,d\tau \tag{8}$$

Note that this is the same as Eq. 3, except that it is evaluated at t_d, since $p(\tau) = 0$ for $t > t_d$. Equation 6 can be used by setting $t = t_d$ within the brackets. Thus,
$$u_2(t) = \frac{p_0}{k}\left\{\sin\omega_n t\left[\frac{1}{\omega_n t_d}(1 - \cos\omega_n t_d)\right] - \cos\omega_n t\left(1 - \frac{1}{\omega_n t_d}\sin\omega_n t_d\right)\right\}$$

Finally, the response ratio for $t > t_d$ is given by
$$R_2(t) = \frac{1}{\omega_n t_d}[(1 - \cos\omega_n t_d)\sin\omega_n t - (\omega_n t_d - \sin\omega_n t_d)\cos\omega_n t] \qquad \text{Ans. (b)} \tag{9}$$

(The values of $R_{\max}$ versus f_n are plotted in Section 5.5.)

From Example 5.2 you can appreciate that although the response of an SDOF system to an arbitrary input may, in principle, be obtained by use of the Duhamel integral, the work involved in evaluating the integrals may be tedious. Also, it may be necessary to obtain the response to an input that is known graphically but not in analytical form: for example, a plot of ground acceleration versus time. In cases such as these, a numerical procedure is needed. Numerical procedures for computing dynamic response are discussed in Chapter 6.

5.5 RESPONSE SPECTRA

In the previous discussions of response calculations it has been assumed that an SDOF system has been defined (i.e., k, c, and m are known) and that the response to a specified input is desired. The problem that is frequently encountered in design, particularly in preliminary design, is to select one or more of the system parameters in a manner that limits a certain response quantity (e.g., the maximum absolute displacement or the

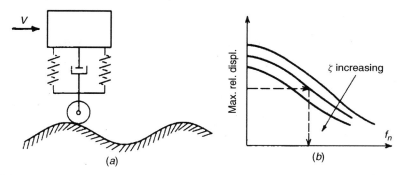

Figure 5.9 Design problem amenable to solution using a response spectrum: (*a*) vehicle moving over a specified bump; (*b*) relative-displacement response spectra.

maximum stress) to a specified range when the system is subjected to a given input. Response spectra have been employed for this purpose in the design of buildings to withstand earthquake excitation, in the design of shock isolators, and in other design studies.

A *response spectrum* is a plot of the maximum "response" (e.g., displacement, stress, acceleration) of SDOF systems to a given input versus some system parameter, generally the undamped natural frequency. A set of such curves, for example, curves plotted for various levels of system damping, may be referred to as *response spectra*.

Figure 5.9 illustrates how response spectra may be used in design. Figure 5.9*b* shows relative-displacement response spectra for a system whose base excitation has a form such as the road profile illustrated in Fig. 5.9*a*. Given the maximum permissible relative displacement and the damping level, the appropriate system natural frequency can be selected from Fig. 5.9*b* in the manner indicated by the dashed arrows. This value of natural frequency, together with the given mass, can be used to determine the required spring stiffness, k.

Example 5.3 (a) Using the response-ratio expressions determined in Example 5.2, plot the response ratio versus t/t_d for $t_d/T_n = 0.25, 0.50, 1.0$, and 1.5 for $0 \leq t/t_d \leq 2.0$. (b) Determine expressions for the maximum response for $t \leq t_d$ and for $t \geq t_d$. Using this information, plot a response spectrum in the form of R_{max} versus $f_n t_d$.

SOLUTION (a) The four plots of the response ratio versus t/t_d are shown in Fig. 1.

(b) First determine an expression for R_{max} for $0 \leq t \leq t_d$. (From the curves shown in Fig. 1 it can be seen that R_{max} occurs in this time interval except for short-duration loading, e.g., the $t_d/T_n = 0.25$ curve above.)

From Eq. 7 of Example 5.2,

$$R_1(t) = 1 - \cos \omega_n t - \frac{t}{t_d} + \frac{1}{\omega_n t_d} \sin \omega_n t \qquad (1)$$

Then,

$$\dot{R}_1(t) = \frac{1}{t_d}(\omega_n t_d \sin \omega_n t - 1 + \cos \omega_n t) \qquad (2)$$

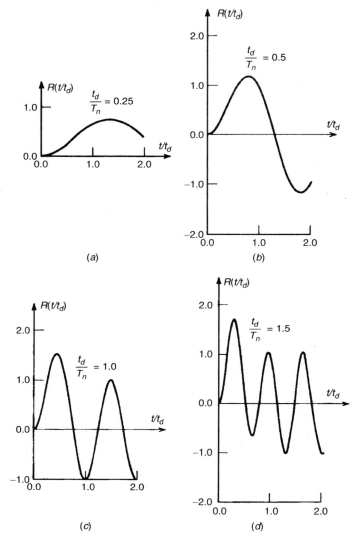

Figure 1 Response-ratio plots.

Setting $\dot{R}_1 = 0$, we get the following expression for the times t_m at which R_1 reaches maxima and minima:

$$\omega_n t_d \sin \omega_n t_m = 1 - \cos \omega_n t_m \qquad (3)$$

The following trigonometric identities are useful here:

$$1 - \cos 2\alpha = 2 \sin^2 \alpha$$
$$\sin 2\alpha = 2 \sin \alpha \cos \alpha \qquad (4)$$
$$\cos 2\alpha = 2 \cos^2 \alpha - 1$$

Then Eq. 3 can be simplified to

$$\tan \frac{\omega_n t_m}{2} = \omega_n t_d \tag{5}$$

From curves (b) through (d) in Fig. 1 we can see that the maximum response occurs at the first occurrence of $\dot{R}_1 = 0$. Thus, $\omega_n t_m/2$ lies in the first quadrant with

$$\omega_n t_m = 2 \tan^{-1} \omega_n t_d \tag{6}$$

and from Eq. 1,

$$(R_1)_{\max} = R_1(t_m) = 1 - \cos \omega_n t_m - \frac{\omega_n t_m}{\omega_n t_d} + \frac{1}{\omega_n t_d} \sin \omega_n t_m \qquad \textbf{Ans.} \tag{7}$$

Next, determine an expression for $R_{\max}$ for $t > t_d$. From Eq. 9 of Example 5.2,

$$R_2(t) = \frac{1}{\omega_n t_d}[(1 - \cos \omega_n t_d) \sin \omega_n t - (\omega_n t_d - \sin \omega_n t_d) \cos \omega_n t] \tag{8}$$

Thus, as in Eq. 3.25a, the amplitude of this combination of $\sin \omega_n t$ and $\cos \omega_n t$ responses is

$$(R_2)_{\max} = \frac{1}{\omega_n t_d}[(1 - \cos \omega_n t_d)^2 + (\omega_n t_d - \sin \omega_n t_d)^2]^{1/2} \qquad \textbf{Ans.} \tag{9}$$

Equation 7 is plotted as the solid curve in Fig. 2, and Eq. 9 is plotted as a dashed curve. For $f_n t_d \leq 0.371$, Eq. 9 gives the larger value of $R_{\max}$, whereas for $f_n t_d \geq 0.371$ the maximum response occurs during the forced-vibration era and is governed by Eq. 7. The response spectrum plotted in Fig. 2 is the composite maximum curve, or maximax curve.

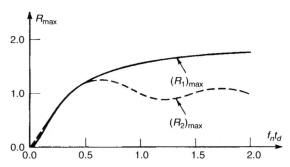

Figure 2 Response spectrum for a triangular pulse.

In general, it is not an easy task to construct a response spectrum, since determination of the time at which the maximum occurs may involve solution of a transcendental equation and careful selection of the maximum from among many local maxima. Consequently, determination of a response spectrum may require numerical evaluation of the response for various values of the system parameters. A number of response spectra for both linear and nonlinear SDOF systems may be found in Ref. [5.1]. In Example 5.4 we consider an application of the response spectrum derived in Example 5.3.

Example 5.4 A building that is subjected to blast forces is modeled as an SDOF system, as shown in Fig. 1, where $k = 9.0$ GN/m and $m = 10$ Mg. Determine the maximum blast force p_0 that can be sustained if the displacement is to be limited to 5 mm if (a) $t_d = 0.4$ s and (b) $t_d = 0.04$ s.

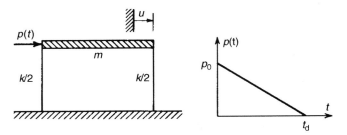

Figure 1 SDOF building subjected to a triangular blast force.

SOLUTION 1. Determine the natural frequency of the system.

$$\omega_n = \left(\frac{k}{m}\right)^{1/2} = \left(\frac{9.0 \text{ GN/m}}{10 \text{ Mg}}\right)^{1/2} = 30 \text{ rad/s} \tag{1}$$

$$f_n = \frac{\omega_n}{2\pi} = 4.77 \text{ Hz}$$

2. Determine the maximum response ratio from Fig. 2 in Example 5.3. For case (a),

$$\omega_n t_d = (30 \text{ rad/s})(0.4 \text{ s}) = 12.0 \text{ rad}$$
$$f_n t_d = (4.77 \text{ Hz})(0.4 \text{ s}) = 1.910 \tag{2a}$$

From Eqs. 6 and 7 and the response spectrum of Example 5.3,

$$(R_1)_{\max} = 1.75 \tag{2b}$$

For case (b),

$$\omega_n t_d = (30 \text{ rad/s})(0.04 \text{ s}) = 1.2 \text{ rad}$$
$$f_n t_d = (4.77 \text{ Hz})(0.04 \text{ s}) = 0.1910 \tag{3a}$$

From Eq. 9 and the response spectrum of Example 5.3,

$$(R_2)_{\max} = 0.58 \tag{3b}$$

3. For each $R_{\max}$, determine the corresponding maximum force. From the definition of response ratio,

$$R_{\max} = \frac{u_{\max}}{p_0/k} \tag{4}$$

or

$$p_0 = \frac{k u_{\max}}{R_{\max}} \tag{5}$$

Thus, for case (a), to limit the displacement to 5 mm, the maximum force must not exceed

$$(p_0)_1 = \frac{(9.0 \text{ GN/m})(5 \text{ mm})}{1.75} = 25.7 \text{ MN} \qquad \textbf{Ans. (6)}$$

and for case (b),

$$(p_0)_2 = \frac{(9.0 \text{ GN/m})(5 \text{ mm})}{0.58} = 77.6 \text{ MN} \qquad \textbf{Ans. (7)}$$

Since many applications of response spectra involve relative motion, it useful to determine appropriate response quantities for this case. Figure 5.10 shows the prototype relative-motion SDOF system. As in Example 2.2, let the relative displacement be

$$w = u - z \qquad (5.39)$$

Then the equation of motion can be written

$$m\ddot{w} + c\dot{w} + kw = -m\ddot{z} \qquad (5.40)$$

A definition of response ratio was given in Eq. 5.6. The response ratio corresponding to relative motion should therefore be defined as

$$R(t) \equiv \frac{kw(t)}{-m\ddot{z}_{max}} = -\frac{\omega_n^2 w(t)}{\ddot{z}_{max}} \qquad (5.41)$$

By letting $\ddot{z} = \ddot{z}_{max} f_a(t)$ and referring to Eqs. 5.32 and 5.34, we can express $R(t)$ in Duhamel integral form. For $w(0) = \dot{w}(0) = 0$,

$$R(t) = \frac{\omega_n^2}{\omega_d} \int_0^t f_a(\tau) e^{-\zeta \omega_n (t-\tau)} \sin \omega_d (t - \tau) \, d\tau \qquad (5.42)$$

for damped systems. Similarly, Eqs. 5.32 and 5.33 lead to the following response ratio for undamped systems:

$$R(t) = \omega_n \int_0^t f_a(\tau) \sin \omega_n (t - \tau) \, d\tau \qquad (5.43)$$

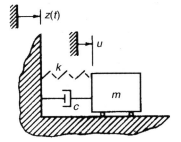

Figure 5.10 Prototype relative-motion SDOF system.

From Eq. 5.41 the maximum relative displacement is given by

$$w_{max} = \frac{R_{max}\ddot{z}_{max}}{\omega_n^2} \tag{5.44}$$

A second quantity that is of interest is the maximum absolute acceleration, $\ddot{u}_{max}$. Equation 5.40 can be written

$$m\ddot{u} + c\dot{w} + kw = 0 \tag{5.45}$$

For an undamped system $\ddot{u}_{max}$ can easily be determined from Eq. 5.45:

$$\ddot{u}_{max} = \omega_n^2 w_{max} \quad \text{(with } c = 0\text{)} \tag{5.46}$$

Then, from Eqs. 5.44 and 5.46,

$$\ddot{u}_{max} = R_{max}\ddot{z}_{max} \quad \text{(with } c = 0\text{)} \tag{5.47}$$

Example 5.5 illustrates a response spectrum for base motion and also indicates a typical form of response spectrum when the spectrum is plotted to logarithmic scales.

Example 5.5 An undamped SDOF system similar to the damped one in Fig. 5.9a experiences a base acceleration as shown in Fig. 1. The initial conditions are all zero. Develop expressions for w_{max} and $\ddot{u}_{max}$, and produce a log-log plot of w_{max} versus f_n.

$$\ddot{z}(t) = \ddot{z}_{max} f_a(t)$$

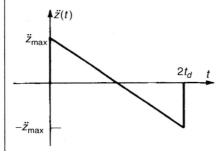

Figure 1 Time history of base acceleration.

where

$$f_a(t) = \begin{cases} 1 - \dfrac{t}{t_d}, & 0 \le t \le 2t_d \\ 0, & 2t_d < t \end{cases}$$

SOLUTION 1. Determine the response for $0 \le t \le 2t_d$. Since the excitation for $0 \le t \le 2t_d$ has the same form as that of Example 5.2, $R_1(t)$ will be the same as that given in Eq. 7 of Example 5.2: namely,

$$R_1(t) = 1 - \cos\omega_n t - \frac{t}{t_d} + \frac{1}{\omega_n t_d}\sin\omega_n t \tag{1}$$

2. Determine the response for $2t_d \leq t$. The most direct way to express this free vibration response is to use Eq. 3.17 in the form

$$R_2(t) = R_1(2t_d) \cos \omega_n(t - 2t_d) + \frac{\dot{R}_1(2t_d)}{\omega_n} \sin \omega_n(t - 2t_d) \qquad (2)$$

3. Determine the maximum response occurring in the time interval $0 \leq t \leq 2t_d$, which is the forced-vibration era.

$$\dot{R}_1(t) = \frac{1}{t_d}(\omega_n t_d \sin \omega_n t - 1 + \cos \omega_n t) \qquad (3)$$

As in Example 5.3, setting $\dot{R}_1(t) = 0$, we are led to

$$\omega_n t_m = 2 \tan^{-1} \omega_n t_d \qquad (4)$$

Unlike Example 5.3, however, the excitation now extends to negative values of $f_a(t)$, and hence the maximum absolute value of $R_1(t)$ may occur near $t = 0$ or near $t = 2t_d$. In the former case,

$$\omega_n t_m = 2 \tan^{-1} \omega_n t_d \qquad (5)$$

with $\omega_n t_m/2$ lying in the first quadrant. In the latter case,

$$\omega_n t_m = 2(\tan^{-1} \omega_n t_d + p\pi) \qquad (6)$$

where p is the largest integer for which $\omega_n t_m < 2\omega_n t_d$, and $\tan^{-1} \omega_n t_d$ is taken in the first quadrant. Then

$$(R_1)_{\max} = R_1(t_m) = 1 - \cos \omega_n t_m - \frac{\omega_n t_m}{\omega_n t_d} + \frac{1}{\omega_n t_d} \sin \omega_n t_m \qquad \textbf{Ans. (7)}$$

4. Determine the maximum occurring in the residual-vibration era, $2t_d < t$. From Eq. 2 this is simply

$$(R_2)_{\max} = \left\{ [R_1(2t_d)]^2 + \left[\frac{\dot{R}_1(2t_d)}{\omega_n} \right]^2 \right\}^{1/2} \qquad \textbf{Ans. (8)}$$

where $R_1(2t_d)$ and $\dot{R}_1(2t_d)$ are based on Eq. 1.

5. Determine expressions for $w_{\max}$ and $\ddot{u}_{\max}$. From Eq. 5.44,

$$w_{\max} = \frac{R_{\max} \ddot{z}_{\max}}{\omega_n^2} \qquad \textbf{Ans. (9)}$$

and from Eq. 5.47,

$$\ddot{u}_{\max} = R_{\max} \ddot{z}_{\max} \qquad \textbf{Ans. (10)}$$

where $R_{\max}$ is the maximax response, that is, the larger of $(R_1)_{\max}$ and $(R_2)_{\max}$.

6. Plot $w_{\max}$ versus f_n using log-log scales. It will be most convenient to plot the nondimensional response $w_{\max}/\ddot{z}_{\max} t_d^2$ versus the nondimensionalized natural frequency $f_n t_d$ (Fig. 2).

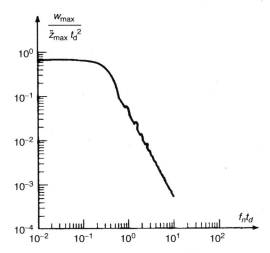

Figure 2 Relative-displacement response spectrum.

It is convenient to present response spectra in log-log form as in Example 5.5 because the spectrum can frequently be represented by two or three straight lines connected by short transition regions. For example, note from Eq. 9 of Example 5.5 that if $R_{max} = $ const, $w_{max} \sim 1/f_n^2$, which corresponds to a slope of -2 on the log-log plot. For values of $f_n t_d$ greater than about $f_n t_d = 0.5$, the log-log plot in Fig. 2 of Example 5.5 approximates this behaviour.

Although creation of response spectra for a given input can be a tedious job, response spectra are useful tools when many systems are to be designed on the basis of their response to a given transient input, for example, a given earthquake time history. Response spectra are discussed further in Chapter 20, *Introduction to Earthquake Response of Structures*.

5.6 SYSTEM RESPONSE BY THE LAPLACE TRANSFORM METHOD: SYSTEM TRANSFER FUNCTION

Section 5.4 describes use of the Duhamel integral to determine the response of linear SDOF systems to general time-dependent inputs. In some cases the response can conveniently be determined analytically by use of the Laplace transform, which is described in this section. However, numerical integration procedures such as those described in Chapters 6 and 16 are required to solve the more complex structural dynamics response problems.

In this book you will be introduced to two important transforms, the Laplace transform (in this section) and the Fourier transform (in Chapter 7). As you will see, the purpose of mathematical transforms is to simplify the solution of certain dynamics problems. The statement of the problem is *transformed* from one domain (i.e., the time domain) to another domain (e.g., the Laplace domain or the Fourier domain) where solution of the problem is simpler. Once that solution has been obtained, it is *inverse-transformed* back to the original domain. The *Laplace Transform Method* is particularly

5.6 System Response by the Laplace Transform Method: System Transfer Function

useful if the forward and inverse transforms can be found directly in a table of Laplace transforms or can be converted to forms that can be obtained by table lookup.[2]

5.6.1 One-Sided Laplace Transform

The Laplace transform of time-dependent variable $u(t)$ is written symbolically as $U(s) \equiv \mathcal{L}[u(t)]$, where s is a complex variable referred to as the *Laplace variable*. The *one-sided Laplace transform* of $u(t)$ is defined by the definite integral

$$U(s) \equiv \mathcal{L}[u(t)] = \int_0^\infty e^{-st} u(t)\, dt \qquad (5.48)$$

Because this is a definite integral with t as the variable of integration, the resulting transform is a function of the Laplace variable s. It is said that the function $u(t)$ in the *time domain* is transformed into the function $U(s)$ in the *Laplace domain*.

The following example illustrates the use of Eq. 5.48 to compute the Laplace transforms of three important simple functions.

Example 5.6 Compute the Laplace transforms of the following three functions: (a) the unit impulse $\delta(t)$, (b) the unit step function beginning at $t=0$, and (c) $\sin \omega t$ starting at $t=0$.

SOLUTION (a) *Unit impulse function:*

$$p_1(t) = \delta(t) \qquad (1)$$

$$P_1(s) = \int_0^\infty e^{-st} \delta(t)\, dt = \int_0^\infty \delta(t)\, dt \qquad (2)$$

since $\delta(t)$ is zero except at $t=0$. Therefore,

$$P_1(s) = 1 \qquad \textbf{Ans. (a)} \quad (3)$$

(b) *Unit step function:*

$$P_2(s) = \int_0^\infty e^{-st}(1)\, dt = -\frac{1}{s} e^{-st} \Big|_0^\infty \qquad (4)$$

$$= \frac{1}{s} \qquad \textbf{Ans. (b)} \quad (5)$$

(c) *Sinusoidal function:*

$$p_3(t) = \sin \omega t \qquad (6)$$

$$P_3(s) = \int_0^\infty e^{-st} \sin \omega t\, dt = \frac{e^{-st}(-s \sin \omega t - \omega \cos \omega t)}{(-s)^2 + \omega^2} \Big|_0^\infty \qquad (7)$$

$$= \frac{\omega}{s^2 + \omega^2} \qquad \textbf{Ans. (c)} \quad (8)$$

Additional Laplace transform pairs are listed in Table C.1.

---

[2] See Appendix C for further background on the Laplace Transform Method.

5.6.2 Laplace Transform Solution of Linear Differential Equations

We wish to use Laplace transforms to solve the prototype SDOF equation of motion (Eq. 3.1):

$$m\ddot{u} + c\dot{u} + ku = p(t) \tag{5.49}$$

with given *excitation* $p(t)$ and subject to specified *initial displacement* $u(0)$ and *initial velocity* $\dot{u}(0)$, respectively. Therefore, we need expressions for the Laplace transforms of the derivatives du/dt and d^2u/dt^2. By using the Laplace transform definition, Eq. 5.48, and performing simple integration by parts, we obtain the following Laplace transform of the first time derivative:

$$\mathcal{L}\left[\frac{du(t)}{dt}\right] = \int_0^\infty e^{-st}\frac{du(t)}{dt}dt = e^{-st}u(t)\Big|_0^\infty + s\int_0^\infty e^{-st}u(t)\,dt \tag{5.50}$$
$$= sU(s) - u(0)$$

Similarly, it is straightforward to show that the Laplace transform of the second time derivative is

$$\mathcal{L}\left[\frac{d^2u(t)}{dt^2}\right] = \int_0^\infty e^{-st}\frac{d^2u(t)}{dt^2}dt = s^2U(s) - su(0) - \dot{u}(0) \tag{5.51}$$

The Laplace transform of the excitation $p(t)$ is simply

$$P(s) \equiv \mathcal{L}[p(t)] = \int_0^\infty e^{-st}p(t)\,dt \tag{5.52}$$

Transforming both sides of the time-domain equation of motion and rearranging, we obtain the following *equation of motion in the Laplace domain*:

$$\boxed{(ms^2 + cs + k)U(s) = P(s) + m\dot{u}(0) + (ms + c)u(0)} \tag{5.53}$$

Note that the Laplace transform process has converted a differential equation in the *time domain* into an algebraic equation in the *Laplace domain*.

Conceptually, each of the three terms on the right-hand side of Eq. 5.53 corresponds to an "input." Therefore, to simplify the following presentation, we consider only the first term and will solve the equation

$$Z(s)U(s) = P(s) \tag{5.54}$$

The function $Z(s)$ is called the *system impedance function*. For the viscous-damped SDOF vibrating system,

$$Z(s) = ms^2 + cs + k = m(s^2 + 2\zeta\omega_n s + \omega_n^2) \tag{5.55}$$

The *transformed solution* $U(s)$ can be written in the form

$$\boxed{U(s) = H(s)P(s)} \tag{5.56}$$

5.6 System Response by the Laplace Transform Method: System Transfer Function

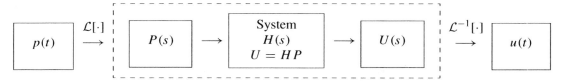

Figure 5.11 Laplace transform procedure for solving linear ordinary differential equations.

where $H(s)$ is the *system transfer function*, or simply the *transfer function*, given here by

$$H(s) = \frac{1}{Z(s)} = \frac{1}{m(s^2 + 2\zeta\omega_n s + \omega_n^2)} \quad (5.57)$$

The *inverse Laplace transform*, designated by the operator $\mathcal{L}^{-1}$, transforms the solution $U(s)$ in the Laplace domain into the solution $u(t)$ in the time domain. Therefore, the entire Laplace transformation procedure can be represented schematically as shown in Fig. 5.11. The dashed box outlines the Laplace domain part of the procedure.

The functions $p(t)$ and $P(s)$ and $u(t)$ and $U(s)$ are said to form *Laplace transform pairs*. Table C.1 lists several important Laplace transform pairs. The inverse Laplace transform can be computed analytically from a line integral in the Laplace domain (i.e., the complex s domain).[5.2] However, it is beyond the scope of this book to discuss this tedious inverse Laplace transform procedure, since structural dynamics response problems can either be solved by the use of Laplace transform pairs that are listed in Table C.1, or else they require numerical solution by one of the methods discussed in Chapters 6 and 16.

As a simple example of the Laplace transform procedure for solving differential equations, let us reconsider the unit impulse response of an undamped SDOF system, which was discussed previously in Section 5.3.

Example 5.7 Determine an expression for the response of an underdamped SDOF system to a unit impulse input at $t = 0$. Let the initial conditions be $u(0) = \dot{u}(0) = 0$. Use Table C.1 to obtain the necessary inverse Laplace transform.

SOLUTION The differential equation is

$$m\ddot{u} + c\dot{u} + ku = p(t) = \delta(t) \quad (1)$$

From Example 5.6(a), the transformed input is

$$P(s) = 1 \quad (2)$$

Combining Eq. 2 with Eqs. 5.56 and 5.57, we obtain the transformed response

$$U(s) = H(s)P(s) = \frac{1}{m(s^2 + 2\zeta\omega_n s + \omega_n^2)} \quad (3)$$

The inverse Laplace transform of $U(s)$ may be obtained from transform pair 14 of Table C.1. Then

$$u(t) = \frac{1}{m\omega_d} e^{-\zeta\omega_n t} \sin \omega_d t \quad \text{Ans.} \quad (4)$$

where, as before, the damped natural frequency is $\omega_d = \omega_n\sqrt{1-\zeta^2}$. This is the same unit impulse response function, $h(t)$, as that obtained in Eq. 5.30.

In many cases, the necessary transform pair is not listed in readily available Laplace transform tables, but it is possible to use the *Method of Partial Fractions*[3] to reduce the transformed response $U(s)$ to a sum of terms, each of which is in the table. Example 5.8 illustrates this procedure.

Example 5.8 A undamped spring–mass oscillator is subjected to a step input of magnitude p_0 starting at $t = 0$. The initial conditions are $u(0) = \dot{u}(0) = 0$. Use partial-fraction expansion and table lookup to determine an expression for the total response, $u(t)$, of this system.

SOLUTION The differential equation is

$$m\ddot{u} + ku = p_0 \qquad \text{for } t > 0 \tag{1}$$

From transform pair 2 of Table C.1, the transformed step input is

$$P(s) = \frac{p_0}{s} \tag{2}$$

Combining Eq. 2 with Eqs. 5.56 and 5.57, we obtain the transformed response

$$U(s) = H(s)P(s) = \frac{p_0}{ms(s^2 + \omega_n^2)} = \frac{p_0}{k}\frac{\omega_n^2}{s(s^2 + \omega_n^2)} \tag{3}$$

The inverse Laplace transform of $U(s)$ has the form

$$u(t) = \frac{p_0}{k}\mathcal{L}^{-1}\left[\frac{\omega_n^2}{s(s^2 + \omega_n^2)}\right] \tag{4}$$

Since there is no $F(s)$ in Table C.1 from which to obtain the inverse transform needed in Eq. 4, let us write this function in the *partial-fraction form*

$$F(s) = \frac{\omega_n^2}{s(s^2 + \omega_n^2)} = \frac{c_1}{s - \lambda_1} + \frac{c_2}{s - \lambda_2} + \frac{c_3}{s - \lambda_3} \tag{5}$$

where the simple poles λ_1 through λ_3 are the roots of the *characteristic equation*

$$s(s^2 + \omega_n^2) = (s - \lambda_1)(s - \lambda_2)(s - \lambda_3) = 0 \tag{6}$$

These roots are

$$\lambda_1 = 0, \qquad \lambda_2 = +i\omega_n, \qquad \lambda_3 = -i\omega_n \tag{7}$$

Equation C.24 in Appendix C.5 can be used to compute the three coefficients:

$$c_k = [(s - \lambda_k)F(s)]|_{s=s_k} \tag{8}$$

---

[3]The Method of Partial Fractions is discussed in Section C.5.

5.6 System Response by the Laplace Transform Method: System Transfer Function

The resulting coefficients are

$$c_1 = \frac{\omega_n^2}{(-i\omega_n)i\omega_n} = 1$$

$$c_2 = \frac{\omega_n^2}{(i\omega_n)2\,i\omega_n} = -\frac{1}{2} \quad (9)$$

$$c_3 = \frac{\omega_n^2}{(-i\omega_n)(-2\,i\omega_n)} = -\frac{1}{2}$$

Combining Eqs. 4, 5, and 9, we get the transformed response in partial-fraction form

$$U(s) = \frac{p_0}{k}\left[\frac{1}{s} - \frac{1}{2(s - i\omega_n)} - \frac{1}{2(s + i\omega_n)}\right] \quad (10)$$

The inverse transforms that are required may be found as transform pairs 2 and 5 of Table C.1. Then the solution $u(t)$ becomes

$$u(t) = \frac{p_0}{k}\left[1 - \frac{1}{2}(e^{i\omega_n t} + e^{-i\omega_n t})\right] \quad (11)$$

But from the Euler formula of Eq. 3.14,

$$\tfrac{1}{2}(e^{i\omega_n t} + e^{-i\omega_n t}) = \cos\omega_n t \quad (12)$$

Therefore, the response of the undamped SDOF system to a step input can be written in the form

$$u(t) = \frac{p_0}{k}(1 - \cos\omega_n t) \qquad \textbf{Ans. (13)}$$

which is the solution given previously in Eq. 5.8.

In Example C.1 in Appendix C.5, the example problem above is solved by use of a partial-fraction expansion where one term has a quadratic denominator of the form

$$F(s) = \frac{As + B}{as^2 + bs + c} \quad (5.58)$$

Next we consider an example that leads to a higher-order pole.

Example 5.9 Consider the same undamped spring–mass oscillator as in Example 5.8, but subjected to a sinusoidal input $p_0 \sin\omega_n t$ starting at $t = 0$. (Note that the excitation frequency is the natural frequency of the system.) The initial conditions are $u(0) = \dot{u}(0) = 0$. Use partial-fraction expansion to determine an expression for the total response, $u(t)$, of this system.

SOLUTION The differential equation is

$$m\ddot{u} + ku = p_0 \sin\omega_n t \qquad \text{for } t > 0 \quad (1)$$

From transform pair 6 of Table C.1, the transformed sinusoidal excitation is

$$P(s) = \frac{p_0 \omega_n}{s^2 + \omega_n^2} \quad (2)$$

Combining Eq. 2 with Eqs. 5.56 and 5.57, we obtain the transformed response

$$U(s) = H(s)P(s) = \frac{p_0 \omega_n}{m(s^2 + \omega_n^2)^2} = \frac{p_0}{k} \frac{\omega_n^3}{(s^2 + \omega_n^2)^2} \quad (3)$$

The inverse Laplace transform of $U(s)$ has the form

$$u(t) = \frac{p_0}{k} \mathcal{L}^{-1} \left[\frac{\omega_n^3}{(s^2 + \omega_n^2)^2} \right] \quad (4)$$

From transform pair 10 in Table C.1, the inverse transform for Eq. 4 is

$$\mathcal{L}^{-1} \left[\frac{\omega_n^3}{(s^2 + \omega_n^2)^2} \right] = \frac{1}{2}(\sin \omega_n t - \omega_n t \cos \omega_n t) \quad (5)$$

Finally,

$$u(t) = \frac{p_0}{2k}(\sin \omega_n t - \omega_n t \cos \omega_n t) \quad \textbf{Ans.} \quad (6)$$

Clearly, as in Fig. 4.3, this solution grows without bound if the excitation continues for more than a few cycles.

In the discussion of experimental modal analysis in Chapter 19 we make extensive use of frequency-response functions. The *frequency-response function* $H(\omega)$ for an underdamped SDOF system is related to the *system transfer function* $H(s)$ in Eq. 5.57 by

$$H(\omega) \equiv H(s)|_{s=i\omega} = \frac{1/m}{(i\omega - \lambda_1)(i\omega - \lambda_1^*)} \quad (5.59)$$

where

$$\lambda_1, \lambda_1^* = -\zeta \omega_n \pm \omega_n \sqrt{1 - \zeta^2} \quad (5.60)$$

are the roots of the characteristic equation. In Chapter 19 it will be useful to express frequency-response functions in partial-fraction format, or *pole-residue format*,

$$H(\omega) = \frac{A}{i\omega - \lambda_1} + \frac{A^*}{i\omega - \lambda_1^*} \quad (5.61)$$

where λ_1, λ_1^* are the *poles* and A, A^* are the corresponding *residues*.

REFERENCES

[5.1] J. M. Biggs, *Introduction to Structural Dynamics*, McGraw-Hill, New York, 1964.

[5.2] J. L. Schiff, *The Laplace Transform: Theory and Applications*, Springer-Verlag, New York, 1999.

PROBLEMS

> For problems whose number is preceded by a **C**, you are to write a computer program and use it to produce the plot(s) requested. *Note*: MATLAB .m-files for many of the plots in this book may be found on the book's website.

C 5.1 Consider an undamped SDOF system that starts from rest at $t = 0$ and is subjected to the ideal step input shown in Fig. 5.1. **(a)** Starting with the equation of motion, Eq. 5.1 (with $c = 0$), and with the initial conditions given by Eqs. 5.2, derive an expression for the response ratio $R(t)$ for an undamped SDOF system subjected to an ideal step input. Show all steps in your derivation. **(b)** Plot five complete cycles of the response ratio $R(t)$ that you obtained in part (a) versus the nondimensional time parameter $(\omega_n t)$; that is, plot $R(t)$ versus $\omega_n t$ for $0 \leq \omega_n t \leq 10\pi$.

5.2 A very simplified model is to be used to study the landing impact of a light aircraft. The airplane is modeled as shown in Fig. P5.2, that is, as a lumped mass with a linear spring representing the landing gear. The mass m has a vertical descent speed of V when the bottom of the spring touches the ground. Call the time of contact $t = 0$, and let $u(0) = 0$. **(a)** Derive the equation of motion for the mass with displacement $u(t)$ measured positive downward from the position of the mass when the spring first comes in contact with the ground. Show your free-body diagram. **(b)** Determine an expression for the vertical position of the mass as a function of time during the time that the spring remains in contact with the ground. **(c)** Determine the time at which the spring loses contact with the ground upon rebound.

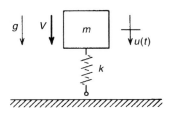

Figure P5.2

5.3 An instrument is being shipped in a container with foam packaging (Fig. P5.3) that has an effective vertical stiffness and provides an effective damping factor. Model this package system as a mass (instrument) on a spring (k = stiffness of foam packaging) as in Fig. P5.2, with an added viscous dashpot (ζ = damping factor due to foam packaging). Gravity g acts downward. **(a)** Derive the equation of motion for the mass, with displacement $u(t)$ measured positive downward from the position of the mass when the spring and dashpot first come in contact with the ground. Show your free-body diagram. **(b)** Determine an expression for the vertical position of the mass as a function of time during the time that the spring remains in contact with the ground. **(c)** If the instrument weighs 40 lb, $k = 100$ lb/in., and $\zeta = 0.05$, and if the container and its contents hit the ground with a vertical speed $V = 150$ in./sec, determine the maximum total force exerted on the instrument by the packaging, that is, the maximum value of the combined force of the "spring" and the "dashpot."

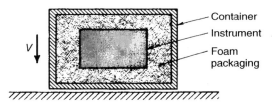

Figure P5.3

5.4 Determine an expression for the transient response of an underdamped $(0 < \zeta < 1)$ SDOF system with initial conditions $u(0) = 0$ and $\dot{u}(0) = v_0$, subjected to an ideal step input, as shown in Fig. 5.1.

Problem Set 5.2

5.5 Determine expressions for the time t_m at which the maximum response occurs for an undamped SDOF system subjected to a rectangular pulse of duration t_d, as shown in Fig. 5.3a. Express your answers in terms of the pulse duration t_d and the undamped natural period of the system, T_n. There should be one expression for $t_d/T_n < 0.5$ and a second expression for $t_d/T_n \geq 0.5$. (See Fig. 5.4b.)

5.6 An undamped SDOF system that is initially at rest is subjected to a ramp forcing function of the form shown in Fig. 5.5. For the case $0 \leq t_r \leq T_n$, determine expressions for **(a)** the time t_m at which the maximum response occurs, and **(b)** the maximum response ratio, $R_{\max}$. Express your answers in terms of the ramp time t_r and the undamped period T_n. (See Fig. 5.6 for an example.)

5.7 An undamped SDOF system is initially at rest. Determine expressions for the response of the system to the forcing function shown in Fig. P5.7 for each of the following time intervals: (a) $0 \leq t \leq t_1$, (b) $t_1 \leq t \leq t_2$, and (c) $t_2 \leq t$.

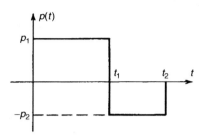

Figure P5.7

Problem Set 5.4

In Problems 5.8 through 5.12, use the *Duhamel Integral Method* to determine expressions for the response of an undamped SDOF system over the time intervals stated. Let $u(0) = \dot{u}(0) = 0$.

5.8 For the excitation shown in Fig. P5.8, determine $u(t)$ for (a) $0 \leq t \leq t_d$, and (b) $t_d \leq t$.

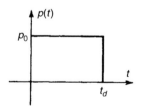

Figure P5.8

5.9 For the excitation shown in Fig. P5.9, determine $u(t)$ for (a) $0 \leq t \leq t_d$, and (b) $t_d \leq t$.

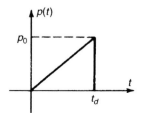

Figure P5.9

5.10 For the excitation shown in Fig. P5.10, determine $u(t)$ for (a) $0 \leq t \leq t_d$, and (b) $t_d \leq t$. Assume that $\omega_n \neq \pi/t_d$.

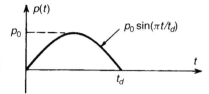

Figure P5.10

5.11 For the excitation shown in Fig. P5.11, determine $u(t)$ for (a) $0 \leq t \leq t_d$, and (b) $t_d \leq t$. Assume that $\omega_n \neq \pi/2t_d$.

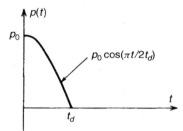

Figure P5.11

5.12 For the excitation shown in Fig. P5.12, determine $u(t)$ for (a) $0 \leq t \leq t_d/2$, (b) $t_d/2 \leq t \leq t_d$, and (c) $t_d \leq t$.

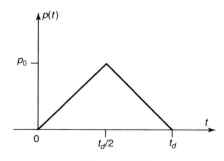

Figure P5.12

C 5.13 Consider an undamped SDOF system that starts from rest at $t = 0$ and is subjected to the ideal step input shown in Fig. 5.1. (a) Use the *Duhamel Integral*

Method to determine an expression for the response of this system for $t \geq 0$. **(b)** Using the expression for $u(t)$ that you obtained in part (a), plot five complete cycles of the response ratio $R(t) = ku(t)/p_0$ versus the nondimensional time parameter $\omega_n t$; that is, plot $R(t)$ versus $\omega_n t$ for $0 \leq \omega_n t \leq 10\pi$.

5.14 When fully loaded, the vehicle in Example 4.3 runs over a half-sine bump at 100 km/h. The length of the bump is 2 m, as shown in Fig. P5.14. Carry out the following steps that would be needed to determine the maximum force exerted on the mass m by the springs: **(a)** Determine t_d, the time it would take for the vehicle to pass over the bump, and write an expression for $z(t)$, assuming that the vehicle reaches the left edge of the bump at $t = 0$. That is, write an expression for $z(t)$ for $0 \leq t \leq t_d$. **(b)** Determine the equation of motion for the mass expressed in terms of the relative motion $w(t) = u(t) - z(t)$. **(c)** Using the *Duhamel Integral Method*, determine expressions for $w_1(t)$ for $0 \leq t \leq t_d$ and $w_2(t)$ for $t_d \leq t$. *Note*: Since this is a damped system, you may leave integrals of the form $\int e^{a\tau}(\sin b\tau)(\cos c\tau)\,d\tau$ in your answers.

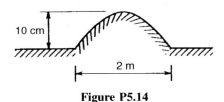

Figure P5.14

Problem Set 5.5

C 5.15 (a) Solve Problem 5.9 by following the steps in Example 5.2. Express your answers in response-ratio form. **(b)** Following Example 5.3, and using the two response-ratio expressions that you obtained as answers to part (a), plot the response ratios versus nondimensional time t/t_d for the time interval $0 \leq t/t_d \leq 4.0$ for input durations $t_d/T_n = 0.25, 0.50, 1.0$, and 1.5. Your plots should be similar to the four plots in Fig. 1 of Example 5.3, but extended to $t/t_d = 4.0$. **(c)** From the data used to create the four plots in part (b), determine the maximum and minimum values of each response ratio and the time at which each occurs.

C 5.16 (a) Solve Problem 5.10 by following the steps in Example 5.2. Express your answers in response-ratio form. **(b)** Also obtain response-ratio expressions that are valid when $\omega_n t_d = \pi$, that is, when $t_d/T_n = 0.5$. **(c)** Following Example 5.3, and using the response-ratio expressions that you obtained as answers to parts (a) and (b), plot the response ratio versus nondimensional time t/t_d for $0 \leq t/t_d \leq 4.0$ for pulse durations $t_d/T_n = 0.25, 0.50, 1.0$, and 1.5. Your plots should be similar to the four plots as in Fig. 1 of Example 5.3, but extended to $t/t_d = 4.0$. **(d)** From the data used to create the four plots in part (c), determine the maximum and minimum values of each response ratio and the time at which each occurs.

Problem Set 5.6

5.17 Using the definition of the Laplace transform in Eq. 5.48, determine the Laplace transforms $F_i(s)$ of the following functions: **(a)** $f_1(t) = t$, **(b)** $f_2(t) = \cos \omega t$, and **(c)** $f_3(t) = e^{\sigma t}\sin \omega t$ with $\Re(s - \sigma) > 0$.

5.18 Using the definition of the Laplace transform in Eq. 5.48, determine an expression for the Laplace transform of $d^2 f(t)/dt^2$. Let $\mathcal{L}[f(t)] \equiv F(s)$.

5.19 Using the definition of the Laplace transform in Eq. 5.48, determine expressions for the Laplace transforms of the following functions: **(a)** $f_1(t) = e^{\sigma t}\sinh \omega t$, and **(b)** $f_2(t) = e^{\sigma t}\cosh \omega t$ with $\Re(s - \sigma) > 0$.

5.20 Using the shifting theorem of Section C.4 and the table of Laplace transforms in Section C.6, determine expressions for the Laplace transforms of the following functions: **(a)** $f_1(t) = e^{\sigma t}\sinh \omega t$, and **(b)** $f_2(t) = e^{\sigma t}\cosh \omega t$ with $\Re(s - \sigma) > 0$.

5.21 An underdamped SDOF system (ω_n, ζ) is given the following initial conditions: $u(0) = u_0$, $\dot{u}(0) = 0$. **(a)** Use the *Laplace Transform Method* to determine the transformed response $U(s)$. **(b)** Use Table C.1 to solve for the response $u(t)$. Express your answers in terms of u_0, ω_n, and ζ.

5.22 An underdamped SDOF system (ω_n, ζ) is at rest $[u(0) = 0, \dot{u}(0) = 0]$ when excitation $p(t) = p_0 \cos \Omega t$, $\Omega \neq \omega_n$ begins. **(a)** Starting with Eq. 4.16, use the *Laplace Transform Method* to determine the transformed response $U(s)$. **(b)** Solve for the total response $u(t)$. Express your answer in the form

$$\frac{ku(t)}{p_0} = C_1 \sin \Omega t + C_2 \cos \Omega t + \frac{C_3}{\omega_d}e^{-\zeta\omega_n t}\sin \omega_d t$$
$$+ C_4 e^{-\zeta \omega_n t}\left(\cos \omega_d t + \frac{\zeta \omega_n}{\omega_d}\sin \omega_d t\right)$$

Hint: See Example 5.8 and use the *Partial-Fractions Method*, with Table C.1, to determine expressions for coefficients C_1 through C_4. Express your answer

$ku(t)/p_0$ in terms of $\Omega, \omega_n, \zeta, r = \Omega/\omega_n$, and $\omega_d = \omega_n(1-\zeta^2)^{1/2}$.

5.23 An overdamped SDOF system ($\omega_n, \zeta > 1$) is given the following initial conditions: $u(0) = 0, \dot{u}(0) = v_0$. **(a)** Determine the transformed response $U(s)$. You may use any method discussed in Section 5.6 or any method discussed in Appendix C. **(b)** Solve for the response $u(t)$. Express your answers in terms of v_0, ω_n, ζ, and $\omega^* = \omega_n(\zeta^2-1)^{1/2}$.

6

Numerical Evaluation of the Dynamic Response of SDOF Systems

In Section 5.4, Duhamel integral expressions were obtained for the response of undamped and underdamped linear SDOF systems subject to arbitrary excitation. For simple forms of excitation, these expressions can be evaluated in closed form. For more complex excitation, a numerical solution is required. Numerical approximation methods used for simulation of structural dynamics transient response problems fall into two broad categories: integration methods for first- and second-order ordinary differential equations. The two different categories of algorithms have become popular for quite different reasons. The second-order integration methods exploit the structure of the ordinary differential equations so common in structural dynamics. For the most part, the goal of this class of methods is efficiency and accuracy in the numerical integration of the second-order ordinary differential equations of motion. Variants of the second-order methods have been derived for some types of nonlinear ordinary differential equations and are discussed in Section 6.3.2. On the other hand, the first-order methods are typically derived assuming a priori that the underlying equations are nonlinear. They are readily able to accommodate structural dynamics models with nonlinearities. In addition, the first-order forms of the governing equations are more amenable to the study of applications in control theory. Over the past few decades, modern structural dynamics has become closely related to many topics studied in control theory.

Upon completion of this chapter you should be able to:

- Use, and in some cases derive, a number of popular second-order integration methods for applications in structural dynamics. These methods should include:
  - Approximation of the excitation via piecewise-constant interpolation
  - Approximation of the excitation via piecewise-linear interpolation
  - The Average Acceleration Method
- Use, and in some cases derive, a number of first-order integration methods for applications in structural dynamics. These methods should include:
  - Direct Taylor Series Methods
  - Runge–Kutta Methods
  - Multistep Methods

Finally, it is important to note that the collection of techniques summarized above represent only a sample of the useful and viable integration methods for applications in structural dynamics. These methods are, however, representatives of classes of methods that are frequently encountered. This chapter, along with the additional techniques discussed in Chapter 16, should provide an excellent foundation on which the reader can build a more specialized study.

6.1 INTEGRATION OF SECOND-ORDER ORDINARY DIFFERENTIAL EQUATIONS

In this section a number of numerical integration methods are discussed that exploit the structure of the prototypical linear ordinary differential equation

$$m\ddot{u} + c\dot{u} + ku = p(t)$$

that arises so frequently in structural dynamics. As will be shown in this chapter and in Chapter 16, the algorithms described in this section can be applied to a host of practical problems in structural dynamics.

6.1.1 Numerical Solution Based on Interpolation of the Excitation Function

In many practical structural dynamics problems the excitation function, $p(t)$, is not known in the form of an analytical expression, but rather, is given by a set of discrete values $p_i \equiv p(t_i)$ for $i = 0$ to N. These may be presented in "tabular form" as a computer file or presented graphically as in Fig. 6.1. The time interval

$$\Delta t_i = t_{i+1} - t_i \tag{6.1}$$

is frequently taken to be a constant, Δt. One approach to obtaining the response to this excitation is to use numerical quadrature formulas (e.g., the trapezoidal rule or Simpson's one-third rule) to evaluate the integrals appearing in the Duhamel integral expressions of Eqs. 5.36 and 5.37.[6.1] This involves interpolation of the integrands appearing in the Duhamel integrals.

A more direct and efficient procedure involves interpolation of the excitation function $p(t)$ and exact solution of the resulting linear response problem using results

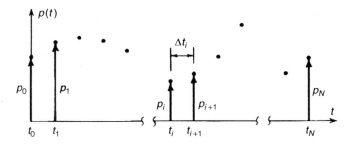

Figure 6.1 Excitation specified at discrete times.

6.1 Integration of Second-Order Ordinary Differential Equations   149

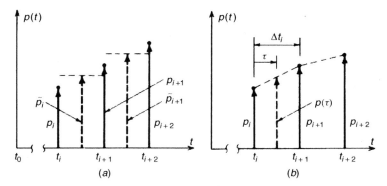

Figure 6.2 (a) Piecewise-constant and (b) piecewise-linear interpolation of excitation functions.

from Chapter 5. Figure 6.2 shows *piecewise-constant* and *piecewise-linear* interpolation of excitation functions. For piecewise-constant interpolation, the value of the force in the interval t_i to t_{i+1} is $\tilde{p}_i$, which could be taken to be the value p_i at the beginning of the interval, the value p_{i+1} at the end of the interval, or the average value $\tilde{p}_i = 0.5(p_i + p_{i+1})$. For piecewise-linear interpolation the interpolated force is given by

$$p(\tau) = p_i + \frac{\Delta p_i}{\Delta t_i}\tau \tag{6.2}$$

where

$$\Delta p_i = p_{i+1} - p_i \tag{6.3}$$

Consider the response of an undamped system. For piecewise-constant interpolation the forced response can be obtained from Eq. 5.8 and the response due to nonzero initial conditions from Eq. 3.17. Thus,

$$u(\tau) = u_i \cos\omega_n\tau + \frac{\dot{u}_i}{\omega_n}\sin\omega_n\tau + \frac{\tilde{p}_i}{k}(1 - \cos\omega_n\tau) \tag{6.4a}$$

$$\frac{\dot{u}(\tau)}{\omega_n} = -u_i\sin\omega_n\tau + \frac{\dot{u}_i}{\omega_n}\cos\omega_n\tau + \frac{\tilde{p}_i}{k}\sin\omega_n\tau \tag{6.4b}$$

Evaluating these expressions at time t_{i+1} (i.e., at $\tau = \Delta t_i$), we get

$$u_{i+1} = u_i\cos(\omega_n\,\Delta t_i) + \frac{\dot{u}_i}{\omega_n}\sin(\omega_n\,\Delta t_i) + \frac{\tilde{p}_i}{k}[1 - \cos(\omega_n\,\Delta t_i)] \tag{6.5a}$$

$$\frac{\dot{u}_{i+1}}{\omega_n} = -u_i\sin(\omega_n\,\Delta t_i) + \frac{\dot{u}_i}{\omega_n}\cos(\omega_n\,\Delta t_i) + \frac{\tilde{p}_i}{k}\sin(\omega_n\,\Delta t_i) \tag{6.5b}$$

Equations 6.5 are the *recurrence formulas* for evaluating the dynamic state $(u_{i+1}, \dot{u}_{i+1})$ at time t_{i+1} given the state $(u_i, \dot{u}_i)$ at time t_i.

In contrast to piecewise-constant interpolation of the excitation, piecewise-linear interpolation permits a closer approximation. Equation 6.2 can be used to derive recurrence formulas based on piecewise-linear interpolation of the excitation. For an undamped system, we get

$$u_{i+1} = u_i \cos(\omega_n \Delta t_i) + \frac{\dot{u}_i}{\omega_n} \sin(\omega_n \Delta t_i)$$
$$+ \frac{p_i}{k}[1 - \cos(\omega_n \Delta t_i)] \qquad (6.6a)$$
$$+ \frac{\Delta p_i}{k} \frac{1}{\omega_n \Delta t_i}[\omega_n \Delta t_i - \sin(\omega_n \Delta t_i)]$$

$$\frac{\dot{u}_{i+1}}{\omega_n} = -u_i \sin(\omega_n \Delta t_i) + \frac{\dot{u}_i}{\omega_n} \cos(\omega_n \Delta t_i)$$
$$+ \frac{p_i}{k} \sin(\omega_n \Delta t_i) \qquad (6.6b)$$
$$+ \frac{\Delta p_i}{k} \frac{1}{\omega_n \Delta t_i}[1 - \cos(\omega_n \Delta t_i)]$$

Recurrence formulas such as Eqs. 6.5 and 6.6 may be expressed conveniently in the following form:

$$\boxed{\begin{aligned} u_{i+1} &= A p_i + B p_{i+1} + C u_i + D \dot{u}_i \\ \dot{u}_{i+1} &= A' p_i + B' p_{i+1} + C' u_i + D' \dot{u}_i \end{aligned}} \qquad (6.7)$$

Table 6.1 gives expressions for the coefficients A through D' for the linear force interpolation for the underdamped case ($\zeta < 1$). Coefficients can also be derived for critically damped and overdamped cases.

Table 6.1 Coefficents for Recurrence Formulas for Underdamped SDOF Systems

$$A = \frac{1}{k\omega_d h}\left\{e^{-\beta h}\left[\left(\frac{\omega_d^2 - \beta^2}{\omega_n^2} - \beta h\right)\sin \omega_d h - \left(\frac{2\omega_d \beta}{\omega_n^2} + \omega_d h\right)\cos \omega_d h\right] + \frac{2\beta \omega_d}{\omega_n^2}\right\}$$

$$B = \frac{1}{k\omega_d h}\left[e^{-\beta h}\left(-\frac{\omega_d^2 - \beta^2}{\omega_n^2}\sin \omega_d h + \frac{2\omega_d \beta}{\omega_n^2}\cos \omega_d h\right) + \omega_d h - \frac{2\beta \omega_d}{\omega_n^2}\right]$$

$$C = e^{-\beta h}\left(\cos \omega_d h + \frac{\beta}{\omega_d}\sin \omega_d h\right)$$

$$D = \frac{1}{\omega_d}e^{-\beta h}\sin \omega_d h$$

$$A' = \frac{1}{k\omega_d h}\{e^{-\beta h}[(\beta + \omega_n^2 h)\sin \omega_d h + \omega_d \cos \omega_d h] - \omega_d\}$$

$$B' = \frac{1}{k\omega_d h}[-e^{-\beta h}(\beta \sin \omega_d h + \omega_d \cos \omega_d h) + \omega_d]$$

$$C' = -\frac{\omega_n^2}{\omega_d}e^{-\beta h}\sin \omega_d h$$

$$D' = e^{-\beta h}\left(\cos \omega_d h - \frac{\beta}{\omega_d}\sin \omega_d h\right)$$

where $\beta \equiv \zeta \omega_n$ and $h \equiv \Delta t_i$

If the time step Δt_i is constant, the coefficients A through D' need be calculated only once, which greatly speeds the computations. Since the recurrence formulas in Table 6.1 are based on exact integration of the equation of motion for the various force interpolations, the only restriction on the size of the time step Δt_i is that it permit a close approximation to the excitation and that it provide output $(u_i, \dot{u}_i)$ at the times where output is desired. In the latter regard, if the maximum response is desired, the time step should satisfy $\Delta t_i \leq T_n/10$ so that peaks due to natural response will not be missed. Recall that T_n is the natural period of the system dynamics.

Example 6.1 For the undamped SDOF system in Example 4.1, determine the response $u(t)$ for $0 \leq t \leq 0.2$ sec: (a) by using piecewise-linear interpolation of the force with $\Delta t = 0.02$ sec, and (b) by evaluating the exact expression for $u(t)$ at these time steps.

SOLUTION From Example 4.1, $k = 40$ lb/in., $\omega_n = 20$ rad/sec, and $p(t) = 10\cos(10t)$ lb.

(a) For the linear interpolation method the displacement and velocity can be written in the form of Eqs. 6.7.

$$u_{i+1} = Ap_i + Bp_{i+1} + Cu_i + D\dot{u}_i$$
$$\dot{u}_{i+1} = A'p_i + B'p_{i+1} + C'u_i + D'\dot{u}_i \tag{1}$$

where the coefficients A through D' can be obtained directly from Eqs. 6.6 or by setting $\beta = 0$ and $\omega_d = \omega_n$ in Table 6.1. Thus, with $\Delta t \equiv h$,

$$A = \frac{1}{k\omega_n h}(\sin\omega_n h - \omega_n h \cos\omega_n h) \qquad B = \frac{1}{k\omega_n h}(\omega_n h - \sin\omega_n h)$$

$$C = \cos\omega_n h \qquad\qquad D = \frac{1}{\omega_n}\sin\omega_n h \tag{2}$$

$$A' = \frac{1}{kh}(\omega_n h \sin\omega_n h + \cos\omega_n h - 1) \qquad B' = \frac{1}{kh}(1 - \cos\omega_n h)$$

$$C' = -\omega_n \sin\omega_n h \qquad\qquad D' = \cos\omega_n h$$

Thus,
$$\omega_n h = 20(0.02) = 0.4 \text{ rad}$$
$$\sin\omega_n h = 0.38942$$
$$\cos\omega_n h = 0.92106$$
$$A = \frac{0.38942 - 0.4(0.92106)}{40(0.4)} = 1.312 \times 10^{-3}$$
$$B = \frac{0.4 - 0.38942}{40(0.4)} = 6.613 \times 10^{-4}$$
$$C = 9.211 \times 10^{-1} \tag{3}$$

(continued on p. 153)

Table 1 Numerical Solution Based on Piecewise-Linear Interpolation of the Excitation (Exponent in Parentheses)

i	t_i	p_i	Ap_i	Bp_{i+1}	Cu_i	$D\dot{u}_i$	$\dot{u}_i$	$A'p_i$	$B'p_{i+1}$	$C'u_i$	$D'\dot{u}_i$	$\ddot{u}_i$
0	0	10.0000	1.312(−2)	6.481(−3)	0.000	0.000	0.000	9.604(−1)	9.671(−1)	0.000	0.000	0.000
1	0.02	9.8007	1.286(−2)	6.091(−3)	1.805(−2)	3.753(−2)	1.960(−2)	9.413(−1)	9.089(−1)	−1.527(−1)	1.775	1.928
2	0.04	9.2106	1.208(−2)	5.458(−3)	6.865(−2)	6.762(−2)	7.453(−2)	8.846(−1)	8.144(−1)	−5.805(−1)	3.199	3.473
3	0.06	8.2534	1.083(−2)	4.607(−3)	1.417(−1)	8.406(−2)	1.538(−1)	7.927(−1)	6.875(−1)	−1.198	3.977	4.318
4	0.08	6.9671	9.141(−3)	3.573(−3)	2.221(−1)	8.293(−2)	2.412(−1)	6.691(−1)	5.332(−1)	−1.878	3.923	4.259
5	0.10	5.4030	7.089(−3)	2.396(−3)	2.927(−1)	6.322(−2)	3.178(−1)	5.189(−1)	3.576(−1)	−2.475	2.991	3.247
6	0.12	3.6236	4.754(−3)	1.124(−3)	3.366(−1)	2.711(−2)	3.654(−1)	3.480(−1)	1.677(−1)	−2.846	1.283	1.392
7	0.14	1.6997	2.230(−3)	−1.931(−4)	3.404(−1)	−2.040(−2)	3.696(−1)	1.632(−1)	−2.881(−2)	−2.878	−9.649(−1)	−1.048
8	0.16	−0.2920	−3.831(−4)	−1.502(−3)	2.966(−1)	−7.221(−2)	3.221(−1)	−2.804(−2)	−2.242(−1)	−2.508	−3.416	−3.709
9	0.18	−2.2720	−2.981(−3)	−2.752(−3)	2.050(−1)	−1.203(−1)	2.226(−1)	−2.182(−1)	−4.107(−1)	−1.733	−5.689	−6.177
10	0.20	−4.1615	—	—	—	—	7.900(−2)	—	—	—	—	−8.051

$$D = \frac{0.38942}{20} = 1.947 \times 10^{-2}$$

$$A' = \frac{0.4(0.38942) + 0.92106 - 1}{40(0.02)} = 9.604 \times 10^{-2}$$

$$B' = \frac{1.0 - 0.92106}{40(0.02)} = 9.868 \times 10^{-2} \qquad \text{(3 cont.)}$$

$$C' = -20.0(0.38942) = -7.788$$

$$D' = 9.211 \times 10^{-1}$$

The results are tabulated in Table 1.

(b) The exact solution is given by Eq. 9 of Example 4.1. For discrete times t_i this can be written

$$u_i = \tfrac{1}{3}[\cos(10t_i) - \cos(20t_i)] \qquad (4)$$

The results are tabulated in Table 2. By comparing the solution based on piecewise-linear interpolation of the excitation with the exact solution, we see that there is good agreement. Since the period of the excitation,

$$T = \frac{2\pi}{\Omega} = 0.628 \text{ sec} \qquad (5)$$

is approximately 30 times Δt, the piecewise-linear approximation of $\cos \Omega t$ should be quite good.

Table 2 Numerical Solution Based on the Exact Response Function (Exponent in Parentheses)

i	t_i	$u_i = \tfrac{1}{3}[\cos(10t_i) - \cos(20t_i)]$
0	0	0
1	0.02	1.967(−2)
2	0.04	7.478(−2)
3	0.06	1.543(−1)
4	0.08	2.419(−1)
5	0.10	3.188(−1)
6	0.12	3.665(−1)
7	0.14	3.707(−1)
8	0.16	3.230(−1)
9	0.18	2.232(−1)
10	0.20	7.916(−2)

For hand calculation of the response, the tabular form of Table 1 is convenient. Recurrence formulas of the form given in Eqs. 6.7 are easily programmed for a digital computer or programmable calculator.

The next example illustrates how the use of MATLAB can provide efficient, easily interpreted implementations of the methods based on exact integration of the governing

SDOF equation using either piecewise-constant or piecewise-linear interpolation of the forcing function.[1]

Example 6.2 Comparison of Piecewise-Constant and Piecewise-Linear Algorithms
Consider again Example 4.1, where $k = 40$ lb/in., $\omega_n = 20$ rad/sec, and $p(t) = 10\cos(10t)$ lb. Use MATLAB to obtain plots of the response based on piecewise-constant interpolation of the forcing function and the response based on piecewise-linear interpolation of the forcing function.

SOLUTION   In Example 6.1 we derived the constants A, B, C, D, A', B', C', and D' that define a simple step-by-step recursion summarized in Eqs. 6.7a and b. This recursion is easy to implement in a few lines of MATLAB code. The .m-files that encode the approximation process are:

- mat-ex-6-1.m: MATLAB example for piecewise-constant approximation of forcing function
- mat-ex-6-2.m: MATLAB example for piecewise-linear approximation of forcing function

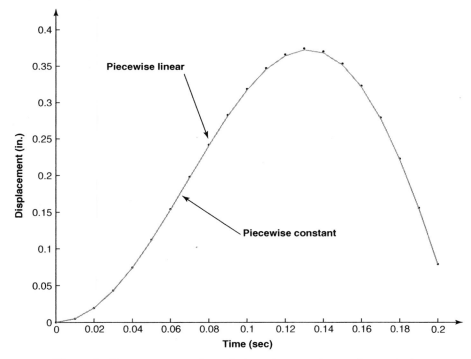

Figure 1   Response comparison of piecewise-exact integration methods.

---

[1] The MATLAB .m-files that are listed in Examples 6.2, 6.4, and 6.6 may be found on the book's website.

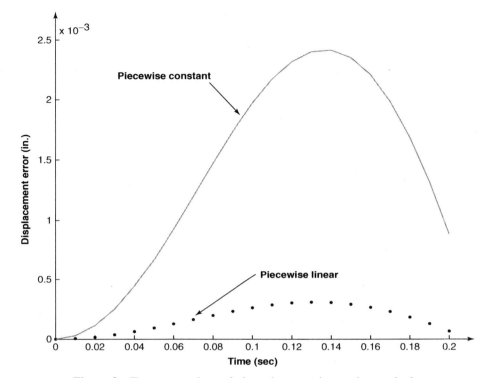

Figure 2 Error comparison of piecewise-exact integration methods.

- pc-exact-int.m: MATLAB function for calculating piecewise-constant exact integration
- plin-exact-int.m: MATLAB function for calculating piecewise-linear exact integration

In this example, the time step has been selected to be $\Delta t = 0.01$ sec. The right-hand-side function $p(t) = 10\cos(10t)$ lb has been approximated by piecewise-constant and piecewise-linear interpolation of the forcing function. The results are depicted in Figs. 1 and 2. Figure 1 clearly shows that there is little difference between the two methods as to the transient response predicted. Figure 2 shows that the error is on the order of 10^{-3} for each method.

6.1.2 Average Acceleration Method

In Section 6.1.1, recurrence relations based on interpolation of the excitation were obtained. These permit the response of a linear SDOF system to be obtained at discrete times t_i. An alternative approach, which may be used for determining the response of both linear and nonlinear systems, is to approximate the derivatives appearing in the system equation of motion and to generate a step-by-step solution using time steps Δt_i. Although many such procedures are available for carrying out this numerical integration

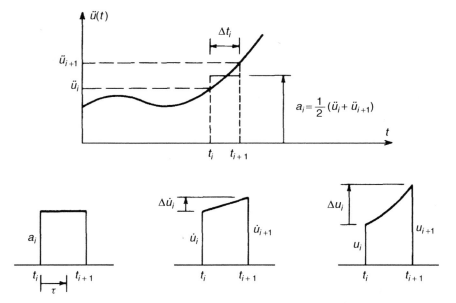

Figure 6.3 Numerical integration using the Average Acceleration Method.

of second-order differential equations (see, e.g., Refs. [6.2] to [6.4]), one of the most common is the *Average Acceleration Method*,[2] which we consider next.

The equation of motion to be integrated is

$$m\ddot{u} + c\dot{u} + ku = p(t) \tag{6.8}$$

with initial conditions $u(0) = u_0$ and $\dot{u}(0) = v_0$ given. The acceleration is approximated as shown in Fig. 6.3. The acceleration in the time interval t_i to t_{i+1} is taken to be the average of the initial and final values of acceleration, that is,

$$\ddot{u}(\tau) = \tfrac{1}{2}(\ddot{u}_i + \ddot{u}_{i+1}) \tag{6.9}$$

Integration of Eq. 6.9 twice gives

$$\dot{u}_{i+1} = \dot{u}_i + \frac{\Delta t_i}{2}(\ddot{u}_i + \ddot{u}_{i+1}) \tag{6.10}$$

and

$$u_{i+1} = u_i + \dot{u}_i\,\Delta t_i + \frac{\Delta t_i^2}{4}(\ddot{u}_i + \ddot{u}_{i+1}) \tag{6.11}$$

In setting up the computational algorithm for this numerical integration problem, it is convenient to employ previously computed values $\dot{u}_i$ and $\ddot{u}_i$, and the incremental

---

[2]This method is also referred to as the *Newmark* $\beta = \tfrac{1}{4}$ *Method*, the *Trapezoidal Rule*, and the *Constant-Average-Acceleration Method*.

quantities Δp_i, Δu_i, $\Delta \dot{u}_i$, where $\Delta p_i = p_{i+1} - p_i$, and so on.[3] Then Eq. 6.11 can be solved for $\Delta \ddot{u}_i$ and Eqs. 6.10 and 6.11 combined to give $\Delta \dot{u}_i$ as follows:

$$\Delta \ddot{u}_i = \frac{4}{\Delta t_i^2}(\Delta u_i - \dot{u}_i \Delta t_i) - 2\ddot{u}_i \tag{6.12}$$

and

$$\Delta \dot{u}_i = \frac{2}{\Delta t_i}\Delta u_i - 2\dot{u}_i \tag{6.13}$$

Since Eq. 6.8 is satisfied at both t_i and t_{i+1}, we can form the following incremental equation of motion:

$$m\,\Delta \ddot{u}_i + c\,\Delta \dot{u}_i + k\,\Delta u_i = \Delta p_i \tag{6.14}$$

Equations 6.12 through 6.14 can be combined to give the equation

$$\boxed{k_i^*\,\Delta u_i = \Delta p_i^*} \tag{6.15}$$

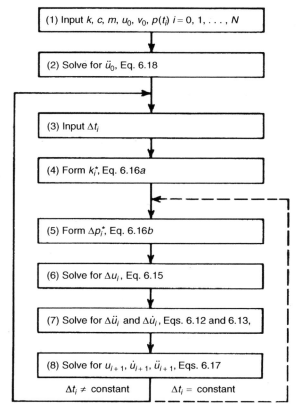

Figure 6.4 Flowchart for step-by-step numerical integration based on the *Average Acceleration Method*.

---

[3]The incremental form presented here is used in the nonlinear analysis of Section 6.3.2. An alternative "operator formulation" is presented in Chapter 16.

where

$$k_i^* = k + \frac{2c}{\Delta t_i} + \frac{4m}{\Delta t_i^2} \qquad (6.16a)$$

and

$$\Delta p_i^* = \Delta p_i + \left(\frac{4m}{\Delta t_i} + 2c\right)\dot{u}_i + 2m\ddot{u}_i \qquad (6.16b)$$

Once Δu_i has been determined from Eq. 6.15, $\Delta \dot{u}_i$ can be obtained from Eq. 6.13 and $\Delta \ddot{u}_i$ from Eq. 6.12, and the updated values of u, $\dot{u}$, and $\ddot{u}_i$ determined from

$$u_{i+1} = u_i + \Delta u_i, \qquad \dot{u}_{i+1} = \dot{u}_i + \Delta \dot{u}_i, \qquad \ddot{u}_{i+1} = \ddot{u}_i + \Delta \ddot{u}_i \qquad (6.17)$$

The acceleration at time t_i can be obtained from the equation of motion,

$$\ddot{u}_i = \frac{1}{m}(p_i - c\dot{u}_i - ku_i) \qquad (6.18)$$

Therefore, $\ddot{u}_{i+1}$ can be obtained from this equation rather than from Eqs. 6.12 and 6.17c. This equation is also used to obtain $\ddot{u}_0$.

A computational algorithm based on the equations above is summarized in the flowchart shown in Fig. 6.4. If Δt_i is constant, steps 3 and 4 can be removed from the inner loop. We discuss numerical integration methods further, in Chapter 16, including the Average Acceleration Method.

Example 6.3 For the undamped SDOF system in Examples 4.1 and 6.1, use the Average Acceleration Method to determine the response $u(t)$ for $0 \le t \le 0.2$ sec. Use a constant time step $\Delta t = 0.02$ sec and also $\Delta t = 0.005$ sec. Compare the results with the exact solution of Example 6.1.

SOLUTION From Example 4.1, $k = 40$ lb/in., $c = 0$, $m = 0.1$ lb-sec²/in., $\omega_n = 20$ rad/sec, and $p(t) = 10\cos(10t)$ lb. The response results are shown in Table 1.

Table 1 Response $u(t_i)$ via the Average Acceleration Method (Exponent in Parentheses)

t_i	Exact	$\Delta t = 0.02$ sec	$\Delta t = 0.005$ sec
0	0	0	0
0.02	1.967(−2)	1.904(−2)	1.963(−2)
0.04	7.478(−2)	7.247(−2)	7.464(−2)
0.06	1.543(−1)	1.498(−1)	1.540(−1)
0.08	2.419(−1)	2.356(−1)	2.416(−1)
0.10	3.188(−1)	3.116(−1)	3.184(−1)
0.12	3.665(−1)	3.602(−1)	3.662(−1)
0.14	3.707(−1)	3.673(−1)	3.705(−1)
0.16	3.230(−1)	3.243(−1)	3.231(−1)
0.18	2.232(−1)	2.303(−1)	2.237(−1)
0.20	7.916(−2)	9.221(−2)	8.002(−2)

A rule of thumb that Δt should satisfy $\Delta t \leq T/10$, where T is the smallest period in the excitation (see Chapter 16) or the natural period T_n, whichever is smaller, is frequently stated. In Example 6.3 the natural period, $T_n = 2\pi/\omega_n = 0.314$ sec, is smaller than the excitation period, $T = 0.628$ sec. The time step $\Delta t = 0.02$ sec satisfies the rule of thumb and, as seen in Example 6.3, gives satisfactory accuracy. The shorter time step, $\Delta t = 0.005$ sec, reproduces the "exact" solution almost identically. Chapter 16 gives further information on the relationship of time step to accuracy of solution.

6.2 INTEGRATION OF FIRST-ORDER ORDINARY DIFFERENTIAL EQUATIONS

In this section we present several techniques for integration of ordinary differential equations in first-order form. As in Section 6.1, the primary purpose of the chapter is the discussion of numerical methods for approximation of the solution of the prototypical equation of structural dynamics:

$$m\ddot{u} + c\dot{u} + ku = p(t), \qquad u(0) = u_0, \quad \dot{u}(0) = v_0 \qquad (6.19)$$

The application of the methods in this section require that Eqs. 6.19 be recast in first-order, or state-variable, form. For the second-order system of Eqs. 6.19, the *state* of the system consists of two variables, the displacement $u(t)$ and the velocity $\dot{u}(t)$. In this book, we use two distinct first-order differential equations that are equivalent to Eqs. 6.19 referred to as the *state-space form* and *generalized state-space form* of the governing equations. The latter form is introduced in Section 10.4 and is applied to experimental structural dynamics in Chapter 18.

To derive the state-space form of the governing equations, a new dependent variable $x(t) \in \mathbb{R}^2$ is introduced.[4]

$$x(t) \equiv \left\{ \begin{array}{c} x_1(t) \\ x_2(t) \end{array} \right\} \triangleq \left\{ \begin{array}{c} u(t) \\ \dot{u}(t) \end{array} \right\} \in \mathbb{R}^2 \qquad (6.20)$$

It is easy to see that Eqs. 6.19 are equivalent to the system

$$\boxed{\begin{array}{l} \dot{x}(t) = f(x(t), t) \\ x(0) = x_0 \end{array}} \qquad (6.21)$$

where

$$f(x(t), t) = \mathbf{A}x(t) + p(t)$$

$$\mathbf{A} = \begin{bmatrix} 0 & 1 \\ -m^{-1}k & -m^{-1}c \end{bmatrix} \in \mathbb{R}^{2 \times 2}$$

$$p(t) = \left\{ \begin{array}{c} 0 \\ m^{-1}p(t) \end{array} \right\} \in \mathbb{R}^2 \qquad (6.22)$$

$$x_0 = \left\{ \begin{array}{c} u_0 \\ v_0 \end{array} \right\} \in \mathbb{R}^2$$

---

[4]It is an accepted practice to use the symbol $\mathbb{R}$ to denote the set of all real numbers, and vectors of length N whose elements are extracted from $\mathbb{R}$ are denoted as belonging to $\mathbb{R}^N$.

A close inspection of Eq. 6.21a for the case when the underlying second-order ODE is *linear*, as in Eq. 6.19, reveals that the new first-order equation becomes

$$\dot{x}(t) = \mathbf{A}x(t) + p(t) \tag{6.23}$$

Thus, the first-order equation is linear in the state $x(t)$. However, the analysis presented in this section is based on the more general nonlinear form, as given in Eq. 6.21a.

6.2.1 Direct Taylor Series Methods

Perhaps the simplest means to achieve numerical integration of the system of first-order ordinary differential equations, Eqs. 6.21, uses the following Taylor series expansion of $x(t)$ in a neighborhood of x_0:

$$x(t) = x_0 + \frac{dx(t_0)}{dt}(t - t_0) + \frac{d^2 x(t_0)}{dt^2}\frac{(t - t_0)^2}{2!} + \cdots \tag{6.24}$$

The evaluation of the right-hand side of Eq. 6.24 requires calculation of the derivatives

$$\frac{dx(t_0)}{dt}, \quad \frac{d^2 x(t_0)}{dt^2}, \quad \frac{d^3 x(t_0)}{dt^3}, \quad \cdots \tag{6.25}$$

But Eqs. 6.21 can be used to obtain each of these quantities. For example, the first and the second derivatives can be calculated as follows:

$$\frac{dx(t_0)}{dt} = f(x_0, t_0) \tag{6.26}$$

By the chain rule of differentiation, we have

$$\begin{aligned}\frac{d^2 x(t_0)}{dt^2} &= \left[\frac{\partial f}{\partial x}\dot{x} + \frac{\partial f}{\partial t}\right]\bigg|_{t_0} \\ &= \frac{\partial f}{\partial x}(x_0, t_0) f(x_0) + \frac{\partial f}{\partial t}(x_0, t_0)\end{aligned} \tag{6.27}$$

The entries of this equation must be interpreted by context. We have the matrix derivative

$$\frac{\partial f}{\partial x} \equiv \begin{bmatrix} \frac{\partial f_1}{\partial x_1} & \frac{\partial f_1}{\partial x_2} \\ \frac{\partial f_2}{\partial x_1} & \frac{\partial f_2}{\partial x_2} \end{bmatrix} \triangleq \left[\frac{\partial f_i}{\partial x_j}\right] \in \mathbb{R}^{2 \times 2}$$

and the vectors

$$f \equiv \begin{Bmatrix} f_1(x_1, x_2; t) \\ f_2(x_1, x_2; t) \end{Bmatrix} \triangleq \{f_i\} \in \mathbb{R}^{2 \times 1}$$

$$\frac{\partial f}{\partial t} = \left\{\frac{\partial f_i}{\partial t}\right\} \in \mathbb{R}^{2 \times 1}$$

where the subscripts i and j denote the row and column, respectively, of the matrix expressions. Higher-order derivatives can be calculated in an identical manner, although the notation becomes increasingly complex.

Table 6.2 Algorithm for Direct Taylor Series Method

Step 1	Choose a partitioning $\{t_k\}_{k=0}^{N}$ of the time interval of interest $[t_0, t_N]$.
Step 2	Loop for all the time steps $i = 1 \ldots N$.
	$h = t_i - t_{i-1}$
	$x_i = x_{i-1} + h T_{k,h}(x_{i-1}, t_{i-1})$
Step 3	Return to step 2.

Similar Taylor series expansions can be performed at later time steps t_i about the current state x_i. With this observation, the development of a numerical integration procedure is straightforward. Define the operator $T_{k,h}(x(t), t)$ to be

$$T_{k,h}(x(t), t) = f(x(t), t) + \frac{h}{2!} \frac{d}{dt}[f(x(t), t)]$$
$$+ \cdots + \frac{h^{k-1}}{k!} \frac{d^{(k-1)}}{dt^{k-1}}[f(x(t), t)] \quad (6.28)$$

Recall that each of the terms on the right-hand side can be constructed from the chain rule, as in Eq. 6.27. An algorithm to achieve the numerical integration of the governing equations is given in Table 6.2.

6.2.2 Runge–Kutta Methods

The derivation in the preceding section showed that higher-order accuracy can be obtained, in principle, using the Taylor series approach for integrating first-order ordinary differential equations. However, it is often not practical to evaluate analytically the higher-order derivatives of f on the right-hand side of Eq. 6.28. One of the most popular methods to integrate the equations of motion numerically in first-order form relies on Taylor series approximations but does not require that analytical derivatives be derived for higher-order derivatives. These methods are referred to as the *Runge–Kutta methods*.

Runge–Kutta Method of Order 2 Runge–Kutta methods include integration formulas of arbitrarily high single-step accuracy. The nature of these methods can be illustrated by considering one of the lower-order versions. Extension to higher-order accuracy follows the same arguments. In the preceding section we studied the vector-valued system of first-order differential equations, Eqs. 6.21. To simplify our notation, we derive the approximate integration methods in the next few sections for the scalar first-order equation

$$\dot{x}(t) = f(x(t), t), \quad x(0) = x_0 \quad (6.29)$$

We leave it as an exercise for the reader (Problem 6.12) to show that the discrete integration methods derived for Eqs. 6.29 hold for Eqs. 6.21 with only the straightforward change to vector notation.

Suppose that we make the assumption that successive approximations x_n and x_{n+1} to the solution $x(t)$ of Eq. 6.29 at times t_n and t_{n+1} are related by the expression

$$x_{n+1} = x_n + c_1 h f(x_n, t_n) + c_2 h f(x_n + c_4 h f(x_n, t_n), t_n + c_3 h) \quad (6.30)$$

In this expression, the constants c_1, c_2, c_3, and c_4 are, as yet, unknown constants that characterize the numerical method. Note that all the right-hand-side terms of Eq. 6.30 can be evaluated once the constants are determined, and (x_n, t_n) are also known. From our discussion of Taylor series approximations in Section 6.2.1, we can calculate an approximation of x_{n+1} via

$$x_{n+1} \equiv x(t_n + h) = x(t_n) + h \left.\frac{dx}{dt}\right|_{t_n} + \frac{h^2}{2!} \left.\frac{d^2x}{dt^2}\right|_{t_n} + \frac{h^3}{3!} \left.\frac{d^3x}{dt^3}\right|_{t_n} + O(h^4) \quad (6.31)$$

We can write

$$\left.\frac{dx}{dt}\right|_{t_n} = f(x_n, t_n) \stackrel{\Delta}{=} f_n$$

$$\left.\frac{d^2x}{dt^2}\right|_{t_n} = \left.\frac{\partial f}{\partial t}\right|_{t_n} + \left.\frac{\partial f}{\partial x}\right|_{t_n} \dot{x}|_{t_n} \stackrel{\Delta}{=} (f_t + f_x f)_n \quad (6.32)$$

$$\left.\frac{d^3x}{dt^3}\right|_{t_n} = (f_{tt} + f_{tx}f + f_{xt}f + f_{xx}f^2 + f_x f_t + f_x^2 f)_n$$

$$= (f_{tt} + 2ff_{tx} + f_{xx}f^2 + f_x f_t + f_x^2 f)_n$$

When we substitute these expressions into the Taylor series in Eq. 6.31, the series becomes

$$x_{n+1} = x_n + h(f_n) + \frac{h^2}{2!}(f_t + ff_x)_n$$
$$+ \frac{h^3}{3!}(f_{tt} + 2ff_{tx} + f_{xx}f^2 + f_x f_t + f_x^2 f)_n + O(h^4) \quad (6.33)$$

On the other hand, we can expand the last term in Eq. 6.30 in a Taylor series to obtain

$$f(x_n + c_4 h f_n, t_n + c_3 h) = f(x_n, t_n) + \left.\frac{\partial f}{\partial t}\right|_n c_3 h + \left.\frac{\partial f}{\partial x}\right|_n c_4 h f_n$$
$$+ \left.\frac{\partial^2 f}{\partial t^2}\right|_n \frac{(c_3 h)^2}{2!} + \left.\frac{\partial^2 f}{\partial t \partial x}\right|_n (c_3 h)(c_4 h f_n)$$
$$+ \left.\frac{\partial^2 f}{\partial x^2}\right|_n \frac{(c_4 h f_n)^2}{2!} + O(h^3)$$

so

$$f(x_n + c_4 h f_n, t_n + c_3 h) = f_n + (c_3 f_t + c_4 f_x f)_n h$$
$$+ (\tfrac{1}{2}c_3^2 f_{tt} + c_3 c_4 f_{tx} f + \tfrac{1}{2}c_4^2 f_{xx} f^2)h^2 + O(h^3) \quad (6.34)$$

Finally, we can substitute Eq. 6.34 into Eq. 6.30, which gives

$$x_{n+1} = x_n + (c_1 + c_2) f_n h + c_2 (c_3 f_t + c_4 f_x f)_n h^2$$
$$+ c_2 \left(\tfrac{1}{2}c_3^2 f_{tt} + c_3 c_4 f f_{tx} + \tfrac{1}{2}c_4^2 f_{xx} f^2\right)_n h^3 + O(h^4) \quad (6.35)$$

We now compare Eqs. 6.33 and 6.35. The goal is to choose the constants c_1, c_2, c_3, and c_4 so that the assumption in Eq. 6.30 is as accurate as possible. That is, we choose the constants so that the expressions match, insofar as this is possible. We see that we must choose

$$c_1 + c_2 = 1, \qquad c_2 c_3 = \tfrac{1}{2}, \qquad c_2 c_4 = \tfrac{1}{2} \tag{6.36}$$

to obtain expressions that match through $O(h^2)$. We have obtained three equations in the four unknowns, and it follows that there are many solutions to this system of equations. A review of the literature suggests that the most popular choice of constants is

$$c_1 = c_2 = \tfrac{1}{2}, \qquad c_3 = c_4 = 1 \tag{6.37}$$

The resulting recursion formula,

$$\boxed{x_{n+1} = x_n + \tfrac{1}{2}[hf_n + hf(x_n + hf_n, t_n + h)]} \tag{6.38}$$

is called the *Runge–Kutta Method of Order 2* often referred to as the *RK2 Method*. Other solutions to Eqs. 6.36 exist, as illustrated by Problem 6.13, but these do not yield the symmetric form of Eq. 6.38.

Example 6.4 Taylor Series and Runge–Kutta Method of Order 2 In this example, we use the Runge–Kutta Method of Order 2 and the Taylor Series Method of Order 2 to approximate the transient response of the SDOF system in Example 4.1. A fixed

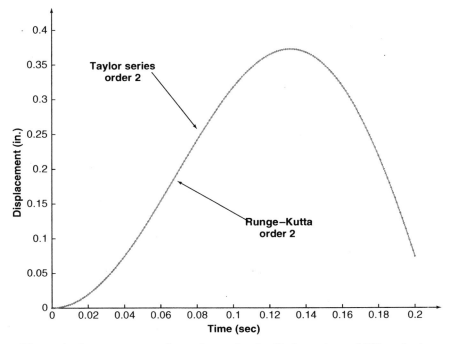

Figure 1 Response comparison of second-order Taylor series and RK methods.

164 Numerical Evaluation of the Dynamic Response of SDOF Systems

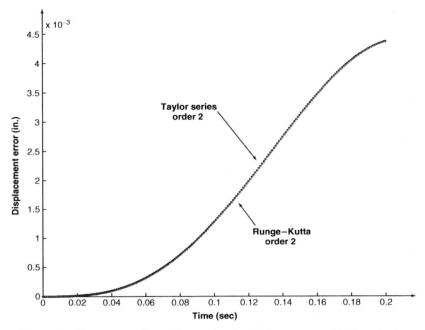

Figure 2 Error comparison of second-order Taylor series and RK methods.

time step of $\Delta t = 0.001$ is used in this simulation. The examples have been written using MATLAB, and the .m-files that encode the approximation process are:

- mat-ex-RKO2.m: MATLAB example of Runge–Kutta Method of Order 2
- f2ndO1stO.m: a function to convert second-order form to first-order form
- mat-ex-tay.m: MATLAB example of Taylor's Method of Order 2

As shown in Fig. 1, the approximations of the transient response provided by these two functions are nearly indistinguishable. Even the error plots in Fig. 2 are nearly overlaid.

Runge–Kutta Method of Order 4 The general procedure outlined in the preceding section can be carried out to obtain higher-order approximations, but the algebraic manipulations become increasingly tedious. For completeness, we simply note here that perhaps the most popular choice among Runge–Kutta methods is the Runge–Kutta Method of Order 4. At time step n, evaluate the following constants:

$$\begin{aligned} f_{n1} &\triangleq hf(x_n, t_n) \\ f_{n2} &\triangleq hf\left(x_n + \tfrac{1}{2}f_{n1}, t_n + \tfrac{1}{2}h\right) \\ f_{n3} &\triangleq hf\left(x_n + \tfrac{1}{2}f_{n2}, t_n + \tfrac{1}{2}h\right) \\ f_{n4} &\triangleq hf(x_n + f_{n3}, t_n + h) \end{aligned} \quad (6.39)$$

Then the *Runge–Kutta Method of Order 4*, commonly known as the *RK4 Method*, uses the following recursion formula:

$$x_{n+1} = x_n + \tfrac{1}{6}(f_{n1} + 2f_{n2} + 2f_{n3} + f_{n4})$$ (6.40)

to generate the approximation $\{x_k\}_{k\in\mathbb{N}}$ to the values of the true solution $\{x(t_k)\}_{k\in\mathbb{N}}$.

In reviewing the Runge–Kutta methods discussed in this section, the following observations should be kept in mind:

- The computational cost of the Runge–Kutta methods is usually summarized in terms of the number of "function evaluations," that is, evaluations of the right-hand side $f(x, t)$ in one time step of the algorithm.
- The Runge–Kutta Method of Order 2 requires two function evaluations:

$$f(x_n, t_n)$$
$$f(x_n + c_4 h f(x_n, t_n), t_n + c_3 h)$$

  at each time step t_n to advance to t_{n+1}. Similarly, the Runge–Kutta Method of Order 4 requires four function evaluations at each time step. This is a generic property of all Runge–Kutta methods: the algorithm of order N requires N function evaluations.
- The Runge–Kutta methods are "self-starting" in the sense that knowledge of the time and state (t_n, x_n) at time step n is all that is required to progress to the next time step. As we will see in the next section, some techniques require several past values of the time and corresponding state.

6.2.3 Linear Multistep Methods

In the Runge–Kutta methods in the preceding section, the propagation of the approximate state from one time step to the next began by using only the current time and state (t_n, x_n). In contrast, the class of *Linear Multistep Methods* begin with the assumption that the next approximate state can be written as a linear combination of the states and derivatives at a number of different time steps. That is, the underlying assumption is that

$$x_{n+1} = \sum_{i=0}^{N} a_i x_{n-i} + \sum_{i=-1}^{M} b_i \dot{x}_{n-i}$$ (6.41)

Several observations should be made at this point.

- The name of these methods derives from the assumption that the time propagation expression relies *linearly* on the past states and functional evaluations. It should be emphasized that the methodology is applicable to *nonlinear* ordinary differential equations of the general class we are studying in Eqs. 6.21.
- The first summation involves only the past states $x_n, x_{n-1}, \ldots, x_{n-N}$, which are known at time step t_n.
- The second summation involves the derivatives of the states $\dot{x}_{n+1}, \dot{x}_n, \dot{x}_{n-1}, \ldots, \dot{x}_{n-M}$. Note that $\dot{x}_{n+1}$ is unknown, being a future quantity, at time step t_n. Thus,

if $b_{-1} \equiv 0$, the entire right-hand side depends only on past values of the states. This class of methods is known as the *explicit, forward, or closed* linear multistep methods.

- On the other hand, if $b_{-1} \neq 0$, the second summation depends on $\dot{x}_{n+1}$. This quantity is not known at time step t_n. This class of methods is known as *implicit, backward, or open* linear multistep methods.

Explicit Multistep Methods In general, the derivation of numerical integration methods from Eq. 6.41 begins by integrating the original equation of motion

$$\dot{x}(t) = f(x(t), t), \qquad x(0) = x_0 \tag{6.29}$$

from t_n to t_{n+1}, thereby achieving

$$\int_{t_n}^{t_{n+1}} \dot{x}(s)\, ds = \int_{t_n}^{t_{n+1}} f(x(s), s)\, ds$$

or

$$x_{n+1} = x_n + \int_{t_n}^{t_{n+1}} f(x(s), s)\, ds \tag{6.42}$$

We approximate $f(x(s), s)$ by a polynomial $\mathcal{P}_M(s)$ that interpolates $f(x(s), s)$ at the sequence of points $\{f_n, f_{n-1}, \ldots, f_{n-M}\}$. It is a classical result of numerical analysis that the approximation will correspond to a polynomial passing through these points provided that we choose

$$\mathcal{P}_M(s) = \sum_{i=0}^{M} (-1)^i \binom{-\xi(s)}{i} \Delta^i f_{n-i} \tag{6.43}$$

where $\xi(t) = (t - t_n)/h$ and Δ^i is the difference operator that is defined recursively via

$$\Delta^i f_k = \begin{cases} f_k, & i = 0 \\ \Delta^{i-1} f_{k+1} - \Delta^{i-1} f_k, & i > 0 \end{cases} \tag{6.44}$$

The notation $\binom{a}{b}$ denotes the binomial function, which is defined by the formula

$$\binom{a}{b} = \begin{cases} 1, & b = 0 \\ \dfrac{a(a-1)\cdots(a-b+1)}{b(b-1)\cdots 1}, & b > 0 \end{cases}$$

Sometimes the last expression is easier to remember when written in terms of the factorial notation

$$\frac{a(a-1)\cdots(a-b+1)}{b(b-1)\cdots 1} = \frac{a!}{b!\,(a-b)!}.$$

Equation 6.43 is known as the *Newton Backward Formula* for the interpolating polynomial of order M passing through the $M+1$ points $f_n, f_{n-1}, \ldots, f_{n-M}$. Substitute Eq. 6.43 into Eq. 6.42 to get

$$x_{n+1} = x_n + \int_{t_n}^{t_{n+1}} \sum_{i=0}^{M} (-1)^i \binom{-\xi(s)}{i} \Delta^i f_{n-i}\, ds \tag{6.45}$$

or

$$x_{n+1} = x_n + \sum_{i=0}^{M} (-1)^i \int_{t_n}^{t_{n+1}} \binom{-\xi(s)}{i} ds \, \Delta^i f_{n-i} \tag{6.46}$$

This last expression can be rewritten as

$$x_{n+1} = x_n + \sum_{i=0}^{M} \alpha_i \, \Delta^i f_{n-i} \tag{6.47}$$

where

$$\boxed{\alpha_i = (-1)^i \int_{t_n}^{t_{n+1}} \binom{-\xi(s)}{i} ds} \tag{6.48}$$

By combining terms, we have found an expression precisely of the form shown in Eq. 6.41.

$$\boxed{x_{n+1} = x_n + \underbrace{\alpha_0 f_n + \alpha_1 \Delta^1 f_{n-1} + \cdots + \alpha_M \Delta^M f_{n-M}}_{\sum_{i=0}^{M} b_i f_{n-i}}} \tag{6.49}$$

The coefficients $\{b_i\}_{i=0}^{M}$ in Eq. 6.41 can be calculated directly from the definition of the difference operator and the coefficients defined in Eq. 6.48. The coefficients can be calculated by hand or by using a symbolic manipulation computer program. Carefully note that *all* of the terms on the right-hand side of Eq. 6.49 are known at time step t_n. Hence, x_{n+1} can be calculated immediately. This property gives rise to the name of these techniques: *Explicit Multistep Methods*.

Example 6.5 Adams–Bashforth Method In Section 6.2.2 we derived the Runge–Kutta Method of Order 2, and in Eq. 6.40 we stated the Runge–Kutta Method of Order 4. In this example we develop the form of the Fourth-Order *Explicit* Linear Multistep Method as defined by Eq. 6.49.

SOLUTION We can directly calculate (or use a symbolic computation program) to obtain the following constants:

$$\alpha_0 = 1, \quad \alpha_1 = \tfrac{1}{2}, \quad \alpha_2 = \tfrac{5}{12}, \quad \alpha_3 = \tfrac{3}{8}, \quad \alpha_4 = \tfrac{251}{720} \tag{1}$$

For example, from the definition of α_i in Eq. 6.48, we can determine $\alpha_0, \alpha_1, \ldots$ as follows:

$$\alpha_i = (-1)^i \int_{t_n}^{t_{n+1}} \binom{-\xi(s)}{i} ds$$

where

$$\xi(s) = \frac{s - t_n}{n}$$

By a simple change of variables, the integral can be written as

$$\alpha_i = (-1)^i \int_0^1 \binom{-\xi}{i} d\xi$$

For $i = 0, 1, 2$, we have

$$\binom{-\xi}{0} = 1, \quad \binom{-\xi}{1} = \frac{-\xi}{1!}, \quad \binom{-\xi}{2} = \frac{(-\xi)(-\xi-1)}{2!}$$

and the corresponding integral expressions are

$$\alpha_0 = (-1)^0 \int_0^1 d\xi = 1$$

$$\alpha_1 = (-1)^1 \int_0^1 -\xi \, d\xi = \frac{1}{2}$$

$$\alpha_2 = (-1)^2 \int_0^1 \frac{\xi^2 + \xi}{2} d\xi = \frac{1}{2}\left(\frac{1}{3} + \frac{1}{2}\right) = \frac{5}{12}$$

By definition of the difference operators, Eq. 6.44, we have

$$\Delta^1 f_{n-1} = \Delta^0 f_n - \Delta^0 f_{n-1}$$
$$= f_n - f_{n-1} \tag{2a}$$

$$\Delta^2 f_{n-2} = \Delta^1 f_{n-1} - \Delta^1 f_{n-2}$$
$$= (f_n - f_{n-1}) - (f_{n-1} - f_{n-2}) \tag{2b}$$
$$= f_n - 2f_{n-1} + f_{n-2}$$

$$\Delta^3 f_{n-3} = \Delta^2 f_{n-2} - \Delta^2 f_{n-3}$$
$$= (f_n - 2f_{n-1} + f_{n-2}) - (f_{n-1} - 2f_{n-2} + f_{n-3}) \tag{2c}$$
$$= f_n - 3f_{n-1} + 3f_{n-2} - f_{n-3}$$

When we substitute the difference formulas above into Eq. 6.49, the following recursion formula is obtained:

$$x_{n+1} = x_n + (1)f_n + \tfrac{1}{2}(f_n - f_{n-1})$$
$$+ \tfrac{5}{12}(f_n - 2f_{n-1} + f_{n-2}) \tag{3}$$
$$+ \tfrac{3}{8}(f_n - 3f_{n-1} + 3f_{n-2} - f_{n-3})$$

or, in final compact form,

$$x_{n+1} = x_n + \tfrac{1}{24}(55 f_n - 59 f_{n-1} + 37 f_{n-2} - 9 f_{n-3}) \quad \text{Ans.} \quad (4)$$

Equation 4 is known as the *Adams–Bashforth Method of Order 4*.

6.2 Integration of First-Order Ordinary Differential Equations

Implicit Multistep Methods Suppose, in contrast to selecting a polynomial $\mathcal{P}_M(s)$ that interpolates the following set of values $\{f_n, f_{n-1}, \ldots f_{n-M}\}$, we had instead chosen a polynomial $\mathcal{P}_{M+1}$ that interpolates the set of values $\{f_{n+1}, f_n, \ldots f_{n-M}\}$ in Eq. 6.42. Such a polynomial can be written in terms of the backward difference formula introduced in the preceding subsection. Thus, from Eq. 6.43,

$$\mathcal{P}_{M+1}(s) = \sum_{i=0}^{M+1} (-1)^i \binom{-\eta(s)}{i} \Delta^i f_{n+1-i} \tag{6.50}$$

where $\eta(t) = (t - t_{n+1})/h$. This equation can be rewritten in terms of the previously introduced function $\xi(t)$ by noting the identity

$$\eta(t) = \frac{t - t_{n+1}}{h} = \frac{t - (t_n + h)}{h} = \xi(t) - 1$$

The polynomial $\mathcal{P}_{M+1}$ is then

$$\mathcal{P}_{M+1}(s) = \sum_{i=0}^{M+1} (-1)^i \binom{1 - \xi(s)}{i} \Delta^i f_{n+1-i}$$

When we substitute this equation into Eq. 6.42, a new expression for the multistep formula is obtained:

$$\begin{aligned}x_{n+1} &= x_n + \int_{t_n}^{t_{n+1}} \sum_{i=0}^{M+1} (-1)^i \binom{1 - \xi(s)}{i} \Delta^i f_{n+1-i}\, ds \\ &= x_n + \sum_{i=0}^{M+1} (-1)^i \int_{t_n}^{t_{n+1}} \binom{1 - \xi(s)}{i} ds\, \Delta^i f_{n+1-i}\end{aligned} \tag{6.51}$$

or

$$x_{n+1} = \sum_{i=0}^{M+1} \beta_i \Delta^i f_{n+1-i} \tag{6.52}$$

where

$$\boxed{\beta_i = (-1)^i \int_{t_n}^{t_{n+1}} \binom{1 - \xi(s)}{i} ds} \tag{6.53}$$

In other words, we have derived a multistep formula that has the form

$$\boxed{x_{n+1} = x_n + \underbrace{\{\beta_0 f_{n+1} + \beta_1 \Delta^1 f_n + \cdots + \beta_{M+1} \Delta^{M+1} f_{n-M}\}}_{\sum_{i=-1}^{M} b_i f_{n-i}}} \tag{6.54}$$

The coefficients $\{b_k\}_{k=0}^{M+1}$ can be calculated by hand or by using a symbolic calculation program. In comparing the summation above to the explicit multistep formulas of the preceding section, it is important to note that the summation above includes the term f_{n+1}, which is not known at time step n. This term gives rise to the name of these techniques: *Implicit Multistep Methods*.

Example 6.6 Adams–Moulton Method In Example 6.5 we derived an explicit fourth-order multistep method known as the Adams–Bashforth Method. In this example, we derive a corresponding fourth-order multistep method that is *implicit*.

SOLUTION We can show that the constants β_i given by Eq. 6.53, are

$$\beta_0 = 1, \quad \beta_1 = -\tfrac{1}{2}, \quad \beta_2 = -\tfrac{1}{12}, \quad \beta_3 = -\tfrac{1}{24}, \quad \beta_4 = -\tfrac{10}{720}$$

By definition of the difference operators, Eq. 6.44, we have

$$\Delta^1 f_n = \Delta^0 f_{n+1} - \Delta^0 f_n \tag{1}$$
$$= f_{n+1} - f_n$$
$$\Delta^2 f_{n-1} = \Delta^1 f_n - \Delta^1 f_{n-1}$$
$$= (f_{n+1} - f_n) - (f_n - f_{n-1}) \tag{2}$$
$$= f_{n+1} - 2f_n + f_{n-1}$$
$$\Delta^3 f_{n-2} = \Delta^2 f_{n-1} - \Delta^2 f_{n-2}$$
$$= (f_{n+1} - 2f_n + f_{n-1}) - (f_n - 2f_{n-1} + f_{n-2}) \tag{3}$$
$$= f_{n+1} - 3f_n + 3f_{n-1} - f_{n-2}$$

The *Adams–Moulton Method* is defined by substituting the definitions above into the implicit multistep formula, Eq. 6.54. Then

$$x_{n+1} = x_n + (1)hf_{n+1} + \left(-\tfrac{1}{2}\right)h(f_{n+1} - f_n)$$
$$+ \left(-\tfrac{1}{12}\right)h(f_{n+1} - 2f_n + f_{n-1}) \tag{4}$$
$$+ \left(-\tfrac{1}{24}\right)h(f_{n+1} - 3f_n + 3f_{n-1} - f_{n-2})$$

Finally,

$$x_{n+1} = x_n + \frac{h}{24}(9f_{n+1} + 19f_n - 5f_{n-1} + f_{n-2}) \quad \textbf{Ans.} \tag{5}$$

Overview of First-Order Numerical Integration In summary, then, we can make the following general observations that will guide us in the selection of an integration method introduced in Section 6.2 for first-order ordinary differential equations.

- The Direct Taylor Series Method provides a good theoretical foundation and conceptual bridge to more advanced methods.
- The Direct Taylor Series Method is seldom used in practice. The right-hand-side function $f(x(t), t)$ must be simple to differentiate to use this method in applications.
- Runge–Kutta methods have the generic advantage that they are *self-starting*. That is, only the state at the current time step is required to propagate a numerical approximation of the state at the next time step.
- Runge–Kutta methods measure their computational costs in terms of the *number of evaluations of the right-hand-side function* $f(x(t), t)$ at each time step.

- Runge–Kutta Methods of Order K require K evaluations of the function $f(x(t), t)$ at each time step.
- Linear Multistep Methods are not, generally speaking, self-starting.
- Explicit Linear Multistep Methods of Order K require one evaluation of the function $f(x(t), t)$ per time step.
- Explicit Linear Multistep Methods of Order K require the storage of roughly $(K-1)$ previous evaluations of the function $f(x(t), t)$ at each time step.

In general, the Runge–Kutta methods are often preferred because they are self-starting. For very long simulations, however, the Linear Multistep Methods can be more efficient. The most popular variants of these methods use error estimators to adapt the time step size or the order of the method to achieve a given accuracy. Other popular techniques combine the Explicit and Implicit Multistep Methods to create *Predictor–Corrector Methods*. These topics are beyond the scope of this book. The interested reader is referred to Refs. [6.5] to [6.8].

6.3 NONLINEAR SDOF SYSTEMS

Both of the numerical approaches summarized in Sections 6.1 and 6.2, those derived for second- and first-order ODEs, can be useful in approximating the response of nonlinear systems. Before considering the numerical solution of the nonlinear equation of motion, let us consider some of the physical phenomena that lead to nonlinearities in the equation of motion of a SDOF system.[5] The equation of motion of a linear SDOF system is, of course,

$$m\ddot{u} + c\dot{u} + ku = p(t) \tag{6.55}$$

The equation of motion of a *nonlinear SDOF system* can often be written in the form

$$m\ddot{u} + f(u, \dot{u}, t) = 0 \tag{6.56}$$

Since this is not a linear differential equation, the principle of superposition does not hold. Thus, for example, the Duhamel integral cannot be used to obtain the solution for the nonlinear response $u(t)$.

In Section 4.7 it was indicated that damping can lead to nonlinear terms in the equation of motion and that equivalent viscous damping may sometimes be employed to approximate the actual nonlinear damping in a system. The damping force of a fluid acting on an object moving through it, and Coulomb damping, were cited as common nonlinear damping mechanisms. Two important classes of nonlinearities related to the displacement $u(t)$ are *geometric nonlinearity* and *material nonlinearity*. These will be illustrated briefly.

Consider the taut string (neglect bending) with attached lumped mass m shown in Fig. 6.5. The tension in the string is given by

$$T = T_0 + \frac{AE}{L}\delta \tag{6.57}$$

---

[5] References [6.9] to [6.11] and many other references are available which go into far greater detail on nonlinear structural dynamics than is possible within the scope of this book. References [6.12] and [6.13], in particular, treat nonlinearities in structures subjected to earthquakes.

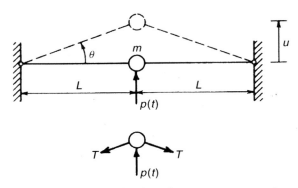

Figure 6.5 Large deflection of a mass on a taut string.

where T_0 is the tension in the undeflected string and δ is the elongation of the string given by

$$\delta = (L^2 + u^2)^{1/2} - L \tag{6.58}$$

From the free-body diagram in Fig. 6.5, the equation of motion is found to be

$$\sum F_u = m\ddot{u} \tag{6.59}$$

or

$$p - 2T \sin\theta = m\ddot{u} \tag{6.60}$$

But

$$\sin\theta = \frac{u}{(L^2 + u^2)^{1/2}} \tag{6.61}$$

Thus, the nonlinear equation of motion is

$$m\ddot{u} + 2\left\{T_0 + \frac{AE}{L}[(L^2 + u^2)^{1/2} - L]\right\} \frac{u}{(L^2 + u^2)^{1/2}} = p(t) \tag{6.62}$$

This exact nonlinear differential equation has the form of Eq. 6.56, with the nonlinear term being a *geometric nonlinearity*; that is, the nonlinearity depends only on geometric quantities, length and displacement, and not on any material properties. For "small" displacements, that is, $u \ll L$, Eq. 6.62 can be approximated by the linear equation

$$m\ddot{u} + \frac{2T_0}{L}u = p(t) \tag{6.63}$$

For somewhat larger displacements,

$$\delta \approx L\left[1 + \frac{1}{2}\left(\frac{u}{L}\right)^2\right] - L = \frac{1}{2L}u^2$$

$$\sin\theta \approx \frac{u}{L}$$

so Eq. 6.62 can be approximated by

$$m\ddot{u} + \frac{2T_0}{L}u + \frac{AE}{L^3}u^3 = p(t) \tag{6.64}$$

The cubic term in Eq. 6.64, preceded by a plus sign, leads to what is referred to as a *hardening spring*. Under harmonic excitation such a system exhibits a frequency-response behavior characterized by the solid curves in Fig. 6.6. The dashed curve in Fig. 6.6 represents the locus of resonant frequencies, indicating that the resonant frequency increases with increasing amplitude of response. Note also that at some values of Ω/ω_n, three different amplitudes are possible. References [6.9], [6.10], and others give detailed studies of the response of systems with a hardening or softening spring to harmonic excitation.

Next, let us consider an example of a *material nonlinearity*. Figure 6.7a represents an idealized steel frame whose columns are assumed to be much more flexible than the horizontal "roof" member. If the load p is increased slowly, the load–deflection curve of Fig. 6.7b results. The load–deflection behavior is linear up to point B, where yielding begins at the outer fibers at the points labeled "plastic hinges," where the moment is the greatest. The load is increased further to point C. Thereafter the load is reduced, and the load–deflection curve from C to E follows a straight line parallel to the original

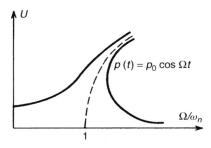

Figure 6.6 Frequency-response curve for a SDOF system with a hardening spring.

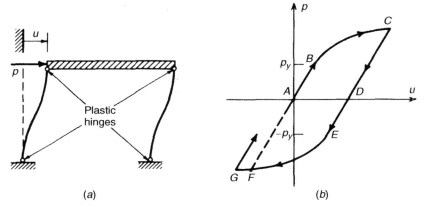

Figure 6.7 (*a*) Steel frame under loading that produces yielding of columns; (*b*) load–deflection diagram.

slope of portion AB. At point D the direction of loading is reversed. The closed loop $ACFA$ is called a *hysteresis loop*. The area inside the loop is the energy dissipated as a result of cyclic plastic deformation. The force–displacement relationship of Fig. 6.7b and other models of structural nonlinearities are discussed in Refs. [6.11] to [6.13]. A simpler model, the elastic–perfectly plastic model, is illustrated in Example 6.8 in Section 6.3.2.

6.3.1 First-Order Formulation of Nonlinear SDOF Systems

The numerical approximation methods derived for first-order ordinary differential equations are directly applicable to the simulation of the response of geometrically nonlinear SDOF systems. For example, the prototypical equation of motion for many nonlinear SDOF systems can be written as

$$m\ddot{u} + f(u, \dot{u}, t) = 0 \tag{6.65}$$

with initial conditions

$$u(0) = u_0, \qquad \dot{u}(0) = v_0 \tag{6.66}$$

Just as in our study of linear SDOF systems, the nonlinear equation of motion can readily be converted to first-order form. We define a *state vector*

$$\boldsymbol{x}(t) \in \mathbb{R}^{2 \times 1} \tag{6.67}$$

$$\boldsymbol{x}(t) \equiv \left\{ \begin{array}{c} x_1(t) \\ x_2(t) \end{array} \right\} \triangleq \left\{ \begin{array}{c} u(t) \\ \dot{u}(t) \end{array} \right\} \tag{6.68}$$

The corresponding first-order equation of motion has the form

$$\dot{\boldsymbol{x}}(t) = \boldsymbol{f}(\boldsymbol{x}(t), t) \tag{6.69}$$

Then Eqs. 6.65 and 6.66 take the state-space form

$$\dot{\boldsymbol{x}}(t) = \left\{ \begin{array}{c} x_2(t) \\ -(1/m)f(x_1(t), x_2(t), t) \end{array} \right\}$$

$$\boldsymbol{x}(0) = \left\{ \begin{array}{c} u_0 \\ v_0 \end{array} \right\} \tag{6.70}$$

We have studied several methods for approximating the transient response of systems having this form. These methods include, for example, the family of Runge–Kutta Methods and the family of Multistep Methods.

In Sections 6.2.2 and 6.2.3 we derived the fourth-order Runge–Kutta Method, fourth-order (Explicit) Adam–Bashforth Multistep Method, and fourth-order (Implicit) Adams–Moulton Multistep Method. The Runge–Kutta and Multistep Methods are so popular that nearly all commercially available numerical integration packages contain some representatives of these classes. In particular, MATLAB has several built-in functions that encode Runge–Kutta and Multistep Methods. In Example 6.7 we show how to use the general framework that has been constructed by the authors of MATLAB to employ their numerical integration packages. Of course, we cannot present the architecture of the MATLAB integration libraries in all their generality. The interested reader is referred to the numerous texts that provide details of the MATLAB environment.

Example 6.7 Runge–Kutta and Multistep Methods for Nonlinear Response In this example we illustrate the use of two specific functions in MATLAB: ODE45 and ODE113. If the reader understands how these two standard packages in MATLAB can be used to integrate ordinary differential equations, it will be clear how to use many of the other numerical integration algorithms in the MATLAB package.

This example computes a simulation of the response of the string–mass geometrically nonlinear system depicted in Fig. 6.5. The MATLAB .m-files that have been written to approximate the transient response of this system include:

- mat-ex-6-4.m: the master .m-file that invokes the MATLAB built-in functions ODE45 and ODE113 to obtain the solution via two different integration methods
- geomnon.m: an .m-file that returns the nonlinear function $f(x(t), t)$ that is defined in the equations of motion having the form $\dot{x}(t) = f(x(t), t)$
- geomnon-forcing.m: an .m-file that returns the right-hand-side forcing function acting on the string–mass system at time t

The computed response $u(t)$ is plotted in Fig. 1.

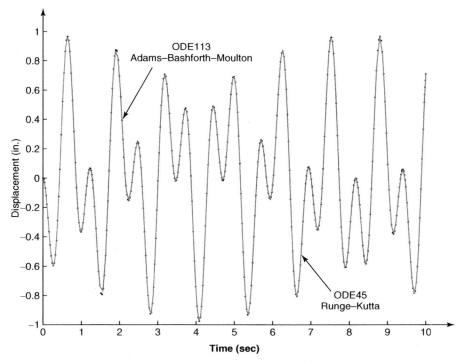

Figure 1 Response based on Runge–Kutta of Order 4 and Multistep Methods.

6.3.2 Second-Order Formulation of Nonlinear SDOF Systems

Incremental, or step-by-step, procedures are the foundation of those numerical integration methods that work directly with the second-order form of the governing nonlinear ordinary differential equations. Most of these methods have a similar philosophy, so that the derivation of a single representative from this class yields a good background for understanding the overall incremental formulation.

As an example, the average acceleration procedure employed in Section 6.1.2 for determining the response of linear systems may be adapted for determining the response of nonlinear systems. Figure 6.8a shows an SDOF system with possible nonlinear "spring" and "damping" elements. Figure 6.8b illustrates a typical nonlinear force–displacement curve for f_S expressed as a function of displacement u. It is assumed that the mass remains constant, although this could easily be generalized to be a known function of time. (The symbols f_S and f_D will be used whether referring to functions of time or to functions of displacement and velocity.) Then the equation of motion may be written at times t_i and $t_{i+1} = t_i + \Delta t_i$ as follows:

$$m\ddot{u}_i + f_{Di} + f_{Si} = p_i$$
$$m\ddot{u}_{i+1} + f_{D(i+1)} + f_{S(i+1)} = p_{i+1} \quad (6.71)$$

Subtracting Eq. 6.71a from Eq. 6.71b, we get

$$m\,\Delta\ddot{u}_i + \Delta f_{Di} + \Delta f_{Si} = \Delta p_i \quad (6.72)$$

where

$$\Delta\ddot{u}_i = \ddot{u}_{i+1} - \ddot{u}_i, \qquad \Delta f_{Di} = f_{D(i+1)} - f_{Di}$$
$$\Delta f_{Si} = f_{S(i+1)} - f_{Si}, \qquad \Delta p_i = p_{i+1} - p_i \quad (6.73)$$

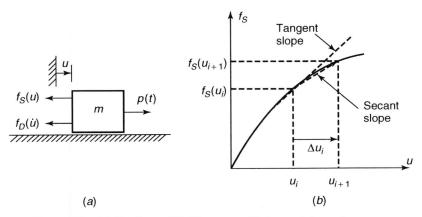

Figure 6.8 (a) Nonlinear SDOF system; (b) force–deformation curve.

Equation 6.72 differs from Eq. 6.14, which was used in the step-by-step integration algorithm, only in the nonlinear f_D and f_S terms. An approximation to the Δf_{Si} term is

$$\Delta f_{Si} \approx k_i \, \Delta u_i \tag{6.74a}$$

where k_i is the tangent slope at u_i as shown in Fig. 6.8b. Similarly, an approximation to the f_D term can be written in the form

$$\Delta f_{Di} \approx c_i \, \Delta \dot{u}_i \tag{6.74b}$$

where c_i is the tangent slope of the f_D versus $\dot{u}$ curve at $\dot{u}_i$. Then Eq. 6.72 becomes

$$m \, \Delta \ddot{u}_i + c_i \, \Delta \dot{u}_i + k_i \, \Delta u_i = \Delta p_i \tag{6.75}$$

This replaces Eq. 6.14 in the step-by-step integration algorithm of Section 6.1.2. Thus, the response of a nonlinear system can be computed using the steps outlined in the flowchart of Fig. 6.4, with c and k being replaced by c_i and k_i evaluated at the beginning of each time step. By using Eq. 6.71b to compute the acceleration at the end of the present time step (and the beginning of the next time step), dynamic equilibrium is enforced at each time step. Due to the approximations in Eqs. 6.74, if $\ddot{u}_i$ were computed using Eqs. 6.12 and 6.17c, dynamic equilibrium would not be satisfied at time t_{i+1} unless c and k remained constant over the time step. In general, the time step Δt_i must be small enough that the difference in a linear solution and nonlinear solution over one time step is not great.

The following example illustrates the use of the average acceleration step-by-step integration algorithm for calculating the response of a nonlinear SDOF system.

Example 6.8 The frame shown in Fig. 6.7a has an elastic, perfectly plastic force–deformation behavior, as shown in Fig. 1a. This is an idealization of the behavior

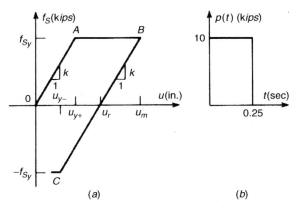

Figure 1 (a) Elastic–plastic material behavior and (b) rectangular pulse loading history for the elastic–plastic frame example.

shown in Fig. 6.7b. The frame, with $m = 0.2$ kip-sec^2/in., $k = 30$ kips/in. (total), and $f_{sy} = 15$ kips, is subjected to a rectangular pulse loading as shown in Fig. 1b. Compare this problem with the corresponding linear problem in Section 5.2. The frame is at rest at $t = 0$. Use the Average Acceleration Method to compute the response $u(t)$ for $t = 0$ to $t = 0.55$ sec. Use $\Delta t = 0.05$ sec (constant).

SOLUTION The flowchart of Fig. 6.4 can be adapted by using the following equations:

$$\ddot{u}_i = \frac{1}{m}(p_i - f_{Si}) = 5.0(p_i - f_{Si}) \tag{1}$$

$$k_i^* = k_i + \frac{4m}{\Delta t^2} = k_i + 320.0 \tag{2}$$

$$\Delta p_i^* = \Delta p_i + \frac{4m}{\Delta t}\dot{u}_i + 2m\ddot{u}_i$$
$$= \Delta p_i + 16.0\dot{u}_i + 0.4\ddot{u}_i \tag{3}$$

$$\Delta u_i = \frac{\Delta p_i^*}{k_i^*} \tag{4}$$

$$\Delta \dot{u}_i = \frac{2}{\Delta t}\Delta u_i - 2\dot{u}_i = 40.0\,\Delta u_i - 2\dot{u}_i \tag{5}$$

$$u_{i+1} = u_i + \Delta u_i \tag{6}$$

$$\dot{u}_i = \dot{u}_i + \Delta \dot{u}_i \tag{7}$$

For the present problem, k_i has the value $k = 30$ kips/in. or zero depending on the deformation history. Other relevant values are:

$$u_y = \frac{f_{Sy}}{k} = \frac{15}{30} = 0.5 \text{ in.}$$

$u_m = $ displacement at which $\dot{u}$ switches from $(+)$ to $(-)$ \hfill (8)

$u_r = $ residual (inelastic) displacement $= u_m - u_y = u_m - 0.5$

$$f_{Si} = \begin{cases} ku_i & \text{for segment } OA \\ f_{Sy} & \text{for segment } AB \\ f_{Sy} - k(u_m - u_i) & \text{for segment } BC \\ \text{etc.} & \end{cases} \tag{9}$$

Tables 1 and 2 outline the calculation of the response of the elastic–plastic system and of an elastic system (assuming that $f_y > \max|f_s|$, i.e., ignoring yielding) using the average acceleration algorithm. Figure 2 shows the elastic–plastic response.

In Tables 1 and 2, note that the expressions for k_i^* and f_{Si} change at the †-symbols; first, where u_i first exceeds u_y, and next, where Δp_i^* changes sign. If the computations were to continue, a further change would be required.

Table 1 Numerical Solution for Elastic–Plastic Response Calculated by the Average Acceleration Method

(1) i	(2) t_i	(3) p_i	(4) f_{Si} Eqs. 9	(5) $\ddot{u}_i$ 5[(3) − (4)]	(6) Δp_i	(7) Δp_i^*	(8) k_i^* Eq. 2	(9) Δu_i (7)/(8)	(10) $\Delta \dot{u}_i$ 40(9) − 2(12)	(11) u_i (11) + (9)	(12) $\dot{u}_i$ (12) + (10)
0	0.00	10.0	0	50.0000	0.0	20.0000	350.0	0.05714	2.2857	0.00000	0.0000
1	0.05	10.0	1.143	41.4286	0.0	53.1429	350.0	0.15184	1.5020	0.05714	2.2857
2	0.10	10.0	6.2694	18.6531	0.0	68.0653	350.0	0.19447	0.2034	0.20898	3.7878
3	0.15	10.0	12.1036	−10.5178	0.0	59.6511	350.0	0.17043	−1.1650	0.40345	3.9911
4	0.20	10.0	15.0000	−25.0000	0.0	35.2181	320.0	0.11006	−1.2500	0.57388†	2.8261
5	0.25	10.0	15.0000	−25.0000	0.0	15.2181	320.0	0.04756	−1.2500	0.68394	1.5761
6	0.30	0.0	15.0000	−75.0000	−10.0	−34.7819†	350.0	−0.09938	−4.6273	0.73150	0.3261
7	0.35	0.0	12.0186	−60.0929	0.0	−92.8565	350.0	−0.26530	−2.0098	0.63212	−4.3012
8	0.40	0.0	4.0595	−20.2973	0.0	−109.0943	350.0	−0.31170	0.1540	0.36682	−6.3110
9	0.45	0.0	−5.2915	26.4574	0.0	−87.9284	350.0	−0.25122	2.2650	0.05512	−6.1570
10	0.50	0.0	−12.8282	64.1410	0.0	−36.6156	350.0	−0.10462	3.5994	−0.19611	−3.8920
11	0.55	0.0	—	—	—	—	—	—	—	−0.30072	−0.2926

Table 2 Numerical Solution for Elastic Response Calculated by the Average Acceleration Method

(1) i	(2) t_i	(3) p_i	(4) $f_{Si} = ku_i$	(5) $\ddot{u}_i$	(6) Δp_i	(7) Δp_i^*	(8) k_i^*	(9) Δu_i	(10) $\Delta \dot{u}_i$	(11) u_i	(12) $\dot{u}_i$
			30(11)	5[(3) − (4)]			Eq. 2	(7)/(8)	40(9) − 2(12)	(11) + (9)	(12) + (10)
0	0.00	10.0	0.0000	50.0000	0.0	20.0000	350.0	0.05714	2.2857	0.00000	0.0000
1	0.05	10.0	1.7143	41.4286	0.0	53.1429	350.0	0.15184	1.5020	0.05714	2.2857
2	0.10	10.0	6.2694	18.6531	0.0	68.0653	350.0	0.19447	0.2034	0.20898	3.7878
3	0.15	10.0	12.1036	−10.5178	0.0	59.6511	350.0	0.17043	−1.1650	0.40345	3.9911
4	0.20	10.0	17.2165	−36.0825	0.0	30.7851	350.0	0.08796	−2.1340	0.57388	2.8261
5	0.25	10.0	19.8552	−49.2761	0.0	−8.6359	350.0	−0.02467	−2.3713	0.66184	0.6922
6	0.30	0.0	19.1150	−95.5751	−10.0	−75.0959	350.0	−0.21456	−5.2242	0.63717	−1.6791
7	0.35	0.0	12.6782	−63.3911	0.0	−135.8088	350.0	−0.38803	−1.7145	0.42261	−6.9033
8	0.40	0.0	1.0375	−5.1873	0.0	−139.9586	350.0	−0.39988	1.2402	0.03458	−8.6177
9	0.45	0.0	−10.9590	54.7949	0.0	−96.1227	350.0	−0.27464	3.7696	−0.36530	−7.3775
10	0.50	0.0	−19.1981	95.9904	0.0	−19.3304	350.0	−0.05523	5.0066	−0.63994	−3.6079
11	0.55	0.0	—	—	—	—	—	—	—	−0.69517	1.3987

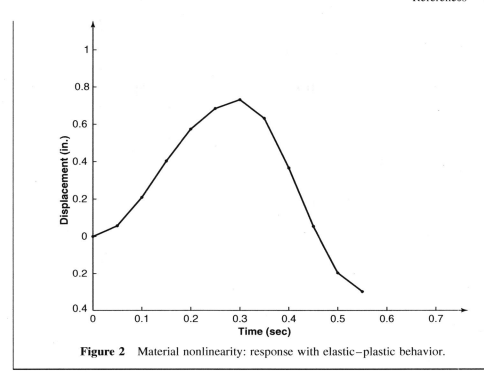

Figure 2 Material nonlinearity: response with elastic–plastic behavior.

Example 6.8 illustrates briefly the complex nature of numerical solution of nonlinear structural dynamics problems. Numerical solution of nonlinear MDOF problems is discussed briefly in Section 16.3.3.

REFERENCES

[6.1] R. W. Clough and J. Penzien, *Dynamics of Structures*, 2nd ed., McGraw-Hill, New York, 1993.

[6.2] N. M. Newmark, "A Method of Computation for Structural Dynamics," *Transactions of ASCE*, Vol. 127, Pt. 1, pp. 1406–1435, 1962.

[6.3] K. J. Bathe and E. L. Wilson, "Stability and Accuracy Analysis of Direct Integration Methods," *Earthquake Engineering and Structural Dynamics,* Vol. 1, pp. 283–291, 1973.

[6.4] H. M. Hilber, T. J. R. Hughes, and R. L. Taylor, "Improved Numerical Dissipation for Time Integration Algorithms in Structural Dynamics," *Earthquake Engineering and Structural Dynamics,* Vol. 5, pp. 283–292, 1977.

[6.5] C. W. Gear, *Numerical Initial Value Problems in Ordinary Differential Equations*, Prentice-Hall, Englewood Cliffs, NJ, 1973.

[6.6] A. Ralston and P. Rabinowitz, *A First Course in Numerical Analysis,* McGraw-Hill, New York, 1978.

[6.7] S. D. Conte and C. de Boor, *Elementary Numerical Analysis: An Algorithmic Approach*, McGraw-Hill, New York, 1980.

[6.8] T. J. R. Hughes, *The Finite Element Method: Linear Static and Dynamic Finite Element Analysis*, Dover, New York, 2000.

[6.9] J. J. Stoker, *Nonlinear Vibration,* Wiley, New York, 1950.

[6.10] W. J. Cunningham, *Introduction to Nonlinear Analysis,* McGraw-Hill, New York, 1958.

[6.11] M. A. Sozen, "Hysteresis in Structural Elements," *Applied Mechanics in Earthquake Engineering,* AMD Vol. 8, ASME, New York, pp. 63–98, 1974.

[6.12] A. K. Chopra, *Dynamics of Structures: Theory and Applications to Earthquake Engineering,* 2nd ed., Prentice Hall, Upper Saddle River, NJ, 2001.

[6.13] G. C. Hart and K. Wong, *Structural Dynamics for Structural Engineers,* Wiley, New York, 2000.

PROBLEMS

For problems whose number is preceded by a **C**, you are to write a computer program and use it to produce the plot(s) requested. *Note:* MATLAB .m-files for many of the plots in this book may be found on the book's website.

Problem Set 6.1

6.1 For the undamped SDOF system in Example 4.1, determine the response $u(t)$ for $0 \leq t \leq 0.4$ s. Use piecewise-constant interpolation of the force, with the magnitude of each impulse determined by the force at the beginning of each time step $\Delta t = 0.02$ s (Fig. P6.1).

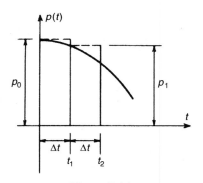

Figure P6.1

6.2 Repeat Problem 6.1, but with the magnitude of each impulse being determined by the force at the middle of each time step $\Delta t = 0.02$ s.

6.3 Repeat Problem 6.1, but with the magnitude of each impulse being determined by the force at the end of each time step $\Delta t = 0.02$ s.

6.4 **(a)** Repeat part (a) of Example 6.1, but double the time step to $\Delta t = 0.04$ s. **(b)** Tabulate your results from part (a), and compare them with the exact solution and with the numerical solution of Problem 6.1.

C 6.5 Write a MATLAB program to approximate the response $u(t)$ for the undamped system in Example 4.1. Use the numerical integration technique derived in Problem 6.1. Choose the time step to be an input variable Δt, and select $\Delta t = 0.02$ s in the simulation. Plot your result $u(t)$ versus t.

C 6.6 Write a MATLAB program to approximate the response $u(t)$ for the undamped system in Example 4.1. Use the numerical integration technique derived in Problem 6.2. Choose the time step to be an input variable Δt, and select $\Delta t = 0.02$ s in the simulation. Plot your result $u(t)$ versus t.

C 6.7 Write a MATLAB program to approximate the response $u(t)$ for the undamped system in Example 4.1. Use the numerical integration technique derived in Problem 6.3. Choose the time step to be an input variable Δt, and select $\Delta t = 0.02$ s in the simulation. Plot your result $u(t)$ versus t.

C 6.8 Repeat the numerical simulations in Problems 6.5, 6.6, and 6.7 with the selection of time steps sizes $\Delta t = 0.1, 0.2, 0.3,$ and 0.4 s, respectively. For each integration method (derived in Problems 6.5, 6.6, and 6.7, respectively), create a graph that contains four plots of error versus time for each of the time following time steps: $\Delta t = 0.1, 0.2, 0.3,$ and 0.4 s.

6.9 The Newmark β method, referred to in footnote 1 in Section 6.1.2, is embodied in the equations

$$\dot{u}_{i+1} = \dot{u}_i + \frac{\Delta t_i}{2}(\ddot{u}_i + \ddot{u}_{i+1})$$

$$u_{i+1} = u_i + \dot{u}_i \,\Delta t_i + \left(\tfrac{1}{2} - \beta\right) \ddot{u}_i (\Delta t_i)^2 + \beta \ddot{u}_{i+1}(\Delta t_i)^2$$

(a) Show that $\beta = \tfrac{1}{6}$ corresponds to the *Linear Acceleration Method,* wherein the acceleration approximation in Fig. 6.3 is replaced by the linear interpolation shown in Fig. P6.9. **(b)** Determine expressions for k_i^* and Δp_i^*

for the Linear Acceleration Method similar to those for the Average Acceleration Method as given by Eqs. 6.16a and b.

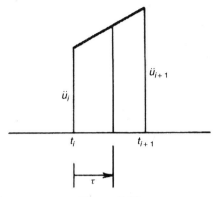

Figure P6.9

6.10 Repeat Example 6.3 using the Linear Acceleration Method as derived in Problem 6.9 instead of the Average Acceleration Method.

C 6.11 Write a MATLAB program that encodes the Newmark β Method for a constant input time step Δt and fixed input choice of β. Obtain a numerical approximation to Example 6.3 using the Linear Acceleration Method (i.e., $\beta = \frac{1}{6}$).

Problem Set 6.2

6.12 In the text we noted that the numerical integration methods we derived for the first-order scalar ordinary differential equation,

$$\dot{x}(t) = f(x(t), t)$$
$$x(0) = x_0$$

can be implemented without change, except notation, to systems of ordinary differential equations having the (vector) form

$$\dot{\boldsymbol{x}}(t) = \boldsymbol{f}(\boldsymbol{x}(t), t)$$
$$\boldsymbol{x}(0) = \boldsymbol{x}_0$$

Argue that this is true. Start with a simple low-dimensional system. Choose, say,

$$\boldsymbol{x}(t) \equiv \begin{Bmatrix} x_1(t) \\ x_2(t) \end{Bmatrix} \in \mathbb{R}^2$$

Extend your argument to higher dimensions.

6.13 In Section 6.2.2 the Runge–Kutta Method of Order 2 was derived, and a popular selection for the four constants is given in Eq. 6.37. The resulting recursion formula is given in Eq. 6.38. (**a**) Determine an alternative set of constants if $c_1 = \frac{1}{4}$ is one of them. (**b**) What is the resulting recursion formula?

C 6.14 Repeat Example 6.4, but generate a comparison of the accuracy of (**a**) the Runge–Kutta Method of Order 2, (**b**) the Taylor Series Method of Order 2, and (**c**) the Runge–Kutta Method of Order 4 as the time step varies from $\Delta t = 0.001$ to $\Delta t = 0.0001$.

6.15 Explicit and Implicit Multistep Methods are frequently used in tandem to generate *Predictor–Corrector Methods*. First, an explicit method is used to generate a *prediction* of the state at the next time step, denoted $x_{n+1}^{(p)}$ from Eq. 6.49. Next, *correction* of the approximate value of the state at the next time step $x_{n+1}^{(c)}$ is generated from Eq. 6.54. The unknown function evaluation f_{n+1} in Eq. 5 of Example 6.6 is approximated by the evaluation

$$f_{n+1} \approx f(x_{n+1}^{(p)}, t_{n+1})$$

Using the results from Examples 6.5 and 6.6, write a detailed flowchart for the fourth-order Adams–Bashforth–Moulton Predictor–Corrector Algorithm. This algorithm is the specific predictor–corrector method that uses the explicit Adams–Bashforth formula in the prediction phase of the algorithm and uses the implicit Adams–Moulton formula during the correction phase.

Problem Set 6.3

C 6.16 Write an Adams–Bashforth–Moulton Predictor–Corrector Algorithm in MATLAB as it is outlined in Problem 6.15. Repeat Example 6.7 using the fixed-time-step predictor–corrector algorithm that you have written. Compare the consistency of this fixed-time-step method with the adaptive time step (and adaptive order) Runge–Kutta and Adams–Bashforth–Moulton Methods that are encoded in the built-in MATLAB integration functions ODE45 and ODE113, respectively.

6.17 Repeat Example 6.8 using the Linear Acceleration Method as derived in Problem 6.9 instead of the Average Acceleration Method.

7

Response of SDOF Systems to Periodic Excitation: Frequency-Domain Analysis

In Chapter 4 you studied the response of SDOF systems to harmonic excitation and became familiar with important concepts such as resonance. You also learned how to simplify the analysis of viscous-damped systems by using complex frequency response. These concepts from Chapter 4 are now extended to determine the response of SDOF systems to periodic excitation.

Upon completion of this chapter you should be able to:

- Determine the Fourier series representation of a periodic function using the real form of the Fourier series.
- Determine the Fourier series representation of a periodic function using the complex form of the Fourier series.
- Determine the steady-state response of an SDOF system to periodic excitation using either the real form or the complex form.
- Use the basic definition to determine the Fourier transform of a transient function $p(t)$.
- Show that the impulse-response function for a linear system and the frequency-response function for the system form a Fourier transform pair.
- Apply the FFT algorithm to compute the Fourier transform of a periodic function $p(t)$, and plot the magnitude and phase of the Fourier transform computed.

7.1 RESPONSE TO PERIODIC EXCITATION: REAL FOURIER SERIES

Forces acting on structures are frequently periodic, or can be approximated closely by periodic forces. For example, the forces exerted on an automobile traveling at constant speed over certain roadway surfaces can be considered to be periodic. Figure 7.1 shows a *periodic function* with period T_1, that is,

$$p(t + T_1) = p(t) \tag{7.1}$$

A periodic function can be separated into its harmonic components by means of a Fourier series expansion. In this section we consider real Fourier series. In Section 7.2

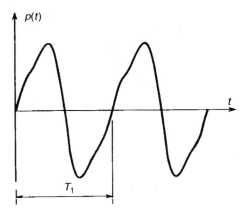

Figure 7.1 Periodic function with period T_1.

complex Fourier series are introduced. The complex form is very useful when combined with the complex frequency-response function of Chapter 4 to study the steady-state response of damped systems.

7.1.1 Real Fourier Series

The periodic function $p(t)$ may be separated into its harmonic components by means of a Fourier series expansion. The *real Fourier series expansion* of $p(t)$ may be defined as

$$p(t) = a_0 + \sum_{n=1}^{\infty} a_n \cos n\Omega_1 t + \sum_{n=1}^{\infty} b_n \sin n\Omega_1 t \qquad (7.2)$$

where

$$\Omega_1 = \frac{2\pi}{T_1} \qquad (7.3)$$

is the *fundamental frequency* (in rad/s), and a_n and b_n are the coefficients of the nth harmonic. The coefficients a_0, a_n, and b_n are related to $p(t)$ by the equations

$$\begin{aligned} a_0 &= \frac{1}{T_1} \int_{\tau}^{\tau+T_1} p(t)\, dt = \text{average value of } p(t) \\ a_n &= \frac{2}{T_1} \int_{\tau}^{\tau+T_1} p(t) \cos n\Omega_1 t\, dt, \qquad n = 1, 2, \ldots \\ b_n &= \frac{2}{T_1} \int_{\tau}^{\tau+T_1} p(t) \sin n\Omega_1 t\, dt, \qquad n = 1, 2, \ldots \end{aligned} \qquad (7.4)$$

where τ is an arbitrary time.

Although theoretically, a Fourier series representation of $p(t)$ may require an infinite number of terms, in actual practice $p(t)$ can generally be approximated with sufficient accuracy by a relatively small number of terms. Example 7.1 illustrates the Fourier series representation of a square wave.

Example 7.1 (a) Determine expressions for the coefficients of a real Fourier series representation of the square wave shown in Fig. 1. Write the Fourier series representation of $p(t)$. (b) Plot truncated series employing, respectively one, two, and three terms of the Fourier series.

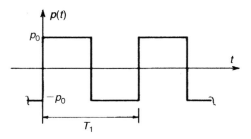

Figure 1 Square wave.

SOLUTION (a) The integrals in Eqs. 7.4 can be evaluated over the period $-T_1/2 < t < T_1/2$ and thus written in the form

$$a_0 = \frac{1}{T_1} \int_{-T_1/2}^{T_1/2} p(t)\, dt \tag{1}$$

$$a_n = \frac{2}{T_1} \int_{-T_1/2}^{T_1/2} p(t) \cos n\Omega_1 t\, dt \tag{2}$$

$$b_n = \frac{2}{T_1} \int_{-T_1/2}^{T_1/2} p(t) \sin n\Omega_1 t\, dt \tag{3}$$

where

$$p(t) = \begin{cases} -p_0 & -T_1/2 \le t < 0 \\ p_0 & 0 \le t < T_1/2 \end{cases} \tag{4}$$

Substituting Eqs. 4 into Eqs. 1 and 2, we get

$$a_0 = a_n = 0 \qquad \textbf{Ans. (a)} \tag{5}$$

This results from the fact that $p(t)$ is an <u>odd</u> function of t [i.e., $p(t) = -p(-t)$], whereas a_0 and a_n are coefficients of <u>even</u> terms in the Fourier series. The coefficient for the odd terms is

$$b_n = \frac{4p_0}{T_1} \int_0^{T_1/2} \sin n\Omega_1 t\, dt \tag{6}$$

so

$$b_n = \frac{4p_0}{T_1} \frac{-1}{n\Omega_1} \cos n\Omega_1 t \Big|_0^{T_1/2} \tag{7}$$

But $\Omega_1 T_1 = 2\pi$, so

$$b_n = -\frac{2p_0}{n\pi}(\cos n\pi - 1) \quad (8)$$

or

$$b_n = \frac{4p_0}{n\pi}, \quad n = 1, 3, 5, \ldots \quad \text{Ans. (a) (9)}$$

The Fourier series representation of the square wave is thus

$$p(t) = \frac{4p_0}{\pi} \sum_{n=1,3,\ldots} \frac{1}{n} \sin n\Omega_1 t \quad \text{Ans. (a) (10)}$$

(b) The plots in Fig. 2 show the contributions over one period of the first three nonzero terms of the Fourier series representation of the square wave in Fig. 1.

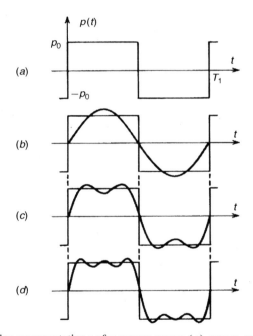

Figure 2 Fourier-series representations of a square wave: (a) square wave; (b) $\sin(\Omega_1 t)$ term; (c) two-term representation; (d) three-term representation.

7.1.2 Steady-State Response to Periodic Excitation

Having determined the response of SDOF systems to harmonic excitation in Chapter 4, and having determined how to represent a periodic function in terms of its harmonic components, we can now determine the response of SDOF systems to periodic excitation. In Example 7.2, an undamped SDOF system is subjected to the square-wave excitation of Example 7.1.

Example 7.2 The undamped SDOF system in Fig. 1 is subjected to a square-wave excitation $p(t)$ like the one in Example 7.1. Determine a Fourier series expression for the steady-state response of the system if $\omega_n = 6\Omega_1$.

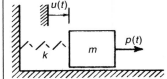

Figure 1 Spring–mass oscillator.

SOLUTION From Eq. 4.9, the steady-state response of an undamped SDOF system to cosine excitation $p_0 \cos \Omega t$ is

$$u = \frac{p_0/k}{1 - r^2} \cos \Omega t \tag{1}$$

where $r = \Omega/\omega_n = \Omega\sqrt{m/k}$.

From Example 7.1 we can write the square-wave excitation $p(t)$ in the form

$$p(t) = \sum_{n=1}^{\infty} P_n \sin n\Omega_1 t \tag{2}$$

where

$$P_n = \begin{cases} \dfrac{4p_0}{n\pi} & n = \text{odd} \\ 0 & n = \text{even} \end{cases} \tag{3}$$

Hence, the steady-state response can be written in the form

$$u(t) = \sum_{n=1}^{\infty} U_n \sin n\Omega_1 t \tag{4}$$

From Eq. 1, the coefficients U_n of the steady-state-response series have the form

$$U_n = \frac{P_n}{k(1 - r_n^2)} \tag{5}$$

where

$$r_n = \frac{n\Omega_1}{\omega_n} \tag{6}$$

Finally, combining Eqs. 3, 4, and 5, we get the following Fourier series expression for the steady-state response:

$$u(t) = \frac{4p_0}{k\pi} \sum_{n=1,3,\ldots} \frac{1}{n[1 - (n/6)^2]} \sin n\Omega_1 t \qquad \textbf{Ans.} \tag{7}$$

since, for the present problem, $\omega_n = 6\Omega_1$.

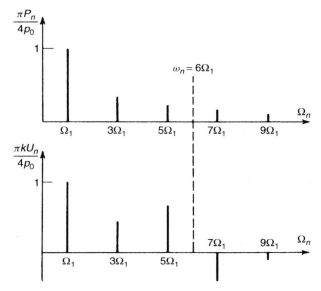

Figure 7.2 Excitation and response spectra based on Example 7.2.

It is very convenient to visualize periodic functions in terms of their *spectra*, that is, plots of the amplitude of each harmonic component versus frequency. The spectra of the periodic excitation $p(t)$ and resulting steady-state response $u(t)$ from Examples 7.1 and 7.2 are plotted in Fig. 7.2. From Eqs. 3 and 5 of Example 7.2, the following nondimensional forms are obtained:

$$\frac{P_n}{p_0} = \begin{cases} \dfrac{4}{n\pi} & n = \text{odd} \\ 0 & n = \text{even} \end{cases} \qquad (7.5a)$$

$$\frac{U_n}{p_0/k} = \begin{cases} \dfrac{4}{n\pi[1 - (n/6)^2]} & n = \text{odd} \\ 0 & n = \text{even} \end{cases} \qquad (7.5b)$$

and

$$\Omega_n = n\Omega_1 \qquad (7.5c)$$

Note that for the particular frequency ratio $\omega_n = 6\Omega_1$, some of the Fourier components of the excitation are at frequencies below resonance, whereas others are above resonance. The response would have a very large Fourier component if $n\Omega_1$ were to fall close to ω_n for some value of n.

7.2 RESPONSE TO PERIODIC EXCITATION: COMPLEX FOURIER SERIES

In Section 4.3, the complex frequency-response function $\overline{H}(\Omega)$ was introduced as a convenient means of representing the response of a viscous-damped system to harmonic excitation. A complex Fourier series representation of periodic excitation and response functions will also prove to be very useful.

7.2.1 Complex Fourier Series

Let the periodic function $p(t)$ be separated into its harmonic components by means of a *complex Fourier-series expansion*, given by[1]

$$\boxed{p(t) = \sum_{n=-\infty}^{\infty} \overline{P}_n(\Omega) e^{i(n\Omega_1 t)}} \tag{7.6}$$

where the fundamental frequency Ω_1 (rad/s) is related to the period of the function by $\Omega_1 T_1 = 2\pi$. The bar over P_n symbolizes the fact that the coefficients of the series may be complex, even though the series as a whole represents a real function of time. To evaluate these components, note that

$$\int_\tau^{\tau+T_1} e^{i(n\Omega_1 t)} e^{-i(m\Omega_1 t)} \, dt = \begin{cases} 0 & n \neq m \\ T_1 & n = m \end{cases} \tag{7.7}$$

So, multiplying Eq. 7.6 by $e^{-i(m\Omega_1 t)}$ and integrating over one period, we get

$$\boxed{\overline{P}_n = \frac{1}{T_1} \int_\tau^{\tau+T_1} p(t) e^{-i(n\Omega_1 t)} \, dt, \qquad n = 0, \pm 1, \ldots} \tag{7.8}$$

Note that

$$\overline{P}_{-n} = \overline{P}_n^* = \text{complex conjugate of } \overline{P}_n \tag{7.9}$$

and that

$$\overline{P}_0 = \frac{1}{T_1} \int_\tau^{\tau+T_1} p(t) \, dt = \text{average value of } p(t) \tag{7.10}$$

[Actually, $\overline{P}_0$ is real-valued since $p(t)$ is real-valued.]

Example 7.3 (a) Show that if $p(t)$ is a real-valued function, the right-hand side of Eq. 7.6 will turn out to be real-valued, as it should be. (b) Show that if $p(t)$ is an odd function, $\overline{P}_n(\Omega)$ is purely imaginary and the coefficients $\overline{P}_n$ and $\overline{P}_{-n}$ are related by the equation

$$\overline{P}_{-n} = -\overline{P}_n$$

SOLUTION (a) From Euler's formula (Eq. 3.14),

$$e^{\pm i\theta} = \cos\theta \pm i\sin\theta \quad \text{and} \quad e^0 = 1 \tag{1}$$

---

[1] By including negative n terms in this series, the Fourier series can represent a real function of time, $p(t)$.

Therefore, Eq. 7.6 can be expanded into the following form:

$$p(t) = \overline{P}_0 + \sum_{n=1}^{\infty} \overline{P}_n (\cos n\Omega_1 t + i \sin n\Omega_1 t)$$
$$+ \sum_{n=1}^{\infty} \overline{P}_{-n} (\cos n\Omega_1 t - i \sin n\Omega_1 t) \quad (2)$$

Since $\overline{P}_n$ is assumed to be complex, it can be expressed in terms of its real and imaginary components as

$$\overline{P}_n = \Re(\overline{P}_n) + i \Im(\overline{P}_n) \quad (3)$$

Then, from Eq. 7.9,

$$\overline{P}_{-n} = \overline{P}_n^* = \Re(\overline{P}_n) - i \Im(\overline{P}_n) \quad (4)$$

Combining Eqs. 2 through 4, we get

$$p(t) = \overline{P}_0 + 2 \sum_{n=1}^{\infty} [\Re(\overline{P}_n) \cos n\Omega_1 t - \Im(\overline{P}_n) \sin n\Omega_1 t] \quad (5)$$

which is real.   Q.E.D.

(b) Making use of Euler's formula, Eq. 1, we can write Eq. 7.8 as

$$\overline{P}_n = \frac{1}{T_1} \int_{\tau}^{\tau+T_1} p(t) (\cos n\Omega_1 t - i \sin n\Omega_1 t) \, dt \quad (6)$$

Since $p(t)$ is said to be an odd function of t, the cosine term drops out and

$$\overline{P}_{+n} = \frac{-i}{T_1} \int_{\tau}^{\tau+T_1} p(t) \sin n\Omega_1 t \, dt \quad (7)$$

Then

$$\overline{P}_{-n} = \frac{-i}{T_1} \int_{\tau}^{\tau+T_1} p(t) \sin(-n\Omega_1 t) \, dt$$
$$= \frac{i}{T_1} \int_{\tau}^{\tau+T_1} p(t) \sin n\Omega_1 t \, dt \quad (8)$$

Hence, from Eq. 7, $\overline{P}_n$ is purely imaginary, and from Eqs. 7 and 8, $\overline{P}_{-n} = -\overline{P}_n$. Q.E.D.

By comparing Eq. 5 of Example 7.3 with Eq. 7.2, you will notice that the coefficients of the real Fourier series and the coefficients of the complex Fourier series are related by the following equations:

$$a_o = \overline{P}_0, \quad a_n = 2\Re(\overline{P}_n) \quad b_n = -2\Im(\overline{P}_n) \quad (7.11)$$

Example 7.4 (a) Determine an expression for the Fourier coefficients $\overline{P}_n$ of the complex Fourier series representation for the square wave of Example 7.1. (b) Sketch spectra of $\Re(\overline{P}_n)$, $\Im(\overline{P}_n)$, and $|\overline{P}_n|$.

SOLUTION (a) Equation 7.8 can be evaluated over the period $0 < t < T_1$, giving

$$\overline{P}_n = \frac{1}{T_1} \int_0^{T_1/2} (p_0) e^{-i(n\Omega_1 t)} \, dt + \frac{1}{T_1} \int_{T_1/2}^{T_1} (-p_0) e^{-i(n\Omega_1 t)} \, dt \tag{1}$$

$$\overline{P}_n = \frac{-p_0}{in\Omega_1 T_1} \left[e^{-i(n\Omega_1 t)} \Big|_0^{T_1/2} - e^{-i(n\Omega_1 t)} \Big|_{T_1/2}^{T_1} \right] \tag{2}$$

But $\Omega_1 T_1 = 2\pi$, so

$$e^{-i(n\Omega_1 T_1/2)} = e^{-i(n\pi)} = \begin{cases} +1 & n = \text{even} \\ -1 & n = \text{odd} \end{cases}$$

$$e^{-i(n\Omega_1 T_1)} = e^{-i(2n\pi)} = 1 \tag{3}$$

Therefore, the complex Fourier coefficients are given by

$$\overline{P}_n = \frac{ip_0}{2\pi n} [2e^{-i(n\pi)} - 1 - e^{-i(2n\pi)}] \tag{4}$$

or

$$\overline{P}_n = \frac{ip_0}{2n\pi} (2e^{-i(n\pi)} - 2) = \begin{cases} 0 & n = \text{even} \\ \dfrac{-2ip_0}{n\pi} & n = \text{odd} \end{cases} \quad \textbf{Ans. (1) (5)}$$

(b) See Fig. 1.

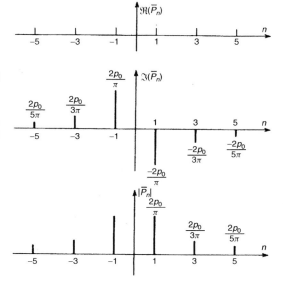

Figure 1 Various spectra for the square wave of Example 7.1.

7.2.2 Complex Frequency Response

The topic of complex frequency response was introduced in Section 4.3. Here we extend that discussion to cover the complex representation for steady-state response of an SDOF system to periodic excitation. From Eqs. 4.30 and 4.33, the steady-state response can be written in complex form as

$$\bar{u}(t) = \overline{U}(\Omega) e^{i\Omega t} = \overline{H}(\Omega) p_0 e^{i\Omega t} \quad (7.12)$$

where the frequency-response function $\overline{H}$ is given by[2]

$$\overline{H}(\Omega) \equiv \overline{H}_{u/p}(\Omega) = \frac{1/k}{[1 - (\Omega/\omega_n)^2] + i[2\zeta(\Omega/\omega_n)]} \quad (7.13)$$

where Ω is the forcing frequency and ω_n is the undamped natural frequency of the SDOF system.

When the excitation is periodic, we can use the complex Fourier series representation of Eq. 7.6, repeated here:

$$p(t) = \sum_{n=-\infty}^{\infty} \overline{P}_n e^{i(n\Omega_1 t)} \quad (7.6)$$

Then the steady-state response can be written as

$$\boxed{u(t) = \sum_{n=-\infty}^{\infty} \overline{U}_n e^{i(n\Omega_1 t)}} \quad (7.14)$$

Noting from Eq. 7.12 that for harmonic excitation $\overline{U} = \overline{H} p_0$, we see that the corresponding expression for periodic response is

$$\boxed{\overline{U}_n = \overline{H}_n \overline{P}_n = |\overline{H}_n||\overline{P}_n| e^{i(\alpha_{H_n} + \alpha_{P_n})}} \quad (7.15)$$

where for a viscous-damped SDOF system,[3]

$$\overline{H}_n \equiv \overline{H}(n\Omega_1) = \frac{1/k}{[1 - (n\Omega_1/\omega_n)^2] + i[2\zeta(n\Omega_1/\omega_n)]} \quad (7.16)$$

Example 7.5 illustrates the use of complex Fourier series in determining the steady-state response of a SDOF system subjected to periodic excitation. The method would be even more beneficial if the system were a damped system.

Example 7.5 (a) Repeat Example 7.2 by determining an expression for the Fourier coefficients $\overline{U}_n$ for the undamped SDOF system subjected to square-wave excitation with $\omega_n = 6\Omega_1$. (b) Sketch magnitude and phase spectra, $|\overline{U}_n|$ and α_{U_n}.

---

[2] Here we use the dimensional form of $\overline{H}$ rather than the nondimensional form given in Eq. 4.33.
[3] The subscript n of the undamped natural frequency ω_n should not be confused with the Fourier series index n.

SOLUTION (a) From Eq. 7.15,

$$\overline{U}_n = \overline{H}_n \overline{P}_n = |\overline{H}_n||\overline{P}_n| e^{i(\alpha_{H_n} + \alpha_{P_n})} \quad (1)$$

and from Eq. 7.16, with $\zeta = 0$,

$$\overline{H}_n = \frac{1/k}{1 - (n\Omega_1/\omega_n)^2} = \frac{1/k}{1 - (n/6)^2} \quad (2)$$

From Eq. 5 of Example 7.4, the Fourier coefficients of the square wave are

$$\overline{P}_n = \begin{cases} 0 & n = \text{even} \\ \dfrac{-2i p_0}{n\pi} & n = \text{odd} \end{cases} \quad (3)$$

Hence, the nth term of the frequency response is

$$\overline{U}_n = \overline{H}_n \overline{P}_n = \begin{cases} 0 & n = \text{even} \\ \dfrac{-i(2p_0)}{nk\pi[1 - (n/6)^2]} & n = \text{odd} \end{cases} \quad \textbf{Ans. (a)} \quad (4)$$

(b) From Eq. 4 we can evaluate expressions for the magnitude and phase angle.

$$|\overline{U}_n| = \frac{2p_0/k\pi}{|n[1 - (n/6)^2]|} \quad (5)$$

Because the nonzero terms in Eq. 4 are pure imaginary, the phase angles will all be $\pm \pi/2$. Hence,

$$\alpha_{U_n} = \begin{cases} -\pi/2 & n = +1, +3, +5, -7, -9, \ldots \\ \pi/2 & n = -1, -3, -5, +7, +9, \ldots \end{cases} \quad (6)$$

Note that sign changes occur as the resonance frequency is passed at $\omega_n = 6\Omega_1$, that is, at $n = 6$.

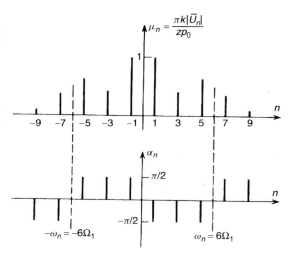

Figure 1 Response spectrum (magnitude and phase).

For sketching purposes, evaluate the nondimensionalized magnitude

$$\mu_n \equiv \frac{(\pi/2)|\overline{U}_n|}{p_0/k} = \frac{|1/n|}{|[1-(n/6)^2]|}, \qquad n = \pm 1, \pm 3, \ldots \tag{7}$$

The values to be plotted are

$$\mu_1 = \mu_{-1} = \frac{36}{35} = 1.029, \qquad \mu_3 = \mu_{-3} = \frac{4}{9} = 0.444, \qquad \mu_5 = \mu_{-5} = \frac{36}{55} = 0.655$$

$$\mu_7 = \mu_{-7} = \frac{36}{91} = 0.396, \qquad \mu_9 = \mu_{-9} = \frac{4}{45} = 0.089$$

The results are shown in Fig. 1.

Compare the sketch of μ_n in Example 7.5 with Fig. 7.2b and note that the complex coefficients μ_n in Example 7.5 have an amplitude that is half that of the corresponding real coefficients, which are plotted in Fig. 7.2b. The contribution of the $-n$ terms in the complex Fourier series accounts for this difference.

It is convenient to think of problems such as Example 7.5 in terms of transformations from the time domain to the frequency domain (spectrum), and from the frequency domain to the time domain. Figure 7.3 illustrates this.

7.3 RESPONSE TO NONPERIODIC EXCITATION: FOURIER INTEGRAL

In previous sections you have seen that a periodic function can be represented by a Fourier series, as in Eq. 7.2 or 7.6. When the function to be represented is not periodic, it can be represented by a Fourier integral. In developing expressions for the Fourier integral transform pair, it will be convenient to employ the complex Fourier series, letting the period T_1 approach infinity.

7.3.1 Fourier Integral Transforms

Equation 7.6, which defines the complex Fourier series, is repeated here:

$$p(t) = \sum_{n=-\infty}^{\infty} \overline{P}_n e^{i(n\Omega_1 t)} \tag{7.6}$$

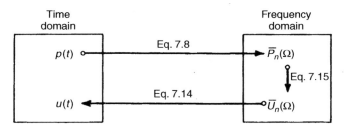

Figure 7.3 Solution of a periodic response problem by transformation to the frequency domain.

where, as given by Eq. 7.8, the Fourier coefficients $\overline{P}_n$ are related to $p(t)$ by

$$\overline{P}_n = \frac{1}{T_1} \int_{\tau}^{\tau+T_1} p(t) e^{-i(n\Omega_1 t)} \, dt, \qquad n = 0, \pm 1, \ldots \tag{7.8}$$

provided that the integral exists.

In letting $T_1 \to \infty$, it will be convenient to introduce the following notation[4]:

$$\Omega_1 = \Delta\Omega, \qquad \Omega_n = n\Omega_1 \tag{7.17}$$

$$\overline{P}(\Omega_n) = T_1 \overline{P}_n = \frac{2\pi}{\Delta\Omega} \overline{P}_n \tag{7.18}$$

Then Eq. 7.6 can be written in the form

$$p(t) = \frac{1}{2\pi} \sum_{n=-\infty}^{\infty} \overline{P}(\Omega_n) e^{i(\Omega_n t)} \Delta\Omega \tag{7.19}$$

where, from Eqs. 7.8 and Eq. 7.18,

$$\overline{P}(\Omega_n) = \int_{-T_1/2}^{T_1/2} p(t) e^{-i(\Omega_n t)} \, dt \tag{7.20}$$

The limits of integration on Eq. 7.20 have been taken as shown so that when $T_1 \to \infty$, the entire time history of $p(t)$ will be included regardless of the specific form of $p(t)$.

As $T_1 \to \infty$, Ω_n becomes the continuous variable Ω, and $\Delta\Omega$ becomes the differential $d\Omega$. Then Eqs. 7.20 and 7.19, respectively, can be written as

$$\overline{P}(\Omega) = \int_{-\infty}^{\infty} p(t) e^{-i\Omega t} \, dt \tag{7.21}$$

$$p(t) = \frac{1}{2\pi} \int_{-\infty}^{\infty} \overline{P}(\Omega) e^{i\Omega t} \, d\Omega \tag{7.22}$$

Equations 7.21 and 7.22 are called a *Fourier transform pair*. $\overline{P}(\Omega)$ is known as the *Fourier transform* of $p(t)$; and $p(t)$ is called the *inverse Fourier transform* of $\overline{P}(\Omega)$. The representation of $p(t)$ by its Fourier transform requires the existence of the integral in Eq. 7.21. Conditions that must be satisfied for the Fourier transform integral to exist are discussed in texts on integral transforms (e.g., Ref. [7.1]). These conditions are met by most physically realizable functions representing forces, displacements, and so on.

Finally, the Fourier transform pair can be written in a more symmetric form if written in terms of the frequency $f = \Omega/2\pi$. Then

$$\boxed{\overline{P}(f) \equiv \mathcal{F}[p(t)] = \int_{-\infty}^{\infty} p(t) e^{-i(2\pi f t)} \, dt} \tag{7.23}$$

$$\boxed{p(t) \equiv \mathcal{F}^{-1}[\overline{P}(f)] = \int_{-\infty}^{\infty} \overline{P}(f) e^{i(2\pi f t)} \, df} \tag{7.24}$$

---

[4] Note that Eq. 7.18 introduces a scaling factor, T_1.

Example 7.6 illustrates the use of straightforward time-domain integration to determine the Fourier transform of a rectangular pulse that is symmetric about $t = 0$. Note that the resulting Fourier transform is a real function of frequency.

Example 7.6 Let $p(t)$ be the rectangular pulse defined by

$$p(t) = \begin{cases} 0 & t < -T \\ p_0 & -T \leq t \leq T \\ 0 & t > T \end{cases}$$

(a) Determine the Fourier transform of this rectangular pulse. Express the transform as a function of the frequency variable Ω. (b) Plot the Fourier transform.

SOLUTION   (a) From Eq. 7.21,

$$\overline{P}(\Omega) = \int_{-\infty}^{\infty} p(t) e^{-i\Omega t}\, dt = \int_{-T}^{T} p_0 e^{-i\Omega t}\, dt \tag{1}$$

Therefore,

$$\overline{P}(\Omega) = \frac{p_0}{-i\,\Omega}(e^{-i\Omega T} - e^{i\Omega T}) \tag{2}$$

which can also be written in terms of the sinc function $(\sin\theta)/\theta$.

$$\overline{P}(\Omega) = 2 p_0 T \frac{\sin \Omega T}{\Omega T} \qquad \textbf{Ans. (a)} \tag{3}$$

(b) The Fourier transform $\overline{P}(\Omega)$ is therefore a real function. It can be plotted versus the frequency variable Ω (Fig. 1) and compared with the discrete Fourier series of Example 7.4.

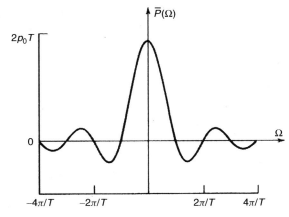

Figure 1   Fourier transform of a symmetric rectangular pulse.

Reference [7.1] contains a table of a number of Fourier transform pairs.

7.3.2 Frequency-Response Functions

In Eq. 7.14 we found that the response of an SDOF system to periodic excitation can be expressed in the form

$$u(t) = \sum_{n=-\infty}^{\infty} \overline{U}_n e^{i(n\Omega_1 t)} \tag{7.14}$$

where, from Eq. 7.15,

$$\overline{U}_n = \overline{H}_n \overline{P}_n \tag{7.15}$$

Following the procedure of Eqs. 7.17 through 7.24, we obtain the following *Fourier transform pair* for the response of an SDOF system:

$$\boxed{\overline{U}(f) = \int_{-\infty}^{\infty} u(t) e^{-i(2\pi f t)} \, dt} \tag{7.25}$$

$$u(t) = \int_{-\infty}^{\infty} \overline{U}(f) e^{i(2\pi f t)} \, df \tag{7.26}$$

where

$$\boxed{\overline{U}(f) = \overline{H}(f)\overline{P}(f)} \tag{7.27}$$

which is the product of the *system frequency-response function*, $\overline{H}(f)$, and the *Fourier transform of the excitation*, $\overline{P}(f)$. Therefore, the response can be expressed by the following inverse Fourier transform:

$$u(t) = \int_{-\infty}^{\infty} \overline{H}(f)\overline{P}(f) e^{i(2\pi f t)} \, df \tag{7.28}$$

In some cases a table of Fourier transform pairs can be used to evaluate this inverse transform.[7.1,7.2] However, evaluation of this definite integral generally involves contour integration in the complex plane, which is beyond the scope of this book.

7.3.3 Parameter Identification

Of great importance (as demonstrated in Chapter 18) is the fact that Eq. 7.27 can be written symbolically in the form

$$\boxed{\overline{H}(f) = \frac{\overline{U}(f)}{\overline{P}(f)}} \tag{7.29}$$

The system frequency-response function, $\overline{H}(f)$, can thus be obtained from Fourier transforms of the measured time histories of excitation $p(t)$ and response $u(t)$. Important system parameters (e.g., the undamped natural frequency ω_n and damping factor ζ of the system) can then be extracted from this system frequency-response function. For example, for a viscous-damped SDOF system, $\overline{H}(f)$ is given by (Eq. 7.13)

$$\overline{H}(f) = \frac{1/k}{[1 - (f/f_n)^2] + i(2\zeta f/f_n)} \tag{7.30}$$

Such parameter identification procedures are discussed in Chapter 18 and in modal analysis references such as Refs. [7.4] and [7.5].

7.4 RELATIONSHIP BETWEEN COMPLEX FREQUENCY RESPONSE AND UNIT IMPULSE RESPONSE

In Section 7.5 we describe how the Fourier integral can be approximated by the discrete Fourier transform (DFT), which, in turn, can be evaluated numerically by use of the fast Fourier transform (FFT) algorithm.

The complex frequency response, $\overline{H}(\Omega)$ or $\overline{H}(f)$, of a linear system describes its response characteristics in the frequency domain, and the unit impulse response $h(t)$ describes the system's response in the time domain. We now show that the unit impulse function and the system frequency-response function form a Fourier transform pair. From Eq. 7.23, the Fourier transform of the unit impulse excitation is

$$\overline{P}(f) = \int_{-\infty}^{\infty} p(t)e^{-i(2\pi ft)}\,dt = 1 \qquad (7.31)$$

Therefore, the response to this unit impulse is, from Eq. 7.28,

$$\boxed{h(t) = \int_{-\infty}^{\infty} \overline{H}(f)e^{i(2\pi ft)}\,df} \qquad (7.32)$$

But from Eq. 7.24, the equation that defines the inverse Fourier transform, this unit impulse response function is just the expression for the inverse Fourier transform of the system frequency-response function $\overline{H}(f)$. Conversely, it follows that the system frequency-response function $\overline{H}(f)$ is the Fourier transform of the unit impulse response function $h(t)$, that is,

$$\boxed{\overline{H}(f) \equiv \mathcal{F}[h(t)] = \int_{-\infty}^{\infty} h(t)e^{-i(2\pi ft)}\,dt} \qquad (7.33)$$

This Fourier transform pair relationship between a system's impulse-response function in the time domain and its frequency-response function in the frequency domain is illustrated in Fig. 7.4. For example, for a viscous-damped SDOF system, $\overline{H}(f)$ is given by Eq. 7.30 and $h(t)$ by Eq. 5.30, both repeated here.

$$\overline{H}(f) = \frac{1/k}{[1 - (f/f_n)^2] + i(2\zeta f/f_n)} \qquad (7.30)$$

$$h(t) = \frac{1}{m\omega_d}e^{-\zeta\omega_n t}\sin\omega_d t, \qquad t > 0 \qquad (5.30)$$

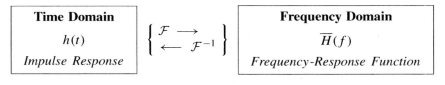

Figure 7.4 Relationship between impulse response and frequency response.

200   Response of SDOF Systems to Periodic Excitation: Frequency-Domain Analysis

It is left as an exercise for the reader to show that these SDOF system functions satisfy Eq. 7.33.

In Chapter 18 it will be noted that some parameter-estimation algorithms work directly with the frequency-response function $\overline{H}(f)$; these are called *frequency-domain algorithms*. Other parameter-estimation algorithms make use of the impulse-response function $h(t)$; these are called *time-domain algorithms*.

7.5 DISCRETE FOURIER TRANSFORM AND FAST FOURIER TRANSFORM

Although the Fourier integral techniques discussed in Section 7.3 provide a means for determining the transient response of a system, numerical implementation of the Fourier integral became a practical reality only with the publication of the Cooley–Tukey algorithm for the fast Fourier transform in 1965.[7.3] Since that date, the FFT has virtually led to a revolution in many areas of technology, including the area of vibration testing (e.g., see Chapter 18 and Refs. [7.4] and [7.5]).

Two steps are involved in the numerical evaluation of Fourier transforms.[7.1] First, discrete Fourier transforms (DFTs), which correspond to Eqs. 7.23 and 7.24, are derived. Then an efficient numerical algorithm, the fast Fourier transform (FFT), is used to the compute the DFTs.

7.5.1 Discrete Fourier Transform Pair

For numerical treatment of the Fourier transform, it is necessary first to define a *discrete Fourier transform pair* corresponding to the continuous Fourier transform pair of Eqs. 7.23 and 7.24. To be transformed, a continuous function must first be sampled at discrete time intervals Δt. Second, due to computer memory and execution-time limitations, only a finite number N of these sampled values can be utilized. Figure 7.5a illustrates a sampled waveform with $N = 16$ samples taken at $\Delta t = 0.25$-sec intervals.

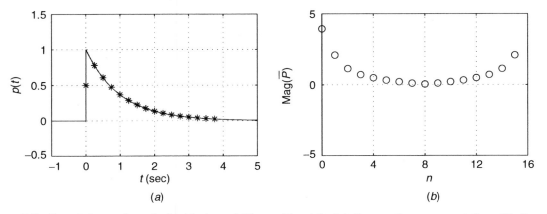

Figure 7.5   Sampled waveform (e^{-t} with $\Delta t = 0.25$ sec, $N = 16$): (*a*) discrete-time representation; (*b*) discrete-frequency representation.

The effects of this *sampling* and *truncation* are to approximate the continuous signal by a periodic signal of period $T_1 = N \Delta t$, sampled at times $t_m = m \Delta t$, $m = 0, 1, \ldots, (N-1)$.

The *total sample time* is T_1, so the fundamental-frequency sinusoid that fits within this sample time has a period T_1. Therefore, the *frequency interval* of the discrete Fourier transform is

$$\Delta f = \frac{1}{T_1} = \frac{1}{N \Delta t} \tag{7.34}$$

The DFT consists of N samples extending up to a maximum frequency of $N \Delta f$. Specific features of sampling and truncation that are illustrated in Fig. 7.5 are discussed following Example 7.7. Sampling and truncation are discussed in greater detail in Section 18.4.

Since a period T_1 consists of N samples, the integral of Eq. 7.23 is replaced by the following finite sum:

$$\overline{P}(f_n) = \sum_{m=0}^{N-1} p(t_m) e^{-i 2\pi (m \Delta t)(n \Delta f)} \Delta t \tag{7.35}$$

Finally, the *discrete Fourier transform* (DFT) can be written

$$\overline{P}(f_n) = \Delta t \sum_{m=0}^{N-1} p(t_m) e^{-i(2\pi m(n/N))}, \qquad n = 0, 1, \ldots, N-1 \tag{7.36}$$

The inverse DFT can be obtained from Eq. 7.24 in a similar manner. Thus,

$$p(t_m) = \sum_{n=0}^{N-1} \overline{P}(f_n) e^{i 2\pi (m \Delta t)(n \Delta f)} \Delta f \tag{7.37}$$

This expression for the *inverse Fourier transform* (IFT) can be written as

$$p(t_m) = \frac{1}{N \Delta t} \sum_{n=0}^{N-1} \overline{P}(f_n) e^{i(2\pi m(n/N))}, \qquad m = 0, 1, \ldots, N-1 \tag{7.38}$$

Equations 7.36 and 7.38 define a *discrete Fourier transform pair* that is consistent with the continuous Fourier transform.

The discrete Fourier transform approximates the continuous Fourier transform at discrete frequencies f_n. The accuracy of a DFT representation depends on the sampling interval Δt and the number of samples, or block size, N. In Section 18.4 we discuss these effects in much greater detail.

Reference [7.1] presents a graphical derivation and a theoretical derivation of the DFT pair. The resulting *discrete Fourier transform* (DFT), written in present notation, is

$$\boxed{\overline{P}(f_n) = \sum_{m=0}^{N-1} p(t_m) e^{-i(2\pi m(n/N))}, \qquad n = 0, 1, \ldots, N-1} \tag{7.39}$$

where $f_n = n\,\Delta f$. The corresponding inverse DFT is

$$p(t_m) = \frac{1}{N}\sum_{n=0}^{N-1}\overline{P}(f_n)e^{i(2\pi m(n/N))}, \qquad m = 0, 1, \ldots, N-1 \qquad (7.40)$$

This form is the one that is utilized in computing FFTs.[7.6] However, a scale factor Δt is required to produce an equivalence between this DFT form and the continuous Fourier transform.

7.5.2 Fast Fourier Transform Algorithm

The *fast Fourier transform* (FFT) is not a new type of transform but rather, an efficient numerical algorithm for evaluating the DFT. Its importance lies in the fact that by eliminating most of the repetition in the calculation of a DFT, it permits much more rapid computation of the DFT.

Either Eq. 7.39 or 7.40 can be cast in the form

$$A_m = \sum_{n=0}^{N-1} B_n W_N^{mn}, \qquad m = 0, 1, \ldots, N-1 \qquad (7.41)$$

where

$$W_N = e^{-i(2\pi/N)} \qquad (7.42)$$

A measure of the amount of computation involved in Eq. 7.41 is the number of complex products implied by the form of the equation and the range of m. It is clear that there are N sums, each of which requires N complex products, or there are N^2 products required for computing all of the A_m's. By taking advantage of the cyclical nature of powers of W_N, the total computational effort can be drastically reduced. Figure 7.6 shows the repetition cycle for W_8^{mn}. The number of complex products for the FFT algorithm is given by $(N/2)\log_2 N$. For example, if $N = 512$, the number of FFT operations is less than 1% of the corresponding number of DFT operations. Signal-processing software products invariably provide for computation of the FFT.[7.6]

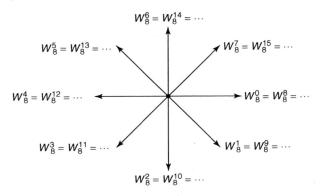

Figure 7.6 Cyclical nature of W_N^{mn} for $N = 8$.

7.5.3 FFT Computations

In Example 7.7 the FFT command in the MATLAB computer program is used to compute the Fourier transform of a square wave, similar to the one treated analytically in Example 7.4 in Section 7.2. Computation of FFTs of nonperiodic signals is discussed in Section 18.4.

***

Example 7.7 Let $p(t)$ be the square wave defined over one period by the function

$$p(t) = \begin{cases} 2 & 0.0 \leq t < 0.5 \text{ sec} \\ 0 & 0.5 \text{ sec} \leq t < 1.0 \text{ sec} \end{cases}$$

represented by 16 samples over the finite interval 0 sec $\leq t < 1$ sec. Use the FFT command in MATLAB to determine the Fourier transform of this square wave, and plot the real and imaginary parts of the resulting Fourier transform.

SOLUTION Relative to an offset average value of 1.0, the square wave is antisymmetric in the time window 0.5 sec $\leq t \leq 0.5$ sec. The 16 discrete-time sample points are shown in Fig. 1a. These 16 data points constitute the input to the FFT algorithm. (See the Comments following this example.)

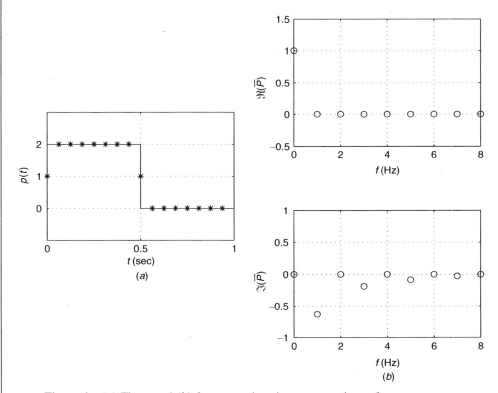

Figure 1 (a) Time- and (b) frequency-domain representations of a square wave.

In Fig. 1*b* the results of a 16-point FFT are plotted versus the discrete frequency f_n for $0 \leq f_n \leq f_{N/2}$. The real part (top right) and the imaginary part (lower right) can be compared with the plots in Fig. 1 of Example 7.4, noting that the square wave in Example 7.4 is an antisymmetric function with an average value of zero, whereas the square wave in this example is an antisymmetric function relative to an average value of 1.0. Other differences are explained in the Comments that follow.

Comments Regarding MATLAB FFT Input/Output

1. The FFT algorithm is executed with the command

$$y = \text{fft}(x, N)$$

   where N is the block size and x is the input sequence, that is, the vector of discrete-time samples of the input function. For structural dynamics applications, x will be a vector of real numbers. It is desirable to let N be a power of 2.

2. The FFT treats the first N numbers in the input sequence as one period of a periodic function. For example, the rectangular pulse in Fig. 1*a* of Example 7.7 will correspond to a square wave that is represented by a Fourier series whose fundamental frequency is 1.0 sec, with the sampled value at $t = 1$ sec repeating the sampled value at $t = 0$ sec. This topic is treated in greater detail in Section 18.4.

3. Where there is a step discontinuity in the input function, the sampled value is taken as the average of the value prior to the jump and the value after the jump, as shown in Fig. 1*a* of Example 7.7 and in Fig. 7.5*a*.

4. Because $\overline{P}(f_n)$ and $\overline{P}(f_{-n})$ are complex conjugates of each other, the second half of the Fourier coefficients are usually not plotted (e.g., as in Fig. 1*b* of Example 7.7). Thus, the complex values in the output vector are sequenced by MATLAB as follows:

$$y(1) = \overline{P}(f_0) = \text{average value of input (real)}$$
$$y(2) = \overline{P}(f_{+1})$$
$$\vdots$$
$$y(N/2 + 1) = \overline{P}(f_{+N/2})$$
$$y(N/2 + 2) = \overline{P}(f_{-N/2+1})$$
$$\vdots$$
$$y(N) = \overline{P}(f_{-1})$$

5. To treat the FFT output values as approximations to the complex Fourier coefficients given by Eq. 7.8, it is necessary to divide the output values $y(n)$ by the block size N.[5]

---

[5] This scaling is necessitated by the scaling factor T_1 introduced in Eq. 7.18, together with the additional scaling factor of Δt between Eqs. 7.36 and 7.39.

6. To convert the input sequence number m to input sample time t_m in seconds, it is necessary to use $t_m = m\Delta t$. Similarly, to convert the frequency scale of the output from coefficient number n to frequency f_n in hertz, one must use $f_n = n \Delta f = n/N \Delta t$.

REFERENCES

[7.1] E. O. Brigham, *The Fast Fourier Transform*, Prentice-Hall, Englewood Cliffs, NJ, 1974.

[7.2] R. A. Gabel and R. A. Roberts, *Signals and Systems*, 3rd ed., Wiley, New York, 1987.

[7.3] J. W. Cooley and J. W. Tukey, "An Algorithm for Machine Calculation of Complex Fourier Series," *Math Computation*, Vol. 19, 1965, pp. 297–301.

[7.4] D. J. Ewins, *Modal Testing: Theory, Practice and Application*, 2nd ed., Research Studies Press, Baldock, Hertfordshire, England, 2000.

[7.5] N. M. M. Maia and J. M. M. Silva, ed., *Theoretical and Experimental Modal Analysis*, Wiley, New York, 1997.

[7.6] *Using MATLAB, Version 6*, The MathWorks, Natick, MA, 2002, pp. 12-41 to 12-48.

PROBLEMS

For problems whose number is preceded by a **C**, you are to write a computer program and use it to produce the plot(s) requested. *Note:* MATLAB .m-files for many of the plots in this book may be found on the book's website.

Problem Set 7.1

7.1 Determine the real Fourier series for the periodic excitation function in Fig. P7.1.

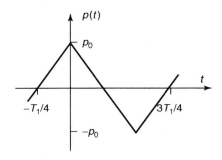

Figure P7.1

7.2 Determine the real Fourier series for the periodic excitation function in Fig. P7.2.

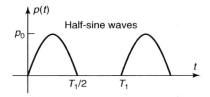

Figure P7.2

C 7.3 (a) Determine the real Fourier series for the square wave shown in Fig. P7.3. (b) By comparing your result from part (a) with the answer in Eq. 10 of Example 7.1, what do you observe to be the effect(s) of the phase shift of the square wave? (c) Run the MATLAB .m-file sd2hw7_3.m, and print out the plots produced. Modify this .m-file to produce the input signal of Example 7.1, and print out the plots produced.

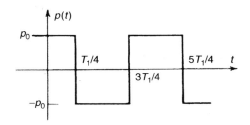

Figure P7.3

7.4 (a) Determine the real Fourier series for the periodic excitation function in Fig. P7.4. (b) By comparing your result from part (a) with the answer in Eq. 10 of Example 7.1, what do you observe to be the effect(s) of the upward shift of the square wave?

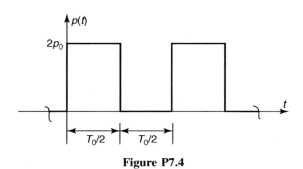

Figure P7.4

7.5 The undamped SDOF system shown in Fig. P7.5 is subjected to the excitation $p(t)$ given in Problem 7.1. (a) Determine a Fourier series expression for the steady-state response of the system if $\omega_n = 4\Omega_1$. (b) Sketch the spectra for P_n and for U_n, similar to those shown in Fig. 7.2.

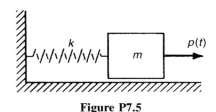

Figure P7.5

7.6 The undamped SDOF system shown in Fig. P7.5 is subjected to the excitation $p(t)$ given in Fig. P7.4. (a) Determine a Fourier series expression for the steady-state response of the system if $\omega_n = 6\Omega_1$. (b) Sketch the spectra for P_n and for U_n similar to those shown in Fig. 7.2.

Problem Set 7.2

7.7 (a) Determine the coefficients of the complex Fourier series that corresponds to the excitation $p(t)$ given in Fig. P7.1. (b) Sketch the real and imaginary parts and the magnitude of the complex coefficients as in Example 7.4.

7.8 (a) Determine the coefficients of the complex Fourier series that corresponds to the excitation $p(t)$ in Fig. P7.3. (b) Sketch the real and imaginary parts and the magnitude of the complex coefficients as in Example 7.4.

7.9 (1) Determine the coefficients of the complex Fourier series that corresponds to the excitation $p(t)$ in Fig. P7.4. (2) Sketch the real and imaginary parts and the magnitude of the complex coefficients as in Example 7.4.

7.10 The undamped SDOF system of Fig. P7.10a is subjected to the sawtooth base motion $z(t)$ illustrated in Fig. 7.10b. Let

$$z(t) = \sum_{n=-\infty}^{\infty} \overline{Z}_n e^{i(n\Omega_1 t)}, \qquad w(t) = \sum_{n=-\infty}^{\infty} \overline{W}_n e^{i(n\Omega_1 t)}$$

where $\overline{W}_n = \overline{H}_n \overline{Z}_n$. (a) Determine $\overline{H}_n$. (b) Determine $\overline{Z}_n$. (These will have the same form as $\overline{P}_n$ of Problem 7.1.) (c) Determine $\overline{W}_n$, and sketch $|\overline{W}_n|/Z$ versus frequency order, as in Example 7.4.

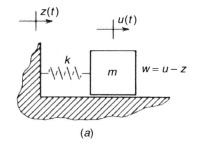

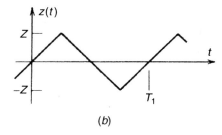

Figure P7.10

Problem Set 7.3

7.11 Determine the continuous Fourier transform for the rectangular pulse in Fig. P7.11. (Note that this involves a time shifting of the pulse in Example 7.6.)

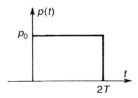

Figure P7.11

7.12 Determine the real and imaginary parts of the continuous Fourier transform of each of the following functions:

(a) $p(t) = p_0 e^{-\alpha |t|}$ *Note:* Absolute value of time t.

(b) $p(t) = \begin{cases} 0, & t < 0 \\ p_0 \sin(2\pi f_0 t), & 0 \leq t \leq \dfrac{1}{2f_0} \\ 0, & t > \dfrac{1}{2f_0} \end{cases}$

Problem Set 7.4

7.13 Using the unit impulse response function $h(t)$ of Eq. 5.30 and the definition in Eq. 7.33 of the continuous Fourier transform $\overline{H}(f)$, show that the Fourier transform of the unit impulse response function is the frequency response function $\overline{H}(f)$ given by Eq. 7.30. That is, using Eq. 7.33, show that $h(t)$ and $\overline{H}(f)$ form a Fourier transform pair, as illustrated in Fig. 7.4.

Problem Set 7.5

C 7.14 (a) Using the input sequence given in Fig. 1a of Example 7.7, use the MATLAB FFT command to compute a 16-point Fourier transform to verify the (real and imaginary) results shown in Figs. 1b. (*Note:* You can find the MATLAB .m-file for Example 7.7 as sd2ex7_7.m on the book's website.) (b) What is the resulting value of the coefficient $(\overline{P}_1)_{16pt}$? Compare this answer with the value given in Eq. 5 of Example 7.4. (c) Repeat part (a) to calculate a 32-point transform and a 64-point transform. Are the values of $(\overline{P}_1)_{16pt}$, $(\overline{P}_1)_{32pt}$, and $(\overline{P}_1)_{64pt}$ converging to the value given in Eq. 5 of Example 7.4?

C 7.15 Let $p(t)$ be the periodic function defined by the 16-point discrete-time sawtooth wave shown in Fig. P7.15. (a) Use the FFT command in MATLAB to determine the 16-point discrete Fourier transform (DFT) of this sawtooth wave. (b) Plot the real and imaginary parts of this DFT transform.

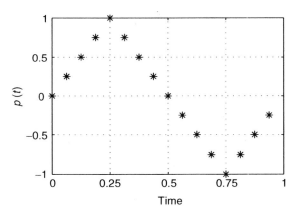

Figure P7.15

PART II

Multiple-Degree-of-Freedom Systems—Basic Topics

8

Mathematical Models of MDOF Systems

In Chapter 2, Newton's Laws and the Principle of Virtual Displacements were employed to derive the equations of motion of single-degree-of-freedom (SDOF) models. Although SDOF models might adequately describe the dynamical behavior of some systems, in most cases it is necessary to employ more complex *multiple-degree-of-freedom* (MDOF) *models*. To illustrate this, consider the axial vibration of a uniform cantilever rod as shown in Fig. 8.1. Only if the axial force $P(t)$ is a "slowly varying" function of time can the dynamics of the rod be described by an SDOF model: for example, by the spring–mass model shown in Fig. 8.1. *Continuous models* (also called *distributed-parameter models*), such as the rod in Fig. 8.1a, are considered in Chapters 12 and 13. While continuous models give valuable insight into the dynamics of a few systems with simple geometry (e.g., uniform bars and beams), the dynamical analysis of most real structures is based on multiple-degree-of-freedom models.

The general form of the equations of motion of a linear N-DOF model of a structure is

$$\boxed{\mathbf{M}\ddot{\mathbf{u}} + \mathbf{C}\dot{\mathbf{u}} + \mathbf{K}\mathbf{u} = \mathbf{p}(t)} \tag{8.1}$$

where $\mathbf{M}$ is the *mass matrix*, $\mathbf{C}$ is the *viscous damping matrix*, and $\mathbf{K}$ is the *stiffness matrix*. These coefficient matrices are all $N \times N$ matrices. The *displacement vector* $\mathbf{u}(t)$, either physical or generalized displacements, and the corresponding *load vector* $\mathbf{p}(t)$ are $N \times 1$ vectors.

In this chapter we explore ways to create *MDOF mathematical models* for use in studying the dynamics of simple structures, and in Chapters 9 through 11 we examine

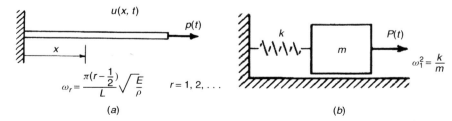

Figure 8.1 Two models of a cantilever uniform rod: (*a*) continuous-system model; (*b*) SDOF model.

212 Mathematical Models of MDOF Systems

the basic solution procedures that apply to MDOF models: namely, eigensolutions and mode-superposition solutions. In Chapters 14 through 18 we take up a number of advanced topics involved in using MDOF models to study the dynamics of real, complex structures, such as airplanes, automobiles, and high-rise buildings, beginning in Chapter 14 with an introduction to *finite element analysis*.

Upon completion of this chapter you should be able to:

- Use Newton's Laws to derive the equations of motion of systems of particles and of rigid bodies in plane motion.
- Use Lagrange's Equations to derive the equations of motion of systems of particles and of rigid bodies in plane motion.
- Use Lagrange's Equations to derive the equations of motion of assumed-modes models of simple continuous systems.

8.1 APPLICATION OF NEWTON'S LAWS TO LUMPED-PARAMETER MODELS

In Chapter 2, SDOF mathematical models were considered. *Lumped-parameter models*, such as systems of connected particles and/or rigid bodies, and rigid bodies in general plane motion, are examples of systems that lead to MDOF models. Several examples of the application of Newton's Laws to determine appropriate mathematical models of such lumped-parameter models are now considered in this section. In view of the fact that the best way to create an MDOF model of a structure is to use some version of the Assumed-Modes Method (introduced in Section 8.4), in the following examples we simply assume that the physical parameters (mass, spring constant, etc.) of lumped-parameter models are known.

Example 8.1   All three of the lumped-parameter systems shown in Fig. 1 have essentially the same set of equations of motion, that is, the same mathematical model. Derive the three equations of motion for the spring–mass system in Fig. 1c.

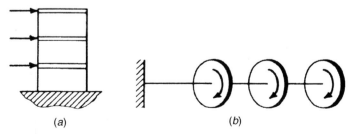

Figure 1   Three 3-DOF models: (*a*) shear-frame building model; (*b*) shaft–disk model; (*c*) spring–mass model (on next page).

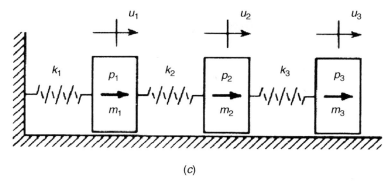

(c)

Figure 1 (continued)

SOLUTION   1. Draw a free-body diagram of each mass in Fig. 1c, and label all unknown forces (Fig. 2).

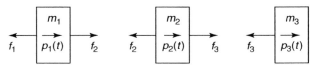

Figure 2 Free-body diagrams.

2. Write Newton's second law for each particle.

$$\overset{+}{\rightarrow} \sum F_1 = m_1 \ddot{u}_1 = p_1 + f_2 - f_1 \tag{1a}$$

$$\overset{+}{\rightarrow} \sum F_2 = m_2 \ddot{u}_2 = p_2 + f_3 - f_2 \tag{1b}$$

$$\overset{+}{\rightarrow} \sum F_3 = m_3 \ddot{u}_3 = p_3 - f_3 \tag{1c}$$

3. For the three springs in Fig. 1c, relate the linearly elastic spring forces shown in Fig. 3 to the displacements u_i of the three masses.

$$f_1 = k_1 e_1 = k_1 u_1 \tag{2a}$$

$$f_2 = k_2 e_2 = k_2(u_2 - u_1) \tag{2b}$$

$$f_3 = k_3 e_3 = k_3(u_3 - u_2) \tag{2c}$$

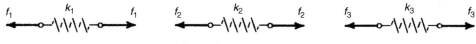

Figure 3 Three springs.

4. Combine and simplify the equations.

$$m_1\ddot{u}_1 + (k_1 + k_2)u_1 - k_2 u_2 = p_1(t) \tag{3a}$$

$$m_2\ddot{u}_2 - k_2 u_1 + (k_2 + k_3)u_2 - k_3 u_3 = p_2(t) \tag{3b}$$

$$m_3\ddot{u}_3 - k_3 u_2 + k_3 u_3 = p_3(t) \tag{3c}$$

These three equations of motion can be written in matrix form, that is,

$$\begin{bmatrix} m_1 & 0 & 0 \\ 0 & m_2 & 0 \\ 0 & 0 & m_3 \end{bmatrix} \begin{Bmatrix} \ddot{u}_1 \\ \ddot{u}_2 \\ \ddot{u}_3 \end{Bmatrix}$$
$$+ \begin{bmatrix} k_1 + k_2 & -k_2 & 0 \\ -k_2 & k_2 + k_3 & -k_3 \\ 0 & -k_3 & k_3 \end{bmatrix} \begin{Bmatrix} u_1 \\ u_2 \\ u_3 \end{Bmatrix} = \begin{Bmatrix} p_1(t) \\ p_2(t) \\ p_3(t) \end{Bmatrix} \quad \textbf{Ans. (4)}$$

or in the symbolic matrix notation of Eq. 8.1,

$$\mathbf{M\ddot{u}} + \mathbf{Ku} = \mathbf{p}(t) \quad \textbf{Ans. (5)}$$

This set of coupled ordinary differential equations constitutes the 3-DOF *mathematical model* (or *math model*) of the lumped-mass system in Fig. 1c. With slightly different notation, the same set of equations constitutes the 3-DOF math model of either of the other two systems in Fig. 1.

The matrix notation employed in Eqs. 4 and 5, although not essential to the solution of this 3-DOF problem, is the "language" used in describing large MDOF systems, particularly when the computer is employed in the dynamical analysis of such systems. The matrix denoted $\mathbf{M}$ is called the *mass matrix*, $\mathbf{K}$ is the *stiffness matrix*, $\mathbf{u}$ is the *displacement vector*, and $\mathbf{p}(t)$ is the *load vector*.

In Eq. 4 of Example 8.1 it should be noted that the mass matrix, $\mathbf{M}$, is a *diagonal matrix*. Since there are off-diagonal terms in the stiffness matrix, $\mathbf{K}$, the system is said to have *stiffness coupling*. In Eqs. 3, where the three equations are written out, the first equation is coupled to the second equation by the presence of the u_2 term in the first equation, the u_1 term in the second equation, and so on.

In Example 8.1 the equations of motion were written in terms of the absolute displacements u_1, u_2, and u_3. As noted in Section 2.2, when the system is subjected to base motion, it is convenient to employ the relative motion between the base and the structural masses. Consider the following *base motion* example involving a viscous-damped 2-DOF lumped-mass system.

Example 8.2 Use Newton's Laws to derive the equations of motion of the system shown in Fig. 1. Express the equations of motion in terms of the displacements of the masses relative to the base.

SOLUTION Let the relative displacements, that is, the displacements of the two masses relative to the moving base, be defined by respectively.

$$w_1 = u_1 - z \qquad (1a)$$
$$w_2 = u_2 - z \qquad (1b)$$

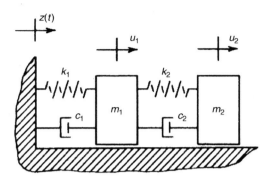

Figure 1 A 2-DOF spring–mass–dashpot system.

1. For each mass in Fig. 1, draw a free-body diagram in Fig. 2, and label all unknown forces.

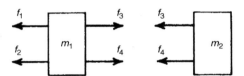

Figure 2 Free-body diagrams.

2. Use Newton's Second Law to write the equations of motion.

$$\overset{+}{\rightarrow} \sum F_1 = m_1 \ddot{u}_1 = m_1(\ddot{z} + \ddot{w}_1) = -f_1 - f_2 + f_3 + f_4 \qquad (2a)$$

$$\overset{+}{\rightarrow} \sum F_2 = m_2 \ddot{u}_2 = m_1(\ddot{z} + \ddot{w}_2) = -f_3 - f_4 \qquad (2b)$$

3. Relate the linearly elastic spring forces (f_1 and f_3) to the displacements, and relate the viscous damping forces (f_2 and f_4) to the velocities.

$$f_1 = k_1(u_1 - z) = k_1 w_1 \qquad (3a)$$
$$f_2 = c_1(\dot{u}_1 - \dot{z}) = c_1 \dot{w}_1 \qquad (3b)$$
$$f_3 = k_2(u_2 - u_1) = k_2(w_2 - w_1) \qquad (3c)$$
$$f_4 = c_2(\dot{u}_2 - \dot{u}_1) = c_2(\dot{w}_2 - \dot{w}_1) \qquad (3d)$$

4. Combine and simplify the equations above.

$$m_1 \ddot{w}_1 + c_1 \dot{w}_1 + k_1 w_1 - c_2(\dot{w}_2 - \dot{w}_1) - k_2(w_2 - w_1) = -m_1 \ddot{z}$$

$$m_2 \ddot{w}_2 + c_2(\dot{w}_2 - \dot{w}_1) + k_2(w_2 - w_1) = -m_2 \ddot{z}$$

Write these equations of motion in matrix form:

$$\begin{bmatrix} m_1 & 0 \\ 0 & m_2 \end{bmatrix} \begin{Bmatrix} \ddot{w}_1 \\ \ddot{w}_2 \end{Bmatrix} + \begin{bmatrix} c_1 + c_2 & -c_2 \\ -c_2 & c_2 \end{bmatrix} \begin{Bmatrix} \dot{w}_1 \\ \dot{w}_2 \end{Bmatrix}$$
$$+ \begin{bmatrix} k_1 + k_2 & -k_2 \\ -k_2 & k_2 \end{bmatrix} \begin{Bmatrix} w_1 \\ w_2 \end{Bmatrix} = \begin{Bmatrix} -m_1 \ddot{z} \\ -m_2 \ddot{z} \end{Bmatrix} \quad \text{Ans. (4)}$$

or in symbolic matrix notation,

$$\mathbf{M}\ddot{\mathbf{w}} + \mathbf{C}\dot{\mathbf{w}} + \mathbf{K}\mathbf{w} = \mathbf{p}_{\text{eff}}(t) \quad \text{Ans. (5)}$$

where the *effective force vector* on the right-hand side of Eq. 5 is related to the base acceleration by the equation

$$\mathbf{p}_{\text{eff}}(t) = \begin{Bmatrix} -m_1 \ddot{z} \\ -m_2 \ddot{z} \end{Bmatrix} \quad (6)$$

Note that writing the equations of motion in terms of relative displacements rather than absolute displacements leads to several advantages: (1) the right-hand side has the form of effective forces that are related directly to the base acceleration $\ddot{z}$, which is simpler to measure than either displacement z or velocity $\dot{z}$; and (2) the relative displacements can be used directly in calculating spring forces and dashpot forces, as shown in Eqs. 3.

The next example treats a system that is modeled as a rigid body in plane motion.

Example 8.3 The motion of a building subjected to earthquake excitation is to be studied by using the lumped-parameter model shown in Fig. 1. Use Newton's Laws (extended to rigid bodies in plane motion) to derive the equations of motion of this system. Consider angle θ to be small, and consider the foundation mass, m, to be a particle undergoing horizontal translation.

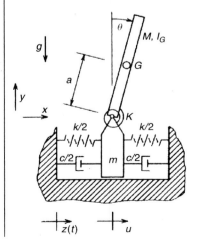

Figure 1 A 2-DOF base-motion example.

SOLUTION 1. Draw free-body diagrams (Fig. 2) of the two components of the idealized lumped-parameter system, and label all unknown forces.

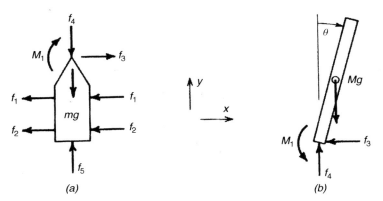

Figure 2 Free-body diagrams.

2. Write the basic equations of motion for each free-body diagram of Fig. 2. For free body (*a*):

$$\stackrel{+}{\rightarrow} \sum F_x = m\ddot{u} = -2f_1 - 2f_2 + f_3 \qquad (1)$$

(The vertical equation of motion and moment equation are not needed since the base is to be treated as a particle.) For free body (*b*):

$$\stackrel{+}{\rightarrow} \sum F_x = M\ddot{x}_G = -f_3 \qquad (2a)$$

$$\uparrow + \sum F_y = M\ddot{y}_G = f_4 - Mg \qquad (2b)$$

$$\stackrel{+}{\circlearrowleft} \sum M_G = I_G \ddot{\theta} = -M_1 + f_4 a \sin\theta + f_3 a \cos\theta \qquad (2c)$$

3. Relate elastic forces to displacements, and relate damping forces to velocities.

$$f_1 = \frac{k}{2}(u - z) \qquad (3a)$$

$$f_2 = \frac{c}{2}(\dot{u} - \dot{z}) \qquad (3b)$$

$$M_1 = K\theta \qquad (3c)$$

4. Use kinematics to relate x_G and y_G to u and θ. For small θ, $\sin\theta \approx \theta$ and $\cos\theta \approx 1$. Therefore, x_G and y_G can be approximated by

$$x_G = u + a\theta \qquad (4a)$$

$$y_G = a \qquad (4b)$$

The resulting two equations of motion, in matrix form, are

$$\begin{bmatrix} M+m & Ma \\ Ma & I_G + Ma^2 \end{bmatrix} \begin{Bmatrix} \ddot{u} \\ \ddot{\theta} \end{Bmatrix} + \begin{bmatrix} c & 0 \\ 0 & 0 \end{bmatrix} \begin{Bmatrix} \dot{u} \\ \dot{\theta} \end{Bmatrix}$$
$$+ \begin{bmatrix} k & 0 \\ 0 & K - Mga \end{bmatrix} \begin{Bmatrix} u \\ \theta \end{Bmatrix} = \begin{Bmatrix} c\dot{z} + kz \\ 0 \end{Bmatrix} \qquad \text{Ans. (5)}$$

or in symbolic matrix notation,

$$\mathbf{M}\ddot{\mathbf{u}} + \mathbf{C}\dot{\mathbf{u}} + \mathbf{K}\mathbf{u} = \mathbf{p}_{\text{eff}}(t) \qquad \text{Ans. (6)}$$

Note that the base motion $z(t)$ leads to the "effective force" terms on the right-hand side of Eq. 5, and note that this system has *inertia coupling* due to the off-diagonal terms in **M**.

As illustrated in Example 8.3, the use of Newton's Laws to derive the equations of motion of systems of coupled rigid bodies can become very tedious. The task of deriving the equations of motion of MDOF systems is greatly simplified in many cases by the use of Lagrange's Equations, discussed in the remainder of this chapter.

8.2 INTRODUCTION TO ANALYTICAL DYNAMICS: HAMILTON'S PRINCIPLE AND LAGRANGE'S EQUATIONS

The study of dynamics may be subdivided into two main categories: *Newtonian Mechanics* and *Analytical Mechanics*. The latter is sometimes referred to as *Variational Principles in Mechanics* or *Energy Methods in Mechanics*.[8.1] The Principle of Virtual Displacements, Hamilton's Principle, and Lagrange's Equations are analytical mechanics methods that are used to derive the equations of motion for various models of dynamical systems. You were introduced to analytical dynamics in Chapter 2, where the Principle of Virtual Displacements was employed in deriving the equations of motion of several SDOF systems. In this section, two very important tools of Analytical Mechanics are introduced: Hamilton's Principle and Lagrange's Equations. These methods are more powerful than the Principle of Virtual Displacements. Since they employ kinetic energy and potential energy, which are scalar quantities, they are also somewhat easier to apply.

We begin by stating the extended form of Hamilton's Principle, followed by the original form of Hamilton's Principle.

8.2.1 Hamilton's Principle

In the formulation of Hamilton's Principle, virtual displacements, or virtual changes of configuration, are employed. Figure 8.2 (also Fig. 2.6) shows a cantilever beam with a virtual change of configuration $\delta v(x, t)$.[1] As noted in Section 2.4, the virtual change of

---

[1] As noted in Section 2.4, $\delta v(x, t) \equiv \delta[v(x, t)]$ is an infinitesimal change in the configuration $v(x, t)$. It is a function of position x, but not of time t.

8.2 Introduction to Analytical Dynamics: Hamilton's Principle and Lagrange's Equations

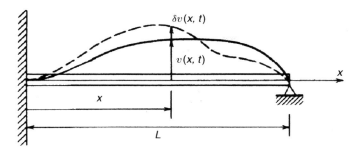

Figure 8.2 End constraints and virtual displacement.

configuration must satisfy all geometric boundary conditions. Hamilton also assumed that the configuration is specified completely at times t_1 and t_2. For the cantilever beam this would imply that $\delta v(x, t_1) = \delta v(x, t_2) = 0$.

The *Extended Hamilton's Principle* may be stated as follows[2]:

The motion of the given system from time t_1 to t_2 is such that

$$\int_{t_1}^{t_2} \delta(\mathcal{T} - \mathcal{V}) \, dt + \int_{t_1}^{t_2} \delta W_{nc} \, dt = 0 \tag{8.2}$$

where

$\mathcal{T}$ = total kinetic energy of the system
$\mathcal{V}$ = potential energy of the system, including the strain energy and the potential energy of conservative external forces
δW_{nc} = virtual work done by nonconservative forces, including damping forces and external forces not accounted for in $\mathcal{V}$
$\delta[\,\cdot\,]$ = symbol denoting the first variation, or virtual change, in the quantity in brackets
t_1, t_2 = times at which the configuration of the system is assumed to be known

For conservative systems $\delta W_{nc} = 0$, so Eq. 8.2 reduces to

$$\int_{t_1}^{t_2} \delta \mathcal{L} \, dt = 0 \tag{8.3}$$

where

$$\mathcal{L} = \mathcal{T} - \mathcal{V} \tag{8.4}$$

is called the *Lagrangian function*. Equation 8.3 is referred to as *Hamilton's Principle*.

Next, the Extended Hamilton's Principle is employed to derive Lagrange's Equations, which are then be used to derive the equations of motion for MDOF models of structures. The Extended Hamilton's Principle is employed directly in Chapter 12 to derive the equations of motion for continuous models of structural elements, such as bars and beams.

---

[2] Some authors refer to this equation simply as *Hamilton's Principle*. For a more detailed discussion of Hamilton's Principle, see Refs. [8.1] to [8.3].

8.2.2 Lagrange's Equations

Some of the difficulties that may arise when Newton's Laws are employed to derive the equations of motion of connected bodies were illustrated in Example 8.3. A separate free-body diagram was required for each component, and the forces of interaction had to be eliminated to arrive at the final set of equations of motion. In Example 2.7 it was shown that use of the Principle of Virtual Displacements eliminates the necessity of using interaction forces directly. Although the Principle of Virtual Displacements can be extended to permit the derivation of equations of motion of MDOF systems, it is far simpler to use Lagrange's Equations for this purpose. This permits use of the scalar quantities—work, potential energy, and kinetic energy—instead of the vector quantities—force and acceleration. Lagrange's Equations can be derived from the Principle of Virtual Displacements or from the Extended Hamilton's Principle. The latter approach is employed here.

Consider Example 8.3 again. Note that although x_G and y_G were required in writing the equations of motion of the mass M, Eqs. 4 permitted x_G and y_G to be related to u and θ, which were the only coordinates appearing in the final equations of motion, Eqs. 5. Kinematical equations such as Eqs. 4 are called *equations of constraint*. The system considered in Example 8.3 is a 2-DOF system, and any arbitrary configuration of the system can be specified by giving the values of u and θ. Note that it is possible to vary θ while holding u constant, and vice versa. Coordinates of this type are called generalized coordinates.

- *Generalized coordinates* are defined as any set of N independent quantities that are sufficient to completely specify the position of every point within an N-DOF system.

In general expressions, the symbol q_i will signify the ith generalized coordinate of a set of N generalized coordinates. In specific examples, other symbols may be used (e.g., $q_1 \to u$ and $q_2 \to \theta$ in Example 8.3).

For most mechanical and structural systems the kinetic energy can be expressed in terms of the generalized coordinates and their first time derivatives, and the potential energy can be expressed in terms of generalized coordinates alone. Also, the virtual work of nonconservative forces, as they act through virtual displacements caused by arbitrary variations in the generalized coordinates, can be expressed as a linear function of those variations. Thus,

$$T = T(q_1, q_2, \ldots, q_N, \dot{q}_1, \dot{q}_2, \ldots, \dot{q}_N, t) \qquad (8.5a)$$

$$V = V(q_1, q_2, \ldots, q_N, t) \qquad (8.5b)$$

$$\delta W_{\text{nc}} = p_1 \delta q_1 + p_2 \delta q_2 + \cdots + p_N \delta q_N \qquad (8.5c)$$

where $p_1, p_2, \ldots, p_N$ are called the *generalized forces*. The generalized forces have units such that each term $p_j \delta q_j$ has the units of work.

Before deriving Lagrange's Equations, we first illustrate how to calculate the foregoing quantities: kinetic energy T, potential energy V, and δW_{nc}, the virtual work of nonconservative forces.

8.2 Introduction to Analytical Dynamics: Hamilton's Principle and Lagrange's Equations

Example 8.4 A particle of mass m slides along a weightless rigid rod as shown in Fig. 1. Write an expression for the kinetic energy of the particle in terms of the generalized coordinates q_1 and q_2 and their time derivatives.

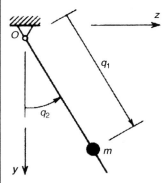

Figure 1 Particle sliding on a pinned rod.

SOLUTION   1. The kinetic energy of the particle is given by

$$T = \tfrac{1}{2}m(\dot{y}^2 + \dot{z}^2) \tag{1}$$

2. From kinematics,

$$y = q_1 \cos q_2 \tag{2a}$$

$$z = q_1 \sin q_2 \tag{2b}$$

Therefore,

$$\dot{y} = \dot{q}_1 \cos q_2 - \dot{q}_2 q_1 \sin q_2 \tag{3a}$$

$$\dot{z} = \dot{q}_1 \sin q_2 + \dot{q}_2 q_1 \cos q_2 \tag{3b}$$

3. Combine and simplify Eqs. 1 and 3 to get the following expression for the kinetic energy:

$$T = \tfrac{1}{2}m[\dot{q}_1^2 + (q_1 \dot{q}_2)^2] \qquad \textbf{Ans. (4)}$$

Note that Eq. 4 has the form of Eq. 8.5a; that is, both generalized coordinates and generalized velocities appear in the kinetic energy expression T.

Example 8.5 As shown in Fig. 1, a force P acts tangent to the path of a particle of weight W, which is attached to a rigid bar of length L. Obtain expressions for the potential energy of weight W and the virtual work done by force P. Also, determine an expression for the generalized force p_θ.

SOLUTION   1. Let the potential energy be zero when $\theta = \pi/2$. Then

$$V = -Wy = -WL\cos\theta \qquad \textbf{Ans. (1)}$$

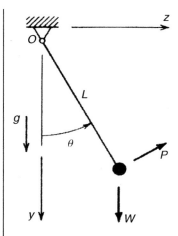

Figure 1 Forces acting on a pinned rod.

2. Determine δW_P due to a variation in θ as shown in Fig. 2. The tangential distance moved by force P is

$$\delta s = L\,\delta\theta \qquad (2)$$

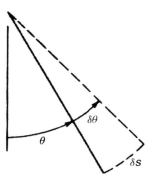

Figure 2 Displacement θ and virtual displacement $\delta\theta$.

Thus, the virtual work done by tangential force P is

$$\delta W_P = P\,\delta s = (PL)\,\delta\theta \qquad \textbf{Ans.} \quad (3)$$

Then, from Eqs. 3 and 8.5c, the generalized force is

$$p_\theta = PL \qquad \textbf{Ans.} \quad (4)$$

Since θ is the generalized coordinate for this problem, Eq. 1 has the form specified by Eq. 8.5b, and Eq. 3 has the form of Eq. 8.5c.

Continuing now with the derivation of Lagrange's Equations, substitute Eqs. 8.5 into the Extended Hamilton's Principle, Eq. 8.2, to give

$$\int_{t_1}^{t_2} \bigg(\frac{\partial T}{\partial q_1} \delta q_1 + \frac{\partial T}{\partial q_2} \delta q_2 + \cdots + \frac{\partial T}{\partial q_N} \delta q_N + \frac{\partial T}{\partial \dot{q}_1} \delta \dot{q}_1 + \frac{\partial T}{\partial \dot{q}_2} \delta \dot{q}_2$$
$$+ \cdots + \frac{\partial T}{\partial \dot{q}_N} \delta \dot{q}_N - \frac{\partial V}{\partial q_1} \delta q_1 - \frac{\partial V}{\partial q_2} \delta q_2 - \cdots - \frac{\partial V}{\partial q_N} \delta q_N \quad (8.6)$$
$$+ p_1 \delta q_1 + p_2 \delta q_2 + \cdots + p_N \delta q_N \bigg) dt = 0$$

The terms involving $\delta \dot{q}_i$ may be integrated by parts in the following manner:

$$\int_{t_1}^{t_2} \frac{\partial T}{\partial \dot{q}_i} \delta \dot{q}_i \, dt = \left[\frac{\partial T}{\partial \dot{q}_i} \delta q_i \right]_{t_1}^{t_2} - \int_{t_1}^{t_2} \frac{d}{dt} \frac{\partial T}{\partial \dot{q}_i} \delta q_i \, dt \quad (8.7)$$

The first term on the right-hand side of Eq. 8.7 is zero for each coordinate, since $\delta q_i(t_1) = \delta q_i(t_2) = 0$ is a basic condition imposed to make Hamilton's Principle valid. When Eq. 8.7 is substituted into Eq. 8.6, the resulting equation can be written in the form

$$\int_{t_1}^{t_2} \left\{ \sum_{i=1}^{N} \left[-\frac{d}{dt} \frac{\partial T}{\partial \dot{q}_i} + \frac{\partial T}{\partial q_i} - \frac{\partial V}{\partial q_i} + p_i \right] \delta q_i \right\} dt = 0 \quad (8.8)$$

Since the variations $\delta q_i (i = 1, 2, \ldots, N)$ must be independent, Eq. 8.8 can be satisfied in general only when the bracketed expression in Eq. 8.8 vanishes for each value of i, that is, when

$$\boxed{\frac{d}{dt} \frac{\partial T}{\partial \dot{q}_i} - \frac{\partial T}{\partial q_i} + \frac{\partial V}{\partial q_i} = p_i(t), \qquad i = 1, 2, \ldots, N} \quad (8.9)$$

The N equations in Eqs. 8.9 are known as *Lagrange's Equations*.

The restrictions imposed in deriving Eqs. 8.9 were that the coordinates q_i be independent and that T, V, and δW_{nc} have the forms shown in Eqs. 8.5. Thus, Eqs. 8.9 are valid for nonlinear systems as well as linear systems.

The remainder of this chapter is devoted to examples of the use of Lagrange's Equations to derive MDOF mathematical models. In Section 8.3, Lagrange's Equations are used to derive mathematical models for discrete-parameter systems modeled as particles and/or rigid bodies. In Section 8.4, Lagrange's Equations are used to develop the Assumed Modes Method for creating MDOF approximations to describe the dynamics of continuous systems. Finally, an extension of Lagrange's Equations to systems represented by constrained (i.e., not independent) coordinates is presented in Section 8.5.

8.3 APPLICATION OF LAGRANGE'S EQUATIONS TO LUMPED-PARAMETER MODELS

The following examples illustrate the application of Lagrange's Equations to systems represented by lumped-parameter models, that is, by particles and/or rigid bodies.

Example 8.6 Symmetrical vibration of an airplane wing–body combination is modeled by a "fuselage" mass M to which are attached "wing" masses m by rigid, weightless beams of length L, as illustrated in Fig. 1. The elastic behavior of the wings is modeled by wing-root springs having spring constant k, which are connected between the fuselage and rigid wings. Use Lagrange's Equations to derive the equations of motion for symmetrical vertical motion of this 2-DOF model. Neglect gravity, and assume that θ remains small (i.e., $\theta \ll 1$).

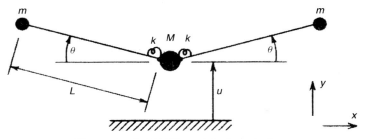

Figure 1 A 2-DOF model of an airplane.

SOLUTION Assign u and θ as the two generalized coordinates; that is, let $q_1 \rightarrow u$, $q_2 \rightarrow \theta$.

1. Write expressions for the kinetic energy T and the potential energy V.

$$T = 2\left(\tfrac{1}{2} m\, \dot{y}_m^2\right) + \tfrac{1}{2} M \dot{u}^2 \tag{1}$$

For small θ,

$$y_m = u + L\theta \tag{2}$$

Therefore, the kinetic energy and potential energy are, respectively,

$$T = m(\dot{u} + L\dot{\theta})^2 + \tfrac{1}{2} M \dot{u}^2 \tag{3}$$

$$V = 2\left(\tfrac{1}{2} k\, \theta^2\right) \tag{4}$$

2. Apply Lagrange's Equations, Eqs. 8.9.

$$\frac{d}{dt}\frac{\partial T}{\partial \dot{q}_i} - \frac{\partial T}{\partial q_i} + \frac{\partial V}{\partial q_i} = p_i, \qquad q_1 \rightarrow u, \quad q_2 \rightarrow \theta \tag{5}$$

The various terms in Eq. 5 are evaluated as follows:

$$\begin{aligned}
\frac{\partial T}{\partial \dot{u}} &= 2m(\dot{u} + L\dot{\theta}) + M\dot{u} \\
\frac{\partial T}{\partial \dot{\theta}} &= 2mL(\dot{u} + L\dot{\theta}) \\
\frac{\partial T}{\partial u} &= \frac{\partial T}{\partial \theta} = 0 \\
\frac{\partial V}{\partial u} &= 0, \qquad \frac{\partial V}{\partial \theta} = 2k\theta \\
p_u &= p_\theta = 0
\end{aligned} \tag{6}$$

Therefore, Eqs. 6 are substituted into Eqs. 5 to give the following two equations of motion for this system:

$$2m(\ddot{u} + L\ddot{\theta}) + M\ddot{u} = 0$$
$$2mL(\ddot{u} + L\ddot{\theta}) + 2k\theta = 0 \tag{7}$$

Equations 7 can be written in matrix form as

$$\begin{bmatrix} M + 2m & 2mL \\ 2mL & 2mL^2 \end{bmatrix} \begin{Bmatrix} \ddot{u} \\ \ddot{\theta} \end{Bmatrix} + \begin{bmatrix} 0 & 0 \\ 0 & 2k \end{bmatrix} \begin{Bmatrix} u \\ \theta \end{Bmatrix} = \begin{Bmatrix} 0 \\ 0 \end{Bmatrix} \quad \textbf{Ans.} \tag{8}$$

Note that the equations of motion above were obtained without the necessity of employing free-body diagrams and interaction forces between the wings and fuselage. Also note that the two equations of motion are inertia-coupled but not stiffness-coupled.

Example 8.7 Use Lagrange's Equations to derive the equations of motion of the system shown in Example 8.3. Assume that θ is small.

SOLUTION Assign u and θ as the two generalized coordinates; that is, let $q_1 \to u$, $q_2 \to \theta$.

1. Write expressions for the kinetic energy T, the potential energy V, and the external virtual work δW_{nc}.

$$T = \tfrac{1}{2}m\dot{u}^2 + \tfrac{1}{2}M(\dot{x}_G^2 + \dot{y}_G^2) + \tfrac{1}{2}I_G\dot{\theta}^2 \tag{1}$$

From kinematics,

$$x_G = u + a\sin\theta, \qquad y_G = a\cos\theta \tag{2a, b}$$

For small θ, $\sin\theta \approx \theta$, $\cos\theta \approx 1 - \tfrac{1}{2}\theta^2$. Therefore, $\dot{x}_G$ and $\dot{y}_G$ can be approximated by

$$\dot{x}_G = \dot{u} + a\dot{\theta}, \qquad \dot{y}_G = 0 \tag{3a, b}$$

respectively, where the nonlinear $\theta\dot{\theta}$ term in $\dot{y}_G$ has been neglected. So the kinetic energy can finally be written as

$$T = \tfrac{1}{2}m\dot{u}^2 + \tfrac{1}{2}M(\dot{u} + a\dot{\theta})^2 + \tfrac{1}{2}I_G\dot{\theta}^2 \tag{4}$$

The potential energy is stored as elastic strain energy in the springs and as gravitational potential energy.

$$V = 2\left[\tfrac{1}{2}\left(\tfrac{k}{2}\right)(u-z)^2\right] + \tfrac{1}{2}K\theta^2 + Mga\cos\theta \tag{5}$$

For small angle θ, the potential energy can be approximated, up to quadratic terms in u and θ, by

$$V = 2\left[\tfrac{1}{2}\left(\tfrac{k}{2}\right)(u-z)^2\right] + \tfrac{1}{2}K\theta^2 + Mga\left(1 - \tfrac{1}{2}\theta^2\right) \tag{6}$$

The nonconservative forces in this problem are the damping forces exerted on the foundation mass. These forces are shown in Fig. 1. (*Note:* This is not a free-body diagram. It is just a sketch that shows the damping forces and how they move.) The virtual work done by the damping forces is

$$\delta W_{nc} = -2[\tfrac{1}{2}c(\dot{u} - \dot{z})]\,\delta u \tag{7}$$

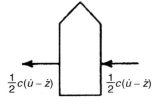

Figure 1 Nonconservative damping forces.

2. Apply Lagrange's Equations, Eqs. 8.9.

$$\frac{d}{dt}\frac{\partial T}{\partial \dot{q}_i} - \frac{\partial T}{\partial q_i} + \frac{\partial V}{\partial q_i} = p_i, \qquad q_1 \to u, \quad q_2 \to \theta \tag{8}$$

The various terms that are needed are

$$\frac{\partial T}{\partial \dot{u}} = m\dot{u} + M(\dot{u} + a\dot{\theta})$$

$$\frac{\partial T}{\partial \dot{\theta}} = Ma(\dot{u} + a\dot{\theta}) + I_G \dot{\theta}$$

$$\frac{\partial T}{\partial u} = \frac{\partial T}{\partial \theta} = 0 \tag{9}$$

$$\frac{\partial V}{\partial u} = k(u - z), \qquad \frac{\partial V}{\partial \theta} = K\theta - Mga\theta$$

$$p_u = -c(\dot{u} - \dot{z}), \qquad p_\theta = 0$$

Therefore, the coupled ordinary differential equations that constitute the mathematical model of this 2-DOF system are

$$(M + m)\ddot{u} + Ma\ddot{\theta} + c\dot{u} + ku = c\dot{z} + kz$$

$$Ma\ddot{u} + (Ma^2 + I_G)\ddot{\theta} + (K - mga)\theta = 0$$

Ans. (10)

It may be observed that Eqs. 10 are the same as Eqs. 5 of Example 8.3.

As a final application of Lagrange's Equations to lumped-parameter systems, consider the following nonlinear problem.

Example 8.8 A particle of mass m is free to slide along a uniform thin rod of mass M and length L as shown in Fig. 1. The rod, in turn, is free to rotate about a pin at its upper end. Use Lagrange's Equations to derive the equations of motion of this 2-DOF system.

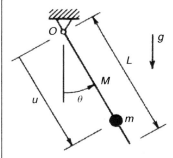

Figure 1 Swinging rod with a sliding mass.

SOLUTION 1. Write expressions for T and V. (The system has no nonconservative forces, so no δW_{nc} is required.) From Example 8.4, the kinetic energy of mass m can be obtained. Then the kinetic energy and potential energy of the rod–mass system are given by

$$T = \tfrac{1}{2}\left(\tfrac{1}{3}ML^2\right)\dot\theta^2 + \tfrac{1}{2}m[\dot u^2 + (u\dot\theta)^2] \tag{1}$$

$$V = -\left(Mg\frac{L}{2} + mgu\right)\cos\theta \tag{2}$$

2. Apply Lagrange's equation, Eq. 8.9.

$$\frac{d}{dt}\frac{\partial T}{\partial \dot q_i} - \frac{\partial T}{\partial q_i} + \frac{\partial V}{\partial q_i} = p_i, \qquad q_1 \to u, \quad q_2 \to \theta \tag{3}$$

The various terms that are needed are

$$\begin{aligned}
\frac{\partial T}{\partial \dot u} &= m\dot u, & \frac{\partial T}{\partial \dot\theta} &= \left(\tfrac{1}{3}ML^2 + mu^2\right)\dot\theta \\
\frac{\partial T}{\partial u} &= m u \dot\theta^2, & \frac{\partial T}{\partial \theta} &= 0 \\
\frac{\partial V}{\partial u} &= -mg\cos\theta & & \\
\frac{\partial V}{\partial \theta} &= \left(Mg\tfrac{L}{2} + mgu\right)\sin\theta & & \\
p_u &= p_\theta = 0 & &
\end{aligned} \tag{4}$$

Therefore, the equations of motion of this 2-DOF system are

$$m\ddot u - m u \dot\theta^2 - mg\cos\theta = 0$$

$$\left(\tfrac{1}{3}ML^2 + mu^2\right)\ddot\theta + 2mu\dot u\dot\theta + \left(Mg\frac{L}{2} + mgu\right)\sin\theta = 0$$

Ans. (5)

Observe that these two equations of motion are highly nonlinear (e.g., the $u\dot\theta^2$ term in the first equation and the $u^2\ddot\theta$ term in the second equation).

8.4 APPLICATION OF LAGRANGE'S EQUATIONS TO CONTINUOUS MODELS: ASSUMED-MODES METHOD

In Section 2.5 the axial displacement of a slender bar was assumed to have the form

$$u(x,t) = \psi(x)q(t) \tag{8.10}$$

This assumption produced a single-DOF model of the continuous system with $q(t)$ being the *generalized coordinate*. To generate an N-DOF model of a continuous system, Eq. 8.10 is expanded to include N *shape functions*, $\psi_i(x)$. Thus, the continuous displacement $u(x,t)$ is approximated by the finite sum

$$\boxed{u(x,t) = \sum_{i=1}^{N} \psi_i(x) q_i(t)} \tag{8.11}$$

where the q_i's are the *generalized coordinates*. The same form is used to approximate the transverse displacement of a beam, $v(x,t)$, the rotation of a torsion bar, $\theta(x,t)$, and so on. The *Assumed-Modes Method* consists of substituting Eq. 8.11 into the appropriate expressions for the kinetic energy T, the strain energy V, and the virtual work of nonconservative forces, δW_{nc}, and then applying Lagrange's Equations to derive equations of motion of the resulting N-DOF model. In this chapter the global form of the Assumed-Modes Method is considered; that is, the functions $\psi_i(x)$ will each represent a displacement shape for the entire structural element under consideration. In Chapter 14 the relationship of the Assumed-Modes Method to the Finite Element Method is noted.

8.4.1 Selection of Shape Functions

It is by choosing the functions $\psi_i(x)$ that the analyst defines the N-DOF *assumed-modes model*. The *shape functions* $\psi_i(x)$:

1. Must form a linearly independent set.
2. Must possess derivatives up to the order appearing in the strain energy V.
3. Must satisfy all *prescribed boundary conditions*, that is, all displacement-type boundary conditions.[3]

Functions that satisfy these three conditions are called *admissible functions*. As an example, consider the propped cantilever beam in Fig. 8.3. The assumed-modes approximation of the transverse displacement of this beam would be written

$$v(x,t) = \sum_{i=1}^{N} \psi_i(x) q_i(t)$$

where each shape function $\psi_i(x)$ must satisfy the boundary conditions

$$\psi_i(0) = \psi_i'(0) = \psi_i(L) = 0$$

---

[3]*Prescribed boundary conditions* are distinct from *natural boundary conditions*, which are force-type boundary conditions.

8.4 Application of Lagrange's Equations to Continuous Models: Assumed-Modes Method

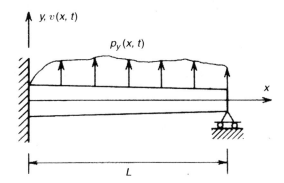

Figure 8.3 Boundary condition example.

since

$$v(0, t) = v'(0, t) = v(L, t) = 0$$

for all t. Since the strain energy expression for a Bernoulli–Euler beam contains $v''(x, t)$, the second derivative of the transverse displacement (e.g., see Section 11.3 of Ref. [8.4]), each $\psi_i(x)$ must be a continuous function of x, and its first derivative with respect to x must be continuous; that is, the beam can have no abrupt changes of displacement or slope.

It is not necessary that the ψ_i's also satisfy the *natural boundary conditions*, that is, force-type boundary conditions, such as the vanishing of moment $M(L, t)$ at the right end of the propped cantilever beam in Fig. 8.3. However, in cases where it is possible to obtain functions $\psi_i(x)$ that satisfy both prescribed and natural boundary conditions, such functions may be used in Eq. 8.11.

8.4.2 Assumed-Modes Method: Axial Deformation

To introduce the Assumed-Modes Method, we apply it first to the problem of determining approximate solutions for the axial vibration of a linearly elastic bar. In this case, $u(x, t)$ represents the axial motion of the plane cross section at x, as shown in Fig. 8.4. Let Eq. 8.11 be written in the form

$$\boxed{u(x, t) = \sum_{i=1}^{N} \psi_i(x) u_i(t)} \quad (8.12)$$

where the generalized coordinates are labeled $u_i(t)$ here to relate them to the axial displacement variable $u(x, t)$.

The strain energy in the bar is given by (e.g., see Section 11.3 of Ref. [8.4])

$$\mathcal{V} = \tfrac{1}{2} \int_0^L EA(u')^2 \, dx \quad (8.13)$$

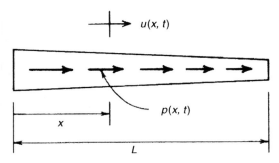

Figure 8.4 Axial motion of a thin bar.

and the kinetic energy of the bar due to the axial displacement $u(x,t)$ is given by

$$\mathcal{T} = \tfrac{1}{2} \int_0^L \rho A (\dot{u})^2 \, dx \qquad (8.14)$$

where $\rho(x)$ is the mass per unit volume and $A(x)$ is the cross-sectional area.

When Eq. 8.12 is substituted into Eq. 8.13, the resulting expression for strain energy may be written in the form

$$\mathcal{V} = \tfrac{1}{2} \sum_{i=1}^{N} \sum_{j=1}^{N} k_{ij} u_i u_j \qquad (8.15)$$

where the *stiffness coefficients* for axial deformation are given by

$$\boxed{k_{ij} = \int_0^L EA \psi_i' \psi_j' \, dx} \qquad (8.16)$$

In other words, $\mathcal{V}$ is a quadratic function of the generalized coordinates, with the coefficients in the quadratic expression being defined by Eq. 8.16. The quadratic expression may be written conveniently in matrix form: namely,

$$\mathcal{V} = \tfrac{1}{2} \mathbf{u}^T \mathbf{K} \mathbf{u} \qquad (8.17)$$

where the *generalized displacement vector* $\mathbf{u}$ and the *stiffness matrix* $\mathbf{K}$ have the respective forms

$$\mathbf{u} = \begin{Bmatrix} u_1 \\ u_2 \\ \vdots \\ u_N \end{Bmatrix}, \qquad \mathbf{K} = \begin{bmatrix} k_{11} & k_{12} & \cdots & k_{1N} \\ k_{21} & k_{22} & \cdots & k_{2N} \\ & & \ddots & \\ k_{N1} & k_{N2} & \cdots & k_{NN} \end{bmatrix} \qquad (8.18)$$

In a similar manner, Eqs. 8.12 and 8.14 may be combined to give the following quadratic expression for the kinetic energy of the bar:

$$\mathcal{T} = \tfrac{1}{2} \sum_{i=1}^{N} \sum_{j=1}^{N} m_{ij} \dot{u}_i \dot{u}_j \qquad (8.19)$$

8.4 Application of Lagrange's Equations to Continuous Models: Assumed-Modes Method

where the m_{ij}'s are given by

$$m_{ij} = \int_0^L \rho A \psi_i \psi_j \, dx \quad (8.20)$$

The kinetic energy can be written in the matrix form

$$T = \tfrac{1}{2} \dot{\mathbf{u}}^T \mathbf{M} \dot{\mathbf{u}} \quad (8.21)$$

When the mass matrix of a system is determined by an expression like Eq. 8.20, the mass matrix is called a *consistent mass matrix*; that is, the mass matrix and stiffness matrix are formed in a consistent manner using the same shape functions.

If the bar is subjected to external forces, as shown in Fig. 8.4, the corresponding generalized forces p_i are determined by employing virtual work. Thus,[4]

$$\delta W = \int_0^L p(x,t)\, \delta u(x,t)\, dx = \sum_{i=1}^N p_i(t)\, \delta u_i \quad (8.22)$$

From Eq. 8.10a, $\delta u(x,t)$ must be approximated by

$$\delta u(x,t) = \sum_{i=1}^N \psi_i(x)\, \delta u_i \quad (8.23)$$

Hence, upon combining Eqs. 8.22 and 8.23, we obtain the following expression for the *generalized forces*:

$$p_i(t) = \int_0^L p(x,t) \psi_i(x)\, dx \quad (8.24)$$

Lagrange's Equations may now be used to determine the equations of motion of the N-DOF model defined by Eq. 8.12. When Eqs. 8.15, 8.19, and 8.22 are substituted into Eq. 8.9, we obtain

$$\sum_{j=1}^N m_{ij} \ddot{u}_j + \sum_{j=1}^N k_{ij} u_j = p_i(t), \quad i = 1, 2, \ldots, N \quad (8.25)$$

which can be written in the familiar matrix form

$$\mathbf{M}\ddot{\mathbf{u}} + \mathbf{K}\mathbf{u} = \mathbf{p}(t) \quad (8.26)$$

8.4.3 Procedure for the Assumed-Modes Method

Although we have gone through Lagrange's Equations to arrive at Eq. 8.26, it may be seen that the only steps that are actually required in using the *Assumed-Modes Method* to arrive at the equations of motion of an N-DOF model of a continuous system are:

1. Select a set of N admissible functions, $\psi_i(x)$.
2. Compute the coefficients k_{ij} of the stiffness matrix by using Eq. 8.16.
3. Compute the coefficients m_{ij} of the mass matrix by using Eq. 8.20.

---

[4] Note that δu is not a function of time, as is $u(x,t)$. See Section 2.5.

4. Determine expressions for the generalized forces $p_i(t)$ corresponding to the applied force $p(x, t)$ by using Eq. 8.24.
5. Form the equations of motion by using Eq. 8.26.

This procedure will now be illustrated.

Example 8.9 Use the Assumed-Modes Method with a polynomial approximation of $u(x, t)$ to obtain a 2-DOF model for axial vibration of a uniform cantilever bar subjected to an end force $P(t)$, as shown in Fig. 1.

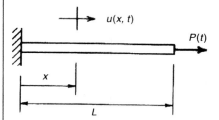

Figure 1 Uniform bar undergoing forced axial vibration.

SOLUTION 1. Select the shape functions, $\psi_i(x)$. For a 2-DOF assumed-modes model, the displacement is to be approximated by

$$u(x, t) = \psi_1(x) u_1(t) + \psi_2(x) u_2(t) \tag{1}$$

[Note that $u_1(t)$ and $u_2(t)$ are generalized coordinates that are the amplitudes of the $\psi_1(x)$ and $\psi_2(x)$ shapes; they are not the motion of two particles or even the motion of two specific points in the bar.] The only prescribed boundary condition is

$$u(0, t) = 0 \tag{2}$$

Thus, the shape functions $\psi_i(x)$ must satisfy

$$\psi_1(0) = \psi_2(0) = 0 \tag{3}$$

Therefore, to make Eq. 1 a two-term polynomial expression in x, let the two shape functions be

$$\psi_1(x) = \frac{x}{L}, \qquad \psi_2(x) = \left(\frac{x}{L}\right)^2 \tag{4}$$

(It is convenient, although not essential, to nondimensionalize the ψ functions, as has been done in Eqs. 4.) These two functions are illustrated in Fig. 2.

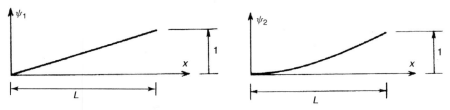

Figure 2 Two shape functions.

8.4 Application of Lagrange's Equations to Continuous Models: Assumed-Modes Method 233

2. Compute the stiffness matrix coefficients k_{ij}. The first derivatives of the two shape functions in Eqs. 4 are

$$\psi'_1(x) = \frac{1}{L}, \qquad \psi'_2(x) = \frac{2}{L}\frac{x}{L} \tag{5}$$

Then, from Eq. 8.16, the stiffness coefficients k_{ij} are

$$k_{11} = \int_0^L EA(\psi'_1)^2 \, dx = \frac{EA}{L} \tag{6a}$$

$$k_{12} = k_{21} = \int_0^L EA\psi'_1\psi'_2 \, dx = \frac{EA}{L} \tag{6b}$$

$$k_{22} = \int_0^L EA(\psi'_2)^2 \, dx = \frac{4EA}{3L} \tag{6c}$$

3. Similarly, compute the mass coefficients, m_{ij}. From Eq. 8.20,

$$m_{11} = \int_0^L \rho A(\psi_1)^2 \, dx = \frac{\rho AL}{3} \tag{7a}$$

$$m_{12} = m_{21} = \int_0^L \rho A\psi_1\psi_2 \, dx = \frac{\rho AL}{4} \tag{7b}$$

$$m_{22} = \int_0^L \rho A(\psi_2)^2 \, dx = \frac{\rho AL}{5} \tag{7c}$$

4. Compute the generalized forces, $p_i(t)$. Equation 8.24 was derived for the distributed forces shown in Fig. 8.4. To determine the generalized forces corresponding to the concentrated end force, $P(t)$, we can go back to virtual work and write

$$\delta W = P(t)\,\delta u(L,t) = p_1\,\delta u_1 + p_2\,\delta u_2 \tag{8}$$

But based on Eq. 1,

$$\delta u(L,t) = \psi_1(L)\,\delta u_1 + \psi_2(L)\,\delta u_2 \tag{9}$$

Therefore,

$$p_1(t) = P(t)\psi_1(L) = P(t), \qquad p_2(t) = P(t)\psi_2(L) = P(t) \tag{10}$$

5. Using the form of Eq. 8.26, assemble the equations of motion in matrix form:

$$\rho AL \begin{bmatrix} \frac{1}{3} & \frac{1}{4} \\ \frac{1}{4} & \frac{1}{5} \end{bmatrix} \begin{Bmatrix} \ddot{u}_1 \\ \ddot{u}_2 \end{Bmatrix} + \frac{EA}{L} \begin{bmatrix} 1 & 1 \\ 1 & \frac{4}{3} \end{bmatrix} \begin{Bmatrix} u_1 \\ u_2 \end{Bmatrix} = \begin{Bmatrix} P(t) \\ P(t) \end{Bmatrix} \qquad \textbf{Ans.} \tag{11}$$

8.4.4 Assumed-Modes Method: Bending of Bernoulli–Euler Beams

In the same manner, the Assumed-Modes Method may be used to derive the equations of motion of other linearly elastic systems. For the Bernoulli–Euler beam the strain energy is given by (e.g., see Section 11.3 of Ref. [8.4])

$$V = \tfrac{1}{2}\int_0^L EI(v'')^2 \, dx \tag{8.27}$$

where $v(x, t)$ refers to transverse displacement, as in Fig. 8.3, and where the flexural rigidity EI may be a function of x. The kinetic energy is given by

$$T = \tfrac{1}{2} \int_0^L \rho A (\dot{v})^2 \, dx \tag{8.28}$$

Let the assumed-modes expression for $v(x, t)$ be

$$\boxed{v(x, t) = \sum_{i=1}^{N} \psi_i(x) v_i(t)} \tag{8.29}$$

with the generalized coordinates labeled $v_i(t)$. When this expression is substituted into Eqs. 8.27 and 8.28, the following expressions for stiffness and mass coefficients are obtained:

$$\boxed{k_{ij} = \int_0^L EI \, \psi_i'' \psi_j'' \, dx} \tag{8.30}$$

$$\boxed{m_{ij} = \int_0^L \rho A \psi_i \psi_j \, dx} \tag{8.31}$$

The generalized forces due to a distributed transverse force $p(x, t)$, shown in Fig. 8.3, are given by

$$\boxed{p_i(t) = \int_0^L p_y(x, t) \psi_i(x) \, dx} \tag{8.32}$$

The next example illustrates how the Assumed-Modes Method can be used to obtain a 2-DOF model to describe the transverse vibration of a beamlike structure.

Example 8.10 An offshore oil-drilling platform is modeled as a uniform flexible beam of length L with a lumped mass M at the top and a rotational spring k at the bottom, as shown in Fig. 1. Use the *Assumed-Modes Method* to derive the equations of motion of a 2-DOF model of this structure. Assume only small rotation of the beam at $x = 0$.

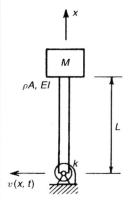

Figure 1 Beam–mass model of an offshore oil-drilling platform.

8.4 Application of Lagrange's Equations to Continuous Models: Assumed-Modes Method

SOLUTION 1. Let the assumed-modes expression for the transverse displacement be

$$v(x, t) = \psi_1(x)v_1(t) + \psi_2(x)v_2(t) \tag{1}$$

and choose the shape functions, $\psi_i(x)$. Since the "beam" is free to rotate at $x = 0$, the beam deflects as illustrated in Fig. 2. So the only prescribed boundary condition is

$$v(0, t) = 0 \tag{2}$$

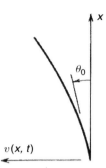

Figure 2 Deflected shape of the beam.

Thus, the shape functions $\psi_i(x)$ must satisfy

$$\psi_1(0) = \psi_2(0) = 0 \tag{3}$$

The simplest functions to use are terms of a polynomial in x. Therefore, as in Example 8.9, choose the following shape functions:

$$\psi_1(x) = \frac{x}{L}, \qquad \psi_2(x) = \left(\frac{x}{L}\right)^2 \tag{4a, b}$$

2. Compute the stiffness coefficients, k_{ij}. When the beam is deflected as shown in Fig. 2, there is potential energy stored in the base spring as well as in the beam. Equation 8.27 must be supplemented by the potential energy in the spring. Thus,

$$\mathcal{V} = \tfrac{1}{2}\int_0^L EI(v'')^2\, dx + \tfrac{1}{2}k\theta_0^2 \tag{5}$$

For small rotation, θ_0, at the base,

$$\theta_0(t) \approx v'(0, t) = \psi_1'(0)v_1(t) + \psi_2'(0)v_2(t) \tag{6}$$

When Eq. 6 is substituted into the second term of Eq. 5, we see that Eq. 8.30 is modified to the form

$$k_{ij} = \int_0^L EI\,\psi_i''\psi_j''\,dx + k\,\psi_i'(0)\psi_j'(0) \tag{7}$$

From Eqs. 4,

$$\psi_1' = \frac{1}{L}, \qquad \psi_1'' = 0 \tag{8a}$$

$$\psi_2' = \frac{2}{L}\frac{x}{L}, \qquad \psi_2'' = \frac{2}{L^2} \tag{8b}$$

Therefore, the stiffness coefficients for this 2-DOF model are

$$k_{11} = 0 + k\frac{1}{L}\frac{1}{L} = \frac{k}{L^2} \tag{9a}$$

$$k_{12} = k_{21} = 0 + 0 = 0 \tag{9b}$$

$$k_{22} = \frac{4EI}{L^3} + 0 = \frac{4EI}{L^3} \tag{9c}$$

3. Compute the mass coefficients, m_{ij}. Due to the presence of the mass M at $x = L$, the expression for the kinetic energy in Eq. 8.28 must be modified, which results in changing Eq. 8.31 to

$$m_{ij} = \int_0^L \rho A \psi_i \psi_j \, dx + M \psi_i(L) \psi_j(L) \tag{10}$$

Thus, the coefficients of the mass matrix for this 2-DOF model are

$$m_{11} = \rho A \int_0^L \left(\frac{x}{L}\right)^2 dx + M(1)(1) = \frac{\rho A L}{3} + M \tag{11a}$$

$$m_{12} = m_{21} = \frac{\rho A L}{4} + M \tag{11b}$$

$$m_{22} = \frac{\rho A L}{5} + M \tag{11c}$$

4. There are no applied external forces, so $p_1(t) = p_2(t) = 0$.
5. Assemble the equations of motion. There are no external forces, so the generalized forces are zero. That is, $p_1 = p_2 = 0$. Finally, the equations of motion for this model are, in matrix form,

$$\begin{bmatrix} \frac{\rho A L}{3} + M & \frac{\rho A L}{4} + M \\ \frac{\rho A L}{4} + M & \frac{\rho A L}{5} + M \end{bmatrix} \begin{Bmatrix} \ddot{v}_1 \\ \ddot{v}_2 \end{Bmatrix} + \begin{bmatrix} \frac{k}{L^2} & 0 \\ 0 & \frac{4EI}{L^3} \end{bmatrix} \begin{Bmatrix} v_1 \\ v_2 \end{Bmatrix} = \begin{Bmatrix} 0 \\ 0 \end{Bmatrix} \quad \text{Ans.} \tag{12}$$

8.4.5 Other Effects: Geometric Stiffness and Distributed Viscous Damping

In a situation like Example 8.10, where a member is subjected to axial loading (e.g., due to the weight of the platform mass M) and also undergoes transverse deflection, the axial load may have a significant effect on the bending stiffness of the member. An expression for the geometric stiffness coefficient for a 1-DOF model was given in Eq. 2.38. For an N-DOF system the corresponding expression for the *geometric stiffness coefficients* k_{Gij} is

$$\boxed{k_{Gij} = \int_0^L N(x) \psi_i'(x) \psi_j'(x) \, dx} \tag{8.33a}$$

If $N(x) = $ constant, Eq. 8.33a becomes

$$k_{Gij} = N \int_0^L \psi_i'(x) \psi_j'(x) \, dx \tag{8.33b}$$

8.4 Application of Lagrange's Equations to Continuous Models: Assumed-Modes Method

Example 8.11 As indicated in Fig. 1, the damping effect of soil on the vertical motion of a foundation beam may be modeled by distributed damping, such that a damping force proportional to the local velocity is exerted on the beam. That is, the distributed damping force has the form

$$p(x, t) = -\xi(x)\dot{v}(x, t)$$

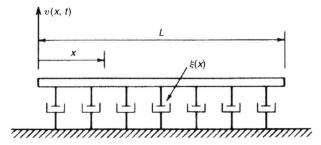

Figure 1 Distributed viscous damping acting on the beam.

where $\xi(x)$ is the distributed viscous-damping coefficient. Determine an expression for the generalized force p_i corresponding to this distributed damping force.

SOLUTION Equation 8.32 can be used directly. Thus,

$$\begin{aligned} p_i &= \int_0^L p(x,t)\psi_i(x)\,dx \\ &= \int_0^L \left[-\xi(x) \sum_{j=1}^N \psi_j(x)\dot{v}_j(t) \right] \psi_i(x)\,dx \end{aligned} \quad (1)$$

where the $\dot{v}_j(t)$ are generalized velocities, or

$$p_i = -\sum_{j=1}^N \dot{v}_j(t) \left[\int_0^L \xi(x)\psi_i(x)\psi_j(x)\,dx \right] \quad \textbf{Ans. (2)}$$

When a system is influenced by viscous damping, as in Example 8.11, it is possible to write the generalized forces corresponding to this as

$$p_i = -\sum_{j=1}^N c_{ij}\dot{v}_j(t) \tag{8.34}$$

where

$$\boxed{c_{ij} = \int_0^L \xi(x)\psi_i(x)\psi_j(x)\,dx} \tag{8.35}$$

Since these generalized forces depend on the unknown generalized velocities, the p_i due to viscous damping may be placed on the left-hand side of Eq. 8.25. Then Eq. 8.26

may be written in the following matrix form:

$$\mathbf{M\ddot{q} + C\dot{q} + Kq = p}(t) \tag{8.36}$$

where **C** is the *damping matrix* and the term $\mathbf{p}(t)$ contains external forces other than the viscous damping forces.

From a practical standpoint, the damping properties of a system are seldom known in the same way that the inertia and stiffness properties of the system are known, so it is generally not practical to seek the coefficients of the damping matrix **C**. Other procedures for treating damping are introduced in Section 10.3.

8.5 CONSTRAINED COORDINATES AND LAGRANGE MULTIPLIERS

In Section 8.2, Lagrange's Equations were derived using a set of N generalized coordinates, designated $q_1, q_2, \ldots, q_N$. In the derivation, a crucial step from Eq. 8.8 to Eq. 8.9 required that the coordinates be independent. Occasionally, it is desirable to employ a set of coordinates that are not independent. Let these be denoted by $g_1, g_2, \ldots, g_M$, where $M > N$. Associated with this set of *constrained coordinates* will be a set of $C = M - N$ constraint equations. Although texts on advanced dynamics handle more general types of constraints[8.3], it will suffice here for us to consider constraint equations that involve only the coordinates g_i (and not derivatives). Let these constraint equations be written in the form

$$f_j(g_1, g_2, \ldots, g_M) = 0, \qquad j = 1, 2, \ldots, C \tag{8.37}$$

Let each coordinate g_r be given a small variation δg_r. Then the first-order change in f_j has the form

$$\delta f_j = \frac{\partial f_j}{\partial g_1} \delta g_1 + \frac{\partial f_j}{\partial g_2} \delta g_2 + \cdots + \frac{\partial f_j}{\partial g_M} \delta g_M = 0 \tag{8.38}$$

or

$$\sum_{i=1}^{M} \frac{\partial f_j}{\partial g_i} \delta g_i = 0, \qquad j = 1, 2, \ldots, C \tag{8.39}$$

Thus, the δg's are not independent but are related to each other by the C linear *equations of constraint*, Eqs. 8.39.

Now let us return to Hamilton's Principle by extending Eq. 8.8 to include the M coordinates, that is,

$$\int_{t_1}^{t_2} \left\{ \sum_{i=1}^{M} \left[-\frac{d}{dt} \frac{\partial \mathcal{T}}{\partial \dot{g}_i} + \frac{\partial \mathcal{T}}{\partial g_i} - \frac{\partial \mathcal{V}}{\partial g_i} + p_i \right] \delta g_i \right\} dt = 0 \tag{8.40}$$

Since the δg's are not independent, we cannot just set the expression in brackets to zero as we did in Eq. 8.9. At this point, however, we can introduce Lagrange multiplier functions, or *Lagrange multipliers*, $\lambda_j(t)$, $j = 1, 2, \ldots, C$. Multiply each of the C equations in Eqs. 8.39 by a corresponding Lagrange multiplier, λ_j, and then sum these.

8.5 Constrained Coordinates and Lagrange Multipliers

That is, form the equation

$$\sum_{j=1}^{C} \lambda_j \sum_{i=1}^{M} \frac{\partial f_j}{\partial g_i} \delta g_i = 0 \tag{8.41}$$

Since this sum is still equal to zero, it may be inserted into Eq. 8.40 to give

$$\int_{t_1}^{t_2} \left\{ \sum_{i=1}^{M} \left[-\frac{d}{dt}\frac{\partial T}{\partial \dot{g}_i} + \frac{\partial T}{\partial g_i} - \frac{\partial \mathcal{V}}{\partial g_i} + p_i + \sum_{j=1}^{C} \lambda_j \frac{\partial f_j}{\partial g_i} \right] \delta g_i \right\} dt = 0 \tag{8.42}$$

Although the δg's are still not independent, we can choose the Lagrange multipliers λ_j so as to make the bracketed expressions for $\delta g_i (i = 1, 2, \ldots, C)$ equal to zero. The remaining $N = M - C$ coordinates can be considered to be independent, so the expression in brackets must also vanish for $\delta g_i (i = C+1, \ldots, M)$. Thus, the bracketed expression must vanish for all δg_i's, giving the following *Lagrange's Equations* modified for constrained coordinates:

$$\boxed{\frac{d}{dt}\frac{\partial T}{\partial \dot{g}_i} - \frac{\partial T}{\partial g_i} + \frac{\partial \mathcal{V}}{\partial g_i} - \sum_{j=1}^{C} \lambda_j \frac{\partial f_j}{\partial g_i} = p_i, \quad i = 1, 2, \ldots, M} \tag{8.43}$$

Alternatively, Lagrange's Equations may be written in the form

$$\frac{d}{dt}\frac{\partial T}{\partial \dot{g}_i} - \frac{\partial T}{\partial g_i} + \frac{\partial \mathcal{V}^*}{\partial g_i} = p_i, \quad i = 1, 2, \ldots, M \tag{8.44}$$

where $\mathcal{V}^*$ is a modified potential energy function given by

$$\mathcal{V}^* = \mathcal{V} - \sum_{j=1}^{C} \lambda_j f_j \tag{8.45}$$

In conclusion, it can be seen that Eqs. 8.37 and 8.43 (or 8.44) provide a set of $M + C$ equations in the $M + C$ unknowns g_i and λ_j. A very simple example of the use of Lagrange multipliers will now be given. A more sophisticated example is given in Section 17.5.

Example 8.12 Use the Assumed-Modes Method with Lagrange multipliers to derive an equation of motion for approximating axial free vibration of the fixed–fixed bar shown in Fig. 1. Begin with the two-term polynomial

$$u(x, t) = \frac{x}{L} g_1 + \left(\frac{x}{L}\right)^2 g_2$$

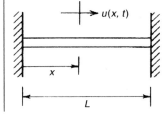

Figure 1 Fixed–fixed bar undergoing axial free vibration.

SOLUTION 1. Establish the constraint equation. The prescribed boundary conditions for the bar are
$$u(0, t) = u(L, t) = 0 \qquad (1)$$
The fixed-end constraint at $x = 0$ is satisfied by each of the given shape functions, x/L and $(x/L)^2$. We must enforce the boundary condition at $x = L$ by setting
$$f(g_1, g_2) \equiv u(L, t) = g_1 + g_2 = 0 \qquad (2)$$
This constraint equation has the form shown in Eq. 8.37.

2. Determine the mass and stiffness coefficients. Since the structure is the same as in Example 8.9, and the two shape functions are the same as the ones used there, we can use the stiffness coefficients, k_{ij}, and the mass coefficients, m_{ij}, derived there.

$$\mathbf{K} = \frac{EA}{L} \begin{bmatrix} 1 & 1 \\ 1 & \frac{4}{3} \end{bmatrix}, \qquad \mathbf{M} = \rho A L \begin{bmatrix} \frac{1}{3} & \frac{1}{4} \\ \frac{1}{4} & \frac{1}{5} \end{bmatrix} \qquad (3)$$

3. Use Eq. 8.43 to obtain the equations of motion. Since there is only one constraint equation, we need only to append $\lambda(\partial f/\partial g_1)$ and $\lambda(\partial f/\partial g_2)$ to their respective equations. But from Eq. 2,
$$\frac{\partial f}{\partial g_1} = \frac{\partial f}{\partial g_2} = 1 \qquad (4)$$
Therefore, the constrained equations of motion, which correspond to Eqs. 8.43, become

$$\rho A L \begin{bmatrix} \frac{1}{3} & \frac{1}{4} \\ \frac{1}{4} & \frac{1}{5} \end{bmatrix} \begin{Bmatrix} \ddot{g}_1 \\ \ddot{g}_2 \end{Bmatrix} + \frac{EA}{L} \begin{bmatrix} 1 & 1 \\ 1 & \frac{4}{3} \end{bmatrix} \begin{Bmatrix} g_1 \\ g_2 \end{Bmatrix} - \begin{Bmatrix} \lambda \\ \lambda \end{Bmatrix} = \begin{Bmatrix} 0 \\ 0 \end{Bmatrix} \qquad (5)$$

The coordinates g_1 and g_2 in Eq. 5 are not independent but are related to each other by the constraint equation, Eq. 2. However, by eliminating g_2 and λ from Eqs. 2 and 5, a single differential equation in g_1 can be obtained. Thus, g_1 becomes the single independent coordinate, and the corresponding differential equation is

$$\frac{\rho A L}{10} \ddot{g}_1 + \frac{EA}{L} g_1 = 0 \qquad \textbf{Ans. (6)}$$

From Eq. 2 we have $g_2 = -g_1$, so $u(x, t)$ has the form
$$u(x, t) = \left[\frac{x}{L} - \left(\frac{x}{L}\right)^2 \right] g_1(t) \qquad (7)$$
with $g_1(t)$ determined by Eq. 6.

REFERENCES

[8.1] H. L. Langhaar, *Energy Methods in Applied Mechanics*, Wiley, New York, 1962.
[8.2] L. Meirovitch, *Fundamentals of Vibrations*, McGraw-Hill, New York, 2001.
[8.3] H. Goldstein, *Classical Mechanics*, Addison-Wesley, Reading, MA, 1950.
[8.4] R. R. Craig, Jr., *Mechanics of Materials*, 2nd ed., Wiley, New York, 2000.

PROBLEMS
Problem Set 8.1

> For *Problems 8.1 through 8.14*, use *Newton's Laws* (with extension to rigid bodies where required) to determine the equations of motion of the MDOF systems below. Show all necessary free-body diagrams for each problem.

8.1 For the two-mass system in Fig. P8.1, derive the equations of motion, and write the equations in matrix form. **(a)** Let coordinate u_1 be the absolute motion of mass 1, and let coordinate u_2 be the absolute motion of mass 2. **(b)** Let coordinate u_1 be the absolute motion of mass 1, and let coordinate u_2 be the motion of mass 2 relative to mass 1. **(c)** Discuss briefly how your answer to part (b) differs from your answer to part (a).

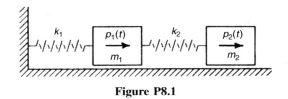

Figure P8.1

8.2 For the mass–pendulum system in Fig. P8.2, derive the equations of motion. Let coordinate u be the absolute motion of mass m_1, and let coordinate θ be the angle that the pendulum makes with the vertical. Assume that angle θ can be "arbitrarily large"; that is, $-(\pi/2) \leq \theta \leq \pi/2$.

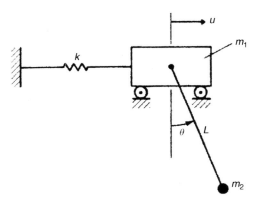

Figure P8.2

8.3 The motion of an airplane wing section of mass m and mass moment of inertia I_G is modeled by the 2-DOF system shown in Fig. P8.3. Use the vertical displacement of point C and the angle of rotation (i.e., the pitch) of the wing as coordinates u and θ, respectively. Assume a small angle of rotation, and neglect gravity. Write the equations of motion in matrix format.

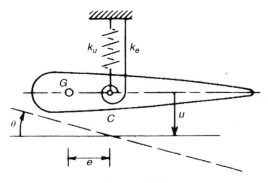

Figure P8.3

8.4 For the bar–spring–mass system in Fig. P8.4, assume that the bar is rigid but its mass is negligible, that the rotation of the bar remains small, and that gravity can be neglected. In parts (a) and (b), derive the equations of motion, and write the equations in matrix form. **(a)** Let coordinates u_1 and u_2 be the vertical displacements of the masses m_1 and m_2, respectively. **(b)** Let coordinates u_1 and u_2 be the vertical displacements of points A and C where springs k_1 and k_2, respectively, are connected to the rigid bar. **(c)** Briefly discuss how the equations of motion are coupled in each case.

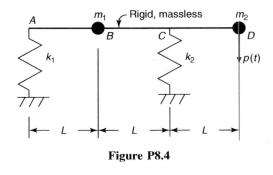

Figure P8.4

8.5 The three-story "shear building" in Fig. P8.5 (the various floors are rigid masses; the columns are flexible, but massless) is subjected to earthquake motion (i.e., base motion) in addition to wind forces. Derive the equations of motion, and write the equations in matrix form. Use the displacements of the concentrated floor masses relative to the ground as the three coordinates $w_1 \triangleq u_1 - z(t)$, $w_2 \triangleq u_2 - z(t)$, and $w_3 \triangleq u_3 - z(t)$.

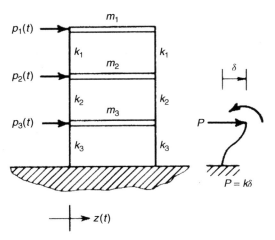

Figure P8.5

8.6 A vehicle is modeled in Fig. P8.6 as a rigid body AB having mass M and mass moment of inertia I_G about its mass center. The mass of the axles and wheels is modeled by lumped masses m, the stiffness of the springs by k_1, and the stiffness of the tires by k_2. The shock absorbers are modeled by viscous dashpots with damping coefficient c, as shown. Use as coordinates u_i, $i = 1$ to 4 the vertical displacements of the smaller masses and the displacements at A and B, as shown in Fig. P8.6. Assume a small angle of rotation of the rigid body AB, and write the equations of motion in matrix format.

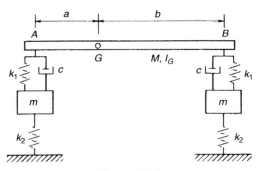

Figure P8.6

8.7 A dynamical system is modeled as two slender, rigid beams AC and DE, having masses M_1 and M_2, respectively. The beams are supported on frictionless pin supports at B and E, respectively, and by the two springs and the dashpot shown in Fig. P8.7. Use as coordinates the beam rotation angles θ_1 and θ_2, with θ_1 being taken positive in a clockwise sense and θ_2 being taken positive in a counterclockwise sense. Assume only small angle of rotation of each of the two beams, and write the equations of motion in matrix format.

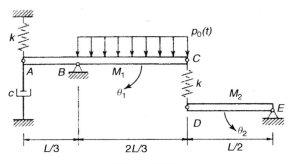

Figure P8.7

8.8 The rotating imbalance of the machine shown in Fig. P8.8 causes it to move vertically on its two spring supports, each of which has a spring constant k_1. The mass of the machine is m_1, and the rotating imbalance has mass m. A mass m_2 attached to the machine mass m_1 by a spring k_2 can be used as a vibration absorber. First determine the form of the vertical force $P(t)$ exerted at C by the arm of the rotating imbalance if the arm rotates counterclockwise at Ω rad/s, as indicated. Then, with vertical displacements u_1 and u_2 as coordinates, determine the equations of motion for vertical motion of masses m_1 and m_2. Neglect gravity.

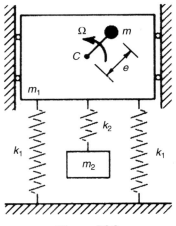

Figure P8.8

8.9 As shown in Fig. P8.9, a coupled-pendulum system is modeled by two lumped masses attached to rigid weightless rods, which are connected by a spring whose stiffness is k. With the angles θ_1 and θ_2 as coordinates, derive the equations of motion, and write them in matrix format.

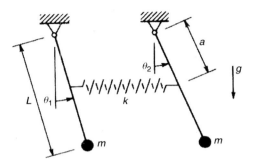

Figure P8.9

8.10 The 2-DOF system in Fig. P8.10 is a model of a high-speed monorail train whose two cars are coupled by a spring and dashpot. Use as coordinates the absolute motion of mass m_1 and the motion of mass m_2 relative to m_1. Write the equations of motion in matrix format.

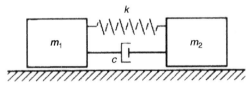

Figure P8.10

8.11 The 3-DOF system in Fig. P8.11 is a model of a high-speed monorail train whose three cars are coupled by springs and dashpots. Derive the equations of motion for this three-car system, and write the equations in matrix format. (a) Use as coordinates u_1 through u_3 the absolute motion of the three respective masses. (b) Use as coordinate u_1 the absolute motion of mass m_1, and for coordinates u_2 and u_3, respectively, the motion of masses m_2 and m_3 relative to m_1.

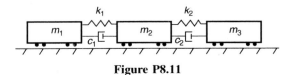

Figure P8.11

8.12 As shown in Fig. P8.12, two gears are attached to a uniform shaft whose torsional rigidity is GJ. Using θ_1 and θ_2, the absolute angles of rotation of the gears, as coordinates, determine the equations of motion of this system. Neglect the rotational inertia of the shaft, and write the two equations of motion in matrix format.

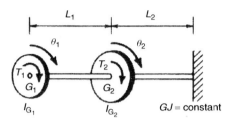

Figure P8.12

8.13 Two gears are attached to a uniform shaft whose torsional rigidity is GJ. The shaft is free to rotate in frictionless bearings, as shown in Fig. P8.13, but there is an external moment T_i applied to each respective gear. Neglect the rotational inertia of the shaft, and write the two equations of motion in matrix format. (a) Using θ_1 and θ_2, the absolute angles of rotation of the gears, as coordinates, determine the equations of motion of this system. (b) Use as coordinate q_1 the absolute rotation (i.e., θ_1) of gear 1, and for coordinate q_2 the rotation of gear 2 relative to gear 1 (i.e., $\theta_2 - \theta_1$).

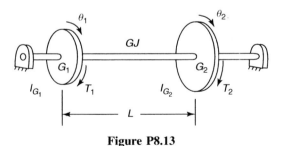

Figure P8.13

8.14 A restaurant and observation deck are situated atop a tall tower. The system is modeled as a lumped mass m supported by a rigid, massless column of length L. The elasticity of the tower and of the soil are modeled by springs, as shown in Fig. P8.14. The base is subjected to horizontal excitation, $z(t)$. Include the gravitational effect of the "restaurant" mass, and employ as coordinates the horizontal motion u of the "foundation" mass M and the angle θ that the column makes with the vertical. Assume that the angle θ remains small.

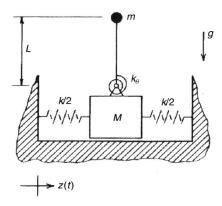

Figure P8.14

Problem Set 8.3

> Write appropriate expressions for the kinetic energy, T, the potential energy, V, and the virtual work of nonconservative forces, δW_{nc}, for the MDOF systems in *Problems 8.15 through 8.26*. Use the set(s) of coordinates specified in the problem statement, and use *Lagrange's Equations* to derive the corresponding equations of motion.

8.15 For the system described in Problem 8.1 and shown in Fig. P8.1, write the equations of motion in matrix format.

8.16 For the system described in Problem 8.2 and shown in Fig. P8.2, write the equations of motion in matrix format.

8.17 For the system described in Problem 8.3 and shown in Fig. P8.3, write the equations of motion in matrix format.

8.18 For the system described in Problem 8.4 and shown in Fig. P8.4, write the equations of motion in matrix format.

8.19 For the system described in Problem 8.5 and shown in Fig. P8.5, write the equations of motion in matrix format.

8.20 For the system described in Problem 8.6 and shown in Fig. P8.6, write the equations of motion in matrix format.

8.21 For the system described in Problem 8.7 and shown in Fig. P8.7, write the equations of motion in matrix format.

8.22 For the system described in Problem 8.10 and shown in Fig. P8.10, write the equations of motion in matrix format.

8.23 For the system described in Problem 8.11 and shown in Fig. P8.11, write the equations of motion in matrix format.

8.24 For the system described in Problem 8.12 and shown in Fig. P8.12, write the equations of motion in matrix format.

8.25 For the system described in Problem 8.13 and shown in Fig. P8.13, write the equations of motion in matrix format.

8.26 For the system described in Problem 8.14 and shown in Fig. P8.14, write the equations of motion in matrix format.

8.27 A double pendulum is modeled by lumped masses and massless, rigid rods, as shown in Fig. P8.27. Neglect friction. The force $P(t)$ remains horizontal. Use θ_1 and θ_2 (not necessarily small) as the system generalized coordinates. See the general instructions for Problem Set 8.3.

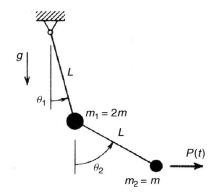

Figure P8.27

Problem Set 8.4

> For *Problems 8.28 through 8.37*, let the mass density be ρ, and let the modulus of elasticity and shear modulus be E and G, respectively.

8.28 The uniform bar in Fig. P8.28 is attached at both ends to rigid supports, and it undergoes axial deformation $u(x, t)$. **(a)** If $u(x, t)$ is approximated by

$$u(x, t) = \psi_1(x)q_1(t) + \psi_2(x)q_2(t)$$

show that the following shape functions satisfy the prescribed boundary conditions:

$$\psi_1(x) = \sin\frac{\pi x}{L}, \qquad \psi_2(x) = \sin\frac{2\pi x}{L}$$

(b) Derive the equations of motion for a 2-DOF assumed-modes model based on the shape functions above. Write the equations of motion in matrix format: $\mathbf{M\ddot{q} + Kq = p}(t)$.

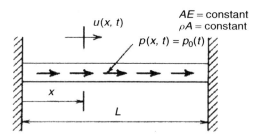

Figure P8.28

8.29 The uniform bar in Fig. P8.28 is attached at both ends to rigid supports, and it undergoes axial deformation $u(x, t)$. Let $u(x, t)$ be approximated by

$$u(x, t) = \psi_1(x)q_1(t) + \psi_2(x)q_2(t)$$

(a) Determine two shape functions $\psi_1(x)$ and $\psi_2(x)$ based on a polynomial in x of the form

$$\psi(x) = a + b\frac{x}{L} + c\left(\frac{x}{L}\right)^2 + d\left(\frac{x}{L}\right)^3$$

and the two prescribed boundary conditions. **(b)** Derive the equations of motion for a 2-DOF assumed-modes model based on the shape functions from part (a), and write the equations of motion in matrix format: $\mathbf{M\ddot{q} + Kq = p}(t)$.

8.30 The uniform bar in Fig. P8.30 is attached to a rigid support at $x = 0$, and it is attached to a point mass and a linear spring at end $x = L$. The bar undergoes axial deformation $u(x, t)$ that is to be approximated by

$$u(x, t) = \psi_1(x)q_1(t) + \psi_2(x)q_2(t)$$

where

$$\psi_1(x) = \frac{x}{L}, \qquad \psi_2(x) = \left(\frac{x}{L}\right)^2$$

Derive the equations of motion for a 2-DOF assumed-modes model based on the shape functions above. Write the equations of motion in matrix format: $\mathbf{M\ddot{q} + Kq = p}(t)$.

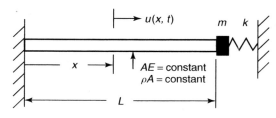

Figure P8.30

8.31 As depicted in Fig. P8.31, a uniform floor beam has a total mass $M = \rho AL$, and it supports a heavy machine of mass m at its midpoint $x = L/2$. Neglect the width of the machine in comparison with length L. Let the transverse deflection $v(x, t)$ be approximated by

$$v(x, t) = \psi_1(x)q_1(t) + \psi_2(x)q_2(t)$$

where the shape functions are

$$\psi_1(x) = \sin\frac{\pi x}{L}, \qquad \psi_2(x) = \sin\frac{2\pi x}{L}$$

Derive the equations of motion for a 2-DOF assumed-modes model based on these shape functions, and write the equations of motion in matrix format: $\mathbf{M\ddot{q} + Kq = p}(t)$.

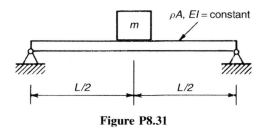

Figure P8.31

8.32 As depicted in Fig. P8.32, a uniform floor beam has a total mass $M = \rho AL$, and it supports a heavy machine of mass m at $x = L/4$. Neglect the width of the machine in comparison with length L. Let the transverse deflection $v(x, t)$ be approximated by

$$v(x, t) = \psi_1(x)q_1(t) + \psi_2(x)q_2(t)$$

where the shape functions are

$$\psi_1(x) = \sin\frac{\pi x}{L}, \qquad \psi_2(x) = \sin\frac{2\pi x}{L}$$

Derive the equations of motion for a 2-DOF assumed-modes model based on these shape functions, and write the equations of motion in matrix format: $\mathbf{M\ddot{q}} + \mathbf{Kq} = \mathbf{p}(t)$.

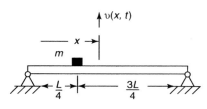

Figure P8.32

8.33 A uniform beam is subjected to a constant horizontal compressive force N and to a uniform, time-dependent downward force $p(x, t) = p_0(t)$, as shown in Fig. P8.33. Using a 2-DOF assumed-modes model based on a polynomial in x, derive the two equations of motion for transverse vibration having the form

$$v(x, t) = \psi_1(x)q_1(t) + \psi_2(x)q_2(t)$$

Write the two equations of motion in matrix format: $\mathbf{M\ddot{q}} + \mathbf{Kq} = \mathbf{p}(t)$.

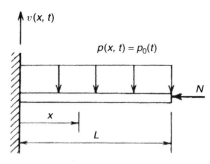

Figure P8.33

8.34 The uniform cantilever shaft in Fig. P8.34 has a cross-sectional area A and a torsional stiffness GJ, and a thin disk of polar moment of inertia I_A is attached to the shaft at end A. Torsional vibration of the shaft–disk system is to be represented by a 2-DOF assumed-modes model of the form

$$\theta(x, t) = \psi_1(x)q_1(t) + \psi_2(x)q_2(t)$$

where the shape functions are assumed to be

$$\psi_1(x) = 1 - \frac{x}{L}, \qquad \psi_2(x) = 1 - \left(\frac{x}{L}\right)^2$$

Derive the equations of motion for a 2-DOF assumed-modes model based on these shape functions, and write the equations of motion in matrix format: $\mathbf{M\ddot{q}} + \mathbf{Kq} = \mathbf{p}(t)$.

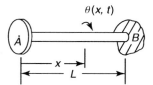

Figure P8.34

8.35 The cantilever shaft in Fig. P8.35 has a torsional stiffness GJ and is subjected to a time-dependent end torque $T_L(t)$. Forced vibration of the shaft is to be represented by a 2-DOF assumed-modes model of the form

$$\theta(x, t) = \psi_1(x)q_1(t) + \psi_2(x)q_2(t)$$

where the shape functions are assumed to be

$$\psi_1(x) = \sin\frac{\pi x}{2L}, \qquad \psi_2(x) = \sin\frac{3\pi x}{2L}$$

Derive the equations of motion for a 2-DOF assumed-modes model based on these shape functions, and write the equations of motion in matrix format: $\mathbf{M\ddot{q}} + \mathbf{Kq} = \mathbf{p}(t)$.

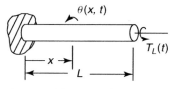

Figure P8.35

8.36 As depicted in Fig. P8.36, a cantilever beam is to be modeled by a 2-DOF assumed-modes model for which the generalized coordinates can be interpreted as the deflection and slope (small) at the free end. That is, $v(x, t)$ is to be approximated by

$$v(x, t) = \psi_1(x)q_1(t) + \psi_2(x)q_2(t)$$

where

$$q_1(t) \equiv v(L, t), \qquad q_2(t) \equiv v'(L, t)$$

The corresponding shape functions should have the shapes shown below. (a) Derive polynomial shape functions $\psi_1(x)$ and $\psi_2(x)$ based on a polynomial in x of the form

$$\psi(x) = a + b\frac{x}{L} + c\left(\frac{x}{L}\right)^2 + d\left(\frac{x}{L}\right)^3$$

(b) Derive the equations of motion for this 2-DOF model, and write the equations of motion in matrix format: $\mathbf{M\ddot{q} + Kq = p}(t)$.

8.37 A large sign is modeled as a uniform rigid rectangular plate of mass M that is welded atop a uniform flexible pole (EI = constant, ρA = constant). Assume that the sign deflects laterally, as depicted in Fig. P8.37. The transverse deflection of the sign is to be approximated by the assumed-modes expression

$$v(x, t) = \psi_1(x)q_1(t) + \psi_2(x)q_2(t)$$

where the shape functions are

$$\psi_1 = 3\left(\frac{x}{L}\right)^2 - 2\left(\frac{x}{L}\right)^3, \quad \psi_2 = -L\left(\frac{x}{L}\right)^2 + L\left(\frac{x}{L}\right)^3$$

With these as the shape functions, the corresponding coordinates are the lateral translation at B, $q_1(t) \equiv v(L, t) = V(t)$, and the (small) slope at B, $q_2(t) \equiv v'(L, t)$. Determine the equations of motion for this 2-DOF model of the sign–pole system, and write the equations of motion in matrix format: $\mathbf{M\ddot{q} + Kq = p}(t)$. *Note*: Since the sign is assumed to be rigid, it is to be treated as a rigid body that rotates in-plane through a (small) angle $\theta(t) \equiv v'(L, t)$, and whose mass center G translates horizontally.

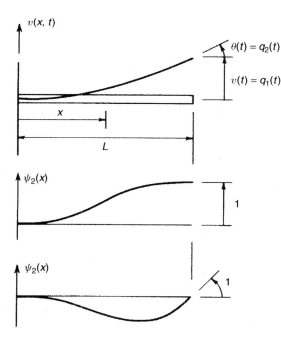

Figure P8.36

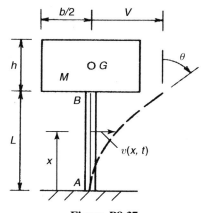

Figure P8.37

9

Vibration of Undamped 2-DOF Systems

Having the equations of motion of MDOF systems, as derived in Chapter 8, we could proceed immediately to a study of the dynamic response of systems with many degrees of freedom. It is very instructive, however, to begin this study with <u>undamped systems having only two degrees of freedom</u> (2 DOF). In Chapter 10 we extend the discussion to general MDOF systems.

The equation of motion of an undamped 2-DOF vibrating system has the form

$$\begin{bmatrix} m_{11} & m_{12} \\ m_{21} & m_{22} \end{bmatrix} \begin{Bmatrix} \ddot{u}_1 \\ \ddot{u}_2 \end{Bmatrix} + \begin{bmatrix} k_{11} & k_{12} \\ k_{21} & k_{22} \end{bmatrix} \begin{Bmatrix} u_1 \\ u_2 \end{Bmatrix} = \begin{Bmatrix} p_1(t) \\ p_2(t) \end{Bmatrix} \quad (9.1)$$

where the coordinates u_i may be physical displacements or may be generalized coordinates. The solution to Eq. 9.1 will consist of a complementary solution plus a particular solution. The complementary solution, obtained by setting the right-hand side of Eq. 9.1 to zero, involves the dynamical properties of the system called *natural frequencies* and *natural modes*. By examining the response of the 2-DOF system to harmonic excitation, you will be introduced to another very important topic in structural dynamics: namely, *mode superposition*.

Upon completion of this chapter you should be able to:

- Calculate the natural frequencies and natural modes of a 2-DOF system.
- Write expressions for the response of a 2-DOF system to nonzero initial conditions.
- Describe the beat phenomenon and discuss why it occurs.
- Calculate the natural frequencies and natural modes of 2- or 3-DOF systems having rigid-body modes.
- Determine the modal matrix, the modal stiffness matrix, and the modal mass matrix of a 2-DOF system.
- Write expressions for the steady-state frequency response of a 2-DOF system both in principal coordinates and in physical coordinates.
- Discuss how an undamped vibration absorber works.

9.1 FREE VIBRATION OF 2-DOF SYSTEMS: NATURAL FREQUENCIES AND MODE SHAPES

We now consider the solution of the equations of motion of a 2-DOF system. The following steps may be followed in obtaining the solution for *free vibration*, that is, the solution of the set of equations

$$\begin{bmatrix} m_{11} & m_{12} \\ m_{21} & m_{22} \end{bmatrix} \begin{Bmatrix} \ddot{u}_1 \\ \ddot{u}_2 \end{Bmatrix} + \begin{bmatrix} k_{11} & k_{12} \\ k_{21} & k_{22} \end{bmatrix} \begin{Bmatrix} u_1 \\ u_2 \end{Bmatrix} = \begin{Bmatrix} 0 \\ 0 \end{Bmatrix} \quad (9.2)$$

1. Assume that the system undergoes *harmonic motion* of the form

$$\begin{aligned} u_1(t) &= U_1 \cos(\omega t - \alpha) \\ u_2(t) &= U_2 \cos(\omega t - \alpha) \end{aligned} \quad (9.3)$$

where U_1 and U_2 are signed constants that determine the amplitudes of the two respective sinusoidal motions.

2. Substitute this assumed solution into the equations of motion to obtain the following *algebraic eigenvalue problem*:

$$\left[\begin{bmatrix} k_{11} & k_{12} \\ k_{21} & k_{22} \end{bmatrix} - \omega^2 \begin{bmatrix} m_{11} & m_{12} \\ m_{21} & m_{22} \end{bmatrix} \right] \begin{Bmatrix} U_1 \\ U_2 \end{Bmatrix} = \begin{Bmatrix} 0 \\ 0 \end{Bmatrix} \quad (9.4)$$

3. Since this is a set of homogeneous linear algebraic equations, the only nontrivial solutions of Eq. 9.4 correspond to values of ω^2 that satisfy the *characteristic equation*

$$\left| \begin{bmatrix} k_{11} & k_{12} \\ k_{21} & k_{22} \end{bmatrix} - \omega_i^2 \begin{bmatrix} m_{11} & m_{12} \\ m_{21} & m_{22} \end{bmatrix} \right| = 0 \quad (9.5)$$

that is, values of ω_i^2 for which the determinant of the coefficients of Eq. 9.4 is equal to zero. This is a polynomial equation of order 2 in ω_i^2.

4. Solve for the two roots of the characteristic equation. Label these roots ω_1^2 and ω_2^2 with $\omega_1 \leq \omega_2$. (In mathematical terminology ω_1^2 and ω_2^2 are called *eigenvalues*.) The parameters ω_1 and ω_2 are called the *circular natural frequencies* (*circular*, since they are expressed in the units rad/sec; *natural*, since there is no excitation of the system; and *frequencies*, since they are the rate of oscillation of the harmonic motion in Eq. 9.3). Although ω_1 and ω_2 are sometimes called the *natural frequencies* of the system, that name is usually reserved for the natural frequencies in hertz (= cycles/s),

$$f_1 = \frac{\omega_1}{2\pi}, \quad f_2 = \frac{\omega_2}{2\pi}$$

5. Substitute ω_1^2 back into the first (or the second, but not both) of the equations in Eq. 9.4 and obtain the ratio $\beta_1 \equiv (U_2/U_1)^{(1)}$. This ratio defines the *natural mode*,

or *mode shape*, corresponding to the natural frequency ω_1. Do the same for ω_2^2 to determine $\beta_2 \equiv (U_2/U_1)^{(2)}$. In mathematical terminology natural modes are called *eigenvectors*. (The natural modes for an undamped system are sometimes referred to as *real modes* to distinguish them from complex modes, which can occur in some damped linear systems. See Section 10.4.)

The following notation is commonly used to characterize the mode shapes of MDOF systems. For example, for a 2-DOF system, the rth mode shape would be designated as $\boldsymbol{\phi}_r$ and written as follows:

$$\boldsymbol{\phi}_r \equiv \begin{Bmatrix} \phi_1 \\ \phi_2 \end{Bmatrix}_r = A_r \begin{Bmatrix} 1 \\ \beta_r \end{Bmatrix}, \qquad r = 1, 2 \tag{9.6}$$

Finally, a sketch is made of the natural modes.

6. Free vibration can occur only at frequencies ω_1 and ω_2 as determined in step 4, and with the respective ratios β_1 and β_2 determined in step 5. The *general solution* of Eq. 9.2 is a linear combination of these two and may be written in the following form:

$$\begin{aligned} u_1(t) &= A_1 \cos(\omega_1 t - \alpha_1) + A_2 \cos(\omega_2 t - \alpha_2) \\ u_2(t) &= \beta_1 A_1 \cos(\omega_1 t - \alpha_1) + \beta_2 A_2 \cos(\omega_2 t - \alpha_2) \end{aligned} \tag{9.7}$$

The constants A_1, A_2, α_1, and α_2 are determined from initial conditions. An alternative form of the general solution is

$$\begin{aligned} u_1(t) &= A_1 \cos \omega_1 t + B_1 \sin \omega_1 t + A_2 \cos \omega_2 t + B_2 \sin \omega_2 t \\ u_2(t) &= \beta_1 A_1 \cos \omega_1 t + \beta_1 B_1 \sin \omega_1 t + \beta_2 A_2 \cos \omega_2 t + \beta_2 B_2 \sin \omega_2 t \end{aligned} \tag{9.8}$$

where the constants A_1, A_2, B_1, and B_2 are determined from initial conditions. Alternatively, in the vector notation of Eq. 9.6, Eq. 9.8 has the form

$$\mathbf{u}(t) \equiv \begin{Bmatrix} u_1(t) \\ u_2(t) \end{Bmatrix} = A_1 \boldsymbol{\phi}_1 \cos \omega_1 t + B_1 \boldsymbol{\phi}_1 \sin \omega_1 t + A_2 \boldsymbol{\phi}_2 \cos \omega_2 t + B_2 \boldsymbol{\phi}_2 \sin \omega_2 t \tag{9.9}$$

where, again, the constants A_1, A_2, B_1, and B_2 are determined from initial conditions.

The determination of the natural frequencies and mode shapes of vibrating structures is, without doubt, the most important exercise in structural dynamics analysis. Example 9.1 illustrates how to calculate frequencies and mode shapes of a 2-DOF system; Example 9.2 then illustrates how to determine the response of a 2-DOF system to *initial conditions*, that is, to nonzero values of one or more of the following: $u_1(0)$, $u_2(0)$, $\dot{u}_1(0)$, and $\dot{u}_2(0)$.

9.1.1 Eigensolution Example: Natural Frequencies and Mode Shapes

Example 9.1 Obtain the natural frequencies and mode shapes of the 2-DOF system shown in Fig. 1. The equations of motion for free vibration are

$$m\ddot{u}_1 + 2ku_1 - ku_2 = 0 \tag{1a}$$

$$m\ddot{u}_2 - ku_1 + 2ku_2 = 0 \tag{1b}$$

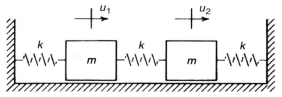

Figure 1 Symmetric 2-DOF spring–mass system.

SOLUTION 1. Assume the harmonic solution

$$u_1(t) = U_1 \cos(\omega_i t - \alpha) \tag{2a}$$

$$u_2(t) = U_2 \cos(\omega_i t - \alpha) \tag{2b}$$

Note that this means that when the 2-DOF system vibrates at natural frequency ω_i, u_1 and u_2 have the same time dependence, and at all times, the amplitude ratio of $u_2/u_1 = U_2/U_1$ has the same value.

2. Substitute Eqs. 2 into Eqs. 1 to get the algebraic eigenvalue problem

$$\begin{bmatrix} 2k - m\omega_i^2 & -k \\ -k & 2k - m\omega_i^2 \end{bmatrix} \begin{Bmatrix} U_1 \\ U_2 \end{Bmatrix} = \begin{Bmatrix} 0 \\ 0 \end{Bmatrix} \tag{3}$$

3. Set up the characteristic equation, which is the determinant of the coefficient matrix in Eq. 3.

$$\begin{vmatrix} 2k - m\omega_i^2 & -k \\ -k & 2k - m\omega_i^2 \end{vmatrix} = 0 \tag{4}$$

Expand the determinant to get the characteristic equation

$$(2k - m\omega_i^2)^2 - k^2 = 0$$

or

$$m^2\omega_i^4 - 4km\omega_i^2 + 3k^2 = 0 \tag{5}$$

4. Obtain the roots of the characteristic equation. In the present case the equation can be factored easily to obtain

$$(m\omega_i^2 - k)(m\omega_i^2 - 3k) = 0$$

so

$$\omega_1^2 = \frac{k}{m}, \qquad \omega_2^2 = \frac{3k}{m} \qquad (6a,b)$$

The (circular) natural frequencies are thus

$$\omega_1 = \left(\frac{k}{m}\right)^{1/2}, \qquad \omega_2 = \left(\frac{3k}{m}\right)^{1/2} \qquad \textbf{Ans. } (7a,b)$$

Note that the lower frequency is labeled ω_1.

5. Substitute the eigenvalues given by Eqs. 6 into the first of Eqs. 3 to obtain the mode-shape ratios.

$$(2k - m\omega_i^2)U_1^{(i)} - kU_2^{(i)} = 0$$

or

$$\beta_i \equiv \left(\frac{U_2}{U_1}\right)^{(i)} = \frac{2k - m\omega_i^2}{k} = 2 - \frac{m\omega_i^2}{k} \qquad (8)$$

Therefore, the mode-shape ratios for the two modes are

$$\beta_1 = 2 - 1 = 1, \qquad \beta_2 = 2 - 3 = -1 \qquad \textbf{Ans. } (9a,b)$$

The mode shapes are sketched in Fig. 2, employing the notation of Eq. 9.6.

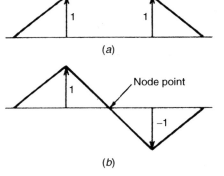

Figure 2 Mode shapes of a symmetric 2-DOF system: (a) mode 1: $\omega_1 = (k/m)^{1/2}$, $\boldsymbol{\phi}_1 = \begin{Bmatrix} 1 \\ 1 \end{Bmatrix}$; (b) mode 2: $\omega_2 = (3k/m)^{1/2}$, $\boldsymbol{\phi}_2 = \begin{Bmatrix} 1 \\ -1 \end{Bmatrix}$.

Notice that the original system is symmetric about the center of the middle spring. For this reason the mode shapes turn out to be a *symmetric mode* (mode 1) and an *antisymmetric mode* (mode 2). This is an important result, since many structures possess such physical symmetry. For example, an airplane is symmetric (nominally) about a vertical plane through the axis of the fuselage.

Also note that mode 2 has a *node point*, that is, a point that always remains stationary.

9.1.2 Free-Vibration Response Example

The general solution for free vibration of an undamped 2-DOF system is given by Eq. 9.7, 9.8, or 9.9. Note that in any case the motion is a combination of motion of the system with mode shape 1 at frequency ω_1 plus motion with mode shape 2 at frequency ω_2. In the following example you can see how the four arbitrary constants in Eqs. 9.8 are determined.

Example 9.2 The following initial conditions are imposed on the 2-DOF system studied in Example 9.1.

$$u_1(0) = \dot{u}_1(0) = \dot{u}_2(0) = 0$$

$$u_2(0) = u_0$$

Determine the subsequent free vibration motion.

SOLUTION Use the general solution form given in Eqs. 9.8.

$$u_1(t) = A_1 \cos\omega_1 t + B_1 \sin\omega_1 t + A_2 \cos\omega_2 t + B_2 \sin\omega_2 t \qquad (1a)$$

$$u_2(t) = \beta_1 A_1 \cos\omega_1 t + \beta_1 B_1 \sin\omega_1 t + \beta_2 A_2 \cos\omega_2 t + \beta_2 B_2 \sin\omega_2 t, \qquad (1b)$$

Differentiate Eqs. 1 to get

$$\dot{u}_1(t) = -A_1\omega_1 \sin\omega_1 t + B_1\omega_1 \cos\omega_1 t - A_2\omega_2 \sin\omega_2 t + B_2\omega_2 \cos\omega_2 t \qquad (2a)$$

$$\dot{u}_2(t) = -\beta_1 A_1\omega_1 \sin\omega_1 t + \beta_1 B_1\omega_1 \cos\omega_1 t - \beta_2 A_2\omega_2 \sin\omega_2 t$$
$$\qquad + \beta_2 B_2\omega_2 \cos\omega_2 t \qquad (2b)$$

From the mode-shape information obtained in Example 9.1,

$$\beta_1 = 1, \qquad \beta_2 = -1 \qquad (3)$$

Therefore, evaluating the initial conditions, we get

$$\begin{aligned} u_1(0) &= A_1 + A_2 = 0, & u_2(0) &= A_1 - A_2 = u_0 \\ \dot{u}_1(0) &= B_1\omega_1 + B_2\omega_2 = 0, & \dot{u}_2(0) &= B_1\omega_1 - B_2\omega_2 = 0 \end{aligned} \qquad (4)$$

Solve these equations to get

$$A_1 = \frac{u_0}{2}, \qquad A_2 = -\frac{u_0}{2}, \qquad B_1 = B_2 = 0 \qquad (5)$$

Therefore, the free-vibration response to the given initial conditions is

$$u_1(t) = \frac{u_0}{2} (\cos\omega_1 t - \cos\omega_2 t) \qquad (6a)$$

Ans.

$$u_2(t) = \frac{u_0}{2} (\cos\omega_1 t + \cos\omega_2 t) \qquad (6b)$$

To see the motion of the two masses when the system is given initial conditions as stated above, take $u_0 = 2$ in. and $\omega_1 = 1$ rad/sec. Then the *fundamental frequency* is $f_1 = \omega_1/2\pi = 1/2\pi = 0.159$ Hz, and the *fundamental period* is $T_1 = 1/f_1 = 6.28$ sec. From Example 9.1, $\omega_2 = \omega_1\sqrt{3}$. Therefore, $\omega_2 = \sqrt{3}$ rad/sec. The frequency of the second mode is $f_2 = \sqrt{3}/2\pi = 0.276$ Hz, and the period of the second mode is $T_2 = 1/f_2 = 3.63$ sec. The resulting motions, $u_1(t)$ and $u_2(t)$, are shown in Fig. 1.

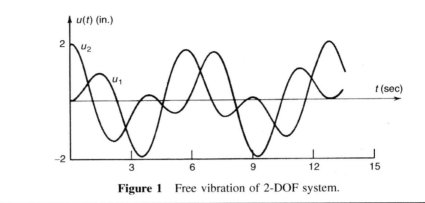

Figure 1 Free vibration of 2-DOF system.

By a proper choice of initial conditions it is possible to initiate free vibration in a single mode.

9.2 BEAT PHENOMENON

Examples 9.1 and 9.2 fully illustrate the basic steps that are required in determining the free vibration of a 2-DOF system. We now continue our study of free-vibration solutions for 2-DOF models by considering the *beat phenomenon*, an important vibration phenomenon that can occur when a system has closely spaced natural frequencies. Two simple examples of such systems are illustrated in Fig. 9.1: two equal masses connected by a weak spring ($k' \ll k$), and two identical pendulums coupled by a weak spring ($k' \ll mgL/a^2$).

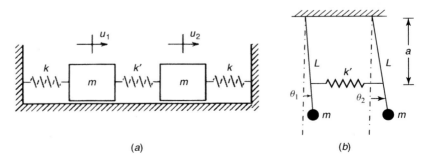

Figure 9.1 Two systems with closely spaced frequencies: (*a*) lightly coupled spring–mass system, $k' \ll k$; (*b*) lightly coupled pendulums, $k' \ll mgL/a^2$.

In the following example we use free vibration of a lightly coupled two-mass system to illustrate the beat phenomenon. To obtain this system, we can modify the spring–mass system of Example 9.1 by connecting the two masses with a coupling spring that is much weaker than the two outer springs, as in Fig. 9.1a.

Example 9.3 In the 2-DOF spring–mass system shown in Fig. 9.1a, let the coupling spring have a spring constant $k' \ll k$. (a) Obtain the natural frequencies and mode shapes of the system. (b) Determine expressions for the motion of the two masses given the initial conditions

$$u_1(0) = u_0, \quad u_2(0) = 0, \quad \dot{u}_1(0) = \dot{u}_2(0) = 0$$

(c) Let system parameters k, k', and m be such that $f_1 = 5.0$ Hz and $f_2 = 5.5$ Hz, and let the initial displacement be $u_0 = 1$. Plot the respective responses of the two masses.

SOLUTION (a) In matrix form, the equations of motion for free vibration of the system in Fig. 9.1a are

$$\begin{bmatrix} m & 0 \\ 0 & m \end{bmatrix} \begin{Bmatrix} \ddot{u}_1 \\ \ddot{u}_2 \end{Bmatrix} + \begin{bmatrix} k+k' & -k' \\ -k' & k+k' \end{bmatrix} \begin{Bmatrix} u_1 \\ u_2 \end{Bmatrix} = \begin{Bmatrix} 0 \\ 0 \end{Bmatrix} \quad (1)$$

Assume the harmonic-motion solution

$$\begin{Bmatrix} u_1(t) \\ u_2(t) \end{Bmatrix} = \begin{Bmatrix} U_1 \\ U_2 \end{Bmatrix} \cos \omega t \quad (2)$$

and let

$$k' = \delta k, \quad \lambda = \frac{\omega^2 m}{k} \quad (3a,b)$$

Then the resulting algebraic eigenproblem can be written in the simplified form

$$\left[\begin{bmatrix} 1+\delta & -\delta \\ -\delta & 1+\delta \end{bmatrix} - \lambda \begin{bmatrix} 1 & 0 \\ 0 & 1 \end{bmatrix} \right] \begin{Bmatrix} U_1 \\ U_2 \end{Bmatrix} = \begin{Bmatrix} 0 \\ 0 \end{Bmatrix} \quad (4)$$

The determinant of coefficients must vanish, so

$$\begin{vmatrix} 1+\delta-\lambda & -\delta \\ -\delta & 1+\delta-\lambda \end{vmatrix} = 0 \quad (5)$$

Equation 5 leads to the characteristic equation

$$\lambda^2 - 2\lambda(1+\delta) + (1+2\delta) = 0 \quad (6)$$

The two roots of Eq. 6 are

$$\begin{Bmatrix} \lambda_1 \\ \lambda_2 \end{Bmatrix} = (1+\delta) \mp \delta = \begin{cases} 1 \\ 1+2\delta \end{cases} \quad (7)$$

From Eqs. 3 and 7 we get the two natural frequencies

$$\omega_1 = \sqrt{\frac{k}{m}}, \qquad \omega_2 = \sqrt{\frac{k}{m}(1+2\delta)} \qquad \textbf{Ans. (a)} \quad (8a,b)$$

From the top equation in matrix Eq. 4, we get the mode-shape ratio

$$\beta_i \equiv \left(\frac{U_2}{U_1}\right)^{(i)} = \frac{1+\delta-\lambda_i}{\delta} \tag{9}$$

$$\beta_1 = 1, \qquad \beta_2 = -1 \tag{10}$$

Therefore, the two modes are given by

$$\boldsymbol{\phi}_1 = \begin{Bmatrix} 1 \\ 1 \end{Bmatrix}, \qquad \boldsymbol{\phi}_2 = \begin{Bmatrix} 1 \\ -1 \end{Bmatrix} \qquad \textbf{Ans. (a)} \quad (11)$$

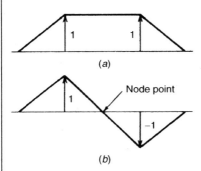

Figure 1 Mode shapes of the 2-DOF system in Fig. 9.1a: (a) mode 1; (b) mode 2.

As we should expect to find for this symmetric system, the fundamental mode is a symmetric mode and the second mode is an antisymmetric mode, as illustrated in Fig. 1.

(b) Following the procedure set out in Example 9.2, we obtain the following free-vibration response to the given initial conditions:

$$u_1(t) = \frac{u_0}{2}(\cos\omega_1 t + \cos\omega_2 t) \tag{12a}$$

$$u_2(t) = \frac{u_0}{2}(\cos\omega_1 t - \cos\omega_2 t) \tag{12b}$$

So far we have not incorporated the fact that if $\delta \ll 1$, the two natural frequencies will be closely spaced and the free vibration solution given by Eqs. 12 will exhibit the beat phenomenon. Using the approximation $(1+\delta)^n \approx (1+n\delta)$, if $\delta \ll 1$, Eqs. 8 become

$$\omega_1 = \sqrt{\frac{k}{m}}, \qquad \omega_2 = \sqrt{\frac{k}{m}(1+2\delta)} \approx (1+\delta)\sqrt{\frac{k}{m}} \tag{13a,b}$$

where the approximation for ω_2 is valid when $\delta \ll 1$. In that case the two natural frequencies will be closely spaced.

Equations 12 can be restated in a form that more clearly points to the beat phenomenon when the two natural frequencies are closely spaced. Using the trigonometric identity

$$\cos(\alpha \pm \beta) = \cos\alpha \cos\beta \mp \sin\alpha \sin\beta$$

and letting

$$\alpha = \frac{\omega_2 + \omega_1}{2} t, \qquad \beta = \frac{\omega_2 - \omega_1}{2} t$$

we can write Eqs. 12 in the following form:

$$u_1(t) = u_0 \left(\cos \frac{\omega_2 - \omega_1}{2} t \; \cos \frac{\omega_2 + \omega_1}{2} t \right) \qquad (14a)$$

$$u_2(t) = u_0 \left(\sin \frac{\omega_2 - \omega_1}{2} t \; \sin \frac{\omega_2 + \omega_1}{2} t \right) \qquad (14b)$$

Let us now examine the beat phenomenon. Let the *beat frequency* ω_B and the *average frequency* ω_{avg} be defined by the following equations:

$$\omega_B = \omega_2 - \omega_1, \qquad \omega_{\text{avg}} = \frac{\omega_2 + \omega_1}{2} \qquad (15)$$

Then Eqs. 14 can be written in the following form:

$$u_1(t) = \left(u_0 \cos \frac{\omega_B t}{2} \right) \cos \omega_{\text{avg}} t \qquad (16a)$$

Ans. (b)

$$u_2(t) = \left(u_0 \sin \frac{\omega_B t}{2} \right) \sin \omega_{\text{avg}} t \qquad (16b)$$

Hence, the displacements $u_1(t)$ and $u_2(t)$ can be considered to be "rapid" harmonic motions at frequency ω_{avg}, with their amplitudes varying slowly with $\cos(\omega_B t/2)$ and $\sin(\omega_B t/2)$, respectively. As the motion of one mass dies down, the motion of the other mass builds up, and vice versa. This behavior, called the *beat phenomenon*, is clearly visible in Fig. 2.

(c) The two natural frequencies of the system are

$$f_1 = \frac{\omega_1}{2\pi} = 5.00 \text{ Hz}, \qquad f_2 = \frac{\omega_2}{2\pi} = 5.50 \text{ Hz} \qquad (17)$$

These two natural frequencies are relatively *closely spaced frequencies*. Let us now see how this fact influences free vibration of this system by using Eqs. 16 to plot $u_1(t)$ and $u_2(t)$. For a system with the parameters given here, the beat frequency and average frequency are

$$\omega_B = \omega_2 - \omega_1 = \pi \text{ rad/sec}, \qquad \omega_{\text{avg}} = \frac{\omega_2 + \omega_1}{2} = 10.5\pi \text{ rad/sec} \qquad (18a,b)$$

The MATLAB .m-file that encodes the following solution can be found on the book's website. Its filename is

- sd2ex9_3f2.m: solution for beat phenomenon illustrated in Fig. 2

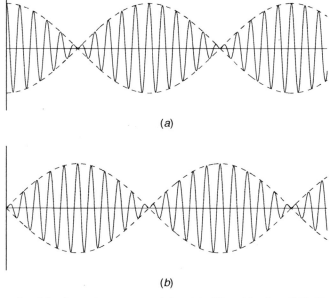

Figure 2 Example of the beat phenomenon: (a) $u_1(t)$; (b) $u_2(t)$. $f_1 = 5.00$ Hz, $f_2 = 5.50$ Hz; $u_1(0) = u_0$, $u_2(0) = 0$, $\dot{u}_1(0) = \dot{u}_2(0) = 0$.

You can see that Fig. 2 clearly illustrates the beat phenomenon due to the relatively closely spaced frequencies $f_1 = 5.00$ Hz and $f_2 = 5.50$ Hz. The envelope curves in Fig. 2a and b, respectively, are

$$(u_1)_{\text{env}} = u_0 \cos \frac{\omega_B t}{2}, \qquad (u_2)_{\text{env}} = u_0 \sin \frac{\omega_B t}{2} \qquad (19a,b)$$

Compare this free vibration with the vibration shown in Fig. 1 of Example 9.2, where the similar 2-DOF system had fairly widely separated natural frequencies of $f_1 = 0.159$ Hz and $f_2 = \sqrt{3} f_1 = 0.276$ Hz. Actually, if Fig. 1 of Example 9.2 were to be plotted in the same format as Fig. 2 above, the beat phenomenon would become more apparent, even for that system with more widely separated natural frequencies.

9.3 ADDITIONAL EXAMPLES OF MODES AND FREQUENCIES OF 2-DOF SYSTEMS: ASSUMED-MODES MODELS

Examples 9.1 and 9.2 have fully illustrated the basic steps that are required in determining the free vibration of a 2-DOF system, and Example 9.3 has demonstrated the beat phenomenon that can occur when a 2-DOF system has closely spaced natural frequencies. This section presents three more examples of modes and frequencies of 2-DOF systems: a model of a vibrating "rigid" beam supported on elastic springs, an assumed-modes model of a uniform cantilever bar, and a lumped-mass model of the same uniform cantilever bar.

9.3 Additional Examples of Modes and Frequencies of 2-DOF Systems: Assumed-Modes Models

Example 9.4 A slender, rigid, uniform beam of mass m and length L is supported at each end by a linearly elastic spring, as shown in Fig. 1. (Note that because the springs are different in stiffness, the system is not symmetric about the middle of the beam.) (a) Using Lagrange's Equations, write the equations of motion in terms of the end displacements u_1 and u_2. Assume that the slope angle of the beam remains small. (b) Determine the two natural frequencies. (c) Determine the two mode shapes. Scale them so that the maximum displacement is 1.0 in each case. Sketch the mode shapes.

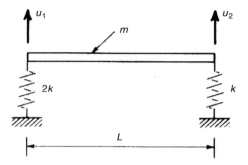

Figure 1 Uniform rigid beam on linearly elastic springs.

SOLUTION (a) Determine the equations of motion for free vibration. Sketch the beam in a displaced configuration (Fig. 2), and define all displacement quantities.

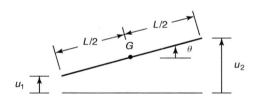

Figure 2 Beam in a displaced configuration.

From Fig. 2, the end displacements u_1 and u_2 are related to the displacement u_G of the mass center and the (small) angle of rotation of the beam, θ, by the following kinematic equations:

$$u_G = \frac{1}{2}(u_1 + u_2), \qquad \theta = \frac{1}{L}(u_2 - u_1) \tag{1}$$

Lagrange's Equation (Eq. 8.9) is

$$\frac{d}{dt}\frac{\partial T}{\partial \dot{q}_i} - \frac{\partial T}{\partial q_i} + \frac{\partial V}{\partial q_i} = (p_i)_{\text{nc}}, \qquad i = 1, 2 \tag{2}$$

where $q_1 \to u_1$ and $q_2 \to u_2$, and where the kinetic energy and potential energy are given by

$$T = \tfrac{1}{2}m\dot{u}_G^2 + \tfrac{1}{2}I_G\dot{\theta}^2, \qquad V = \tfrac{1}{2}2ku_1^2 + \tfrac{1}{2}ku_2^2 \tag{3}$$

Since there are no external forces on the beam, the generalized forces are

$$(p_1)_{nc} = (p_2)_{nc} = 0 \tag{4}$$

The partial derivatives in Eq. 2 are evaluated as follows:

$$\frac{\partial T}{\partial \dot{u}_1} = m\dot{u}_G \frac{\partial \dot{u}_G}{\partial \dot{u}_1} + I_G \dot{\theta} \frac{\partial \dot{\theta}}{\partial \dot{u}_1}$$

$$= m\left(\frac{\dot{u}_1 + \dot{u}_2}{2}\right)\left(\frac{1}{2}\right) + \frac{1}{12}mL^2\left(\frac{\dot{u}_2 - \dot{u}_1}{L}\right)\left(-\frac{1}{L}\right)$$

$$= \frac{m}{6}(2\dot{u}_1 + \dot{u}_2)$$

$$\frac{\partial T}{\partial u_1} = 0, \quad \frac{\partial V}{\partial u_1} = 2ku_1$$

$$\frac{\partial T}{\partial \dot{u}_2} = m\dot{u}_G \frac{\partial \dot{u}_G}{\partial \dot{u}_2} + I_G \dot{\theta} \frac{\partial \dot{\theta}}{\partial \dot{u}_2}$$

$$= m\left(\frac{\dot{u}_1 + \dot{u}_2}{2}\right)\left(\frac{1}{2}\right) + \frac{1}{12}mL^2\left(\frac{\dot{u}_2 - \dot{u}_1}{L}\right)\left(\frac{1}{L}\right)$$

$$= \frac{m}{6}(\dot{u}_1 + 2\dot{u}_2)$$

$$\frac{\partial T}{\partial u_2} = 0, \quad \frac{\partial V}{\partial u_2} = ku_2$$

Substituting the expressions above into Lagrange's Equation, Eq. 2, we get the following matrix equation of motion:

$$\frac{m}{6}\begin{bmatrix} 2 & 1 \\ 1 & 2 \end{bmatrix}\begin{Bmatrix} \ddot{u}_1 \\ \ddot{u}_2 \end{Bmatrix} + k\begin{bmatrix} 2 & 0 \\ 0 & 1 \end{bmatrix}\begin{Bmatrix} u_1 \\ u_2 \end{Bmatrix} = \begin{Bmatrix} 0 \\ 0 \end{Bmatrix} \quad \textbf{Ans. (a)} \tag{5}$$

(b) Determine the two natural frequencies. Assume harmonic motion.

$$\begin{Bmatrix} u_1 \\ u_2 \end{Bmatrix} = \begin{Bmatrix} U_1 \\ U_2 \end{Bmatrix} \cos(\omega t - \alpha) \tag{6}$$

Then, combining Eqs. 5 and 6, we get the following algebraic eigenproblem:

$$\left[k\begin{bmatrix} 2 & 0 \\ 0 & 1 \end{bmatrix} - \frac{\omega^2 m}{6}\begin{bmatrix} 2 & 1 \\ 1 & 2 \end{bmatrix}\right]\begin{Bmatrix} U_1 \\ U_2 \end{Bmatrix} = \begin{Bmatrix} 0 \\ 0 \end{Bmatrix} \tag{7}$$

To obtain the two natural frequencies, we set the determinant of the coefficients in Eq. 7 equal to zero. Let

$$\lambda = \frac{\omega^2 m}{6k} \tag{8}$$

9.3 Additional Examples of Modes and Frequencies of 2-DOF Systems: Assumed-Modes Models

Then the determinant of the coefficients in Eq. 7 becomes

$$\begin{vmatrix} 2 - 2\lambda & -\lambda \\ -\lambda & 1 - 2\lambda \end{vmatrix} = 0$$

which gives the following characteristic equation:

$$(2 - 2\lambda)(1 - 2\lambda) - \lambda^2 = 0$$

whose roots are

$$\lambda = \frac{6 \pm \sqrt{36 - 24}}{6}, \quad \begin{cases} \lambda_1 = 0.4226 \\ \lambda_2 = 1.5774 \end{cases} \tag{9}$$

Finally, combining Eqs. 8 and 9, we get the following two natural frequencies:

$$\omega_1 = 1.592\sqrt{\frac{k}{m}}, \quad \omega_2 = 3.076\sqrt{\frac{k}{m}} \quad \textbf{Ans. (b)} \tag{10}$$

(c) Determine the two mode shapes and sketch them. From the top equation in matrix Eq. 7, we can determine the mode-shape ratios:

$$(2 - 2\lambda)U_1 - \lambda U_2 = 0$$

or

$$\left(\frac{U_2}{U_1}\right)_i = \frac{2 - 2\lambda_i}{\lambda_i} \tag{11}$$

so

$$\left(\frac{U_2}{U_1}\right)_1 = \frac{2 - 2(0.4226)}{0.4226} = 2.7321, \quad \left(\frac{U_2}{U_1}\right)_2 = \frac{2 - 2(1.5774)}{1.5774} = -0.7321$$

We are to scale the mode shapes so that the largest displacement in each is 1.0. Therefore, the two scaled mode shapes are

$$\boldsymbol{\phi}_1 = \begin{Bmatrix} 0.366 \\ 1.000 \end{Bmatrix}, \quad \boldsymbol{\phi}_2 = \begin{Bmatrix} 1.000 \\ -0.732 \end{Bmatrix} \quad \textbf{Ans. (c)} \tag{12}$$

These mode shapes are sketched in Fig. 3.

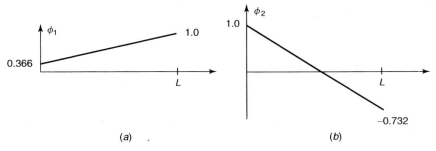

Figure 3 Scaled mode shapes: (*a*) mode 1; (*b*) mode 2.

262 Vibration of Undamped 2-DOF Systems

The next example illustrates how to solve for natural frequencies and mode shapes when the equations of motion represent an assumed-modes model.

Example 9.5 A 2-DOF assumed-modes math model for axial vibration of the uniform cantilever bar in Fig. 1 was obtained in Example 8.9. Solve for the natural frequencies and modes of this model, and sketch the modes.

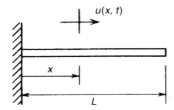

Figure 1 Uniform cantilever bar.

SOLUTION 1. Write the equations of motion for free axial vibration. In Example 8.9 the assumed-modes model of axial vibration was based on the displacement approximation

$$u(x,t) = \psi_1(x)u_1(t) + \psi_2(x)u_2(t) \tag{1}$$

with

$$\psi_1(x) = \frac{x}{L}, \qquad \psi_2(x) = \left(\frac{x}{L}\right)^2 \tag{2}$$

Note that the coordinates u_1 and u_2 in Eq. 1 are generalized coordinates, not physical displacements. From Eq. 11 of Example 8.9, the equations of motion for this 2-DOF model, written in matrix form, are

$$\frac{\rho A L}{60}\begin{bmatrix} 20 & 15 \\ 15 & 12 \end{bmatrix}\begin{Bmatrix} \ddot{u}_1 \\ \ddot{u}_2 \end{Bmatrix} + \frac{EA}{3L}\begin{bmatrix} 3 & 3 \\ 3 & 4 \end{bmatrix}\begin{Bmatrix} u_1 \\ u_2 \end{Bmatrix} = \begin{Bmatrix} 0 \\ 0 \end{Bmatrix} \tag{3}$$

2. Assume harmonic motion.

$$\begin{Bmatrix} u_1 \\ u_2 \end{Bmatrix} = \begin{Bmatrix} U_1 \\ U_2 \end{Bmatrix}\cos(\omega t - \alpha) \tag{4}$$

3. Obtain the algebraic eigenvalue problem

$$\left[\begin{bmatrix} 3 & 3 \\ 3 & 4 \end{bmatrix} - \lambda_i \begin{bmatrix} 20 & 15 \\ 15 & 12 \end{bmatrix}\right]\begin{Bmatrix} U_1 \\ U_2 \end{Bmatrix} = \begin{Bmatrix} 0 \\ 0 \end{Bmatrix} \tag{5}$$

where $\lambda_i = (\rho L^2/20E)\omega_i^2$.

4. From the determinant of the coefficients, obtain the characteristic equation.

$$(3 - 20\lambda_i)(4 - 12\lambda_i) - (3 - 15\lambda_i)^2 = 0$$

or

$$15\lambda_i^2 - 26\lambda_i + 3 = 0 \tag{6}$$

9.3 Additional Examples of Modes and Frequencies of 2-DOF Systems: Assumed-Modes Models

5. Solve for the roots of the characteristic equation.

$$\lambda_i = \frac{26 \mp \sqrt{496}}{30} = \begin{cases} 0.1243 \\ 1.6090 \end{cases} \tag{7}$$

Then, from Eq. 7 and the definition of λ_i,

$$(\omega_1 L)^2 = \frac{20E}{\rho}\lambda_1 = 2.486\frac{E}{\rho} \tag{8a}$$

Ans.

$$(\omega_2 L)^2 = \frac{20E}{\rho}\lambda_2 = 32.18\frac{E}{\rho} \tag{8b}$$

In Section 13.1 you will find that the "exact" natural frequencies of this system, based on the continuous model, are

$$(\omega_1 L)^2_{\text{exact}} = 2.467\frac{E}{\rho}, \qquad (\omega_2 L)^2_{\text{exact}} = 22.21\frac{E}{\rho} \tag{9a,b}$$

Note that in both cases the approximate frequencies are higher than the exact frequencies and that the fundamental (i.e., lowest) frequency is calculated with much greater accuracy than the second frequency. It is characteristic of assumed-modes models that frequencies are too high and that the higher frequencies are inaccurate. These phenomena are discussed further in connection with finite element modeling in Section 14.8.

6. Determine the mode shapes. Substitute λ_i into the first of Eqs. 5.

$$(3 - 20\lambda_i)U_1^{(i)} + (3 - 15\lambda_i)^2 U_2^{(i)} = 0 \tag{10}$$

or

$$\beta_i \equiv \left(\frac{U_2}{U_1}\right)^{(i)} = -\frac{3 - 20\lambda_i}{3 - 15\lambda_i} \tag{11}$$

Therefore,

$$\beta_1 = -\frac{0.514}{1.136} = -0.453, \qquad \beta_2 = -1.381 \qquad \textbf{Ans. (12a,b)}$$

To write the two mode shapes as functions of x so they can be sketched, we must recall from Eqs. 1 and 2 that

$$u(x, t) = \frac{x}{L}u_1(t) + \left(\frac{x}{L}\right)^2 u_2(t) \tag{13}$$

From Eqs. 4, 11, and 13 we find that motion in the ith mode is given by

$$u^{(i)}(x, t) = \phi_i(x)\cos(\omega_i t - \alpha_i) \tag{14}$$

where the approximated mode shapes $\phi_i(x)$ are given by

$$\phi_i(x) = \left[\frac{x}{L} + \beta_i\left(\frac{x}{L}\right)^2\right], \qquad i = 1, 2 \qquad \textbf{Ans. (15)}$$

with the two values of β_i from Eq. 12.

264 Vibration of Undamped 2-DOF Systems

The two approximate mode shapes are sketched in Fig. 2. These approximate mode shapes can be compared with the "exact" modes shown in Fig. 2 of Example 13.1.

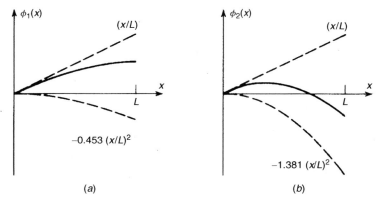

Figure 2 Mode shapes based on 2-DOF assumed-modes model: (*a*) mode 1; (*b*) mode 2.

Example 9.5 indicates that good estimates of natural frequencies are attainable by the use of assumed-modes models. Consider now how the assumed-modes model compares with a 2-DOF lumped-mass model of the same uniform cantilever bar. There are no rigorous guidelines for defining lumped-mass models, but the model presented in Example 9.6 is a logical model to assume.

Example 9.6 The cantilever bar of Example 9.5 is now to be modeled by a massless uniform bar to which are attached two lumped masses representing the mass of the original system. This is "equivalent" to the spring–mass system with $k = 2AE/L$ and $m = \rho AL$. Determine the natural frequencies and mode shapes of this model. (*Note*: The coordinates u_1 and u_2 represent physical displacement of the two respective masses, as shown in Fig. 1.)

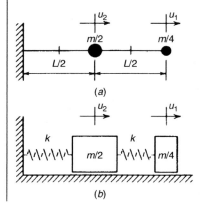

Figure 1 Lumped-mass model of a uniform cantilever bar: (*a*) lumped masses on a massless bar; (*b*) equivalent spring–mass model.

9.3 Additional Examples of Modes and Frequencies of 2-DOF Systems: Assumed-Modes Models

SOLUTION The same steps will be followed as in previous examples.

$$\frac{\rho A L}{4}\begin{bmatrix} 1 & 0 \\ 0 & 2 \end{bmatrix}\begin{Bmatrix} \ddot{u}_1 \\ \ddot{u}_2 \end{Bmatrix} + \frac{2EA}{L}\begin{bmatrix} 1 & -1 \\ -1 & 2 \end{bmatrix}\begin{Bmatrix} u_1 \\ u_2 \end{Bmatrix} = \begin{Bmatrix} 0 \\ 0 \end{Bmatrix} \quad (1)$$

$$\begin{Bmatrix} u_1 \\ u_2 \end{Bmatrix} = \begin{Bmatrix} U_1 \\ U_2 \end{Bmatrix}\cos(\omega t - \alpha) \quad (2)$$

$$\left[\begin{bmatrix} 1 & -1 \\ -1 & 2 \end{bmatrix} - \lambda_i \begin{bmatrix} 1 & 0 \\ 0 & 2 \end{bmatrix}\right]\begin{Bmatrix} U_1 \\ U_2 \end{Bmatrix}^{(i)} = \begin{Bmatrix} 0 \\ 0 \end{Bmatrix} \quad (3)$$

where

$$\lambda_i = \frac{\rho L^2}{8E}\omega_i^2 \quad (4)$$

The characteristic equation is

$$(1-\lambda_i)(2-2\lambda_i) - (-1)^2 = 0$$

or

$$2\lambda_i^2 - 4\lambda_i + 1 = 0 \quad (5)$$

$$\lambda_i = \frac{4 \mp \sqrt{8}}{4} = \begin{cases} 0.2929 \\ 1.707 \end{cases} \quad (6)$$

Then the natural frequencies can be obtained from the following expressions:

$$(\omega_1 L)^2 = 8(0.2929)\frac{E}{\rho} = 2.343\frac{E}{\rho} \quad (7a)$$

Ans.

$$(\omega_2 L)^2 = 8(1.707)\frac{E}{\rho} = 13.66\frac{E}{\rho} \quad (7b)$$

Recall from Example 9.4 that the corresponding "exact" coefficient values are 2.467 and 22.21. Hence, it can be seen that this lumped-mass model gives frequencies that are below the exact value in each case. Further examples are given in Section 14.8.

For the mode shapes, Eq. 3 gives

$$(1 - \lambda_i)U_1^{(i)} - U_2^{(i)} = 0 \quad (8)$$

or

$$\beta_i \equiv \left(\frac{U_2}{U_1}\right)^{(i)} = 1 - \lambda_i \quad (9)$$

Therefore, the mode-shape ratios are

$$\beta_1 = 0.707, \quad \beta_2 = -0.707 \quad (10a,b)$$

and the corresponding scaled modal vectors are

$$\phi_1 = \begin{Bmatrix} 1.000 \\ 0.707 \end{Bmatrix}, \quad \phi_2 = \begin{Bmatrix} 1.000 \\ -0.707 \end{Bmatrix} \quad \textbf{Ans. (11)}$$

The mode shapes can be plotted as shown in Fig. 2. Note that u_1 is the displacement of the right-hand mass in Fig. 1.

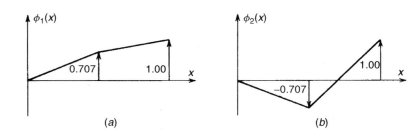

Figure 2 Mode shapes based on the lumped-mass model: (*a*) mode 1; (*b*) mode 2.

By comparing the above 2-DOF lumped-mass example of a vibrating uniform rod (Example 9.6) with the corresponding 2-DOF assumed-modes example (Example 9.5), you can observe that the choice of model is quite important. (See Problem 9.7 for an assumed-modes model of the cantilever bar with added tip mass.) The "exact" solution for free vibration of the uniform cantilever bar is given in Example 13.1.

We will continue to use lumped-mass models in some of the example. However, for modeling real systems, lumped-mass modeling has largely been supplanted by *finite element modeling*, a powerful computer-based version of assumed-modes modeling, which is the subject of Chapters 14 through 17.

From the examples in Sections 9.1 through 9.3, you should now be very familiar with how to calculate modes and frequencies of 2-DOF vibrating systems and how to determine their free-vibration response to given initial conditions. In Chapter 10 these concepts are generalized to multi-degree-of-freedom (MDOF) systems.

9.4 FREE VIBRATION OF SYSTEMS WITH RIGID-BODY MODES

Frequently, systems are encountered that have one or more *rigid-body modes*; that is, they can move about in space without changing their shape. This is true for aerospace vehicles in flight, since the vehicle structure is not attached to any fixed base and is therefore free to translate or rotate as a rigid body without deformation. For example, an airplane wing and body modeled by three lumped masses along a flexible beam has

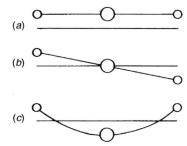

Figure 9.2 Mode shapes of a three-mass model of an airplane wing illustrating rigid-body modes: (*a*) mode 1: translation (plunge) rigid-body mode; (*b*) mode 2: rotation (roll) rigid-body mode; (*c*) mode 3: flexible mode.

the mode shapes shown Fig. 9.2. The first two modes are rigid-body modes (a *plunge mode* and a *roll mode* in airplane terminology). Rigid-body modes have a corresponding natural frequency of zero, as will be seen in Example 9.7.

To illustrate how rigid-body modes are treated, we consider the system in Example 9.7. This system has only one rigid-body mode.

***

Example 9.7 Determine the natural frequencies and mode shapes of the system shown in Fig. 1. (Note that this system is free to translate as a rigid body with the spring remaining unstretched; that is, the masses can move with $u_1 = u_2$ without any external forcing.)

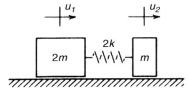

Figure 1 A 2-DOF system with a rigid-body mode.

SOLUTION Proceed in exactly the same manner as before; that is, follow the steps outlined in Section 9.1. The equations of motion are

$$2m\ddot{u}_1 + 2ku_1 - 2ku_2 = 0$$
$$m\ddot{u}_2 - 2ku_1 + 2ku_2 = 0 \qquad (1)$$

1. Assume the harmonic solution

$$u_1(t) = U_1 \cos \omega_i t$$
$$u_2(t) = U_2 \cos \omega_i t \qquad (2)$$

2. Substitute Eqs. 2 into Eqs. 1 to get the algebraic eigenvalue problem.

$$\begin{bmatrix} 2k - 2m\omega_i^2 & -2k \\ -2k & 2k - m\omega_i^2 \end{bmatrix} \begin{Bmatrix} U_1 \\ U_2 \end{Bmatrix} = \begin{Bmatrix} 0 \\ 0 \end{Bmatrix} \qquad (3)$$

3. Set up the characteristic equation, which is the determinant of the coefficient matrix in Eq. 3.

$$(2k - 2m\omega_i^2)(2k - m\omega_i^2) - (2k)^2 = 0$$

or

$$m\omega_i^2(2m\omega_i^2 - 6k) = 0 \qquad (4)$$

4. Solve for the roots of the characteristic equation.

$$\omega_1^2 = 0, \qquad \omega_2^2 = \frac{3k}{m} \qquad \textbf{Ans.} \ (5)$$

Note that the frequency labeled ω_1 is zero, corresponding to the rigid-body mode.

5. Substitute the eigenvalues given by Eqs. 5 into the first of Eqs. 3 to obtain the mode-shape ratios.

$$(2k - 2m\omega_i^2)U_1^{(i)} - 2kU_2^{(i)} = 0$$

Therefore, the mode-shape ratio is

$$\beta_i = \left(\frac{U_2}{U_1}\right)^{(i)} = \frac{2k - 2m\omega_i^2}{2k} = 1 - \frac{m\omega_i^2}{k} \qquad (6)$$

and the mode-shape ratios for the two modes are

$$\beta_1 = 1 - 0 = 1, \qquad \beta_2 = 1 - 3 = -2 \qquad \textbf{Ans.} \ (7)$$

Sketch the mode shapes by letting $(\phi_1)_i \equiv U_1^{(i)} = 1$. Then $(\phi_2)_i \equiv U_2^{(i)} = \beta_i$.

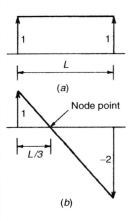

Figure 2 Mode shapes of a 2-DOF system with one rigid-body mode: (a) mode 1 (rigid-body mode): $\omega_1 = 0$, $\boldsymbol{\phi}_1 = \begin{Bmatrix} 1 \\ 1 \end{Bmatrix}$; (b) mode 2: $\omega_2 = \left(\dfrac{3k}{m}\right)^{1/2}$, $\boldsymbol{\phi}_2 = \begin{Bmatrix} 1 \\ -2 \end{Bmatrix}$.

Note that for the rigid-body mode the frequency is zero and the spring is undeformed. In mode 2 the masses move in opposite directions and there is a *node point* (i.e., a point that remains stationary), as shown in Fig. 2b.

An eigenvalue problem that results in one or more zero eigenvalues, as in Example 9.7, is called a *semidefinite eigenvalue problem*. In structural dynamics, zero eigenvalues result when the structure can move as a rigid body. Mathematically, the zero eigenvalues result from the fact that the stiffness matrix is singular; that is, the determinant of the coefficients of the stiffness matrix is zero.

9.5 INTRODUCTION TO MODE SUPERPOSITION: FREQUENCY RESPONSE OF AN UNDAMPED 2-DOF SYSTEM

In Chapter 4 you studied the response of single-DOF systems to harmonic excitation. You learned that when the forcing frequency is equal to the natural frequency of the system (i.e., $\Omega = \omega_n$), resonance occurs, that is, the displacement of the undamped system becomes unbounded. In the preceding sections of this chapter you have seen that MDOF systems possess multiple natural frequencies (e.g., 2-DOF systems have two natural frequencies), so it is reasonable to expect resonance to be an important factor to consider in MDOF systems. In this section you will consider the response of a 2-DOF system to harmonic excitation. In addition, you will be introduced to the *mode-superposition method* for solving for the dynamic response of MDOF systems.

9.5 Introduction to Mode Superposition: Frequency Response of an Undamped 2-DOF System

Key mode-superposition equations in the following example problem are "framed," just as other key equations in the text have been, and this important topic is considered in greater detail in Chapter 11.

Example 9.8 The 2-DOF system shown in Fig. 1 is subjected to a single harmonic force, $p_1(t) = P_1 \cos \Omega t$. Determine the steady-state response of each of the masses as a function of frequency. Use the mode-superposition method in solving this problem. The matrix equation of motion of this undamped 2-DOF system is

$$m \begin{bmatrix} 1 & 0 \\ 0 & 2 \end{bmatrix} \begin{Bmatrix} \ddot{u}_1 \\ \ddot{u}_2 \end{Bmatrix} + k \begin{bmatrix} 2 & -1 \\ -1 & 3 \end{bmatrix} \begin{Bmatrix} u_1 \\ u_2 \end{Bmatrix} = \begin{Bmatrix} P_1 \\ 0 \end{Bmatrix} \cos \Omega t$$

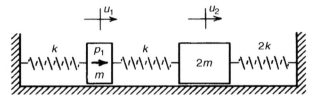

Figure 1 A 2-DOF system with harmonic excitation.

or, in symbolic matrix form,

$$\mathbf{M}\ddot{\mathbf{u}} + \mathbf{K}\mathbf{u} = \mathbf{p}(t)$$

and the natural frequencies and modes are depicted in Fig. 2.

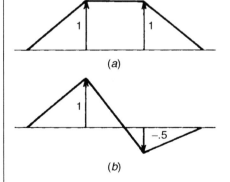

Figure 2 Mode shapes of a 2-DOF lumped-mass system: mode 1: $\omega_1^2 = k/m$, $\boldsymbol{\phi}_1 = \begin{Bmatrix} 1 \\ 1 \end{Bmatrix}$; mode 2: $\omega_2^2 = \frac{5}{2}(k/m)$, $\boldsymbol{\phi}_2 = \begin{Bmatrix} 1 \\ -0.5 \end{Bmatrix}$.

SOLUTION 1. Form the modal matrix. This is a matrix whose respective columns are the natural modes.

$$\boldsymbol{\Phi} = [\boldsymbol{\phi}_1 \; \boldsymbol{\phi}_2] = \begin{bmatrix} \phi_{11} & \phi_{12} \\ \phi_{21} & \phi_{22} \end{bmatrix} = \begin{bmatrix} 1 & 1 \\ 1 & -\frac{1}{2} \end{bmatrix} \quad (1)$$

2. Define the set of *principal coordinates*, $\eta_1(t)$ and $\eta_2(t)$, by the *mode-superposition equation*

$$\mathbf{u}(t) = \boldsymbol{\phi}_1 \eta_1(t) + \boldsymbol{\phi}_2 \eta_2(t) \tag{2a}$$

or

$$\begin{Bmatrix} u_1(t) \\ u_2(t) \end{Bmatrix} = \begin{bmatrix} 1 & 1 \\ 1 & -\frac{1}{2} \end{bmatrix} \begin{Bmatrix} \eta_1(t) \\ \eta_2(t) \end{Bmatrix} = \begin{Bmatrix} \eta_1 + \eta_2 \\ \eta_1 - \frac{1}{2}\eta_2 \end{Bmatrix} \tag{2b}$$

which can be written as the matrix-transformation equation

$$\boxed{\mathbf{u}(t) = \boldsymbol{\Phi}\,\boldsymbol{\eta}(t)} \tag{3}$$

3. Transform the equations of motion to principal coordinates.

$$\boldsymbol{\Phi}^T[\mathbf{M}\ddot{\mathbf{u}} + \mathbf{K}\mathbf{u} = \mathbf{p}(t)] \tag{4}$$

$$\boxed{\mathscr{M}\ddot{\boldsymbol{\eta}} + \mathscr{K}\boldsymbol{\eta} = \mathscr{p}(t)} \tag{5}$$

where the matrices $\mathscr{M}$ and $\mathscr{K}$ and the vector $\mathscr{p}(t)$, defined by the equations

$$\boxed{\mathscr{M} = \boldsymbol{\Phi}^T \mathbf{M} \boldsymbol{\Phi}, \quad \mathscr{K} = \boldsymbol{\Phi}^T \mathbf{K} \boldsymbol{\Phi}, \quad \mathscr{p}(t) = \boldsymbol{\Phi}^T \mathbf{p}(t)} \tag{6}$$

re called the *modal mass matrix*, the *modal stiffness matrix*, and the *modal force vector*, respectively. (Note the difference in the font used for the original mass matrix $\mathbf{M}$ and the modal mass matrix $\mathscr{M}$, etc.) Equation 5 is called the *equation of motion in principal coordinates*, or the *modal equation of motion*, for short.

Use Eqs. 1 and 6 to evaluate $\mathscr{M}$, $\mathscr{K}$, and $\mathscr{p}$.

$$\mathscr{M} = \begin{bmatrix} 1 & 1 \\ 1 & -\frac{1}{2} \end{bmatrix} \begin{bmatrix} m & 0 \\ 0 & 2m \end{bmatrix} \begin{bmatrix} 1 & 1 \\ 1 & -\frac{1}{2} \end{bmatrix} = m \begin{bmatrix} 3 & 0 \\ 0 & \frac{3}{2} \end{bmatrix} \tag{7}$$

$$\mathscr{K} = \begin{bmatrix} 1 & 1 \\ 1 & -\frac{1}{2} \end{bmatrix} \begin{bmatrix} 2k & -k \\ -k & 3k \end{bmatrix} \begin{bmatrix} 1 & 1 \\ 1 & -\frac{1}{2} \end{bmatrix} = k \begin{bmatrix} 3 & 0 \\ 0 & \frac{15}{4} \end{bmatrix} \tag{8}$$

$$\mathscr{p}(t) = \begin{bmatrix} 1 & 1 \\ 1 & -\frac{1}{2} \end{bmatrix} \begin{Bmatrix} P_1 \\ 0 \end{Bmatrix} \cos \Omega t = \begin{Bmatrix} P_1 \\ P_1 \end{Bmatrix} \cos \Omega t \tag{9}$$

Therefore, the equation of motion in principal coordinates is

$$m \begin{bmatrix} 3 & 0 \\ 0 & \frac{3}{2} \end{bmatrix} \begin{Bmatrix} \ddot{\eta}_1 \\ \ddot{\eta}_2 \end{Bmatrix} + k \begin{bmatrix} 3 & 0 \\ 0 & \frac{15}{4} \end{bmatrix} \begin{Bmatrix} \eta_1 \\ \eta_2 \end{Bmatrix} = \begin{Bmatrix} P_1 \\ P_1 \end{Bmatrix} \cos \Omega t \tag{10}$$

Note that the transformation to principal coordinates has *uncoupled the equations of motion*. (Compare Eq. 10 with the original equation of motion.) This is due to the *orthogonality relationships* that are satisfied by the mode shapes, a topic that is discussed further in Section 10.1. This leads to two separate single-DOF equations, which can each be solved by the methods of Chapter 4. Thus, the equations to solve are

$$3m\ddot{\eta}_1 + 3k\eta_1 = P_1 \cos \Omega t$$

$$\tfrac{3}{2} m\ddot{\eta}_2 + \tfrac{15}{4} k\eta_2 = P_1 \cos \Omega t \tag{11a,b}$$

9.5 Introduction to Mode Superposition: Frequency Response of an Undamped 2-DOF System

4. Solve the uncoupled equations of motion above for the two steady-state responses. Assume harmonic motion.

$$\eta_1 = Y_1 \cos \Omega t$$
$$\eta_2 = Y_2 \cos \Omega t \qquad (12a,b)$$

Then, by substituting these into Eqs. 11, we obtain the following *modal response amplitudes*:

$$Y_1 = \frac{P_1}{3k - 3m\Omega^2} = \frac{(1/3k)P_1}{1 - (\Omega/\omega_1)^2}$$

$$Y_2 = \frac{P_1}{\frac{15}{4}k - \frac{3}{2}m\Omega^2} = \frac{(4/15k)P_1}{1 - (\Omega/\omega_2)^2} \qquad (13a,b)$$

5. Transform the responses back to physical coordinates. Equations 3, 12, and 13 can be combined to give the following steady-state responses of the two masses:

$$u_1 = U_1 \cos \Omega t, \qquad u_2 = U_2 \cos \Omega t \qquad (14a,b)$$

where the respective amplitudes are

$$U_1 = \frac{P_1/3k}{1 - (\Omega/\omega_1)^2} + \frac{4P_1/15k}{1 - (\Omega/\omega_2)^2} \qquad (15a)$$

Ans.

$$U_2 = \frac{P_1/3k}{1 - (\Omega/\omega_1)^2} - \frac{1}{2}\left[\frac{4P_1/15k}{1 - (\Omega/\omega_2)^2}\right] \qquad (15b)$$

Figure 3 shows the nondimensionalized *frequency-response functions* $kU_1(\Omega)/P_1$ and $kU_2(\Omega)/P_1$, both plotted versus the nondimensional frequency Ω/ω_1. Note the resonance "peaks" in these figures: the first at $\Omega = \omega_1$ and the second at $\Omega = \omega_2 = \sqrt{5/2}\,\omega_1$. Also note how the responses change sign whenever the excitation frequency crosses one of the natural frequencies.

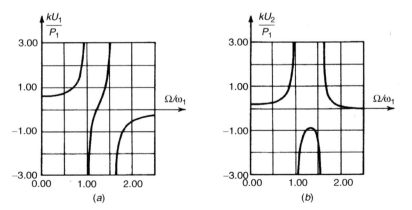

Figure 3 Frequency-response functions for (*a*) mass 1 and (*b*) mass 2.

Example 9.8 could easily have been solved by substituting Eqs. 14 directly into the original equations of motion. However, this would not have been the case with general (i.e., not harmonic) excitation **p**(*t*), as you will see in Chapter 11.

From Example 9.8 you should have discovered two important reasons for solving for the modes and frequencies of MDOF systems:

- The modal matrix can be used to uncouple the equations of motion of undamped (and in some cases, damped) systems. This makes it easy to treat the MDOF system as a collection of SDOF systems.
- The natural frequencies and modes of a system greatly influence how it will respond to dynamic excitation, especially to harmonic excitation near one of the system's natural frequencies.

In Chapter 10 we study some of the important properties of modes and frequencies, and later (in Chapter 15 and in Section 17.8) discuss some important numerical procedures that may be used to solve for the modes and frequencies of systems with very many degrees of freedom.

9.6 UNDAMPED VIBRATION ABSORBER

Figure 9.3a shows a "machine" that consists of an undamped spring–mass system with harmonic excitation at frequency Ω rad/s. This will be referred to as the *primary system*. It is assumed that excitation frequency Ω is a fixed operating frequency and that it is close enough to the primary system's undamped natural frequency to produce an undesirably high amplitude of vibration of the primary system's mass, m_1. That is, the harmonic excitation produces a near-resonance steady-state condition for the primary system.

As shown in Fig. 9.3b, a second spring–mass system, called the *absorber system*, can be attached to the primary mass, and this added SDOF system can be tuned to act as a vibration absorber, reducing the response of the primary mass to zero at the excitation frequency Ω.

Before analyzing this vibration-absorber system, let us review Example 9.8, where a similar undamped 2-DOF system was subjected to harmonic excitation at one of its masses. Note that the frequency-response function in Fig. 3a of Example 9.8 shows that between the first and second natural frequencies of the 2-DOF system, there is an *antiresonance* frequency, that is, a frequency at which the response of the mass where

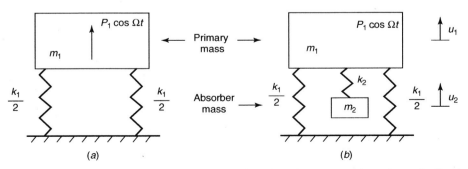

Figure 9.3 (a) Original spring–mass system; (b) original system with undamped vibration absorber.

9.6 Undamped Vibration Absorber

the excitation is applied is zero. The principle behind the *undamped vibration absorber* is to attach a SDOF spring–mass absorber system to the primary SDOF system and to *tune the absorber* so that the antiresonance frequency between the two new natural frequencies ω_1 and ω_2 is at the original excitation frequency Ω.

The equations of motion of the 2-DOF system in Fig. 9.3b, written in matrix form, are

$$\begin{bmatrix} m_1 & 0 \\ 0 & m_2 \end{bmatrix} \begin{Bmatrix} \ddot{u}_1 \\ \ddot{u}_2 \end{Bmatrix} + \begin{bmatrix} k_1 + k_2 & -k_2 \\ -k_2 & k_2 \end{bmatrix} \begin{Bmatrix} u_1 \\ u_2 \end{Bmatrix} = \begin{Bmatrix} P_1 \\ 0 \end{Bmatrix} \cos \Omega t \quad (9.10)$$

We are interested in the frequency response of this system, so let us assume the following steady-state responses:

$$u_1(t) = U_1 \cos \Omega t, \qquad u_2(t) = U_2 \cos \Omega t \quad (9.11)$$

Substitution of Eqs. 9.11 into Eqs. 9.10 leads to two simultaneous algebraic equations, which can be written in the following matrix form:

$$\begin{bmatrix} k_1 + k_2 - \Omega^2 m_1 & -k_2 \\ -k_2 & k_2 - \Omega^2 m_2 \end{bmatrix} \begin{Bmatrix} U_1 \\ U_2 \end{Bmatrix} = \begin{Bmatrix} P_1 \\ 0 \end{Bmatrix} \quad (9.12)$$

Solving these two equations for U_1 and U_2, we get the following *frequency-response* equations:

$$U_1(\Omega) = \frac{(k_2 - \Omega^2 m_2) P_1}{(k_1 + k_2 - \Omega^2 m_1)(k_2 - \Omega^2 m_2) - k_2^2}$$

$$U_2(\Omega) = \frac{k_2 P_1}{(k_1 + k_2 - \Omega^2 m_1)(k_2 - \Omega^2 m_2) - k_2^2} \quad (9.13)$$

It is helpful to introduce the following notation:

$$\omega_p \overset{\Delta}{=} \sqrt{k_1/m_1} = \text{natural frequency of the primary system alone}$$

$$\omega_a \overset{\Delta}{=} \sqrt{k_2/m_2} = \text{natural frequency of the absorber system alone}$$

$$u_{st} \overset{\Delta}{=} P_1/k_1 = \text{static deflection of the primary system}$$

$$\mu \overset{\Delta}{=} m_2/m_1 = \text{ratio of the absorber mass to the primary mass}$$

Then Eqs. 9.13 can be written in the form

$$\frac{U_1}{u_{st}} = \frac{1 - (\Omega/\omega_a)^2}{[1 + \mu(\omega_a/\omega_p)^2 - (\Omega/\omega_p)^2][1 - (\Omega/\omega_a)^2] - \mu(\omega_a/\omega_p)^2}$$

$$\frac{U_2}{u_{st}} = \frac{1}{[1 + \mu(\omega_a/\omega_p)^2 - (\Omega/\omega_p)^2][1 - (\Omega/\omega_a)^2] - \mu(\omega_a/\omega_p)^2} \quad (9.14)$$

From the first of Eqs. 9.14, we can conclude that the steady-state amplitude of the primary mass can be reduced to zero at excitation frequency Ω by attaching an absorber system with frequency $\omega_a = \Omega$. That is, if the stiffness and mass parameters of the undamped absorber system are "tuned" so that $\omega_a \triangleq \sqrt{k_2/m_2} = \Omega$, the amplitude of the primary mass will be reduced to zero at the operating frequency Ω.

But now we have a 2-DOF system whose main mass is not moving but whose absorber mass definitely is moving. From the second of Eqs. 9.14, we get the following response of the absorber mass when $\omega_a = \Omega$:

$$U_2 = -\left(\frac{\omega_p}{\omega_a}\right)^2 \frac{u_{st}}{\mu} = -\frac{P_1}{k_2} \tag{9.15}$$

Since the main mass is not moving when $\omega_a = \Omega$, the force in the absorber spring will be

$$f_2 = k_2 u_2 = k_2 U_2 \cos \Omega t = -P_1 \cos \Omega t \tag{9.16}$$

Therefore, if the absorber is tuned so that its natural frequency ω_a is exactly equal to the excitation frequency Ω, the absorber will exert a force on the main mass that is exactly equal and opposite to the original disturbing force, regardless of the separate values k_2 and m_2.

The MATLAB .m-file that encodes the following solution can be found on the book's website. Its filename is

- sd2fig9_4.m: solution for vibration absorber plot in Fig. 9.4

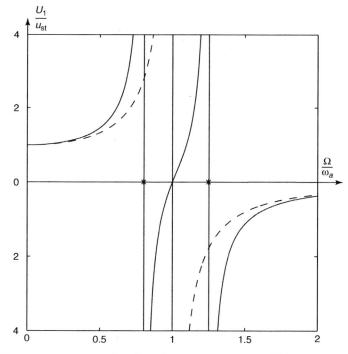

Figure 9.4 Frequency-response plots for primary mass m_1. $\mu = 0.20$, $\omega_p = \omega_a$. Asterisks, natural frequencies with absorber: $\omega_1/\omega_a = 0.8011$, $\omega_2/\omega_a = 1.2483$. Dashed lines, primary system alone; solid lines, primary system with absorber.

Figure 9.4 shows a sample of frequency-response magnitude curves for the primary mass, both for the primary system alone (dashed curves) and for the primary system with absorber system attached to form a 2-DOF system (solid curves).

If the primary system is a machine with a rotating part that causes the harmonic excitation, as the machine's speed increases from zero up to the operating speed Ω, the excitation frequency will pass through the first resonant frequency, ω_1. However, if the excitation does not dwell at that resonant frequency, the amplitudes of vibration of the two masses will not become large. Other design considerations, such as how to limit the amplitude of the absorber mass and the maximum force in the absorber spring, and whether damping should be introduced, are discussed in Ref. [9.1]. Problem 19.1 contrasts the design of a conventional vibration absorber with the design of a piezoelectric vibration absorber.

REFERENCE

[9.1] D. J. Inman, *Engineering Vibration*, 2nd ed., Prentice Hall, Upper Saddle River, NJ, 2001.

PROBLEMS

Problem Set 9.1

9.1 The stiffness and mass matrices of the two-story building shown in Fig. P9.1 are

$$\mathbf{K} = \begin{bmatrix} 1000 & -1000 \\ -1000 & 2000 \end{bmatrix} \text{ kips/in.,}$$

$$\mathbf{M} = \begin{bmatrix} 2 & 0 \\ 0 & 3 \end{bmatrix} \text{ kip-sec}^2\text{/in.}$$

(a) Determine the two natural frequencies (in hertz) of the structure. (b) Determine the two corresponding mode shapes. Scale them so that the maximum displacement is 1.0. Sketch the two modes.

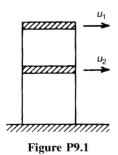

Figure P9.1

9.2 Figure P9.2 shows a lumped-mass model of an airplane wing. The stiffness and mass matrices for this model are

$$\mathbf{K} = k \begin{bmatrix} 4 & -10 \\ -10 & 30 \end{bmatrix}, \qquad \mathbf{M} = m \begin{bmatrix} 2 & 0 \\ 0 & 5 \end{bmatrix}$$

(a) Determine the two natural frequencies of the structure, ω_1 and ω_2, and express them in terms of the stiffness and mass coefficients k and m, respectively.
(b) Determine the two corresponding mode shapes. Scale them so that the maximum displacement is 1.0. Sketch the two modes.

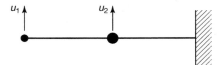

Figure P9.2

9.3 A slender, rigid uniform beam of mass m is supported by two linear springs, as shown in Fig. P9.3. (a) Derive the two equations of motion in terms of the end displacements u_1 and u_2. Write these equations in matrix format. (b) Determine the two natural frequencies of the structure, ω_1 and ω_2, and express them in terms of the stiffness and mass coefficients k and m, respectively. (c) Determine the two corresponding mode shapes. Scale them so that the maximum displacement is 1.0. Sketch the two modes.

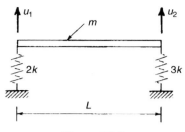

Figure P9.3

9.4 For the bar–spring–mass system in Fig. P9.4, assume that the bar is rigid but its mass is negligible, that the rotation of the bar remains small, and that gravity can be neglected. (a) For the coordinate system below (i.e., u_1 and u_2 are the vertical displacements of the masses m_1 and m_2, respectively), use Newton's Laws to derive the equations of motion, and write the equations in matrix format. (b) Determine the two natural frequencies of the structure, ω_1 and ω_2, and express them in terms of the stiffness and mass coefficients k and m, respectively. (c) Determine the two corresponding mode shapes. Scale each so that its maximum displacement is 1.0. Sketch the two modes.

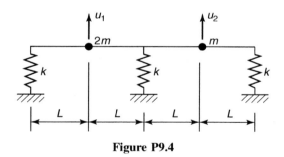

Figure P9.4

9.5 The frequencies and mode shapes of the structure in Fig. P9.1 are

$$\omega_1^2 = \frac{500}{3} \text{ rad}^2/\text{sec}^2, \qquad \omega_2^2 = 1000 \text{ rad}^2/\text{sec}^2$$

$$\phi_1 = \begin{Bmatrix} 1 \\ \frac{2}{3} \end{Bmatrix}, \qquad \phi_2 = \begin{Bmatrix} 1 \\ -1 \end{Bmatrix}$$

If the initial conditions are

$$u_1(0) = 2 \text{ in.}, \quad u_2(0) = 1 \text{ in.}, \quad \dot{u}_1(0) = \dot{u}_2(0) = 0$$

determine expressions for $u_1(t)$ and $u_2(t)$.

Problem Set 9.2

9.6 Replace the spring between masses m_1 and m_2 in Example 9.1 by a weaker coupling spring having a stiffness $k/20$. (a) Determine the two natural frequencies of the system, ω_1 and ω_2, and express them in terms of the stiffness coefficient k and the mass coefficient m. (b) Determine the two corresponding mode shapes, and scale each so that its maximum displacement is 1.0. Sketch the two modes. (c) For the initial conditions

$$u_2(0) = u_0, \qquad u_1(0) = \dot{u}_1(0) = \dot{u}_2(0) = 0$$

determine expressions for the displacements $u_1(t)$ and $u_2(t)$. (d) Show that displacements $u_1(t)$ and $u_2(t)$ can be written in the form

$$u_1(t) = u_0 \left(\sin \frac{\omega_2 + \omega_1}{2} t \right) \left(\sin \frac{\omega_2 - \omega_1}{2} t \right)$$

$$u_2(t) = u_0 \left(\cos \frac{\omega_2 + \omega_1}{2} t \right) \left(\cos \frac{\omega_2 - \omega_1}{2} t \right)$$

Note: This form indicates that the system will exhibit a *beat phenomenon*; that is, the motion appears to be harmonic motion at the faster frequency $(\omega_2 + \omega_1)/2$ with a slowly varying amplitude at frequency $(\omega_2 - \omega_1)/2$.

Problem Set 9.3

> For problems whose number is preceded by a **C**, you are to write a computer program and use it to produce the plot(s) requested. *Note*: MATLAB .m-files for the many of the plots in this book may be found on the book's website.

C 9.7 Axial vibration of the uniform rod and tip mass system in Fig. P9.7 is to be modeled by a 2-DOF system based on the shape functions

$$\psi_1(x) = \frac{x}{L}, \qquad \psi_2(x) = \left(\frac{x}{L}\right)^2$$

The mass of the "particle" at $x = L$ is $m = \rho A L/5$. $AE = $ constant and $\rho A = $ constant. (a) Determine the equations of motion for this 2-DOF model. (b) Determine the two natural frequencies (squared), ω_1^2 and ω_2^2. (c) Determine expressions for the two mode shapes, and write them in the form of $\phi_i(x)$, as given in Eqs. 12 and 15 of Example 9.5. (d) Use MATLAB to plot these two mode shapes, as in Fig. 2 of Example 9.5.

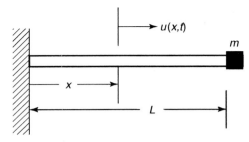

Figure P9.7

C 9.8 Axial vibration of the system in Fig. P9.8 is to be modeled by a 2-DOF system based on the shape functions

$$\psi_1(x) = \frac{x}{L}, \qquad \psi_2(x) = \left(\frac{x}{L}\right)^2$$

The stiffness of the linear spring at $x = L$ is $k = AE/2L$. AE = constant and ρA = constant. (a) Determine the equations of motion for this 2-DOF model. (b) Determine the two natural frequencies (squared), ω_1^2 and ω_2^2. (c) Determine expressions for the two mode shapes, and write them in the form of $\phi_i(x)$, as given in Eqs. 12 and 15 of Example 9.5. (d) Use MATLAB to plot these two mode shapes, as in Fig. 2 of Example 9.5.

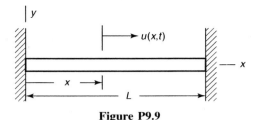

Figure P9.9

C 9.10 Transverse vibration of the uniform cantilever beam in Fig. P9.10 is to be modeled by a 2-DOF system based on the shape functions

$$\psi_1(x) = \left(\frac{x}{L}\right)^2, \qquad \psi_2(x) = \left(\frac{x}{L}\right)^3$$

EI = constant and ρA = constant. (a) Determine the equations of motion for this 2-DOF transverse-vibration model. (b) Determine the two natural frequencies (squared), ω_1^2 and ω_2^2. (c) Determine expressions for the two transverse-vibration mode shapes, and write them in the form of $\phi_i(x)$, in the same manner as given in Eqs. 12 and 15 of Example 9.5 for axial-vibration modes. (d) Use MATLAB to plot these two transverse-vibration mode shapes, in the same manner as the axial-vibration modes shown in Fig. 2 of Example 9.5.

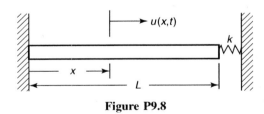

Figure P9.8

C 9.9 Axial vibration of the uniform fixed–fixed bar in Fig. P9.9 is to be modeled by a 2-DOF system based on the shape functions

$$\psi_1(x) = \frac{x}{L}\left[1 - \left(\frac{x}{L}\right)^2\right], \qquad \psi_2(x) = \left(\frac{x}{L}\right)^2\left[1 - \left(\frac{x}{L}\right)^2\right]$$

AE = constant and ρA = constant. (a) Determine the equations of motion for this 2-DOF model. (b) Determine the two natural frequencies (squared), ω_1^2 and ω_2^2. (c) Determine expressions for the two mode shapes and write them in the form of $\phi_i(x)$, as given in Eqs. 12 and 15 of Example 9.5. (d) Use MATLAB to plot these two mode shapes, as in Fig. 2 of Example 9.5.

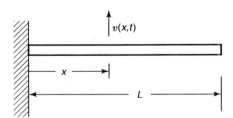

Figure P9.10

C 9.11 Transverse vibration of the uniform cantilever beam with tip mass, shown in Fig. P9.11, is to be modeled by a 2-DOF system based on the shape functions

$$\psi_1(x) = \left(\frac{x}{L}\right)^2, \qquad \psi_2(x) = \left(\frac{x}{L}\right)^3$$

The tip mass is $m = \rho AL/5$. EI = constant and ρA = constant. (a) Determine the equations of motion for this 2-DOF transverse-vibration model. (b) Determine the two natural frequencies (squared), ω_1^2 and ω_2^2. (c) Determine expressions for the two transverse-vibration mode shapes, and write them in the form of $\phi_i(x)$, in the same manner as given in Eqs. 12 and 15 of Example 9.5 for axial-vibration modes. **(d)** Use MATLAB to plot these two transverse-vibration mode shapes, in the same manner as the axial-vibration modes shown in Fig. 2 of Example 9.5.

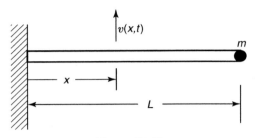

Figure P9.11

C 9.12 The uniform floor beam shown in Fig. P8.31 has a total mass $M = \rho AL$, and it supports a heavy machine of mass m at its center, $x = L/2$. Neglect the width of the machine in comparison with length L. Let the transverse deflection $v(x, t)$ be approximated by

$$v(x, t) = \psi_1(x)v_1(t) + \psi_2(x)v_2(t)$$

where the shape functions are

$$\psi_1(x) = \sin\frac{\pi x}{L}, \qquad \psi_2(x) = \sin\frac{2\pi x}{L}$$

(a) Derive the equations of motion for a 2-DOF assumed-modes model based on these shape functions, and write the equations of motion in matrix format: $\mathbf{M\ddot{q}} + \mathbf{Kq} = \mathbf{p}(t)$. **(b)** Let the magnitude of the machine's mass be $m = \rho AL/5$. Determine the two natural frequencies (squared), ω_1^2 and ω_2^2, for this model. **(c)** Comment on the effect of the added mass m on each of the two natural frequencies. **(d)** Determine expressions for the two transverse-vibration mode shapes, and write them as $\phi_i(x)$, in the same manner as given in Eqs. 12 and 15 of Example 9.5 for axial-vibration modes. **(e)** Use MATLAB to plot these two transverse-vibration mode shapes, in the same manner as the axial-vibration modes shown in Fig. 2 of Example 9.5.

C 9.13 The uniform floor beam shown in Fig. P8.32 has a total mass $M = \rho AL$, and it supports a heavy machine of mass m at $x = L/4$. Neglect the width of the machine in comparison with length L. Let the transverse deflection $v(x, t)$ be approximated by

$$v(x, t) = \psi_1(x)v_1(t) + \psi_2(x)v_2(t)$$

where the shape functions are

$$\psi_1(x) = \sin\frac{\pi x}{L}, \qquad \psi_2(x) = \sin\frac{2\pi x}{L}$$

(a) Derive the equations of motion for a 2-DOF assumed-modes transverse-vibration model based on these shape functions, and write the equations of motion in matrix format: $\mathbf{M\ddot{q}} + \mathbf{Kq} = \mathbf{p}(t)$. **(b)** Let the magnitude of the machine's mass be $m = \rho AL/5$. Determine the two natural frequencies (squared), ω_1^2 and ω_2^2, for this model. **(c)** Determine expressions for the two transverse-vibration mode shapes, and write them as $\phi_i(x)$, in the same manner as given in Eqs. 12 and 15 of Example 9.5 for axial-vibration modes. **(d)** Use MATLAB to plot these two transverse-vibration mode shapes, in the same manner as the axial-vibration modes shown in Fig. 2 of Example 9.5.

9.14 The uniform cantilever shaft in Fig. P8.34 has a cross-sectional area A and polar moment of inertia I_p, and is made of a metal whose shear modulus is G. A thin disk of polar moment of inertia I_A is attached to the shaft at end A. Torsional vibration of the shaft–disk system is to be represented by a 2-DOF assumed-modes torsion model of the form

$$\theta(x, t) = \psi_1(x)q_1(t) + \psi_2(x)q_2(t)$$

where the shape functions that are to be used are

$$\psi_1(x) = 1 - \frac{x}{L}, \qquad \psi_2(x) = 1 - \left(\frac{x}{L}\right)^2$$

(a) Derive the equations of motion for a 2-DOF assumed-modes torsion model based on these shape functions, and write the equations of motion in matrix format: $\mathbf{M\ddot{q}} + \mathbf{Kq} = \mathbf{p}(t)$. **(b)** Let $I_A = I_p L/4$. Determine the two natural frequencies (squared), ω_1^2 and ω_2^2, for this model. **(c)** Determine expressions for the two torsional mode shapes, and write them in the form of $\phi_i(x)$, in the same manner as given in Eqs. 12 and 15 of Example 9.5 for axial-vibration modes.

Problem Set 9.4

> Although all of the problems in Problem Set 9.4 are for 3-DOF systems, each has a rigid-body mode ($\omega_1 = 0$), so each system has only two flexible modes.

9.15 Determine the natural frequencies and mode shapes of the 3-DOF system shown in Fig. P9.15. Let $m_1 = m_2 = m_3 = m$; and let $k_1 = k_2 = k$. Sketch the three mode shapes. (Note that mode 1 is a rigid-body mode; that is, $\omega_1 = 0$.)

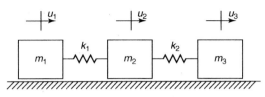

Figure P9.15

9.16 Determine the natural frequencies and mode shapes of the 3-DOF system shown in Fig. P9.15. Let $m_1 = m_3 = m$ and $m_2 = 2m$; and let $k_1 = k_2 = k$. Sketch the three mode shapes. (Note that mode 1 is a rigid-body mode; that is, $\omega_1 = 0$.)

9.17 Determine the natural frequencies and mode shapes of the 3-DOF system shown in Fig. P9.15. Let $m_1 = m_2 = m$ and $m_3 = 2m$; and let $k_1 = k$ and $k_2 = 2k$. Sketch the three mode shapes. (Note that mode 1 is a rigid-body mode; that is, $\omega_1 = 0$.)

9.18 The axial vibration of a free–free uniform bar is to be studied by using a 3-DOF lumped-mass model, as shown in Fig. P9.18. In matrix form, the resulting equations of motion are

$$m \begin{bmatrix} 1 & 0 & 0 \\ 0 & 2 & 0 \\ 0 & 0 & 1 \end{bmatrix} \begin{Bmatrix} \ddot{u}_1 \\ \ddot{u}_2 \\ \ddot{u}_3 \end{Bmatrix} + k \begin{bmatrix} 1 & -1 & 0 \\ -1 & 2 & -1 \\ 0 & -1 & 1 \end{bmatrix} \begin{Bmatrix} u_1 \\ u_2 \\ u_3 \end{Bmatrix} = \begin{Bmatrix} 0 \\ 0 \\ 0 \end{Bmatrix}$$

where $m = \rho A L/4$ and $k = 2AE/L$. (a) Determine the natural frequencies of this 3-DOF lumped-mass model. (Note that mode 1 is a rigid-body mode; that is, $\omega_1 = 0$.)

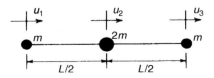

Figure P9.18

(b) Solve for the three mode shapes. Scale each mode so that the maximum displacement is 1.0, and sketch the modes.

9.19 The 3-DOF system shown in Fig. P9.19 has the following equation of motion:

$$m \begin{bmatrix} 1 & 1 & 0 \\ 1 & 5 & 1 \\ 0 & 1 & 1 \end{bmatrix} \begin{Bmatrix} \ddot{w}_1 \\ \ddot{w}_2 \\ \ddot{w}_3 \end{Bmatrix} + k \begin{bmatrix} 1 & 0 & 0 \\ 0 & 0 & 0 \\ 0 & 0 & 1 \end{bmatrix} \begin{Bmatrix} w_1 \\ w_2 \\ w_3 \end{Bmatrix} = \begin{Bmatrix} 0 \\ 0 \\ 0 \end{Bmatrix}$$

where $w_2 \equiv u_2$ and where the relative displacements $w_1 = u_1 - u_2$ and $w_3 = u_3 - u_2$ are used. (Note that this makes u_2 a rigid-body degree of freedom for the system.) (a) Determine the natural frequencies (squared), ω_r^2, of this system. (b) Determine the three mode shapes (eigenvectors) $\boldsymbol{\phi}_{wi}$ of this system, expressed in terms of the **w** coordinates and scaled so that the maximum entry in each of the vectors is 1. (c) Sketch the mode shapes for this system in terms of absolute displacements u_1, u_2, and u_3. Note that you will first need to determine the mode shapes $\boldsymbol{\phi}_{wi}$ in the **w** coordinates and then convert to modes $\boldsymbol{\phi}_{ui}$ in the **u** coordinates by using the coordinate transformation equations above.

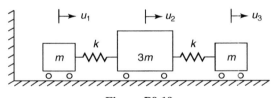

Figure P9.19

Problem Set 9.5

C 9.20 In Example 9.8, which illustrates the method of *mode superposition* for determining the frequency response of a 2-DOF system, the excitation force is applied to mass 1. Repeat *all* of the steps in Example 9.8, replacing the given excitation force with the following excitation force vector:

$$\mathbf{p}(t) \equiv \begin{Bmatrix} p_1(t) \\ p_2(t) \end{Bmatrix} = \begin{Bmatrix} 0 \\ P_2 \end{Bmatrix} \cos \Omega t$$

In particular, determine expressions for the frequency-response functions $kU_1(\Omega)/P_2$ and $kU_2(\Omega)/P_2$ corresponding to Eqs. 15a and b of Example 9.8. Use MATLAB to create plots of these corresponding to the two nondimensionalized plots in Fig. 3 of Example 9.8.

9.21 Example 9.8 illustrates the method of *mode superposition*, which is employed extensively in Chapter 11. A more direct solution for the steady-state response of the system in Example 9.8 is obtained by substituting Eqs. 14a and b directly into the equation of motion, which is given in the problem statement for Example 9.8. Show that this leads to the frequency-response expressions in Eqs. 15a and b of Example 9.8.

9.22 Using information from Problems 9.1 and 9.5, and using the mode-superposition method of Example 9.8, determine the steady-state response of the two-story building of Fig. P9.1 to the following harmonic excitation:

$$\mathbf{p}(t) = \begin{Bmatrix} 10 \\ 0 \end{Bmatrix} \sin 25t \quad \text{kips}$$

9.23 Using the mode-superposition method of Example 9.8, show that antisymmetric harmonic excitation given by

$$\mathbf{p}(t) \equiv \begin{Bmatrix} p_1(t) \\ p_2(t) \end{Bmatrix} = \begin{Bmatrix} P \\ -P \end{Bmatrix} \cos \Omega t$$

produces steady-state response only in the antisymmetric mode for the symmetric system of Example 9.1.

Problem Set 9.6

C 9.24 One method of reducing the vibration amplitude of a SDOF system subjected to harmonic excitation is to attach a tuned vibration absorber, which is a second spring–mass system, as shown in Fig. P9.24. (a) For the resulting 2-DOF system shown below, determine the equations of motion. (b) Let the steady-state response be given by

$$u_1 = U_1 \cos \Omega t, \quad u_2 = U_2 \cos \Omega t$$

and show that

$$U_1(\Omega) = \frac{(k_2 - m_2 \Omega^2) P_1}{D(\Omega)}, \quad U_2(\Omega) = \frac{k_2 P_1}{D(\Omega)}$$

where

$$D(\Omega) = (k_1 + k_2 - m_1 \Omega^2)(k_2 - m_2 \Omega^2) - k_2^2$$

(c) The absorber is tuned so that $k_2/m_2 = k_1/m_1$. Thus, when $\Omega^2 = k_1/m_1$, that is, when the original system is excited at resonance, the response amplitude U_1 is reduced to zero. Let $m_2/m_1 = 0.25$ for a particular "tuned" absorber system. Use MATLAB to plot the frequency-response functions $k_1 U_1/P_1$ and $k_1 U_2/P_1$ versus the frequency ratio $r_a = \Omega/\omega_a$, where $\omega_a = \sqrt{k_2/m_2}$. (See Fig. 3 of Example 9.8 and Fig. 9.4.)

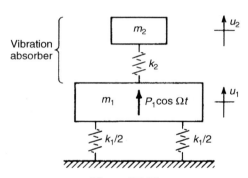

Figure P9.24

10

Vibration Properties of MDOF Systems: Modes, Frequencies, and Damping

In Chapter 8 the equations of motion of MDOF systems were derived, and in Chapter 9 you were introduced to techniques for obtaining natural frequencies and natural modes as you considered free vibration of undamped 2-DOF systems. In the present chapter a number of properties of natural frequencies and natural modes of undamped MDOF systems are discussed. We also introduce ways to characterize damping in MDOF systems and to free vibration of damped MDOF systems.

Upon completion of this chapter you should be able to:

- Define the following terms: positive definite matrix, singular matrix, natural mode, normal mode, modal stiffness, modal mass, orthogonality, modal matrix, eigenvalue matrix, Rayleigh quotient, Ritz vector, modal damping, Rayleigh damping, and complex mode.
- Determine the mode shapes of an undamped MDOF system given the distinct and/or repeated natural frequencies of the system.
- Discuss the condition that leads to rigid-body modes of an undamped system.
- Derive the orthogonality equations that must be satisfied by modes that correspond to distinct frequencies of an undamped MDOF system.
- Obtain an estimate of the fundamental frequency of an undamped MDOF system by using the Rayleigh Method, and state the fundamental property of the Rayleigh quotient.
- Obtain an estimate of $\hat{N}$ frequencies of an undamped N-DOF system ($\hat{N} < N$) by using the Rayleigh–Ritz method. Discuss how these frequency estimates could be improved.
- Obtain an expression for the physical damping matrix $\mathbf{C}$ that corresponds to uncoupled damping with specified modal damping factors ζ_r.
- Write the equations of motion of a viscous-damped N-DOF system in state-variable form, and write the corresponding generalized state algebraic eigenvalue equation.
- Solve a simple state eigenvalue problem and discuss the significance of complex-conjugate eigenvalues and eigenvectors.
- Develop the orthogonality equations for state eigenvectors.

10.1 SOME PROPERTIES OF NATURAL FREQUENCIES AND NATURAL MODES OF UNDAMPED MDOF SYSTEMS

The purpose of the present section is to discuss some of the more important properties of natural frequencies and natural modes of undamped systems. In the interest of conciseness, most of the equations will be written in symbolic matrix form. Before considering properties of frequencies and modes, it will be useful for us to consider some the properties of the stiffness and mass matrices encountered in structural dynamics problems.

10.1.1 Properties of K and M

In Section 8.4, stiffness matrix **K** and mass matrix **M** were related to strain energy and kinetic energy, respectively, by the quadratic forms

$$\mathcal{V} = \tfrac{1}{2}\mathbf{u}^T\mathbf{K}\mathbf{u}, \qquad \mathcal{T} = \tfrac{1}{2}\dot{\mathbf{u}}^T\mathbf{M}\dot{\mathbf{u}} \qquad (10.1)$$

where $\mathbf{u}(t)$ is the displacement vector, and **K** and **M** were seen to be symmetric, that is, $\mathbf{K}^T = \mathbf{K}$ and $\mathbf{M}^T = \mathbf{M}$.[1]

For most structures **K** and **M** are *positive definite matrices*. That is, when arbitrary vectors $\mathbf{u}$ and $\dot{\mathbf{u}}$ are chosen and $\mathcal{V}$ and $\mathcal{T}$ are computed from Eqs. 10.1, the resulting values of $\mathcal{V}$ and $\mathcal{T}$ are positive, except for the trivial cases $\mathbf{u} = \mathbf{0}$ and $\dot{\mathbf{u}} = \mathbf{0}$, respectively. The physical significance of positive definiteness is as follows:

- For any arbitrary displacement of a system with positive definite **K** from its undeformed configuration, the strain energy will be positive.
- For any arbitrary velocity distribution of a system with positive definite **M**, a positive kinetic energy will result.

Exceptions to the positive definiteness of **K** are systems that have rigid-body freedom. Then the displacement **u** can be a rigid-body displacement. In this case **K** is said to be *positive semidefinite*; that is, $\mathcal{V}$ can be either zero (for rigid-body motion) or greater than zero (for motion resulting in deformation, i.e., change of shape, of the structure). When **K** is positive semidefinite, the determinant of the stiffness matrix vanishes; that is, $|\mathbf{K}| \equiv \det(\mathbf{K}) = 0$. A matrix whose determinant vanishes is called a *singular matrix*. Examples of semidefinite systems are given in Examples 10.2 and 10.3.

Exceptions to the positive definiteness of **M** are systems for which there are degrees of freedom that have no associated inertia. Such a model is sometimes generated by the lumped-mass approach, as illustrated in Fig. 10.1 and discussed further in Section 14.2.

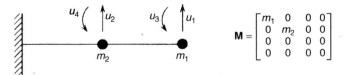

Figure 10.1 Beam with a positive semidefinite mass matrix.

---

[1] The symmetry of stiffness coefficients, that is, $k_{ij} = k_{ji}$, for linear structures is often referred to as *Maxwell's reciprocity relationship*. The form of k_{ij} obtained by the assumed modes method [e.g., Eq. (8.16)] shows that $k_{ij} = k_{ji}$. Problem 10.1 illustrates how a nonsymmetric stiffness matrix or mass matrix might arise.

In Fig. 10.1 mass is associated with the two translational degrees of freedom, but no rotational inertia is associated with the two rotational degrees of freedom, and there is no mass coupling. Hence, the mass matrix **M** in Fig. 10.1 is singular.

10.1.2 Eigensolution

The free-vibration equation of motion of an undamped MDOF system, written in symbolic matrix form, is

$$\mathbf{M\ddot{u} + Ku = 0} \tag{10.2}$$

where **M** and **K** are $N \times N$ matrices and $\mathbf{u}(t)$ is the corresponding $N \times 1$ vector of physical or generalized displacement coordinates.

Harmonic motion, given by

$$\mathbf{u}(t) = \mathbf{U}\cos(\omega t - \alpha) \tag{10.3}$$

may be substituted into Eq. 10.2 to give the Nth-order *algebraic eigenvalue problem*

$$[\mathbf{K} - \omega^2 \mathbf{M}]\mathbf{U} = \mathbf{0} \tag{10.4}$$

For there to be a nontrivial solution of Eq. 10.4 it is necessary that

$$\det(\mathbf{K} - \omega^2 \mathbf{M}) = 0 \tag{10.5}$$

This is called the *characteristic equation*. When the determinant of Eq. 10.5 is expanded, there results a polynomial equation of degree N in ω^2, whose roots are the *eigenvalues*, or squared *natural frequencies*, ω_r^2. These can be ordered from lowest to highest:

$$0 \leq \omega_1^2 \leq \omega_2^2 \leq \cdots \leq \omega_r^2 \leq \cdots \leq \omega_N^2 \tag{10.6}$$

Corresponding to each eigenvalue ω_r^2 there will be an *eigenvector*, or *natural mode*, $\mathbf{U}_r$, where

$$\mathbf{U}_r = \begin{Bmatrix} U_1 \\ U_2 \\ \vdots \\ U_N \end{Bmatrix}_r, \qquad r = 1, 2, \ldots, N \tag{10.7}$$

The modes are determined only to within a constant multiplier. Thus, modes may be scaled in any convenient manner.

10.1.3 Scaling (Normalizing) the Modes

If the value of one of the N elements of a natural mode vector $\mathbf{U}_r$ is assigned a specified value, the remaining $N - 1$ elements are determined uniquely. Thus, we say that the *mode shape* is determined uniquely, but not the mode's amplitude. The process of scaling a natural mode so that each of its elements has a unique value is called *normalization*, and the resulting modal vectors are called *normal modes*.

We denote by $\boldsymbol{\phi}_r$ a mode that has been scaled to make its amplitude unique, and we assume $\boldsymbol{\phi}_r$ to be dimensionless; that is, an arbitrary modal vector corresponding to ω_r^2 can be written in the form

$$\mathbf{U}_r = c_r \boldsymbol{\phi}_r \tag{10.8}$$

where c_r is a scaling constant whose units are such that $\boldsymbol{\phi}_r^T \mathbf{M} \boldsymbol{\phi}_r$ has the dimensions of mass.[2]

There are three commonly employed procedures for normalizing modes:

1. Scale the rth mode so that $(\phi_i)_r = 1$ at a specified coordinate i.
2. Scale the rth mode so that $(\phi_i)_r = 1$ where $|(\phi_i)_r| = \max_j |(\phi_j)_r|$; that is, the maximum displacement is at coordinate i.
3. Scale the rth mode so that its generalized mass, or *modal mass*, defined by

$$\boxed{M_r = \boldsymbol{\phi}_r^T \mathbf{M} \boldsymbol{\phi}_r} \tag{10.9}$$

has a specified value. The value $M_r = 1$ is used frequently. As noted previously, it is convenient to scale $\boldsymbol{\phi}_r$ so that the product $\boldsymbol{\phi}_r^T \mathbf{M} \boldsymbol{\phi}_r$ has the units of mass. Thus, $M_r = 1$ kg (or 1 slug) is implied when we set $M_r = 1$.

The generalized stiffness, or *modal stiffness*, for the rth mode is defined as

$$\boxed{K_r = \boldsymbol{\phi}_r^T \mathbf{K} \boldsymbol{\phi}_r} \tag{10.10}$$

If Eq. 10.4 is written for rth mode and premultiplied by $\boldsymbol{\phi}_r^T$, we obtain

$$\boldsymbol{\phi}_r^T \mathbf{K} \boldsymbol{\phi}_r = \omega_r^2 (\boldsymbol{\phi}_r^T \mathbf{M} \boldsymbol{\phi}_r)$$

Therefore, the square of the natural frequency is related to the modal stiffness and modal mass by

$$\boxed{\omega_r^2 = \frac{K_r}{M_r}} \tag{10.11}$$

10.1.4 Mode Shapes: Distinct Frequency Cases

If ω_r is a distinct eigenvalue, the modal vector $\boldsymbol{\phi}_r$ can be determined in the following manner. Let the *dynamic matrix* be defined by the equation

$$\mathbf{D}(\omega_r) \triangleq \mathbf{K} - \omega_r^2 \mathbf{M} \tag{10.12}$$

Assume that coordinate 1 is not a node point of the rth mode (i.e., it is not a point whose displacement is zero) and partition Eq. 10.4 as follows:

$$\begin{bmatrix} \mathbf{D}_{aa}(\omega_r) & \mathbf{D}_{ab}(\omega_r) \\ \mathbf{D}_{ba}(\omega_r) & \mathbf{D}_{bb}(\omega_r) \end{bmatrix} \begin{Bmatrix} 1 \\ \boldsymbol{\phi}_b \end{Bmatrix}_r = \begin{Bmatrix} 0 \\ \mathbf{0}_b \end{Bmatrix} \tag{10.13}$$

---

[2] The reason for this particular definition of *dimensionless* is that some vectors $\mathbf{U}$ may contain a mixture of types of coordinates, for example, translations and rotations. Thus, it would not be possible to make all components of $\boldsymbol{\phi}$ simultaneously dimensionless in the usual sense of the word.

10.1 Some Properties of Natural Frequencies and Natural Modes of Undamped MDOF Systems

where $\boldsymbol{\phi}_r$ has been scaled by setting $(\phi_1)_r = 1$, where $\{\boldsymbol{\phi}_b\}_r$ has the form

$$\{\boldsymbol{\phi}_b\}_r \equiv \begin{Bmatrix} \phi_2 \\ \phi_3 \\ \vdots \\ \phi_N \end{Bmatrix}_r \tag{10.14}$$

and where $\mathbf{0}_b$ is a vector of $N - 1$ zeros. Since ω_r is a natural frequency, Eq. 10.4 states that $\det[\mathbf{D}(\omega_r)] = 0$; that is, $\mathbf{D}(\omega_r)$ is singular. The *rank* of a matrix is defined as the largest submatrix having nonzero determinant. If ω_r is a distinct natural frequency, the rank of $\mathbf{D}(\omega_r)$ is $N - 1$, and the coordinates can be labeled so that $\mathbf{D}_{bb}(\omega_r)$ is nonsingular. Therefore, $[\mathbf{D}_{bb}(\omega_r)]^{-1}$ exists, and the lower partition of Eq. 10.13 can be solved for the remainder of the mode shape $\boldsymbol{\phi}_r$. That is,

$$\{\boldsymbol{\phi}_b\}_r = -[\mathbf{D}_{bb}(\omega_r)]^{-1}\mathbf{D}_{ba}(\omega_r) \tag{10.15}$$

The example that follows illustrates the characteristic polynomial and the use of Eq. 10.15.

Example 10.1 (a) For the system shown in Fig. 1, determine the values of ω_r^2 for $r = 1, 2, 3$, and sketch the characteristic polynomial for $0 \leq \omega \leq 2.0$. (b) Solve for the mode shape corresponding to ω_2.

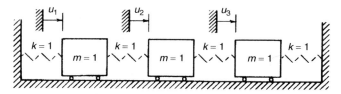

Figure 1 A 3-DOF spring–mass system.

The stiffness matrix and mass matrix for this system are

$$\mathbf{K} = \begin{bmatrix} 2 & -1 & 0 \\ -1 & 2 & -1 \\ 0 & -1 & 2 \end{bmatrix}, \quad \mathbf{M} = \begin{bmatrix} 1 & 0 & 0 \\ 0 & 1 & 0 \\ 0 & 0 & 1 \end{bmatrix} \tag{1}$$

SOLUTION (a) The dynamic matrix is

$$\mathbf{D}(\omega) = \begin{bmatrix} 2 - \omega^2 & -1 & 0 \\ -1 & 2 - \omega^2 & -1 \\ 0 & -1 & 2 - \omega^2 \end{bmatrix} \tag{2}$$

Expanding $\det[\mathbf{D}(\omega)]$ using the first row (or first column), we get

$$\det[\mathbf{D}(\omega)] = (2 - \omega^2) \begin{vmatrix} 2 - \omega^2 & -1 \\ -1 & 2 - \omega^2 \end{vmatrix} + 1 \begin{vmatrix} -1 & -1 \\ 0 & 2 - \omega^2 \end{vmatrix}$$

so the characteristic polynomial is

$$\det[\mathbf{D}(\omega)] = (2 - \omega^2)(\omega^4 - 4\omega^2 + 2) \tag{3}$$

and its three roots, the eigenvalues, are

$$\omega_1^2 = 2 - \sqrt{2}, \qquad \omega_2^2 = 2, \qquad \omega_3^2 = 2 + \sqrt{2} \qquad \textbf{Ans. (a)} \tag{4}$$

The characteristic polynomial can be computed for values of ω in the range $0 \leq \omega \leq 2.0$ and plotted as shown in Fig. 2. Note that the characteristic polynomial crosses zero at each of the three natural frequencies.

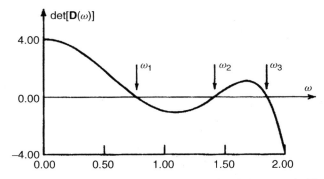

Figure 2 Characteristic polynomial for the 3-DOF system in Fig. 1.

(b) To solve for the second mode shape, we can partition the dynamic matrix $\mathbf{D}(\omega_2)$ as in Eq. (10.13):

$$\begin{bmatrix} \mathbf{D}_{aa}(\omega_r) & \mathbf{D}_{ab}(\omega_r) \\ \mathbf{D}_{ba}(\omega_r) & \mathbf{D}_{bb}(\omega_r) \end{bmatrix} = \begin{bmatrix} 0 & -1 & 0 \\ -1 & 0 & -1 \\ 0 & -1 & 0 \end{bmatrix} \tag{5}$$

$$\mathbf{D}_{bb}(\omega_2) = \begin{bmatrix} 0 & -1 \\ -1 & 0 \end{bmatrix}, \qquad \mathbf{D}_{ba}(\omega_2) = \begin{Bmatrix} -1 \\ 0 \end{Bmatrix} \tag{6}$$

The inverse of $\mathbf{D}_{bb}(\omega_2)$ is

$$[\mathbf{D}_{bb}(\omega_2)]^{-1} = \begin{bmatrix} 0 & -1 \\ -1 & 0 \end{bmatrix} \tag{7}$$

and from Eq. 10.15 we get

$$\{\boldsymbol{\phi}_b\}_2 = -[\mathbf{D}_{bb}(\omega_2)]^{-1}\mathbf{D}_{ba}(\omega_2)$$

$$= -\begin{bmatrix} 0 & -1 \\ -1 & 0 \end{bmatrix}\begin{Bmatrix} -1 \\ 0 \end{Bmatrix} = \begin{Bmatrix} 0 \\ -1 \end{Bmatrix} \tag{8}$$

10.1 Some Properties of Natural Frequencies and Natural Modes of Undamped MDOF Systems

So since $(\phi_1)_2 = 1$, mode 2 becomes

$$\phi_2 = \begin{Bmatrix} 1 \\ 0 \\ -1 \end{Bmatrix} \qquad \textbf{Ans. (b)} \quad (9)$$

as shown in Fig. 3.

Figure 3 Mode 2 of the 3-DOF system in Fig. 1.

In the next two examples, some features of semidefinite systems, that is, systems with rigid-body modes, are illustrated.

Example 10.2 The Assumed-Modes Method of Section 8.4 (or the finite element method, which is discussed in Chapter 14) may be used to derive the following 3-DOF model of a uniform bar that undergoes unrestrained axial motion. As shown in Fig. 1, the bar has a total length of $2L$, and the displacement coordinates u_1, u_2, and u_3 are the physical displacements of the left end, center, and right end of the bar, respectively. $\rho A = $ constant and $EA = $ constant.

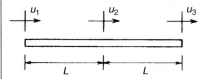

Figure 1 Axial motion of an unrestrained uniform bar.

The matrix equation of motion for this assumed-modes model is

$$\frac{\rho A L}{6} \begin{bmatrix} 2 & 1 & 0 \\ 1 & 4 & 1 \\ 0 & 1 & 2 \end{bmatrix} \begin{Bmatrix} \ddot{u}_1 \\ \ddot{u}_2 \\ \ddot{u}_3 \end{Bmatrix} + \frac{EA}{L} \begin{bmatrix} 1 & -1 & 0 \\ -1 & 2 & -1 \\ 0 & -1 & 1 \end{bmatrix} \begin{Bmatrix} u_1 \\ u_2 \\ u_3 \end{Bmatrix} = \begin{Bmatrix} 0 \\ 0 \\ 0 \end{Bmatrix}$$

(a) Show that the stiffness matrix is singular. (b) Solve for the natural frequencies, and thus show that $\omega_1 = 0$. (c) Use the procedure of Eq. 10.15 to determine the corresponding rigid-body mode.

SOLUTION (a) Evaluate the determinant of the stiffness matrix, $\det(\mathbf{K})$.

$$\det(\mathbf{K}) = \frac{EA}{L} \begin{vmatrix} 1 & -1 & 0 \\ -1 & 2 & -1 \\ 0 & -1 & 1 \end{vmatrix}$$

Therefore, expanding on the first row of the determinant on the right-hand side, we get

$$\det(\mathbf{K}) = \frac{EA}{L}\left[\begin{vmatrix} 2 & -1 \\ -1 & 1 \end{vmatrix} + \begin{vmatrix} -1 & -1 \\ 0 & 1 \end{vmatrix}\right] = 0 \quad \text{Ans. (a)} \quad (1)$$

Since $\det(\mathbf{K}) = 0$, $\mathbf{K}$ is a singular matrix. This is consistent with the fact that the bar can move axially as a rigid body.

(b) Set up the characteristic polynomial and determine its roots. To simplify the notation, let

$$\lambda \equiv \omega^2 \frac{\rho L^2}{6E} \quad (2)$$

Then $\mathbf{D}(\omega)$ can be written in partitioned form as

$$\mathbf{D}(\lambda) = \begin{bmatrix} 1-2\lambda & -1-\lambda & 0 \\ -1-\lambda & 2(1-2\lambda) & -1-\lambda \\ 0 & -1-\lambda & 1-2\lambda \end{bmatrix} \quad (3)$$

Expanding $\det[\mathbf{D}(\lambda)]$, we obtain the characteristic polynomial

$$\det[\mathbf{D}(\lambda)] = \lambda(1-2\lambda)(-2+\lambda) = 0 \quad (4)$$

The roots of the characteristic polynomial are thus

$$\lambda_1 = 0, \quad \lambda_2 = \tfrac{1}{2}, \quad \lambda_3 = 2 \quad (5)$$

so combining Eqs. 2 and 5, we get

$$\omega_1^2 = 0, \quad \omega_2^2 = 3\frac{E}{\rho L^2}, \quad \omega_3^2 = 12\frac{E}{\rho L^2} \quad \text{Ans. (b)} \quad (6)$$

(c) For the rigid-body mode, $\omega_1^2 = 0$. Therefore, use Eq. 10.13 to determine $\boldsymbol{\phi}_1$, which is a rigid-body mode. From Eq. 3, with $\lambda = 0$,

$$\mathbf{D}_{bb}(\omega_1) = \begin{bmatrix} 2 & -1 \\ -1 & 1 \end{bmatrix}, \quad \mathbf{D}_{ba}(\omega_1) = \begin{Bmatrix} -1 \\ 0 \end{Bmatrix} \quad (7)$$

The inverse of $\mathbf{D}_{bb}(\omega_1)$ is

$$[\mathbf{D}_{bb}(\omega_1)]^{-1} = \begin{bmatrix} 1 & 1 \\ 1 & 2 \end{bmatrix} \quad (8)$$

From Eq. 10.15 we get

$$\{\boldsymbol{\phi}_b\}_1 \equiv \begin{Bmatrix} \phi_2 \\ \phi_3 \end{Bmatrix}_1 = -[\mathbf{D}_{bb}(\omega_1)]^{-1}\mathbf{D}_{ba}(\omega_1)$$

$$= -\begin{bmatrix} 1 & 1 \\ 1 & 2 \end{bmatrix}\begin{Bmatrix} -1 \\ 0 \end{Bmatrix} = \begin{Bmatrix} 1 \\ 1 \end{Bmatrix} \quad (9)$$

Finally, since $(\phi_1)_1 = 1$, mode 1 becomes

$$\phi_1 = \begin{Bmatrix} 1 \\ 1 \\ 1 \end{Bmatrix} \qquad \text{Ans. (c)} \quad (10)$$

This is the expected rigid-body mode shape, since Eq. 10 says that all points on the bar move the same distance in this mode. Thus, there is no deformation along the bar.

10.1.5 Orthogonality

The most important property of natural modes is the *orthogonality property*. We begin by writing Eq. 10.4 for mode r and premultiplying the equation by ϕ_s^T to get

$$(\phi_s^T \mathbf{K} \phi_r) - \omega_r^2 (\phi_s^T \mathbf{M} \phi_r) = 0 \qquad (10.16)$$

Then, writing Eq. 10.4 for the sth mode and premultiplying it by ϕ_r^T, we get

$$(\phi_r^T \mathbf{K} \phi_s) - \omega_s^2 (\phi_r^T \mathbf{M} \phi_s) = 0 \qquad (10.17)$$

Since $\mathbf{K}$ and $\mathbf{M}$ are symmetric, Eq. 10.17 can be transposed and written

$$(\phi_s^T \mathbf{K} \phi_r) - \omega_s^2 (\phi_s^T \mathbf{M} \phi_r) = 0 \qquad (10.18)$$

Equation 10.18 may be subtracted from Eq. 10.16 to give

$$(\omega_s^2 - \omega_r^2)(\phi_s^T \mathbf{M} \phi_r) = 0 \qquad (10.19)$$

For modes with distinct frequencies, that is, if $\omega_r \neq \omega_s$, it is necessary that[3]

$$\boxed{\phi_s^T \mathbf{M} \phi_r = 0 \qquad \text{if } \omega_r \neq \omega_s} \qquad (10.20)$$

The rth and sth modes are said to be *orthogonal with respect to the mass matrix*. Equation 10.20 can be substituted into Eq. 10.16 to show that the rth and sth modes are also *orthogonal with respect to the stiffness matrix*; that is,

$$\boxed{\phi_s^T \mathbf{K} \phi_r = 0 \qquad \text{if } \omega_r \neq \omega_s} \qquad (10.21)$$

Equations 10.20 and 10.21 constitute the *orthogonality properties* of eigenvectors (mode shapes) of undamped MDOF systems. As you will see, these orthogonality properties are extremely important.

10.1.6 Mode Shapes: Repeated Frequency Case

It frequently happens in complex systems that there are "closely spaced" frequencies, that is, cases in which ω_r and ω_{r+1} differ by less than 1% or so. It occasionally happens that a system has a repeated frequency; that is, $\omega_r = \omega_{r+1} = \cdots = \omega_{r+p-1}$. A theorem of linear algebra[4] states that if the eigenvalue is repeated p times, there will be p *linearly*

---

[3] The repeated-frequency case is discussed next.
[4] The theorem is usually proved for the standard eigenvalue problem $\mathbf{Ax} = \lambda \mathbf{x}$, where $\mathbf{A}$ is symmetric. However, the generalized eigenvalue problem, $\mathbf{K}\phi = \omega^2 \mathbf{M}\phi$, can be cast into the required standard, symmetric form.

independent eigenvectors associated with this repeated eigenvalue. Although it is not necessary that these eigenvectors be orthogonal to each other, it is possible to choose the eigenvectors such that they will, in fact, satisfy the orthogonality relationships of Eqs. 10.20 and 10.21, even though $\omega_r = \omega_s$.

The procedure for determining mode shapes corresponding to a repeated frequency differs slightly from that established in Eq. 10.13 and illustrated in Examples 10.1 and 10.2. The rank of $\mathbf{D}(\omega_r)$ is $N - p$ if the frequency ω_r is repeated p times. Let Eq. 10.4 be written in the partitioned form

$$\begin{bmatrix} \mathbf{D}_{aa}(\omega_r) & \mathbf{D}_{ab}(\omega_r) \\ {}_{p\times p} & {}_{p\times(N-p)} \\ \mathbf{D}_{ba}(\omega_r) & \mathbf{D}_{bb}(\omega_r) \\ {}_{(N-p)\times p} & {}_{(N-p)\times(N-p)} \end{bmatrix} \begin{Bmatrix} \boldsymbol{\phi}_a \\ \boldsymbol{\phi}_b \end{Bmatrix}_r = \begin{Bmatrix} \mathbf{0}_a \\ \mathbf{0}_b \end{Bmatrix} \tag{10.22}$$

where $\mathbf{D}_{bb}$ is nonsingular, and where

$$\{\boldsymbol{\phi}_a\} \equiv \begin{Bmatrix} \phi_1 \\ \phi_2 \\ \vdots \\ \phi_p \end{Bmatrix}_r, \quad \{\boldsymbol{\phi}_b\} \equiv \begin{Bmatrix} \phi_{p+1} \\ \phi_{p+2} \\ \vdots \\ \phi_N \end{Bmatrix}_r \tag{10.23}$$

The lower partition of Eq. 10.22 can be solved for $\{\boldsymbol{\phi}_b\}_r$, giving

$$\boxed{\{\boldsymbol{\phi}_b\}_r = -[\mathbf{D}_{bb}(\omega_r)]^{-1}\mathbf{D}_{ba}(\omega_r)\{\boldsymbol{\phi}_a\}_r} \tag{10.24}$$

According to the theorem stated above, there will be p linearly independent vectors corresponding to the repeated frequency ω_r. Hence, we must pick p linearly independent vectors $\{\boldsymbol{\phi}_a\}_r, \{\boldsymbol{\phi}_a\}_{(r+1)}, \ldots, \{\boldsymbol{\phi}_a\}_{(r+p-1)}$, and Eq. 10.24 will determine the remaining components of the modal vectors $\{\boldsymbol{\phi}\}_r$, and so on. The following set of p linearly independent vectors form a convenient set of vectors to use as $\{\boldsymbol{\phi}_a\}$ vectors in Eq. 10.24:

$$\{\boldsymbol{\phi}_a\}_r \equiv \begin{Bmatrix} \phi_1 \\ \phi_2 \\ \phi_3 \\ \vdots \\ \phi_p \end{Bmatrix}_r = \begin{Bmatrix} 1 \\ 0 \\ 0 \\ \vdots \\ 0 \end{Bmatrix}, \quad \{\boldsymbol{\phi}_a\}_{(r+1)} = \begin{Bmatrix} 0 \\ 1 \\ 0 \\ \vdots \\ 0 \end{Bmatrix}, \quad \ldots, \quad \{\boldsymbol{\phi}_a\}_{(r+p-1)} = \begin{Bmatrix} 0 \\ 0 \\ 0 \\ \vdots \\ 1 \end{Bmatrix}$$

(10.25)

Example 10.3 illustrates how $\mathbf{D}(\omega)$ is partitioned for a system with $\omega_1 = \omega_2 = 0$; that is, $p = 2$ for the frequency $\omega = 0$.

Example 10.3 A model of a uniform beam is created by lumping mass at three nodes,[5] as shown in Fig. 1. The resulting equation of motion is

$$\frac{\rho AL}{4}\begin{bmatrix} 1 & 0 & 0 \\ 0 & 2 & 0 \\ 0 & 0 & 1 \end{bmatrix}\begin{Bmatrix} \ddot{v}_1 \\ \ddot{v}_2 \\ \ddot{v}_3 \end{Bmatrix} + \frac{12EI}{L^3}\begin{bmatrix} 1 & -2 & 1 \\ -2 & 4 & -2 \\ 1 & -2 & 1 \end{bmatrix}\begin{Bmatrix} v_1 \\ v_2 \\ v_3 \end{Bmatrix} = \begin{Bmatrix} 0 \\ 0 \\ 0 \end{Bmatrix}$$

---

[5] The names *grid point* and *node* are used for locations on a structure where displacement coordinates, (e.g., v_1, v_2) are assigned. This use of *node* should not be confused with a node of a mode shape, as noted in Example 10.1.

10.1 Some Properties of Natural Frequencies and Natural Modes of Undamped MDOF Systems

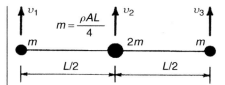

Figure 1 Free–free 3-DOF beam–mass system.

(a) Solve for the three natural frequencies of this system. (b) Solve for the three normal modes of the system. (c) Evaluate the orthogonality relationships $\boldsymbol{\phi}_1^T \mathbf{M} \boldsymbol{\phi}_2$ and $\boldsymbol{\phi}_1^T \mathbf{M} \boldsymbol{\phi}_3$.

SOLUTION (a) Set up and solve the characteristic equation. Let

$$\lambda = \omega^2 \frac{\rho A L^4}{48 EI} \tag{1}$$

Then the eigenvalue equation

$$[\mathbf{K} - \omega^2 \mathbf{M}]\boldsymbol{\phi} = \mathbf{0} \tag{2}$$

can be written in the form

$$\mathbf{D}(\lambda)\boldsymbol{\phi} = \begin{bmatrix} 1-\lambda & -2 & 1 \\ -2 & 2(2-\lambda) & -2 \\ 1 & -2 & 1-\lambda \end{bmatrix} \begin{Bmatrix} \phi_1 \\ \phi_2 \\ \phi_3 \end{Bmatrix} = \begin{Bmatrix} 0 \\ 0 \\ 0 \end{Bmatrix} \tag{3}$$

The characteristic equation is $\det[\mathbf{D}(\lambda)] = 0$, which simplifies to

$$\lambda^2(\lambda - 4) = 0 \tag{4}$$

so

$$\lambda_1 = \lambda_2 = 0, \qquad \lambda_3 = 4 \tag{5}$$

Then the three natural frequencies (squared) are

$$\omega_1^2 = \omega_2^2 = 0, \qquad \omega_3^2 = 4\frac{48 EI}{\rho A L^4} \qquad \textbf{Ans. (a)} \tag{6}$$

(b) Solve first for mode 3, which corresponds to the unique frequency ω_3. Use the same procedure as in Examples 10.1 and 10.2. Partition the matrix $[\mathbf{D}(\lambda_3)]$ of Eq. 3 as in Eq. 10.13 by isolating the top row and the leftmost column. Then

$$\begin{bmatrix} \mathbf{D}_{aa}(\lambda_3) & \mathbf{D}_{ab}(\lambda_3) \\ \mathbf{D}_{ba}(\lambda_3) & \mathbf{D}_{bb}(\lambda_3) \end{bmatrix} \begin{Bmatrix} 1 \\ \boldsymbol{\phi}_b \end{Bmatrix}_3 = \begin{bmatrix} -3 & -2 & 1 \\ -2 & -4 & -2 \\ 1 & -2 & -3 \end{bmatrix} \begin{Bmatrix} 1 \\ \phi_2 \\ \phi_3 \end{Bmatrix}_3 \tag{7}$$

Now, $\mathbf{D}_{bb}$ is nonsingular, and its inverse is

$$[\mathbf{D}_{bb}(\lambda_3)]^{-1} = \frac{1}{8}\begin{bmatrix} -3 & 2 \\ 2 & -4 \end{bmatrix} \tag{8}$$

Then, from Eq. 10.15,

$$\{\boldsymbol{\phi}_b\}_3 \equiv \left\{\begin{array}{c} \phi_2 \\ \phi_3 \end{array}\right\}_3 = -[\mathbf{D}_{bb}(\lambda_3)]^{-1}\mathbf{D}_{ba}(\lambda_3)$$

$$= -\frac{1}{8}\begin{bmatrix} -3 & 2 \\ 2 & -4 \end{bmatrix}\left\{\begin{array}{c} -2 \\ 1 \end{array}\right\} = \left\{\begin{array}{c} -1 \\ 1 \end{array}\right\} \qquad (9)$$

Finally, from Eqs. 7 and 9, mode 3 is

$$\boldsymbol{\phi}_3 = \left\{\begin{array}{c} 1 \\ -1 \\ 1 \end{array}\right\} \qquad \textbf{Ans. (b)} \quad (10)$$

Mode 3, the only flexible mode of this 3-DOF model, is sketched in Fig. 2.

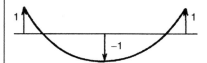

Figure 2 Mode 3 of the 3-DOF beam–mass system.

(b cont.) Solve for the two modes corresponding to $\omega = 0$. For $\lambda = 0$, $\mathbf{D}(\lambda)$ becomes

$$\mathbf{D}(\lambda_1) = \mathbf{D}(\lambda_2) = \mathbf{D}(0) = \begin{bmatrix} 1 & -2 & 1 \\ -2 & 4 & -2 \\ 1 & -2 & 1 \end{bmatrix} \qquad (11)$$

It is easily seen that each row (column) is a multiple of the first row (column), and that there is no 2×2 submatrix of $\mathbf{D}(0)$ that has a nonzero determinant. Thus, the rank of $\mathbf{D}(0)$ is $N - p = 3 - 2 = 1$. For $\lambda = 0$, let Eq. 3 be written in the partitioned form

$$\begin{bmatrix} \mathbf{D}_{aa}(\lambda_3) & \mathbf{D}_{ab}(\lambda_3) \\ \mathbf{D}_{ba}(\lambda_3) & \mathbf{D}_{bb}(\lambda_3) \end{bmatrix}\left\{\begin{array}{c} \boldsymbol{\phi}_a \\ \boldsymbol{\phi}_b \end{array}\right\} = \begin{bmatrix} 1 & -2 & 1 \\ -2 & 4 & -2 \\ \hline 1 & -2 & 1 \end{bmatrix}\left\{\begin{array}{c} \phi_1 \\ \phi_2 \\ \phi_3 \end{array}\right\} = \left\{\begin{array}{c} 0 \\ 0 \\ 0 \end{array}\right\} \qquad (12)$$

From Eq. 10.24 (or from the bottom partition of Eq. 12),

$$(\phi_3)_r = -1[\,1\ \ -2\,]\left\{\begin{array}{c} \phi_1 \\ \phi_2 \end{array}\right\}_r, \qquad r = 1, 2 \qquad (13)$$

Choose the linearly independent vectors

$$\left\{\begin{array}{c} \phi_1 \\ \phi_2 \end{array}\right\}_1 = \left\{\begin{array}{c} 1 \\ 0 \end{array}\right\} \quad \text{and} \quad \left\{\begin{array}{c} \phi_1 \\ \phi_2 \end{array}\right\}_2 = \left\{\begin{array}{c} 0 \\ 1 \end{array}\right\} \qquad (14)$$

From Eqs. 13 and 14,

$$(\phi_3)_1 = [\,-1\ \ 2\,]\left\{\begin{array}{c} 1 \\ 0 \end{array}\right\} = -1 \qquad (15a)$$

$$(\phi_3)_2 = [\,-1\ \ 2\,]\left\{\begin{array}{c} 0 \\ 1 \end{array}\right\} = 2 \qquad (15b)$$

10.1 Some Properties of Natural Frequencies and Natural Modes of Undamped MDOF Systems

Finally, the two modes corresponding to the repeated frequency $\omega = 0$ are

$$\boldsymbol{\phi}_1 = \begin{Bmatrix} 1 \\ 0 \\ -1 \end{Bmatrix}, \quad \boldsymbol{\phi}_2 = \begin{Bmatrix} 0 \\ 1 \\ 2 \end{Bmatrix} \qquad \text{Ans. (b) (16)}$$

which are the rigid-body modes sketched in Fig. 3.

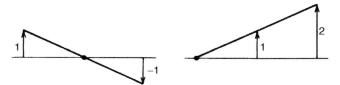

Figure 3 Rigid-body modes of the 3-DOF beam–mass system.

Note that these two rigid-body modes are not unique; that is, any linear combination of the two vectors in Eq. 16 is also a valid rigid-body mode. However, the procedure that we have employed guarantees that we obtain p linearly independent modes.

(c) Evaluate the orthogonality relationships $\boldsymbol{\phi}_1^T \mathbf{M} \boldsymbol{\phi}_2$ and $\boldsymbol{\phi}_1^T \mathbf{M} \boldsymbol{\phi}_3$.

$$\boldsymbol{\phi}_1^T \mathbf{M} \boldsymbol{\phi}_2 = \begin{bmatrix} 1 & 0 & -1 \end{bmatrix} \begin{bmatrix} m & 0 & 0 \\ 0 & 2m & 0 \\ 0 & 0 & m \end{bmatrix} \begin{Bmatrix} 0 \\ 1 \\ 2 \end{Bmatrix} = -2m \neq 0 \qquad (17)$$

Therefore, the two rigid-body modes given in Eqs. 16, although linearly independent, are not orthogonal with respect to the mass matrix. Let $\hat{\boldsymbol{\phi}}_2 = \boldsymbol{\phi}_1 + \boldsymbol{\phi}_2$, which is a particular linear combination of the rigid-body modes above. Then, check the orthogonality relationship

$$\boldsymbol{\phi}_1^T \mathbf{M} \hat{\boldsymbol{\phi}}_2 = \begin{bmatrix} 1 & 0 & -1 \end{bmatrix} \begin{bmatrix} m & 0 & 0 \\ 0 & 2m & 0 \\ 0 & 0 & m \end{bmatrix} \begin{Bmatrix} 1 \\ 1 \\ 1 \end{Bmatrix} = 0 \qquad (18)$$

Thus, the modes $\boldsymbol{\phi}_1$ and $\hat{\boldsymbol{\phi}}_2$ form a pair of rigid-body modes that are orthogonal with respect to the mass matrix.

Now evaluate $\boldsymbol{\phi}_1^T \mathbf{M} \boldsymbol{\phi}_3$.

$$\boldsymbol{\phi}_1^T \mathbf{M} \boldsymbol{\phi}_3 = \begin{bmatrix} 1 & 0 & -1 \end{bmatrix} \begin{bmatrix} m & 0 & 0 \\ 0 & 2m & 0 \\ 0 & 0 & m \end{bmatrix} \begin{Bmatrix} 1 \\ -1 \\ 1 \end{Bmatrix} = 0 \qquad (19)$$

Thus, orthogonality is satisfied, since $\omega_1 \neq \omega_3$.

Although the procedure given in Eqs. 10.22 through 10.25 guarantees that p linearly independent modes are computed when frequency ω_r is repeated p times, Example 10.3 shows that it is also possible to use this set of p linearly independent modes to create a set of p modes that are orthogonal with respect to the mass and stiffness matrices. The

Gram–Schmidt Procedure, a formal procedure for creating a set of orthogonal vectors, is introduced in Section 15.2.3.

As noted in Example 10.3, systems that possess repeated frequencies, or even very closely spaced frequencies, require special consideration when natural modes are to be determined.

10.1.7 Modal Matrix and Eigenvalue Matrix

It will be convenient to assume that all N modes of an N-DOF system are orthogonal, that is, Eqs. 10.20 and 10.21 hold for $r \neq s$, including cases where $\omega_r = \omega_s$.

The *modal matrix* of an N-DOF system is a matrix whose columns are the respective normal modes, that is,

$$\boxed{\boldsymbol{\Phi} \equiv [\boldsymbol{\phi}_1 \ \boldsymbol{\phi}_2 \ \cdots \ \boldsymbol{\phi}_N]} \tag{10.26}$$

The algebraic eigenproblem, Eq. 10.4, can be written for all N modes as follows:

$$\mathbf{K}\boldsymbol{\Phi} = \mathbf{M}\boldsymbol{\Phi}\boldsymbol{\Lambda} \tag{10.27}$$

where $\boldsymbol{\Lambda}$ is the *eigenvalue matrix*, which is a diagonal matrix of eigenvalues. That is,

$$\boxed{\boldsymbol{\Lambda} \equiv \mathrm{diag}(\omega_1^2, \omega_2^2, \ldots, \omega_N^2)} \tag{10.28}$$

Note that the $\boldsymbol{\Lambda}$ matrix postmultiplies the $\boldsymbol{\Phi}$ matrix in Eq. 10.27 so that each ω_r^2 multiplies the proper modal vector $\boldsymbol{\phi}_r$.

10.1.8 Generalized Mass Matrix and Generalized Stiffness Matrix

By using the definitions of modal mass and modal stiffness given in Eqs. 10.9 and 10.10 together with the orthogonality equations, Eqs. 10.20 and 10.21, we obtain a diagonal *modal mass matrix* M and a diagonal *modal stiffness matrix* K, given by

$$\boxed{M = \boldsymbol{\Phi}^T \mathbf{M} \boldsymbol{\Phi} = \mathrm{diag}(M_1, M_2, \ldots, M_N)} \tag{10.29}$$

and

$$\boxed{K = \boldsymbol{\Phi}^T \mathbf{K} \boldsymbol{\Phi} = \mathrm{diag}(K_1, K_2, \ldots, K_N)} \tag{10.30}$$

If the natural modes are normalized so that $M_r = 1$, $r = 1, 2, \ldots, N$, they are said to form a set of *orthonormal vectors*, since they are both orthogonal and normalized. Then the modal mass matrix M becomes the $N \times N$ unit matrix; that is,

$$\boldsymbol{\Phi}^T \mathbf{M} \boldsymbol{\Phi} = \mathbf{I} \tag{10.31}$$

and the corresponding modal stiffness matrix becomes

$$\boldsymbol{\Phi}^T \mathbf{K} \boldsymbol{\Phi} = \boldsymbol{\Lambda} \tag{10.32}$$

where $\boldsymbol{\Lambda}$ is given by Eq. 10.28.

10.1.9 Expansion Theorem

An *expansion theorem* can be stated for MDOF systems. The normal modes $\boldsymbol{\phi}_r$, $r = 1, 2, \ldots, N$, form a mutually orthogonal set of N-dimensional vectors. It can be shown (see Problem 10.8) that the N modes also form a linearly independent set of N-dimensional vectors. Hence, an arbitrary vector $\mathbf{u}$ can be expressed as a linear combination of the normal modes by the equation

$$\mathbf{u} = \sum_{r=1}^{N} c_r \boldsymbol{\phi}_r \tag{10.33}$$

where the c_r's are determined by the equation

$$c_r = \frac{1}{M_r} \boldsymbol{\phi}_r^T \mathbf{M} \mathbf{u} = \frac{\boldsymbol{\phi}_r^T \mathbf{M} \mathbf{u}}{\boldsymbol{\phi}_r^T \mathbf{M} \boldsymbol{\phi}_r} \tag{10.34}$$

The next example illustrates how a given N-dimensional vector can be represented in terms of the modes of an N-DOF system.

Example 10.4 A uniform beam was modeled in Example 10.3 by three lumped masses on a massless beam, as repeated in Fig. 1. The model is symmetric about its central mass. The mutually orthogonal normal modes can be identified as two symmetric modes and one antisymmetric mode, as shown in Fig. 2. Show that the expansion of any arbitrary symmetric deflection $\mathbf{v}$ using Eq. 10.33 will involve only the symmetric modes of the structure.

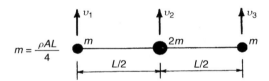

Figure 1 A 3-DOF model of a free–free beam.

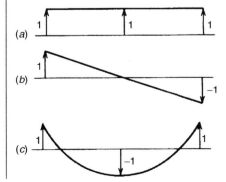

Figure 2 Three normal modes of a 3-DOF free–free beam: (*a*) mode 1, symmetric rigid-body mode; (*b*) mode 2, antisymmetric rigid-body mode; (*c*) mode 3, symmetric flexible mode.

SOLUTION We need to show that for an arbitrary symmetric vector of the form

$$\mathbf{v} = \begin{Bmatrix} a \\ b \\ a \end{Bmatrix} \tag{1}$$

the coefficient c_2 of the antisymmetric mode vanishes in the expansion

$$\mathbf{v} = \sum_{r=1}^{3} c_r \boldsymbol{\phi}_r \tag{2}$$

Therefore, evaluate c_2 using Eq. 10.34.

$$c_2 = \frac{\boldsymbol{\phi}_2^\mathrm{T} \mathbf{M} \mathbf{v}}{\boldsymbol{\phi}_2^\mathrm{T} \mathbf{M} \boldsymbol{\phi}_2} \tag{3}$$

But

$$\boldsymbol{\phi}_2^\mathrm{T} \mathbf{M} \mathbf{v} = [1\ 0\ -1] \begin{bmatrix} m & 0 & 0 \\ 0 & 2m & 0 \\ 0 & 0 & m \end{bmatrix} \begin{Bmatrix} a \\ b \\ a \end{Bmatrix} = [1\ 0\ -1] \begin{Bmatrix} ma \\ 2mb \\ ma \end{Bmatrix} = 0 \tag{4}$$

Therefore, $c_2 = 0$, and the expansion of any symmetric deflection involves only the two symmetric modes of the system.

10.1.10 Mode Superposition Employing Modes of the Undamped Structure

Because the exact nature of damping in complex structural systems is seldom known, it is often assumed that the damping of the structure can be represented by viscous damping. That is, the MDOF equations of motion are assumed to be

$$\mathbf{M}\ddot{\mathbf{u}} + \mathbf{C}\dot{\mathbf{u}} + \mathbf{K}\mathbf{u} = \mathbf{p}(t) \tag{10.35}$$

where matrix $\mathbf{C}$ is the system viscous damping matrix in physical (or generalized) coordinates $\mathbf{u}$. Figure 10.2 outlines the three strategies for integrating MDOF equations of motion to obtain dynamic response: (1) mode superposition using modes of the undamped system; (2) mode superposition using the complex modes of the damped system, and (3) direct integration of the coupled equations of motion.

Equation 10.33 forms the basis of what is known as the *normal-mode method* or the *mode-superposition method* for solving the MDOF equations of motion. Mode superposition was introduced in Section 9.5 and illustrated in Example 9.8. Let $(\omega_r, \boldsymbol{\phi}_r)$ be the eigenpairs corresponding to the undamped free-vibration problem. The modal matrix

Mode superposition using real modes of the undamped system	or	*Mode superposition* using complex modes of the damped system	or	*Direct integration* of the coupled equations of motion

Figure 10.2 Three strategies for solving for the response of MDOF systems.

10.1 Some Properties of Natural Frequencies and Natural Modes of Undamped MDOF Systems

$\boldsymbol{\Phi}$ is given by Eq. 10.26 and the corresponding eigenvalue matrix $\boldsymbol{\Lambda}$ by Eq. 10.28. A mode-superposition solution based on the free-vibration modes of the undamped structure has the form

$$\mathbf{u}(t) = \sum_{r=1}^{N} \boldsymbol{\phi}_r \eta_r(t) = \boldsymbol{\Phi}\boldsymbol{\eta}(t) \tag{10.36}$$

where the coordinates η_r are called *modal coordinates*, or *principal coordinates*.[6] Equation 10.36 is substituted into Eq. 10.35, and the resulting equation is multiplied by $\boldsymbol{\Phi}^T$ to give the *equation of motion in principal coordinates*: namely,

$$\boldsymbol{M}\ddot{\boldsymbol{\eta}} + \boldsymbol{C}\dot{\boldsymbol{\eta}} + \boldsymbol{K}\boldsymbol{\eta} = \boldsymbol{\Phi}^T \mathbf{p}(t) \tag{10.37}$$

where

$$\begin{aligned} \boldsymbol{M} &= \boldsymbol{\Phi}^T \mathbf{M} \boldsymbol{\Phi} = \text{modal mass matrix (diag.)} \\ \boldsymbol{C} &= \boldsymbol{\Phi}^T \mathbf{C} \boldsymbol{\Phi} = \text{generalized damping matrix} \\ \boldsymbol{K} &= \boldsymbol{\Phi}^T \mathbf{K} \boldsymbol{\Phi} = \text{modal stiffness matrix (diag.)} \\ \boldsymbol{\Phi}^T \mathbf{p}(t) &= \text{modal force vector} \end{aligned} \tag{10.38}$$

Mode-superposition solutions employing modes of the undamped system are discussed at length in Chapter 11.

10.1.11 Generalized Damping Matrix

Typically, in those cases where it is possible to define the physical damping matrix $\mathbf{C}$, the generalized damping matrix

$$\boldsymbol{C} = \boldsymbol{\Phi}^T \mathbf{C} \boldsymbol{\Phi} \tag{10.39}$$

is not diagonal. Section 10.3 describes several important forms of damping (e.g., modal damping and Rayleigh damping) that do lead to a diagonal generalized damping matrix $\boldsymbol{C}$; and Chapter 11 presents several examples that illustrate mode-superposition solutions for systems with uncoupled modal damping.

There are a number of situations in which damping cannot properly be represented by an uncoupled modal-damping model. One such instance is the modeling of structures with concentrated energy absorbers, such as automobile shock absorbers. Another example is a building and the surrounding soil. The damping level for the building on a rigid foundation is less than that for the soil, and when a combined soil-structure model is generated and system modes are used to define the principal coordinates, the resulting damping matrix in principal coordinates contains off-diagonal coupling terms. For such systems with general viscous damping, there are two options: mode superposition using complex modes of the damped system, and direct integration of the coupled equations of motion. The former approach is discussed in Sections 10.4 and 10.5. Procedures for incorporating damping when direct integration is used to solve the coupled MDOF equations of motion are discussed in Section 16.1. Special techniques that have recently been developed for solving for the frequency response of very large systems with localized damping mechanisms are mentioned in Section 17.8.

---

[6]Some authors refer to these as *natural coordinates* or as *normal coordinates*. Some authors reserve the name *normal coordinates* to use when modes are normalized so that $M_r = 1$ for every mode.

10.2 MODEL REDUCTION: RAYLEIGH, RAYLEIGH–RITZ, AND ASSUMED-MODES METHODS

10.2.1 Rayleigh Quotient

The *Rayleigh quotient* for an N-DOF system is defined by

$$\boxed{\omega_R^2 \equiv \mathcal{R}(\mathbf{v}) = \frac{\mathbf{v}^T \mathbf{K} \mathbf{v}}{\mathbf{v}^T \mathbf{M} \mathbf{v}}} \qquad (10.40)$$

where $\mathbf{v}$ is any N-dimensional vector. Let $\mathbf{v}$ be expanded in a series of the orthonormal modes of the N-DOF system, that is, N modes $\boldsymbol{\phi}_r$, which have been scaled so that $M_r = 1$ for every mode. Then

$$\mathbf{v} = \sum_{r=1}^{N} c_r \boldsymbol{\phi}_r \qquad (10.41)$$

Due to the orthogonality properties of the modes and Eq. 10.11, $\mathcal{R}(\mathbf{v})$ has the form

$$\mathcal{R}(\mathbf{v}) = \frac{\omega_1^2 c_1^2 + \omega_2^2 c_2^2 + \cdots + \omega_N^2 c_N^2}{c_1^2 + c_2^2 + \cdots + c_N^2} \qquad (10.42)$$

Using Eq. 10.42, we can prove that

$$\omega_1^2 \leq \mathcal{R}(\mathbf{v}) \leq \omega_N^2 \qquad (10.43)$$

If $\omega_1 \neq 0$, we can write Eq. 10.42 in the form

$$\mathcal{R}(\mathbf{v}) = \omega_1^2 \frac{1 + (c_2/c_1)^2 (\omega_2/\omega_1)^2 + \cdots + (c_N/c_1)^2 (\omega_N/\omega_1)^2}{1 + (c_2/c_1)^2 + \cdots + (c_N/c_1)^2} \qquad (10.44)$$

Since $\omega_1 \leq \omega_2 \cdots \leq \omega_N$, each term in the numerator is greater than or equal to the corresponding term in the denominator. Hence,

$$\mathcal{R}(\mathbf{v}) \geq \omega_1^2 \qquad (10.45)$$

A similar procedure can be employed to show that $\mathcal{R}(\mathbf{v}) \leq \omega_N^2$, and hence Eq. 10.43 is proved.

10.2.2 Rayleigh Method for an N-DOF System

The *Rayleigh Method* is a procedure for approximating a continuous system by an SDOF system, and using energy conservation to calculate an approximate fundamental frequency, which can be denoted ω_R. A similar procedure can be used to reduce an N-DOF system to a SDOF system that approximates the fundamental mode of that system. Let the displacement vector of the N-dimensional system be approximated by

$$\mathbf{u}(t) = \mathbf{v} \cos \omega_R t = \boldsymbol{\psi} \hat{v} \cos \omega_R t \qquad (10.46)$$

where $\boldsymbol{\psi}$ is any N-dimensional vector that approximates the expected shape of the fundamental mode of the system, and $\hat{v}$ is the amplitude. Then

$$\dot{\mathbf{u}}(t) = -\omega_R \mathbf{v} \sin \omega_R t = -\omega_R \boldsymbol{\psi} \hat{v} \sin \omega_R t \qquad (10.47)$$

10.2 Model Reduction: Rayleigh, Rayleigh–Ritz, and Assumed-Modes Methods

Since the potential energy and kinetic energy of the system are given by

$$\mathcal{V} = \tfrac{1}{2}\mathbf{u}^T\mathbf{K}\mathbf{u}, \qquad \mathcal{T} = \tfrac{1}{2}\dot{\mathbf{u}}^T\mathbf{M}\dot{\mathbf{u}} \tag{10.48}$$

respectively, their maximum values are given by

$$\mathcal{V}_{max} = \tfrac{1}{2}\widehat{k}\,\widehat{v}^2, \qquad \mathcal{T}_{max} = \tfrac{1}{2}\widehat{m}\,\overline{v}^2 \tag{10.49}$$

where the generalized stiffness $\widehat{k}$ and the generalized mass $\widehat{m}$ are given by

$$\widehat{k} = \boldsymbol{\psi}^T\mathbf{K}\boldsymbol{\psi}, \qquad \widehat{m} = \boldsymbol{\psi}^T\mathbf{M}\boldsymbol{\psi} \tag{10.50}$$

For energy conservation to hold, $\mathcal{V}_{max} = \mathcal{T}_{max}$, which gives

$$\mathcal{R}(\mathbf{v}) = \omega_R^2 = \frac{\widehat{k}}{\widehat{m}} \tag{10.51}$$

Then, from Eqs. 10.43 and 10.51,

$$\omega_1^2 \le \omega_R^2 \le \omega_N^2 \tag{10.52}$$

If the assumed mode shape $\boldsymbol{\psi}$ closely approximates the shape of the true fundamental mode $\boldsymbol{\phi}_1$, ω_R will give a close upper bound to the fundamental frequency ω_1.

10.2.3 Rayleigh–Ritz Method for MDOF Systems

The *Rayleigh–Ritz Method* for N-DOF systems permits approximate values of the frequencies of $\widehat{N}$ modes ($\widehat{N} < N$) to be computed. The displacement is assumed to be harmonic in time

$$\mathbf{u}(t) = \mathbf{v}\cos(\omega t - \alpha) \tag{10.53}$$

The shape vector $\mathbf{v}$ is assumed to be given by the series expansion

$$\mathbf{v} = \sum_{i=1}^{\widehat{N}} \boldsymbol{\psi}_i \widehat{v}_i = \widehat{\boldsymbol{\Psi}}\,\widehat{\mathbf{v}} \tag{10.54}$$

where

$$\widehat{\boldsymbol{\Psi}} = [\boldsymbol{\psi}_1\,\boldsymbol{\psi}_2\,\cdots\,\boldsymbol{\psi}_{\widehat{N}}] \tag{10.55}$$

The $\boldsymbol{\psi}_i$'s are preselected linearly independent assumed-mode vectors. Setting $\mathcal{V}_{max} = \mathcal{T}_{max}$ leads to the following form for the *Rayleigh quotient*:

$$\mathcal{R}(\mathbf{v}) \equiv \widehat{\omega}^2 = \frac{\widehat{\mathbf{v}}^T\widehat{\mathbf{K}}\widehat{\mathbf{v}}}{\widehat{\mathbf{v}}^T\widehat{\mathbf{M}}\widehat{\mathbf{v}}} = \frac{\sum_{i=1}^{\widehat{N}}\sum_{j=1}^{\widehat{N}}\widehat{v}_i\widehat{v}_j\widehat{k}_{ij}}{\sum_{i=1}^{\widehat{N}}\sum_{j=1}^{\widehat{N}}\widehat{v}_i\widehat{v}_j\widehat{m}_{ij}} \tag{10.56}$$

where

$$\widehat{\mathbf{K}} = \widehat{\boldsymbol{\Psi}}^T\mathbf{K}\widehat{\boldsymbol{\Psi}}, \qquad \widehat{\mathbf{M}} = \widehat{\boldsymbol{\Psi}}^T\mathbf{M}\widehat{\boldsymbol{\Psi}} \tag{10.57}$$

Comparing Eq. 10.51 with Eq. 10.56, we see that a definite frequency ω_R^2 is established by the former, whereas the value of $\widehat{\omega}^2$ in Eq. 10.56 depends on the values of the amplitude coefficients $\widehat{v}_i$, $i = 1, 2, \ldots, \widehat{N}$.

Ritz proposed that the coefficients $\widehat{v}_i$ be chosen to make $\mathcal{R}(\mathbf{v})$ stationary, that is, to make

$$\frac{\partial \mathcal{R}(\mathbf{v})}{\partial \widehat{v}_i} = 0, \qquad i = 1, 2, \ldots, \widehat{N} \tag{10.58}$$

Let

$$\mathcal{R}(\mathbf{v}) \equiv \frac{\mathcal{N}(\mathbf{v})}{\mathcal{D}(\mathbf{v})} \tag{10.59}$$

Then Eq. 10.58 gives

$$\mathcal{N}(\mathbf{v}) \frac{\partial \mathcal{D}(\mathbf{v})}{\partial \widehat{v}_i} - \mathcal{D}(\mathbf{v}) \frac{\partial \mathcal{N}(\mathbf{v})}{\partial \widehat{v}_i} = 0 \tag{10.60}$$

But since $\widehat{k}_{ij} = \widehat{k}_{ji}$,

$$\frac{\partial \mathcal{N}(\mathbf{v})}{\partial \widehat{v}_i} = 2 \sum_{j=1}^{\widehat{N}} \widehat{k}_{ij} \widehat{v}_j \tag{10.61}$$

and similarly for the derivative of $\mathcal{D}(\mathbf{v})$. Combining Eqs. 10.56, 10.60, and 10.61, we get

$$\sum_{j=1}^{\widehat{N}} (\widehat{k}_{ij} - \widehat{\omega}^2 \widehat{m}_{ij}) \widehat{v}_j = 0, \qquad i = 1, 2, \ldots, \widehat{N} \tag{10.62}$$

or, in matrix form, we get the *reduced-order eigenproblem*,

$$\boxed{[\widehat{\mathbf{K}} - \widehat{\omega}^2 \widehat{\mathbf{M}}] \widehat{\mathbf{v}} = \mathbf{0}} \tag{10.63}$$

This eigenvalue problem leads to a set of $\widehat{N}$ approximate frequencies $\widehat{\omega}_r$ and corresponding modes $\widehat{\mathbf{v}}_r$. The relationship of the $\widehat{N}$ approximate frequencies to the N exact frequencies is shown later in this section.

10.2.4 Assumed-Modes Method for Model Reduction of MDOF Systems

The Rayleigh–Ritz procedure described above is a model order-reduction method that applies specifically to free vibration (Eq. 10.53). This earlier method can be considered to be special case of applying the *Assumed-Modes Method* to reduce an N-DOF system to a $\widehat{N}$-DOF system by assuming that

$$\boxed{\mathbf{u}(t) = \sum_{i=1}^{\widehat{N}} \boldsymbol{\psi}_i \widehat{u}_i(t) = \widehat{\boldsymbol{\Psi}} \widehat{\mathbf{u}}(t)} \tag{10.64}$$

This approximation can be substituted into expressions for strain energy $\mathcal{V}$ and kinetic energy $\mathcal{T}$ in Eqs. 10.48, and the resulting expressions substituted into Lagrange's equation, Eq. 8.9, to give

$$\widehat{\mathbf{M}}\ddot{\hat{\mathbf{u}}} + \widehat{\mathbf{K}}\hat{\mathbf{u}} = \mathbf{0} \tag{10.65}$$

for undamped free vibration. For forced vibration of a viscous-damped system, the reduced-order equations of motion are given by

$$\widehat{\mathbf{M}}\ddot{\hat{\mathbf{u}}} + \widehat{\mathbf{C}}\dot{\hat{\mathbf{u}}} + \widehat{\mathbf{K}}\hat{\mathbf{u}} = \hat{\mathbf{p}}(t) \tag{10.66}$$

where

$$\widehat{\mathbf{C}} = \widehat{\mathbf{\Psi}}^T \mathbf{C}\widehat{\mathbf{\Psi}}, \qquad \hat{\mathbf{p}}(t) = \widehat{\mathbf{\Psi}}^T \mathbf{p}(t) \tag{10.67}$$

Several specific procedures for creating reduced-order models by selecting appropriate assumed-mode vectors $\boldsymbol{\psi}_i$ are discussed in Section 14.6.

10.2.5 Eigenvalue Separation Property

A very interesting and useful property of eigenvalues (natural frequencies) is the *eigenvalue separation property*.[10.1] Let the original N-DOF eigenvalue problem be stated as

$$[\mathbf{K} - \lambda \mathbf{M}]\mathbf{v} = \mathbf{0} \tag{10.68}$$

where $\lambda = \omega^2$, and let

$$[\mathbf{K}^{(m)} - \lambda^{(m)}\mathbf{M}^{(m)}]\mathbf{v}^{(m)} = \mathbf{0}, \qquad m = 0, 1, \ldots, N-1 \tag{10.69}$$

be the mth constrained eigenvalue problem, where $\mathbf{K}^{(m)}$ and $\mathbf{M}^{(m)}$ are obtained by deleting the last m rows and columns of $\mathbf{K}$ and $\mathbf{M}$, respectively. By definition, $\mathbf{K}^{(0)} \equiv \mathbf{K}$ and $\mathbf{M}^{(0)} \equiv \mathbf{M}$. Then the *eigenvalue separation theorem* states that

$$\boxed{\lambda_1^{(m)} \leq \lambda_1^{(m+1)} \leq \lambda_2^{(m)} \leq \lambda_2^{(m+1)} \leq \cdots \leq \lambda_{(N-m)}^{(m)} \qquad \text{for } m = 0, 1, 2, \ldots, N-2} \tag{10.70}$$

That is, the eigenvalues of the $(m+1)$st problem separate the eigenvalues of the mth problem, as illustrated in Fig. 10.3, where

$$p^{(m)}(\mu) \equiv \det[\mathbf{K}^{(m)} - \mu \mathbf{M}^{(m)}] \tag{10.71}$$

Although, as illustrated in Fig. 10.3, the constraints that reduce the model order appear as physical constraints that reduce the rows and columns of the mass and stiffness matrices of a system, the eigenvalue-separation theorem applies to other forms of "constraint." For example, if the Assumed-Modes Method is employed to create an N-DOF model of a continuous system (e.g., a beam), one "constraint" is placed on the model if one of the shape functions is omitted so that the assumed-modes model is just an $(N-1)$-DOF model. This is illustrated in Section 14.8.

The eigenvalue separation theorem of Eq. 10.70 can be employed directly to show the convergence properties of frequencies obtained by the Rayleigh–Ritz method. The result is shown in Table 10.1. Thus, each of the $\widehat{N} < N$ eigenvalues produced by a Rayleigh–Ritz approximation to an N-DOF system is an upper bound to the corresponding exact eigenvalue, and the eigenvalues approach the exact values from above as the number of degrees of freedom, $\widehat{N}$, increases.

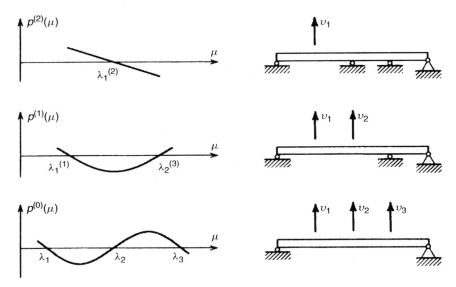

Figure 10.3 Eigenvalue-separation property.

Table 10.1 Convergence Properties of Rayleigh–Ritz Frequencies

DOF $= \hat{N} =$	1	2	3	$\cdots$	$N-1$	N
"Constraints" $= m =$	$N-1$	$N-2$	$N-3$	$\cdots$	1	0
First eigenvalue	$\lambda_1^{(N-1)} \geq$	$\lambda_1^{(N-2)} \geq$	$\lambda_1^{(N-3)} \geq$	$\cdots$	$\lambda_1^{(1)} \geq$	λ_1
Second eigenvalue		$\lambda_2^{(N-2)} \geq$	$\lambda_2^{(N-3)} \geq$	$\cdots$	$\lambda_2^{(1)} \geq$	λ_2
Third eigenvalue			$\lambda_3^{(N-3)} \geq$	$\cdots$	$\lambda_3^{(1)} \geq$	λ_3
$\vdots$						$\vdots$
Nth eigenvalue						λ_N

10.3 UNCOUPLED DAMPING IN MDOF SYSTEMS

In the discussion of mode superposition near the end of Section 10.1, it was noted that it is often assumed that the damping of the structure can be represented by viscous damping. That is, the MDOF equations of motion are assumed to be

$$\mathbf{M}\ddot{\mathbf{u}} + \mathbf{C}\dot{\mathbf{u}} + \mathbf{K}\mathbf{u} = \mathbf{p}(t) \tag{10.72}$$

where matrix $\mathbf{C}$ is the system viscous damping matrix in physical (or generalized) coordinates $\mathbf{u}$. A *mode-superposition solution* based on the free-vibration modes of the undamped structure led to the *equation of motion in principal coordinates*: namely,

$$\mathbf{M}\ddot{\boldsymbol{\eta}} + \mathbf{C}\dot{\boldsymbol{\eta}} + \mathbf{K}\boldsymbol{\eta} = \boldsymbol{\Phi}^T \mathbf{p}(t) \tag{10.73}$$

where the coordinates η_r are called *modal coordinates*, or *principal coordinates*; and the *modal mass matrix* and *modal stiffness matrix* were defined in Eqs. 10.29 and 10.30, respectively:

$$M = \mathbf{\Phi}^T\mathbf{M}\mathbf{\Phi} = \text{diag}(M_r), \qquad K = \mathbf{\Phi}^T\mathbf{K}\mathbf{\Phi} = \text{diag}(K_r) = \text{diag}(\omega_r^2 M_r) \qquad (10.74)$$

Typically, in those cases where it is possible to define the damping matrix **C**, the *generalized damping matrix*

$$C = \mathbf{\Phi}^T\mathbf{C}\mathbf{\Phi} \qquad (10.75)$$

is not diagonal. In principle, the damping matrix could be derived by finite element procedures analogous to those employed in Section 14.2 for deriving stiffness and mass matrices. However, material damping properties may not be well defined enough to permit this, and furthermore, much damping in structures results from joints and from nonstructural elements (e.g., partitions, etc.). Therefore, damping of a structure is usually defined at the system level rather than in terms of individual element properties. In this section we discuss four ways in which a system's damping can lead to a diagonal generalized damping matrix **C**.

10.3.1 Raleigh Damping

One procedure for defining a system viscous damping matrix **C** that leads to a diagonal generalized damping matrix C is to employ *Rayleigh damping*, which is defined by

$$\boxed{\mathbf{C} = a_0\mathbf{M} + a_1\mathbf{K}} \qquad (10.76)$$

This is also called *proportional damping*, since the damping matrix is proportional to a linear combination of the mass matrix and the stiffness matrix. The constants a_0 and a_1 can be chosen to produce specified modal damping factors for two selected modes.

Combining Eqs. 10.74, 10.75, and 10.76, we get

$$C = \mathbf{\Phi}^T\mathbf{C}\mathbf{\Phi} = \text{diag}(C_r) = \text{diag}(a_0 + a_1\omega_r^2)M_r = \text{diag}(2\zeta_r\omega_r M_r) \qquad (10.77)$$

where the last term defines the *modal damping factor* ζ_r in the same way it was defined in Chapter 3 for viscous-damped SDOF systems. Therefore, the N modal damping factors are related to the two Rayleigh damping coefficients by

$$\boxed{\zeta_r = \frac{1}{2}\left(\frac{a_0}{\omega_r} + a_1\omega_r\right)} \qquad (10.78)$$

Thus, Rayleigh damping is easy to define by choosing ζ_r for two modes and solving for the corresponding Rayleigh damping coefficients a_0 and a_1. The damping in the remaining modes is then determined by Eq. 10.78. The mass-proportional contribution in Eq. 10.76 gives a contribution to ζ_r in Eq. 10.78 that is inversely proportional to ω_r. The stiffness-proportional term, on other hand, leads to a contribution to ζ_r that increases linearly with ω_r.

10.3.2 Modal Damping

The special type of damping that is most frequently used in structural dynamics computations is referred to in the literature as *modal damping*, or sometimes proportional damping or classical damping. Modal damping is assumed to satisfy orthogonality, so that Eq. 10.75 becomes

$$\boxed{C = \Phi^T C \Phi = \text{diag}(C_r) = \text{diag}(2\zeta_r \omega_r M_r)} \qquad (10.79)$$

With this diagonalized modal damping, the coupled equations of motion, Eq. 10.72, are transformed into the following set of N *uncoupled equations of motion* in modal coordinates:

$$\boxed{M_r \ddot{\eta}_r + 2M_r \omega_r \zeta_r \dot{\eta}_r + \omega_r^2 M_r \eta_r = \phi_r^T p(t), \qquad r = 1, 2, \ldots, N} \qquad (10.80)$$

Unlike Rayleigh damping, where Eq. 10.78 indicates that all of the damping factors are determined by just two coefficients, in modal damping all N damping factors may be given distinct values. The *modal damping factor* values ζ_r are assumed on the basis of providing damping that is characteristic of the type of structure under consideration. Typical values lie in the range $0.01 \leq \zeta_r \leq 0.1$. In Chapter 18, experimental methods for determining suitable damping factors are discussed. Mode-superposition solutions based on modal damping are illustrated in Section 11.2.

10.3.3 Damping Matrix C for Modal Damping

We now take up the topic, "How can damping be represented when a damping matrix **C** in physical coordinates is required, as when direct integration (Chapter 16) is used to solve Eq. 10.72?" The Rayleigh Method, discussed above, and the following two methods answer this question. The obvious disadvantage of Rayleigh damping is that since it is defined by just two coefficients, it does not permit realistic damping to be defined for all the modes of interest. The method that follows permits a damping matrix to be generated, which leads to modal damping with specified damping factors for a given number of modes. The remaining modes are undamped.

Assume that it is necessary to construct a physical (or generalized) viscous damping matrix **C** that corresponds to modal damping in all N modes of a structure. That is, we want to construct the **C** matrix such Eq. 10.79 is satisfied. Then, assuming that we have the complete $N \times N$ modal matrix Φ, the physical damping matrix **C** is given by

$$C = \Phi^{-T} C \Phi^{-1} \qquad (10.81)$$

A convenient expression for Φ^{-1} can be developed from the orthogonality property of the modes. Recall that

$$M = \Phi^T M \Phi = \text{diag}(M_r) \qquad (10.82a)$$

Then

$$I = M^{-1} M = (M^{-1} \Phi^T M) \Phi = \Phi^{-1} \Phi \qquad (10.82b)$$

Therefore,

$$\Phi^{-1} = M^{-1} \Phi^T M \qquad (10.82c)$$

Equations 10.81 and 10.82c can be combined to give

$$\mathbf{C} = (\mathbf{M}\boldsymbol{\Phi}M^{-1})C(M^{-1}\boldsymbol{\Phi}^{\mathrm{T}}\mathbf{M}) \tag{10.83}$$

Since M and C are diagonal, Eq. 10.83 can be written in the form

$$\mathbf{C} = \sum_{r=1}^{N} \frac{2\zeta_r \omega_r}{M_r}(\mathbf{M}\boldsymbol{\phi}_r)(\mathbf{M}\boldsymbol{\phi}_r)^{\mathrm{T}} \tag{10.84}$$

Due to orthogonality of modes it can be seen that Eq. 10.84 gives

$$\boldsymbol{\phi}_s^{\mathrm{T}}\mathbf{C}\boldsymbol{\phi}_s = 2\zeta_s \omega_s M_s \tag{10.85}$$

so that the modes for which a nonzero value of ζ_r is specified in Eq. 10.84 will have that damping present in $\mathbf{C}$, while there will be no damping of those modes for which ζ_r is set to zero in Eq. 10.84.

10.3.4 Damping Matrix C for Augmented Modal Damping

If a limited number of the lower-frequency modes are considered to be important in the response calculations, a truncated form of Eq. 10.84 can be employed, as follows:

$$\mathbf{C} = \sum_{r=1}^{N_c} \frac{2\zeta_r \omega_r}{M_r}\mathbf{M}\boldsymbol{\phi}_r(\mathbf{M}\boldsymbol{\phi}_r)^{\mathrm{T}} \tag{10.86}$$

This produces a damping matrix $\mathbf{C}$ that yields no damping in the modes $N_c + 1, N_c + 2, \ldots, N$. It may, however, be desirable to provide damping in these higher modes.

It is possible to modify Eq. 10.86 such that the modes $r = 1, 2, \ldots, N_c$, have specified damping ratios, and the modes $N_c + 1, N_c + 2, \ldots, N$ have damping greater than that in mode N_c. This is possible by letting

$$\mathbf{C} = a_1 \mathbf{K} + \sum_{r=1}^{N_c-1} \frac{2\hat{\zeta}_r \omega_r}{M_r}\mathbf{M}\boldsymbol{\phi}_r(\mathbf{M}\boldsymbol{\phi}_r)^{\mathrm{T}} \tag{10.87}$$

where

$$a_1 = \frac{2\zeta_{N_c}}{\omega_{N_c}}, \qquad \hat{\zeta}_r = \zeta_r - \zeta_{N_c}\frac{\omega_r}{\omega_{N_c}} \tag{10.88}$$

Then

$$\zeta_r = \begin{cases} \text{specified value}, & r = 1, 2, \ldots, N_c \\ \zeta_{N_c}\dfrac{\omega_r}{\omega_{N_c}}, & r = N_c + 1, N_c + 2, \ldots, N \end{cases} \tag{10.89}$$

Whenever direct integration requires use of a physical damping matrix, Rayleigh damping defined by Eq. 10.76 or modal damping defined by Eq. 10.84 or Eq. 10.86 or Eq. 10.87 may be employed to approximate the damping.

Example 10.5 (a) Use Eq. 10.87 to define a physical damping matrix **C** for the four-story building in Fig. 1. Assign two damping factors: $\zeta_1 = \zeta_2 = 0.01$. (b) Determine the resulting damping ratios ζ_3 and ζ_4.

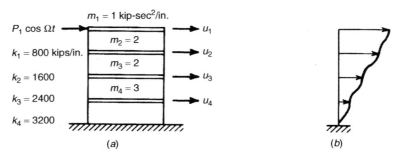

Figure 1 (a) Four-story shear building; (b) mode 1.

The stiffness matrix and mass matrix for the building are

$$\mathbf{K} = 800 \begin{bmatrix} 1 & -1 & 0 & 0 \\ -1 & 3 & -2 & 0 \\ 0 & -2 & 5 & -3 \\ 0 & 0 & -3 & 7 \end{bmatrix} \text{ kips/in.}, \quad \mathbf{M} = \begin{bmatrix} 1 & 0 & 0 & 0 \\ 0 & 2 & 0 & 0 \\ 0 & 0 & 2 & 0 \\ 0 & 0 & 0 & 3 \end{bmatrix} \text{ kip-sec}^2/\text{in.}$$

and from an eigensolution, the following modal data were obtained:

$$\boldsymbol{\phi}_1 = \begin{Bmatrix} 1.00000 \\ 0.77910 \\ 0.49655 \\ 0.23506 \end{Bmatrix}, \quad \boldsymbol{\omega} = \begin{Bmatrix} 13.294 \\ 29.660 \\ 41.079 \\ 55.882 \end{Bmatrix} \text{ rad/sec}$$

SOLUTION (a) From Eq. 10.87, with $N_c = 2$,

$$\mathbf{C} = a_1 \mathbf{K} + \frac{2\hat{\zeta}_1 \omega_1}{M_1}(\mathbf{M}\boldsymbol{\phi}_1)(\mathbf{M}\boldsymbol{\phi}_1)^T \tag{1}$$

where

$$a_1 = \frac{2\zeta_2}{\omega_2}, \quad \hat{\zeta}_1 = \zeta_1 - \zeta_2 \frac{\omega_1}{\omega_2} \tag{2}$$

Thus,

$$a_1 = \frac{2(0.01)}{29.660} = 6.7431 \times 10^{-4}$$

$$\hat{\zeta}_1 = 0.01 - 0.01\left(\frac{13.294}{29.660}\right) = 5.5179 \times 10^{-3} \tag{3}$$

$$\mathbf{M}\boldsymbol{\phi}_1 = \begin{bmatrix} 1 & 0 & 0 & 0 \\ 0 & 2 & 0 & 0 \\ 0 & 0 & 2 & 0 \\ 0 & 0 & 0 & 3 \end{bmatrix} \begin{Bmatrix} 1.00000 \\ 0.77910 \\ 0.49655 \\ 0.23506 \end{Bmatrix} = \begin{Bmatrix} 1.00000 \\ 1.55820 \\ 0.99310 \\ 0.70518 \end{Bmatrix} \tag{4}$$

From Eq. 10.29,

$$M_1 = \boldsymbol{\phi}_1^T \mathbf{M} \boldsymbol{\phi}_1 = 2.87288 \text{ kip-sec}^2/\text{in.} \tag{5}$$

Finally, Eq. (5) may be substituted into Eq. 1 to give the physical damping matrix

$$\mathbf{C} = \begin{bmatrix} 0.59051 & -0.45988 & 0.05071 & 0.03601 \\ -0.45988 & 1.74233 & -0.99987 & 0.05611 \\ 0.05071 & -0.99987 & 2.74760 & -1.58258 \\ 0.03601 & 0.05611 & -1.58258 & 3.80153 \end{bmatrix} \text{ kip-sec/in.} \quad \textbf{Ans. (a)} \tag{6}$$

(b) From Eq. 10.89b,

$$\zeta_r = \zeta_2 \frac{\omega_r}{\omega_2}, \qquad r = 3, 4 \tag{7}$$

Therefore, the damping factors for the two higher modes are

$$\zeta_3 = 0.01 \left(\frac{41.079}{29.660} \right) = 0.0138$$

$$\zeta_4 = 0.01 \left(\frac{55.882}{29.660} \right) = 0.0188$$

$\textbf{Ans. (b)}$ (8)

10.4 STRUCTURES WITH ARBITRARY VISCOUS DAMPING: COMPLEX MODES

Because the exact nature of damping in complex structural systems is seldom known, it is often assumed that the damping of the structure can be represented by viscous damping. That is, the N-DOF matrix equation of motion is assumed to be

$$\mathbf{M}\ddot{\mathbf{u}} + \mathbf{C}\dot{\mathbf{u}} + \mathbf{K}\mathbf{u} = \mathbf{p}(t) \tag{10.90}$$

where matrix $\mathbf{C}$ is the system viscous damping matrix in physical (or generalized) coordinates $\mathbf{u}$. In Section 10.3.2 we introduced a very special form of damping, where the normal modes of the undamped structure diagonalize (or are assumed to diagonalize) the viscous damping matrix. Mode-superposition solutions that utilize this form of damping are treated in Chapter 11. One procedure for handling systems with more general viscous damping is to transform the N second-order equations of motion, matrix Eq. 10.90, into $2N$ first-order equations.

10.4.1 State-Space Form of the Equations of Motion

To discuss modes and frequencies of N-DOF systems with general viscous damping, we first expand Eq. 10.90 into state-space form as follows. Let the *state vector* corresponding to displacement vector $\mathbf{u}$ be defined as the $2N$ vector:

$$\mathbf{z}(t) \equiv \begin{Bmatrix} \mathbf{u}(t) \\ \mathbf{v}(t) \end{Bmatrix} \tag{10.91}$$

308 Vibration Properties of MDOF Systems: Modes, Frequencies, and Damping

where vector $\mathbf{v}(t)$ is defined as the effective velocity vector by the momentum equality

$$\mathbf{M}\mathbf{v}(t) = \mathbf{M}\dot{\mathbf{u}}(t) \tag{10.92}$$

By incorporating Eq. 10.92, we can rewrite Eq. 10.90 in the form

$$\mathbf{M}\dot{\mathbf{v}} + \mathbf{C}\dot{\mathbf{u}} + \mathbf{K}\mathbf{u} = \mathbf{p}(t) \tag{10.93}$$

Then Eqs. 10.92 and 10.93 can be combined in *generalized state-space form*

$$\boxed{\mathbf{A}\dot{\mathbf{z}}(t) + \mathbf{B}\mathbf{z}(t) = \mathbf{F}(t)} \tag{10.94}$$

where

$$\boxed{\mathbf{A} = \begin{bmatrix} \mathbf{C} & \mathbf{M} \\ \mathbf{M} & \mathbf{0} \end{bmatrix}, \quad \mathbf{B} = \begin{bmatrix} \mathbf{K} & \mathbf{0} \\ \mathbf{0} & -\mathbf{M} \end{bmatrix}, \quad \mathbf{F}(t) = \begin{Bmatrix} \mathbf{p}(t) \\ \mathbf{0} \end{Bmatrix}} \tag{10.95}$$

are the two $2N \times 2N$ constant coefficient matrices and the $2N$ *state forcing vector*, respectively. Note that this produces coefficient matrices $\mathbf{A}$ and $\mathbf{B}$ that are symmetric, since the constituent matrices $\mathbf{M}$, $\mathbf{C}$, and $\mathbf{K}$ are all symmetric.[7]

10.4.2 State-Space Eigenvalue Problem

Starting with the generalized state equation of motion, it is possible to solve for eigenvalues and eigenvectors and to determine appropriate orthogonality equations so that mode-superposition solutions become possible for systems with general viscous damping. For free vibration of the damped system, Eq. 10.94 becomes

$$\mathbf{A}\dot{\mathbf{z}}(t) + \mathbf{B}\mathbf{z}(t) = \mathbf{0} \tag{10.96}$$

Because this is a homogeneous set of ordinary differential equations with constant coefficients, its solution has the form

$$\mathbf{z}(t) = \boldsymbol{\theta}\, e^{\lambda t} \equiv \begin{Bmatrix} \boldsymbol{\theta}_u \\ \boldsymbol{\theta}_v \end{Bmatrix} e^{\lambda t} = \begin{Bmatrix} \boldsymbol{\theta}_u \\ \lambda \boldsymbol{\theta}_u \end{Bmatrix} e^{\lambda t} \tag{10.97}$$

where λ is a scalar and $\boldsymbol{\theta}$ is a $2N$ vector. When Eq. 10.97 is substituted into Eq. 10.96 and the result divided by $e^{\lambda t}$, we obtain the following *generalized algebraic eigenvalue equation*:

$$\boxed{[\lambda \mathbf{A} + \mathbf{B}]\boldsymbol{\theta} = \mathbf{0}} \tag{10.98}$$

The solution of this eigenvalue problem consists of $2N$ *eigenvalues* $\lambda_i, i = 1, 2, \ldots, 2N$, and $2N$ corresponding *eigenvectors* $\boldsymbol{\theta}_i, i = 1, 2, \ldots, 2N$. The eigenvalues must satisfy the *characteristic equation*

$$\det(\lambda \mathbf{A} + \mathbf{B}) = 0 \tag{10.99}$$

---

[7]This form may be found in Ref. [10.3]. Some authors (e.g., Chapters 6, 15, and 16 of this book and Ref. [10.2] and [10.4]) utilize a standard state-space form. See Section D.4.3.

The following properties of the eigensolution can be stated[10.4]:

- Because the coefficient matrices are real, the $2N$ eigenvalues must either be real or they must occur in complex conjugate pairs. Since real eigenvalues indicate very high damping leading to overdamped modes, most structures will have N complex conjugate pairs of eigenvalues and corresponding eigenvectors.
- The eigenvector $\boldsymbol{\theta}_r$ belonging to a complex eigenvalue λ_r is also complex; and the eigenvector belonging to the complex conjugate of λ_r is the complex conjugate of its eigenvector $\boldsymbol{\theta}_r$. These are also called *complex modes*.

10.4.3 Orthogonality Equations for Complex Modes

It is possible to develop orthogonality equations for complex modes similar to the orthogonality equations for modes of an undamped system. First, let us write Eq. 10.85 for the ith eigenpair,

$$[\lambda_i \mathbf{A} + \mathbf{B}]\boldsymbol{\theta}_i = \mathbf{0} \tag{10.100}$$

and also for the jth eigenpair

$$[\lambda_j \mathbf{A} + \mathbf{B}]\boldsymbol{\theta}_j = \mathbf{0} \tag{10.101}$$

Now, premultiply the left and right sides of Eq. 10.100 by $\boldsymbol{\theta}_j^T$, and premultiply the left and right sides of Eq. 10.101 by $\boldsymbol{\theta}_i^T$. The resulting equations are

$$\boldsymbol{\theta}_j^T [\lambda_i \mathbf{A} + \mathbf{B}]\boldsymbol{\theta}_i = 0 \tag{10.102}$$

and

$$\boldsymbol{\theta}_i^T [\lambda_j \mathbf{A} + \mathbf{B}]\boldsymbol{\theta}_j = 0 \tag{10.103}$$

Since $\mathbf{A}$ and $\mathbf{B}$ are both symmetric matrices, Eq. 10.102 can be transposed to give

$$\boldsymbol{\theta}_i^T [\lambda_i \mathbf{A} + \mathbf{B}]\boldsymbol{\theta}_j = 0 \tag{10.104}$$

Since the $\mathbf{B}$ terms in these last two equations are identical, when we subtract Eq. 10.104 from Eq. 10.103 we get

$$(\lambda_j - \lambda_i)\boldsymbol{\theta}_i^T \mathbf{A}\boldsymbol{\theta}_j = 0 \tag{10.105}$$

But if $\lambda_j \neq \lambda_i$, Eq. 10.105 is satisfied only if

$$\boxed{\boldsymbol{\theta}_i^T \mathbf{A}\boldsymbol{\theta}_j = 0, \quad \lambda_j \neq \lambda_i, \quad i,j = 1, 2, \ldots, 2N} \tag{10.106}$$

This *orthogonality equation* states that *any two eigenvectors $\boldsymbol{\theta}_i$ and $\boldsymbol{\theta}_j$ are orthogonal with respect to $\mathbf{A}$ when the two eigenvectors correspond to distinct eigenvalues*. This orthogonality equation can be substituted back into Eq. 10.102, which gives

$$\boxed{\boldsymbol{\theta}_j^T \mathbf{B}\boldsymbol{\theta}_i = 0, \quad \lambda_j \neq \lambda_i, \quad i,j = 1, 2, \ldots, 2N} \tag{10.107}$$

This second *orthogonality equation* states that *any two eigenvectors $\boldsymbol{\theta}_i$ and $\boldsymbol{\theta}_j$ are orthogonal with respect to $\mathbf{B}$ when the two eigenvectors correspond to distinct eigenvalues.*

Let the *complex modal matrix* be defined by

$$\Theta = [\theta_1 \quad \theta_2 \quad \cdots \quad \theta_{2N}] \tag{10.108}$$

These orthogonality equations can be combined with definitions of diagonal terms *modal a_r* and *modal b_r* in the following equations:

$$\Theta^T A \Theta = \text{diag}(a_r), \qquad \Theta^T B \Theta = \text{diag}(b_r) \tag{10.109}$$

By writing Eq. 10.98 for the rth eigenvalue/eigenvector and premultiplying that equation by θ_r^T, it is easy to show that

$$\lambda_r = -\frac{b_r}{a_r} \tag{10.110}$$

Although a_r and b_r play roles similar to the modal mass M_r and modal stiffness K_r associated with the modes of undamped structures (Eqs. 10.9 and 10.10), these are generally complex numbers.

10.4.4 Interpretation of State-Space Eigenvalues

The solution of the state-space eigenproblem, Eq. 10.98, produces $2N$ eigenvalues λ_r and $2N$ corresponding eigenvectors θ_r. With notation similar to the notation used for viscous-damped SDOF systems, let us identify the real and imaginary parts of λ_r by the following notation:

$$\lambda_r \equiv \alpha_r + i\beta_r = -\zeta_r \omega_r + i\omega_r \sqrt{1 - \zeta_r^2} = -\zeta_r \omega_r + i\omega_{dr} \tag{10.111}$$

Then the *natural frequency* ω_r and the *damping factor* ζ_r are given by

$$\omega_r = \sqrt{\alpha_r^2 + \beta_r^2}, \qquad \zeta_r = \frac{-\alpha_r}{\omega_r} \tag{10.112}$$

It should be noted that ω_r is here called the *natural frequency*, not the undamped natural frequency. For viscous-damped systems, the natural frequency given by Eq. 10.112a is equal to the undamped natural frequency only if the system has proportional damping.[10.3]

10.4.5 Interpretation of State-Space Eigenvectors: Scaling and Rotating of Complex Eigenvectors

Since the eigenvalues and eigenvectors can be complex, let the eigensolution in Eq. 10.97 be expressed in terms of real and imaginary parts by

$$\theta e^{\lambda t} = \begin{Bmatrix} \theta_u \\ \lambda \theta_u \end{Bmatrix} e^{\lambda t} \tag{10.113}$$

where
$$\boldsymbol{\theta}_u = \mathbf{x} + i\mathbf{y}, \qquad \lambda = \alpha + i\beta \qquad (10.114)$$

The following example is presented to give you a physical feeling for the results of state-vector eigensolutions. The spring–mass system was used in Example 9.8. Since $N = 2$ for this problem, there will be four eigenvalues and eigenvectors of the state eigenproblem, Eq. 10.98. The eigenvalues will be listed as a column vector λ, and the eigenvectors in a 4×4 matrix $\boldsymbol{\Theta}$ or as two 4×2 matrices. The example illustrates four distinct damping cases: the undamped system, a system with dashpots that produces **M**-proportional damping, an underdamped system with a single local damper, and an overdamped system with a single local damper. You should study these examples carefully and compare them against each other so that you will get some feeling for how damping affects the dynamic behavior of real systems.

Example 10.6 For the four listed versions of the 2-DOF spring–mass–dashpot system in Fig. 1, use state-space eigensolutions to determine the natural frequencies, damping factors, and mode shapes. Discuss your solutions. The four systems are: (a) the undamped system, (b) a system with dashpots that produces $\mathbf{C} = 0.5\mathbf{M}$, (c) an underdamped system with a single local damper $c_3 = 20$, and (d) an overdamped system with a single local damper $c_3 = 200$.

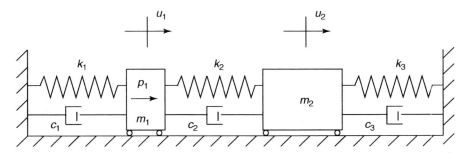

Figure 1 A 2-DOF spring–mass–dashpot system.

For all four systems, let $k_1 = 1600$, $k_2 = 600$, $k_3 = 3200$, $m_1 = 1$, $m_2 = 2$. Then the stiffness matrix and mass matrix are

$$\mathbf{K} = \begin{bmatrix} k_1 + k_2 & -k_2 \\ -k_2 & k_2 + k_3 \end{bmatrix} = \begin{bmatrix} 2200 & -600 \\ -600 & 3800 \end{bmatrix}, \qquad \mathbf{M} = \begin{bmatrix} m_1 & 0 \\ 0 & m_2 \end{bmatrix} = \begin{bmatrix} 1 & 0 \\ 0 & 2 \end{bmatrix}$$

The viscous damping matrix **C** has the form

$$\mathbf{C} \equiv \begin{bmatrix} c_{11} & c_{12} \\ c_{21} & c_{22} \end{bmatrix} = \begin{bmatrix} c_1 + c_2 & -c_2 \\ -c_2 & c_2 + c_3 \end{bmatrix}$$

which has enough flexibility to permit Rayleigh (proportional) damping matrices to be implemented.

SOLUTION For all four cases, state matrices **A** and **B** have the form given by Eqs. 10.95a and b, respectively.

$$\mathbf{A} = \begin{bmatrix} \mathbf{C} & \mathbf{M} \\ \mathbf{M} & \mathbf{0} \end{bmatrix}, \quad \mathbf{B} = \begin{bmatrix} \mathbf{K} & \mathbf{0} \\ \mathbf{0} & -\mathbf{M} \end{bmatrix} \quad (1)$$

The MATLAB .m-files that encode the following solutions can be found on the book's website. Their filenames are:

- sd2ex10_6a.m: solution for part (a), undamped system
- sd2ex10_6b.m: solution for part (b), system with **M**-proportional damping
- sd2ex10_6d.m: solution for part (c), underdamped system with local damping
- sd2ex10_6e.m: solution for part (d), overdamped system with local damping

(a) *Undamped system.* The modes and frequencies of the undamped system are given in Example 9.8. The results of the corresponding state-space solution are

$$\lambda = \begin{Bmatrix} 0 + 50.0i \\ 0 - 50.0i \\ 0 + 40.0i \\ 0 - 40.0i \end{Bmatrix} \quad \text{Ans. (a) (2)}$$

$$\Theta = \begin{bmatrix} 0 + 0.020i & 0 - 0.020i & 0 + 0.025i & 0 - 0.025i \\ 0 - 0.010i & 0 + 0.010i & 0 + 0.025i & 0 - 0.025i \\ \hline -1.00 & -1.00 & -1.00 & -1.00 \\ 0.50 & 0.50 & -1.00 & -1.00 \end{bmatrix}$$

Ans. (a) (3)

From the data above, we can make the following observations:

1. The eigensolver lists the higher-frequency (50 rad/s) mode first, then the lower-frequency (40 rad/s) mode. Therefore, the user must sort the eigenvalues and associated eigenvectors if they are required to be listed in order of increasing frequency. For the four cases presented in this example, the eigenvalues and eigenvectors will be left unsorted.
2. There are two sets of complex-conjugate eigenvalues and two sets of associated complex-conjugate eigenvectors. The vertical lines in Θ delineate the four eigenvector columns.
3. The horizontal line in Θ separates the (upper) θ_u parts of the eigenvectors from the (lower) θ_v parts. Clearly, the latter have the correct form, $\theta_v = \lambda \theta_u$, for the respective eigenvectors.
4. Each of the eigenvalues is pure imaginary, which indicates that there is no damping.
5. Each of the eigenvectors is a pure imaginary vector. So the two elements of each eigenvector are in phase or $180°$ out of phase with each other.

6. Let us add the top parts of the third and fourth eigenvectors as follows:

$$\boldsymbol{\theta}_{u3}e^{\lambda_3 t} + \boldsymbol{\theta}_{u4}e^{\lambda_4 t} = \begin{Bmatrix} 0.025\,i \\ 0.025\,i \end{Bmatrix} e^{(40.0\,i)t} + \begin{Bmatrix} -0.025\,i \\ -0.025\,i \end{Bmatrix} e^{(-40.0\,i)t}$$

$$= \begin{Bmatrix} 0.025\,i \\ 0.025\,i \end{Bmatrix} [\cos(40.0\,t) + i\sin(40.0\,t)]$$

$$+ \begin{Bmatrix} -0.025\,i \\ -0.025\,i \end{Bmatrix} [\cos(40.0\,t) - i\sin(40.0\,t)] \qquad (4)$$

$$= -\begin{Bmatrix} 0.050 \\ 0.050 \end{Bmatrix} \sin(40.0\,t)$$

Clearly, the combination of the two complex-conjugate eigenvectors leads to the real motion of the symmetric mode (i.e., first mode) at its undamped natural frequency of $\omega_1 = 40$ rad/s. This illustrates why complex eigenvectors must occur in complex-conjugate pairs. It also illustrates how state eigenvectors that are pure imaginary correspond to modes of undamped systems, which are often called *real normal modes*.

(b) *System with **M**-proportional damping with* $\mathbf{C} = 0.5\mathbf{M}$. The results of the MATLAB state-space solution for this system are

$$\boldsymbol{\lambda} = \begin{Bmatrix} -0.2500 + 49.9994\,i \\ -0.2500 - 49.9994\,i \\ -0.2500 + 39.9992\,i \\ -0.2500 - 39.9992\,i \end{Bmatrix}, \quad \boldsymbol{\omega} = \begin{Bmatrix} 50.0 \\ 50.0 \\ 40.0 \\ 40.0 \end{Bmatrix}, \quad \boldsymbol{\zeta} = \begin{Bmatrix} 0.0050 \\ 0.0050 \\ 0.0063 \\ 0.0063 \end{Bmatrix}$$

Ans. (b) (5)

The eigenvectors, in rectangular form, are listed as two complex-conjugate pairs:

$$[\boldsymbol{\theta}_1 \mid \boldsymbol{\theta}_2] = \begin{bmatrix} 0.0004 + 0.0197\,i & 0.0004 - 0.0197\,i \\ -0.0002 - 0.0098\,i & -0.0002 + 0.0098\,i \\ -0.9846 + 0.0154\,i & -0.9846 - 0.0154\,i \\ 0.4923 - 0.0077\,i & 0.4923 + 0.0077\,i \end{bmatrix}$$

(6)

$$[\boldsymbol{\theta}_3 \mid \boldsymbol{\theta}_4] = \begin{bmatrix} -0.0006 - 0.0245\,i & -0.0006 + 0.0245\,i \\ -0.0006 - 0.0245\,i & -0.0006 + 0.0245\,i \\ 0.9809 - 0.0191\,i & 0.9809 + 0.0191\,i \\ 0.9809 - 0.0191\,i & 0.9809 + 0.0191\,i \end{bmatrix}$$

It is much more instructive to list the eigenvectors in polar (magnitude, phase angle) form. The phase angles are given in degrees.

$$[\boldsymbol{\theta}_1 \mid \boldsymbol{\theta}_2] = \begin{bmatrix} 0.0197 \; \angle 88.8190 & 0.0197 \; \angle -88.8190 \\ 0.0098 \; \angle -91.1810 & 0.0098 \; \angle 91.1810 \\ 0.9847 \; \angle 179.1054 & 0.9847 \; \angle -179.1054 \\ 0.4924 \; \angle -0.8946 & 0.4924 \; \angle 0.8946 \end{bmatrix}$$

Ans. (b) (7)

$$[\boldsymbol{\theta}_3 \mid \boldsymbol{\theta}_4] = \begin{bmatrix} 0.0245 \; \angle -91.4762 & 0.0245 \; \angle 91.4762 \\ 0.0245 \; \angle -91.4762 & 0.0245 \; \angle 91.4762 \\ 0.9810 \; \angle -1.1181 & 0.9810 \; \angle 1.1181 \\ 0.9810 \; \angle -1.1181 & 0.9810 \; \angle 1.1181 \end{bmatrix}$$

From the data above, we can make the following observations that hold true for viscous-damped systems with proportional damping. The same hold for systems with modal damping, which does not couple the modes.

1. The natural frequencies, listed in the ω vector, are the same as the natural frequencies of the undamped system [given in part (a)].
2. The damping factors satisfy Eq. 10.78 for **M**-proportional Rayleigh damping, that is,

$$\zeta_r = \frac{a_0}{2\omega_r}, \quad \text{e.g.,} \quad \zeta_1 = \frac{0.5}{2(50.0)} = 0.0050, \quad \zeta_3 = \frac{0.5}{2(40.0)} = 0.00625$$

3. From the eigenvector magnitude and phase information in Eq. 7, it can be seen that the two masses are $180°$ out of phase ($88.8190 + 91.1810 = 180.0000$) in the first pair (the antisymmetric mode), and that the two masses are in phase in the second pair (the symmetric mode). When the damping is proportional, all masses of a system will be either in phase or $180°$ out of phase, just as for an undamped system. Therefore, like the eigenvectors of the undamped system in part (a), the eigenvectors for this type of proportionally damped system are sometimes called *real normal modes*.

(c) *Underdamped system with local damping with* $c_3 = 20$. The results of the MATLAB state-space solution for this system are

$$\lambda = \begin{Bmatrix} -1.4958 + 49.3641\,i \\ -1.4958 - 49.3641\,i \\ -3.5042 + 40.3448\,i \\ -3.5042 - 40.3448\,i \end{Bmatrix}, \quad \omega = \begin{Bmatrix} 49.3868 \\ 49.3868 \\ 40.4967 \\ 40.4967 \end{Bmatrix}, \quad \zeta = \begin{Bmatrix} 0.0303 \\ 0.0303 \\ 0.0865 \\ 0.0865 \end{Bmatrix}$$

Ans. (c) (8)

The eigenvectors are stated below in polar (magnitude, phase angle) form. The phase angles are given in degrees.

$$[\theta_1 \mid \theta_2] = \begin{bmatrix} 0.0165 & \angle 72.9389 & 0.0165 & \angle -72.9389 \\ 0.0076 & \angle -74.8685 & 0.0076 & \angle 74.8685 \\ 0.8138 & \angle 164.6745 & 0.8138 & \angle -164.6745 \\ 0.3760 & \angle 16.8671 & 0.3760 & \angle -16.8671 \end{bmatrix}$$

Ans. (c) (9)

$$[\theta_3 \mid \theta_4] = \begin{bmatrix} 0.0177 & \angle 89.8007 & 0.0177 & \angle -89.8007 \\ 0.0191 & \angle 63.9883 & 0.0191 & \angle -63.9883 \\ 0.7149 & \angle -175.2353 & 0.7149 & \angle 175.2353 \\ 0.7737 & \angle 158.9524 & 0.7737 & \angle -158.9524 \end{bmatrix}$$

From the data in Eqs. 8 and 9, we can make the following observations for this underdamped system with nonproportional damping.

1. Since the system is underdamped, the eigenvalues and eigenvectors occur in two complex-conjugate pairs.
2. The natural frequencies listed in the ω vector in Eq. 8 are <u>not</u> exactly the same as the natural frequencies of the undamped system [given in part (a)]. However, since this is a fairly lightly damped system, the natural frequencies are close to the respective undamped natural frequencies.

3. The eigenvectors are sketched in Fig. 2. As can be seen, the masses are not in phase or 180° out of phase as they were in part (b) for the system with proportional damping. However, since this system is lightly damped, it has *complex modes* that are only "moderately" complex.

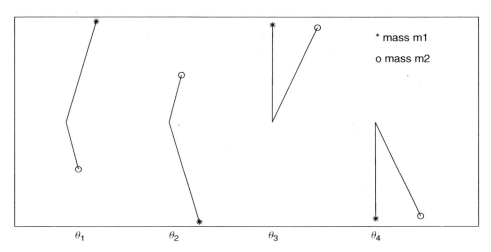

Figure 2 Eigenvectors for an underdamped system with local damping.

The results for this underdamped system should be compared with the following overdamped system, which has the same form of local damping.

(d) *Overdamped system with local damping with $c_3 = 200$.* (This system becomes overdamped when $c_3 > 172$.) The results of the MATLAB state-space solution for this system are

$$\lambda = \begin{Bmatrix} -0.4097 + 46.9362\,i \\ -0.4097 - 46.9362\,i \\ -74.9604 \\ -24.2203 \end{Bmatrix}, \quad \omega = \begin{Bmatrix} 46.9380 \\ 46.9380 \\ -- \\ -- \end{Bmatrix}, \quad \zeta = \begin{Bmatrix} 0.0087 \\ 0.0087 \\ -- \\ -- \end{Bmatrix}$$

Ans. (d) (10)

The eigenvectors are stated below in polar (magnitude, phase angle) form. The phase angles are given in degrees.

$$[\boldsymbol{\theta}_1 \mid \boldsymbol{\theta}_2] = \begin{bmatrix} 0.0209 \angle -89.4497 & 0.0209 \angle 89.4497 \\ 0.0013 \angle 176.3264 & 0.0013 \angle -176.3264 \\ 0.9822 \angle 1.0504 & 0.9822 \angle -1.0504 \\ 0.0631 \angle -93.1735 & 0.0631 \angle 93.1735 \end{bmatrix}$$

Ans. (d) (11)

$$[\boldsymbol{\theta}_3 \mid \boldsymbol{\theta}_4] = \begin{bmatrix} 0.0010 \angle 180.0000 & 0.0089 \angle 0 \\ 0.0133 \angle 180.0000 & 0.0413 \angle 0 \\ 0.0767 \angle 0 & 0.2153 \angle 180.0000 \\ 1.0000 \angle 0 & 1.0000 \angle 180.0000 \end{bmatrix}$$

From the data in Eqs. 10 and 11, we can make the following observations for this overdamped system with nonproportional damping.

1. There are two large negative real eigenvalues, (λ_3 and λ_4). That is, the system is overdamped to the extent that there is only one pair of complex conjugate eigenvalues and associated eigenvectors.
2. The natural frequencies ω_1 and ω_2 are <u>not</u> exactly the same as natural frequencies of the undamped system [given in part (a) above].
3. As can be seen from the eigenvectors (displacement part), the masses are not in phase or 180° out of phase as they were in part (b) for the system with proportional damping. And since this system is heavily damped, it has "very" complex modes compared with the complex modes in part (c).

Special procedures are required if the damped system has rigid-body modes. This is the subject of the next section.

10.5 NATURAL FREQUENCIES AND MODE SHAPES OF DAMPED STRUCTURES WITH RIGID-BODY MODES

Special treatment is required in determining the state eigenvectors of systems that have rigid-body modes. It is necessary to incorporate *generalized eigenvectors* and the *Jordan form* of the eigenvalue matrix.[8]

10.5.1 Generalized Eigenvectors: Jordan Form

An $N \times N$ matrix **D** that fails to have a linearly independent set of N eigenvectors is said to be a *defective matrix*. This may occur when **D** has a repeated eigenvalue. It is then not possible to transform **D** into diagonal form; that is, there exists no eigenvector matrix $\mathbf{\Phi}$ such that

$$\mathbf{D\Phi} = \mathbf{\Phi\Lambda}$$

where $\mathbf{\Lambda}$ is a diagonal matrix. But it is possible to find a linearly independent set of *generalized eigenvectors* that transform **D** into almost-diagonal *Jordan form*

$$\mathbf{DQ} = \mathbf{QJ} \qquad (10.115)$$

Reference [10.6] defines these concepts and shows, for example, that when **D** has a repeated eigenvalue λ_2 of multiplicity 3 and an eigenvalue λ_3 of multiplicity 2, the Jordan matrix will have the form (zeros in off-diagonal blocks omitted for clarity)

$$\mathbf{J} = \begin{bmatrix} \lambda_1 & & & & & \\ & \lambda_2 & 1 & 0 & & \\ & 0 & \lambda_2 & 1 & & \\ & 0 & 0 & \lambda_2 & & \\ & & & & \lambda_3 & 1 \\ & & & & 0 & \lambda_3 \\ & & & & & & \lambda_4 \end{bmatrix} \qquad (10.116)$$

---

[8]This section is based on material originally published in Ref. [10.5], copyright © by the American Institute of Aeronautics and Astronautics, Inc. Reprinted with permission.

where the repeated eigenvalues lead to Jordan blocks having the eigenvalue on the diagonal and ones on the superdiagonal.

It will now be shown that systems that have rigid-body modes and are described by the state-variable equation of the form Eq. 10.94 require the use of generalized eigenvectors.

10.5.2 Undamped Systems with Rigid-Body Modes

Before considering damped systems, let us examine solution of the state eigenvalue problem for undamped MDOF systems. Consider an undamped N-DOF system described by the equation of motion

$$\mathbf{M}\ddot{\mathbf{u}} + \mathbf{K}\mathbf{u} = \mathbf{0} \tag{10.117}$$

and let this system have N_r rigid-body modes; that is, $\mathbf{K}$ is of rank $N - N_r$. Then the N_r rigid-body displacement modes can be obtained from the equation

$$\mathbf{K}\boldsymbol{\Theta}_{ur} = \mathbf{0}_{nr} \tag{10.118}$$

where $\boldsymbol{\Theta}_{ur}$ is an $N \times N_r$ matrix of rigid-body displacement modes.

Now let the same undamped system be described by the state-variable equation (Eq. 10.96)

$$\mathbf{A}\dot{\mathbf{z}}(t) + \mathbf{B}\mathbf{z}(t) = \mathbf{0} \tag{10.119}$$

where $\mathbf{A}$ and $\mathbf{B}$ are given by Eqs. 10.95a and b, respectively, with, of course, $\mathbf{C} = \mathbf{0}$ for undamped systems. The corresponding state eigenproblem is (Eq. 10.98)

$$[\lambda \mathbf{A} + \mathbf{B}]\boldsymbol{\theta} = \mathbf{0} \tag{10.120}$$

and the corresponding characteristic equation is (Eq. 10.99)

$$\det(\lambda \mathbf{A} + \mathbf{B}) = 0 \tag{10.121}$$

If the state rigid-body modes are defined as those state vectors that satisfy Eq. 10.120 with $\lambda = 0$, these state rigid-body modes must satisfy the following equation:

$$\mathbf{B}\boldsymbol{\theta} \equiv \begin{bmatrix} \mathbf{K} & \mathbf{0} \\ \mathbf{0} & -\mathbf{M} \end{bmatrix} \begin{Bmatrix} \boldsymbol{\theta}_u \\ \boldsymbol{\theta}_v \end{Bmatrix} = \begin{Bmatrix} \mathbf{K}\boldsymbol{\theta}_u \\ -\mathbf{M}\boldsymbol{\theta}_v \end{Bmatrix} = \begin{Bmatrix} \mathbf{0} \\ \mathbf{0} \end{Bmatrix} \tag{10.122}$$

If $\mathbf{M}$ is nonsingular, $\boldsymbol{\theta}_v$ must be zero, and the only state rigid-body modes will have the form

$$\boldsymbol{\theta}'_r = \begin{Bmatrix} \boldsymbol{\theta}_u \\ \mathbf{0} \end{Bmatrix} \tag{10.123}$$

and there will be only N_r such modes. On the other hand, the eigenvalue $\lambda = 0$ will occur as a root of multiplicity $2N_r$ of Eq. 10.121. Thus, the generalized eigenproblem of Eq. 10.120 is defective. It possesses one set of N_r *regular state rigid-body modes* given by

$$\boldsymbol{\Theta}'_r = \begin{bmatrix} \boldsymbol{\Theta}_{ur} \\ \mathbf{0}_{nr} \end{bmatrix}_{2N \times N_r} \tag{10.124}$$

where $\boldsymbol{\Theta}_{ur}$ is given by Eq. 10.118.

For each column of $\boldsymbol{\Theta}_{ur}$; that is, for each regular state rigid-body mode corresponding to $\lambda = 0$, there will be a corresponding *generalized state rigid-body mode* $\boldsymbol{\theta}''_r$, defined by a generalization of Eq. 10.115: namely,

$$\mathbf{A}[\boldsymbol{\theta}'_r \ \boldsymbol{\theta}''_r] \begin{bmatrix} 0 & 1 \\ 0 & 0 \end{bmatrix} + \mathbf{B}[\boldsymbol{\theta}'_r \ \boldsymbol{\theta}''_r] = [\mathbf{0} \ \mathbf{0}] \tag{10.125}$$

or

$$\mathbf{B}\boldsymbol{\Theta}'_r = \mathbf{0}_{sr} \tag{10.126}$$

$$\mathbf{B}\boldsymbol{\Theta}''_r = -\mathbf{A}\boldsymbol{\Theta}'_r \tag{10.127}$$

where the subscript s stands for the $2N$ rows of a state vector. In expanded form, Eq. 10.127 becomes

$$\begin{bmatrix} \mathbf{K} & \mathbf{0} \\ \mathbf{0} & -\mathbf{M} \end{bmatrix} \begin{bmatrix} \boldsymbol{\Theta}''_{ur} \\ \boldsymbol{\Theta}''_{vr} \end{bmatrix} = -\begin{bmatrix} \mathbf{0} & \mathbf{M} \\ \mathbf{M} & \mathbf{0} \end{bmatrix} \begin{bmatrix} \boldsymbol{\Theta}_{ur} \\ \mathbf{0}_{nr} \end{bmatrix} \tag{10.128}$$

The row partitions of Eq. 10.128 are

$$\mathbf{K}\boldsymbol{\Theta}''_{ur} = \mathbf{0}_{nr} \tag{10.129}$$

$$-\mathbf{M}[\boldsymbol{\Theta}''_{vr} - \boldsymbol{\Theta}_{ur}] = \mathbf{0}_{nr} \tag{10.130}$$

When $\mathbf{M}$ is nonsingular, Eq. 10.130 requires that

$$\boldsymbol{\Theta}''_{vr} = \boldsymbol{\Theta}_{ur} \tag{10.131}$$

Equation 10.129 states that either $\boldsymbol{\Theta}''_{ur}$ satisfies the same equation as $\boldsymbol{\Theta}_{ur}$ or that it is zero. Since the regular state rigid-body mode already contains this displacement representation, it is sufficient to just set $\boldsymbol{\Theta}''_{ur} = \mathbf{0}_{nr}$. Thus, the complete set of $2N_r$ *state rigid-body modes* for an undamped system is given by

$$\boxed{\boldsymbol{\Theta}_r \equiv [\boldsymbol{\Theta}'_r \ \boldsymbol{\Theta}''_r] = \begin{bmatrix} \boldsymbol{\Theta}_{ur} & \mathbf{0}_{nr} \\ \mathbf{0}_{nr} & \boldsymbol{\Theta}_{ur} \end{bmatrix}} \tag{10.132}$$

where $\boldsymbol{\Theta}_{ur}$ is given by Eq. 10.118.

Example 10.7 As an example of the preceding theory for undamped systems, consider the 2-DOF spring–mass system shown in Fig. 1. Determine the full set of four linearly independent state eigenvectors.

10.5 Natural Frequencies and Mode Shapes of Damped Structures with Rigid-Body Modes

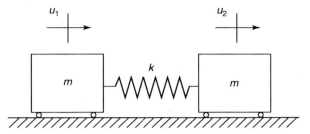

Figure 1 A 2-DOF spring–mass system with rigid-body freedom. $k = m = 1$.

SOLUTION   The state matrices are

$$\mathbf{A} = \begin{bmatrix} 0 & 0 & 1 & 0 \\ 0 & 0 & 0 & 1 \\ 1 & 0 & 0 & 0 \\ 0 & 1 & 0 & 0 \end{bmatrix}, \quad \mathbf{B} = \begin{bmatrix} 1 & -1 & 0 & 0 \\ -1 & 1 & 0 & 0 \\ 0 & 0 & -1 & 0 \\ 0 & 0 & 0 & -1 \end{bmatrix} \quad (1a,b)$$

The eigenvalues are $\lambda = 0, 0, i\sqrt{2}, -i\sqrt{2}$. Since the rank of $[\lambda_1 \mathbf{A} + \mathbf{B}]$ is 3, there will only be one regular state eigenvector corresponding to $\lambda_1 = 0$. This is given by

$$\mathbf{B}\boldsymbol{\theta}_1 = \mathbf{0} \tag{2}$$

Therefore, the regular state rigid-body mode is

$$\boldsymbol{\theta}_1^T = [1 \ 1 \ 0 \ 0] \tag{3}$$

From Eq. 10.117, the generalized state rigid-body mode is given by

$$\mathbf{B}\boldsymbol{\theta}_2 = -\mathbf{A}\boldsymbol{\theta}_1 \tag{4}$$

which gives

$$\boldsymbol{\theta}_2^T = [0 \ 0 \ 1 \ 1] \tag{5}$$

It is left to the reader to determine the regular state modes corresponding to $\boldsymbol{\theta}_3$ and $\boldsymbol{\theta}_4$.

Finally, the full set of linearly independent state eigenvectors (modes) for this 2-DOF system with rigid-body freedom is

$$\boldsymbol{\Theta} = \begin{bmatrix} 1 & 0 & 1 & 1 \\ 1 & 0 & -1 & -1 \\ 0 & 1 & i\sqrt{2} & -i\sqrt{2} \\ 0 & 1 & -i\sqrt{2} & i\sqrt{2} \end{bmatrix} \quad \text{Ans. (6)}$$

and the corresponding Jordan matrix is

$$\mathbf{J} = \left[\begin{array}{cc|cc} 0 & 1 & 0 & 0 \\ 0 & 0 & 0 & 0 \\ \hline 0 & 0 & i\sqrt{2} & 0 \\ 0 & 0 & 0 & -i\sqrt{2} \end{array} \right] \quad \text{Ans. (7)}$$

Note the Jordan block corresponding to the repeated root $\lambda = 0$.

> A physical significance can be attributed to the first two columns of $\boldsymbol{\Theta}$: namely, that rigid-body motion with arbitrary velocity can be expressed as a linear superposition of these two columns.

10.5.3 Viscous-Damped Systems with Rigid-Body Modes

Consider a viscous-damped system whose stiffness matrix is of rank $N - N_r$. Then the eigenproblem

$$[\lambda^2 \mathbf{M} + \lambda \mathbf{C} + \mathbf{K}]\boldsymbol{\theta}_u = \mathbf{0} \tag{10.133}$$

has N_r eigenvalues $\lambda = 0$, with rigid-body displacement modes $\boldsymbol{\Theta}_{ur}$ given by Eq. 10.118. For a system with viscous damping, the equation for determining the generalized eigenvectors of a defective system takes the form

$$\begin{bmatrix} \mathbf{K} & \mathbf{0} \\ \mathbf{0} & -\mathbf{M} \end{bmatrix} \begin{bmatrix} \boldsymbol{\Theta}''_{ur} \\ \boldsymbol{\Theta}''_{vr} \end{bmatrix} = - \begin{bmatrix} \mathbf{C} & \mathbf{M} \\ \mathbf{M} & \mathbf{0} \end{bmatrix} \begin{bmatrix} \boldsymbol{\Theta}_{ur} \\ \mathbf{0}_{nr} \end{bmatrix} \tag{10.134}$$

The row partitions of Eq. 10.134 are

$$\mathbf{K}\boldsymbol{\Theta}''_{ur} = -\mathbf{C}\boldsymbol{\Theta}_{ur} \tag{10.135}$$

$$-\mathbf{M}[\boldsymbol{\Theta}''_{vr} - \boldsymbol{\Theta}_{ur}] = \mathbf{0}_{nr} \tag{10.136}$$

Since $\mathbf{M}$ is assumed to be nonsingular, Eq. 10.136 requires that

$$\boldsymbol{\Theta}''_{vr} = \boldsymbol{\Theta}_{ur} \tag{10.137}$$

as in the undamped case. Since $\mathbf{K}$ is singular, Eq. 10.135 will have a solution only for certain forms of $\mathbf{C}$. For example, if any of the columns of $\mathbf{C}\boldsymbol{\Theta}_{ur}$ are zero, there will be a solution for the corresponding column of $\mathbf{K}\boldsymbol{\Theta}''_{ur}$, just as for the undamped case. However, for those columns of $\mathbf{C}\boldsymbol{\Theta}_{ur}$ that are not zero, there will not be a solution for the corresponding columns of $\mathbf{K}\boldsymbol{\Theta}''_{ur}$. Hence, the assumption that $\lambda = 0$ is a double root leading to both regular and generalized state rigid-body modes is not valid. Both of these cases are illustrated in the following example.

> **Example 10.8** As an example of the preceding theory for damped systems, consider the 2-DOF spring–mass–dashpot system shown in Fig. 1, where $k_1 = k_2 = m_1 = m_2 = m_3 = 1$. Determine the $\lambda = 0$ eigenvectors of the following systems: (a) *a case for which* $\mathbf{C}\boldsymbol{\theta}_{ur} = \mathbf{0}$; (b) *a case for which* $\mathbf{C}\boldsymbol{\theta}_{ur} \neq \mathbf{0}$.
>
>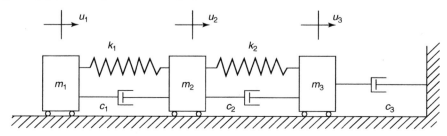
>
> **Figure 1** A 3-DOF spring–mass–dashpot system with rigid-body freedom.

10.5 Natural Frequencies and Mode Shapes of Damped Structures with Rigid-Body Modes

SOLUTION For this system,

$$\mathbf{M} = \begin{bmatrix} 1 & 0 & 0 \\ 0 & 1 & 0 \\ 0 & 0 & 1 \end{bmatrix}, \quad \mathbf{K} = \begin{bmatrix} 1 & -1 & 0 \\ -1 & 2 & -1 \\ 0 & -1 & 1 \end{bmatrix} \quad (1)$$

(a) *Case for which* $\mathbf{C}\boldsymbol{\theta}_{ur} = \mathbf{0}$. Let $c_1 = 0.2$, $c_2 = 0.4$, and $c_3 = 0$. That is, the 3-DOF system is not connected to ground by either a spring or a dashpot. Then, for this system,

$$\mathbf{C} = \begin{bmatrix} 0.2 & -0.2 & 0.0 \\ -0.2 & 0.6 & -0.4 \\ 0.0 & -0.4 & 0.4 \end{bmatrix} \quad (2)$$

The characteristic equation for this system is

$$\lambda^2 (\lambda^4 + 1.2\lambda^3 + 4.24\lambda^2 + 1.8\lambda + 3) = 0 \quad (3)$$

so $\lambda = 0$ is a root of multiplicity 2. The rigid-body displacement mode based on Eq. 10.118 is

$$\boldsymbol{\theta}_{u1} = [\,1\ 1\ 1\,]^T \quad (4)$$

Even though $\mathbf{C}$ is not proportional to $\mathbf{K}$, it is seen that $\mathbf{C}\boldsymbol{\theta}_{u1} = \mathbf{0}$. This is consistent with the fact that since there is no direct connection between the system and ground, $\lambda = 0$ is a root of multiplicity 2. Consequently, the system will have both a regular state rigid-body mode and a generalized state rigid-body mode. Therefore, from Eq. 10.133, the state rigid-body mode matrix will be

$$[\,\boldsymbol{\theta}_1\ \boldsymbol{\theta}_2\,] = \begin{bmatrix} 1 & 0 \\ 1 & 0 \\ 1 & 0 \\ 0 & 1 \\ 0 & 1 \\ 0 & 1 \end{bmatrix} \quad \textbf{Ans. (a)} \quad (5)$$

(b) *Case for which* $\mathbf{C}\boldsymbol{\theta}_{ur} \neq \mathbf{0}$. Let $c_1 = 0.2$, $c_2 = 0.4$, and $c_3 = 0.2$. That is, this 3-DOF system is connected to ground by a dashpot but not by a spring. Then, for this system,

$$\mathbf{C} = \begin{bmatrix} 0.2 & -0.2 & 0.0 \\ -0.2 & 0.6 & -0.4 \\ 0.0 & -0.4 & 0.6 \end{bmatrix} \quad (6)$$

The characteristic equation for this system is

$$\lambda (\lambda^5 + 1.4\lambda^4 + 4.4\lambda^3 + 2.416\lambda^2 + 3.12\lambda + 0.2) = 0 \quad (7)$$

so $\lambda = 0$ is a single root. The rigid-body displacement mode is the same as for case A, but in the present case, $\mathbf{C}\boldsymbol{\theta}_{u1} \neq \mathbf{0}$. Therefore, the system is nondefective, and there is no generalized eigenvector corresponding to $\lambda = 0$.

Reference [10.5] gives, as an additional example of damped MDOF systems, a four-element finite element model of a beam with nonproportional damping. This example requires both regular and generalized rigid-body modes.

REFERENCES

[10.1] K.-J. Bathé and E. L. Wilson, *Numerical Methods in Finite Element Analysis*, Prentice-Hall, Englewood Cliffs, NJ, 1976.

[10.2] D. J. Inman, *Engineering Vibration*, 2nd ed., Prentice Hall, Upper Saddle River, NJ, 2000.

[10.3] D. J. Ewins, *Modal Testing: Theory, Practice and Application*, Research Studies Press, Baldock, Hertfordshire, England, 2000.

[10.4] L. Meirovitch, *Fundamentals of Vibrations*, McGraw-Hill, New York, 2001.

[10.5] R. R. Craig, Jr., T.-J. Su, and Z. Ni, "State-Variable Models of Structures Having Rigid-Body Modes," *AIAA Journal of Guidance, Navigation and Control*, Vol. 13, No. 6, November–December 1990, pp. 1157–1160.

[10.6] C.-T. Chen, *Linear System Theory and Design*, Holt, Rinehart and Winston, New York, 1984.

PROBLEMS

Problem Set 10.1

For problems whose number is preceded by a **C**, you are to write a computer program and use it to produce the plot(s) requested. *Note:* MATLAB .m-files for many of the plots in this book may be found on the book's website.

10.1 This problem illustrates the fact that the symmetry of **K** and **M** depends on the coordinates employed. (a) For the 2-DOF spring-mass system in Fig. P10.1, derive the equations of motion by using Newton's Laws and using as coordinates u_1 and u_2 the absolute motions of masses m_1 and m_2, respectively. Write the equations of motion in matrix format. (b) Repeat part (a) but use as coordinates the absolute motion, u_1, of mass m_1 and the displacement of mass m_2 relative to m_1, that is, $u_r = u_2 - u_1$. (c) Using the coordinates of part (b), use Lagrange's equations to derive the equations of motion for this system. What conclusions about symmetry of the coefficient matrices can you draw by comparing these equations with the ones that were derived by Newton's Laws in part (a)?

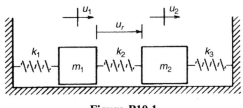

Figure P10.1

C 10.2 For the 3-DOF spring–mass system shown in Fig. P10.2, the values of the masses m_i and stiffnesses k_i are as shown. (a) Derive the equations of motion and determine the characteristic polynomial

$$\mathcal{P}(\omega^2) \equiv \det(\mathbf{K} - \omega^2 \mathbf{M})$$

(b) Use MATLAB or another computer program to plot $\mathcal{P}(\omega^2)$ from $\omega^2 = 0$ to $\omega^2 = 3.2$. (c) Solve $\mathcal{P}(\omega_r^2) = 0$ for ω_1^2, ω_2^2, and ω_3^2. (d) Determine the three mode shapes and scale them so that $M_r = 1$.

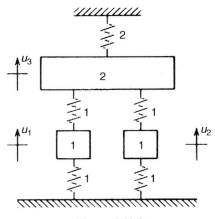

Figure P10.2

10.3 (a) Using the method of Eq. 10.13, solve for modes $\boldsymbol{\phi}_1$ and $\boldsymbol{\phi}_3$ of the 3-DOF spring–mass system in Example 10.1. (b) Scale modes $\boldsymbol{\phi}_1$ and $\boldsymbol{\phi}_3$ so that $M_1 = M_3 = m = 1$. (c) Show that $\boldsymbol{\phi}_1$ and $\boldsymbol{\phi}_3$ are orthogonal with respect to the mass matrix and also with respect to the stiffness matrix.

10.4 (a) Using the method of Eq. 10.13, solve for modes $\boldsymbol{\phi}_2$ and $\boldsymbol{\phi}_3$ of the 3-DOF assumed-modes model of the bar in Example 10.2. (b) Scale modes $\boldsymbol{\phi}_2$ and $\boldsymbol{\phi}_3$ so that $M_2 = M_3 = \rho A L$. (c) Show that $\boldsymbol{\phi}_2$ is orthogonal (with respect to the mass matrix) to both $\boldsymbol{\phi}_1$ and $\boldsymbol{\phi}_3$.

10.5 The 3-DOF spring–mass system in Fig. P10.5 has a repeated frequency $\omega_2 = \omega_3$. The values of the masses m_i and stiffnesses k_i are shown on the figure. (a) Determine the three frequencies (squared), ω_r^2. (b) Using the method of Eq. 10.22, determine mode shapes $\boldsymbol{\phi}_2$ and $\boldsymbol{\phi}_3$. (c) Are the two modes that you have obtained orthogonal with respect to the mass matrix?

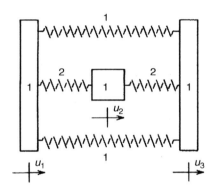

Figure P10.5

10.6 Equation 9.9 for the free-vibration response of an undamped 2-DOF system can be extended to an undamped N-DOF system by using the equation

$$\mathbf{u}(t) = \sum_{i=1}^{N} \boldsymbol{\phi}_i (a_i \cos \omega_i t + b_i \sin \omega_i t) \quad (1)$$

where the $2N$ coefficients a_i and b_i are determined from the $2N$ initial conditions $\mathbf{u}(0)$ and $\dot{\mathbf{u}}(0)$. Starting with Eq. 1, show that these coefficients can be determined by using the following equation for the a_i coefficients:

$$a_i = \frac{\boldsymbol{\phi}_i^T \mathbf{M} \mathbf{u}(0)}{\boldsymbol{\phi}_i^T \mathbf{M} \boldsymbol{\phi}_i}$$

and a similar equation for the b_i coefficients.

10.7 Equation 9.9 for the free-vibration response of an undamped 2-DOF system can be extended to an undamped N-DOF system with N_r rigid-body modes by using the following equation:

$$\mathbf{u}(t) = \sum_{i=1}^{N_r} \boldsymbol{\phi}_i (a_i + b_i t) + \sum_{i=N_r+1}^{N} \boldsymbol{\phi}_i (a_i \cos \omega_i t + b_i \sin \omega_i t)$$

(a) If the rigid-body modes $\boldsymbol{\phi}_i$, $i = 1, 2, \ldots, N_r$ form a mutually orthogonal set of modes, determine expressions for the a_i and b_i coefficients for the rigid-body modes $i = 1, 2, \ldots, N_r$. (See Problem 10.6 for a hint regarding the desired form of answer for this problem.) (b) The modes $\boldsymbol{\phi}_1$, $\boldsymbol{\phi}_2$, and $\boldsymbol{\phi}_3$ shown in Fig. 2 of Example 10.4 form a mutually orthogonal set of modes for the 3-DOF beam system whose mass matrix and stiffness matrix are given in Example 10.3. Determine a complete expression for the free-vibration response $\mathbf{v}(t)$ if the initial displacement vector and the initial velocity vector are

$$\mathbf{v}(0) = \begin{Bmatrix} 0 \\ 0 \\ 1 \end{Bmatrix}, \quad \dot{\mathbf{v}}(0) = \begin{Bmatrix} 0 \\ 0 \\ 0 \end{Bmatrix}$$

respectively.

10.8 Prove that since the N normal modes $\boldsymbol{\phi}_i$ of an N-DOF system are mutually orthogonal, they form a set of N linearly independent N-dimensional vectors.

Problem Set 10.2

10.9 Show that the eigenvalue separation theorem of Eq. 10.70 holds true for the 3-DOF system in Example 10.1 by restraining, respectively, mass 3 and then masses 2 and 3, and computing the natural frequencies (squared) of these two restrained systems. For the original 3-DOF system's natural frequencies, you may use the answer given in Example 10.1.

Problem Set 10.3

10.10 A three-story building is modeled as shown in Fig. P10.10. The stiffness matrix and the mass matrix for the structure are, respectively,

$$\mathbf{K} = \begin{bmatrix} 800 & -800 & 0 \\ -800 & 2400 & -1600 \\ 0 & -1600 & 4000 \end{bmatrix} \text{ kips/in.},$$

$$\mathbf{M} = \begin{bmatrix} 1 & 0 & 0 \\ 0 & 2 & 0 \\ 0 & 0 & 2 \end{bmatrix} \text{ kip-sec}^2/\text{in.}$$

(a) Determine the Rayleigh damping coefficients a_0 and a_1 in Eq. 10.76 such that the 3-DOF system has damping factors $\zeta_1 = \zeta_2 = 0.01$ in its first two modes. (b) What is the resulting value of the damping factor ζ_3 of the third mode?

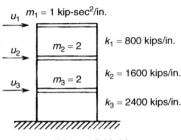

Figure P10.10

10.11 A 2-DOF spring–mass–dashpot system is modeled as shown in Fig. 1 of Example 10.6. As given in Example 10.6, the stiffness matrix and the mass matrix for the structure are, respectively,

$$\mathbf{K} = \begin{bmatrix} 2200 & -600 \\ -600 & 3800 \end{bmatrix}, \quad \mathbf{M} = \begin{bmatrix} 1 & 0 \\ 0 & 2 \end{bmatrix}$$

(a) Determine the two undamped natural frequencies, ω_1 and ω_2, of this system. (b) Determine the Rayleigh damping coefficients a_0 and a_1 in Eq. 10.76 such that the 2-DOF system will have damping factors $\zeta_1 = \zeta_2 = 0.1$ in its two modes. (c) What is the resulting damping matrix $\mathbf{C}$?

Problem Set 10.4

C 10.12 Example 10.6 illustrates the generalized state-variable approach to determining the natural frequencies, damping factors, and complex modes of viscous-damped MDOF systems. (a) Execute the MATLAB computer program sd2ex10_6c.m that may be found on the book's website. Use damping having the form $\mathbf{C} = a_1 \mathbf{K}$ with the scale factor a_1 that you select according to Eq. 10.78 such that the damping in mode 1 is $\zeta_1 = 0.01$. Obtain a complete printout of the results from this run. (b) Discuss how the natural frequencies, damping factors, and complex modes from this system with $\mathbf{K}$-proportional viscous damping differ from the results for $\mathbf{M}$-proportional that are presented in part (b) of Example 10.6.

C 10.13 Example 10.6 illustrates the generalized state-variable approach to determining the natural frequencies, damping factors, and complex modes of viscous-damped MDOF systems. (a) Execute the MATLAB computer program sd2ex10_6d.m that may be found on the book's website. Use local viscous damping between the two masses having the form

$$\mathbf{C} = \begin{bmatrix} c_2 & -c_2 \\ -c_2 & c_2 \end{bmatrix}$$

Select the value of c_2 such that the damping in mode 2 is $\zeta_2 = 0.01$. Obtain a complete printout of the results from this run. (b) Discuss how the natural frequencies, damping factors, and complex modes from this system differ from the results for the underdamped system presented in part (c) of Example 10.6.

11

Dynamic Response of MDOF Systems: Mode-Superposition Method

In Section 9.5 you were introduced to the subject of dynamic response of MDOF systems by considering the response of an undamped 2-DOF system to harmonic excitation. You were also introduced to the very important mode-superposition method for solving for the dynamic response of MDOF systems. In this chapter you will make a more detailed study of the mode-superposition method and will learn when and how to apply it. In Chapter 16 we cover the dynamic response of systems for which the mode-superposition method is not appropriate.

Upon completion of this chapter you should be able to:

- Transform the equations of motion of an N-DOF system to principal coordinates.
- Determine the initial conditions in principal coordinates.
- Generate complex frequency response plots for N-DOF systems having uncoupled modal damping, and describe the type of curves that can be expected if (a) the system has widely spaced frequencies and light damping, or (b) the system has closely spaced frequencies and light damping.
- Determine the displacement and internal stress response of an N-DOF system to harmonic or transient excitation by using the mode-displacement method.
- Determine modal static displacements and describe the significance of the distribution of the input forces on the modal responses.
- Determine the displacement and internal stress response of an N-DOF system to harmonic or transient excitation using the mode-acceleration method if (a) the system has no rigid-body modes, or (b) the system has rigid-body modes.

11.1 MODE-SUPERPOSITION METHOD: PRINCIPAL COORDINATES

The equations of motion of a linear MDOF system are

$$\mathbf{M}\ddot{\mathbf{u}} + \mathbf{C}\dot{\mathbf{u}} + \mathbf{K}\mathbf{u} = \mathbf{p}(t) \tag{11.1}$$

In general, the coefficient matrices in Eq. 11.1 (**M**, **C**, and **K**) have nonzero coupling terms (e.g., $k_{ij} = k_{ji} \neq 0$), so that to solve Eq. 11.1 in its present form would require simultaneous solution of N equations in N unknowns, where N could even exceed 1 million. In this section we illustrate use of the *mode-superposition method*, or *normal-mode method*, by which such a set of coupled equations can be transformed into a set of uncoupled equations through use of the normal modes of the undamped system. In Sections 11.1 through 11.4, examples are presented for undamped systems and for systems with *modal damping*, which was introduced in Section 10.3. Response of systems with more general viscous damping is considered in Chapter 16.

Equation 11.1 is the original set of coupled equations of motion for an N-DOF system, where $\mathbf{u}(t)$ may be physical or generalized coordinates. The response of the system to the excitation $\mathbf{p}(t)$ and to the initial conditions

$$\mathbf{u}(0) = \mathbf{u}_0, \qquad \dot{\mathbf{u}}(0) = \mathbf{v}_0 \tag{11.2}$$

is sought.

11.1.1 Mode-Superposition Procedure Employing Modes of the Undamped Structure

The first step in a mode-superposition solution is to obtain the natural frequencies and natural modes of the undamped system. (We assume here that all N modes are to be used. The important topic of truncation, that is, using fewer than N modes, is discussed at length in Sections 11.2 through 11.6.) The natural frequencies and modes of the undamped system satisfy the algebraic eigenproblem

$$[\mathbf{K} - \omega_r^2 \mathbf{M}]\boldsymbol{\phi}_r = \mathbf{0} \tag{11.3}$$

giving the *eigenpairs* $(\omega_r^2, \boldsymbol{\phi}_r)$, $r = 1, 2, \ldots, N$. The modes $\boldsymbol{\phi}_r$ are assumed to have been normalized by one of the schemes listed in Section 10.1.3 and the *modal masses*, M_r, and *modal stiffnesses*, K_r, calculated using

$$M_r = \boldsymbol{\phi}_r^T \mathbf{M} \boldsymbol{\phi}_r, \qquad K_r = \boldsymbol{\phi}_r^T \mathbf{K} \boldsymbol{\phi}_r = \omega_r^2 M_r \tag{11.4}$$

We will assume that if there are any repeated frequencies, the associated modes have been orthogonalized so that the orthogonality equations

$$\boldsymbol{\phi}_r^T \mathbf{M} \boldsymbol{\phi}_s = \boldsymbol{\phi}_r^T \mathbf{K} \boldsymbol{\phi}_s = 0 \tag{11.5}$$

are satisfied for all $r \neq s$. (See the discussion of repeated frequencies in Section 10.1.6.) The modes are then collected to form the *modal matrix*

$$\boldsymbol{\Phi} = [\boldsymbol{\phi}_1 \quad \boldsymbol{\phi}_2 \quad \cdots \quad \boldsymbol{\phi}_N] \tag{11.6}$$

Then Eqs. 11.4 and 11.5 can be combined and written in the following form:

$$\mathcal{M} = \boldsymbol{\Phi}^T \mathbf{M} \boldsymbol{\Phi} = \text{diag}(M_r), \qquad \mathcal{K} = \boldsymbol{\Phi}^T \mathbf{K} \boldsymbol{\Phi} = \text{diag}(K_r) = \text{diag}(\omega_r^2 M_r) \tag{11.7}$$

The key step in the *mode-superposition procedure* is to introduce the coordinate transformation

$$\boxed{\mathbf{u}(t) = \boldsymbol{\Phi}\boldsymbol{\eta}(t) = \sum_{r=1}^{N} \boldsymbol{\phi}_r \eta_r(t)} \tag{11.8}$$

The coordinates $\eta_r(t)$ will be referred to as *principal coordinates* or *modal coordinates*.[1] Equation 11.8 is substituted into Eq. 11.1, and the resulting equation is multiplied by $\mathbf{\Phi}^T$ to give the *equation of motion in principal coordinates*: namely,

$$\mathbf{M}\ddot{\boldsymbol{\eta}} + \mathbf{C}\dot{\boldsymbol{\eta}} + \mathbf{K}\boldsymbol{\eta} = \mathbf{f}(t) \tag{11.9}$$

where

$$\begin{aligned}\mathbf{M} &= \mathbf{\Phi}^T\mathbf{M}\mathbf{\Phi} = \text{modal mass matrix} = \text{diag}(M_r)\\ \mathbf{C} &= \mathbf{\Phi}^T\mathbf{C}\mathbf{\Phi} = \text{modal damping matrix}\\ \mathbf{K} &= \mathbf{\Phi}^T\mathbf{K}\mathbf{\Phi} = \text{modal stiffness matrix} = \text{diag}(\omega_r^2 M_r)\\ \mathbf{f}(t) &= \mathbf{\Phi}^T\mathbf{p}(t) = \text{modal force vector}\end{aligned} \tag{11.10}$$

11.1.2 Modal Damping: Uncoupled Equations of Motion

As indicated in Eq. 11.7, modal matrices $\mathbf{M}$ and $\mathbf{K}$ are diagonal matrices, so the equations of motion in modal coordinates, Eq. 11.9, are coupled only through nonzero off-diagonal coefficients in the (coupled) generalized damping matrix, $\mathbf{C}$.

The type of damping that is most frequently used in structural dynamics computations is referred to in the literature as *modal damping*, or sometimes, proportional damping or classical damping. This form of uncoupled modal damping was introduced in Section 10.3. Modal damping is assumed to satisfy orthogonality, so that Eq. 11.10b becomes

$$\mathbf{C} = \mathbf{\Phi}^T\mathbf{C}\mathbf{\Phi} = \text{diag}(C_r) = \text{diag}(2\zeta_r\omega_r M_r) \tag{11.11}$$

The *modal damping factor* values ζ_r are assumed on the basis of providing damping typical of the type of structure under consideration. Typical values lie in the range $0.01 \leq \zeta_r \leq 0.1$. In Chapter 18, experimental methods for determining suitable damping factors are discussed.

With this diagonalized modal damping, the coupled equations of motion, Eq. 11.1, are reduced to the following set of N uncoupled *modal equations of motion*:

$$\boxed{M_r\ddot{\eta}_r + 2M_r\omega_r\zeta_r\dot{\eta}_r + \omega_r^2 M_r\eta_r = f_r(t), \qquad r = 1, 2, \ldots, N} \tag{11.12}$$

11.1.3 Initial Conditions in Modal Coordinates

The total response $\boldsymbol{\eta}(t)$ can be obtained as a superposition of the *response due to initial conditions alone* and the *response due to the excitation alone*. From Eq. 11.8, the mode-superposition expressions for the displacement vector and the velocity vector at time $t = 0$ are, respectively,

$$\mathbf{u}(0) = \mathbf{\Phi}\boldsymbol{\eta}(0), \qquad \dot{\mathbf{u}}(0) = \mathbf{\Phi}\dot{\boldsymbol{\eta}}(0) \tag{11.13}$$

Multiplying these equations by $\mathbf{\Phi}^T\mathbf{M}$, we get

$$\mathbf{\Phi}^T\mathbf{M}\mathbf{u}(0) = \mathbf{M}\boldsymbol{\eta}(0), \qquad \mathbf{\Phi}^T\mathbf{M}\dot{\mathbf{u}}(0) = \mathbf{M}\dot{\boldsymbol{\eta}}(0) \tag{11.14}$$

---

[1] Some authors refer to these as *natural coordinates* or as *normal coordinates*. Some authors reserve the name *normal coordinates* to use when modes are normalized so that $M_r = 1$ for every mode.

Since $\mathbf{M}$ is diagonal, Eqs. 11.14 can be solved for the *modal initial conditions*, giving

$$\left. \begin{array}{l} \eta_r(0) = \dfrac{1}{M_r}\boldsymbol{\phi}_r^T \mathbf{M}\mathbf{u}(0) \\[6pt] \dot{\eta}_r(0) = \dfrac{1}{M_r}\boldsymbol{\phi}_r^T \mathbf{M}\dot{\mathbf{u}}(0) \end{array} \right\} \quad r = 1, 2, \ldots, N \qquad (11.15)$$

These formulas apply to both rigid-body modes, if any, and flexible modes.

An example of mode superposition for free vibration is given next. In Section 9.5 we introduced mode superposition with an example, Example 9.8, that discussed frequency response of an undamped 2-DOF system. In Section 11.2 we continue discussion of this important topic of frequency response by looking at examples where the modal damping matrix is diagonal. We consider additional mode-superposition solutions for undamped systems in Sections 11.3 through 11.6. Mode superposition for MDOF systems that have rigid-body modes is treated in Section 11.6.

11.1.4 Mode-Superposition Solution for Free Vibration of an Undamped MDOF System

Referring to Eq. 10.3 and to the 2-DOF free-vibration solution in Section 9.1, we see that the free vibration of an undamped N-DOF system vibrating in its rth flexible (i.e., not rigid-body) mode is given by

$$\mathbf{u}_r(t) = c_r \boldsymbol{\phi}_r \cos(\omega_r t - \alpha_r) \qquad (11.16)$$

Thus, the general solution for free vibration may be written in the form

$$\mathbf{u}(t) = \sum_{r=1}^{N} c_r \boldsymbol{\phi}_r \cos(\omega_r t - \alpha_r) \qquad (11.17)$$

or in the form that was employed in Example 9.2,

$$\mathbf{u}(t) = \sum_{r=1}^{N} \boldsymbol{\phi}_r (a_r \cos \omega_r t + b_r \sin \omega_r t) \qquad (11.18)$$

The $2N$ coefficients (c_r, α_r) or (a_r, b_r) are determined by the initial conditions $\mathbf{u}(0)$ and $\dot{\mathbf{u}}(0)$. The coefficients a_r and b_r can be determined by using orthogonality as in Eqs. 10.33 and 10.34. Thus, from Eq. 11.18 and its derivative evaluated at time $t = 0$,

$$\mathbf{u}(0) = \sum_{r=1}^{N} \boldsymbol{\phi}_r a_r, \qquad \dot{\mathbf{u}}(0) = \sum_{r=1}^{N} \boldsymbol{\phi}_r \omega_r b_r \qquad (11.19a, b)$$

Multiplying Eqs. 11.19 by $(\boldsymbol{\phi}_s^T \mathbf{M})$ and observing the mass orthogonality of the modes, we obtain

$$a_r = \frac{\boldsymbol{\phi}_r^T \mathbf{M}\mathbf{u}(0)}{M_r}, \qquad b_r = \frac{\boldsymbol{\phi}_r^T \mathbf{M}\dot{\mathbf{u}}(0)}{M_r \omega_r} \qquad (11.20a, b)$$

11.1 Mode-Superposition Method: Principal Coordinates

This provides a straightforward procedure for determining the modal coefficients of **u** in Eq. 11.18 and hence for determining the free vibration due to initial conditions **u**(0) and **u̇**(0).

Example 11.1 Using Eqs. 11.18 and 11.20, determine expressions for the free vibration response of the 2-DOF system of Examples 9.1 and 9.2 if the initial conditions are

$$\mathbf{u}(0) = \begin{Bmatrix} 0 \\ u_0 \end{Bmatrix}, \qquad \dot{\mathbf{u}}(0) = \begin{Bmatrix} 0 \\ 0 \end{Bmatrix}$$

The mass matrix and mode shapes for the system are

$$\mathbf{M} = \begin{bmatrix} m & 0 \\ 0 & m \end{bmatrix}, \qquad \boldsymbol{\phi}_1 = \begin{Bmatrix} 1 \\ 1 \end{Bmatrix}, \qquad \boldsymbol{\phi}_2 = \begin{Bmatrix} 1 \\ -1 \end{Bmatrix}$$

SOLUTION Determine the modal masses M_r. From Eq. 11.4a,

$$M_r = \boldsymbol{\phi}_r^T \mathbf{M} \boldsymbol{\phi}_r \tag{1}$$

so the modal masses are

$$M_1 = \begin{bmatrix} 1 & 1 \end{bmatrix} \begin{bmatrix} m & 0 \\ 0 & m \end{bmatrix} \begin{Bmatrix} 1 \\ 1 \end{Bmatrix} = 2m$$

$$M_2 = \begin{bmatrix} 1 & -1 \end{bmatrix} \begin{bmatrix} m & 0 \\ 0 & m \end{bmatrix} \begin{Bmatrix} 1 \\ -1 \end{Bmatrix} = 2m \tag{2}$$

Determine the coefficients a_r and b_r from Eqs. 11.20.

$$a_r = \frac{\boldsymbol{\phi}_r^T \mathbf{M} \mathbf{u}(0)}{M_r}, \qquad b_r = \frac{\boldsymbol{\phi}_r^T \mathbf{M} \dot{\mathbf{u}}(0)}{M_r \omega_r} \tag{3}$$

$$\mathbf{M}\mathbf{u}(0) = \begin{bmatrix} m & 0 \\ 0 & m \end{bmatrix} \begin{Bmatrix} 0 \\ u_0 \end{Bmatrix} = \begin{Bmatrix} 0 \\ mu_0 \end{Bmatrix}$$

$$\mathbf{M}\dot{\mathbf{u}}(0) = \begin{bmatrix} m & 0 \\ 0 & m \end{bmatrix} \begin{Bmatrix} 0 \\ 0 \end{Bmatrix} = \begin{Bmatrix} 0 \\ 0 \end{Bmatrix} \tag{4}$$

Therefore,

$$a_1 = \frac{\boldsymbol{\phi}_1^T \mathbf{M} \mathbf{u}(0)}{M_1} = \frac{1}{2m} \begin{bmatrix} 1 & 1 \end{bmatrix} \begin{Bmatrix} 0 \\ mu_0 \end{Bmatrix} = \frac{u_0}{2}$$

$$a_2 = \frac{\boldsymbol{\phi}_2^T \mathbf{M} \mathbf{u}(0)}{M_2} = \frac{1}{2m} \begin{bmatrix} 1 & -1 \end{bmatrix} \begin{Bmatrix} 0 \\ mu_0 \end{Bmatrix} = -\frac{u_0}{2} \tag{5}$$

Finally, from Eq. 11.18,

$$\mathbf{u}(t) = \sum_{r=1}^{2} a_r \boldsymbol{\phi}_r \cos \omega_r t \qquad (6)$$

so

$$\begin{Bmatrix} u_1(t) \\ u_2(t) \end{Bmatrix} = \frac{u_0}{2} \begin{Bmatrix} 1 \\ 1 \end{Bmatrix} \cos \omega_1 t - \frac{u_0}{2} \begin{Bmatrix} 1 \\ -1 \end{Bmatrix} \cos \omega_2 t$$

or

$$u_1(t) = \frac{u_0}{2}(\cos \omega_1 t - \cos \omega_2 t)$$

$$u_2(t) = \frac{u_0}{2}(\cos \omega_1 t + \cos \omega_2 t)$$

Ans. (7)

which is the same result determined by a different procedure in Example 9.2.

Although the procedure of Example 9.2 works well for a 2-DOF problem, the method employed in Example 11.1 is systematic and can be easily applied in a computer solution for the free vibration response of a system of any size. And as noted earlier, this solution for the response to initial conditions can be added to the solution for the response due to external excitation.

11.2 MODE-SUPERPOSITION SOLUTIONS FOR MDOF SYSTEMS WITH MODAL DAMPING: FREQUENCY-RESPONSE ANALYSIS

In this section we consider mode-superposition solutions for the response of systems with viscous damping distributed in such a way that the modal equations of motion are uncoupled. We will also use this opportunity to consider the topic of *frequency response* of viscous-damped systems, that is, the response of damped systems to harmonic excitation. Frequency-response analysis plays a very important role in the testing of structures to determine their dynamical properties, as discussed in Chapter 18.

Consider an N-DOF system with viscous damping that satisfies the orthogonality conditions of modal damping given by Eq. 11.11. Then Eq. 11.12 gives the uncoupled modal equations of motion, which may be written in the form

$$\ddot{\eta}_r + 2\zeta_r \omega_r \dot{\eta}_r + \omega_r^2 \eta_r = \frac{1}{M_r} f_r(t) = \frac{1}{M_r} \boldsymbol{\phi}_r^T \mathbf{p}(t) \qquad \text{for } r = 1, 2, \ldots, N \qquad (11.21)$$

11.2.1 Mode-Superposition Solution for Frequency Response of a Viscous-Damped System

A general mode-superposition solution is presented in Section 11.3. Consider now the mode-superposition solution for the steady-state response of a viscous-damped system subjected to harmonic excitation $\mathbf{p}(t) = \mathbf{P} \cos \Omega t$. Then, from Eq. 11.21, the modal equation of motion takes the form

$$\ddot{\eta}_r + 2\zeta_r \dot{\eta}_r + \omega_r^2 \eta_r = \frac{1}{M_r} F_r \cos \Omega t \qquad (11.22)$$

11.2 Frequency-Response Analysis for MDOF Systems with Modal Damping

where

$$F_r = \boldsymbol{\phi}_r^T \mathbf{P} \tag{11.23}$$

As in the case of a SDOF system, we can solve Eq. 11.22 by using complex frequency-response techniques.

$$\ddot{\overline{\eta}}_r + 2\zeta_r \omega_r \dot{\overline{\eta}}_r + \omega_r^2 \overline{\eta}_r = \omega_r^2 \frac{F_r}{K_r} e^{i\Omega t} \tag{11.24}$$

Then the steady-state solution $\overline{\eta}_r$ can be written in the form

$$\overline{\eta}_r = \overline{H}_{\eta_r/F_r}(\Omega) F_r e^{i\Omega t} \tag{11.25}$$

where $\overline{H}_{\eta_r/F_r}(\Omega)$ is the *complex frequency-response function in principal coordinates*, given by

$$\boxed{\overline{H}_{\eta_r/F_r}(\Omega) = \frac{1/K_r}{(1-r_r^2) + i(2\zeta_r r_r)}} \tag{11.26}$$

where r_r is the *modal frequency ratio*, given by

$$r_r = \frac{\Omega}{\omega_r} \tag{11.27}$$

As in Section 4.3, we can determine the magnitude and phase of $\overline{H}_{\eta_r/F_r}(\Omega)$. Then the *modal response* $\eta_r(t)$ becomes

$$\boxed{\eta_r(t) = \frac{F_r/K_r}{\sqrt{(1-r_r^2)^2 + (2\zeta_r r_r)^2}} \cos(\Omega t - \alpha_r)} \tag{11.28a}$$

where the *phase angle* α_r is given by

$$\boxed{\tan \alpha_r = \frac{2\zeta_r r_r}{1 - r_r^2}} \tag{11.28b}$$

The complex frequency response in physical (or generalized) coordinates $\mathbf{u}$ can be obtained by writing Eq. 11.8 in complex form as

$$\overline{\mathbf{u}}(t) = \boldsymbol{\Phi} \overline{\boldsymbol{\eta}}(t) = \sum_{r=1}^{N} \boldsymbol{\phi}_r \overline{\eta}_r(t) \tag{11.29}$$

Combining Eqs. 11.23, 11.25, 11.26, and 11.29, we get

$$\overline{\mathbf{u}}(t) = \sum_{r=1}^{N} \frac{\boldsymbol{\phi}_r \boldsymbol{\phi}_r^T \mathbf{P}}{K_r} \frac{1}{(1-r_r^2) + i(2\zeta_r r_r)} e^{i\Omega t} \tag{11.30}$$

The *complex frequency-response function* (FRF) *in physical coordinates*, $\overline{H}_{ij}(\Omega)$, gives the response at coordinate u_i due to unit harmonic excitation at p_j. Thus, from Eq. 11.30,

$$\overline{H}_{ij}(\Omega) \equiv \overline{H}_{u_i/p_j}(\Omega) = \sum_{r=1}^{N} \frac{\phi_{ir}\phi_{jr}}{K_r} \frac{1}{(1 - r_r^2) + i(2\zeta_r r_r)} \quad (11.31)$$

where ϕ_{ir} is the element in row i of the rth mode $\boldsymbol{\phi}_r$, that is, the element in row i and column r of the modal matrix $\boldsymbol{\Phi}$. The *steady-state response* $\mathbf{u}(t)$ can also be obtained from Eq. 11.30. It is

$$\mathbf{u}(t) = \sum_{r=1}^{N} \frac{\boldsymbol{\phi}_r \boldsymbol{\phi}_r^T \mathbf{P}}{K_r} \frac{1}{\sqrt{(1 - r_r^2)^2 + (2\zeta_r r_r)^2}} \cos(\Omega t - \alpha_r) \quad (11.32)$$

where the phase angle α_r is again given by Eq. 11.28b.

A plot of Eq. 11.31 in the complex plane (Argand plane) is referred to as a *Nyquist FRF plot* or *complex frequency-response plot*. In this case, the forcing frequency f (or Ω) is a parameter, and the imaginary part $\Im(\overline{H})$ is plotted versus the real part $\Re(\overline{H})$. On the other hand, it is frequently convenient to display $\Re(\overline{H})$ versus the frequency in hertz, $f = \Omega/2\pi$, and the imaginary part $\Im(\overline{H})$ versus frequency f. Expressions for these can be obtained from Eq. 11.31. Thus, the real part is given by

$$\Re(\overline{H}_{ij}) = \sum_{r=1}^{N} \frac{\phi_{ir}\phi_{jr}}{K_r} \frac{1 - r_r^2}{(1 - r_r^2)^2 + (2\zeta_r r_r)^2} \quad (11.33a)$$

and the imaginary part by

$$\Im(\overline{H}_{ij}) = \sum_{r=1}^{N} \frac{\phi_{ir}\phi_{jr}}{K_r} \frac{-2\zeta_r r_r}{(1 - r_r^2)^2 + (2\zeta_r r_r)^2} \quad (11.33b)$$

Complex frequency-response functions (sometimes called *transfer functions*), as given by Eq. 11.31, are frequently employed in determining the vibrational characteristics of a system experimentally, as discussed in Chapter 18.

Two examples are now presented which show the nature of complex frequency-response functions. Example 11.2 treats a 2-DOF system with light damping and widely separated natural frequencies, and Example 11.3 indicates the problems that arise when a system has two natural frequencies that are very close.

Example 11.2 For the 2-DOF system in Fig. 1, where $p_1(t) = P_1 \cos \Omega t$, $k = 987$, $k' = 217$, $m = 1$, $c = 0.6284$, and $c' = 0.0628$: (a) Determine the system's natural

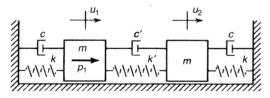

Figure 1 A 2-DOF spring–mass–dashpot system with harmonic excitation.

11.2 Frequency-Response Analysis for MDOF Systems with Modal Damping

frequencies f_1 and f_2 and the corresponding mode shapes $\boldsymbol{\phi}_1$ and $\boldsymbol{\phi}_2$. (b) Determine the modal mass matrix $\boldsymbol{M}$, the modal damping matrix $\boldsymbol{C}$, and the modal stiffness matrix $\boldsymbol{K}$. (c) Determine the modal damping factors ζ_1 and ζ_2. (d) Determine the complex frequency-response functions $\overline{H}_{11}$ and $\overline{H}_{21}$. (e) Plot $\overline{H}_{11}$ and $\overline{H}_{21}$ in the complex plane for the frequency range $f = 4$ to 7 Hz. (f) Plot $\Re(\overline{H}_{11})$ and $\Im(\overline{H}_{11})$ versus frequency f for the frequency range $f = 4$ to 7 Hz.

SOLUTION (a) Determine the frequencies and mode shapes using the methods of Chapter 9. The complete equation of motion is

$$\begin{bmatrix} m & 0 \\ 0 & m \end{bmatrix} \begin{Bmatrix} \ddot{u}_1 \\ \ddot{u}_2 \end{Bmatrix} + \begin{bmatrix} c+c' & -c' \\ -c' & c+c' \end{bmatrix} \begin{Bmatrix} \dot{u}_1 \\ \dot{u}_2 \end{Bmatrix} + \begin{bmatrix} k+k' & -k' \\ -k' & k+k' \end{bmatrix} \begin{Bmatrix} u_1 \\ u_2 \end{Bmatrix} = \begin{Bmatrix} p_1(t) \\ 0 \end{Bmatrix} \quad (1)$$

while, for undamped free vibration, the equation of motion is

$$\begin{bmatrix} m & 0 \\ 0 & m \end{bmatrix} \begin{Bmatrix} \ddot{u}_1 \\ \ddot{u}_2 \end{Bmatrix} + \begin{bmatrix} k+k' & -k' \\ -k' & k+k' \end{bmatrix} \begin{Bmatrix} u_1 \\ u_2 \end{Bmatrix} = \begin{Bmatrix} 0 \\ 0 \end{Bmatrix} \quad (2)$$

Let

$$\mathbf{u} = \boldsymbol{\phi} \cos \omega t \quad (3)$$

Then the algebraic eigenvalue equation can be written

$$\left[\begin{bmatrix} k+k' & -k' \\ -k' & k+k' \end{bmatrix} - \omega^2 \begin{bmatrix} m & 0 \\ 0 & m \end{bmatrix} \right] \begin{Bmatrix} \phi_1 \\ \phi_2 \end{Bmatrix} = \begin{Bmatrix} 0 \\ 0 \end{Bmatrix} \quad (4)$$

The undamped natural frequencies (squared) can be found to be

$$\omega_1^2 = \frac{k}{m}, \quad \omega_2^2 = \frac{k+2k'}{m} \quad (5)$$

and the corresponding mode shapes are

$$\boldsymbol{\phi}_1 = \begin{Bmatrix} 1 \\ 1 \end{Bmatrix}, \quad \boldsymbol{\phi}_2 = \begin{Bmatrix} 1 \\ -1 \end{Bmatrix} \quad \text{Ans. (a)} \quad (6)$$

For the specific values of k, k', and m above,

$$\omega_1^2 = \frac{987}{1} = 987, \quad \omega_2^2 = \frac{1421}{1} = 1421$$

$$\omega_1 = 31.42 \text{ rad/s}, \quad \omega_2 = 37.70 \text{ rad/s}$$

Therefore, the natural frequencies in hertz are

$$f_1 = \frac{\omega_1}{2\pi} = 5.00 \text{ Hz}, \quad f_2 = \frac{\omega_2}{2\pi} = 6.00 \text{ Hz} \quad \text{Ans. (a)} \quad (7)$$

(b) Determine M, C, and K. Based on the mode shapes given in Eq. 6, the modal matrix is

$$\Phi = \begin{bmatrix} 1 & 1 \\ 1 & -1 \end{bmatrix}$$

$$M = \Phi^T M \Phi = \begin{bmatrix} 1 & 1 \\ 1 & -1 \end{bmatrix} \begin{bmatrix} 1 & 0 \\ 0 & 1 \end{bmatrix} \begin{bmatrix} 1 & 1 \\ 1 & -1 \end{bmatrix} = \begin{bmatrix} 2 & 0 \\ 0 & 2 \end{bmatrix} \quad \text{Ans. (b) (8a)}$$

$$C = \Phi^T C \Phi = \begin{bmatrix} 1 & 1 \\ 1 & -1 \end{bmatrix} \begin{bmatrix} 0.6912 & -0.0628 \\ -0.0628 & 0.6912 \end{bmatrix} \begin{bmatrix} 1 & 1 \\ 1 & -1 \end{bmatrix}$$

$$= \begin{bmatrix} 1.2568 & 0 \\ 0 & 1.5080 \end{bmatrix} \quad \text{Ans. (b) (8b)}$$

$$K_1 = \omega_1^2 M_1 = 987(2) = 1974, \qquad K_2 = \omega_2^2 M_2 = 1421(2) = 2842$$

Therefore,

$$K = \begin{bmatrix} 1974 & 0 \\ 0 & 2842 \end{bmatrix} \quad \text{Ans. (b) (8c)}$$

(c) Determine the damping factors ζ_1 and ζ_2. From Eq. 11.11,

$$\zeta_r = \frac{C_r}{2M_r \omega_r} \quad (9)$$

$$\zeta_1 = \frac{1.2568}{2(2)(31.42)} = 0.0100, \qquad \zeta_2 = \frac{1.5080}{2(2)(37.70)} = 0.0100 \quad \text{Ans. (c) (10)}$$

(d) Determine $\overline{H}_{11}(\Omega)$ and $\overline{H}_{21}(\Omega)$. From Eq. 11.31,

$$\overline{H}_{ij}(\Omega) = \sum_{r=1}^{2} \frac{\phi_{ir}\phi_{jr}}{K_r} \frac{1}{1 - (\Omega/\omega_r)^2 + i(2\zeta_r \Omega/\omega_r)} \quad (11)$$

Then

$$\overline{H}_{11}(\Omega) = \frac{1(1)}{1974} \left[\frac{1}{1 - (\Omega/31.42)^2 + i[2(0.01)\Omega/31.42]} \right]$$

$$+ \frac{1(1)}{2842} \left[\frac{1}{1 - (\Omega/37.70)^2 + i[2(0.01)\Omega/37.70]} \right]$$

Finally, the complex frequency-response function $\overline{H}_{11}$ is

$$\overline{H}_{11}(\Omega) = \frac{5.066 \times 10^{-4}}{1 - (\Omega/31.42)^2 + i(0.02\Omega/31.42)}$$

$$+ \frac{3.519 \times 10^{-4}}{1 - (\Omega/37.70)^2 + i(0.02\Omega/37.70)} \quad \text{Ans. (d) (12a)}$$

11.2 Frequency-Response Analysis for MDOF Systems with Modal Damping

Similarly, from Eq. 11, the complex frequency-response function $\overline{H}_{21}$ is

$$\overline{H}_{21}(\Omega) = \frac{1(1)}{1974}\left[\frac{1}{1-(\Omega/31.42)^2 + i\,[2(0.01)\Omega/31.42\,]}\right]$$

$$+ \frac{(-1)(1)}{2842}\left[\frac{1}{1-(\Omega/37.70)^2 + i\,[2(0.01)\Omega/37.70\,]}\right]$$

Note the effect of the minus sign in the term for mode 2. The result is

$$\overline{H}_{21}(\Omega) = \frac{5.066 \times 10^{-4}}{1-(\Omega/31.42)^2 + i\,(0.02\Omega/31.42\,)}$$
$$- \frac{3.519 \times 10^{-4}}{1-(\Omega/37.70)^2 + i\,(0.02\Omega/37.70\,)} \quad \textbf{Ans. (d)} \;(12b)$$

(e) Plot $\overline{H}_{11}$ and $\overline{H}_{21}$ in the Argand plane. We will need to plot $\Im(\overline{H}_{ij})$ versus $\Re(\overline{H}_{ij})$. Equations 11.33a and b give

$$\Re(\overline{H}_{ij}) = \sum_{r=1}^{2} \frac{\phi_{ir}\phi_{jr}}{K_r}\frac{1-r_r^2}{(1-r_r^2)^2 + (2\zeta_r r_r)^2} \quad (13a)$$

and

$$\Im(\overline{H}_{ij}) = \sum_{r=1}^{2} \frac{\phi_{ir}\phi_{jr}}{K_r}\frac{-2\zeta_r r_r}{(1-r_r^2)^2 + (2\zeta_r r_r)^2} \quad (13b)$$

Numerical values needed in Eqs. 13 for plotting $\overline{H}_{11}$ and $\overline{H}_{21}$ can be taken from Eqs. 12a and b and are not repeated here. (*Note:* Different scales are used for the two plots in Fig. 2.) Note that $f_2 = 1.2 f_1$, so that the two natural frequencies of this system can be considered to be "widely separated."

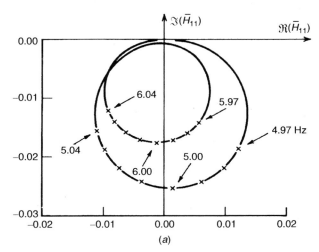

Figure 2 Nyquist FRF plots for a 2-DOF system with widely separated frequencies: (*a*) $\overline{H}_{11}$; (*b*) $\overline{H}_{21}$.

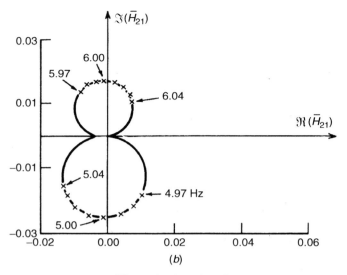

Figure 2 (*continued*)

(f) Equations 13 can be used to plot $\Re(\bar{H}_{11})$ and $\Im(\bar{H}_{11})$ as functions of the excitation frequency $f = \Omega/2\pi$, as shown in Fig. 3.

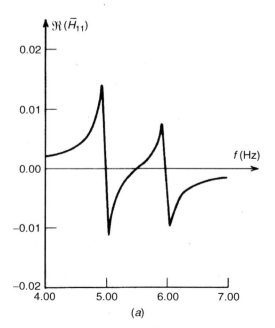

Figure 3 (*a*) Real and (*b*) imaginary FRF plots for a 2-DOF system with widely separated frequencies.

11.2 Frequency-Response Analysis for MDOF Systems with Modal Damping

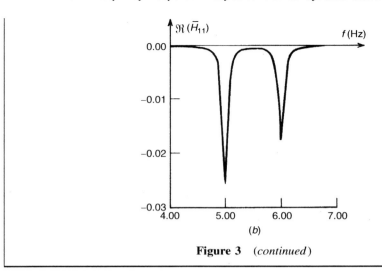

(b)

Figure 3 (*continued*)

From the complex frequency-response plot of $\overline{H}_{11}$ in Fig. 2a of Example 11.2, it can be seen that:

1. The response resembles two distinct SDOF complex frequency-response plots as seen in Fig. 4.9; that is, the response of this 2-DOF system resembles that of two independent SDOF systems.
2. Frequency spacing is greatest near the undamped natural frequencies $f_1 = 5.00$ Hz and $f_2 = 6.00$ Hz.
3. $\Re(\overline{H}_{11})$ is approximately zero at these natural frequencies.
4. $\Im(\overline{H}_{11})$ has its minimum/maximum at these natural frequencies.

Conclusions 3 and 4 can also be deduced from the plots of $\Re(\overline{H}_{11})$ and $\Im(\overline{H}_{11})$, respectively. The complex frequency-response plot of $\overline{H}_{21}$ leads to similar conclusions, but it also exhibits a phase change due to the sign difference between modes 1 and 2.

Real structures of any complexity at all generally have closely spaced frequencies. That is, a structure may have well over 100 natural frequencies between 0 and 50 Hz. In Section 10.1.6, techniques for treating systems with repeated frequencies were discussed and it was pointed out that repeated frequencies do not possess unique corresponding mode shapes. As noted in Example 11.2, the two natural frequencies of that system were widely separated. In contrast, in Example 11.3 we consider a system with "closely spaced" frequencies. This example illustrates how frequency response plots of such a system may differ from those of a system with more widely spaced frequencies, as illustrated by Example 11.2.

Example 11.3 Using the following system parameters for the 2-DOF system in Fig. 1 of Example 11.2:

$$k = 987, \quad k' = 10, \quad m = 1, \quad c = 0.6284, \quad c' = 0.0031$$

repeat the tasks listed in Example 11.2. (*Note:* These parameters were chosen to give 1% damping and to give natural frequencies separated by only 1%.)

SOLUTION (a) The equation of motion has the same form as in Example 11.2, so the frequencies will be given by

$$\omega_1^2 = \frac{k}{m}, \qquad \omega_2^2 = \frac{k + 2k'}{m} \qquad (1)$$

and the mode shapes will be the same as in Example 11.2. For the specific values of k, k', and m above,

$$\omega_1^2 = \frac{987}{1} = 987, \qquad \omega_2^2 = \frac{1007}{1} = 1007$$

$$\omega_1 = 31.42 \text{ rad/s}, \qquad \omega_2 = 31.73 \text{ rad/s}$$

Therefore, the natural frequencies in hertz are

$$f_1 = \frac{\omega_1}{2\pi} = 5.00 \text{ Hz}, \qquad f_2 = \frac{\omega_2}{2\pi} = 5.05 \text{ Hz} \qquad \textbf{Ans. (a) } (2a)$$

and the modal matrix is

$$\mathbf{\Phi} = \begin{bmatrix} 1 & 1 \\ 1 & -1 \end{bmatrix} \qquad \textbf{Ans. (a) } (2b)$$

as in Example 11.2.

(b) Determine $\mathbf{M}$, $\mathbf{C}$, and $\mathbf{K}$. From Example 11.2, the modal mass matrix is

$$\mathbf{M} = \mathbf{\Phi}^T \mathbf{M} \mathbf{\Phi} = \begin{bmatrix} 2 & 0 \\ 0 & 2 \end{bmatrix} \qquad \textbf{Ans. (b) } (3a)$$

The modal damping matrix is

$$\mathbf{C} = \mathbf{\Phi}^T \mathbf{C} \mathbf{\Phi} = \begin{bmatrix} 1 & 1 \\ 1 & -1 \end{bmatrix} \begin{bmatrix} 0.6315 & -0.0031 \\ -0.0031 & 0.6315 \end{bmatrix} \begin{bmatrix} 1 & 1 \\ 1 & -1 \end{bmatrix}$$

$$= \begin{bmatrix} 1.2568 & 0 \\ 0 & 1.2692 \end{bmatrix} \qquad \textbf{Ans. (b) } (3b)$$

The modal stiffness coefficients are given by

$$K_1 = \omega_1^2 M_1 = 987(2) = 1974, \qquad K_2 = \omega_2^2 M_2 = 1007(2) = 2014$$

so the modal stiffness matrix is

$$\mathbf{K} = \begin{bmatrix} 1974 & 0 \\ 0 & 2014 \end{bmatrix} \qquad \textbf{Ans. (b) } (3c)$$

(c) Determine the damping factors ζ_1 and ζ_2. From Eq. 11.11,

$$\zeta_r = \frac{C_r}{2 M_r \omega_r} \qquad (4)$$

$$\zeta_1 = \frac{1.2568}{2(2)(31.42)} = 0.0100, \qquad \zeta_2 = \frac{1.2692}{2(2)(31.73)} = 0.0100 \qquad \textbf{Ans. (c) } (5)$$

11.2 Frequency-Response Analysis for MDOF Systems with Modal Damping

(d) Determine $\overline{H}_{11}(\Omega)$ and $\overline{H}_{21}(\Omega)$. Using Eq. 11 of Example 11.2, together with numerical values from the problem statement above, we get

$$\overline{H}_{11}(\Omega) = \frac{1(1)}{1974} \left[\frac{1}{1 - (\Omega/31.42)^2 + i\,[2(0.01)\Omega/31.42]} \right]$$
$$+ \frac{1(1)}{2014} \left[\frac{1}{1 - (\Omega/31.73)^2 + i\,[2(0.01)\Omega/31.73]} \right]$$

Finally, the complex frequency-response function $\overline{H}_{11}$ is

$$\overline{H}_{11}(\Omega) = \frac{5.066 \times 10^{-4}}{1 - (\Omega/31.42)^2 + i\,(0.02\Omega/31.42)}$$
$$+ \frac{4.965 \times 10^{-4}}{1 - (\Omega/31.73)^2 + i\,(0.02\Omega/31.73)} \qquad \text{Ans. (d) } (6a)$$

Similarly, from Eq. 11 of Example 11.2, the complex frequency-response function $\overline{H}_{21}$ is

$$\overline{H}_{21}(\Omega) = \frac{5.066 \times 10^{-4}}{1 - (\Omega/31.42)^2 + i\,(0.02\Omega/31.42)}$$
$$- \frac{4.965 \times 10^{-4}}{1 - (\Omega/31.73)^2 + i\,(0.02\Omega/31.73)} \qquad \text{Ans. (d) } (6b)$$

(e) The plots of $\overline{H}_{11}$ and $\overline{H}_{21}$ in Figs. 1 and in 2 are based on the formulas in Eqs. 11.33, with numerical values from Eqs. 6a and b. (*Note:* Different scales are used for the two plots in Fig. 1.) Note that $f_2 = 1.01 f_1$, so that the two natural frequencies of this system are considered to be "closely spaced."

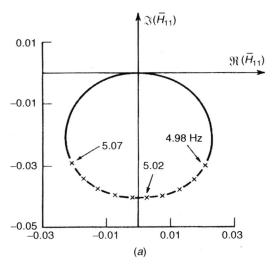

Figure 1 FRF Nyquist plots for a 2-DOF system with closely spaced frequencies: (*a*) $\overline{H}_{11}$; (*b*) $\overline{H}_{21}$.

340 Dynamic Response of MDOF Systems: Mode-Superposition Method

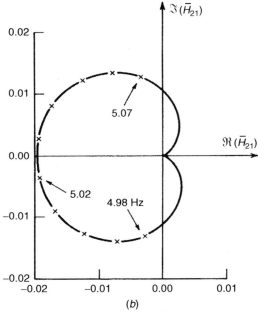

(b)

Figure 1 (*continued*)

(f) The real and imaginary plots, $\Re(\overline{H}_{11})$ and $\Im(\overline{H}_{11})$, respectively, are obtained as in Example 11.2 and are shown in Fig. 2.

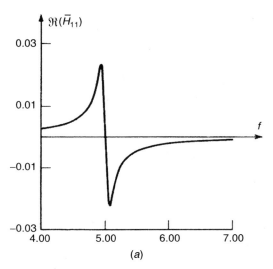

(a)

Figure 2 (*a*) Real and (*b*) imaginary FRF plots for a 2-DOF system with closely spaced frequencies.

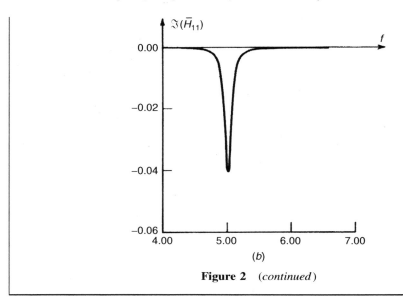

(b)

Figure 2 (*continued*)

By comparing the frequency-response plots of Examples 11.2 and 11.3, you can see that the complex frequency-response plot of $\overline{H}_{11}$ of the system with closely spaced frequencies gives little indication that two modes are present. The same is true of the plot of $\overline{H}_{11}$ versus the forcing frequency. On the other hand, the plot of $\overline{H}_{21}$ for the system with closely spaced modes does indicate a rapid phase change, and this indicates the presence of two modes whose responses at this coordinate have opposite signs.

Great difficulty is encountered in experimentally separating modes of systems with closely spaced frequencies or with heavy damping. Figure 11.1 is an inertance Bode plot (i.e., a log-log plot of acceleration/force as a function of frequency) of the response of a complex system with excitation at one point and acceleration response measured at another point.

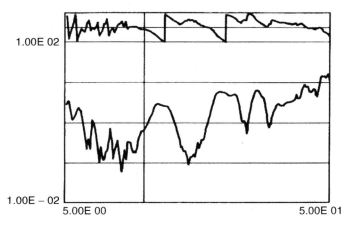

Figure 11.1 Inertance Bode plot for a complex structure.

Because of the importance of understanding frequency-response behavior of MDOF systems, that topic has been presented as the first example of mode-superposition solutions for damped structures. In the context of system identification, Section 18.2 presents an extensive discussion of mode-superposition solutions for FRFs of MDOF systems, including FRFs of velocity/force and acceleration/force, and mode superposition using complex modes. Other mode-superposition solutions are presented in Sections 11.3 through 11.6.

11.3 MODE-DISPLACEMENT SOLUTION FOR THE RESPONSE OF MDOF SYSTEMS

In Eq. 11.8 the coordinate transformation relating physical (or generalized) coordinates **u** and modal coordinates $\boldsymbol{\eta}$ includes all N system modes. If the system is assumed to have a diagonal modal damping matrix, the modal coordinates are governed by Eq. 11.12, the N uncoupled modal equations of motion. The complete solution of Eq. 11.12 can be written in the same form as the SDOF solution given in Eq. 5.37; that is, for underdamped modes ($\zeta_r < 1$),

$$\eta_r(t) = \frac{1}{M_r \omega_{dr}} \int_0^t f_r(\tau) e^{-\zeta_r \omega_r (t-\tau)} \sin \omega_{dr}(t - \tau) \, d\tau \\ + e^{-\zeta_r \omega_r t} \left[\eta_r(0) \cos \omega_{dr} t + \frac{\dot{\eta}_r(0) + \zeta_r \omega_n \eta_r(0)}{\omega_{dr}} \sin \omega_d t \right] \quad (11.34)$$

where the *damped modal natural frequency* is

$$\omega_{dr} = \omega_r \sqrt{1 - \zeta_r^2} \quad (11.35)$$

and the modal initial conditions $\eta_r(0)$ and $\dot{\eta}_r(0)$ are as given by Eqs. 11.15. Frequently, in the absence of more definitive information about damping, modal damping as given by Eq. 11.11 is simply assumed to be valid, and "reasonable" values of ζ_r are assumed.

If the N solutions $\eta_r(t)$ from Eq. 11.34 are substituted back into Eq. 11.8, the response **u**(t) is determined. However, in many cases only a subset of the modes of the system is available, and the mode-superposition solution then involves only a fraction of the N modes of the original MDOF model. Hence, we examine modal truncation and determine the factors that should be considered in deciding how many modes to include in a mode-superposition solution in order to produce results of acceptable accuracy.

11.3.1 Mode-Displacement and Mode-Acceleration Solutions

If only $\widehat{N}$ modes are retained in the solution, and the following truncated form of Eq. 11.8 is used, we have a *mode-displacement solution*:

$$\widehat{\mathbf{u}}(t) = \widehat{\boldsymbol{\Phi}} \, \widehat{\boldsymbol{\eta}}(t) = \sum_{r=1}^{\widehat{N}} \boldsymbol{\phi}_r \eta_r(t) \quad (11.36)$$

11.3 Mode-Displacement Solution for the Response of MDOF Systems

where the truncated modal matrix is given by

$$\widehat{\boldsymbol{\Phi}} = [\boldsymbol{\phi}_1 \quad \boldsymbol{\phi}_2 \quad \cdots \quad \boldsymbol{\phi}_{\widehat{N}}] \tag{11.37}$$

Equation 11.36 is the coordinate transformation for mode superposition, with truncation of modes from N to $\widehat{N} < N$ (e.g., $N = 10,000$, $\widehat{N} = 100$). This solution ignores completely the contribution of the modes not included in the set $\widehat{\boldsymbol{\Phi}}$, but the set of "kept" modes is not restricted to the $\widehat{N}$ lowest-frequency modes if some modes of higher frequency are available and are considered to be important.

A more accurate *mode-acceleration solution* can be obtained for these viscous-damped systems. This can be obtained by writing Eq. 11.1 in the form[2]

$$\mathbf{u} = \mathbf{K}^{-1}[\mathbf{p}(t) - \mathbf{C}\dot{\mathbf{u}} - \mathbf{M}\ddot{\mathbf{u}}] \tag{11.38}$$

and then incorporating Eq. 11.8 on the right-hand side,

$$\mathbf{u} = \mathbf{K}^{-1}[\mathbf{p}(t) - \mathbf{C}\boldsymbol{\Phi}\dot{\boldsymbol{\eta}} - \mathbf{M}\boldsymbol{\Phi}\ddot{\boldsymbol{\eta}}] \tag{11.39}$$

When the damping matrix $\mathbf{C}$ satisfies the conditions for modal damping, Eq. 11.11, the last two terms can be simplified (see Problem 11.7). Finally, let the velocity and acceleration terms be approximated by their $\widehat{N}$ mode-displacement approximations. Then the *mode-acceleration approximation* is given by

$$\boxed{\tilde{\mathbf{u}}(t) = \mathbf{K}^{-1}\mathbf{p}(t) - \sum_{r=1}^{\widehat{N}} \frac{2\zeta_r}{\omega_r}\boldsymbol{\phi}_r \dot{\eta}_r(t) - \sum_{r=1}^{\widehat{N}} \frac{1}{\omega_r^2}\boldsymbol{\phi}_r \ddot{\eta}_r(t)} \tag{11.40}$$

The first term on the right side of Eq. 11.40 is the *pseudostatic response*, while the third term gives the method its name, the *mode-acceleration method*. Reference [11.2] gives some interesting numerical convergence studies based on Eq. 11.40.

11.3.2 Mode-Displacement Solution for the Response of an Undamped MDOF System

For an undamped system, the uncoupled equations of motion in modal coordinates, Eq. 11.12, reduce to the following $\widehat{N}$ uncoupled *modal equations of motion*:

$$\boxed{M_r\ddot{\eta}_r + \omega_r^2 M_r \eta_r = f_r(t), \qquad r = 1, 2, \ldots, \widehat{N}} \tag{11.41}$$

where the modal mass M_r and the modal stiffness K_r are given by Eqs. 11.4 and where the modal force is given by[3]

$$f_r(t) = \boldsymbol{\phi}_r^T \mathbf{p}(t) \tag{11.42}$$

The total response of the rth mode can be expressed as a superposition of $\widehat{N}$ responses due to modal initial conditions, which are given by Eqs. 11.15, and the $\widehat{N}$ modal response due to $f_r(t)$. The Duhamel integral can be used to represent this

---
[2] See Refs. [11.2] to [11.4] for additional discussion of improved mode-superposition solution techniques.
[3] The modal force $f_r(t)$ should not be confused with the natural frequency (f_r, hertz) of the rth mode.

response symbolically. Equation 5.36 for the response of an undamped SDOF system can be written in the following form for the *modal responses* $\eta_r(t)$:

$$\begin{aligned} \eta_r(t) &= \eta_r(0)\cos\omega_r t + \frac{1}{\omega_r}\dot{\eta}_r(0)\sin\omega_r t \\ &\quad + \frac{1}{M_r\omega_r}\int_0^t \boldsymbol{\phi}_r^T \mathbf{p}(\tau)\sin\omega_r(t-\tau)\,d\tau \end{aligned} \tag{11.43}$$

Any of the special response solutions from Chapters 4 or 5 or the numerical techniques of Chapter 6 can, of course, be employed for evaluating the modal time histories, $\eta_r(t)$.

The process of computing modal responses and substituting these back into Eq. 11.36 to obtain the approximate system response is called the *mode-displacement method*.

11.3.3 Mode-Displacement Solution for the Steady-State Response of an Undamped MDOF System

Consider the steady-state response of an undamped MDOF system with harmonic excitation given by

$$\mathbf{p}(t) = \mathbf{P}\cos\Omega t \tag{11.44}$$

The resulting rth modal force is

$$f_r(t) = \boldsymbol{\phi}_r^T \mathbf{p}(t) = F_r \cos\Omega t \tag{11.45a}$$

where

$$F_r(t) = \boldsymbol{\phi}_r^T \mathbf{P} \tag{11.45b}$$

Substituting Eq. 11.45a into Eq. 11.41 and using Eqs. 4.2 and 4.3, we obtain the *steady-state modal response*

$$\eta_r(t) = \frac{F_r}{K_r}\frac{1}{1-(\Omega/\omega_r)^2}\cos\Omega t \tag{11.46}$$

Then, combining Eqs. 11.36 and 11.46, we obtain the following *mode-displacement solution for the steady-state response* in the original coordinates:

$$\widehat{\mathbf{u}}(t) = \sum_{r=1}^{\widehat{N}} \boldsymbol{\phi}_r \frac{F_r}{K_r}\left[\frac{1}{1-(\Omega/\omega_r)^2}\right]\cos\Omega t \tag{11.47}$$

where $\widehat{N} \leq N$. The quality of the approximation depends on how many modes, $\widehat{N}$, are used in the summation; the solution is exact if all N modes are used.

Note from Eq. 11.47 that the steady-state response in mode r becomes infinite if the excitation frequency, Ω, is equal to the rth undamped natural frequency, ω_r. Now consider an example steady-state response solution that will permit a cursory look at the effect of modal truncation.

11.3 Mode-Displacement Solution for the Response of MDOF Systems

Example 11.4 The four-story shear building in Fig. 1 has the natural frequencies and modal matrix shown. Harmonic excitation $P_1 \cos \Omega t$ is applied at the top story. Consider steady-state response only in parts (c) through (e). (a) Determine the four modal masses M_r and the four modal stiffnesses K_r. (b) Determine the four modal forces $f_r(t)$. (c) Determine an expression for the steady-state responses $\eta_r(t)$. (d) Using Eq. 11.47, determine an expression for $\widehat{u}_1(t)$. (e) Prepare a table showing the amplitude of $\widehat{u}_1(t)$ using $\widehat{N} = 1$, $\widehat{N} = 2$, and $\widehat{N} = 3$ for excitation frequencies $\Omega = 0$, $\Omega = 0.5\omega_1$, and $\Omega = 1.3\omega_3$. Also, determine $u_1(t)$ using $N = 4$. (f) From the results tabulated in part (e), what conclusions can you draw concerning truncation to one mode, to two modes, or to three modes?

$$\mathbf{K} = 800 \begin{bmatrix} 1 & -1 & 0 & 0 \\ -1 & 3 & -2 & 0 \\ 0 & -2 & 5 & -3 \\ 0 & 0 & -3 & 7 \end{bmatrix} \text{ kips/in.}, \quad \mathbf{M} = \begin{bmatrix} 1 & 0 & 0 & 0 \\ 0 & 2 & 0 & 0 \\ 0 & 0 & 2 & 0 \\ 0 & 0 & 0 & 3 \end{bmatrix} \text{ kip-sec}^2/\text{in.}$$

$$\omega^2 = \begin{Bmatrix} 0.17672 \\ 0.87970 \\ 1.68746 \\ 3.12279 \end{Bmatrix} \times 10^3, \quad \omega = \begin{Bmatrix} 13.294 \\ 29.660 \\ 41.079 \\ 55.882 \end{Bmatrix} \text{ rad/sec}$$

$$\boldsymbol{\Phi} = \begin{bmatrix} 1.00000 & 1.00000 & -0.90145 & 0.15436 \\ 0.77910 & -0.09963 & 1.00000 & -0.44817 \\ 0.49655 & -0.53989 & -0.15859 & 1.00000 \\ 0.23506 & -0.43761 & -0.70797 & -0.63688 \end{bmatrix}$$

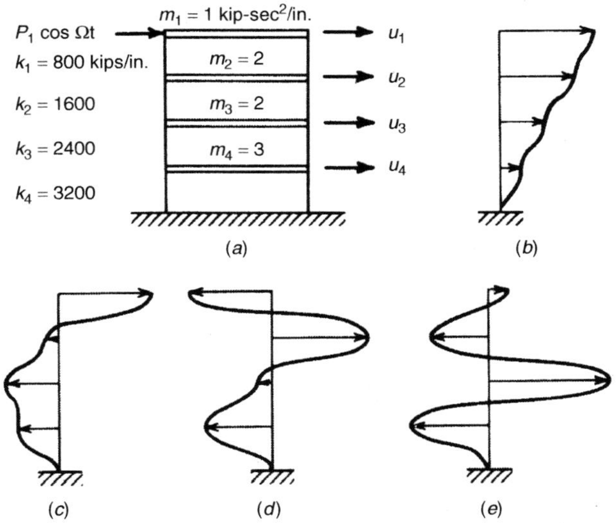

Figure 1 (a) Undamped four-story shear building; (b) mode 1; (c) mode 2; (d) mode 3; (e) mode 4.

SOLUTION (a) Determine modal masses from Eq. 11.4a and modal stiffnesses from Eq. 11.4b.

$$M_r = \boldsymbol{\phi}_r^T \mathbf{M} \boldsymbol{\phi}_r, \qquad K_r = \omega_r^2 M_r \qquad (1)$$

For example,

$$M_1 = \begin{Bmatrix} 1.00000 \\ 0.77910 \\ 0.49655 \\ 0.23506 \end{Bmatrix}^T \begin{bmatrix} 1 & 0 & 0 & 0 \\ 0 & 2 & 0 & 0 \\ 0 & 0 & 2 & 0 \\ 0 & 0 & 0 & 3 \end{bmatrix} \begin{Bmatrix} 1.00000 \\ 0.77910 \\ 0.49655 \\ 0.23506 \end{Bmatrix} = 2.87290 \text{ kip-sec}^2/\text{in.}$$

$$K_1 = \omega_1^2 M_1 = 176.72(2.87288) = 507.691 \text{ kips/in.}$$

The remaining modal masses and modal stiffnesses are calculated in a similar manner, giving

$$M_1 = 2.87290 \text{ kip-sec}^2/\text{in.}, \qquad K_1 = 507.691 \text{ kips/in.}$$
$$M_2 = 2.17732 \text{ kip-sec}^2/\text{in.}, \qquad K_2 = 1915.39 \text{ kips/in.}$$
$$M_3 = 4.36660 \text{ kip-sec}^2/\text{in.}, \qquad K_3 = 7368.45 \text{ kips/in.}$$
$$M_4 = 3.64239 \text{ kip-sec}^2/\text{in.}, \qquad K_4 = 11,374.4 \text{ kips/in.}$$

Ans. (a) (2)

(Note that the mode shapes are normalized with the largest element in each mode equal to 1. In the sketches of the four mode shapes, it can be seen that as the mode number increases, the deformation of the columns increases. Since the modal stiffness K_r is a measure of the strain energy stored in mode r, the increase in deformation with mode number is evidenced by the corresponding increase in modal stiffness with mode number. There is no similar correlation of modal masses with mode number.)

(b) The modal forces are calculated from Eq. 11.45b. Since the physical-coordinate force vector is

$$\mathbf{p}(t) = \mathbf{P} \cos \Omega t = \begin{Bmatrix} P_1 \\ 0 \\ 0 \\ 0 \end{Bmatrix} \cos \Omega t \qquad (3)$$

and

$$F_r = \boldsymbol{\phi}_r^T \mathbf{P} \qquad (4)$$

the modal force amplitudes are

$$F_1 = P_1$$
$$F_2 = P_1$$
$$F_3 = -0.90145 P_1$$
$$F_4 = 0.15436 P_1$$

Ans. (b) (5)

11.3 Mode-Displacement Solution for the Response of MDOF Systems

(c) Since this is harmonic excitation, Eq. 11.46 gives

$$\eta_r(t) = \frac{F_r}{K_r} \frac{1}{1-(\Omega/\omega_r)^2} \cos \Omega t \qquad \textbf{Ans. (c) (6)}$$

(d) The mode-displacement approximation $\widehat{\mathbf{u}}(t)$ is given by Eq. 11.36. For $\widehat{u}_1(t)$ we take the first component of each $\boldsymbol{\phi}_r$ to get

$$\widehat{u}_1(t) = \sum_{r=1}^{\widehat{N}} \phi_{1r} \eta_r(t) \qquad \textbf{Ans. (d) (7)}$$

(e) We can write the expression for $u_1(t)$ without truncation and then indicate the terms corresponding to mode-displacement solutions with $\widehat{N} = 1$, $\widehat{N} = 2$, and $\widehat{N} = 3$. Thus,

$$u_1(t) = \frac{1.0(P_1 \cos \Omega t)}{507.691[1-(\Omega^2/176.72)]} \left.\vphantom{\int}\right] \widehat{N}=1$$
$$+ \frac{1.0(P_1 \cos \Omega t)}{1915.39[1-(\Omega^2/879.70)]} \left.\vphantom{\int}\right] \widehat{N}=2$$
$$+ \frac{-0.90145(-0.90145\, P_1 \cos \Omega t)}{7368.45[1-(\Omega^2/1687.46)]} \left.\vphantom{\int}\right] \widehat{N}=3$$
$$+ \frac{0.15436(0.15436\, P_1 \cos \Omega t)}{11{,}374.4[1-(\Omega^2/3122.79)]}$$

Ans. (e) (8)

For
$\Omega = 0.5\,\omega_1$, $\quad \Omega = 6.6468$ rad/sec, $\quad \Omega^2 = 44.179$ (rad/sec)^2
$\Omega = 1.3\,\omega_1$, $\quad \Omega = 53.402$ rad/sec, $\quad \Omega^2 = 2851.80$ (rad/sec)^2

Constant C in $u_1(t) = CP_1 \cos \Omega t$:

	$\widehat{N}=1$	$\widehat{N}=2$	$\widehat{N}=3$	$\widehat{N}=4$
$\Omega = 0$	$1.970(10^{-3})$	$2.492(10^{-3})$	$2.602(10^{-3})$	$2.604(10^{-3})$
$\Omega = 0.5\,\omega_1$	$2.626(10^{-3})$	$3.176(10^{-3})$	$3.289(10^{-3})$	$3.291(10^{-3})$
$\Omega = 1.3\,\omega_1$	$-1.301(10^{-3})$	$-3.630(10^{-3})$	$-5.228(10^{-3})$	$-4.987(10^{-3})$

(f) From the data in the foregoing table, the following conclusions can be reached:

1. A one-mode solution is not accurate at any of the three frequencies.
2. A three-mode solution is quite accurate for $\Omega = 0$ and for $\Omega = 0.5\,\omega_1$, but since $\Omega = 1.3\,\omega_1$ is almost equal to ω_4, an important contribution to $u_1(t)$ at this frequency is the mode 4 contribution. A truncated solution is useless at this frequency.

Equation 8 of Example 11.4 illustrates the importance that the distribution and the frequency of the input have on the response, as is also shown in Eq. 11.47. The importance of the distribution of the input forces is also illustrated in Example 11.5. By comparing Eqs. 4.9 and 11.46, we see that the *modal static displacement*, defined by

$$D_r \equiv \frac{F_r}{K_r} = \frac{\boldsymbol{\phi}_r^T \mathbf{P}}{K_r} \tag{11.48}$$

plays the same role as the static displacement U_0 of a SDOF system (Eq. 4.4).

Example 11.5 For the four-story building of Example 11.4, compute (a) F_r and (b) D_r for the following two load distributions:

$$\mathbf{P}_a = \begin{Bmatrix} 1 \\ 0 \\ 0 \\ 0 \end{Bmatrix} \text{ kips}, \quad \mathbf{P}_b = \begin{Bmatrix} 1 \\ 1 \\ 1 \\ 1 \end{Bmatrix} \text{ kips}$$

SOLUTION (a) Using the mode shapes from Example 11.4, we obtain

$$F_{1a} = \boldsymbol{\phi}_1^T \mathbf{P}_a = \begin{Bmatrix} 1.00000 \\ 0.77910 \\ 0.49655 \\ 0.23506 \end{Bmatrix}^T \begin{Bmatrix} 1 \\ 0 \\ 0 \\ 0 \end{Bmatrix} = 1.00000 \text{ kip} \tag{1a}$$

$$F_{1b} = \boldsymbol{\phi}_1^T \mathbf{P}_b = \begin{Bmatrix} 1.00000 \\ 0.77910 \\ 0.49655 \\ 0.23506 \end{Bmatrix}^T \begin{Bmatrix} 1 \\ 1 \\ 1 \\ 1 \end{Bmatrix} = 2.51071 \text{ kips} \tag{1b}$$

The other modal forces are determined in a similar manner:

$$\begin{array}{llll}
F_{1a} = & 1.00000 \text{ kip}, & F_{1b} = & 2.51071 \text{ kips} \\
F_{2a} = & 1.00000 \text{ kip}, & F_{2b} = & -0.07713 \text{ kip} \\
F_{3a} = & -0.90145 \text{ kip}, & F_{3b} = & -0.76801 \text{ kip} \\
F_{4a} = & 0.15436 \text{ kip}, & F_{4b} = & 0.06931 \text{ kip}
\end{array} \quad \text{Ans. (a) (2)}$$

(b) From Eq. 11.48,

$$D_r \equiv \frac{F_r}{K_r} = \frac{\boldsymbol{\phi}_r^T \mathbf{P}}{K_r} \tag{3}$$

Using this equation together with the results of part (a) and the generalized stiffnesses calculated in Example 11.4, we get

$$\begin{array}{llll}
D_{1a} = & 1.9697(10^{-3}) \text{ in.}, & D_{1b} = & 4.9453(10^{-3}) \text{ in.} \\
D_{2a} = & 5.2209(10^{-4}) \text{ in.}, & D_{2b} = & -0.4027(10^{-4}) \text{ in.} \\
D_{3a} = & -1.2234(10^{-4}) \text{ in.}, & D_{3b} = & -1.0423(10^{-4}) \text{ in.} \\
D_{4a} = & 1.3571(10^{-5}) \text{ in.}, & D_{4b} = & 0.6094(10^{-5}) \text{ in.}
\end{array} \quad \text{Ans. (b) (4)}$$

Note that although the total force applied in case (b) is four times that in case (a), the mode shape together with the force distribution determines the value of F_r and hence the excitation force level of each mode. Also note that when modes are normalized so that the largest component is 1.0, the modal stiffness K_r increases with mode number. This influences the modal static deflections D_r.

11.4 MODE-ACCELERATION SOLUTION FOR THE RESPONSE OF UNDAMPED MDOF SYSTEMS

As noted in Example 11.4, a mode-displacement solution may fail to give an accurate solution even when static loading is applied. Frequently, the convergence is slow and many modes would be needed to give an accurate mode-displacement solution. This difficulty can be alleviated to some extent by use of the mode-acceleration method, which was introduced in Eq. 11.40. Because of the improved convergence properties of this method, $\widehat{N}$ can be reduced, and fewer natural frequencies and modes are required from the eigensolution.

For an undamped MDOF system, the mode-acceleration solution, Eq. 11.40, reduces to[4]

$$\tilde{\mathbf{u}}(t) = \mathbf{K}^{-1}\mathbf{p}(t) - \sum_{r=1}^{\widehat{N}} \frac{1}{\omega_r^2} \boldsymbol{\phi}_r \ddot{\eta}_r(t) \tag{11.49}$$

The first term in the equation above is the *pseudostatic response*, while the second term gives the method its name, the *mode-acceleration method*. The presence of ω_r^2 in the denominator of this term improves the convergence of the method as compared to the mode-displacement method.[11.5] Williams[11.1] is credited with first suggesting the mode-acceleration method.

Example 11.6 For the four-story building of Example 11.4: (a) Use Eq. 11.49 to determine an expression for $\tilde{u}_1(t)$. (b) Prepare a table showing the amplitude of $\tilde{u}_1(t)$ using $\widehat{N} = 1$, $\widehat{N} = 2$, and $\widehat{N} = 3$ for excitation frequencies $\Omega = 0$, $\Omega = 0.5\omega_1$, and $\Omega = 1.3\omega_3$. (c) By comparing the table prepared in part (b) with the mode-displacement table of Example 11.4, state any conclusions you can reach about convergence of the two methods.

The flexibility matrix is

$$\mathbf{A} \equiv \mathbf{K}^{-1} = \begin{bmatrix} 2.60417 & 1.35417 & 0.72917 & 0.31250 \\ 1.35417 & 1.35417 & 0.72917 & 0.31250 \\ 0.72917 & 0.72917 & 0.72917 & 0.31250 \\ 0.31250 & 0.31250 & 0.31250 & 0.31250 \end{bmatrix} (10^3) \text{ in./kip}$$

SOLUTION   (a) As in part (e) of Example 11.4, we can write an expression for $u_1(t)$ without truncation and then indicate the terms corresponding to $\widehat{N} = 1$, $\widehat{N} = 2$, and

---

[4] Since $\mathbf{K}^{-1}$ doesn't exist for systems with rigid-body modes, Eq. 11.49 does not apply to such systems. Systems with rigid-body modes are treated in Section 11.6.

$\widehat{N} = 3$. Thus, from Eq. 11.49,

$$\tilde{u}_1 = a_{11} P_1 \cos \Omega t - \sum_{r=1}^{\widehat{N}} \frac{1}{\omega_r^2} \phi_{1r} \ddot{\eta}_r \qquad (1)$$

which, when combined with Eq. 11.46, gives

$$\tilde{u}_1 = a_{11} P_1 \cos \Omega t + \sum_{r=1}^{\widehat{N}} \frac{\Omega_r^2}{\omega_r^2} \phi_{1r} \frac{F_r}{K_r} \frac{1}{1 - (\Omega/\omega_r)^2} \cos \Omega t \qquad (2)$$

Thus, we can write the expression for $u_1(t)$ without truncation and then indicate the terms corresponding to mode-acceleration solutions with $\widehat{N} = 1$, $\widehat{N} = 2$, and $\widehat{N} = 3$.

$$\left. \begin{array}{l} u_1(t) = 2.60417(10^{-3})(P_1 \cos \Omega t) \\[4pt] \left. \begin{array}{l} + \dfrac{(\Omega^2/176.72)(1.0)(P_1 \cos \Omega t)}{507.695[1 - (\Omega^2/176.72)]} \\[10pt] + \dfrac{(\Omega^2/879.70)(1.0)(P_1 \cos \Omega t)}{1915.39[1 - (\Omega^2/879.70)]} \end{array} \right\} \widehat{N} = 2 \\[20pt] + \dfrac{(\Omega^2/1687.46)(-0.90145)(-0.90145\, P_1 \cos \Omega t)}{7368.43[1 - (\Omega^2/1687.46)]} \\[10pt] + \dfrac{(\Omega^2/3122.79)(0.15436)(0.15436\, P_1 \cos \Omega t)}{11,374.4[1 - (\Omega^2/3122.79)]} \end{array} \right\} \widehat{N} = 3 \quad \textbf{Ans. (a)} \ (3)$$

(b) As in Example 11.4, the three frequencies of interest give

$$\Omega^2 = 0, \qquad \Omega^2 = 44.179 \ (\text{rad/sec})^2, \qquad \Omega^2 = 2851.80 \ (\text{rad/sec})^2$$

Constant C in $u_1(t) = C P_1 \cos \Omega t$:

	$\widehat{N} = 1$	$\widehat{N} = 2$	$\widehat{N} = 3$	$\widehat{N} = 4$
$\Omega = 0$	$2.604(10^{-3})$	$2.604(10^{-3})$	$2.604(10^{-3})$	$2.604(10^{-3})$
$\Omega = 0.5\,\omega_1$	$3.261(10^{-3})$	$3.288(10^{-3})$	$3.291(10^{-3})$	$3.291(10^{-3})$
$\Omega = 1.3\,\omega_1$	$5.044(10^{-3})$	$-2.506(10^{-3})$	$-5.207(10^{-3})$	$-4.987(10^{-3})$

(c) From the foregoing table we can conclude that:

1. The exact static solution is produced at $\Omega = 0$ without any contribution from normal modes.
2. At the low frequency of $\Omega = 0.5\,\omega_1$, even a one-term solution is fairly accurate, and the mode-acceleration solution is accurate to two places for $\widehat{N} = 2$, as compared with $\widehat{N} = 3$ for the mode-displacement solution.
3. Since the forcing frequency $\Omega = 1.3\,\omega_3$ lies between ω_3 and ω_4, a truncated mode-acceleration solution is not any better than a truncated mode-displacement solution—the fourth mode is needed in either case.

As noted in Example 11.6, the mode-acceleration method is particularly useful in improving the static or low-frequency-response convergence. Whether the mode-displacement method or the mode-acceleration method is used, it is always important to include every mode whose natural frequency is "in vicinity of" any excitation frequency.

For undamped systems, Eq. 11.43 can be used to obtain a general expression for $\ddot{\eta}_r(t)$ to substitute into the mode-acceleration equation, Eq. 11.49. Thus,

$$\ddot{\eta}_r(t) = -\omega_r^2 \eta_r(0) \cos \omega_r t - \omega_r \dot{\eta}_r(0) \sin \omega_r t + \frac{f_r(t)}{M_r} - \frac{\omega_r}{M_r} \int_0^t f_r(\tau) \sin \omega_r (t - \tau) \, d\tau \quad (11.50)$$

Numerical procedures described in Chapter 6 can also be employed to determine modal acceleration histories.

In Section 11.3 and the present section it has generally been implied that $\widehat{N} < N$ modes could be employed for mode superposition, with the first $\widehat{N}$ modes (i.e., starting from lowest frequency and counting the $\widehat{N}$ lowest-frequency modes out of the set of N modes). Engineering judgment is required in determining how many modes to include. Examples 11.4 through 11.6 are intended to point out the importance of frequency ratio and mode shape in determining how many modes are important.

11.5 DYNAMIC STRESSES BY MODE SUPERPOSITION

In Sections 11.3 and 11.4 the basic techniques for determining the displacement history $\mathbf{u}(t)$ by using the mode-displacement method and the mode-acceleration method were described. In addition to determining the displacement history, a dynamic analysis usually includes the determination of stress histories, or at least the determination of maximum values of stress at specified locations in the structure. Here the word *stress* refers to any internal-force-related quantity (e.g., bending moment, transverse shear force, axial force, or axial stress).

The following are symbolic representations of stresses obtained by mode superposition. For the mode-displacement method, the internal stresses are given by

$$\widehat{\boldsymbol{\sigma}}(t) = \sum_{r=1}^{\widehat{N}} \mathbf{s}_r \eta_r(t) \quad (11.51)$$

where $\mathbf{s}_r$ is the contribution to the stress vector $\boldsymbol{\sigma}$ due to a unit displacement of the rth mode, that is, for $\eta_r = 1$. For the mode-acceleration method for an undamped system, the displacement approximation in Eq. 11.49 leads to the stress approximation

$$\widetilde{\boldsymbol{\sigma}}(t) = \boldsymbol{\sigma}_{\text{pseudostatic}} - \sum_{r=1}^{\widehat{N}} \frac{1}{\omega_r^2} \mathbf{s}_r \ddot{\eta}_r(t) \quad (11.52)$$

The mode-acceleration method is particularly beneficial in speeding convergence of the internal stresses.

Example 11.7 For the shear building of Example 11.4, write an expression for the shear force at the ith story, σ_i, corresponding to mode r.

SOLUTION The story shears are determined by the relative displacements between floors. Thus,

$$\sigma_1 = k_1(u_1 - u_2), \qquad \sigma_2 = k_2(u_2 - u_3)$$
$$\sigma_3 = k_3(u_3 - u_4), \qquad \sigma_4 = k_4 u_4 \qquad (1)$$

or in matrix form,

$$\begin{Bmatrix} \sigma_1 \\ \sigma_2 \\ \sigma_3 \\ \sigma_4 \end{Bmatrix} = \begin{bmatrix} k_1 & -k_1 & 0 & 0 \\ 0 & k_2 & -k_2 & 0 \\ 0 & 0 & k_3 & -k_3 \\ 0 & 0 & 0 & k_4 \end{bmatrix} \begin{Bmatrix} u_1 \\ u_2 \\ u_3 \\ u_4 \end{Bmatrix} \qquad (2)$$

Thus, when the displacement shape is $\boldsymbol{\phi}_r$, Eq. 2 leads to the following expression for the modal shears $\mathbf{s}_r$:

$$\begin{Bmatrix} s_1 \\ s_2 \\ s_3 \\ s_4 \end{Bmatrix}_r = \begin{bmatrix} k_1 & -k_1 & 0 & 0 \\ 0 & k_2 & -k_2 & 0 \\ 0 & 0 & k_3 & -k_3 \\ 0 & 0 & 0 & k_4 \end{bmatrix} \begin{Bmatrix} \phi_1 \\ \phi_2 \\ \phi_3 \\ \phi_4 \end{Bmatrix}_r \qquad \text{Ans. (3)}$$

Using the numerical values of $\mathbf{K}$ and $\boldsymbol{\phi}_r$ from Example 11.4, we can use Eq. 3 to evaluate the respective modal shear forces:

$$\mathbf{s}_1 = \begin{Bmatrix} 176.72 \\ 452.08 \\ 627.58 \\ 752.19 \end{Bmatrix} \text{kips/in.}, \qquad \mathbf{s}_2 = \begin{Bmatrix} 879.70 \\ 704.42 \\ -245.47 \\ -1400.35 \end{Bmatrix} \text{kips/in.}$$

$$\mathbf{s}_3 = \begin{Bmatrix} -1521.16 \\ 1853.74 \\ 1318.51 \\ -2265.50 \end{Bmatrix} \text{kips/in.}, \qquad \mathbf{s}_4 = \begin{Bmatrix} 482.02 \\ -2317.07 \\ 3928.51 \\ -2038.02 \end{Bmatrix} \text{kips/in.}$$

Ans. (4)

Note that while the modes $\boldsymbol{\phi}_r$ given in Example 11.4 and used in Example 11.7 to determine the modal shears $\mathbf{s}_r$ are scaled such that the maximum displacement in each mode is $+1$, the magnitudes of the modal shears tend to increase with increasing mode number. It is for this reason that convergence of stresses is slower than the convergence of displacements, and hence the mode acceleration method proves to be especially beneficial when stresses are to to be computed.

11.6 MODE SUPERPOSITION FOR UNDAMPED SYSTEMS WITH RIGID-BODY MODES

Mode superposition may be employed to determine the response of systems with rigid-body modes. In this section we consider the application of the mode-displacement method and the mode-acceleration method to undamped systems with rigid-body modes. Extension of the methods presented here to systems with modal damping is straightforward.

For systems with rigid-body modes, the basic equation of mode superposition, Eq. 11.8, can be written in the form

$$\mathbf{u}(t) = \mathbf{u}_R(t) + \mathbf{u}_E(t) = \boldsymbol{\Phi}_R \boldsymbol{\eta}_R(t) + \boldsymbol{\Phi}_E \boldsymbol{\eta}_E(t) \qquad (11.53)$$

where $\boldsymbol{\Phi}_R$ and $\boldsymbol{\Phi}_E$ are modal matrices containing the N_R rigid-body modes and the N_E elastic-deformation modes, respectively ($N = N_R + N_E$). Substituting Eq. 11.53 into the equation of motion, Eq. 11.1, with $\mathbf{C} = 0$, we get the following modal equations of motion:

$$\boldsymbol{M}_R \ddot{\boldsymbol{\eta}}_R = \boldsymbol{\Phi}_R^T \mathbf{p}(t) \qquad (11.54a)$$

and

$$\boldsymbol{M}_E \ddot{\boldsymbol{\eta}}_E + \boldsymbol{K}_E \boldsymbol{\eta}_E = \boldsymbol{\Phi}_E^T \mathbf{p}(t) \qquad (11.54b)$$

From Eq. 11.54a, the rigid-body coordinates are obtained by double integration; that is,

$$\boxed{\eta_r(t) = \int_0^t \int_0^\tau \frac{1}{M_r} f_r(\xi)\, d\xi\, d\tau + t\, \dot{\eta}_r(0) + \eta_r(0), \qquad r = 1, 2, \ldots, N_R} \qquad (11.55)$$

For the N_E elastic-mode coordinates, $\eta_r(t)$ is given by Eq. 11.43, as before.

11.6.1 Mode-Displacement Method for Systems with Rigid-Body Modes

To determine the system displacements by the *mode-displacement method*, Eq. 11.53 is employed with all rigid-body modes included but with the number of elastic modes truncated to $\widehat{N}_E$. Thus,

$$\boxed{\widehat{\mathbf{u}}(t) = \boldsymbol{\Phi}_R \boldsymbol{\eta}_R(t) + \widehat{\boldsymbol{\Phi}}_E \widehat{\boldsymbol{\eta}}_E(t)} \qquad (11.56)$$

where $\widehat{\boldsymbol{\Phi}}_E$ contains the $\widehat{N}_E$ retained elastic modes. These may be the $\widehat{N}_E$ lowest-frequency modes, or they may be a set of $\widehat{N}_E$ elastic modes that are deemed to be most significant for the response calculation. Equations 11.55 and 11.43 are used to determine the N_R rigid-body modal displacement histories and the $\widehat{N}_E$ elastic modal displacement histories, respectively.

To determine stresses by the mode-displacement method, Eq. 11.51 is simply modified to account for the fact that rigid-body displacements do not give rise to internal stresses. Hence,

$$\widehat{\boldsymbol{\sigma}}(t) = \widehat{\mathbf{S}}_E \widehat{\boldsymbol{\eta}}_E = \sum_{r=1}^{\widehat{N}_E} \mathbf{s}_r \eta_r(t) \qquad (11.57)$$

where the columns of $\widehat{\mathbf{S}}_E$ are the internal stress vectors corresponding to unit values of each of the $\widehat{N}_E$ retained elastic modes.

11.6.2 Mode-Acceleration Method for Systems with Rigid-Body Modes

Since the stiffness matrix for a system with rigid-body modes is singular and therefore cannot be inverted, the mode-acceleration method cannot be employed in the straightforward manner indicated by Eq. 11.49.

Let us begin by solving Eq. 11.54b for $\boldsymbol{\eta}_E(t)$ and substituting this into Eq. 11.53. We get

$$\mathbf{u}(t) = \boldsymbol{\Phi}_R \boldsymbol{\eta}_R(t) + (\boldsymbol{\Phi}_E \boldsymbol{K}_E^{-1} \boldsymbol{\Phi}_E^T)\mathbf{p}(t) - (\boldsymbol{\Phi}_E \boldsymbol{K}_E^{-1} \boldsymbol{M}_E)\ddot{\boldsymbol{\eta}}_E \qquad (11.58)$$

The rigid-body term is determined just as for the mode-displacement equation; that is, all rigid-body modes are retained. To cast Eq. 11.58 into mode-acceleration form similar to Eq. 11.49, we can observe that truncating the last term in Eq. 11.58 leads to the same expression as the last term in Eq. 11.49. Thus, let us truncate the number of elastic modes and rewrite Eq. 11.58 in the form

$$\tilde{\mathbf{u}}(t) = \boldsymbol{\Phi}_R \boldsymbol{\eta}_R(t) + \mathbf{A}_E \mathbf{p}(t) - (\widehat{\boldsymbol{\Phi}}_E \widehat{\boldsymbol{K}}_E^{-1} \widehat{\boldsymbol{M}}_E)\ddot{\widehat{\boldsymbol{\eta}}}_E \qquad (11.59)$$

where the only term leading to difficulty is the *pseudostatic deflection term*, which we have written as $\mathbf{A}_E \mathbf{p}(t)$.

One expression for the *elastic flexibility matrix* $\mathbf{A}_E$ is

$$\mathbf{A}_E = \boldsymbol{\Phi}_E \boldsymbol{K}_E^{-1} \boldsymbol{\Phi}_E^T \qquad (11.60)$$

which is found in Eq. 11.58. We must include the flexibility of all N_E flexible modes. However, we must do this without solving for all N_E of the elastic modes, as we would have to do if we used Eq. 11.60. First, let us substitute Eq. 11.53 into the equation of motion, noting that $\mathbf{K}\mathbf{u}_R = \mathbf{0}$. Then the remaining terms in the equation of motion are

$$\mathbf{M}\ddot{\mathbf{u}}_E + \mathbf{K}\mathbf{u}_E = \mathbf{p}_E(t) \qquad (11.61a)$$

where

$$\mathbf{p}_E(t) = \mathbf{p}(t) - \mathbf{M}\ddot{\mathbf{u}}_R \qquad (11.61b)$$

Thus, in determining the elastic displacements, we use a self-equilibrated force system of applied forces and rigid-body inertia forces. Since, from Eqs. 11.53 and 11.54a,

$$\ddot{\mathbf{u}}_R = \boldsymbol{\Phi}_R \ddot{\boldsymbol{\eta}}_R = \boldsymbol{\Phi}_R \boldsymbol{M}_R^{-1} \boldsymbol{\Phi}_R^T \mathbf{p}(t) \qquad (11.62)$$

we can write the elastic force vector $\mathbf{p}_E$ as

$$\mathbf{p}_E(t) = \mathbf{R}\mathbf{p}(t) \qquad (11.63a)$$

where

$$\mathbf{R} = \mathbf{I} - \mathbf{M}\boldsymbol{\Phi}_R \boldsymbol{M}_R^{-1} \boldsymbol{\Phi}_R^T \qquad (11.63b)$$

is called the *inertia-relief matrix*.

11.6 Mode Superposition for Undamped Systems with Rigid-Body Modes

Since the original system has N_R rigid-body modes, in order to calculate a flexibility matrix we need to impose N_R arbitrary constraints. Then let $\mathbf{A}_R$ be the flexibility matrix of the system relative to these statically determinate constraints, with zeros filling in the N_R rows and columns corresponding to the constraints. Then let

$$\mathbf{w} = \mathbf{A}_R \mathbf{p}_E \tag{11.64}$$

be the pseudostatic elastic displacement of the system relative to the statically determinate constraints imposed. But the deflection $\mathbf{w}$ may contain some component of the rigid-body modes, so a new vector $\mathbf{w}_E$ is created that is orthogonal to the rigid-body modes.

$$\mathbf{w}_E = \mathbf{w} - \mathbf{\Phi}_R \mathbf{c}_R \tag{11.65}$$

where $\mathbf{c}_R$ is such that $\mathbf{\Phi}_R^T \mathbf{M} \mathbf{w}_E = \mathbf{0}$. Then

$$\mathbf{c}_R = M_R^{-1} \mathbf{\Phi}_R^T \mathbf{M} \mathbf{w} \tag{11.66}$$

Then

$$\mathbf{w}_E = (\mathbf{I} - \mathbf{\Phi}_R M_R^{-1} \mathbf{\Phi}_R^T \mathbf{M}) \mathbf{w} = \mathbf{R}^T \mathbf{w} \tag{11.67}$$

In conclusion, we get

$$\mathbf{w}_E = \mathbf{A}_E \mathbf{p} \tag{11.68}$$

where $\mathbf{A}_E$, the symmetric *total elastic flexibility matrix*, has the same meaning as in Eq. 11.60 but is now given by

$$\mathbf{A}_E = \mathbf{R}^T \mathbf{A}_R \mathbf{R} \tag{11.69}$$

Finally, with $\widehat{N}_R$ rigid-body modes and $\widehat{N}_E$ retained elastic modes, and with the inertia-relief matrix $\mathbf{R}$ given by Eq. 11.63b, the equation

$$\boxed{\tilde{\mathbf{u}}(t) = \mathbf{\Phi}_R \boldsymbol{\eta}_R(t) + \mathbf{R}^T \mathbf{A}_R \mathbf{R} \mathbf{p}(t) - (\widehat{\mathbf{\Phi}}_E \widehat{K}_E^{-1} \widehat{M}_E) \ddot{\widehat{\boldsymbol{\eta}}}_E(t)} \tag{11.70}$$

is used to calculate the displacement vector by the *mode-acceleration method* for systems having rigid-body modes.[5]

11.6.3 Stresses in Truncated Models of Systems with Rigid-Body Modes

The stresses in a system having rigid-body modes can be determined by the mode-acceleration expression

$$\tilde{\boldsymbol{\sigma}}(t) = \boldsymbol{\sigma}_{\text{pseudostatic}} - \sum_{r=1}^{\widehat{N}_E} \frac{1}{\omega_r^2} \mathbf{s}_r \ddot{\eta}_r(t) \tag{11.71}$$

where $\boldsymbol{\sigma}_{\text{pseudostatic}}$ is the vector of internal stresses resulting from the self-equilibrated elastic force vector $\mathbf{p}_E$ given by Eq. 11.63a. Determination of stress by using

---

[5] A similar treatment of rigid-body modes is required in Section 17.3.

Eq. 11.71 is considerably simpler than determining displacements through the use of Eq. 11.70. Example 11.8 illustrates the use the mode-displacement method and the mode-acceleration method to determine internal stresses.

Example 11.8 Use the mode-displacement method and the mode-acceleration method to determine expressions for the maximum force in each of the two springs shown in Fig. 1 due to application of a step force $p_3(t) = p_0$, $t \geq 0$. Compare the convergence of the two methods. The system is at rest at $t = 0$.

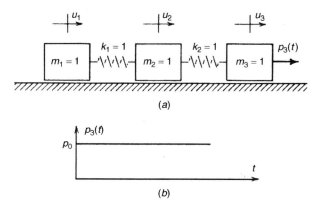

Figure 1 (a) Unrestrained 3-DOF spring–mass system; (b) step force input at mass 3.

SOLUTION   1. Solve for the modes and frequencies.

$$\begin{bmatrix} 1 & 0 & 0 \\ 0 & 1 & 0 \\ 0 & 0 & 1 \end{bmatrix} \begin{Bmatrix} \ddot{u}_1 \\ \ddot{u}_2 \\ \ddot{u}_3 \end{Bmatrix} + \begin{bmatrix} 1 & -1 & 0 \\ -1 & 2 & -1 \\ 0 & -1 & 1 \end{bmatrix} \begin{Bmatrix} u_1 \\ u_2 \\ u_3 \end{Bmatrix} = \begin{Bmatrix} 0 \\ 0 \\ 0 \end{Bmatrix} \quad (1)$$

Let
$$\mathbf{u} = \boldsymbol{\phi} \cos \omega t \quad (2)$$

Then the algebraic eigenvalue equation is

$$\begin{bmatrix} 1-\omega^2 & -1 & 0 \\ -1 & 2-\omega^2 & -1 \\ 0 & -1 & 1-\omega^2 \end{bmatrix} \begin{Bmatrix} \phi_1 \\ \phi_2 \\ \phi_3 \end{Bmatrix} = \begin{Bmatrix} 0 \\ 0 \\ 0 \end{Bmatrix} \quad (3)$$

Setting the determinant of coefficients of Eq. 3 to zero, we get the characteristic equation

$$\omega^2(\omega^4 - 4\omega^2 + 3) = 0 \quad (4)$$

whose roots are

$$\omega_1^2 = 0, \quad \omega_2^2 = 1, \quad \omega_3^2 = 3 \quad (5)$$

The corresponding modes are

$$\boldsymbol{\phi}_1 = \begin{Bmatrix} 1 \\ 1 \\ 1 \end{Bmatrix}, \quad \boldsymbol{\phi}_2 = \begin{Bmatrix} 1 \\ 0 \\ -1 \end{Bmatrix}, \quad \boldsymbol{\phi}_3 = \begin{Bmatrix} 1 \\ -2 \\ 1 \end{Bmatrix} \quad (6)$$

Note that mode $\boldsymbol{\phi}_1$ is the rigid-body mode of this system.

The modal masses and modal stiffnesses are given by

$$M_r = \boldsymbol{\phi}_r^\mathrm{T} \mathbf{M} \boldsymbol{\phi}_r = \boldsymbol{\phi}_r^\mathrm{T} \boldsymbol{\phi}_r, \quad K_r = \omega_r^2 M_r \quad (7)$$

so

$$\begin{aligned} M_1 &= 3, & M_2 &= 2, & M_3 &= 6 \\ K_1 &= 0, & K_2 &= 2, & K_3 &= 18 \end{aligned} \quad (8)$$

2. Since the system is initially at rest,

$$\eta_r(0) = \dot{\eta}_r(0) = 0, \quad r = 1, 2, 3 \quad (9)$$

3. The generalized forces are given by

$$f_r(t) = \boldsymbol{\phi}_r^\mathrm{T} \mathbf{p}(t) \quad (10)$$

so

$$f_1(t) = p_3(t) = p_0, \quad f_2(t) = -p_3(t) = -p_0, \quad f_3(t) = p_3(t) = p_0 \quad (11)$$

4. The equations of motion in modal coordinates are thus

$$\left. \begin{aligned} 3\ddot{\eta}_1 &= p_0 \\ 2\ddot{\eta}_2 + 2\eta_2 &= -p_0 \\ 6\ddot{\eta}_3 + 18\eta_3 &= p_0 \end{aligned} \right\} \quad t \geq 0 \quad (12)$$

Then the solutions that satisfy these ordinary differential equations and the initial conditions of Eqs. 9 are

$$\begin{aligned} \eta_1 &= \frac{p_0 t^2}{6} \\ \eta_2 &= \frac{-p_0}{2}(1 - \cos \omega_2 t) \\ \eta_3 &= \frac{p_0}{18}(1 - \cos \omega_3 t) \end{aligned} \quad (13)$$

where the latter two can be obtained from Eq. 5.8. From Eqs. 13b and c,

$$\begin{aligned} \ddot{\eta}_2 &= -\frac{p_0}{2} \cos \omega_2 t \\ \ddot{\eta}_3 &= \frac{p_0}{6} \cos \omega_3 t \end{aligned} \quad (14)$$

5. Determine the *modal stresses*, that is, the spring forces due to the elastic modes $\boldsymbol{\phi}_2$ and $\boldsymbol{\phi}_3$.

$$\mathbf{s}_r = \left\{ \begin{array}{c} s_1 \\ s_2 \end{array} \right\}_r = \left[\begin{array}{ccc} -1 & 1 & 0 \\ 0 & -1 & 1 \end{array} \right] \left\{ \begin{array}{c} \phi_1 \\ \phi_2 \\ \phi_3 \end{array} \right\}_r \tag{15}$$

Therefore, the *modal stress vectors* are

$$\mathbf{s}_2 = \left\{ \begin{array}{c} -1 \\ -1 \end{array} \right\}, \quad \mathbf{s}_3 = \left\{ \begin{array}{c} -3 \\ 3 \end{array} \right\} \tag{16}$$

6. The mode-displacement approximation to the spring forces (internal stresses) is given by Eq. 11.57; that is,

$$\widehat{\boldsymbol{\sigma}}_{\text{(one-mode)}} = \mathbf{s}_2 \eta_2(t) \qquad \text{Ans. (17}a\text{)}$$

if one elastic mode is employed, or

$$\widehat{\boldsymbol{\sigma}}_{\text{(two-mode)}} = \boldsymbol{\sigma}(t) = \mathbf{s}_2 \eta_2(t) + \mathbf{s}_3 \eta_3(t) \qquad \text{Ans. (17}b\text{)}$$

if both elastic modes are included. Then

$$\widehat{\boldsymbol{\sigma}}_{\text{(one-mode)}} = \left\{ \begin{array}{c} \widehat{\sigma}_1 \\ \widehat{\sigma}_2 \end{array} \right\} = \left\{ \begin{array}{c} -1 \\ -1 \end{array} \right\} \frac{-p_0}{2} (1 - \cos \omega_2 t)$$

$$= \frac{p_0}{2} \left\{ \begin{array}{c} 1 \\ 1 \end{array} \right\} (1 - \cos \omega_2 t) \qquad \text{Ans. (18}a\text{)}$$

$$\widehat{\boldsymbol{\sigma}}_{\text{(two-mode)}} = \boldsymbol{\sigma} \equiv \left\{ \begin{array}{c} \sigma_1 \\ \sigma_2 \end{array} \right\} = \frac{p_0}{2} \left\{ \begin{array}{c} 1 \\ 1 \end{array} \right\} (1 - \cos \omega_2 t)$$

$$+ \frac{p_0}{6} \left\{ \begin{array}{c} -1 \\ 1 \end{array} \right\} (1 - \cos \omega_3 t) \qquad \text{Ans. (18}b\text{)}$$

7. The mode-acceleration solution for internal stresses is based on Eq. 11.71. The self-equilibrated loading $\mathbf{p}_E$, which is based on Eq. 11.61b, is shown in Fig. 2. Therefore, by taking free-body diagrams, we get the pseudostatic spring forces

$$\boldsymbol{\sigma}_{\text{pseudostatic}} = \left\{ \begin{array}{c} \sigma_1 \\ \sigma_2 \end{array} \right\}_{\text{pseudostatic}} = \left\{ \begin{array}{c} \dfrac{p_0}{3} \\ \dfrac{2p_0}{3} \end{array} \right\} \tag{19}$$

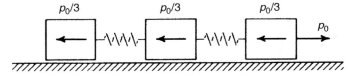

Figure 2 Self-equilibrating force system due to rigid-body motion.

Then the mode-acceleration solution based on one elastic mode is

$$\tilde{\sigma}(t) = \sigma_{\text{pseudostatic}} - \frac{1}{\omega_2^2}\mathbf{s}_2\ddot{\eta}_2(t) \tag{20a}$$

and based on both elastic modes is

$$\tilde{\sigma}(t) = \sigma = \sigma_{\text{pseudostatic}} - \frac{1}{\omega_2^2}\mathbf{s}_2\ddot{\eta}_2(t) - \frac{1}{\omega_3^2}\mathbf{s}_3\ddot{\eta}_3(t) \tag{20b}$$

Finally, the spring forces (internal stresses) based on the mode-acceleration method are

$$\tilde{\sigma}_{\text{(one-mode)}} = \begin{Bmatrix} \tilde{\sigma}_1 \\ \tilde{\sigma}_2 \end{Bmatrix} = \frac{p_0}{3}\begin{Bmatrix} 1 \\ 2 \end{Bmatrix} - \frac{p_0}{2}\begin{Bmatrix} 1 \\ 1 \end{Bmatrix}\cos\omega_2 t \quad \textbf{Ans. (21a)}$$

and

$$\tilde{\sigma}_{\text{(two-mode)}} = \sigma \equiv \begin{Bmatrix} \sigma_1 \\ \sigma_2 \end{Bmatrix} = \frac{p_0}{3}\begin{Bmatrix} 1 \\ 2 \end{Bmatrix}$$
$$- \frac{p_0}{2}\begin{Bmatrix} 1 \\ 1 \end{Bmatrix}\cos\omega_2 t + \frac{p_0}{6}\begin{Bmatrix} 1 \\ -1 \end{Bmatrix}\cos\omega_3 t \quad \textbf{Ans. (21b)}$$

As should be expected, the two-mode mode-displacement solution, Eq. 18b, and the two-mode mode-acceleration solution, Eq. 21b, are identical, since this system has only two elastic modes.

	Mode-displacement one-mode	Mode-acceleration one-mode	Exact[a]
σ_1/p_0	1.000000	0.833333	0.999933
σ_1/p_0	1.000000	1.166667	1.333241

[a] "Exact" values computed by evaluating σ_1 and σ_2 from Eq. 18b at 1° intervals to $\omega t = 100\pi$.

Example 11.8 is too small to indicate improved "convergence" of the mode-acceleration method over the mode-displacement method. It does however, illustrate the procedures used in evaluating internal stresses by the two methods. A more extensive application of the two methods to a system with rigid-body modes may be found in Ref. [11.5].

REFERENCES

[11.1] D. Williams, *Dynamic Loads in Aeroplanes Under Given Impulsive Loads with Particular Reference to Landing and Gust Loads on a Large Flying Boat*, Great Britain RAE Reports SME 3309 and 3316, 1945.

[11.2] R. E. Cornwell, R. R. Craig, Jr., and C. P. Johnson, "On the Application of the Mode-Acceleration Method to Structural Engineering Problems," *Earthquake Engineering and Structural Dynamics*, Vol. 11, 1983, pp. 679–688.

[11.3] C. J. Camarda, et al., "An Evaluation of Higher-Order Modal Methods for Calculating Transient Structural Response," *Computers & Structures*, Vol. 27, No. 1, 1987, pp. 89–101.

[11.4] D. J. Rixen, "Generalized Mode Acceleration Methods and Modal Truncation Augmentation," presented at the 42nd AIAA/ASME/ASCE/AHS/ASC Structures, Structural Dynamics, and Materials Conference, Seattle, WA, April 2001.

[11.5] R. L. Bisplinghoff, H. Ashley, and R. L. Halfman, *Aeroelasticity*, Addison-Wesley, Reading, MA, 1955.

PROBLEMS
Problem Set 11.1

11.1 The system in Fig. P11.1 has the following equation of motion:

$$m\begin{bmatrix} 3 & 0 \\ 0 & 2 \end{bmatrix}\begin{Bmatrix} \ddot{u}_1 \\ \ddot{u}_2 \end{Bmatrix} + k\begin{bmatrix} 5 & -2 \\ -2 & 2 \end{bmatrix}\begin{Bmatrix} u_1 \\ u_2 \end{Bmatrix} = \begin{Bmatrix} 0 \\ 0 \end{Bmatrix}$$

(a) Solve for the natural frequencies and mode shapes for this system. Scale each of the modes so that its largest component is equal to 1 and write down the modal matrix. Sketch the two mode shapes. (b) Using the mode shapes from part (a), determine the modal mass matrix M and the modal stiffness matrix K. Write down the modal equations of motion for free vibration with the mode shapes scaled as in part (a). (c) Using the results of parts (a) and (b), determine expressions for the modal initial conditions $\eta_r(0)$ and $\dot{\eta}_r(0)$ ($r = 1, 2$) if the initial conditions in physical coordinates are

$$\mathbf{u}(0) = \begin{Bmatrix} u_1(0) \\ u_2(0) \end{Bmatrix}, \qquad \dot{\mathbf{u}}(0) = \begin{Bmatrix} \dot{u}_1(0) \\ \dot{u}_2(0) \end{Bmatrix}$$

(d) Normalize the eigenvectors so that the modal masses are equal to m. Write down the modal equations of motion with the eigenvectors scaled in this way.

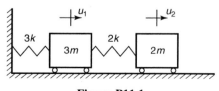

Figure P11.1

11.2 The system in Fig. P11.2 has the following equation of motion:

$$m\begin{bmatrix} 4 & 0 \\ 0 & 3 \end{bmatrix}\begin{Bmatrix} \ddot{u}_1 \\ \ddot{u}_2 \end{Bmatrix} + k\begin{bmatrix} 5 & -3 \\ -3 & 3 \end{bmatrix}\begin{Bmatrix} u_1 \\ u_2 \end{Bmatrix} = \begin{Bmatrix} 0 \\ p_2(t) \end{Bmatrix}$$

where $m = 1$ kg, $k = 1$ N/mm, and the modal matrix and the initial conditions are given by

$$\Phi = \begin{bmatrix} 3 & 1 \\ 4 & -1 \end{bmatrix}, \qquad \mathbf{u}(0) = \begin{Bmatrix} 2 \\ 2 \end{Bmatrix} \text{ mm},$$

$$\dot{\mathbf{u}}(0) = \begin{Bmatrix} 0 \\ -1 \end{Bmatrix} \text{ mm/s}$$

(a) Determine the natural frequencies of this system. (b) Verify the given mode shapes and use them to determine the modal mass matrix M and the modal stiffness matrix K. (c) Determine the modal initial conditions $\eta_r(0)$ and $\dot{\eta}_r(0)$ ($r = 1, 2$) that correspond to the physical-coordinate initial conditions given above. (d) If $p_2(t)$ is a step force of magnitude 5 N starting at $t = 0$, determine the modal forces $f_r(t) = \boldsymbol{\phi}_r^T \mathbf{p}(t)$ ($r = 1, 2$). (e) Write down the modal equations of motion. Solve for the modal responses $\eta_r(t)$ for $t > 0$. (*Hint:* See Section 5.1 for the response of a SDOF system that is initially at rest to a step input force. Add the response due to the initial conditions.) (f) Finally, determine expressions for the physical responses $u_1(t)$ and $u_2(t)$ for $t > 0$.

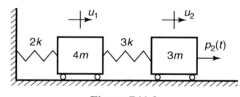

Figure P11.2

11.3 The system shown in Fig. P11.3 has the following equation of motion in **w** coordinates:

$$m\begin{bmatrix} 1 & 1 & 0 \\ 1 & 4 & 1 \\ 0 & 1 & 1 \end{bmatrix}\begin{Bmatrix} \ddot{w}_1 \\ \ddot{w}_2 \\ \ddot{w}_3 \end{Bmatrix} + k\begin{bmatrix} 2 & 0 & 0 \\ 0 & 0 & 0 \\ 0 & 0 & 2 \end{bmatrix}\begin{Bmatrix} w_1 \\ w_2 \\ w_3 \end{Bmatrix} = \begin{Bmatrix} 0 \\ 0 \\ 0 \end{Bmatrix}$$

where $w_2 \equiv u_2$ and where the relative displacements $w_1 = u_1 - u_2$ and $w_3 = u_3 - u_2$ are used. **(a)** Determine the natural frequencies, ω_r^2, of this system. **(b)** Determine the three mode shapes (eigenvectors) of this system, expressed in terms of the **w** coordinates and scaled so that the maximum entry in each of the vectors is 1. **(c)** Sketch the three mode shapes in absolute (i.e., **u**) coordinates. **(d)** If the initial velocities are all zero and the initial displacements are such that only the modal displacement in the first (i.e., lowest-frequency) mode is nonzero, describe the resulting motion. What is the potential energy associated with this motion? Explain why.

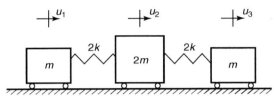

Figure P11.3

Problem Set 11.2

11.4 The system in Fig. P11.4 has the following equation of motion:

$$m \begin{bmatrix} 2 & 0 \\ 0 & 3 \end{bmatrix} \begin{Bmatrix} \ddot{u}_1 \\ \ddot{u}_2 \end{Bmatrix} + k \begin{bmatrix} 3 & -2 \\ -2 & 7 \end{bmatrix} \begin{Bmatrix} u_1 \\ u_2 \end{Bmatrix} = \begin{Bmatrix} 0 \\ p_2(t) \end{Bmatrix}$$

where $m = 1$ lb-sec^2/in., $k = 1$ lb/in., and the modal matrix and the initial conditions are given by

$$\boldsymbol{\Phi} = \begin{bmatrix} 2 & -3 \\ 1 & 4 \end{bmatrix}, \quad \mathbf{u}(0) = \begin{Bmatrix} -1 \\ 3 \end{Bmatrix} \text{ in., } \quad \dot{\mathbf{u}}(0) = \mathbf{0}$$

(a) Determine the natural frequencies of this system. **(b)** Verify the given mode shapes and use them to determine the modal mass matrix M and the modal stiffness matrix K. **(c)** Determine the modal initial conditions $\eta_r(0)$ and $\dot{\eta}_r(0)$ ($r = 1, 2$) that correspond to the physical-coordinate initial conditions given above. **(d)** If $p_2(t)$ is the harmonic force $p_2(t) = (5 \text{ lb}) \cos 20t$, determine the modal forces $f_r(t) = \boldsymbol{\phi}_r^T \mathbf{p}(t)$ for ($r = 1, 2$). **(e)** Write down the modal equations of motion. Solve these equations for the (total) modal responses $\eta_r(t)$ for $t > 0$, that is, the sum of the steady-state responses due to $\mathbf{p}(t)$ and the responses due to the initial conditions. **(f)** Finally, determine expressions for the physical responses $u_1(t)$ and $u_2(t)$ for $t > 0$.

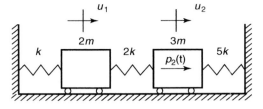

Figure P11.4

For problems whose number is preceded by a **C**, you are to write a computer program and use it to produce the plot(s) requested. *Note*: MATLAB .m-files for many of the plots in this book may be found on the book's website.

C 11.5 For the viscous-damped 2-DOF system in Example 11.2, change the stiffness coupling coefficient to $k' = 40.25$ and the damping coupling coefficient to $c' = 0.0125$. **(a)** Repeat all six steps in Example 11.2. Use MATLAB (or another computer program) to plot the required FRFs in parts (e) and (f). **(b)** Compare your results with those of Examples 11.2 and 11.3. Would you classify the frequencies in this problem to be "closely spaced"? Explain your answer. **(c)** Could the two natural frequencies be identified accurately from peaks in the plot of $\Im(\overline{H}_{11})$ versus frequency f? Could the two natural frequencies be identified accurately from peaks in the plot of $\Im(\overline{H}_{21})$ versus frequency f?

C 11.6 The viscous-damped 3-DOF system in Fig. P11.6 has dashpots that produce modal damping such that the damping factors are $\zeta_1 = \zeta_2 = \zeta_3 = 0.01$. **(a)** Determine the undamped natural frequencies

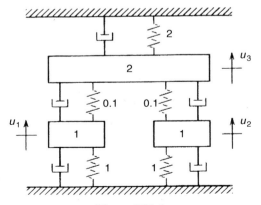

Figure P11.6

and mode shapes of this 3-DOF system. Scale the modes such that the largest element in each modal vector is 1.0. (*Hint:* You should be able to factor the characteristic equation for this 3-DOF system quite easily.) (**b**) Determine expressions for FRFs $\overline{H}_{11}$ and $\overline{H}_{31}$. (**c**) Use MATLAB (or another computer program) to create Nyquist plots of the FRFs obtained in part (b). (**d**) Use MATLAB (or another computer program) to create plots of $\Im(\overline{H}_{11})$ versus frequency f and $\Im(\overline{H}_{31})$ versus frequency f. (**e**) In a brief paragraph, describe the important features of the steady-state response that you can observe in the plots in parts (c) and (d).

Problem Set 11.3

11.7 (**a**) Starting with Eq. 11.7b, which defines the modal stiffness matrix, show that the flexibility matrix $\mathbf{K}^{-1}$ can be calculated by use of the equation

$$\mathbf{K}^{-1} = \mathbf{\Phi} K^{-1} \mathbf{\Phi}^T$$

(**b**) Show that when the damping matrix $\mathbf{C}$ satisfies the modal damping orthogonality conditions of Eq. 11.11, the damping term $\mathbf{K}^{-1}\mathbf{C}\mathbf{\Phi}\dot{\eta}$ in Eq. 11.39 reduces to the form $\sum_{r=1}^{N} 2\zeta_r/\omega_r \boldsymbol{\phi}_r \dot{\eta}_r(t)$, which is approximated by $\widehat{N}$ terms in Eq. 11.40.

11.8 The three-story building in Fig. P11.8 has the following stiffness and mass matrices, natural frequencies, and mode shapes:

$$\mathbf{K} = \begin{bmatrix} 600 & -600 & 0 \\ -600 & 1800 & -1200 \\ 0 & -1200 & 3000 \end{bmatrix} \text{ kips/in.,}$$

$$\mathbf{M} = \begin{bmatrix} 1 & 0 & 0 \\ 0 & 2 & 0 \\ 0 & 0 & 2 \end{bmatrix} \text{ kip-sec}^2/\text{in.}$$

$\omega_1^2 = 188.32, \qquad \omega_2^2 = 900.0, \qquad \omega_3^2 = 1911.7$

$$\mathbf{\Phi} = \begin{bmatrix} 1.00000 & 1.00000 & 0.31386 \\ 0.68614 & -0.50000 & -0.68614 \\ 0.31386 & -0.50000 & 1.00000 \end{bmatrix}$$

(**a**) Determine the modal mass matrix M and the modal stiffness matrix K. (**b**) If the force acting on the building is the sinusoidal force

$$\mathbf{p}(t) = \begin{Bmatrix} 100 \\ 100 \\ 100 \end{Bmatrix} \cos \Omega t \quad \text{kips}$$

determine the modal forces $f_r(t)$ ($r = 1, 2, 3$). (**c**) Determine expressions for the steady-state responses $\eta_r(t)$. (**d**) Determine the response $u_1(t)$ of the top mass by using the mode-superposition equation, Eq. 11.8. Clearly indicate the contribution from each mode as is done in Example 11.4. (**e**) Form a table, as in Example 11.4, using $\Omega = 0$, $\Omega = 0.5\omega_1$, and $\Omega = \frac{1}{2}(\omega_1 + \omega_2)$.

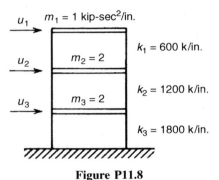

Figure P11.8

11.9 For the three-story building in Problem 11.8: (**a**) Determine F_r and D_r ($r = 1, 2, 3$) for the force distribution

$$\mathbf{p}(t) = \begin{Bmatrix} P_1 \\ 0 \\ 0 \end{Bmatrix} \cos \Omega t$$

(**b**) If the base of the building structure in Fig. P11.8 has harmonic motion $z(t) = Z \cos \Omega t$, determine the effective force $\mathbf{p}_{\text{eff}}(t)$. Using this effective force vector, determine the corresponding expressions for F_r and D_r ($r = 1, 2, 3$).

11.10 The horizontal stabilizer of a light aircraft is modeled as a 3-DOF lumped-mass system as shown in Fig. P11.10a. The stiffness and mass matrices and the natural frequencies and mode shapes are given. The airplane hits a sudden gust that produces a step force

$$\mathbf{p}(t) = \begin{Bmatrix} 500 \\ 100 \\ 100 \end{Bmatrix} f(t) \quad \text{lb}$$

where $f(t)$ is the unit-step force shown in Fig. P11.10b. (**a**) Determine expressions for the modal responses $\eta_r(t)$ to the gust if the initial conditions are: $\mathbf{v}(0) = \dot{\mathbf{v}}(0) = \mathbf{0}$. (**b**) Using the mode-superposition equation (Eq. 11.8),

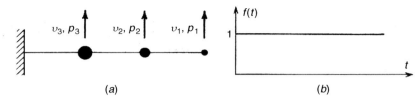

Figure P11.10

determine an expression for the tip response $v_1(t)$. Clearly indicate the contribution of each of the three modes.

The horizontal stabilizer has the following stiffness and mass matrices, natural frequencies, and mode shapes:

$$\mathbf{K} = \begin{bmatrix} 0.0656 & -0.1538 & 0.1220 \\ -0.1538 & 0.4797 & -0.5843 \\ 0.1220 & -0.5843 & 1.2593 \end{bmatrix}(10^5) \text{ lb/in.},$$

$$\mathbf{M} = \begin{bmatrix} 4 & 0 & 0 \\ 0 & 6 & 0 \\ 0 & 0 & 8 \end{bmatrix}\left(\frac{1}{386}\right) \text{ lb-sec}^2/\text{in.}$$

$\omega_1^2 = 5.96(10^4), \quad \omega_2^2 = 1.33(10^6), \quad \omega_3^2 = 8.40(10^6)$

$$\mathbf{\Phi} = \begin{bmatrix} 8.31 & -4.96 & 1.70 \\ 4.08 & 5.36 & -4.35 \\ 1.10 & 3.80 & 5.71 \end{bmatrix}$$

Problem Set 11.4

11.11 The three-story building in Problem 11.8 has the following flexibility matrix:

$$\mathbf{A} = \mathbf{K}^{-1} = \begin{bmatrix} 3.0556 & 1.3889 & 0.5556 \\ 1.3889 & 1.3889 & 0.5556 \\ 0.5556 & 0.5556 & 0.5556 \end{bmatrix}(10^{-3}) \text{ in./kip}$$

(a) Using the expressions for $\eta_r(t)$ obtained in part (c) of Problem 11.8, and using Eq. 11.49, determine the mode-acceleration expression for $u_1(t)$. Include all three modes, and indicate clearly the portion of the solution corresponding to $\widehat{N} = 1$, $\widehat{N} = 2$, and $\widehat{N} = 3$, as in Example 11.6. (b) Form a table, as in Example 11.6, using the following excitation frequencies: $\Omega = 0$, $\Omega = 0.5\omega_1$, and $\Omega = (\omega_1 + \omega_2)/2$.

Problem Set 11.5

11.12 (a) Using the modal story shears s_r given in Eq. 4 of Example 11.7 and the modal steady-state responses $\eta_r(t)$ from Example 11.4, determine expressions for the dynamic shears in the four-story building by the mode-displacement method, Eq. 11.51. (b) Using the story shears again, determine expressions for the dynamic shears by the mode-acceleration method, Eq. 11.52. Use free-body diagrams to solve for the pseudostatic shear forces. (c) Let the bottom story shear, $\sigma_4(t)$, be written in the form

$$\sigma_4(t) = CP_1 \cos \Omega t$$

Construct tables like the ones in Examples 11.4 and 11.6 that show the convergence of the amplitude of $\sigma_4(t)$ when the expressions developed in parts (a) and (b) are evaluated at the following excitation frequencies: $\Omega = 0$, $\Omega = 0.5\omega_1$, and $\Omega = 1.3\omega_3$.

Problem Set 11.6

11.13 Axial vibration of a free-free uniform bar was modeled in Problem 9.18 as a 3-DOF lumped-mass system. The system is shown in Fig. P11.13 with a force $P_1(t)$ applied to the left end. (a) Using the results of Problem 9.18, determine the modal axial forces in the two "springs" due to the two elastic modes; that is, determine

$$\mathbf{s}_2 = \begin{Bmatrix} s_1 \\ s_2 \end{Bmatrix}_2 \quad \text{and} \quad \mathbf{s}_3 = \begin{Bmatrix} s_1 \\ s_2 \end{Bmatrix}_3$$

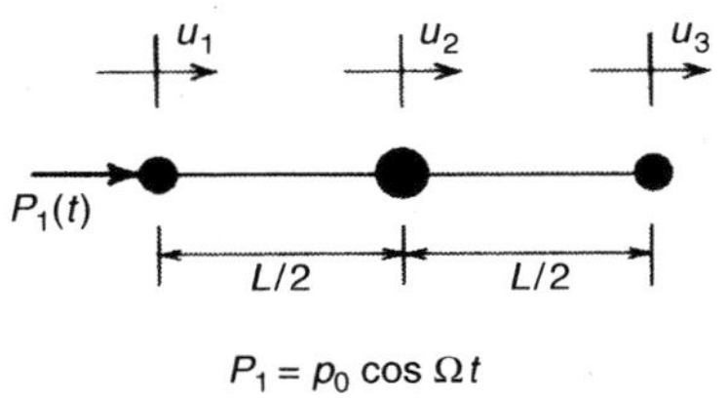

$P_1 = p_0 \cos \Omega t$

Figure P11.13

where s_1 is the axial force in the rod joining masses m_1 and m_2, and s_2 is the axial force in the rod joining masses m_2 and m_3. **(b)** Let

$$\sigma(t) = \begin{Bmatrix} \sigma_1(t) \\ \sigma_2(t) \end{Bmatrix}$$

be the dynamic axial force in the two rods. Use the mode-displacement method, Eq. 11.57, to compute the steady-state force vector $\sigma(t)$. Retain both elastic modes and indicate the contribution from each elastic mode. **(c)** Use the mode-acceleration method, Eq. 11.71, to compute the steady-state force vector $\sigma(t)$. Retain both elastic modes, and indicate the contribution from each elastic mode.

PART III

Continuous Systems

12

Mathematical Models of Continuous Systems

All structures are actually three-dimensional solid bodies, and every point in such a body, unless restrained, can displace along three mutually perpendicular directions, say x, y, and z. In Chapter 1 the distinction was made between discrete-parameter models and continuous models (also called distributed-parameter models) of structures. In Chapters 2 through 11 we considered SDOF and MDOF discrete-parameter models and their responses to various types of excitation. In the present chapter and in Chapter 13, we consider briefly structures represented by continuous models, that is, by partial-differential-equation models. The information presented in these two chapters is intended to provide you with "exact" solutions for simple structures against which you can compare solutions based on a variety of MDOF models, including the finite element models presented in Chapter 14.

Upon completion of this chapter you should be able to:

- Use Newton's Laws to derive the equation(s) of motion of "one-dimensional" bodies undergoing axial deformation, torsional deformation, bending deformation, or a combination of these.
- State appropriate mathematical boundary conditions for axial, transverse, or torsional free vibration of members having various boundary conditions.
- Use the Extended Hamilton's Principle to derive the equation(s) of motion of "one-dimensional" bodies undergoing axial deformation, torsional deformation, bending deformation, or a combination of these.

12.1 APPLICATIONS OF NEWTON'S LAWS: AXIAL DEFORMATION AND TORSION

As you study the derivation for axial deformation in Section 12.1.1 and the derivation for torsional deformation in Section 12.1.2, note the similarities in the derivations and in the resulting equations of motion and boundary conditions.

12.1.1 Axial Deformation

We consider first the axial deformation of a long, thin member, a portion of which is shown in Fig. 12.1a. To derive the equation of motion for axial vibration we isolate a

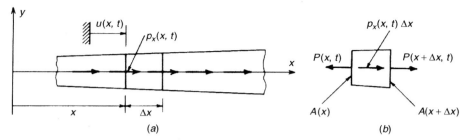

Figure 12.1 (a) Portion of a member undergoing axial deformation; (b) free-body diagram.

free-body diagram of an element of length Δx. Let $u(x, t)$ be the displacement of the cross section along the axial direction, and let $p_x(x, t)$ be the externally applied axial force per unit length with resultant lying along the centroidal axis of the member. The cross-sectional area is A, and ρ is the mass density (i.e., the mass per unit volume). It is assumed that either the member is prismatic (i.e., it has constant cross section) or that its cross section varies only slowly with x, as indicated in Fig. 12.1a.

The *axial deformation assumptions*, as treated in elementary solid mechanics (mechanics of materials, strength of materials) are (see, e.g., Chapter 3 of Ref. [12.1]):

1. The axis of the member remains straight.
2. Cross sections remain plane and remain perpendicular to the axis of the member.
3. The material is linearly elastic.
4. The material properties (E, ρ) are constant at a given cross section, but may vary with x.

Based on these assumptions, the following three equations are obtained. From the kinematics of axial deformation (assumptions 1 and 2), we have the *strain-displacement equation*, which relates the extensional strain, $\epsilon(x, t)$, to the axial displacement, $u(x, t)$.[1]

$$\epsilon(x, t) = \frac{\partial u(x, t)}{\partial x} \tag{12.1}$$

From Hooke's Law for linearly elastic materials, the normal stress σ is related to the extensional strain ϵ by the *constitutive equation*

$$\sigma = E\epsilon \tag{12.2}$$

where E is the *modulus of elasticity*, also called *Young's modulus*. The axial force, $P(x, t)$, on the cross section (Fig. 12.1b) is defined by

$$P(x, t) = \iint_A \sigma \, dA = A\sigma \tag{12.3}$$

Equations 12.1 through 12.3 can be combined to give

$$\boxed{\frac{\partial u}{\partial x} = \frac{P}{AE}} \tag{12.4}$$

---

[1] Since we are dealing with functions of two variables, a spatial coordinate x and time t, we must use partial derivatives rather than ordinary derivatives. The dependence on x and t is made explicit in Eq. 12.1 but is dropped in many of the equations that follow.

12.1 Applications of Newton's Laws: Axial Deformation and Torsion

If we apply Newton's Second Law,

$$\overset{+}{\rightarrow} \sum F_x = \Delta m a_x \qquad (12.5)$$

to the free-body diagram in Fig. 12.1b, we get

$$p_x \Delta x + P(x + \Delta x, t) - P(x, t) = \rho A \Delta x \frac{\partial^2 u}{\partial t^2} \qquad (12.6)$$

where because of the limit process to follow, the first and last terms above are taken at x rather than as averages over Δx. Dividing Eq. 12.6 by Δx and taking the limit as $\Delta x \to 0$ gives

$$\lim_{\Delta x \to 0} \frac{P(x + \Delta x, t) - P(x, t)}{\Delta x} + p_x(x, t) = \rho A \frac{\partial^2 u}{\partial t^2}$$

or

$$\frac{\partial P}{\partial x} + p_x = \rho A \frac{\partial^2 u}{\partial t^2} \qquad (12.7)$$

Equation 12.4 may be substituted into Eq. 12.7 to give

$$\boxed{\frac{\partial}{\partial x}\left(AE \frac{\partial u}{\partial x}\right) + p_x(x, t) = \rho A \frac{\partial^2 u}{\partial t^2}, \qquad 0 < x < L} \qquad (12.8)$$

This is the *differential equation of motion* for axial vibration of a linearly elastic bar.

Two common *boundary conditions*, or end conditions, are the *fixed end* and the *force-free end*. The appropriate mathematical boundary condition for a fixed end at $x = x_e$ is

$$\boxed{u(x_e, t) = 0, \qquad \text{fixed end}} \qquad (12.9a)$$

For a force-free end at $x = x_e$, the appropriate mathematical boundary condition is

$$P(x_e, t) = 0, \qquad \text{force-free end} \qquad (12.9b)$$

By using Eq. 12.4, we can write Eq. 12.9b as

$$\boxed{\left.\frac{\partial u}{\partial x}\right|_{x_e} = 0, \qquad \text{force-free end}} \qquad (12.9c)$$

Only one of the equations above, that is, Eq. 12.9a or 12.9c, may be enforced at a given end. Example 12.1 illustrates other possible types of boundary conditions.

Example 12.1 Determine the appropriate axial-deformation boundary conditions at $x = 0$ for the two members shown in Fig. 1: (a) a point mass at $x = 0$ (Fig. 1a), and (b) a linear spring at $x = 0$ (Fig. 1b).

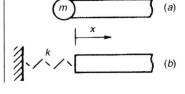

Figure 1 Two types of axial-deformation boundary condition at end $x = 0$: (a) tip mass; (b) linear spring end restraint.

SOLUTION (a) For the tip-mass problem, draw a free-body diagram of the tip mass, exposing the force exerted on the mass by the end of the bar. This is Fig. 2. (Note that end $x = 0$ is the left end of the bar. The derivations below would have different signs if we were dealing with a right end at cross section x_e.)

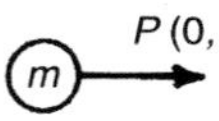

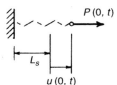

Figure 2 Free-body diagram of a tip mass at $x = 0$.

From Newton's Second Law, the equation of motion for the tip mass is

$$\overset{+}{\rightarrow} \sum F_x = P(0, t) = m \left. \frac{\partial^2 u}{\partial t^2} \right|_{x=0} \qquad (1)$$

From Eq. 12.4 we get the following expression for the force on the end of the linearly elastic bar at $x = 0$:

$$P(0, t) = \left(AE \frac{\partial u}{\partial x} \right)\bigg|_{x=0} \qquad (2)$$

Thus, the end condition for a bar with mass m at $x = 0$ is

$$\left(AE \frac{\partial u}{\partial x} \right)\bigg|_{x=0} = m \left. \frac{\partial^2 u}{\partial t^2} \right|_{x=0} \qquad \text{Ans. (a)} \quad (3)$$

which reduces to Eq. 12.8c when $m = 0$.

(b) For the bar with a linear spring at $x = 0$, sketch the spring stretched by the (tensile) force $P(0, t)$ that the spring exerts on the bar at end $x = 0$. This is Fig. 3.

Figure 3 Extension of a linear spring at $x = 0$.

From the force–elongation equation of the linear spring,

$$P(0, t) = k u(0, t) \qquad (4)$$

Incorporating Eq. 2 for the force on the end of the linearly elastic bar at $x = 0$, we get

$$\left(AE \frac{\partial u}{\partial x} \right)\bigg|_{x=0} = k u(0, t) \qquad \text{Ans. (b)} \quad (5)$$

To satisfy the axial deformation assumptions, the line of action of restraints at the ends (e.g., fixed or pinned end, mass-loaded end, spring-restrained end) must pass through the centroid of the cross section.

12.1.2 Torsional Deformation of Rods with Circular Cross Section

Newton's Laws will now be used to derive the differential equation of motion and the boundary conditions for a cantilevered rod undergoing torsional deformation. (In Section 12.3, Hamilton's Principle is used to obtain the same results.) This derivation applies only to rods with circular cross section. You will see that the results of this torsion derivation are very similar to the results just obtained in the derivation for axial deformation.

Consider the tapered rod in Fig. 12.2, which has a time-dependent externally applied torque $T_L(t)$ at end $x = L$ and a distributed externally applied torque $t_\theta(x, t)$ along its length. The angle of rotation (or twist angle) of the cross section at x is denoted by $\theta(x, t)$, positive in the sense shown in Fig. 12.2a.

The *torsional-deformation assumptions*, as treated in elementary solid mechanics (mechanics of materials, strength of materials), are (see, e.g., Section 4.2 of Ref. [12.1]):

1. The axis of the member, which is labeled the x axis, remains straight.
2. Cross sections remain plane and remain perpendicular to the axis of the member.
3. Radial lines in each cross section remain straight and radial as the cross section rotates through angle θ about the axis.
4. The material is linearly elastic. That is, $\tau = G\gamma$, where τ is *shear stress*, γ is *shear strain*, and G is the *shear modulus of elasticity*.
5. The shear modulus is constant at a given cross section but may vary with x.

The assumptions above are combined to give the *torque–twist equation* for static loading (see, e.g., Sections 4.2 and 4.3 of Ref. [12.1]):

$$\boxed{\frac{\partial \theta}{\partial x} = \frac{T}{GI_p}} \qquad (12.10)$$

where $T(x, t)$ is the twisting moment on the cross section at x, as shown in Fig. 12.2b. The polar moment of inertia of area at section x is labeled $I_p(x)$. Note the similarity in the form of Eqs. 12.10 and 12.4. Consequently, we can follow the same steps as those used above in deriving the equation of motion and boundary conditions for axial deformation.

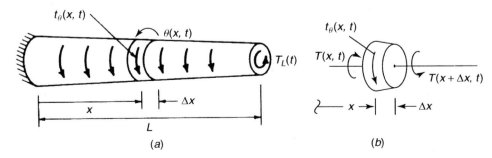

Figure 12.2 (a) Rod undergoing torsional deformation; (b) free-body diagram.

First, we apply Newton's Law for moments about the axis of the bar (Eq. 2.8b) to the free-body diagram in Fig. 12.2b.

$$\sum M_x = (\rho I_p \, \Delta x) \frac{\partial^2 \theta}{\partial t^2} \tag{12.11}$$

Here I_p is the area polar moment of inertia of the cross section of the rod and ρ is the mass density of the rod:

$$t_\theta(x, t) \, \Delta x + T(x + \Delta x, t) - T(x, t) = (\rho I_p \, \Delta x) \frac{\partial^2 \theta}{\partial t^2} \tag{12.12}$$

where, because of the limit process to follow, the first and last terms above are taken at x rather than as averages over Δx. Taking the limit as $\Delta x \to 0$ gives

$$\lim_{\Delta x \to 0} \frac{T(x + \Delta x, t) - T(x, t)}{\Delta x} + t_\theta(x, t) = \rho I_p \frac{\partial^2 \theta}{\partial t^2}$$

or

$$\frac{\partial T}{\partial x} + t_\theta = \rho I_p \frac{\partial^2 \theta}{\partial t^2} \tag{12.13}$$

Equation 12.10 may be substituted into Eq. 12.13 to give

$$\boxed{\frac{\partial}{\partial x}\left(GI_p \frac{\partial \theta}{\partial x}\right) + t_\theta(x, t) = \rho I_p \frac{\partial^2 \theta}{\partial t^2}, \qquad 0 < x < L} \tag{12.14}$$

This is the *differential equation of motion* for torsional vibration of a linearly elastic rod with circular cross section.

Two common *boundary conditions*, or end conditions, are the *fixed end* and the *torque-loaded end*. The appropriate mathematical boundary condition for a fixed end at $x = x_e$ is

$$\boxed{\theta(x_e, t) = 0, \qquad \text{fixed end}} \tag{12.15a}$$

For a torque-loaded end at $x = x_e$, with external torque $T_e(t)$ taken to be positive in the same sense as the internal torque $T(x, t)$, the appropriate mathematical boundary condition is

$$T(x_e, t) = T_e(t), \qquad \text{torque-loaded end} \tag{12.15b}$$

By using Eq. 12.10, we can write Eq. 12.15b as

$$\boxed{\left(GI_p \frac{\partial \theta}{\partial x}\right)\bigg|_{x_e} = T_e(t), \qquad \text{torque-loaded end}} \tag{12.15c}$$

Only one of the equations above, that is, Eq. 12.15a or c, may be enforced at a given end. The effect of replacing the end torque $T_L(t)$ at end $x = L$ on the rod in Fig. 12.2a by a thin disk is treated in the same manner as the tip mass in Example 12.1.

Example 12.2 Determine the appropriate torsion boundary condition at $x = L$ for the rod–disk system shown in Fig. 1.

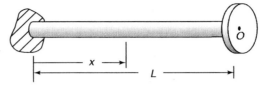

Figure 1 Torsion rod with a thin, heavy disk at end $x = L$.

SOLUTION   For the rod–disk problem, draw a free-body diagram of the disk, exposing the torque exerted on it by the end of the bar. This is Fig. 2.

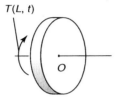

Figure 2 Free-body diagram of the disk at $x = L$.

From Newton's Law for a rotating rigid body (Eq. 2.8*b*), the equation of motion for the disk is

$$\sum M_x = -T(L, t) = I_o \ddot{\theta}(L, t) \tag{1}$$

where I_o is the mass polar moment of inertia of the disk about its center.

From Eq. 12.10 we get the following expression for the force on the end of the linearly elastic bar at $x = 0$:

$$T(L, t) = \left(GI_p \frac{\partial \theta}{\partial x} \right) \bigg|_{x=L} \tag{2}$$

Thus, the end condition for a rod with disk I_o at $x = L$ is

$$\left(GI_p \frac{\partial \theta}{\partial x} \right) \bigg|_{x=L} = -I_o \ddot{\theta}(L, t) \qquad \textbf{Ans. (3)}$$

Note the difference in signs on the right-hand side of Eq. 3 here and Eq. 3 of Example 12.1. The difference in sign is because the disk is at $x = L$, whereas the tip mass in Example 12.1 is at $x = 0$. Having a correct free-body diagram is a great aid in getting the signs right!

12.2 APPLICATION OF NEWTON'S LAWS: TRANSVERSE VIBRATION OF LINEARLY ELASTIC BEAMS (BERNOULLI–EULER BEAM THEORY)

The equation of motion of long, thin members undergoing transverse vibration may also be derived using Newton's Second Law. Figure 12.3a shows a portion of a member undergoing transverse motion (i.e., motion in the y direction), and Fig. 12.3b shows an appropriate free-body diagram. The *transverse displacement* of the point $(x, 0)$ on the neutral axis of the beam is labeled $v(x, t)$, with positive v in the $+y$ direction. The *bending moment* at section x is $M(x, t)$, the *transverse shear force* is $S(x, t)$, and the external *transverse force per unit length* is $p_y(x, t)$, with the sign convention for these specified in Fig 12.3b.

The *Bernoulli–Euler assumptions of elementary beam theory* are (see, e.g., Section 6.3 of Ref. [12.1]):

1. The x–y plane is a principal plane of the beam, and it remains plane as the beam deforms in the y direction.
2. There is an axis of the beam, which undergoes no extension or contraction. This is called the *neutral axis*, and it is labeled the x axis. The original xz plane is called the *neutral surface*.
3. Cross sections, which are perpendicular to the neutral axis in the undeformed beam, remain plane and remain perpendicular to the deformed neutral axis; that is, transverse shear deformation is neglected.
4. The material is linearly elastic, with modulus of elasticity $E(x)$; that is, the beam is homogeneous at any cross section. (Generally, E = constant throughout the beam.)
5. Stresses σ_y and σ_z are negligible compared to σ_x.

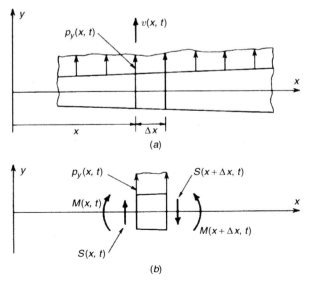

Figure 12.3 (*a*) Portion of a member undergoing transverse vibration; (*b*) free-body diagram.

12.2 Application of Newton's Laws: Transverse Vibration of Linearly Elastic Beams

In addition to the assumptions above, which are made in the static analysis of beams, the following dynamics assumptions will be made:

6. The rotatory inertia of the beam may be neglected in the moment equation.
7. The mass density, $\rho(x)$, is constant at each cross section, so that the mass center coincides with the centroid of the cross section. (Generally, $\rho =$ constant throughout the beam.)

From kinematics based on assumptions 2 and 3 (see, e.g., Section 6.2 of Ref. [12.1]), the extensional strain, $\epsilon\ (x, y, t)$, can be related to the curvature, $1/\mu(x, t)$, of the beam by[2]

$$\epsilon = -\frac{y}{\mu} \tag{12.16}$$

where y is distance in the cross section measured from the neutral surface. Then, for a linearly elastic beam whose properties are independent of position in the cross section, the bending moment can be related to the curvature by the *moment–curvature equation* (see, e.g., Section 6.3 of Ref. [12.1])

$$M(x, t) = \frac{EI}{\mu} \tag{12.17}$$

where I is the area moment of inertia of the cross section.

The equations of motion for the mass Δm in the free-body diagram in Fig. 12.3b may be derived by use of Newton's Laws. Thus,

$$\uparrow + \sum F_y = \Delta m a_y \tag{12.18}$$

and

$$\circlearrowleft + \sum M_G = \Delta I_G \alpha \tag{12.19}$$

where G is the mass center of Δm and α is the angular acceleration. However, it was stated in assumption (6) that rotatory inertia would be neglected in the moment equation, so Eq. 12.19 reduces to

$$\circlearrowleft + \sum M_G = 0 \tag{12.20}$$

Applying Eq. 12.18 to the free-body diagram in Fig. 12.3b, we obtain

$$S(x, t) - S(x + \Delta x, t) + p_y(x, t)\,\Delta x = \rho A\,\Delta x\,\frac{\partial^2 v}{\partial t^2} \tag{12.21}$$

By dividing Eq. 12.21 by Δx and taking the limit as $\Delta x \to 0$, we obtain

$$-\frac{\partial S}{\partial x} + p_y(x, t) = \rho A\,\frac{\partial^2 v}{\partial t^2} \tag{12.22}$$

From the moment equation, Eq. 12.20, we obtain, in a similar manner, the equation

$$S = \frac{\partial M}{\partial x} \tag{12.23}$$

---

[2] In most elementary mechanics of solids books the symbol that is used for the radius of curvature of a beam is the Greek letter ρ. In this structural dynamics book, however, the radius of curvature is called μ, since the Greek letter ρ is used for mass density.

If the slope of the beam remains small, that is, if $\partial v/\partial x \ll 1$, the curvature may be approximated by $\partial^2 v/\partial x^2$ (see, e.g., Section 7.2 of Ref. [12.1]). Then the moment–curvature equation, Eq. 12.17, becomes

$$M(x, t) = EI \frac{\partial^2 v}{\partial x^2} \quad (12.24)$$

Combining Eqs. 12.22, 12.23, and 12.24, we get

$$\boxed{\frac{\partial^2}{\partial x^2}\left(EI \frac{\partial^2 v}{\partial x^2}\right) + \rho A \frac{\partial^2 v}{\partial t^2} = p_y(x, t), \quad 0 < x < L} \quad (12.25)$$

This is the *differential equation of motion* governing transverse vibration of a beam that satisfies the assumptions stated above. It is valid only for beams that are relatively long and thin. In Example 12.4 the derivation above is modified account for an axial preload on the beam. In Section 12.4 shear deformation and rotary inertia are added.

The *boundary conditions* most frequently encountered in analyzing vibration of Bernoulli–Euler beams are the following:

1. Fixed end at $x = x_e$: The displacement and slope vanish at a fixed end; that is,

$$\boxed{v(x_e, t) = 0, \quad \text{and} \quad \left.\frac{\partial v}{\partial x}\right|_{x=x_e} = 0} \quad \text{Fixed end} \quad (12.26a, b)$$

2. Simply supported end at $x = x_e$: The displacement and bending moment vanish at a simply supported end; that is,[3]

$$v(x_e, t) = 0 \quad \text{and} \quad M(x_e, t) = 0 \quad (12.27a, b)$$

Using Eq. 12.24, we can express the bending moment boundary condition in terms of v. Therefore, the two boundary conditions for a simply supported end are

$$\boxed{v(x_e, t) = 0 \quad \text{and} \quad \left.\frac{\partial^2 v}{\partial x^2}\right|_{x=x_e} = 0} \quad \begin{array}{l}\text{Simply}\\\text{supported}\\\text{end}\end{array} \quad (12.28a, b)$$

3. Force-free end at $x = x_e$: The transverse shear and the bending moment vanish at a free end, so

$$S(x_e, t) = 0 \quad \text{and} \quad M(x_e, t) = 0 \quad (12.29a, b)$$

Equations 12.23 and 12.24 may be combined with Eqs. 12.29a and b to give the following two boundary conditions for a force-free end:

$$\boxed{\left.\frac{\partial}{\partial x}\left(EI \frac{\partial^2 v}{\partial x^2}\right)\right|_{x=x_e} = 0 \quad \text{and} \quad \left.\frac{\partial^2 v}{\partial x^2}\right|_{x=x_e} = 0} \quad \text{Free end} \quad (12.30a, b)$$

---

[3]The moment is not necessarily zero at a simply supported end, but this is generally the case. Otherwise, $M_{(x_e, t)} = M_e(t) =$ given moment.

12.2 Application of Newton's Laws: Transverse Vibration of Linearly Elastic Beams

In summary, at each end of a beam two end conditions are required. Different end conditions arise when, for example, a lumped mass or a spring is attached to the end of the beam. Because signs are so important, the best procedure to employ in such cases is to draw careful free-body diagrams employing the sign convention of Fig. 12.3b. This is illustrated by the following example.

Example 12.3 Determine the appropriate transverse-vibration boundary conditions if a point mass m is attached to the end of the beam at $x = L$, as indicated in Fig. 1.

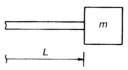

Figure 1 Beam with the tip mass at $x = L$.

SOLUTION Isolate a free-body diagram of the mass m (Fig. 2). Use the sign convention of Fig. 12.3b.

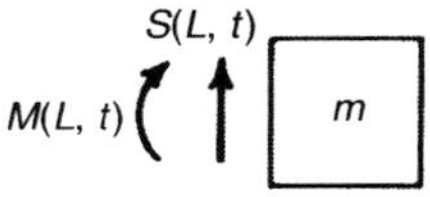

Figure 2 Free-body diagram of the tip mass at $x = L$.

Apply Newton's Second Law to this "particle."

$$\uparrow + \sum F_y = ma_y \tag{1}$$

From the free-body diagram,

$$S(L, t) = m \left. \frac{\partial^2 v}{\partial t^2} \right|_{x=L} \tag{2}$$

Combine Eqs. 2, 12.23, and 12.24 to get

$$\left. \frac{\partial}{\partial x} \left(EI \frac{\partial^2 v}{\partial x^2} \right) \right|_{x=L} = m \left. \frac{\partial^2 v}{\partial t^2} \right|_{x=L} \qquad \text{Ans. (3)}$$

Since the particle has no rotatory inertia,

$$\circlearrowleft + \sum M_G = 0 \tag{4}$$

Then

$$M(L, t) = 0 \tag{5}$$

or

$$\left. \frac{\partial^2 v}{\partial x^2} \right|_{x=L} = 0 \qquad \text{Ans. (6)}$$

Equations 3 and 6 are the appropriate end conditions for a beam with a mass m at end $x = L$, with the rotational inertia I_G of mass m neglected.

Example 12.4 Determine the equation of motion for a beam that is subjected to a compressive end load N that remains parallel to the x axis, as shown in Fig. 1. Neglect axial strain. (*Note:* Coupled axial-bending motion is considered in Problem 12.3.)

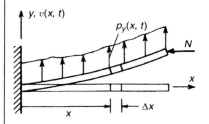

Figure 1 Beam with a tip force at $x = L$.

SOLUTION   1. Draw a free-body diagram of a beam element of length Δx. To account for the effect of the constant axial preload N, it will be necessary to draw a free-body diagram of the deformed element, as shown in Fig. 2.

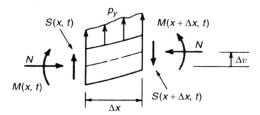

Figure 2 Free-body diagram of a deformed beam element.

2. Write the equations of motion. As in the previous derivation in Example 12.3 for a beam without axial pre-load,

$$\uparrow + \sum F_y = ma_y \tag{1}$$

$$\circlearrowleft + \sum M_G = 0 \tag{2}$$

As in Eq. 12.22, Eq. 1 leads to

$$-\frac{\partial S}{\partial x} + p_y(x, t) = \rho A \frac{\partial^2 v}{\partial t^2} \tag{3}$$

Equation 2 gives

$$M(x + \Delta x, t) - M(x, t) + N[v(x + \Delta x, t) - v(x, t)] - S(x + \Delta x, t)\Delta x = 0 \tag{4}$$

Dividing Eq. 4 by Δx and taking the limit as $\Delta x \to 0$, we get

$$\frac{\partial M}{\partial x} + N \frac{\partial v}{\partial x} = S \tag{5}$$

Equations 3 and 5 may now be combined to give

$$\frac{\partial^2 M}{\partial x^2} + N\frac{\partial^2 v}{\partial x^2} + \rho A \frac{\partial^2 v}{\partial t^2} = p_y(x,t) \quad (6)$$

Finally, Eq. 12.24 can be incorporated into Eq. 6 giving the final equation of motion for a beam with constant axial force.

$$\frac{\partial^2}{\partial x^2}\left(EI\frac{\partial^2 v}{\partial x^2}\right) + N\frac{\partial^2 v}{\partial x^2} + \rho A \frac{\partial^2 v}{\partial t^2} = p_y(x,t) \quad \text{Ans. (7)}$$

12.3 APPLICATION OF HAMILTON'S PRINCIPLE: TORSION OF A ROD WITH CIRCULAR CROSS SECTION

The study of dynamics is frequently subdivided into *Newtonian Mechanics* and *Analytical Mechanics*. The latter is also referred to as the study of *Energy Methods*. The Principle of Virtual Displacements, Hamilton's Principle, and Lagrange's Equations are topics treated in Analytical Mechanics. In Sections 2.4 and 2.5, the Principle of Virtual Displacements was employed in deriving the equations of motion of various SDOF systems. The Extended Hamilton's Principle was introduced in Section 8.2, where it was used to derive Lagrange's Equations. Since it employs kinetic energy and potential energy, which are scalar quantities, the Extended Hamilton's Principle provides a powerful method that can be used directly to derive the equations of motion of continuous systems. Use of the Extended Hamilton's Principle provides an important bonus: namely, all boundary conditions are handled systematically in the process of obtaining the equation(s) of motion.[4]

In the formulation of the Extended Hamilton's Principle, virtual displacements, or virtual changes of configuration, are employed. Figure 8.2 shows a cantilever beam with a virtual change of configuration $\delta v(x,t) \equiv \delta[v(x,t)]$. As indicated in Sections 2.4 and 8.2, a virtual displacement is <u>not a function of time</u>, it is an imaginary, infinitesimal change in the quantity that describes the configuration of a system at some time [e.g., a change in $v(x,t)$ or in $\theta(t)$]. We abbreviate the notation from $\delta[v(x,t)]$ to $\delta v(x,t)$, as indicated. As noted in Section 8.2, the virtual change of configuration must satisfy all geometric boundary conditions. Hamilton also assumed that the configuration is specified at times t_1 and t_2. For example, for transverse vibration of a beam this would imply that $\delta v(x,t_1) = \delta v(x,t_2) = 0$.

The *Extended Hamilton's Principle* may be stated as

$$\boxed{\int_{t_1}^{t_2} \delta(\mathcal{T} - \mathcal{V})\,dt + \int_{t_1}^{t_2} \delta\mathcal{W}_{nc}\,dt = 0} \quad (12.31)$$

---

[4] See Ref. [12.2], pp. 168–170, for a discussion of the important role played by energy methods in early attempts to formulate a theory for the bending of flat plates.

where

T = total kinetic energy of the system.
V = potential energy of the system, including the strain energy and the potential energy of conservative external forces.
δW_{nc} = virtual work done by nonconservative forces, including damping forces and external forces not accounted for in V.
$\delta[\,\cdot\,]$ = symbol denoting the first variation, or virtual change, in the quantity in brackets.
t_1, t_2 = times at which the configuration of the system is assumed to be known.

In this section, use of the Extended Hamilton's Principle is illustrated by a derivation of the differential equation of motion and the boundary conditions for a cantilevered rod undergoing torsional deformation. (Newton's Laws were used in Section 12.1 to obtain essentially the same results.) The tapered rod in Fig. 12.4 is undergoing torsional deformation due to distributed externally applied torque $t_\theta(x, t)$. In addition, a torsional moment $T_L(t)$ is applied through a pulley that is attached to the end of the rod at $x = L$. The rotation of the cross section at x is denoted by $\theta(x, t)$. As was the case for the torsion equations developed in Section 12.1, the torsion equations developed here apply only to rods with circular cross section.

The rod is attached to a rigid wall at $x = 0$, but it is free to rotate at end $x = L$. Therefore, the only *geometric boundary condition* (i.e., displacement-type boundary condition) is

$$\boxed{\theta(0, t) = 0} \tag{12.32}$$

The potential energy and the kinetic energy for a member undergoing torsion are given by

$$V = \tfrac{1}{2} \int_0^L GJ(x)[\theta'(x, t)]^2 \, dx \tag{12.33}$$

and

$$T = T_{\text{rod}} + T_{\text{pulley}} = \tfrac{1}{2} \int_0^L \rho I_p(x)[\dot\theta(x, t)]^2 \, dx + \tfrac{1}{2} I_0 [\dot\theta(L, t)]^2 \tag{12.34}$$

respectively, where the spatial partial derivative and the time partial derivative are symbolized by

$$(\,\cdot\,)' \equiv \frac{\partial(\,\cdot\,)}{\partial x}, \qquad (\dot{\,\cdot\,}) \equiv \frac{\partial(\,\cdot\,)}{\partial t}$$

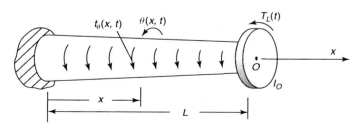

Figure 12.4 Rod undergoing torsional deformation.

12.3 Application of Hamilton's Principle: Torsion of a Rod with Circular Cross Section

respectively, and where I_O is the mass polar moment of inertia of the pulley. For the rod with circular cross section, the polar moment of inertia of the cross-sectional area is[5]

$$I_p(x) = \iint r^2 \, dA = \frac{\pi R^4}{2} \tag{12.35}$$

where $R(x)$ is the radius of the rod at section x. For the torsion rod in Fig. 12.4, the virtual work of the external torques is

$$\delta W_{nc} = \int_0^L t_\theta(x,t) \, \delta\theta(x,t) \, dx + T_L \, \delta\theta(L,t) \tag{12.36}$$

Substituting Eqs. 12.33, 12.34, and 12.36 into Eq. 12.31, we get the following expression for the Extended Hamilton's Principle:

$$\int_{t_1}^{t_2} \delta \left[\frac{1}{2} \int_0^L \rho I_p(\dot\theta)^2 \, dx + \frac{1}{2} I_O [\dot\theta(L,t)]^2 - \frac{1}{2} \int_0^L GJ(\theta')^2 \, dx \right] dt \\ + \int_{t_1}^{t_2} \left[\int_0^L t_\theta(x,t) \, \delta\theta(x,t) \, dx + T_L \, \delta\theta(L,t) \right] dt = 0 \tag{12.37}$$

As in Section 2.4, the virtual change in strain energy can be written

$$\delta V = \int_0^L GJ\theta' \, \delta\theta' \, dx \tag{12.38}$$

and similarly for the kinetic energy,

$$\delta T = \int_0^L \rho I_p \dot\theta \, \delta\dot\theta \, dx + I_O \dot\theta(L,t) \, \delta\dot\theta(L,t) \tag{12.39}$$

It is necessary to remove the $(\cdot)'$ and $(\dot{\cdot})$ from the $\delta\theta'$ and $\delta\dot\theta$ terms. This is done by integration by parts. Integrating by parts with respect to x, we get

$$\int_{t_1}^{t_2} \delta V \, dt = \int_{t_1}^{t_2} \left(\int_0^L GJ\theta' \, \delta\theta' \, dx \right) dt \\ = \int_{t_1}^{t_2} \left[(GJ\theta') \, \delta\theta \Big|_0^L - \int_0^L (GJ\theta')' \, \delta\theta \, dx \right] dt \tag{12.40}$$

and integrating by parts with respect to t gives

$$\int_{t_1}^{t_2} \delta T \, dt = \int_0^L \left(\int_{t_1}^{t_2} \rho I_p \dot\theta \, \delta\dot\theta \, dt \right) dx + \int_{t_1}^{t_2} I_O \dot\theta(L,t) \, \delta\dot\theta(L,t) \, dt \\ = \int_0^L \left[(\rho I_p \dot\theta \, \delta\theta) \Big|_{t_1}^{t_2} - \int_{t_1}^{t_2} (\rho I_p \ddot\theta \, \delta\theta) \, dt \right] dx \\ + [I_O \dot\theta(L,t) \, \delta\theta(L,t)] \Big|_{t_1}^{t_2} - I_O \ddot\theta(L,t) \, \delta\theta(L,t) \tag{12.41}$$

---

[5] For a circular cross section, the torsion constant J is equal to the polar moment of inertia I_p, that is, $J = I_p$. For other cross-sectional shapes $I_p = \iint r^2 \, dA \neq J$. The reader should consult texts on advanced strength of materials or theory of elasticity to obtain expressions for the torsional constant J for noncircular cross-sectional areas.

In Eq. 12.41 the terms evaluated at t_1 and t_2 involve $\delta\theta(x, t_1)$, $\delta\theta(x, t_2)$, $\delta\theta(L, t_1)$, and $\delta\theta(L, t_2)$. In the formulation of the Extended Hamilton's Principle, it is assumed that the initial and final configurations are known, that is, $\theta(x, t_1)$, and so on, are known. Thus, $\delta\theta(x, t_1) = \delta\theta(x, t_2) = \delta\theta(L, t_1) = \delta\theta(L, t_2) = 0$, so the integrated terms on the right-hand side of Eq. 12.41 vanish. Combining Eqs. 12.37, 12.40, and 12.41, we obtain

$$\int_{t_1}^{t_2} \int_0^L [(GJ\theta')' - \rho I_p \ddot{\theta} + t_\theta(x,t)]\delta\theta \, dx \, dt$$
$$+ \int_{t_1}^{t_2} \{[(GJ\theta')\delta\theta]_{x=0} - [(GJ\theta' - T_L)\delta\theta]_{x=L} - I_o\ddot{\theta}(L,t)\,\delta\theta(L,t)\}\,dt = 0 \quad (12.42)$$

Because of the geometric boundary condition in Eq. 12.32, we must set $\delta\theta(0,t) = 0$, so the first boundary condition term in Eq. 12.42 vanishes. Since $\theta(L,t)$ is not prescribed, $\delta\theta(L,t)$ is completely arbitrary. Therefore, collecting the boundary terms that are evaluated at $x = L$, we obtain the natural boundary condition

$$\boxed{(GJ\theta')_{x=L} + I_o\ddot{\theta}(L,t) = T_L(t)} \quad (12.43)$$

Compare this with Eq. 3 in Example 12.2, where there was no external moment applied to the disk.

A *natural boundary condition* is a boundary condition that occurs when no geometric boundary condition is prescribed on a portion of the boundary. Note how the natural boundary condition arises automatically as part of the integration-by-parts evaluation of boundary terms when the Extended Hamilton's Principle is used to derive equations of motion. Natural boundary conditions provide boundary conditions for force-type quantities, like the net end torque $[T_L(t) - I_o\ddot{\theta}(L,t)]$ in Eq. 12.43.

Over the interval $0 < x < L$, $\delta\theta(x,t)$ is arbitrary. Hence, the first term in Eq. 12.42 leads to the *partial differential equation of motion* of the torsion rod.

$$\boxed{-(GJ\theta')' + \rho I_p \ddot{\theta} = t_\theta(x,t), \qquad 0 < x < L} \quad (12.44)$$

In summary, then, the motion of the torsion member in Fig. 12.4 is governed by its *partial differential equation of motion* (Eq. 12.44), together with a *geometric boundary condition* at $x = 0$ (Eq. 12.32) and a *natural boundary condition* at $x = L$ (Eq. 12.43).

12.4 APPLICATION OF THE EXTENDED HAMILTON'S PRINCIPLE: BEAM FLEXURE INCLUDING SHEAR DEFORMATION AND ROTATORY INERTIA (TIMOSHENKO BEAM THEORY)

In Section 12.2 the equation of motion was derived for transverse vibration of a long, thin beam, for which it is valid to ignore shear deformation and rotatory inertia. In this section the Extended Hamilton's Principle is used in deriving the equations of motion and boundary conditions which include these effects, so that vibration of shorter, stubbier beams can be analyzed.

Figure 12.5 shows the kinematics of deformation of a beam that undergoes shear deformation in addition to pure bending. The transverse displacement of the neutral

12.4 Application of the Extended Hamilton's Principle: Timoshenko Beam Theory

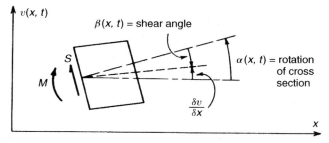

Figure 12.5 Kinematics of deformation of a beam including shear deformation.

axis of the beam is designated $v(x, t)$, and $\alpha(x, t)$ is the angle of rotation of the cross section at x. From Fig. 12.5, the following kinematical relationship is obtained:

$$\beta = \alpha - \frac{\partial v}{\partial x} \tag{12.45}$$

where β is the shear angle. From elementary beam theory (e.g., see Sections 6.2 and 6.3 of Ref. [12.1]), the *moment–curvature equation* is

$$M = EI\alpha' \tag{12.46}$$

and the corresponding *bending strain energy* is

$$V_b = \tfrac{1}{2} \int_0^L EI(\alpha')^2 \, dx \tag{12.47}$$

The *shear strain energy* can be expressed as

$$V_s = \tfrac{1}{2} \int_0^L \kappa GA\beta^2 \, dx \tag{12.48}$$

where the shear coefficient κ can be obtained by computing the strain energy due to shear using

$$V_s = \tfrac{1}{2} \int_0^L \left(\iint_A \tau \gamma \, dA \right) dx \tag{12.49}$$

For rectangular beams, the shear coefficient is $\kappa = \tfrac{5}{6}$.

The kinetic energy of the beam is

$$T = \tfrac{1}{2} \int_0^L \rho A (\dot v)^2 \, dx + \tfrac{1}{2} \int_0^L \rho I (\dot\alpha)^2 \, dx \tag{12.50}$$

The first term is the kinetic energy due to translation, and the second term is the *rotatory inertia term*. For a beam with distributed transverse load $p_y(x, t)$, the virtual work due to the transverse load is given by

$$\delta W_{nc} = \int_0^L p_y(x, t) \, \delta v(x, t) \, dx \tag{12.51}$$

It seems logical that geometric boundary conditions can be imposed independently on the translational displacement, $v(x, t)$, and on the rotation of the cross section, $\alpha(x, t)$. Hence, v and α will be retained as the unknown displacement variables, and Eq. 12.45 can be used to eliminate the shear angle, $\beta(x, t)$. Then Eqs. 12.47, 12.48, 12.50, and 12.51 can be substituted into the Extended Hamilton's Principle equation, Eq. 12.31, to give

$$\frac{1}{2}\int_{t_1}^{t_2}\int_0^L \delta[\rho A(\dot{v})^2 + \rho I(\dot{\alpha})^2 - EI(\alpha')^2 - \kappa GA(\alpha - v')^2]\,dx\,dt$$
$$+ \int_{t_1}^{t_2}\int_0^L p_y\,\delta v\,dx\,dt = 0 \tag{12.52}$$

Integrating by parts, as was done in Section 12.3, and noting that $\delta v(x, t_1) = \delta v(x, t_2) = \delta\alpha(x, t_1) = \delta\alpha(x, t_2) = 0$, we obtain

$$\int_{t_1}^{t_2}\int_0^L \{-\rho A\ddot{v} - [\kappa GA(\alpha - v')]' + p_y\}\,\delta v\,dx\,dt$$
$$+ \int_{t_1}^{t_2}\int_0^L [-\rho I\ddot{\alpha} + (EI\alpha')' - \kappa GA(\alpha - v')]\,\delta\alpha\,dx\,dt \tag{12.53}$$
$$+ \int_{t_1}^{t_2} [\kappa GA(\alpha - v')\,\delta v]\Big|_0^L\,dt - \int_{t_1}^{t_2} [(EI\alpha')\,\delta\alpha]\Big|_0^L\,dt = 0$$

Since δv and $\delta\alpha$ are arbitrary except where geometric boundary conditions are prescribed, where they are zero, Eq. 12.53 leads to the following two *partial differential equations of motion*:

$$\boxed{\begin{aligned}[\kappa GA(\alpha - v')]' + \rho A\ddot{v} &= p_y(x, t) \\ \kappa GA(\alpha - v') - (EI\alpha')' + \rho I\ddot{\alpha} &= 0\end{aligned}} \tag{12.54}$$

The following *generalized boundary conditions* are also obtained from Eq. 12.53:

$$\boxed{\begin{aligned}(\kappa GA\beta)\,\delta v &= 0 & \text{at } x &= 0 \\ (\kappa GA\beta)\,\delta v &= 0 & \text{at } x &= L \\ (EI\,\alpha')\,\delta\alpha &= 0 & \text{at } x &= 0 \\ (EI\,\alpha')\,\delta\alpha &= 0 & \text{at } x &= L\end{aligned}} \tag{12.55}$$

Each of the four boundary conditions above must be satisfied either as a *geometric boundary condition*, that is, with v (or α) specified so that $\delta v = 0$ (or $\delta\alpha = 0$), or as a *natural boundary condition*, that is, with $\beta = 0$ (or $\alpha' = 0$).

If the beam has uniform cross-sectional properties, Eqs. 12.54 may be combined to give a single equation in v. From Eq. 12.54a we get

$$\alpha' = v'' + \frac{1}{\kappa GA}(p_y - \rho A\ddot{v}) \tag{12.56}$$

Differentiating Eq. 12.54b and substituting Eq. 12.56 into the resulting equation, we get

$$\underbrace{EI\frac{\partial^4 v}{\partial x^4} - \left(p_y - \rho A \frac{\partial^2 v}{\partial t^2}\right)}_{\text{Bernoulli–Euler theory}} \underbrace{- \rho I \frac{\partial^4 v}{\partial x^2 \partial t^2}}_{\text{principal rotatory inertia term}}$$

$$+ \underbrace{\frac{EI}{\kappa GA}\frac{\partial^2}{\partial x^2}\left(p_y - \rho A \frac{\partial^2 v}{\partial t^2}\right)}_{\text{principal shear deformation term}} \underbrace{- \frac{\rho I}{\kappa GA}\frac{\partial^2}{\partial t^2}\left(p_y - \rho A \frac{\partial^2 v}{\partial t^2}\right)}_{\text{combined rotatory inertia and shear deformation}} = 0 \quad (12.57)$$

From Eq. 12.57, it is possible to identify the terms of Bernoulli–Euler theory, Eq. 12.25, and the additional correction terms that account for shear deformation and rotatory inertia.

REFERENCES

[12.1] R. R. Craig, Jr., *Mechanics of Materials*, 2nd ed., Wiley, New York, 2000.
[12.2] H. L. Langhaar, *Energy Methods in Applied Mechanics*, Wiley, New York, 1962.

PROBLEMS

Problem Set 12.1

12.1 The bar system shown in Fig. P12.1a consists of two long "planks" of length L and cross-sectional dimensions $b \times h$ that are securely bonded together at their interface. The materials have moduli of elasticity E_1 and E_2 and mass densities ρ_1 and ρ_2, respectively. (**a**) By carrying out a complete derivation similar to the one in Section 12.1, show that axial deformation (the axis remains straight; cross sections remain plane and remain perpendicular to the axis) is possible if $E_1/E_2 = \rho_1/\rho_2$. *Note*: You will need to replace Eq. 12.3 with the two equations

$$P_1 = A_1 \sigma_1, \qquad P_2 = A_2 \sigma_2$$

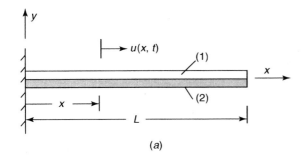

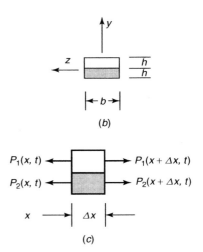

Figure P12.1 (*a*) Two-material bar undergoing axial deformation; (*b*) cross section of the bar; (*c*) free-body diagram of the bar segment from x to $(x + \Delta x)$.

In addition to the $\sum F$ equation in Eq. 12.5, you will need to use the free-body diagram in Fig. P12.1b and write an equation for $\sum M$. In these equations, include the inertia terms from the upper block and the lower block as two separate terms. (**b**) Give the equation of motion for axial deformation of this system. For homogeneous bars, your answer should reduce to Eq. 12.8.

12.2 A shaft consists of two identical cylindrical segments, each of length L, that are fixed to rigid "walls" at their ends and are welded in the middle to a disk of mass M and radius R (Fig. P12.2). Refer to the left-hand segment of shaft as segment 1, and the right-hand segment as segment 2. For each segment of shaft there will be a torsional equation of motion like Eq. 12.14, one in terms of $\theta_1(x_1, t)$ and the other in terms of $\theta_2(x_2, t)$. **(a)** State the fixed-end boundary condition for shaft 1 at $x_1 = 0$, and state the fixed-end boundary condition for shaft 2 at $x_2 = L$. **(b)** State the displacement-type boundary condition where the disk is attached to the two shafts. That is, relate θ_1 at $x_1 = L$ and θ_2 at $x_2 = 0$. **(c)** Finally, one more boundary condition is required. Sketch a free-body diagram of the disk, and derive the torque-type (natural) boundary condition of the type that is given for a single torsion rod by Eq. 3 of Example 12.2. Give special attention to the proper sign conventions for torques, rotation angles, and so on, as you draw your free-body diagram of the disk.

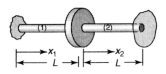

Figure P12.2

Problem Set 12.2

12.3 The effect of a tip mass on the boundary conditions for axial vibration was discussed in Example 12.1 and for bending in Example 12.3. If the mass is located eccentrically, as in Fig. P12.3, it couples the axial vibration and the bending vibration. There will be an equation of motion for axial vibration, such as Eq. 12.8, and there will also be an equation of motion for transverse vibration, such as Eq. 12.25. Using an appropriate free-body diagram of the mass m and the bar that attaches it to the end of the beam, determine the boundary conditions that couple axial and transverse motion. You will also need to draw a kinematical diagram of the mass m and the bar that attaches it to the end of the beam, and write one or more kinematical equations that relate the x and y motions and slope of the beam at $x = L$ to the x and y motions of the mass m.

12.4 A uniform, thin, rigid bar AB having mass M and length b is attached to a uniform flexible beam BC. Assume a small transverse displacement $v(x, t)$. Using appropriate diagrams (free-body and kinematics), determine the boundary conditions for the beam at B and C, that is, at $x = 0$ and at $x = L$.

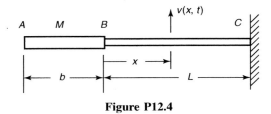

Figure P12.4

Problem Set 9.3

12.5 A small mass, $m = \rho AL/4$, is attached to the tip of a uniform beam by a rigid, weightless arm of length e as shown in Fig. P12.3. Consider vibration in the x–y plane only. Use *Hamilton's Principle* to derive the differential equations of motion (axial and transverse) and the boundary conditions for the system.

12.6 An airplane wing with a tip tank is modeled as a uniform cantilever beam (EI = constant, ρA = constant) with a lumped mass M attached at the axis of the beam at $x = L$ (Fig. P12.6). Use *Hamilton's Principle* to derive the differential equation of motion and the boundary conditions for transverse vibration of the wing (beam).

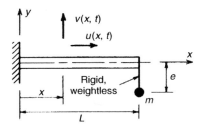

Figure P12.3

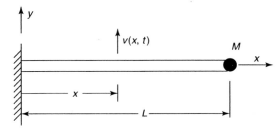

Figure P12.6

12.7 The beam shown in Fig. P12.7 consists of two long "planks" of length L and cross-sectional dimensions ($b \times h$) that are securely bonded together at their interface, which is taken to be the x–z plane. The materials have moduli of elasticity E_1 and E_2 and mass densities ρ_1 and ρ_2, respectively. (**a**) Use *Hamilton's Principle* to carry out a complete derivation of the two (coupled) differential equations of motion for axial motion $u(x, t)$ and transverse motion $v(x, t)$, and of the boundary conditions at $x = 0$ and at $x = L$. In your derivation you will need to carry out integrations with respect to y as well as with respect to x, since the properties of the beam are not constant throughout the depth of the beam. (**b**) Show that axial motion and transverse motion are uncoupled if $E_1/E_2 = \rho_1/\rho_2$.

12.8 The uniform cantilever rod in Fig. P12.8 has torsional stiffness GJ, vertical bending stiffness EI, mass per unit length ρA, and torsional mass moment of inertia ρI_p. At end $x = L$ a rigid, weightless bar BC is attached parallel to the y-axis. At C a concentrated mass m is attached to bar BC, and also at C a load $P(t)$ acts parallel to the z-axis. Assume that rod AB bends in the vertical (i.e., x–z) plane with deflection $v(x, t)$ and twists about its axis with angle of twist $\theta(x, t)$. Assume all deflections to be small (e.g., m moves approximately vertically if θ is small), and neglect gravity. Use the *Extended Hamilton's Principle* to derive the equations of motion and the boundary conditions for this system.

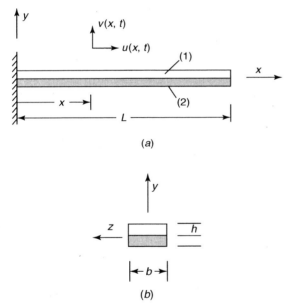

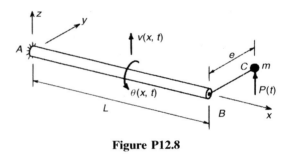

Figure P12.8

Figure P12.7 (*a*) Two-material beam undergoing axial and bending deformation; (*b*) cross section of the beam.

12.9 Use *Hamilton's Principle* to derive the equation of motion and the boundary conditions for the flexible beam BC in Problem 12.4.

13
Free Vibration of Continuous Systems

In Chapter 12 the equations of motion and boundary conditions were derived for several continuous models of "one-dimensional" structures, that is, structures whose deformation depends on one spatial variable, labeled x, and time, t. In this chapter the important concepts of natural frequency and mode shape are discussed. You will find that these are essentially the same concepts of natural frequency and mode shape that were obtained for undamped multi-degree-of-freedom models in Chapters 9 and 10. The results that are obtained in this chapter for uniform members will serve as "exact solutions" against which the MDOF results obtained in Chapters 9, 10, and 14 can be compared. Forced vibration of continuous systems is not treated in this book, since very few practical structures can be conveniently modeled as uniform continuous members. Instead, finite element models, which are introduced in Chapter 14, are employed to determine the dynamic behavior of complex structures.

Upon completion of this chapter you should be able to:

- Solve for the natural frequencies and mode shapes for axial, transverse, or torsional free vibration of uniform members with specified boundary conditions.
- Discuss conditions under which shear deformation and/or rotatory inertia should be included in considering transverse vibration of beams.
- Discuss the relationship of frequencies of axial vibration and transverse vibration of a uniform beam.
- Use the Rayleigh method to approximate the fundamental frequency of a member, and state the relationship of the approximate frequency to the exact frequency.
- Derive the orthogonality equations for axial, transverse, or torsional vibration of bars.
- Describe three procedures for normalizing mode shapes. Define generalized stiffness and generalized mass.

13.1 FREE AXIAL AND TORSIONAL VIBRATION

As noted in Chapter 12, axial free vibration and torsional free vibration are governed by virtually identical equations.

13.1.1 Free Axial Vibration

The equation of motion for axial vibration of a bar with distributed axial load is given by Eq. 12.8. For free vibration this reduces to

$$(AEu')' - \rho A \ddot{u} = 0 \tag{13.1}$$

Assume harmonic motion given by the equation

$$u(x,t) = U(x)\cos(\omega t - \alpha) \tag{13.2}$$

Substitute this into Eq. 13.1 to obtain the *eigenvalue differential equation*

$$(AEU')' + \omega^2(\rho A U) = 0 \tag{13.3}$$

For a given set of boundary conditions this ordinary differential equation possesses solutions only for certain values ω_r and corresponding functions $U_r(x)$.

Consider a uniform bar; that is, A = constant, E = constant, and ρ = constant. Then Eq. 13.3 reduces to

$$\frac{d^2 U}{dx^2} + \frac{\rho \omega^2}{E} U = 0 \tag{13.4}$$

which may be simplified by writing it in the form

$$\frac{d^2 U}{dx^2} + \lambda^2 U = 0 \tag{13.5}$$

where the *eigenvalue* λ is defined by[1]

$$\lambda = \omega \sqrt{\frac{\rho}{E}} \tag{13.6}$$

The general solution of Eq. 13.5 is

$$U(x) = A_1 \cos \lambda x + A_2 \sin \lambda x \tag{13.7}$$

The end conditions for free vibration are obtained by substituting Eq. 13.2 into Eqs. 12.9a and c to obtain

(a) Fixed end:
$$U = 0 \tag{13.8a}$$

(b) Free end:
$$\frac{dU}{dx} = 0 \tag{13.8b}$$

Example 13.1 Determine expressions for the natural frequencies and mode shapes for a uniform cantilever bar in axial motion, as shown in Fig. 1

SOLUTION *Eigenvalues (natural frequencies).* The boundary conditions are

$$U(0) = \left.\frac{dU}{dx}\right|_{x=L} = 0 \tag{1}$$

---

[1] The constants ω_r and ω_r^2 are also frequently called *eigenvalues*.

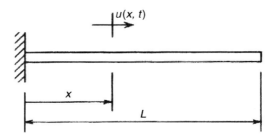

Figure 1 Uniform bar undergoing axial free vibration.

Differentiating Eq. 13.7 with respect to x gives

$$\frac{dU}{dx} = -A_1 \lambda \sin \lambda x + A_2 \lambda \cos \lambda x \tag{2}$$

Evaluating the two boundary conditions of Eq. 1, we obtain

$$U(0) = A_1 = 0 \tag{3}$$

$$\left.\frac{dU}{dx}\right|_{x=L} = A_2 \lambda \cos \lambda L = 0 \tag{4}$$

Since $A_1 = A_2 = 0$ would represent a trivial solution, and since Eq. 3 requires that $A_1 = 0$, we must choose λ in Eq. 4 such that

$$\cos \lambda L = 0 \tag{5}$$

This is the *characteristic equation* for this eigenvalue problem. The roots of the characteristic equation are

$$\lambda L = \frac{\pi}{2}, \frac{3\pi}{2}, \ldots, \left(r - \frac{1}{2}\right)\pi, \ldots \tag{6}$$

Using Eq. 13.6, we get

$$\omega_r = \lambda_r \left(\frac{E}{\rho}\right)^{1/2} = \frac{(\lambda L)_r}{L}\left(\frac{E}{\rho}\right)^{1/2} \tag{7}$$

so that the *natural frequencies* (in rad/s) are given by

$$\omega_r = \frac{(2r-1)\pi}{2L}\left(\frac{E}{\rho}\right)^{1/2} \qquad \text{Ans. (8)}$$

Eigenfunctions (mode shapes). The corresponding *mode shapes* are obtained by combining Eqs. 13.7 and 3 to get

$$U_r = C \sin \lambda_r x \tag{9}$$

or

$$U_r = C \sin\left(\frac{2r-1}{2}\frac{\pi x}{L}\right) \tag{10}$$

where C is an arbitrary scaling factor. (Section 13.5 discusses scaling.) Scaling the modes so that the maximum displacement is one, we might call the *modes* $\phi_r(x)$ and rewrite Eq. 10 setting $C = 1$.

$$\phi_r(x) = \sin\left(\frac{2r-1}{2}\frac{\pi x}{L}\right) \quad \textbf{Ans. (11)}$$

Hence, the mode shapes are sine curves, starting with the quarter-sine-wave mode 1 and the three-quarter-sine-wave mode 2, as shown in Fig. 2. *Note:* The vibratory motion $u(x, t)$ is actually along the axial direction of the bar, but the modes are plotted transverse to the bar for clarity.

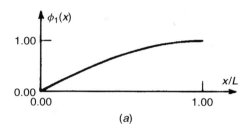

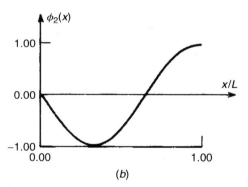

Figure 2 Two axial free-vibration modes of a uniform cantilever bar: (*a*) mode 1; (*b*) mode 2.

In mathematical terminology Eq. 13.5 is an *eigenvalue (differential) equation*, λ_r is an *eigenvalue*, and $\phi_r(x)$ is the corresponding *eigenfunction*.

Note that the curve of $\phi_1(x)$ in Example 13.1 exhibits no crossing of the x axis, while $\phi_2(x)$ has one point (at about $x = 0.6L$) for which $\phi_2(x) = 0$. Such a point is called a *node point*. As the *mode number* r increases, the number of node points increases accordingly, for example, $\phi_5(x)$ would have four node points. Note the difference in the meaning of the two similar-sounding words *mode* and *node*.

13.1.2 Free Torsional Vibration

Because the differential equation of motion for rods with circular cross section undergoing torsional vibration, Eq. 12.14, has the same form as the differential equation governing axial vibration, Eq. 12.8, and there are corresponding boundary conditions in both cases, the solution for free vibration in torsion can be obtained directly from the axial vibration solution of Section 13.1.1 by making the following simple changes in notation:

$$u(x) \longrightarrow \theta(x), \qquad E \longrightarrow G, \qquad A \longrightarrow J$$

13.2 FREE TRANSVERSE VIBRATION OF BERNOULLI–EULER BEAMS

Transverse vibration of Bernoulli–Euler beams is governed by Eq. 12.25. For free vibration this reduces to

$$(EIv'')'' + \rho A \ddot{v} = 0 \tag{13.9}$$

Again, assume harmonic motion given by the equation

$$v(x, t) = V(x) \cos(\omega t - \alpha) \tag{13.10}$$

and substitute this into Eq. 13.9 to obtain the *eigenvalue equation*

$$(EIV'')'' - \rho A \omega^2 V = 0 \tag{13.11}$$

Since closed-form solutions are not available for this equation with variable coefficients, we will restrict our attention to free vibration of uniform beams. Free transverse vibration of uniform beams has been studied extensively, and the mode shapes of uniform beams have frequently been used in conjunction with the Assumed-Modes Method, as discussed in Section 8.4.

For free vibration of a uniform beam, Eq. 13.11 reduces to

$$\frac{d^4 V}{dx^4} - \lambda^4 V = 0 \tag{13.12}$$

where

$$\lambda^4 = \omega^2 \frac{\rho A}{EI} \tag{13.13}$$

[Note that Eq. 13.12 is a fourth-order ordinary differential equation, whereas axial free vibration and torsional free vibration (Section 13.1) are governed by second-order ODEs.]

The general solution of Eq. 13.12 may be written in the form

$$V(x) = A_1 e^{\lambda x} + A_2 e^{-\lambda x} + A_3 e^{i\lambda x} + A_4 e^{-i\lambda x} \tag{13.14a}$$

Two useful alternative forms are

$$V(x) = B_1 e^{\lambda x} + B_2 e^{-\lambda x} + B_3 \sin \lambda x + B_4 \cos \lambda x \tag{13.14b}$$

and

$$V(x) = C_1 \sinh \lambda x + C_2 \cosh \lambda x + C_3 \sin \lambda x + C_4 \cos \lambda x \tag{13.14c}$$

13.2 Free Transverse Vibration of Bernoulli–Euler Beams

There are five constants in the general solution: the four amplitude constants and the eigenvalue λ. The end (boundary) conditions are used in evaluating the eigenvalue and three of the amplitude constants, as illustrated in the examples below; the remaining amplitude constant remains arbitrary. For free vibration of uniform beams, Eq. 13.10 can be substituted into the end-condition equations of Section 12.2 to give the following:

(a) Fixed end:
$$V = \frac{dV}{dx} = 0 \qquad (13.15)$$

(b) Simply supported end:
$$V = \frac{d^2V}{dx^2} = 0 \qquad (13.16)$$

(e) Free end:
$$\frac{d^2V}{dx^2} = \frac{d^3V}{dx^3} = 0 \qquad (13.17)$$

Example 13.2 Determine the natural frequencies and natural modes of a uniform beam that is simply supported at both ends, as shown in Fig. 1.

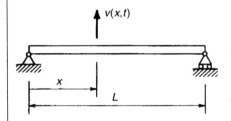

Figure 1 Simply supported uniform beam.

SOLUTION Use the general solution given in Eq. 13.14c.
$$V(x) = C_1 \sinh \lambda x + C_2 \cosh \lambda x + C_3 \sin \lambda x + C_4 \cos \lambda x \qquad (1)$$

The boundary conditions are, at $x = 0$:
$$V(0) = 0, \qquad \left.\frac{d^2V}{dx^2}\right|_{x=0} = 0 \qquad (2a, b)$$

and at $x = L$:
$$V(L) = 0, \qquad \left.\frac{d^2V}{dx^2}\right|_{x=L} = 0 \qquad (3a, b)$$

From Eq. 1,
$$\frac{d^2V}{dx^2} = \lambda^2(C_1 \sinh \lambda x + C_2 \cosh \lambda x - C_3 \sin \lambda x - C_4 \cos \lambda x) \qquad (4)$$

Eigenvalues (natural frequencies). Evaluating the boundary conditions at $x = 0$, we obtain

$$C_2 + C_4 = 0, \qquad \lambda^2(C_2 - C_4) = 0 \qquad (5a, b)$$

Thus, $C_2 = C_4 = 0$. Evaluating the remaining boundary conditions, we get

$$C_1 \sinh \lambda L + C_3 \sin \lambda L = 0 \qquad (6a)$$

$$\lambda^2 (C_1 \sinh \lambda L - C_3 \sin \lambda L) = 0 \qquad (6b)$$

Since this is a pair of homogeneous, linear algebraic equations in C_1 and C_3, a nontrivial solution exists only if the determinant of the coefficients vanishes, namely

$$\begin{vmatrix} \sinh \lambda L & \sin \lambda L \\ \lambda^2 \sinh \lambda L & -\lambda^2 \sin \lambda L \end{vmatrix} = 0 \qquad (7)$$

which simplifies to give

$$\sinh \lambda L \sin \lambda L = 0 \qquad (8)$$

Since $\sinh \lambda L = 0$ only if $\lambda L = 0$, the only nontrivial solutions of Eq. 8 are obtained if

$$\sin \lambda L = 0 \qquad (9)$$

Equation 9 is the *characteristic equation* for this problem, that is, for free transverse vibration of a simply supported uniform beam. It determines the eigenvalues λ_r. From Eq. 9 the eigenvalues are given by

$$\lambda_1 = \pi/L, \quad \lambda_2 = 2\pi/L, \quad \ldots, \quad \lambda_r = r\pi/L, \quad \ldots \qquad (10)$$

The *natural frequency*, ω_r, can be obtained by combining Eqs. 13.13 and 10 to get

$$\omega_r = \left(\frac{r\pi}{L}\right)^2 \left(\frac{EI}{\rho A}\right)^{1/2} \qquad \text{Ans. (11)}$$

Eigenfunctions (mode shapes). If Eq. 9 is substituted back into Eq. 6a (or 6b), we get

$$C_1 = 0 \qquad (12)$$

Thus, $C_1 = C_2 = C_4 = 0$, so the *mode shapes* of the beam are given by

$$V_r(x) = C \sin \lambda_r x \qquad (13)$$

where λ_r is determined from Eq. 10 and where C is an arbitrary amplitude factor. When Eq. 10 is substituted into Eq. 13, the equation for the mode shapes becomes

$$\phi_r(x) = \sin \frac{r\pi x}{L} \qquad \text{Ans. (14)}$$

where the mode has been renamed ϕ_r, since it has been normalized by setting $C = 1$ in Eq. 13.

Three of these sine-wave mode shapes are plotted in Fig. 2. Note that these modes are, respectively, symmetric, antisymmetric, and symmetric about the center of the beam. This grouping of symmetric modes and antisymmetric modes is typical of structures that have a point of symmetry, a line of symmetry, or a plane of symmetry.

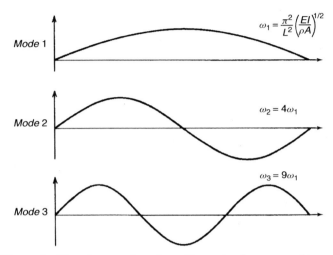

Figure 2 First three modes of a uniform simply supported beam.

Example 13.3 Determine the natural frequencies and natural modes of the uniform cantilever beam shown in Fig. 1.

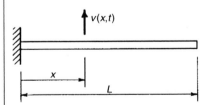

Figure 1 Uniform cantilever beam.

SOLUTION Use the general solution given in Eq. 13.14c.

$$V(x) = C_1 \sinh \lambda x + C_2 \cosh \lambda x + C_3 \sin \lambda x + C_4 \cos \lambda x \qquad (1)$$

Eigenvalues (natural frequencies). The boundary conditions are, at $x = 0$:

$$V(0) = 0, \qquad \left.\frac{dV}{dx}\right|_{x=0} = 0 \qquad (2a, b)$$

and at $x = L$:

$$\left.\frac{d^2 V}{dx^2}\right|_{x=L} = 0, \qquad \left.\frac{d^3 V}{dx^3}\right|_{x=L} = 0 \qquad (3a, b)$$

From Eq. 1,

$$\frac{dV}{dx} = \lambda(C_1 \cosh \lambda x + C_2 \sinh \lambda x + C_3 \cos \lambda x - C_4 \sin \lambda x) \tag{4a}$$

$$\frac{d^2V}{dx^2} = \lambda^2(C_1 \sinh \lambda x + C_2 \cosh \lambda x - C_3 \sin \lambda x - C_4 \cos \lambda x) \tag{4b}$$

$$\frac{d^3V}{dx^3} = \lambda^3(C_1 \cosh \lambda x + C_2 \sinh \lambda x - C_3 \cos \lambda x + C_4 \sin \lambda x) \tag{4c}$$

Substituting Eqs. 1 and 4 into the boundary condition equations, Eqs. 2 and 3, we obtain the following four equations, expressed in matrix format:

$$\begin{bmatrix} 0 & 1 & 0 & 1 \\ \lambda & 0 & \lambda & 0 \\ \lambda^2 \sinh \lambda L & \lambda^2 \cosh \lambda L & -\lambda^2 \sin \lambda L & -\lambda^2 \cos \lambda L \\ \lambda^3 \cosh \lambda L & \lambda^3 \sinh \lambda L & -\lambda^3 \cos \lambda L & \lambda^3 \sin \lambda L \end{bmatrix} \begin{Bmatrix} C_1 \\ C_2 \\ C_3 \\ C_4 \end{Bmatrix} = \begin{Bmatrix} 0 \\ 0 \\ 0 \\ 0 \end{Bmatrix} \tag{5}$$

For this set of homogeneous equations to have a nontrivial solution, the determinant of the coefficients must vanish. This leads to the *characteristic equation*

$$\cos \lambda L \cosh \lambda L + 1 = 0 \tag{6}$$

whose roots are the eigenvalues λ_r, times the length L. No simple expression for the roots of the characteristic equation is available, so a numerical solution of Eq. 6 is required. Values of $(\lambda_r L)$ may be found in tables in Ref. [13.2]. A few values of $(\lambda_r L)$ are listed below:

$$\begin{aligned} \lambda_1 L &= 1.8751, & \lambda_2 L &= 4.6941, \\ \lambda_3 L &= 7.8548, & \lambda_4 L &= 10.996 \end{aligned} \tag{7}$$

From Eq. 13.13, the *natural frequency* (in rad/s) is given by

$$\omega_r = \frac{(\lambda_r L)^2}{L^2} \left(\frac{EI}{\rho A}\right)^{1/2} \tag{8}$$

so

$$\omega_1 = \frac{3.516}{L^2} \left(\frac{EI}{\rho A}\right)^{1/2}$$

$$\omega_2 = \frac{22.03}{L^2} \left(\frac{EI}{\rho A}\right)^{1/2} \qquad \textbf{Ans.} \tag{9}$$

$$\omega_3 = \frac{61.70}{L^2} \left(\frac{EI}{\rho A}\right)^{1/2}$$

Eigenfunctions (mode shapes). We determine the mode shape by employing any three of the four equations in Eq. 5 to express three of the constants in terms of the

13.2 Free Transverse Vibration of Bernoulli–Euler Beams

fourth, which remains arbitrary. The first two equations of Eq. 5 say that

$$C_4 = -C_2$$
$$C_3 = -C_1 \tag{10}$$

The third equation says that

$$C_1 \sinh \lambda_r L + C_2 \cosh \lambda_r L - C_3 \sin \lambda_r L - C_4 \cos \lambda_r L = 0 \tag{11}$$

which can be combined with Eqs. 10 to give

$$C_1(\sinh \lambda_r L + \sin \lambda_r L) + C_2(\cosh \lambda_r L + \cos \lambda_r L) = 0 \tag{12}$$

or

$$C_1 = -C_2 \frac{\cosh \lambda_r L + \cos \lambda_r L}{\sinh \lambda_r L + \sin \lambda_r L} \equiv -k_r C_2 \tag{13}$$

Finally, Eqs. 10 and 13 can be combined with Eq. 1 to give the following expression for the *mode shapes*:

$$V_r(x) = C[(\cosh \lambda_r x - \cos \lambda_r x) - k_r (\sinh \lambda_r x - \sin \lambda_r x)] \qquad \textbf{Ans.} \tag{14}$$

where k_r is given in Eq. 13, and C is an arbitrary amplitude constant. Three mode shapes of a uniform cantilever beam are sketched in Fig. 2.

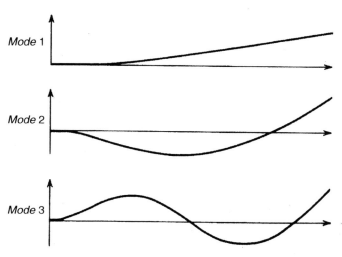

Figure 2 First three mode shapes of a uniform cantilever beam.

In the examples above it may be noted that the natural frequencies (eigenvalues) are determined by solving for the roots of a transcendental equation. It should also be noted that there is an arbitrary amplitude associated with each mode shape. These and other properties of modes and frequencies are discussed further in Section 13.5.

13.3 RAYLEIGH'S METHOD FOR APPROXIMATING THE FUNDAMENTAL FREQUENCY OF A CONTINUOUS SYSTEM

In Section 10.2 you were introduced to the Rayleigh Method for obtaining a SDOF approximation of an MDOF system. This is a variant of the method used by Lord Rayleigh to estimate the fundamental frequency of an undamped continuous system, which is described in this section. Consider free transverse vibration of a beam, like the one in Fig. 13.1. Lord Rayleigh observed that for undamped free vibration, the motion is simple harmonic motion. Thus,

$$v(x, t) = V(x) \cos \omega_R t = C \psi(x) \cos \omega_R t \tag{13.18}$$

where ω_R designates the Rayleigh approximation to the fundamental frequency and where $\psi(x)$ is an assumed *shape function*, which must satisfy the conditions prescribed in Section 2.5 for an admissible function. Rayleigh also observed that energy is conserved. Hence, since energy is conserved, the maximum kinetic energy is equal to the maximum potential energy, that is,

$$\mathcal{T}_{\max} = \mathcal{V}_{\max} \tag{13.19}$$

For the beam in Fig. 13.1 the strain energy is given by

$$\mathcal{V} = \tfrac{1}{2} \int_0^L EI(v'')^2 \, dx + \tfrac{1}{2} k_i v_i^2 \tag{13.20}$$

and the kinetic energy is given by

$$\mathcal{T} = \tfrac{1}{2} \int_0^L \rho A(\dot{v})^2 \, dx + \tfrac{1}{2} m_s \dot{v}_s^2 \tag{13.21}$$

Substitution of Eq. 13.18 into these produces

$$\mathcal{V}_{\max} = \tfrac{1}{2} k C^2 \tag{13.22}$$

and

$$\mathcal{T}_{\max} = \tfrac{1}{2} \omega_R^2 m C^2 \tag{13.23}$$

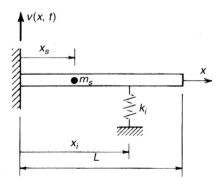

Figure 13.1 Transverse vibration of a cantilever beam.

13.3 Rayleigh's Method for Approximating the Fundamental Frequency of a Continuous System

where, as in Eqs. 2.37 and 2.35, respectively, in Section 2.5,

$$k = \int_0^L EI(\psi'')^2\, dx + k_i[\psi(x_i)]^2 \tag{13.24}$$

and

$$m = \int_0^L \rho A \psi^2\, dx + m_s[\psi(x_s)]^2 \tag{13.25}$$

Finally, from Eqs. 13.19, 13.22, and 13.23 we obtain the *Rayleigh quotient*:

$$\boxed{\mathcal{R}(V) = \omega_R^2 = \frac{k}{m}} \tag{13.26}$$

Note that this same free-vibration result would be obtained by using the assumed-modes generalized stiffness and generalized mass of Section 2.5 in the expression for ω^2 for an undamped SDOF system (Eq. 3.4a). Thus, when the same shape function $\psi(x)$ is used, the Rayleigh Method and the Assumed-Modes Method lead to the same expression for the approximate frequency of an SDOF model of a continuous system.

Example 13.4 Use the Rayleigh Method to obtain an approximate fundamental frequency of a uniform cantilever beam. Use

$$\psi(x) = \left(\frac{x}{L}\right)^2$$

as the shape function, and compare the resulting frequency with the "exact" frequency obtained in Example 13.3.

SOLUTION

$$\psi''(x) = \frac{2}{L^2} \tag{1}$$

$$k = \int_0^L EI(\psi'')^2\, dx = \frac{4EI}{L^3} \tag{2}$$

$$m = \int_0^L \rho A \psi^2\, dx = \frac{\rho A L}{5} \tag{3}$$

From Eq. 13.26,

$$\omega_R^2 = \frac{k}{m} = \frac{20 EI}{\rho A L^4} \tag{4}$$

Then

$$\omega_R = \frac{\sqrt{20}}{L^2}\left(\frac{EI}{\rho A}\right)^{1/2} \tag{5}$$

or

$$\omega_R = \frac{4.472}{L^2}\left(\frac{EI}{\rho A}\right)^{1/2} \qquad \textbf{Ans.} \tag{6}$$

This value compares with the "exact" value of ω_1 given in Example 13.3:

$$\omega_1 = \frac{3.516}{L^2}\left(\frac{EI}{\rho A}\right)^{1/2}$$

It will be shown in Section 13.5 that the Rayleigh Method always gives an upper bound to the exact fundamental frequency, as is the case in Example 13.4.

13.4 FREE TRANSVERSE VIBRATION OF BEAMS INCLUDING SHEAR DEFORMATION AND ROTATORY INERTIA

The equations of motion and generalized boundary conditions for a Timoshenko beam were derived in Section 12.4. To gain some insight into the effect of shear deformation and rotatory inertia on the vibration of beams, consider a uniform, simply supported beam like the one in Example 13.2. The geometric boundary conditions for a simply supported beam are

$$v(0, t) = v(L, t) = 0 \tag{13.27}$$

Since the rotation α is not specified at the ends, Eqs. 12.55c and d give the following natural boundary conditions:

$$\alpha'(0, t) = \alpha'(L, t) = 0 \tag{13.28}$$

It will be convenient to use Eqs. 12.56 and 12.57 as the equations of motion, although Eqs. 12.54 could equally well have been chosen. For free vibration, the equations of motion reduce to

$$\alpha' = v'' - \frac{\rho}{\kappa G}\ddot{v} \tag{13.29}$$

and

$$EI\frac{\partial^4 v}{\partial x^4} + \rho A\frac{\partial^2 v}{\partial t^2} - \rho A r_G^2 \left(1 + \frac{E}{\kappa G}\right)\frac{\partial^4 v}{\partial x^2 \partial t^2} + \frac{\rho^2 A r_G^2}{\kappa G}\frac{\partial^4 v}{\partial t^4} = 0 \tag{13.30}$$

where r_G is the radius of gyration, $r_G^2 = I/A$.

Assume harmonic motion of the form

$$v(x, t) = V(x)\cos\omega t \tag{13.31}$$

Equation 13.29 then becomes

$$\alpha' = \left(V'' - \frac{\rho\omega^2}{\kappa G}V\right)\cos\omega t \tag{13.32}$$

so the boundary conditions of Eqs. 13.27 and 13.28 reduce, respectively, to

$$V(0) = V(L) = 0 \tag{13.33a}$$

and

$$V''(0) = V''(L) = 0 \tag{13.33b}$$

which are the same as those of simple beam theory. Substituting Eq. 13.31 into Eq. 13.30, we obtain the spatial form of the equation of motion,

$$V^{iv} - \lambda^4 V + \lambda^4 r_G^2 \left(1 + \frac{E}{\kappa G}\right) V'' + \lambda^8 r_G^4 \frac{E}{\kappa G} V = 0 \tag{13.34}$$

where λ^4 is given by Eq. 13.13.

The simply supported beam mode shape

$$V_r(x) = C \sin \frac{r\pi x}{L} \tag{13.35}$$

satisfies both the boundary conditions, Eqs. 13.33, and the equation of motion, Eq. 13.34. Therefore, it provides a solution for the present problem. Substitution of this solution into Eq. 13.34 gives

$$\underbrace{\left(\frac{r\pi}{L}\right)^4 - \lambda^4}_{(a)} + \underbrace{\lambda^4 r_G^2 \left(\frac{r\pi}{L}\right)^2}_{(b)} + \underbrace{\lambda^4 r_G^2 \frac{E}{\kappa G} \left(\frac{r\pi}{L}\right)^2}_{(c)} + \underbrace{\lambda^8 r_G^4 \frac{E}{\kappa G}}_{(d)} = 0 \tag{13.36}$$

The terms denoted (a) are those of the simple Bernoulli–Euler beam theory characteristic equation. To estimate the relative importance of terms (c) and (d), approximate λ in term (d) by the simple beam solution $(r\pi/L)^4$. Then term (d) is approximately

$$\underbrace{\lambda^8 r_G^4 \frac{E}{\kappa G}}_{(d)} \approx \underbrace{\lambda^4 r_G^2 \frac{E}{\kappa G} \left(\frac{r\pi}{L}\right)^2}_{(c)} \left(\frac{r\pi r_G}{L}\right)^2$$

Thus, when $r\pi r_G/L \ll 1$, term (d) will be small compared with term (c). To compare terms (b) and (c), observe that $E/\kappa G \approx 3$ for a rectangular beam made of typical construction materials. Hence, the shear correction term (c) will be about three times as important as the rotatory inertia correction term (b).

Neglecting term (d), we can approximate the solution of the characteristic equation, Eq. 13.36, by

$$\lambda^4 = \left(\frac{r\pi}{L}\right)^4 \frac{1}{1 + r_G^2(r\pi/L)^2(1 + E/\kappa G)} \tag{13.37}$$

Thus, the correction due to shear and rotatory inertia increases as the mode number r increases, and it decreases as the slenderness ratio L/r_G increases. It is important to note that although a beam may be physically "slender," that is, have a large value of L/r_G, it is the *effective slenderness ratio* based on the wavelength L/r, not based on the physical length L, that governs whether corrections for shear deformation and rotatory inertia are needed.

13.5 SOME PROPERTIES OF NATURAL MODES OF CONTINUOUS SYSTEMS

In Sections 13.1 and 13.2 we present typical examples of the determination of modes and frequencies of continuous members. In the present section we consider the following properties associated with the modes: scaling (or normalization), orthogonality, the expansion theorem, and the Rayleigh quotient. These are illustrated by using the Bernoulli–Euler beam equations. However, the following discussion is not limited to uniform beams, and the properties that are described apply to other continuous systems as well. This section parallels the treatment of properties of modes and frequencies of MDOF systems presented in Section 10.1.

From Eq. 13.11, the eigenvalue equation for a uniform Bernoulli–Euler beam is

$$(EIV_r'')'' - \rho A \omega_r^2 V_r = 0 \tag{13.38}$$

Homogeneous *generalized boundary conditions* can be stated as

$$(EIV'')' \, \delta V = 0 \quad \text{at } x = 0 \tag{13.39a}$$

$$(EIV'') \, \delta V' = 0 \quad \text{at } x = 0 \tag{13.39b}$$

$$(EIV'')' \, \delta V = 0 \quad \text{at } x = L \tag{13.39c}$$

$$(EIV'') \, \delta V' = 0 \quad \text{at } x = L \tag{13.39d}$$

From these, any combination of fixed end (Eqs. 13.15), simply supported end (Eqs. 13.16), or free end (Eqs. 13.17) can be obtained.

As noted previously, Eqs. 13.38 and 13.39 determine a set of *natural frequencies*, ω_r ($r = 1, 2, \ldots$), and corresponding *natural modes*, $V_r(x)$ ($r = 1, 2, \ldots$), where the latter are determined only to within a constant multiplier, that is, the *mode shape* is determined but not the amplitude. Thus, modes can be scaled in any convenient manner.

13.5.1 Scaling (Normalization)

Modes that have been scaled so that they have a unique amplitude will be denoted by $\phi_r(x)$. Let

$$V_r(x) = c_r \phi_r(x) \tag{13.40}$$

The process of rendering the amplitude of a mode to be unique is called *normalization*, and the resulting modes, $\phi_r(x)$, are called *normal modes*. It is convenient to consider $\phi_r(x)$ to be dimensionless and to let c_r carry the units of $V(x)$.

Three procedures that are employed for scaling modes are:

1. Scale the mode so that the amplitude at a particular location in the structure has a specified value; for example, make $\phi_r(x_s) = 1$ at a specified location x_s.[2]
2. Scale the mode so that the value of $\phi_r(x)$ at the point where $|\phi_r(x)|$ is a maximum is a specified value, that is, so that $\max_x |\phi_r(x)| = 1$.
3. Scale the mode so that the *generalized mass*, or *modal mass*, defined by

$$\boxed{M_r = \int_0^L \rho A \phi_r^2 \, dx} \tag{13.41}$$

---

[2] This scaling procedure is sometimes used where the point x_s has some particular significance: for example, the location of the main deck of an offshore platform, the location of the top story of a building, and so on. Although this method is frequently used to scale modes, it is subject to one limitation. Namely, the selected location x_s could be a node point for some mode, and hence $V_r(x_s) = 0$. Therefore, this scaling procedure could not be employed for such a mode.

has a specified value. Generally, the value $M_r = 1$ is used. Since it is convenient to consider ϕ_r to be dimensionless, the units of mass ($\rho A L$) should be associated with M_r; for example, $M_r = 1$ slug.[3]

The *generalized stiffness*, or *modal stiffness*, for the rth mode is defined as

$$K_r = \int_0^L EI\,(\phi_r'')^2\,dx \tag{13.42}$$

K_r and M_r can be related to each other in the following manner. Let Eq. 13.38, be multiplied by ϕ_r and integrated from 0 to L. Then

$$\int_0^L (EI\,\phi_r'')''\phi_r\,dx - \omega_r^2 \int_0^L \rho A \phi_r^2\,dx = 0 \tag{13.43}$$

Noting that K_r contains the term $(\phi_r'')^2$, we can integrate the first term of Eq. 13.43 by parts twice to get

$$(EI\,\phi_r'')'\phi_r \Big|_0^L - (EI\,\phi_r'')\phi_r' \Big|_0^L + \int_0^L EI\,(\phi_r'')^2\,dx - \omega_r^2 \int_0^L \rho A \phi_r^2\,dx = 0 \tag{13.44}$$

From Eqs. 13.39 it will be recognized that all of the boundary terms in Eq. 13.44 vanish due to either geometric boundary conditions or natural boundary conditions, regardless of the type of support at each end. From the definitions of M_r and K_r, then, Eq. 13.44 reduces to

$$\omega_r^2 = \frac{K_r}{M_r} \tag{13.45}$$

13.5.2 Orthogonality

A very important property of natural modes is the property of *orthogonality*. To derive the orthogonality property, we multiply Eq. 13.38 by ϕ_s ($\neq \phi_r$) and integrate from 0 to L.

$$\int_0^L (EI\,\phi_r'')''\phi_s\,dx - \omega_r^2 \int_0^L \rho A \phi_r \phi_s\,dx = 0 \tag{13.46}$$

Integrating by parts and noting that the same boundary conditions apply to modes r and s, we can reduce Eq. 13.46 to

$$\int_0^L EI\,(\phi_r'')\phi_s''\,dx - \omega_r^2 \int_0^L \rho A \phi_r \phi_s\,dx = 0 \tag{13.47}$$

Starting with Eq. 13.38 written for mode s, multiplying by ϕ_r, and integrating from 0 to L, we get

$$\int_0^L EI\,(\phi_s'')\phi_r''\,dx - \omega_s^2 \int_0^L \rho A \phi_r \phi_s\,dx = 0 \tag{13.48}$$

---

[3] Some authors use the term *normal mode* in a more general sense to refer to any free-vibration mode shape. Others restrict the term to those modes normalized so that $M_r = 1$.

Subtracting Eq. 13.47 from Eq. 13.48, we obtain

$$(\omega_r^2 - \omega_s^2) \int_0^L \rho A \phi_r \phi_s \, dx = 0 \qquad (13.49)$$

For two modes having distinct frequencies, therefore, the *orthogonality property*

$$\boxed{\int_0^L \rho A \phi_r \phi_s \, dx = 0, \qquad \omega_r \neq \omega_s} \qquad (13.50)$$

holds true. The modes are said to be *orthogonal with respect to the mass distribution*.[4] By substituting Eq. 13.50 into either Eq. 13.47 or 13.48, we see that modes ϕ_r and ϕ_s are also *orthogonal with respect to the stiffness distribution*, as defined by

$$\boxed{\int_0^L EI \phi_r'' \phi_s'' \, dx = 0, \qquad \omega_r \neq \omega_s} \qquad (13.51)$$

Mode sets ϕ_r for which the modal mass satisfies the equation

$$M_r = \int_0^L \rho A \phi_r^2 \, dx = 1 \qquad (13.52)$$

are said to be *normalized*, and the two *orthogonality conditions*

$$\int_0^L \rho A \phi_r \phi_s \, dx = 0, \qquad \int_0^L EI \phi_r'' \phi_s'' \, dx = 0 \qquad (13.53)$$

also hold for these modes. Therefore, this set of modes $\phi_r(x), r = 1, 2, \ldots$ is said to form a *set of orthonormal modes*.

13.5.3 Expansion Theorem

An *expansion theorem* for continuous systems can be stated as follows: Any function $V(x)$ that satisfies the same boundary conditions as are satisfied by a given set of orthonormal modes, $\phi_r (r = 1, 2, \ldots)$, and is such that $(EIV'')''$ is a continuous function, can be represented by an absolutely and uniformly convergent series of the form

$$\boxed{V(x) = \sum_{r=1}^{\infty} c_r \phi_r(x)} \qquad (13.54)$$

where the expansion coefficients c_r are given by

$$c_r = \frac{\int_0^L \rho A V \phi_r \, dx}{\int_0^L \rho A \phi_r^2 \, dx} \qquad (13.55a)$$

---

[4]This equation holds for transverse vibration of Bernoulli–Euler beams with homogeneous boundary conditions, that is, boundary conditions that satisfy Eqs. 13.39. See the Problems for the derivation of the appropriate form of the orthogonality equations when there are springs or masses attached at the boundaries.

or if the modes are normalized (i.e., $M_r = 1$),

$$c_r = \int_0^L \rho A V \phi_r \, dx \qquad (13.55b)$$

In deriving Eqs. 13.55 from Eq. 13.54, the orthogonality equation, Eq. 13.50, was employed.

13.5.4 Rayleigh Quotient

Let $V(x)$ be an arbitrary function satisfying the boundary and continuity conditions mentioned in the expansion theorem above. Then the *Rayleigh quotient* is defined as

$$\boxed{\mathcal{R}(V) = \frac{k}{m} = \frac{\int_0^L EI(V'')^2 \, dx}{\int_0^L \rho A V^2 \, dx}} \qquad (13.56)$$

for a beam with no lumped springs or masses. The Rayleigh quotient was used in Section 13.3 to obtain an approximate value of the fundamental frequency. Equation 13.45 shows that $\mathcal{R}(V) = \omega_r^2$ when the function V is any of the natural mode shapes, that is, when $V = c_r \phi_r(x)$. Let Eq. 13.54 be substituted into Eq. 13.56. Then, using the orthonormality conditions, we get

$$\mathcal{R}(V) = \frac{c_1^2 \omega_1^2 + c_2^2 \omega_2^2 + c_3^2 \omega_3^2 + \cdots}{c_1^2 + c_2^2 + c_3^2 + \cdots}$$

which can be written in the form

$$\mathcal{R}(V) = \omega_1^2 \frac{1 + (c_2/c_1)^2 (\omega_2/\omega_1)^2 + (c_3/c_1)^2 (\omega_3/\omega_1)^2 + \cdots}{1 + (c_2/c_1)^2 + (c_3/c_1)^2 + \cdots} \qquad (13.57)$$

By comparing the numerator and denominator term by term we see that since ω_1 is the lowest frequency, the numerator is larger than the denominator. Hence,

$$\boxed{\mathcal{R}(V) \geq \omega_1^2} \qquad (13.58)$$

the equality holding when $c_2 = c_3 = \cdots = 0$. The Rayleigh quotient thus provides an *upper bound to the fundamental frequency*, as shown in Example 13.4. Therefore, if a function $V(x)$ is selected that has a shape that is very similar to that of the fundamental mode shape, the Rayleigh quotient can be used to obtain a close estimate to the fundamental natural frequency.

13.6 FREE VIBRATION OF THIN FLAT PLATES

In Chapter 12 the equations of motion were derived for several one-dimensional structural members, and so far in Chapter 13 solutions have been obtained for the modes and frequencies of these one-dimensional members. Flat plates and thin shells are common types of structures that have one dimension that is much smaller than the other

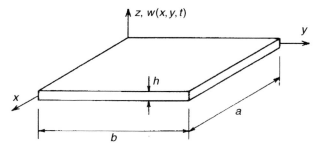

Figure 13.2 Thin rectangular plate.

two dimensions. Figure 13.2 shows a flat rectangular plate, the plate thickness h being much smaller than the major dimensions a and b.

Although a thorough discussion of the vibration of plates and shells is beyond the scope of this book, a brief introduction is presented here so that you can take note of some of the features that arise in vibration of multidimensional structures. Excellent surveys of the topics of vibration of plates and vibration of shells have been compiled by Leissa.[13.4,13.5]

The *classical small-deflection theory of flat plates*, also called *Kirchhoff–Love theory*, is based on the following assumptions [13.6–13.8]:

1. During bending of the plate, the middle plane of the plate does not undergo in-plane deformation. Therefore, the middle plane is the *neutral plane* of the plate.
2. The transverse deflection of points in the neutral plane is small in comparison with the thickness of the plate. That is, $|w(x, y, t)| < h$.
3. Straight lines that are normal to the middle plane before deformation remain straight and inextensional, and they remain normal to the deformed neutral plane.
4. The plate is made of material that is homogeneous, isotropic, and linearly elastic.
5. The normal stress in the direction transverse to the plate, that is, σ_z, is negligible in comparison with the in-plane stresses σ_x and σ_y.

Based on the assumptions above, the equation of motion of a flat plate is

$$D\nabla^4 w + \rho h w_{tt} = 0 \tag{13.59}$$

where subscripts indicate partial differentiation and where D is the *flexural rigidity* of the plate, as defined by

$$D = \frac{Eh^3}{12(1-\nu^2)} \tag{13.60}$$

where h is the (constant) thickness of the plate, $w(x, y, t)$ is the transverse displacement of the point $(x, y, z = 0)$ at time t, and ∇^4 is the biharmonic operator, given by

$$\nabla^4(\cdot) = (\cdot)_{xxxx} + 2(\cdot)_{xxyy} + (\cdot)_{yyyy} \tag{13.61}$$

Figure 13.2 shows a rectangular plate. On an edge $y = $ constant, the following generalized boundary conditions pertain:

$$[w_{yyy} + (2 - \nu)w_{xxy}]\delta w = 0 \tag{13.62a}$$

$$[w_{yy} + \nu w_{xx}]\delta w_y = 0 \tag{13.62b}$$

Conditions for an edge $x = $ constant are obtained by permutation of x and y.

Exact solutions for free vibration are only available for a rectangular plate with a pair of opposite edges simply supported. The case of a rectangular plate with all edges simply supported leads to a particularly easy solution and is presented here to illustrate some features of plate vibration. The boundary conditions for the simply supported rectangular plate in Fig. 13.2 are:

$$w(0, y, t) = w(a, y, t) = w(x, 0, t) = w(x, b, t) = 0 \tag{13.63a}$$

$$w_{xx}(0, y, t) = w_{xx}(a, y, t) = w_{yy}(x, 0, t) = w_{yy}(x, b, t) = 0 \tag{13.63b}$$

Equations 13.63a are *geometric boundary conditions* and Eqs. 13.63b are *natural boundary conditions* obtained from Eq. 13.62b.

The equation of motion admits a solution that is harmonic in time, that is,

$$w(x, y, t) = W(x, y)\cos(\omega t - \alpha) \tag{13.64}$$

Then the equation of motion, Eq. 13.59, becomes the eigenvalue equation

$$\nabla^4 W - \lambda^4 W = 0 \tag{13.65}$$

where

$$\lambda^4 = \frac{\rho h \omega^2}{D} \tag{13.66}$$

The spatial function

$$W_{(m,n)}(x, y) = C \sin\frac{m\pi x}{a} \sin\frac{n\pi y}{b} \tag{13.67}$$

satisfies all of the boundary conditions, and when substituted into Eq. 13.65, it gives

$$\lambda^2_{(m,n)} = \left(\frac{m\pi}{a}\right)^2 + \left(\frac{n\pi}{b}\right)^2 \tag{13.68}$$

Using Eqs. 13.66 and 13.68, we can express the natural frequency $\omega_{(m,n)}$ in the form

$$\omega_{(m,n)} = \left(\frac{\pi^4 D}{\rho h a^4}\right)^{1/2}\left[m^2 + n^2\left(\frac{a}{b}\right)^2\right] \tag{13.69a}$$

or

$$\omega_{(m,n)} = \left(\frac{\pi^4 D}{\rho h b^4}\right)^{1/2}\left[m^2\left(\frac{b}{a}\right)^2 + n^2\right] \tag{13.69b}$$

For a two-dimensional structure such as the rectangular plate, it is convenient to use the parameters (m, n) to describe the mode rather than a single parameter as was used for one-dimensional structural members. For the simply supported plate, these designate the number of half-sine waves in the x and y directions, respectively (Eq. 13.67).

Modes of two-dimensional structures are frequently visualized by sketching node lines, as in Fig. 13.3. These are the lines along which $W_{(m,n)}(x, y, t) = 0$ for all time. The $\oplus$ and $\ominus$ indicate the relative sign of $W_{(m,n)}(x, y)$ in various portions of the plate. Note from Eq. 13.69a that if $a > b$, the mode $(2, 1)$ has a lower frequency than the mode $(1, 2)$. Referring to the $(2, 1)$ and $(1, 2)$ figures in Fig. 13.3, we can interpret this as saying that the plate will seek to divide itself into "squares."

One feature that is easily demonstrated on simply supported square plates (i.e., $a = b$), but that can also occur in other structures, is the phenomenon of *repeated frequencies*; that is, $\omega_{(m,n)} = \omega_{(n,m)}$. Let $W_{(m,n)}$ and $W_{(n,m)}$ be distinct spatial modes corresponding to $\lambda_{(m,n)} = \lambda_{(n,m)}$. Then any linear combination $W_{(m,n)} + cW_{(n,m)}$ satisfies the eigenvalue equation

$$\nabla^4 W - \lambda^4 W = 0 \qquad (13.70)$$

This means that there are not two unique mode shapes, but there is a two-dimensional *subspace of modes* of the form

$$W = A_{(1,4)} W_{(1,4)} + A_{(4,1)} W_{(4,1)} \qquad (13.71)$$

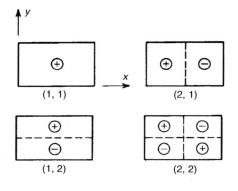

Figure 13.3 Node lines of modes $W_{(m,n)}(x, y)$ of a thin rectangular plate.

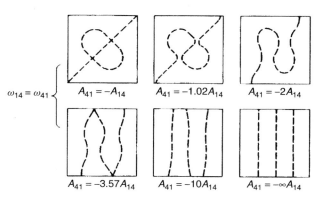

Figure 13.4 Nodal patterns for plates having a natural frequency $\omega_{(1,4)} = \omega_{(4,1)}$. (From Ref. [13.4].)

Figure 13.4 shows some interesting nodal patterns obtained for a square plate, where $W_{(1,4)}$ and $W_{(4,1)}$ are defined by Eq. 13.67.

REFERENCES

[13.1] R. R. Craig, Jr., *Mechanics of Materials*, 2nd ed., Wiley, New York, 2000.

[13.2] T.-C. Chang and R. R. Craig, Jr., "Normal Modes of Uniform Beams," *Proceedings of ASCE*, Vol. 95, No. EM4, 1969, pp. 1025–1031.

[13.3] G. Temple and W. G. Bickley, *Rayleigh's Principle*, Oxford University Press, London, 1933. Reprinted by Dover, New York, 1956.

[13.4] A. W. Leissa, *Vibration of Plates*, NASA SP-160, National Aeronautics and Space Administration, Washington, DC, 1969.

[13.5] A. W. Leissa, *Vibration of Shells*, NASA SP-288, National Aeronautics and Space Administration, Washington, DC, 1973.

[13.6] S. Timoshenko and S. Woinowsky-Krieger, *Theory of Plates and Shells*, McGraw-Hill, New York, 1959.

[13.7] H. L. Langhaar, *Energy Methods in Applied Mechanics*, Wiley, New York, 1962.

[13.8] L. Meirovitch, *Principles and Techniques of Vibrations*, Prentice Hall, Upper Saddle River, NJ, 1997.

PROBLEMS

For problems whose number is preceded by a **C**, you are to write a computer program and use it to produce the plot(s) requested. *Note:* MATLAB .m-files for many of the plots in this book may be found on the book's website.

Problem Set 13.1

C 13.1 The uniform bar shown in Fig. P9.9 has uniform cross-sectional area A, modulus of elasticity E, and mass density ρ. It is attached to rigid supports at its ends. **(a)** Obtain the characteristic equation from which eigenvalues for axial free vibration can be determined. Your answer should contain the parameters λ and L, where λ is defined by Eq. 13.6. **(b)** Determine an expression for the fundamental axial frequency of the bar (ω_1 rad/s). Your answer should contain the parameters E, ρ, and L. **(c)** Determine an expression for the fundamental axial mode shape, $\phi_1(x)$, and use MATLAB, or another computer program, to plot the fundamental mode. Scale the mode so that its maximum value is 1.0.

C 13.2 Consider axial free vibration of the uniform bar (AE = constant, ρA = constant) in Fig. P9.7. The mass "particle" at $x = L$ is $m = \rho AL/5$. **(a)** Determine the boundary condition at end $x = L$. **(b)** Obtain the characteristic equation from which eigenvalues can be determined. Your answer should contain the parameters λ and L, where λ is defined by Eq. 13.6. **(c)** Determine an expression for the fundamental axial frequency of the bar (ω_1 rad/s). **(d)** Determine an expression for the fundamental axial mode shape, $\phi_1(x)$, and use MATLAB, or another computer program, to plot the fundamental mode. Scale the mode so that its maximum value is 1.0.

C 13.3 Consider axial free vibration of the uniform bar (AE = constant, ρA = constant) in Fig. P9.8. The stiffness of the linear spring at $x = L$ is $k = AE/2L$; ignore its mass. **(a)** Determine the boundary condition at end $x = L$. **(b)** Obtain the characteristic equation from which eigenvalues can be determined. Your answer should contain the parameters λ and L, where λ is defined by Eq. 13.6. **(c)** Determine an expression for the fundamental axial frequency of the bar (ω_1 rad/s). **(d)** Determine an expression for the fundamental axial mode shape, $\phi_1(x)$, and use MATLAB, or another computer program, to plot the fundamental mode. Scale the mode so that its maximum value is 1.0.

C 13.4 The axial frequencies of a free–free structure are sometimes of great importance, as in the "Pogo" stability problem experienced by some rockets. Consider axial motion of the uniform free–free bar in Fig. P13.4. (a) Show that the system has a zero-frequency, or rigid-body, mode. (b) Obtain the characteristic equation from which the nonzero eigenvalues can be determined. Your answer should contain the parameters λ and L, where λ is defined by Eq. 13.6. (c) Determine an expression for the fundamental (flexible) axial frequency of the bar (ω_1 rad/s). (d) Determine an expression for the fundamental (flexible) axial mode shape, $\phi_1(x)$, and use MATLAB, or another computer program, to plot the fundamental mode. Scale the mode so that its maximum value is 1.0.

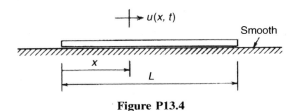

Figure P13.4

Problem Set 13.2

C 13.5 Consider transverse vibration of the uniform clamped–clamped beam shown in Fig. P13.5. (a) Determine the characteristic equation for transverse vibration. Your answer should contain the parameters λ and L, where λ is defined by Eq. 13.13. (b) Determine an expression for the fundamental transverse-vibration frequency of the beam (ω_1 rad/s). (c) Determine an expression for the fundamental transverse-vibration mode shape, $\phi_1(x)$, and use MATLAB, or another computer program, to plot the fundamental mode. Scale the mode so that its maximum value is 1.0. (d) Using the results of Example 13.2, compare the fundamental frequency of the clamped–clamped beam with that of a simply supported beam. What is the effect of adding the slope constraints to the beam at $x = 0$ and $x = L$?

C 13.6 Flight vehicles such as airplanes and rockets are unconstrained, and therefore they can undergo rigid-body motion. The free–free uniform beam gives some insight into the behavior of such free–free structures. Consider transverse vibration of the uniform beam (EI = constant, ρA = constant) shown in Fig. P13.6. (a) Show that the beam has two zero-frequency (i.e., rigid-body) modes, one mode in translation and one in rotation. (b) Determine the characteristic equation governing the nonzero frequencies. Your answer should contain the parameters λ and L, where λ is defined by Eq. 13.13. (Note that this is the same equation as obtained in Problem 13.5 for a uniform clamped–clamped beam.) (c) Determine an expression for the fundamental nonzero frequency of the beam (ω_3 rad/s). (*Note:* For the rigid-body modes, $\omega_1 = \omega_2 = 0$.) (d) Determine an expression for the fundamental flexible transverse-vibration mode shape, $\phi_3(x)$, and use MATLAB, or another computer program, to plot the fundamental flexible mode. Scale the mode so that its maximum value is 1.0.

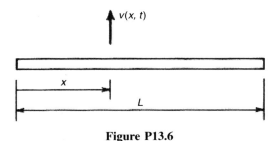

Figure P13.6

13.7 Most structural beams are neither clamped–clamped nor simply supported but can be considered to have partial end fixity. At $x = 0$ and at $x = L$ the uniform beam shown in Fig. P13.7 cannot displace laterally; that is, $v(0, t) = v(L, t) = 0$. However, at each end, rotation through (slope) angle θ is permitted, but rotation is resisted by a spring that induces a bending moment at the end of the beam that is proportional to the slope of the end of beam. Determine the characteristic equation for the beam shown in Fig. P13.7, where $0 \leq \beta \leq \infty$ is a dimensionless parameter controlling the amount of rotational restraint ($\beta = 0$ corresponds to a simply supported end; $\beta = \infty$ corresponds to a fixed end).

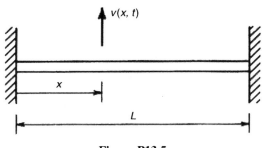

Figure P13.5

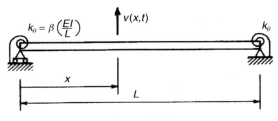

Figure P13.7

13.8 The statement "neglect axial motion" is frequently encountered in modeling frame structures such as the one shown in Fig. P13.8a. The justification for this lies in the difference in the natural frequency values associated with axial vibration and those associated with transverse vibration. Using the results of Examples 10.1 and 10.3, tabulate the first three axial frequencies and the first three bending frequencies of a cantilever beam. Use the physical parameters of the 10-ft **W** 10×30 steel beam shown in Fig. P13.8b ($I = 170$ in^4, $A = 8.84$ in^2, $\gamma = 0.284$ lb/in^3).

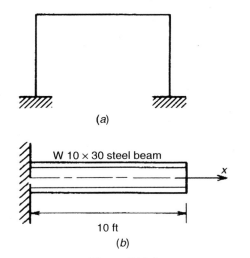

Figure P13.8

13.9 Obtain the characteristic equation for determining the frequencies of the coupled bending-axial motion of the system in Fig. P12.3. You may start with the axial vibration general solution given in Eq. 13.7 and the transverse vibration general solution given in Eqs. 13.14. You will need to derive the boundary conditions that couple the axial vibration and transverse vibration (see Problem 12.3).

13.10 Consider transverse vibration, $v(x, t)$ of the uniform cantilever beam ($EI = $ constant, $\rho A = $ constant) with tip mass, as shown in Fig. P9.11. The mass "particle" at $x = L$ is $m = \rho A L/5$. (**a**) Determine the two boundary conditions at end $x = L$. (**b**) Obtain the characteristic equation from which the eigenvalues can be determined. Your answer should contain the parameters λ and L, where λ is defined by Eq. 13.13. (*Hint:* See Eq. 6 of Example 13.3.) (**c**) Solve the characteristic equation [part (b)] for the lowest eigenvalue, λ_1, and then determine the fundamental transverse-bending frequency, ω_1.

Problem Set 13.3

13.11 In Example 13.4 the assumed mode $\psi(x) = (x/L)^2$ gave only a crude estimate of the fundamental frequency of a uniform cantilever beam. Here, a different function will be used for the shape function $\psi(x)$. (**a**) Determine an expression for the static deflection curve of a uniform beam of length L carrying a uniformly distributed load w per unit length. Put this deflection curve into dimensionless form by using x/L and by dropping physical parameters such as w and EI (**b**) Use this nondimensionalized static deflection curve as $\psi(x)$ in estimating the fundamental frequency of a uniform cantilever beam by the Rayleigh Method. Compare your answer with the exact fundamental frequency and the Rayleigh estimate obtained in Example 13.4.

13.12 Consider transverse vibration of the beam with tip mass shown in Fig. P9.11. As in Problem 9.11, the tip mass is $m = \rho A L/5$. Use the Rayleigh Method to estimate the fundamental bending frequency of this cantilever beam with tip mass. Let $\psi(x)$ be the nondimensionalized static deflection curve of a uniform cantilever beam with concentrated tip force as shown in Fig. P13.12. (The "exact" fundamental frequency was determined in Problem 13.10.)

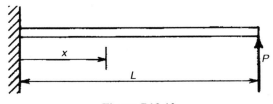

Figure P13.12

13.13 A heavy air conditioner is located at the center of a uniform simply supported roof beam ($EI = $ constant,

ρA = constant), as shown in Fig. P13.13. Neglect the width of the air conditioner's contact with the beam; that is, assume that it is a point mass at $x = L/2$. Use the Rayleigh Method to obtain an expression for estimating the effect of the air-conditioner mass on the fundamental beam frequency. For $\psi(x)$ use the fundamental mode shape of the uniform beam alone, as found in Example 13.2. Express your answer in terms of the mass ratio $M/\rho AL$.

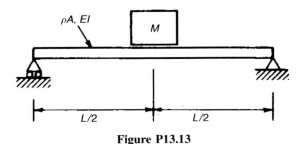

Figure P13.13

13.14 A massless linear spring is located at the center of a simply supported beam (EI = constant, ρA = constant), as shown in Fig. P13.14a. (a) Determine an expression for the static deflection curve of a uniform beam of length L with a concentrated load P at its center, as shown in Fig. P13.14b. Since the deflection curve will be symmetric about the center of the beam, determine only the expression that is valid for the left half of the beam, that is, for $0 \leq x \leq L/2$. Put this deflection curve into dimensionless form by using x/L and by dropping physical parameters such as P and EI. (b) Use the Rayleigh Method to obtain an expression for estimating the effect of the spring on the fundamental beam frequency, ω_1. For the shape function $\psi(x)$ use the nondimensionalized static deflection curve determined in part (a). Express your answer in terms of the stiffness ratio $k/(EI/L^3)$.

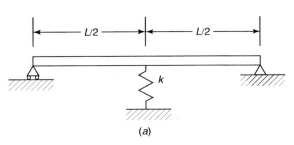

(a)

Figure P13.14

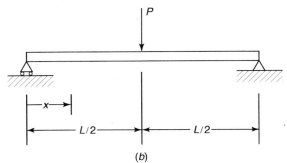

(b)

Figure P13.14 (*continued*)

Problem Set 13.5

13.15 Consider axial vibration of the uniform bar with tip mass, shown in Fig. P9.7. Starting with the eigenvalue equation

$$(EA\phi'_r)' + \rho A \omega_r^2 \phi_r = 0$$

show that the rth natural frequency, ω_r, is given by Eq. 13.45 when the generalized mass M_r is defined by

$$M_r = \int_0^L \rho A \phi_r^2 \, dx + m[\phi_r(L)]^2$$

and the generalized stiffness K_r is defined by

$$K_r = \int_0^L EA \, (\phi'_r)^2 \, dx$$

respectively.

13.16 Consider axial vibration of the uniform bar with spring, as shown in Fig. P9.8. Starting with the eigenvalue equation

$$(EA\phi'_r)' + \rho A \omega_r^2 \phi_r = 0$$

show that the rth natural frequency, ω_r, is given by Eq. 13.45 when the generalized mass M_r is defined by

$$M_r = \int_0^L \rho A \phi_r^2 \, dx$$

and the generalized stiffness K_r is defined by

$$K_r = \int_0^L EA \, (\phi'_r)^2 \, dx + k[\phi_r(L)]^2$$

respectively.

13.17 Consider transverse vibration of the uniform beam with tip mass, shown in Fig. P9.11. Starting with the eigenvalue equation

$$(EI\phi_r'')'' - \rho A \omega_r^2 \phi_r = 0$$

show that the rth natural frequency, ω_r, is given by Eq. 13.45 when the generalized mass M_r is defined by

$$M_r = \int_0^L \rho A \phi_r^2 \, dx + m[\phi_r(L)]^2$$

and the generalized stiffness K_r is defined by

$$K_r = \int_0^L EI \, (\phi_r'')^2 \, dx$$

respectively.

13.18 Consider the bar–mass system in Fig. P9.7. (a) Starting with the eigenvalue equation

$$(EA\phi_r')' - \rho A \omega_r^2 \phi_r = 0$$

derive the equation that axial-vibration modes ϕ_r and ϕ_s must satisfy in order to be "orthogonal with respect to the mass distribution." (b) Derive the equation that they must satisfy to be "orthogonal with respect to the stiffness distribution."

13.19 Consider the beam–mass system in Fig. P9.11. (a) Starting with the eigenvalue equation

$$(EI\phi_r'')'' - \rho A \omega_r^2 \phi_r = 0$$

derive the equation that transverse-vibration modes ϕ_r and ϕ_s must satisfy to be "orthogonal with respect to the mass distribution." (b) Derive the equation that they must satisfy to be "orthogonal with respect to the stiffness distribution."

13.20 Consider the beam–spring system in Fig. P13.7. (a) Starting with the eigenvalue equation

$$(EI\phi_r'')'' - \rho A \omega_r^2 \phi_r = 0$$

derive the equation that transverse-vibration modes ϕ_r and ϕ_s must satisfy to be "orthogonal with respect to the mass distribution." (b) Derive the equation that they must satisfy to be "orthogonal with respect to the stiffness distribution." (c) Finally, determine appropriate expressions for the generalized stiffness K_r and the generalized mass M_r for this system.

PART IV

Computational Methods in Structural Dynamics

14

Introduction to Finite Element Modeling of Structures

In Section 8.4 you were introduced to the Assumed-Modes Method for generating a finite-DOF model of a continuous system. The procedures described there may be referred to as the *Global Assumed-Modes Method* because the assumed modes described deflection patterns throughout the entire structure [e.g., $\psi_i(x), 0 \leq x \leq L$ for a beam of length L]. A far more powerful approximation method, which is a version of the Assumed-Modes Method, is the Finite Element Method. In this chapter we provide an introduction to the use of finite element models in structural dynamics analyses, and in Chapters 15 to 17 treat computational methods for solving large FE problems.

Upon completion of this chapter you should be able to:

- Derive expressions for shape functions for an axial element or a Bernoulli–Euler bending element.
- Generate the element stiffness matrix, mass matrix, or load vector referred to element axes for axial, bending, and torsion elements.
- Derive transformation matrices for truss and frame elements, and transform element matrices to the global reference frame.
- Assemble system stiffness and mass matrices and load vector by the "direct stiffness" method, and enforce boundary conditions.
- Reduce the number of degrees-of-freedom (DOF) of a lumped-mass model by using static condensation.
- Reduce the number of degrees-of-freedom of a consistent-mass finite element model by using Guyan–Irons Reduction.
- Generate a Ritz Transformation matrix corresponding to a stated set of displacement constraint equations.
- Use the *ISMIS* computer program, or some other computer program, to determine modes and frequencies of small structures using consistent mass or lumped-mass FE models.
- Describe the relationship of frequencies of simple structures based on consistent-mass models and on lumped-mass models with frequencies determined from continuum (i.e., partial differential equations) models.

418 Introduction to Finite Element Modeling of Structures

14.1 INTRODUCTION TO THE FINITE ELEMENT METHOD

In Section 8.4, finite-DOF approximations to continuous systems were created using the *Global Assumed-Modes Method*, that is, by approximating the displacement function for the continuous system by an expression of the form

$$v(x,t) = \sum_{i=1}^{N} \psi_i(x) v_i(t) \qquad (14.1)$$

where each $\psi_i(x)$ describes a deflected shape of the entire structure. As noted in Section 8.4, an N-DOF mathematical model of the structure is obtained, with generalized stiffness and mass coefficients and generalized forces being obtained from integrals involving the ψ_i's and their derivatives. The advent of the digital computer facilitated evaluation of these integrals, but it failed to remedy other serious drawbacks of the Global Assumed Modes Method: (1) It is extremely difficult for an analyst to choose a set of ψ_i's for a structure that has complex geometry; (2) the equations that result from applying the global assumed modes procedure are usually highly coupled, and this requires more computer time and memory than is required if the coefficient matrices are sparsely populated; and (3) there is little carryover from one problem to the next, that is, for each new geometry a new set of ψ_i's must be selected. The *Finite Element Method* (FEM), which is introduced in this chapter, overcomes these difficulties and has become the principal computational tool for structural dynamics analysis.

Finite element modeling of a structure may be considered to be an application of the Assumed-Modes Method, wherein the ψ_i's represent deflection shapes that are limited

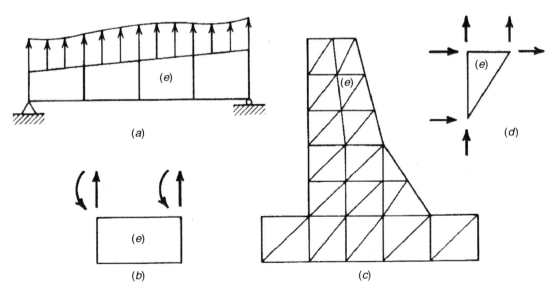

Figure 14.1 Structures represented by finite elements.

to a portion (finite element) of the structure, with the elements being assembled to form the structural system. Figure 14.1a and b show side views of a tapered beam and a typical beam element of uniform cross section, respectively. Figure 14.1c and d show a typical section of a gravity dam modeled by triangular elements. The elements are joined together at nodes, or joints,[1] and displacement compatibility is enforced at these joints. Although the Finite Element Method has a very wide range of applicability, only aspects of FE theory that are particularly applicable to structural dynamics are treated here. To simplify the presentation, emphasis is placed on one-dimensional elements such as the beam element of Fig 14.1b.

In Section 14.2 stiffness and mass matrices and force vectors are derived for several finite element types. Section 14.3 treats the transformation of these element matrices so that all forces and displacements are referred to a global reference frame. Finally, in Section 14.4 the assembling of elements to form a structural system is described. Sections 14.5 through 14.7 discuss the enforcement of boundary conditions and other forms of constraints, including constraints imposed to reduce the number of active degrees of freedom used to model a system. Finally, Section 14.8 presents several small FE structural dynamics examples solved by use of the *ISMIS* computer program.

14.2 ELEMENT STIFFNESS AND MASS MATRICES AND ELEMENT FORCE VECTOR

In this section we develop stiffness and mass matrices and load vectors for uniform one-dimensional elements subjected to axial deformation, bending, and torsion, or to a combination of these. In this chapter, element stiffness matrices and mass matrices and displacement and load vectors are represented by lowercase symbols (e.g., **k**, **p**). System matrices and system vectors are designated by capital letters (e.g., **K**, **U**).

14.2.1 Axial Motion

Consider a uniform bar element of length L, mass density ρ, elastic modulus E, and cross-sectional area A. Choose a reference frame as shown in Fig. 14.2. This is referred to as an *element reference frame* because of the alignment of one axis, labeled the x axis, along the centerline of the element. The simplest approximation of axial displacement within the element employs linear interpolation between the displacements at the two ends and is given by

$$u(x, t) = \psi_i(x)u_1(t) + \psi_2(x)u_2(t) \tag{14.2}$$

---

[1] The term *node* is generally used in finite element literature, but the term *joint* is used in this book because of the alternative meaning that the word *node* has in structural dynamics and vibrations.

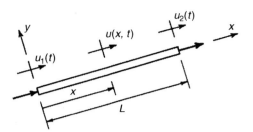

Figure 14.2 Uniform bar element undergoing axial deformation.

Since the two *element displacements referred to element coordinates*, $u_1(t)$ and $u_2(t)$, are defined by $u_1(t) = u(0, t)$ and $u_2(t) = u(L, t)$, the *shape functions* (assumed modes) $\psi_1(x)$ and $\psi_2(x)$ must satisfy the following boundary conditions:

$$\psi_1(0) = 1, \quad \psi_1(L) = 0 \\ \psi_2(0) = 0, \quad \psi_2(L) = 1 \tag{14.3}$$

The shape functions can be derived by considering axial deformation under static loads that would produce the boundary conditions of Eqs. 14.3. Specializing Eq. 12.8 to static deformation gives

$$(AEu')' = 0 \tag{14.4}$$

For the uniform element $AE = $ constant, so

$$u(x) = c_1 + c_2 \frac{x}{L} \tag{14.5}$$

The linear term is nondimensionalized to x/L so that c_1 and c_2 will both carry the dimensions of $u(x)$. From Eqs. 14.3 and 14.5, we obtain the two shape functions that are used to approximate axial displacement along an element of length L. These shape functions are shown in Fig. 14.3.

$$\boxed{\psi_1(x) = 1 - \frac{x}{L}, \quad \psi_2(x) = \frac{x}{L}} \tag{14.6}$$

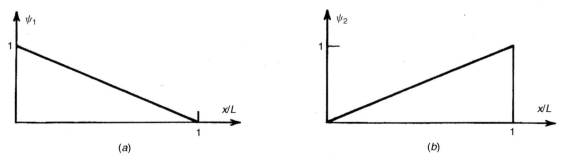

Figure 14.3 Shape functions (a) $\psi_1(x)$ and (b) $\psi_2(x)$ for an axial-deformation element.

14.2 Element Stiffness and Mass Matrices and Element Force Vector

Expressions for the stiffness coefficients k_{ij}, the mass coefficients m_{ij}, and the generalized forces $p_i(t)$ for axial motion are given by Eqs. 8.16, 8.20, and 8.24, respectively. Thus,

$$k_{ij} = \int_0^L EA\psi_i'\psi_j' \, dx, \quad m_{ij} = \int_0^L \rho A \psi_i \psi_j \, dx, \quad p_i(t) = \int_0^L p_x(x,t)\psi_i \, dx \tag{14.7}$$

Substituting Eqs. 14.6 into 14.7a and b we obtain the following stiffness and mass matrices for a uniform element undergoing axial deformation:

$$\mathbf{k} = \frac{AE}{L}\begin{bmatrix} 1 & -1 \\ -1 & 1 \end{bmatrix}, \quad \mathbf{m} = \frac{\rho AL}{6}\begin{bmatrix} 2 & 1 \\ 1 & 2 \end{bmatrix} \tag{14.8}$$

Use of Eq. 14.7c to determine generalized forces is illustrated in Example 14.1.

14.2.2 Transverse Motion: Bernoulli–Euler Beam Theory

Consider a uniform beam element of length L, mass density ρ, elastic modulus E, cross-sectional area A, and moment of inertia I. Let the finite element displacement coordinates for transverse motion be the end displacements and slopes numbered as shown in Fig. 14.4.

Let the displacement $v(x, t)$ be interpolated over $0 \leq x \leq L$ by the equation

$$v(x, t) = \sum_{i=1}^{4} \psi_i(x) v_i(t) \tag{14.9}$$

where the four *shape functions* $\psi_i(x)$ must satisfy the following boundary conditions:

$$\begin{aligned}
\psi_1(0) &= 1, & \psi_1'(0) &= \psi_1(L) = \psi_1'(L) = 0 \\
\psi_2'(0) &= 1, & \psi_2(0) &= \psi_2(L) = \psi_2'(L) = 0 \\
\psi_3(L) &= 1, & \psi_3(0) &= \psi_3'(0) = \psi_3'(L) = 0 \\
\psi_4'(L) &= 1, & \psi_4(0) &= \psi_4'(0) = \psi_4(L) = 0
\end{aligned} \tag{14.10}$$

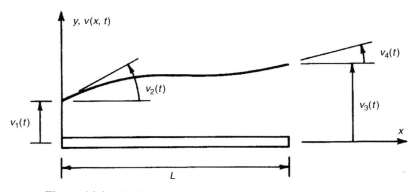

Figure 14.4 Uniform element undergoing transverse deflection.

Appropriate shape functions can easily be derived by considering the beam element in Fig. 14.4 to be loaded statically by end shears and bending moments to produce the various static deflection shapes that satisfy Eqs. 14.10. Thus, for a beam loaded only at its ends, the equilibrium equation is

$$(EI v'')'' = 0 \tag{14.11}$$

The general solution of Eq. 14.11 for a uniform beam (EI = constant) is a cubic polynomial of the form

$$v(x) = c_1 + c_2 \frac{x}{L} + c_3 \left(\frac{x}{L}\right)^2 + c_4 \left(\frac{x}{L}\right)^3 \tag{14.12}$$

Note that x/L is again used instead of x, so that all c_i's will have the same dimension. Substituting the four sets of boundary conditions of Eqs. 14.10 into Eq. 14.12 we obtain the following shape functions for transverse deflection of a beam element of length L:

$$\boxed{\begin{aligned}
\psi_1(x) &= 1 - 3\left(\frac{x}{L}\right)^2 + 2\left(\frac{x}{L}\right)^3 \\
\psi_2(x) &= x - 2L\left(\frac{x}{L}\right)^2 + L\left(\frac{x}{L}\right)^3 \\
\psi_3(x) &= 3\left(\frac{x}{L}\right)^2 - 2\left(\frac{x}{L}\right)^3 \\
\psi_4(x) &= -L\left(\frac{x}{L}\right)^2 + L\left(\frac{x}{L}\right)^3
\end{aligned}} \tag{14.13}$$

These shape functions are illustrated in Fig. 14.5. With these cubic-polynomial shape functions, both displacement and slope can be made compatible at both ends of beam elements.

Note that although it appears that the shape functions defined by Eqs. 14.13 lead to deflections in Fig. 14.5 that are "large deflections," in Eq. 14.9 each shape function is multiplied by its corresponding amplitude $v_i(t)$, and these amplitudes will remain small enough so that the overall deflection of the beam remains "small" in comparison with the depth of the beam. Although these functions are <u>exact</u> displacement shapes <u>for uniform beams that are statically loaded only by end shears and moments</u>, they will be used to <u>approximate dynamic behavior</u>, even of nonuniform beam elements that may be loaded transversely between the two ends of the element.

Equations 8.30 through 8.32 are expressions for k_{ij}, m_{ij}, and $p_i(t)$ for Bernoulli–Euler beams. Thus,

$$k_{ij} = \int_0^L EI \psi_i'' \psi_j'' \, dx \tag{14.14a}$$

$$m_{ij} = \int_0^L \rho A \psi_i \psi_j \, dx \tag{14.14b}$$

$$p_i(t) = \int_0^L p_y(x, t) \psi_i \, dx \tag{14.14c}$$

14.2 Element Stiffness and Mass Matrices and Element Force Vector

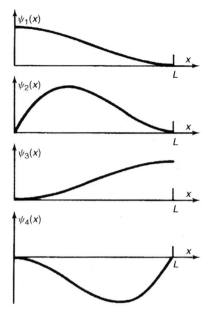

Figure 14.5 Shape functions for transverse deformation of a beam element.

Substituting Eqs. 14.13 into Eqs. 14.14a and b, we obtain the following stiffness and mass matrices for a Bernoulli–Euler beam element:

$$\mathbf{k} = \frac{EI}{L^3} \begin{bmatrix} 12 & 6L & -12 & 6L \\ & 4L^2 & -6L & 2L^2 \\ & & 12 & -6L \\ \text{symm.} & & & 4L^2 \end{bmatrix}$$

$$\mathbf{m} = \frac{\rho A L}{420} \begin{bmatrix} 156 & 22L & 54 & -13L \\ & 4L^2 & 13L & -3L^2 \\ & & 156 & -22L \\ \text{symm.} & & & 4L^2 \end{bmatrix} \quad (14.15)$$

Example 14.1 Determine the generalized load vector $\mathbf{p}(t)$ for a beam element subjected to a uniform transverse load $f(t)$, as depicted in Fig. 1.

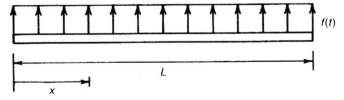

Figure 1 Beam element with a uniformly distributed time-dependent transverse loading.

SOLUTION The beam element displacement coordinates referred to element axes are shown in Fig. 14.4 and Fig. 2, and the shape functions are given in Eqs. 14.13. Thus,

$$\psi_1(x) = 1 - 3\left(\frac{x}{L}\right)^2 + 2\left(\frac{x}{L}\right)^3$$
$$\psi_2(x) = x - 2L\left(\frac{x}{L}\right)^2 + L\left(\frac{x}{L}\right)^3$$
$$\psi_3(x) = 3\left(\frac{x}{L}\right)^2 - 2\left(\frac{x}{L}\right)^3 \quad (1)$$
$$\psi_4(x) = -L\left(\frac{x}{L}\right)^2 + L\left(\frac{x}{L}\right)^3$$

Figure 2 Beam element displacement coordinates.

The equation for the generalized force components is Eq. 14.14c; that is,

$$p_i(t) = \int_0^L p_y(x,t)\psi_i(x)\,dx \quad (2)$$

Combining Eqs. 1 and 2, we get

$$\begin{aligned}
p_1(t) &= f(t)\int_0^L \psi_1(x)\,dx \\
&= f(t)\int_0^L \left[1 - 3\left(\frac{x}{L}\right)^2 + 2\left(\frac{x}{L}\right)^3\right]dx \\
&= f(t)L\left[\left(\frac{x}{L}\right) - \left(\frac{x}{L}\right)^3 + \left(\frac{2}{4}\right)\left(\frac{x}{L}\right)^4\right]_0^L \\
&= \frac{f(t)L}{2}
\end{aligned} \quad (3)$$

The remaining generalized forces are determined in a similar fashion, giving

$$\mathbf{p}(t) = \begin{Bmatrix} \frac{1}{2}f(t)L \\ \frac{1}{12}f(t)L^2 \\ \frac{1}{2}f(t)L \\ -\frac{1}{12}f(t)L^2 \end{Bmatrix} \quad \textbf{Ans.} \ (4)$$

14.2.3 Torsion

In Section 12.1.2 the equation of motion and boundary conditions for torsional deformation of a circular rod were obtained. We wish now to derive expressions for stiffness and mass coefficients and generalized forces (torsional moments) for a uniform torsion element of length L, as shown in Fig. 14.6.

Let the local x axis be the centroidal axis. Let I_p be the area polar moment of inertia about the centroidal axis and let GJ be the torsional stiffness.[2] Then the strain and kinetic energies for pure torsion are given by

$$\mathcal{V} = \tfrac{1}{2} \int_0^L GJ(\theta')^2 \, dx, \qquad \mathcal{T} = \tfrac{1}{2} \int_0^L \rho I_p (\dot{\theta})^2 \, dx \qquad (14.16)$$

The rotation along the element is to be given by the assumed-modes form

$$\theta(x, t) = \psi_1(x)\theta_1(t) + \psi_2(x)\theta_2(t) \qquad (14.17)$$

where the appropriate boundary conditions for the shape functions are given by Eqs. 14.3. For a torsion member loaded statically by end torques, the equilibrium equation, obtained from Eq. 12.14, is

$$(GJ\theta')' = 0 \qquad (14.18)$$

Since this equation has the same form as the equilibrium equation for axial deformation, Eq. 14.4, and since the torsion shape functions must satisfy the same boundary conditions as the axial shape functions, the shape functions for the torsion element and axial element are the same; that is, ψ_1 and ψ_2 are given by Eqs. 14.6.

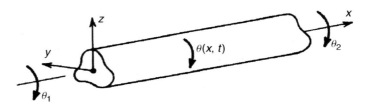

Figure 14.6 Uniform element undergoing torsional deformation.

---

[2] For a circular cross section, $J = I_p$. For other cross-sectional shapes the reader may consult texts on advanced strength of materials or elasticity for expressions for J. For noncircular members the axis of twist and the centroidal axis frequently do not coincide. This leads to a coupling of bending and torsion. An example of this coupling, vibration of an airplane wing, has been discussed by numerous authors.[14.1] Hence, great care should be exercised in modeling any structural dynamics problem where member torsion is involved.

Using the procedures employed in Chapter 8, we can combine Eqs. 14.16 and 14.17 to obtain the expressions

$$k_{ij} = \int_0^L GJ \psi_i' \psi_j' \, dx$$

$$m_{ij} = \int_0^L \rho I_p \psi_i \psi_j \, dx \quad (14.19)$$

$$p_i(t) = \int_0^L t_\theta(x,t) \psi_i \, dx$$

where $t_\theta(x,t)$ is a distributed torque per unit length. Inserting the shape functions of Eqs. 14.6 into Eqs. 14.19, we obtain the following stiffness matrix and mass matrix for a uniform torsion element:

$$\boxed{\mathbf{k} = \frac{GJ}{L} \begin{bmatrix} 1 & -1 \\ -1 & 1 \end{bmatrix}, \quad \mathbf{m} = \frac{\rho I_p L}{6} \begin{bmatrix} 2 & 1 \\ 1 & 2 \end{bmatrix}} \quad (14.20)$$

As noted in footnote 2 accompanying Eqs. 14.16, it is again emphasized that treating the dynamics of torsion members that have a noncircular cross section (e.g., structural channel sections) requires great care and is beyond the scope of this book.

14.2.4 Three-Dimensional Frame Element

The stiffness and mass matrices for a slender three-dimensional frame element can be obtained from the axial, bending, and torsion elements discussed above. In particular, Bernoulli–Euler beam theory was used in deriving the bending element. Przemieniecki[14.2] presents derivations for the stiffness and mass matrices of a three-dimensional frame element including shear deformation and rotatory inertia effects.

Figure 14.7 shows the local reference frame and the displacement coordinates for a three-dimensional frame element. The x axis is along the line of centroids of the cross sections and the y and z axes are principal axes in the cross section. The area moments of inertia of the cross section are I_y and I_z, and I_p is the area polar moment of inertia about the x axis.

The stiffness and mass coefficients associated with axial displacements u_1 and u_7 are given by Eqs. 14.8; those associated with u_2, u_6, u_8, and u_{12} are based on Eqs. 14.15 for bending in the x–y plane, while u_9, u_{11}, u_3, and u_5 are for bending in the x–z plane. The order of these bending degrees of freedom preserves the sense of the rotations shown in Fig. 14.4. Finally, the coefficients associated with torsional degrees of freedom u_4 and u_{10} are obtained from Eqs. 14.20. Finally, the 12-DOF stiffness matrix and mass matrix for a uniform three-dimensional element are given in Eqs. 14.21a and b, respectively.

$$\mathbf{k} = \begin{bmatrix}
\frac{EA}{L} & 0 & 0 & 0 & 0 & 0 & -\frac{EA}{L} & 0 & 0 & 0 & 0 & 0 \\
0 & \frac{12EI_z}{L^3} & 0 & 0 & 0 & \frac{6EI_z}{L^2} & 0 & -\frac{12EI_z}{L^3} & 0 & 0 & 0 & \frac{6EI_z}{L^2} \\
0 & 0 & \frac{12EI_y}{L^3} & 0 & -\frac{6EI_y}{L^2} & 0 & 0 & 0 & -\frac{12EI_y}{L^3} & 0 & -\frac{6EI_y}{L^2} & 0 \\
0 & 0 & 0 & \frac{GJ}{L} & 0 & 0 & 0 & 0 & 0 & -\frac{GJ}{L} & 0 & 0 \\
0 & 0 & -\frac{6EI_y}{L^2} & 0 & \frac{4EI_y}{L} & 0 & 0 & 0 & \frac{6EI_y}{L^2} & 0 & \frac{2EI_y}{L} & 0 \\
0 & \frac{6EI_z}{L^2} & 0 & 0 & 0 & \frac{4EI_z}{L} & 0 & -\frac{6EI_z}{L^2} & 0 & 0 & 0 & \frac{2EI_z}{L} \\
-\frac{EA}{L} & 0 & 0 & 0 & 0 & 0 & \frac{EA}{L} & 0 & 0 & 0 & 0 & 0 \\
0 & -\frac{12EI_z}{L^3} & 0 & 0 & 0 & -\frac{6EI_z}{L^2} & 0 & \frac{12EI_z}{L^3} & 0 & 0 & 0 & -\frac{6EI_z}{L^2} \\
0 & 0 & -\frac{12EI_y}{L^3} & 0 & \frac{6EI_y}{L^2} & 0 & 0 & 0 & \frac{12EI_y}{L^3} & 0 & \frac{6EI_y}{L^2} & 0 \\
0 & 0 & 0 & -\frac{GJ}{L} & 0 & 0 & 0 & 0 & 0 & \frac{GJ}{L} & 0 & 0 \\
0 & 0 & -\frac{6EI_y}{L^2} & 0 & \frac{2EI_y}{L} & 0 & 0 & 0 & \frac{6EI_y}{L^2} & 0 & \frac{4EI_y}{L} & 0 \\
0 & \frac{6EI_z}{L^2} & 0 & 0 & 0 & \frac{2EI_z}{L} & 0 & -\frac{6EI_z}{L^2} & 0 & 0 & 0 & \frac{4EI_z}{L}
\end{bmatrix} \quad (14.21a)$$

$$\mathbf{m} = \frac{\rho A L}{420} \begin{bmatrix} 140 & & & & & & 70 & & & & & \\ & 156 & & & & 22L & & 54 & & & & -13L \\ & & 156 & & -22L & & & & 54 & & 13L & \\ & & & \dfrac{140 I_p}{A} & & & & & & \dfrac{70 I_p}{A} & & \\ & & -22L & & 4L^2 & & & & -13L & & -3L^2 & \\ & 22L & & & & 4L^2 & & 13L & & & & -3L^2 \\ 70 & & & & & & 140 & & & & & \\ & 54 & & & & 13L & & 156 & & & & -22L \\ & & 54 & & -13L & & & & 156 & & 22L & \\ & & & \dfrac{70 I_p}{A} & & & & & & \dfrac{140 I_p}{A} & & \\ & & 13L & & -3L^2 & & & & 22L & & 4L^2 & \\ & -13L & & & & -3L^2 & & -22L & & & & 4L^2 \end{bmatrix}$$

$$(14.21b)$$

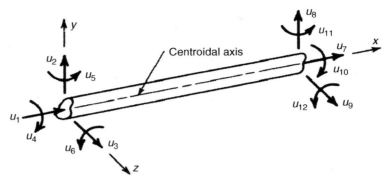

Figure 14.7 Notation for a three-dimensional frame element.

Stiffness and mass matrices for other finite elements, including two- and three-dimensional elements, may be found in reference texts and journal articles on finite elements, for example, Ref. [14.2].

14.2.5 Lumped-Mass Matrix for Beam Elements

The mass matrices derived above are referred to as *consistent-mass matrices* because the same shape functions are used to derive the coefficients of the mass matrix as are used for the stiffness matrix. This leads, for example, to coupling between translational coordinates and rotations in the mass matrix for a beam, Eq. 14.15b. A simpler model of the inertia properties of a beam is the *lumped-mass model*. Figure 14.8 shows a beam divided into finite elements with the mass of elements lumped at the element ends and reassembled to form a lumped-mass model of the beam. For elements of equal length

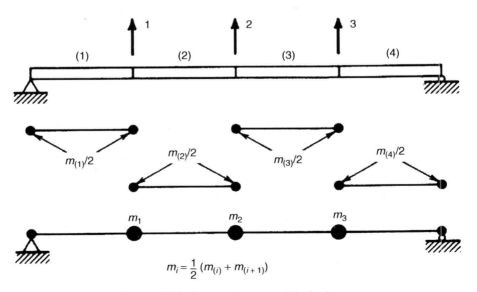

Figure 14.8 Lumped-mass model of a beam.

the lumped-mass matrix for the beam of Fig. 14.8 would be

$$\mathbf{M} = \frac{\rho A L}{4} \begin{bmatrix} 1 & 0 & 0 \\ 0 & 1 & 0 \\ 0 & 0 & 1 \end{bmatrix} \qquad (14.22)$$

Archer[14.3], Leckie and Lindberg[14.4], and Tong et al.,[14.5] have compared eigensolutions based on lumped- and consistent-mass formulations. We return to this topic in Section 14.6 when we discuss static condensation.

Thus far, expressions have been obtained for element stiffness and mass matrices and for element load vectors. Some procedures for handling damping in structures were discussed in Section 10.3.

14.3 TRANSFORMATION OF ELEMENT MATRICES

In Section 14.1 it was indicated that the finite element method consists of locally approximating displacements and then tying elements together through displacement compatibility conditions at the joints. This approach is called the *displacement method*, because it is the displacements within an element that are directly approximated. Texts on the finite element method also discuss the *force method*, wherein stresses are approximated directly, and *mixed methods*. We pursue only the displacement method in this book because it is the one that is most commonly employed in structural dynamics analyses.

In Section 14.2, stiffness and mass matrices were derived for several uniform elements, with the displacement coordinates referred to an *element reference frame*, labeled the *xyz* frame. Truss and frame structures frequently have members that are not aligned with a common set of axes, that is, with a *global reference frame* or *XYZ* frame. Such is the case for the simple planar truss (or plane truss) in Fig. 14.9, where the ends of elements are labeled i and j, and the local axis sign convention places the origin at the i end with the x axis along the member from i toward j.

In this section we consider the transformation of element matrices (displacement and force vectors, stiffness and mass matrices) to the global reference so that compatibility equations and assembly of system matrices can be treated more directly. We begin with a planar truss element, which is a two-force axial-deformation member.

14.3.1 Displacement Transformation: Planar Truss Element

Figure 14.10a shows the *element displacement coordinates referred to element axes*, which will be designated the *ECE* coordinates for a planar truss element; and Fig. 14.10b shows the *element displacement coordinates referred to global axes*, which will be designated the *ECG* coordinates for this element. The angle θ shown in Fig. 14.10 is the angle measured from the global X axis to the local x axis. In Fig. 14.10a the axial displacements u_1 and u_3 are associated with rigid-body motion along the x-axis direction and axial deformation of the truss element. For static-deformation problems, only these

14.3 Transformation of Element Matrices   431

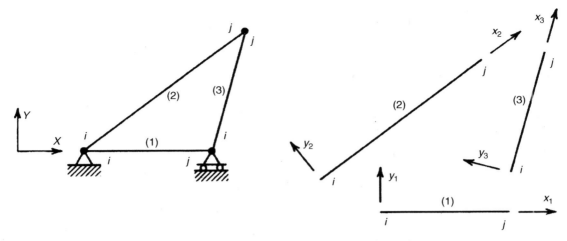

Figure 14.9 Global and element reference frames for a planar truss.

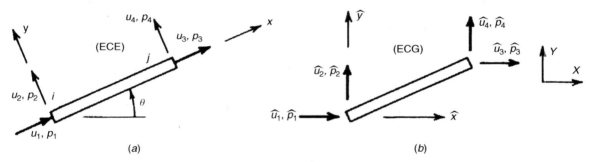

Figure 14.10 Two sets of element displacement coordinates for a planar truss element: (a) element displacements referred to element axes; (b) element displacements referred to global axes.

two axial coordinates are needed. However, for structural dynamics problems, even though all bending of the planar truss element is ignored, the two transverse displacements u_2 and u_4 are still needed to express rigid-body motion of the element in the y direction, including small rotation of the element as a rigid body.

When the truss element is assembled as one member of a planar truss, it is the displacements along the global directions $\hat{x}\ (= X)$ and $\hat{y}\ (= Y)$ that are employed. For the element, then, $\hat{u}_1$ and $\hat{u}_2$ are the displacements in the $\hat{x}$ and $\hat{y}$ directions at the i end, while $\hat{u}_3$ and $\hat{u}_4$ are the displacements in the $\hat{x}$ and $\hat{y}$ directions at the j end of the element.

Figure 14.11a shows the change of configuration of a truss element due to an $\hat{x}$ displacement $\hat{u}_1$, and Fig. 14.11b shows the change of configuration of the element due to a $\hat{y}$ displacement $\hat{u}_2$. In each figure, the contribution to the x displacement u_1 is labeled. From the geometry of these figures, it is easy to see that the sum of these two

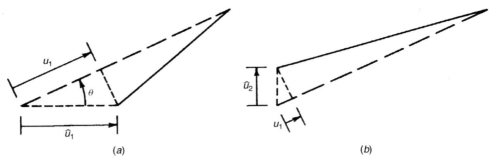

Figure 14.11 Sample displacement transformation for a planar truss element (displacements exaggerated): (a) $\hat{x}$ displacement at end i; (b) $\hat{y}$ displacement at end i. (Long-dashed lines represent the original undeformed element.)

contributions gives

$$u_1 = \hat{u}_1 \cos\theta + \hat{u}_2 \sin\theta \tag{14.23a}$$

Similarly, the projections of $\hat{u}_1$ and $\hat{u}_2$ onto the $\hat{y}$ direction give the total displacement of

$$u_2 = -\hat{u}_1 \sin\theta + \hat{u}_2 \cos\theta \tag{14.23b}$$

It is left as a homework exercise for the reader to show that the projections of $\hat{u}_3$ and $\hat{u}_4$ at the j end of the element lead to the following displacements in the local xy frame:

$$u_3 = \hat{u}_3 \cos\theta + \hat{u}_4 \sin\theta \tag{14.23c}$$

$$u_4 = -\hat{u}_3 \sin\theta + \hat{u}_4 \cos\theta \tag{14.23d}$$

These equations can be combined into a single matrix equation

$$\boxed{\mathbf{u} = \mathbf{T}\hat{\mathbf{u}}} \tag{14.24}$$

where for the planar truss element,

$$\mathbf{u} = \lfloor u_1\ u_2\ u_3\ u_4 \rfloor^T, \qquad \hat{\mathbf{u}} = \lfloor \hat{u}_1\ \hat{u}_2\ \hat{u}_3\ \hat{u}_4 \rfloor^T \tag{14.25}$$

Therefore, the *displacement transformation matrix for a planar truss element* is

$$\mathbf{T} = \begin{bmatrix} \cos\theta & \sin\theta & 0 & 0 \\ -\sin\theta & \cos\theta & 0 & 0 \\ 0 & 0 & \cos\theta & \sin\theta \\ 0 & 0 & -\sin\theta & \cos\theta \end{bmatrix} \tag{14.26}$$

Example 14.2 Three bars are pin-connected to form a planar truss, as shown in Fig. 1. Set up the displacement transformation matrices for elements 1 and 2. Label the ends of the elements as indicated in Fig. 1, and orient the global axes XY as shown on the figure.

14.3 Transformation of Element Matrices

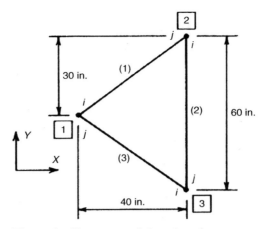

Figure 1 Pin-connected three-bar planar truss.

SOLUTION From Fig. 2a, for element 1,

$$L_1 = \sqrt{(30 \text{ in.})^2 + (40 \text{ in.})^2} = 50 \text{ in.} \tag{1}$$

$$\cos\theta_1 = \tfrac{4}{5}, \qquad \sin\theta_1 = \tfrac{4}{5} \tag{2}$$

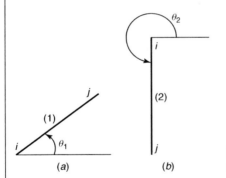

Figure 2 Original configurations of (a) element 1 and (b) element 2.

so the displacement transformation matrix for element 1 is

$$\mathbf{T}_1 = \begin{bmatrix} \cos\theta_1 & \sin\theta_1 & 0 & 0 \\ -\sin\theta_1 & \cos\theta_1 & 0 & 0 \\ 0 & 0 & \cos\theta_1 & \sin\theta_1 \\ 0 & 0 & -\sin\theta_1 & \cos\theta_1 \end{bmatrix} = \frac{1}{5}\begin{bmatrix} 4 & 3 & 0 & 0 \\ -3 & 4 & 0 & 0 \\ 0 & 0 & 4 & 3 \\ 0 & 0 & -3 & 4 \end{bmatrix} \quad \textbf{Ans.} \tag{3}$$

From Fig. 2b, for element 2,

$$\cos\theta_2 = 0, \qquad \sin\theta_2 = -1 \tag{4}$$

(*Note:* Since the direction from end i to end j is downward for element 2, $\sin\theta_2 = -1$.) Finally, the displacement transformation matrix for element 2 is

$$\mathbf{T}_2 = \begin{bmatrix} \cos\theta_2 & \sin\theta_2 & 0 & 0 \\ -\sin\theta_2 & \cos\theta_2 & 0 & 0 \\ 0 & 0 & \cos\theta_2 & \sin\theta_2 \\ 0 & 0 & -\sin\theta_2 & \cos\theta_2 \end{bmatrix} = \begin{bmatrix} 0 & -1 & 0 & 0 \\ 1 & 0 & 0 & 0 \\ 0 & 0 & 0 & -1 \\ 0 & 0 & 1 & 0 \end{bmatrix} \quad \textbf{Ans. (5)}$$

14.3.2 Displacement Transformation: Plane Frame Element

Truss elements can only sustain axial forces, whereas frame elements can sustain transverse shear forces and bending moments as well as axial forces. As for displacement coordinates at the ends of the elements, frame elements have rotational degrees of freedom in addition to translational degrees of freedom (Fig. 14.12), whereas truss elements have only translational degrees of freedom (Fig. 14.10). Thus, the *ECE*-coordinates and *ECG*-coordinates for the planar frame element (or plane frame element)

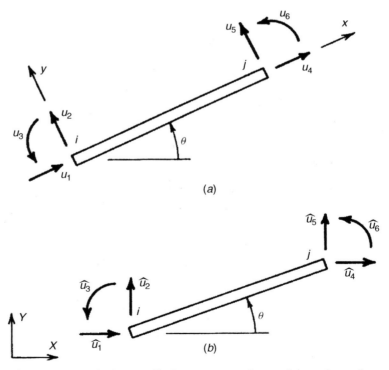

Figure 14.12 Two sets of element displacement coordinates for a planar frame element: (*a*) element displacements referred to element axes; (*b*) element displacements referred to global axes.

are, respectively,

$$\mathbf{u} = \lfloor u_1 \; u_2 \; u_3 \; u_4 \; u_5 \; u_6 \rfloor^T, \qquad \widehat{\mathbf{u}} = \lfloor \widehat{u}_1 \; \widehat{u}_2 \; \widehat{u}_3 \; \widehat{u}_4 \; \widehat{u}_5 \; \widehat{u}_6 \rfloor^T \qquad (14.27)$$

The displacement transformation matrix for frame elements can be deduced directly from Eq. 14.26 by renumbering the translational coordinates and adding the two rotational-DOF equations

$$u_3 = \widehat{u}_3, \qquad u_6 = \widehat{u}_6$$

Again, the displacement transformation has the form of Eq. 14.24, where the displacement coordinates for the planar frame element are defined in Fig. 14.12 and given by Eq. 14.27, and the *displacement transformation matrix for the planar frame element* is

$$\mathbf{T} = \begin{bmatrix} \cos\theta & \sin\theta & 0 & 0 & 0 & 0 \\ -\sin\theta & \cos\theta & 0 & 0 & 0 & 0 \\ 0 & 0 & 1 & 0 & 0 & 0 \\ 0 & 0 & 0 & \cos\theta & \sin\theta & 0 \\ 0 & 0 & 0 & -\sin\theta & \cos\theta & 0 \\ 0 & 0 & 0 & 0 & 0 & 1 \end{bmatrix} \qquad (14.28)$$

14.3.3 Displacement Transformation: Three-Dimensional Truss Element

Figure 14.13 shows a three-dimensional truss element, which has three translational displacement coordinates at each end. Again, the displacement transformation has the form of Eq. 14.24, where the displacement coordinates for the three-dimensional truss

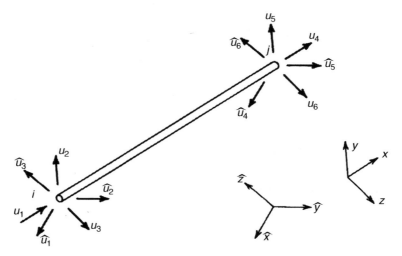

Figure 14.13 Two sets of element displacement coordinates for a three-dimensional truss element.

element are defined in Fig. 14.13 and given by Eq. 14.27. The transformation matrix that transforms $\widehat{\mathbf{u}}$ displacements to $\mathbf{u}$ displacements can be written as a matrix of direction cosines. Thus, the *displacement transformation matrix for a three-dimensional truss element* has the form

$$\mathbf{T} = \begin{bmatrix} \mathbf{T}_c & \mathbf{0} \\ \mathbf{0} & \mathbf{T}_c \end{bmatrix} \tag{14.29a}$$

where $\mathbf{T}_c$ is the following matrix of direction cosines:

$$\mathbf{T}_c = \begin{bmatrix} \cos x\widehat{x} & \cos x\widehat{y} & \cos x\widehat{z} \\ \cos y\widehat{x} & \cos y\widehat{y} & \cos y\widehat{z} \\ \cos z\widehat{x} & \cos z\widehat{y} & \cos z\widehat{z} \end{bmatrix} \tag{14.29b}$$

14.3.4 Transformation of Element Force Vectors and Element Stiffness and Element Mass Matrices

To obtain the transformations for the force vector, the stiffness matrix, and the mass matrix for the various elements, it is convenient to use symbolic equations and to employ virtual work and energy concepts. To transform the force vector $\mathbf{p}$ associated with the displacement vector $\mathbf{u}$ in local element coordinates to force vector $\widehat{\mathbf{p}}$ associated with displacement vector $\widehat{\mathbf{u}}$ in global coordinates, consider the virtual work. The scalar quantity $\mathcal{W}$ should be the same for both representations; that is,

$$\delta \mathcal{W} = \delta \mathbf{u}^T \mathbf{p} = \delta \widehat{\mathbf{u}}^T \widehat{\mathbf{p}} \tag{14.30}$$

Substituting the displacement transformation equation, Eq. 14.24, into the virtual work equation above, we get

$$\delta \mathcal{W} = \delta \mathbf{u}^T \mathbf{p} = (\delta \widehat{\mathbf{u}}^T \mathbf{T}^T) \mathbf{p} = \delta \widehat{\mathbf{u}}^T (\mathbf{T}^T \mathbf{p}) \tag{14.31}$$

By comparing the terms multiplying $\delta\widehat{u}_1$, $\delta\widehat{u}_2$, and so on, in Eqs. 14.30 and 14.31, we can conclude that

$$\boxed{\widehat{\mathbf{p}} = \mathbf{T}^T \mathbf{p}} \tag{14.32}$$

since the $\delta\widehat{u}_i$'s are independent. It is quite straightforward to show that Eq. 14.32 is nothing other than the projection of the *ECE* element force vector onto the *ECG* axes.

To obtain the transformation equations for stiffness and mass matrices, we can employ the strain energy and the kinetic energy, respectively. From Eqs. 8.17 and 8.21 we get

$$\begin{aligned} \mathcal{V} &= \tfrac{1}{2}\mathbf{u}^T\mathbf{k}\mathbf{u} = \tfrac{1}{2}\widehat{\mathbf{u}}^T\widehat{\mathbf{k}}\widehat{\mathbf{u}} \\ \mathcal{T} &= \tfrac{1}{2}\dot{\mathbf{u}}^T\mathbf{m}\dot{\mathbf{u}} = \tfrac{1}{2}\dot{\widehat{\mathbf{u}}}^T\widehat{\mathbf{m}}\dot{\widehat{\mathbf{u}}} \end{aligned} \tag{14.33}$$

Thus, inserting the displacement transformation equation, Eq. 14.24, we get

$$\mathcal{V} = \tfrac{1}{2}\mathbf{u}^T\mathbf{k}\mathbf{u} = \tfrac{1}{2}(\widehat{\mathbf{u}}^T\mathbf{T}^T)\mathbf{k}(\mathbf{T}\widehat{\mathbf{u}}) = \tfrac{1}{2}\widehat{\mathbf{u}}^T(\mathbf{T}^T\mathbf{k}\mathbf{T})\widehat{\mathbf{u}} = \tfrac{1}{2}\widehat{\mathbf{u}}^T\widehat{\mathbf{k}}\widehat{\mathbf{u}} \tag{14.34}$$

Therefore, the *coordinate transformation for the element stiffness matrix* is

$$\widehat{\mathbf{k}} = \mathbf{T}^T \mathbf{k} \mathbf{T} \tag{14.35a}$$

where **k** is the *ECE* element stiffness matrix. This gives the stiffness matrix referred to the global coordinates. Similarly, from the kinetic energy expressions we can obtain the following *coordinate transformation for the element mass matrix*:

$$\widehat{\mathbf{m}} = \mathbf{T}^T \mathbf{m} \mathbf{T} \tag{14.35b}$$

which gives the mass matrix referred to the global coordinates. These are the forms of element matrices that are needed to assemble the stiffness and mass matrices for the assembled structure in global coordinates.

14.3.5 Transformation of Stiffness Matrices and Mass Matrices: Planar Truss

The transformation of the stiffness and mass matrices for a planar truss element requires special consideration. Since this element has stiffness only along its axial direction, the *element stiffness matrix in ECE coordinates* can be obtained from the axial stiffness matrix in Eq. 14.8a, renumbered according to the degrees of freedom in Fig. 14.10a. Thus,

$$\mathbf{k} = \frac{AE}{L} \begin{bmatrix} 1 & 0 & -1 & 0 \\ 0 & 0 & 0 & 0 \\ -1 & 0 & 1 & 0 \\ 0 & 0 & 0 & 0 \end{bmatrix} \tag{14.36a}$$

Combining Eqs. 14.26, 14.35a, and 14.36a, we get the following expression for the *planar truss element stiffness matrix in global (ECG) coordinates*:

$$\widehat{\mathbf{k}} = \frac{AE}{L} \begin{bmatrix} c^2 & cs & -c^2 & -cs \\ & s^2 & -cs & -s^2 \\ & & c^2 & cs \\ \text{symm.} & & & s^2 \end{bmatrix} \tag{14.36b}$$

where $c = \cos\theta$ and $s = \sin\theta$, with θ as defined in Fig. 14.10. Although this stiffness matrix was formed by using the triple matrix product of Eq. 14.35a, it is simpler just to refer directly to Eq. 14.36b when forming element stiffness matrices "by hand" for simple structures.

Whereas the truss element has stiffness only along its axial direction, the element has rigid-body inertia in the transverse directions as well as the inertia associated with the two axial coordinates. The rigid-body translation in the transverse direction can be interpolated by the same linear shape function as the axial deformation. Hence, Eq. 14.8b can be expanded to give the following *element mass matrix in local (ECE) coordinates*:

$$\mathbf{m} = \frac{\rho A L}{6} \begin{bmatrix} 2 & 0 & 1 & 0 \\ 0 & 2 & 0 & 1 \\ 1 & 0 & 2 & 0 \\ 0 & 1 & 0 & 2 \end{bmatrix} \tag{14.37a}$$

Combining Eqs. 14.26, 14.35b, and 14.37a, we get the following expression for the *planar truss element mass matrix in global (ECG) coordinates:*

$$\hat{\mathbf{m}} = \frac{\rho AL}{6} \begin{bmatrix} 2 & 0 & 1 & 0 \\ 0 & 2 & 0 & 1 \\ 1 & 0 & 2 & 0 \\ 0 & 1 & 0 & 2 \end{bmatrix} \quad (14.37b)$$

Note that $\hat{\mathbf{m}} = \mathbf{m}$ regardless of the orientation of the truss element!

14.4 ASSEMBLY OF SYSTEM MATRICES: DIRECT STIFFNESS METHOD

Several processes are involved in arriving at the final set of equations of motion based on a finite element model of a system. So far we have considered processes at the element level: generating element matrices in the local element coordinates and transforming them to global coordinates. There remains the assembly of system matrices, the enforcement of boundary conditions, and the enforcement of other constraints, if any (e.g., reducing the number of coordinates of the system). In many finite element codes, these processes are carried out simultaneously; that is, boundary conditions and other constraints are imposed as the system matrices are assembled. For clarity of presentation, however, we consider these activities separately.

In Section 14.3 the transformation matrix $\mathbf{T}$ was introduced to allow all element matrices to be referred to a common reference frame, the global *XYZ* frame. In this section the direct stiffness method is employed to assemble system matrices $\mathbf{M}$ and $\mathbf{K}$ and system load vector $\mathbf{P}$ from the corresponding element matrices.

When several elements are combined to form a structure, the common set of *system displacement coordinates* is designated by the vector $\mathbf{U}$. Figure 14.14 shows the six system displacements of an unrestrained three-bar truss. The joints are labeled 1, 2,

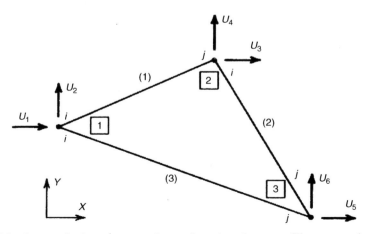

Figure 14.14 System displacement coordinates for a three-bar truss. Element numbers in parentheses, node (joint) numbers in boxes.

14.4 Assembly of System Matrices: Direct Stiffness Method

and 3, and the system displacements are taken in the following order: U_1 is the displacement in the X direction of joint 1, U_2 is the displacement in the Y direction of joint 1, and so on, as shown in Fig. 14.14. Notice that we have now introduced three distinct types of displacement coordinates: *element displacements referred to element axes* (ECEs), *element displacements referred to global axes* (ECGs), and *system displacement coordinates*. In this section we use the relationships between the latter two types in assembling the system matrices.

The *Direct Stiffness Method* for assembling system matrices is based on the fact that work and energy are scalar quantities, so that, for example, the total strain energy of a structure is the sum of the strain energy contributions of all of its elements. In addition, the element displacements referred to global axes can simply be identified with the appropriate system displacements; that is, the displacements $\hat{\mathbf{u}}_e$ of element e can be identified with the appropriate system displacements in $\mathbf{U}$.

For example, consider element 3 of the three-bar truss in Fig. 14.14. The element displacements referred to global axes are identified on Fig. 14.15, where the orientation of element axes is defined by the i, j notation on Fig. 14.14. By comparing Fig. 14.15 with Fig. 14.14, we note that the displacements of element 3 are related by

$$\hat{u}_1 = U_1, \quad \hat{u}_2 = U_2, \quad \hat{u}_3 = U_5, \quad \hat{u}_4 = U_6 \tag{14.38}$$

The element coordinates referred to element axes can be related to system displacement coordinates through *locator matrices* (or label matrices) $\mathbf{L}_e$ such that

$$\boxed{\hat{\mathbf{u}}_e = \mathbf{L}_e \mathbf{U}} \tag{14.39}$$

For example, for element 3 the locator matrix is

$$\mathbf{L}_3 = \begin{bmatrix} 1 & 0 & 0 & 0 & 0 & 0 \\ 0 & 1 & 0 & 0 & 0 & 0 \\ 0 & 0 & 0 & 0 & 1 & 0 \\ 0 & 0 & 0 & 0 & 0 & 1 \end{bmatrix} \tag{14.40}$$

The number of rows of $\mathbf{L}_e$ is, of course, equal to the number of rows of $\hat{\mathbf{u}}_e$, and the number of columns of $\mathbf{L}_e$ is equal to the number of system degrees of freedom. Since $\mathbf{L}_e$ consists of only ones and zeros, it is more efficient in computer applications to store the locator information as a *locator vector* $\mathbf{l}_e$, which lists the system coordinates that

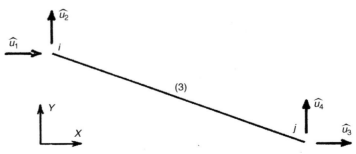

Figure 14.15 Element displacement coordinates for element 3 referred to global axes (*ECG*).

correspond to the respective element coordinates in $\hat{\mathbf{u}}_e$. For example, for element 3,

$$\mathbf{l}_3^T = \lfloor 1 \quad 2 \quad 5 \quad 6 \rfloor \tag{14.41}$$

We can arrive at the direct stiffness assembly procedure by expressing the strain energy of the structure as a sum of the element strain energies; that is,

$$\mathcal{V} = \sum_{e=1}^{N_e} \mathcal{V}_e = \sum_{e=1}^{N_e} (\tfrac{1}{2}\hat{\mathbf{u}}_e^T \hat{\mathbf{k}}_e \hat{\mathbf{u}}_e) = \tfrac{1}{2} \sum_{e=1}^{N_e} \mathbf{U}^T \mathbf{K}_e \mathbf{U} \tag{14.42}$$

where Eq. 14.39 has been used and where the *element stiffness matrix expanded to system DOFs* is

$$\boxed{\mathbf{K}_e = \mathbf{L}_e^T \hat{\mathbf{k}}_e \mathbf{L}_e} \tag{14.43}$$

The system strain energy can also be written as

$$\mathcal{V} = \tfrac{1}{2} \mathbf{U}^T \mathbf{K} \mathbf{U} \tag{14.44}$$

where $\mathbf{K}$ is the *system stiffness matrix*, given by

$$\boxed{\mathbf{K} = \sum_{e=1}^{N_e} \mathbf{K}_e} \tag{14.45}$$

Although the process of assembling the system stiffness matrix from the N_e element matrices $\hat{\mathbf{k}}_e$ appears to involve the two steps given by Eqs. 14.43 and 14.45, the two steps can be combined in an efficient process called the *Direct Stiffness Method*, which will be illustrated for the three-bar truss of Fig. 14.14.

First, we will show that Eq. 14.43 simply locates the elements of the $\hat{\mathbf{k}}_e$ matrix in the proper row and column locations corresponding to the system degrees of freedom of element e. Consider again element 3, for which the locator matrix $\mathbf{L}_3$ is given by Eq. 14.40. Let the element stiffness matrix in global coordinates be represented symbolically by

$$\hat{\mathbf{k}}_3 = \begin{bmatrix} \hat{k}_{11} & \hat{k}_{12} & \hat{k}_{13} & \hat{k}_{14} \\ \hat{k}_{21} & \hat{k}_{22} & \hat{k}_{23} & \hat{k}_{24} \\ \hat{k}_{31} & \hat{k}_{32} & \hat{k}_{33} & \hat{k}_{34} \\ \hat{k}_{41} & \hat{k}_{42} & \hat{k}_{43} & \hat{k}_{44} \end{bmatrix} \tag{14.46}$$

Then

$$\hat{\mathbf{k}}_3 \mathbf{L}_3 = \begin{bmatrix} \hat{k}_{11} & \hat{k}_{12} & 0 & 0 & \hat{k}_{13} & \hat{k}_{14} \\ \hat{k}_{21} & \hat{k}_{22} & 0 & 0 & \hat{k}_{23} & \hat{k}_{24} \\ \hat{k}_{31} & \hat{k}_{32} & 0 & 0 & \hat{k}_{33} & \hat{k}_{34} \\ \hat{k}_{41} & \hat{k}_{42} & 0 & 0 & \hat{k}_{43} & \hat{k}_{44} \end{bmatrix} \tag{14.47}$$

That is, postmultiplication by $\mathbf{L}_e$ simply places coefficients of $\hat{\mathbf{k}}_e$ into the proper columns of $\mathbf{K}_e$. Premultiplication by $\mathbf{L}_e^T$ expands the rows and places the stiffness coefficients in the proper row locations of $\mathbf{K}_e$. Finally, Eq. 14.45 simply takes the stiffness coefficients and sums them into the proper locations of the system stiffness matrix $\mathbf{K}$.

The postmultiplication by $\mathbf{L}_e$ and premultiplication by $\mathbf{L}_e^T$ involves many unnecessary multiplications by ones and zeros, since the final result is simply a moving of each element of $\widehat{\mathbf{k}}_e$ to the proper row and column position for addition into the system stiffness matrix $\mathbf{K}$. Bypassing these multiplications and just directly adding coefficients from $\widehat{\mathbf{k}}_e$ into their proper row–column locations in $\mathbf{K}$ with the aid of locator information from a locator vector $\mathbf{l}_e$, such as the one in Eq. 14.41, is referred to as the *Direct Stiffness Method* of assembling the system stiffness matrix $\mathbf{K}$. This process is demonstrated in Example 14.3.

For dynamics problems the "direct stiffness" procedure described above can be employed to assemble the system mass matrix $\mathbf{M}$ from the individual element mass matrices referred to global coordinates. This can be seen by considering the kinetic energy of the system.

$$T = \sum_{e=1}^{N_e} T_e = \sum_{e=1}^{N_e} (\tfrac{1}{2} \dot{\widehat{\mathbf{u}}}_e^T \widehat{\mathbf{m}}_e \dot{\widehat{\mathbf{u}}}_e) = \tfrac{1}{2} \sum_{e=1}^{N_e} \dot{\mathbf{U}}^T \mathbf{M}_e \dot{\mathbf{U}} = \tfrac{1}{2} \dot{\mathbf{U}}^T \mathbf{M} \dot{\mathbf{U}} \qquad (14.48)$$

where the *element mass matrix expanded to system DOFs* is

$$\boxed{\mathbf{M}_e = \mathbf{L}_e^T \widehat{\mathbf{m}}_e \mathbf{L}_e} \qquad (14.49a)$$

and where $\mathbf{M}$ is the *system mass matrix*, given by

$$\boxed{\mathbf{M} = \sum_{e=1}^{N_e} \mathbf{M}_e} \qquad (14.49b)$$

Example 14.3 Each of the truss members in Example 14.2 has the same cross-sectional area A and modulus of elasticity E. Using the lengths and the element and joint numbering shown in Example 14.2, (a) form the three element stiffness matrices referred to global coordinates, and (b) form the system stiffness matrix.

SOLUTION The numbering of element coordinates and system coordinates is defined by the numbering of joints and elements and by the element ends (i, j) shown in Example 14.2.

(a) Determine the element stiffness matrices $\widehat{\mathbf{k}}_e$, $e = 1, 2, 3$. Since the elements are planar truss elements, Eq. 14.36b can be used in forming the three $\widehat{\mathbf{k}}_e$ matrices.

$$\widehat{\mathbf{k}}_e = \left(\frac{AE}{L}\right)_e \begin{bmatrix} c_e^2 & c_e s_e & -c_e^2 & -c_e s_e \\ & s_e^2 & -c_e s_e & -s_e^2 \\ & & c_e^2 & c_e s_e \\ \text{symm.} & & & s_e^2 \end{bmatrix} \qquad (1)$$

From the geometry of the three truss elements in Fig. 1a,

$$\begin{aligned}
\cos\theta_1 &= \tfrac{4}{5}, & \sin\theta_1 &= \tfrac{3}{5}, & L_1 &= 50 \text{ in.} \\
\cos\theta_2 &= 0, & \sin\theta_2 &= -1, & L_2 &= 60 \text{ in.} \\
\cos\theta_3 &= -\tfrac{4}{5}, & \sin\theta_3 &= \tfrac{3}{5}, & L_3 &= 50 \text{ in.}
\end{aligned} \qquad (2)$$

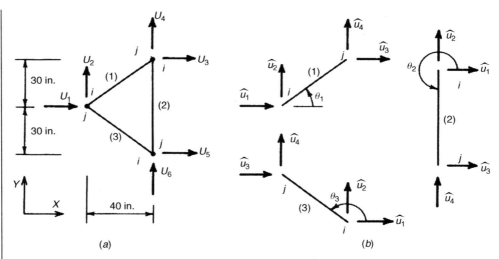

Figure 1 (a) System displacement coordinates and (b) element ECG displacement coordinates for a three-bar truss.

Thus,

$$\widehat{\mathbf{k}}_1 = \frac{AE}{50 \text{ in.}} \frac{1}{25} \begin{bmatrix} 16 & 12 & -16 & -12 \\ 12 & 9 & -12 & -9 \\ -16 & -12 & 16 & 12 \\ -12 & -9 & 12 & 9 \end{bmatrix} \begin{matrix} 1 \\ 2 \\ 3 \\ 4 \end{matrix}$$

$$\begin{matrix} & 1 & 2 & 3 & 4 \end{matrix}$$

$$\widehat{\mathbf{k}}_2 = \frac{AE}{60 \text{ in.}} \begin{bmatrix} 0 & 0 & 0 & 0 \\ 0 & 1 & 0 & -1 \\ 0 & 0 & 0 & 0 \\ 0 & -1 & 0 & 1 \end{bmatrix} \begin{matrix} 3 \\ 4 \\ 5 \\ 6 \end{matrix}$$

$$\begin{matrix} & 3 & 4 & 5 & 6 \end{matrix}$$

Ans. (a) (3)

$$\widehat{\mathbf{k}}_3 = \frac{AE}{50 \text{ in.}} \frac{1}{25} \begin{bmatrix} 16 & -12 & -16 & 12 \\ -12 & 9 & 12 & -9 \\ -16 & 12 & 16 & -12 \\ 12 & -9 & -12 & 9 \end{bmatrix} \begin{matrix} 5 \\ 6 \\ 1 \\ 2 \end{matrix}$$

$$\begin{matrix} & 5 & 6 & 1 & 2 \end{matrix}$$

(b) Using the locator information adjacent to the rows and columns of the $\widehat{\mathbf{k}}_e$ matrices above, assemble the system stiffness matrix **K** by summing the elements of the $\widehat{\mathbf{k}}_e$ matrices into the appropriate rows and columns of **K**.

Equation 14.45 is the basic equation for this direct stiffness assembly procedure.

$$\mathbf{K} = \frac{AE}{25} \begin{bmatrix} \frac{16}{50}+\frac{16}{50} & \frac{12}{50}-\frac{12}{50} & -\frac{16}{50} & -\frac{12}{50} & -\frac{16}{50} & \frac{12}{50} \\ \frac{12}{50}-\frac{12}{50} & \frac{9}{50}+\frac{9}{50} & -\frac{12}{50} & -\frac{9}{50} & \frac{12}{50} & -\frac{9}{50} \\ -\frac{16}{50} & -\frac{12}{50} & \frac{16}{50} & \frac{12}{50} & 0 & 0 \\ -\frac{12}{50} & -\frac{9}{50} & \frac{12}{50} & \frac{9}{50}+\frac{25}{60} & 0 & -\frac{25}{60} \\ -\frac{16}{50} & \frac{12}{50} & 0 & 0 & \frac{16}{50} & -\frac{12}{50} \\ \frac{12}{50} & -\frac{9}{50} & 0 & -\frac{25}{60} & -\frac{12}{50} & \frac{9}{50}+\frac{25}{60} \end{bmatrix} \begin{matrix} 1 \\ 2 \\ 3 \\ 4 \\ 5 \\ 6 \end{matrix}$$

with columns labeled 1, 2, 3, 4, 5, 6.

Ans. (b) (4)

The virtual work done by the loads can be used to obtain the system load vector **P** corresponding to the system displacements **U**. Thus,

$$\delta \mathcal{W} = \sum_{e=1}^{N_e} \delta \widehat{\mathbf{u}}_e^{\mathrm{T}} \widehat{\mathbf{p}}_e = \delta \mathbf{U}^{\mathrm{T}} \mathbf{P} \tag{14.50}$$

Combining Eq. 14.39 with Eq. 14.50, we get

$$\sum_{e=1}^{N_e} \delta \widehat{\mathbf{u}}_e^{\mathrm{T}} \widehat{\mathbf{p}}_e = \delta \mathbf{U}^{\mathrm{T}} \left(\sum_{e=1}^{N_e} \mathbf{L}_e^{\mathrm{T}} \widehat{\mathbf{p}}_e \right) \tag{14.51}$$

Therefore, since the virtual displacements δU_i are independent, we get the following equation that describes the assembly of the *system load vector* from the element load vectors referred to global coordinates:

$$\boxed{\mathbf{P} = \sum_{e=1}^{N_e} \mathbf{L}_e^{\mathrm{T}} \widehat{\mathbf{p}}_e + \mathbf{P}_{\mathrm{ext}}} \tag{14.52}$$

Just as it was unnecessary to carry out the pre- and postmultiplications in Eq. 14.43, it is unnecessary to carry out the multiplications in Eq. 14.52. To assemble the system load vector **P**, it is merely necessary to use the "locator" information to assign the components of each element load vector $\widehat{\mathbf{p}}_e$ to the proper row of the system load vector **P**. As indicated in Eq. 14.52, there may also be some concentrated external loads applied directly at nodes (joints) that must also be added to complete the final system load vector.

Example 14.4 A beam is divided into two elements (Fig. 1). The transverse load on each element is a function of time only. Determine the system load vector **P**.

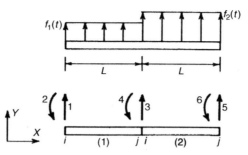

Figure 1 (a) Uniformly distributed loads and (b) system DOF numbering.

SOLUTION Since the elements are aligned with the global X axis, $\widehat{\mathbf{p}}_e = \mathbf{p}_e$. The generalized element load vector for a uniform load distribution on a beam element is given in Example 14.1. Thus,

$$\widehat{\mathbf{p}}_e = \mathbf{p}_e = f_e(t)L_e \left[\frac{1}{2} \quad \frac{L_e}{12} \quad \frac{1}{2} \quad \frac{-L_e}{12} \right]^T \qquad (1)$$

So the two element load vectors are

$$\widehat{\mathbf{p}}_1 = f_1(t)L \begin{Bmatrix} \frac{1}{2} \\ \frac{L}{12} \\ \frac{1}{2} \\ \frac{-L}{12} \end{Bmatrix} \begin{matrix} 1 \\ 2 \\ 3 \\ 4 \end{matrix} \qquad \widehat{\mathbf{p}}_2 = f_2(t)L \begin{Bmatrix} \frac{1}{2} \\ \frac{L}{12} \\ \frac{1}{2} \\ \frac{-L}{12} \end{Bmatrix} \begin{matrix} 3 \\ 4 \\ 5 \\ 6 \end{matrix} \qquad (2)$$

The "locator" information is shown alongside the element load vectors. This gives the system degree of freedom corresponding to the particular element degree of freedom. This information is used in adding the element contributions into the system load vector **P** in accordance with Eq. 14.52.

$$\mathbf{P}(t) = \begin{Bmatrix} \dfrac{f_1(t)L}{2} \\ \dfrac{f_1(t)L^2}{12} \\ \dfrac{f_1(t)L}{2} + \dfrac{f_2(t)L}{2} \\ -\dfrac{f_1(t)L^2}{12} + \dfrac{f_2(t)L^2}{12} \\ \dfrac{f_2(t)L}{2} \\ -\dfrac{f_2(t)L^2}{12} \end{Bmatrix} \begin{matrix} 1 \\ 2 \\ 3 \\ 4 \\ 5 \\ 6 \end{matrix} \qquad \textbf{Ans.} \ (3)$$

14.5 BOUNDARY CONDITIONS

For dynamics problems stated in matrix form it is necessary to incorporate boundary conditions of two types: prescribed forces and prescribed displacements. Equation 14.52 provides for assembly of the system force vector from element force vectors. As noted in Eq. 14.52, to this could be added any concentrated external loads applied at the joints. In this section, therefore, we examine a procedure for enforcing boundary constraints on a finite element model of a structure.

In Section 14.4 the system matrices $\mathbf{K}$ and $\mathbf{M}$ were assembled as though all joints of the structure were unrestrained (e.g., see Fig. 14.14). This leads to a singular stiffness matrix $\mathbf{K}$, and the system has rigid-body freedom. When a number of the joint displacements are prescribed to be zero, the most straightforward procedure for enforcing this condition is to partition the system into $\underline{a}$ctive degrees-of-freedom and $\underline{c}$onstrained degrees of freedom (DOFs). Thus, for an undamped system the partitioned equation of motion takes the form

$$\begin{bmatrix} \mathbf{M}_{aa} & \mathbf{M}_{ac} \\ \mathbf{M}_{ca} & \mathbf{M}_{cc} \end{bmatrix} \begin{Bmatrix} \ddot{\mathbf{U}}_a \\ \ddot{\mathbf{U}}_c \end{Bmatrix} + \begin{bmatrix} \mathbf{K}_{aa} & \mathbf{K}_{ac} \\ \mathbf{K}_{ca} & \mathbf{K}_{cc} \end{bmatrix} \begin{Bmatrix} \mathbf{U}_a \\ \mathbf{U}_c \end{Bmatrix} = \begin{Bmatrix} \mathbf{P}_a(t) \\ \mathbf{P}_c(t) \end{Bmatrix} \quad (14.53)$$

But if $\mathbf{U}_c = \mathbf{0}$ due to total constraint of the c-coordinates, Eq. 14.53 can be written as two separate equations:

$$\mathbf{M}_{aa}\ddot{\mathbf{U}}_a + \mathbf{K}_{aa}\mathbf{U}_a = \mathbf{P}_a(t) \quad (14.54a)$$

$$\mathbf{P}_c(t) = \mathbf{M}_{ca}\ddot{\mathbf{U}}_a + \mathbf{K}_{ca}\mathbf{U}_a \quad (14.54b)$$

Equation 14.54a must first be solved for the active displacement vector $\mathbf{U}_a$. Then the reaction forces $\mathbf{P}_c(t)$ at the constraints can be obtained from Eq. 14.54b. However, since only $\mathbf{M}_{aa}$ and $\mathbf{K}_{aa}$ are required in the solution for the active displacement vector $\mathbf{U}_a$, the remaining portions of the $\mathbf{K}$ and $\mathbf{M}$ matrices need not be assembled from the corresponding element matrices. In this case, the reactions may be obtained from element forces after the latter have been calculated rather than by use of Eq. 14.54b.

Example 14.5 A uniform propped cantilever beam is divided into two elements of length $L_1 = L_2 = L/2$ (Fig. 1). Enforce the system boundary conditions by assembling only the active matrices $\mathbf{K}_{aa}$ and $\mathbf{M}_{aa}$.

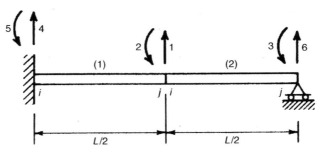

Figure 1 Two-element propped cantilever beam.

SOLUTION The system displacement coordinates are numbered such that the active coordinates are 1, 2, and 3, and the constrained coordinates are 4, 5, and 6. The element matrices are given by Eqs. 14.15.

$$\widehat{\mathbf{k}}_e = \frac{EI}{L_e^3} \begin{matrix} & 4 & 5 & 1 & 2 \\ & 1 & 2 & 6 & 3 \\ \begin{bmatrix} 12 & 6L_e & -12 & 6L_e \\ 6L_e & 4L_e^2 & -6L_e & 2L_e^2 \\ -12 & -6L_e & 12 & -6L_e \\ 6L_e & 2L_e^2 & -6L_e & 4L_e^2 \end{bmatrix} & \begin{matrix} 1 & 4 \\ 2 & 5 \\ 6 & 1 \\ 3 & 2 \end{matrix} \end{matrix} \quad (1)$$

$$\widehat{\mathbf{m}}_e = \frac{\rho A L_e}{420} \begin{matrix} & 4 & 5 & 1 & 2 \\ & 1 & 2 & 6 & 3 \\ \begin{bmatrix} 156 & 22L_e & 54 & -13L_e \\ 22L_e & 4L_e^2 & 13L_e & -3L_e^2 \\ 54 & 13L_e & 156 & -22L_e \\ -13L_e & -3L_e^2 & -22L_e & 4L_e^2 \end{bmatrix} & \begin{matrix} 1 & 4 \\ 2 & 5 \\ 6 & 1 \\ 3 & 2 \end{matrix} \end{matrix} \quad (2)$$

The element locator information is given alongside rows and columns of $\widehat{\mathbf{k}}_e$ and $\widehat{\mathbf{m}}_e$. This information enables $\mathbf{K}_{aa}$ and $\mathbf{M}_{aa}$ to be assembled. The system stiffness matrix for active DOFs is

$$\mathbf{K}_{aa} = \frac{EI}{(L/2)^3} \begin{matrix} & 1 & 2 & 3 \\ \begin{bmatrix} 12+12 & -3L+3L & 3L \\ -3L+3L & L^2+L^2 & L^2/2 \\ 3L & L^2/2 & L^2 \end{bmatrix} & \begin{matrix} 1 \\ 2 \\ 3 \end{matrix} \end{matrix} \quad (3)$$

or

$$\mathbf{K}_{aa} = \frac{4EI}{L^3} \begin{bmatrix} 48 & 0 & 6L \\ 0 & 4L^2 & L^2 \\ 6L & L^2 & 2L^2 \end{bmatrix} \quad \text{Ans. (4)}$$

In as similar manner, the system mass matrix for active DOFs is

$$\mathbf{M}_{aa} = \frac{\rho A(L/2)}{420} \begin{matrix} & 1 & 2 & 3 \\ \begin{bmatrix} 156+156 & -11L+11L & -13L/2 \\ -11L+11L & L^2+L^2 & -3L^2/4 \\ -13L/2 & -3L^2/4 & L^2 \end{bmatrix} & \begin{matrix} 1 \\ 2 \\ 3 \end{matrix} \end{matrix} \quad (3)$$

or

$$\mathbf{M}_{aa} = \frac{\rho AL}{3360} \begin{bmatrix} 1248 & 0 & -26L \\ 0 & 8L^2 & -3L^2 \\ -26L & -3L^2 & L^2 \end{bmatrix} \quad \text{Ans. (5)}$$

14.6 CONSTRAINTS: REDUCTION OF DEGREES OF FREEDOM

In Section 14.5 we treated constraints imposed as boundary conditions on a structure. Frequently, there arises a need for specifying relationships among system displacement coordinates.

14.6.1 Displacement Constraints

Displacement constraints may be enforced through the use of a transformation of coordinates

$$\boxed{\mathbf{U} = \mathbf{T}\widehat{\mathbf{U}}} \tag{14.55}$$

where $\widehat{N} \leq N$ and where $\widehat{\mathbf{U}}$ is a vector of generalized coordinates. Equation 14.55, which was introduced in Section 10.2, may be referred to as a *Ritz transformation*.

Using energy equivalence, as we have done on many previous occasions, we obtain the *reduced-order system equation of motion*

$$\widehat{\mathbf{M}}\ddot{\widehat{\mathbf{U}}} + \widehat{\mathbf{C}}\dot{\widehat{\mathbf{U}}} + \widehat{\mathbf{K}}\widehat{\mathbf{U}} = \widehat{\mathbf{P}}(t) \tag{14.56}$$

where

$$\widehat{\mathbf{M}} = \mathbf{T}^T\mathbf{M}\mathbf{T}, \qquad \widehat{\mathbf{C}} = \mathbf{T}^T\mathbf{C}\mathbf{T}, \qquad \widehat{\mathbf{K}} = \mathbf{T}^T\mathbf{K}\mathbf{T}, \qquad \widehat{\mathbf{P}} = \mathbf{T}^T\mathbf{P} \tag{14.57}$$

The transformation matrix $\mathbf{T}$ may arise as a result of the need to specify a relationship among several system displacement coordinates. The *equation of constraint* may be written in matrix form as

$$\mathbf{R}\mathbf{U} \equiv [\mathbf{R}_{da} \ \mathbf{R}_{dd}] \begin{Bmatrix} \mathbf{U}_a \\ \mathbf{U}_d \end{Bmatrix} = \mathbf{0} \tag{14.58}$$

where $\mathbf{U}_d$ is a vector of N_d <u>d</u>ependent coordinates and $\mathbf{U}_a$ is the vector of independent, or <u>a</u>ctive, coordinates. Equation 14.58 may be solved for $\mathbf{U}_d$, giving

$$\mathbf{U}_d = -\mathbf{R}_{dd}^{-1}\mathbf{R}_{da}\mathbf{U}_a \tag{14.59}$$

Therefore, the original coordinate set $\mathbf{U}$ can be related to the active coordinates by the equation

$$\mathbf{U} \equiv \begin{Bmatrix} \mathbf{U}_a \\ \mathbf{U}_d \end{Bmatrix} = \begin{bmatrix} \mathbf{I}_{aa} \\ \mathbf{T}_{da} \end{bmatrix} \mathbf{U}_a \equiv \mathbf{T}\mathbf{U}_a \tag{14.60}$$

where the active coordinates form the $\widehat{\mathbf{U}}$ reduced set of coordinates and where

$$\mathbf{T}_{da} = -\mathbf{R}_{dd}^{-1}\mathbf{R}_{da} \tag{14.61}$$

A constraint equation such as Eq. 14.58 may arise from a situation where a rigid body is connected to an elastic structure. Example 14.6 treats this problem.

Example 14.6 A rigid flat plate is connected to two joints of a flexible planar truss, as shown in Fig. 1. Write a constraint equation treating the translation coordinates (U_1 and U_2) and the rotation coordinate (U_3) of the plate as the active coordinates. Assume small rotations.

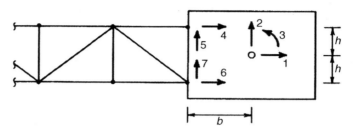

Figure 1 Rigid plate attached to a plane truss structure.

SOLUTION It is easiest to take each of the active displacements (U_1, U_2, U_3) in turn and determine the resulting displacement of each of the four dependent coordinates ($U_4, \ldots, U_7$). For $U_1 \neq 0$,

$$U_4 = U_6 = U_1, \qquad U_5 = U_7 = 0 \tag{1}$$

For $U_2 \neq 0$,

$$U_5 = U_7 = U_2, \qquad U_4 = U_6 = 0 \tag{2}$$

For (small rotation) $U_3 \neq 0$,

$$U_4 = -hU_3, \qquad U_5 = -bU_3, \qquad U_6 = hU_3, \qquad U_7 = -bU_3 \tag{3}$$

Combining Eqs. 1 through 3, we can write the following constraint equation, which relates the dependent coordinates U_4, U_5, U_6, and U_7 to the active coordinates U_1, U_2, and U_3:

$$\begin{bmatrix} -1 & 0 & h & 1 & 0 & 0 & 0 \\ 0 & -1 & b & 0 & 1 & 0 & 0 \\ -1 & 0 & -h & 0 & 0 & 1 & 0 \\ 0 & -1 & b & 0 & 0 & 0 & 1 \end{bmatrix} \begin{Bmatrix} U_1 \\ U_2 \\ U_3 \\ \hline U_4 \\ U_5 \\ U_6 \\ U_7 \end{Bmatrix} = \begin{Bmatrix} 0 \\ 0 \\ 0 \\ 0 \end{Bmatrix} \qquad \text{Ans.} \quad (4)$$

where the matrix partitioning follows the scheme in Eq. 14.58.

14.6.2 Static Condensation

Static condensation is another example of the application of constraint equations to reduce some coordinates. Consider an undamped system whose inertia is represented by a lumped-mass matrix, as described in Section 14.2. The equation of motion may be written in partitioned-matrix form:

$$\begin{bmatrix} \mathbf{M}_{aa} & \mathbf{0}_{ad} \\ \mathbf{0}_{da} & \mathbf{0}_{dd} \end{bmatrix} \begin{Bmatrix} \ddot{\mathbf{U}}_a \\ \ddot{\mathbf{U}}_d \end{Bmatrix} + \begin{bmatrix} \mathbf{K}_{aa} & \mathbf{K}_{ad} \\ \mathbf{K}_{da} & \mathbf{K}_{dd} \end{bmatrix} \begin{Bmatrix} \mathbf{U}_a \\ \mathbf{U}_d \end{Bmatrix} = \begin{Bmatrix} \mathbf{P}_a(t) \\ \mathbf{0}_d \end{Bmatrix} \quad (14.62)$$

The lower partition provides a static constraint equation

$$\mathbf{K}_{da}\mathbf{U}_a + \mathbf{K}_{dd}\mathbf{U}_d = \mathbf{0}_d \quad (14.63)$$

which is of the same form as Eq. 14.58. Then the *static condensation* transformation matrix is

$$\mathbf{T} \equiv \begin{bmatrix} \mathbf{I}_{aa} \\ \mathbf{T}_{da} \end{bmatrix} = \begin{bmatrix} \mathbf{I}_{aa} \\ -\mathbf{K}_{dd}^{-1}\mathbf{K}_{da} \end{bmatrix} \quad (14.64)$$

The transformation matrix of Eq. 14.64 can be used in conjunction with Eqs. 14.57 to produce the equation of motion for the active coordinates $\widehat{\mathbf{U}} \equiv \mathbf{U}_a$. The reduced mass matrix and reduced stiffness matrix are given by

$$\widehat{\mathbf{M}}_{aa} = \mathbf{T}^T \mathbf{M} \mathbf{T} = [\mathbf{I}_{aa} \ \mathbf{T}_{da}^T] \begin{bmatrix} \mathbf{M}_{aa} & \mathbf{0}_{ad} \\ \mathbf{0}_{da} & \mathbf{0}_{dd} \end{bmatrix} \begin{bmatrix} \mathbf{I}_{aa} \\ \mathbf{T}_{da} \end{bmatrix} = \mathbf{M}_{aa} \quad (14.65a)$$

$$\widehat{\mathbf{K}}_{aa} = \mathbf{T}^T \mathbf{K} \mathbf{T} = [\mathbf{I}_{aa} \ \mathbf{T}_{da}^T] \begin{bmatrix} \mathbf{K}_{aa} & \mathbf{K}_{ad} \\ \mathbf{K}_{da} & \mathbf{K}_{dd} \end{bmatrix} \begin{bmatrix} \mathbf{I}_{aa} \\ \mathbf{T}_{da} \end{bmatrix} = \mathbf{K}_{aa} - \mathbf{K}_{da}^T \mathbf{K}_{dd}^{-1} \mathbf{K}_{da} \quad (14.65b)$$

Thus, for a lumped-mass model the problem size may be reduced by using only coordinates having associated inertia. The mass matrix is unchanged, while the stiffness matrix becomes the *reduced stiffness matrix* $\widehat{\mathbf{K}}_{aa}$, defined by Eq. 14.65b.

Example 14.7 Create a 2-DOF lumped-mass model of a uniform cantilever beam of length L.

SOLUTION   One possible model is a cantilever beam having two elements of length $L/2$, as depicted in Fig. 1, with the two rotational degrees of freedom reduced out to produce a 2-DOF model.

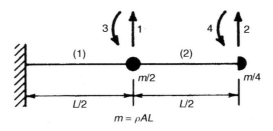

Figure 1 Cantilever beam with two lumped masses.

Since $L_e = L/2$, the element stiffness matrix is, from Eq. 14.15a,

$$\widehat{\mathbf{k}}_e = \mathbf{k}_e = \frac{8EI}{L^3} \begin{bmatrix} \overset{1}{\underset{x}{12}} & \overset{3}{\underset{x}{3L}} & \overset{2}{\underset{1}{-12}} & \overset{4}{\underset{3}{3L}} \\ 3L & L^2 & -3L & L^2/2 \\ -12 & -3L & 12 & -3L \\ 3L & L^2/2 & -3L & L^2 \end{bmatrix} \begin{matrix} x\ 1 \\ x\ 3 \\ 1\ 2 \\ 3\ 4 \end{matrix} \tag{1}$$

Therefore, the assembled system equation of motion for free vibration is

$$\frac{\rho A L}{4} \begin{bmatrix} 2 & 0 & | & 0 & 0 \\ 0 & 1 & | & 0 & 0 \\ \hline 0 & 0 & | & 0 & 0 \\ 0 & 0 & | & 0 & 0 \end{bmatrix} \begin{Bmatrix} \ddot{U}_1 \\ \ddot{U}_2 \\ \ddot{U}_3 \\ \ddot{U}_4 \end{Bmatrix}$$

$$+ \frac{8EI}{L^3} \begin{bmatrix} 24 & -12 & | & 0 & 3L \\ -12 & 12 & | & -3L & -3L \\ \hline 0 & -3L & | & 2L^2 & L^2/2 \\ 3L & -3L & | & L^2/2 & L^2 \end{bmatrix} \begin{Bmatrix} U_1 \\ U_2 \\ U_3 \\ U_4 \end{Bmatrix} = \begin{Bmatrix} 0 \\ 0 \\ 0 \\ 0 \end{Bmatrix} \tag{2}$$

which is in the same form as Eq. 14.62. Therefore, the respective matrix partitions of the unreduced system stiffness matrix $\mathbf{K}$ can be identified as

$$\widehat{\mathbf{K}}_{aa} = \frac{8EI}{L^3} \begin{bmatrix} 24 & -12 \\ -12 & 12 \end{bmatrix}$$

$$\widehat{\mathbf{K}}_{da} = \frac{8EI}{L^3} \begin{bmatrix} 0 & -3L \\ 3L & -3L \end{bmatrix}, \quad \widehat{\mathbf{K}}_{dd} = \frac{8EI}{L} \begin{bmatrix} 2 & \frac{1}{2} \\ \frac{1}{2} & 1 \end{bmatrix} \tag{3}$$

Then

$$\mathbf{K}_{dd}^{-1} = \frac{L}{28EI} \begin{bmatrix} 2 & -1 \\ -1 & 4 \end{bmatrix} \tag{4}$$

and

$$\mathbf{K}_{dd}^{-1} \mathbf{K}_{da} = \frac{L}{28EI} \begin{bmatrix} 2 & -1 \\ -1 & 4 \end{bmatrix} \frac{8EI}{L^3} \begin{bmatrix} 0 & -3L \\ 3L & -3L \end{bmatrix}$$

$$= \frac{-6}{7L} \begin{bmatrix} 1 & 1 \\ -4 & 3 \end{bmatrix} \tag{5}$$

Finally, from Eq. 14.65b, the reduced stiffness matrix for the 2-DOF lumped-mass model is

$$\widehat{\mathbf{K}}_{aa} = \mathbf{K}_{aa} - \mathbf{K}_{da}^{T} \mathbf{K}_{dd}^{-1} \mathbf{K}_{da} = \frac{8EI}{L^3} \begin{bmatrix} 24 & -12 \\ -12 & 12 \end{bmatrix}$$

$$- \frac{8EI}{L^3} \begin{bmatrix} 0 & 3L \\ -3L & -3L \end{bmatrix} \frac{-6}{7L} \begin{bmatrix} 1 & 1 \\ -4 & 3 \end{bmatrix} \tag{6}$$

or

$$\widehat{\mathbf{K}}_{aa} = \frac{48EI}{7L^3} \begin{bmatrix} 16 & -5 \\ -5 & 2 \end{bmatrix} \tag{7}$$

From Eqs. 14.65a and 2, the reduced mass matrix is simply

$$\widehat{\mathbf{M}}_{aa} = \mathbf{M}_{aa} = \frac{\rho AL}{4} \begin{bmatrix} 2 & 0 \\ 0 & 1 \end{bmatrix} \qquad (8)$$

Thus, the system equation in reduced coordinates (1 and 2) is

$$\frac{\rho AL}{4} \begin{bmatrix} 2 & 0 \\ 0 & 1 \end{bmatrix} \begin{Bmatrix} \ddot{U}_1 \\ \ddot{U}_2 \end{Bmatrix} + \frac{48EI}{7L^3} \begin{bmatrix} 16 & -5 \\ -5 & 2 \end{bmatrix} \begin{Bmatrix} U_1 \\ U_2 \end{Bmatrix} = \begin{Bmatrix} 0 \\ 0 \end{Bmatrix} \qquad \text{Ans. (9)}$$

14.6.3 Guyan–Irons Reduction

A technique that has frequently been employed to reduce the number of degrees of freedom of a system is referred to as the *Guyan–Irons Reduction Method* (or, often, just Guyan Reduction).[14.6] Even though inertia may be associated with all coordinates, a subset of the coordinates is selected arbitrarily as the set of *a*ctive (or "master") coordinates, and the remaining coordinates are *d*ependent (or "slave") coordinates. Then the Ritz transformation matrix of Eq. 14.55 is taken to be the same as that obtained for a lumped-mass system in Eq. 14.64. The *Ritz basis vectors*, which are the columns of the Ritz transformation matrix **T**, are the displacement patterns associated with unit displacement of the respective *a*-coordinates, while the *d*-coordinates are released. Anderson et al.,[14.7] discuss the application of this reduction procedure to vibration and buckling analyses and show the effect of the choice of coordinates and number of coordinates retained on the accuracy of the eigenvalues and eigenvectors obtained. An example Guyan Reduction solution is presented in Section 14.8.

Reference [14.8] discusses the use of substructuring as a procedure for selecting the degrees of freedom to be retained and those to be reduced out and discusses a quadratic eigensolution procedure for improving the frequencies and modes obtained by using Guyan Reduction. In Chapter 17 some particular forms of coordinate reduction associated with substructuring are treated.

14.7 SYSTEMS WITH RIGID-BODY MODES

Some algebraic eigensolvers, for example, iteration methods such as those that are discussed in Section 15.2, require the solution of the eigenvalue equation

$$\mathbf{K}^{-1}\mathbf{M}\boldsymbol{\phi} = \frac{1}{\lambda}\boldsymbol{\phi} \qquad (14.66)$$

When a structure has rigid-body modes, the inversion of **K** is not possible, since **K** is singular. A procedure for eliminating the rigid-body freedoms can be expressed in the form of a transformation of coordinates that separates *r*igid-body coordinates from *e*lastic deformation coordinates. Let the original **K** matrix be written

$$\mathbf{K} = \begin{bmatrix} \mathbf{K}_{ee} & \mathbf{K}_{er} \\ \mathbf{K}_{re} & \mathbf{K}_{rr} \end{bmatrix} \qquad (14.67)$$

A Ritz transformation **T** having the form of Eq. 14.55, namely

$$\begin{Bmatrix} \mathbf{U}_e \\ \mathbf{U}_r \end{Bmatrix} = \begin{bmatrix} \mathbf{I}_{ee} & -\mathbf{K}_{ee}^{-1}\mathbf{K}_{er} \\ \mathbf{0}_{re} & \mathbf{I}_{rr} \end{bmatrix} \begin{Bmatrix} \widehat{\mathbf{U}}_e \\ \widehat{\mathbf{U}}_r \end{Bmatrix} \qquad (14.68)$$

together with Eq. 14.57c leads to the generalized stiffness matrix

$$\widehat{\mathbf{K}} = \mathbf{T}^T \mathbf{K} \mathbf{T} = \begin{bmatrix} \mathbf{K}_{ee} & \mathbf{0}_{er} \\ \mathbf{0}_{re} & \mathbf{0}_{rr} \end{bmatrix} \quad (14.69)$$

The transformation given by Eq. 14.68 leads to a particularly simple form of $\widehat{\mathbf{K}}$. It is clear that the coordinates $\widehat{\mathbf{U}}_r \equiv \mathbf{U}_r$ represent rigid-body displacement, since all stiffness submatrices in Eq. 14.69 that are associated with these coordinates are zero. Any linearly independent set of vectors could replace the $\mathbf{I}_{ee}$ matrix in Eq. 14.68, but other choices would not lead to $\widehat{\mathbf{K}}_{ee} = \mathbf{K}_{ee}$ as in Eq. 14.69. Reference [14.9] discusses further numerical aspects of the rigid-body problem.

Example 14.8 illustrates the process of transforming displacement coordinates into a set that separates the rigid-body coordinates from the coordinates that represent elastic deformation.

Example 14.8 The uniform bar in Fig. 1 is to be modeled by two elements of length $L/2$. Let $k = 2AE/L$, and let $U_r = U_3$. (a) Derive the transformation matrix of Eq. 14.68; (b) sketch the shape functions that correspond to the vectors that are the Ritz basis vectors of this transformation; and (c) show that the transformed stiffness matrix has the form given in Eq. 14.69.

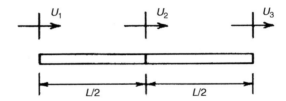

Figure 1 A 3-DOF uniform bar undergoing axial motion.

SOLUTION   (a) Set up the system stiffness matrix in partitioned form as in Eq. 14.67.

$$\mathbf{K} = \begin{bmatrix} \mathbf{K}_{ee} & \mathbf{K}_{er} \\ \mathbf{K}_{re} & \mathbf{K}_{rr} \end{bmatrix} = k \begin{bmatrix} 1 & -1 & 0 \\ -1 & 2 & -1 \\ 0 & -1 & 1 \end{bmatrix} \quad (1)$$

Therefore,

$$\mathbf{K}_{ee}^{-1} \mathbf{K}_{er} = \frac{1}{k} \begin{bmatrix} 2 & 1 \\ 1 & 1 \end{bmatrix} (k) \begin{bmatrix} 0 \\ -1 \end{bmatrix} = \begin{bmatrix} -1 \\ -1 \end{bmatrix} \quad (2)$$

Then, from Eqs. 14.68 and 2,

$$\mathbf{T} = \begin{bmatrix} \mathbf{I}_{ee} & -\mathbf{K}_{ee}^{-1} \mathbf{K}_{er} \\ \mathbf{0}_{re} & \mathbf{I}_{rr} \end{bmatrix} = \begin{bmatrix} 1 & 0 & 1 \\ 0 & 1 & 1 \\ 0 & 0 & 1 \end{bmatrix} \quad \textbf{Ans. (a)} \quad (3)$$

(b) Sketch the shape functions that correspond to the Ritz vectors, that is, to the columns of the transformation matrix **T**. In Fig. 2 these columns are labeled $\mathbf{t}_i$. Note that column 3, that is, $\mathbf{t}_3$, corresponds to rigid-body translation of the bar.

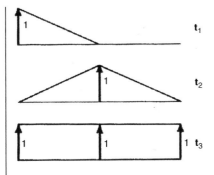

Figure 2 Shape functions representing two elastic modes and one rigid-body mode.

(c) Form the transformed stiffness matrix $\widehat{\mathbf{K}}$ using the matrix triple product in Eqs. 14.69 and 1 and 3.

$$\widehat{\mathbf{K}} = \mathbf{T}^T\mathbf{K}\mathbf{T} = \begin{bmatrix} 1 & 0 & 0 \\ 0 & 1 & 0 \\ 1 & 1 & 1 \end{bmatrix} (k) \begin{bmatrix} 1 & -1 & 0 \\ -1 & 2 & -1 \\ 0 & -1 & 1 \end{bmatrix} \begin{bmatrix} 1 & 0 & 1 \\ 0 & 1 & 1 \\ 0 & 0 & 1 \end{bmatrix} \quad (4)$$

Finally, the transformed system stiffness matrix is

$$\widehat{\mathbf{K}} = k \begin{bmatrix} 1 & -1 & 0 \\ -1 & 2 & 0 \\ 0 & 0 & 0 \end{bmatrix} \quad \textbf{Ans. (c) (5)}$$

Note that $\widehat{\mathbf{K}}$ does indeed have the form indicated in partitioned matrix in Eq. 14.69.

14.8 FINITE ELEMENT SOLUTIONS FOR NATURAL FREQUENCIES AND MODE SHAPES

The purpose of the present section is to present some solutions for modes and frequencies of simple structural members based on FE representations of the members, and to introduce the use of the *ISMIS* computer program for solving small structural dynamics problems. Consistent-mass and lumped-mass FE models are compared with each other and with "exact" solutions, and the reduction of coordinates is demonstrated. The *ISMIS* computer program is a collection of MATLAB files that generate element stiffness and mass matrices and assemble system matrices. *ISMIS*, a small DOF-based structural analysis code, permits direct communication with MATLAB to carry out other matrix operations (e.g., performing an eigensolution and plotting the results). Unlike larger commercial FE codes, which start with generation of the system geometry and nodes and elements, and then proceed to carry out the remaining steps of "canned" solution sequences, *ISMIS* requires the user to perform matrix operations to carry out every step of the FE analysis. A brief *ISMIS User's Manual* and several sample analysis files can be found on the book's website.

In Example 8.9 a 2-DOF assumed-modes model of a uniform cantilever bar undergoing axial vibration was generated, and in Example 9.3 the natural frequencies and modes of this model were obtained. However, in the present section we concentrate on the transverse vibration of a uniform cantilever beam and use it to illustrate the following:

1. A "hand-crank" 2-DOF solution based on a FE model
2. The effect of increasing the number of degrees of freedom of an FE model that is based on a consistent-mass matrix
3. The effect of reducing out freedoms by using the Guyan Reduction Method
4. The comparison of lumped-mass models with consistent-mass models

Consider the following single-element 2-DOF model of a uniform cantilever beam undergoing transverse vibration.

Example 14.9 (a) Using stiffness and mass matrices for a uniform beam finite element, as given in Eqs. 14.15, create a 2-DOF math model of the uniform cantilever beam shown in Fig. 1. (b) Solve for the natural frequencies of this model and compare them with the "exact" frequencies given in Example 13.3.

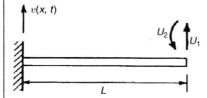

Figure 1 Transverse vibration of a single-element FE model of a uniform cantilever beam.

SOLUTION (a) The stiffness and mass matrices in Eqs. 14.15 correspond to the element coordinate numbering shown in Fig. 2. In creating "system" stiffness and mass matrices based on the two system degrees of freedom at the free end, we can use the following locator information

Element DOF	System DOF
1	constrained
2	constrained
3	1
4	2

Figure 2 Original element coordinate numbering.

Using the entries in this locator table and the mass matrix and stiffness matrix for the beam element from Eqs. 14.15, we get the following system equations of motion:

$$\frac{\rho A L}{420}\begin{bmatrix} 156 & -22L \\ -22L & 4L^2 \end{bmatrix}\begin{Bmatrix} \ddot{U}_1 \\ \ddot{U}_2 \end{Bmatrix} + \frac{EI}{L^3}\begin{bmatrix} 12 & -6L \\ -6L & 4L^2 \end{bmatrix}\begin{Bmatrix} U_1 \\ U_2 \end{Bmatrix} = \begin{Bmatrix} 0 \\ 0 \end{Bmatrix} \quad \textbf{Ans. (a) (1)}$$

(b) Equation 1 has the form

$$\mathbf{M\ddot{U} + KU = 0} \tag{2}$$

To set up the eigenproblem, assume harmonic motion of the form

$$\mathbf{U}(t) = \boldsymbol{\phi} \cos \omega t \tag{3}$$

and let

$$\mu \equiv \frac{\rho A L^4}{420 EI} \tag{4}$$

Then the resulting algebraic eigenvalue problem is

$$\left[\begin{bmatrix} 12 & -6L \\ -6L & 4L^2 \end{bmatrix} - \mu \begin{bmatrix} 156 & -22L \\ -22L & 4L^2 \end{bmatrix} \right] \begin{Bmatrix} \phi_1 \\ \phi_2 \end{Bmatrix} = \begin{Bmatrix} 0 \\ 0 \end{Bmatrix} \tag{5}$$

We set the determinant of the coefficients to zero, getting the characteristic equation

$$(12 - 156\mu)(4L^2 - 4L^2\mu) - (-6L + 22L\mu)^2 = 0 \tag{6}$$

The roots of Eq. 6 are

$$\mu_1 = 2.97147 \times 10^{-2}, \qquad \mu_2 = 2.88457$$

Then, inserting these values into Eq. 4, we get

$$\omega_1^2 = 12.4802 \frac{EI}{\rho A L^4}, \qquad \omega_2^2 = 1211.52 \frac{EI}{\rho A L^4}$$

Finally, the (circular) natural frequencies based on this 2-DOF FE model are

$$\omega_1 = 3.533 \left(\frac{EI}{\rho A L^4} \right)^{1/2}, \qquad \omega_2 = 34.81 \left(\frac{EI}{\rho A L^4} \right)^{1/2} \qquad \textbf{Ans. (b)} \tag{7}$$

From Example 13.3, the corresponding "exact" frequencies for the cantilever beam are

$$\omega_1 = 3.516 \left(\frac{EI}{\rho A L^4} \right)^{1/2}, \qquad \omega_2 = 22.03 \left(\frac{EI}{\rho A L^4} \right)^{1/2} \qquad \textbf{Ans. (b)} \tag{8}$$

From Eqs. 7 and 8 of Example 14.9, it can be seen that the single-element 2-DOF FE model produces upper bounds to the first and second exact frequencies. The fundamental frequency of the 2-DOF model is very accurate, whereas the second frequency is too high to be of any use. These results are in agreement with the information on Rayleigh–Ritz bounds summarized in Table 10.1, where the number of degrees of freedom of the continuum model is infinite, that is, $N = \infty$.

14.8.1 *ISMIS* Finite Element Models

The *ISMIS* computer program will now be used to solve the 2-DOF FE problem of Example 14.9.

Example 14.10 Use the *ISMIS* computer program to solve the problem stated in Example 14.9.

SOLUTION The *ISMIS* computer program has commands <framel> and <framms> that generate the 6 × 6 element stiffness matrix and element mass matrix for a plane frame element whose element coordinates are numbered as shown in Fig. 1a. To obtain the constant that multiplies the factor $(EI/\rho AL^4)^{1/2}$ in Example 14.9, $E = \rho = I = A = L = 1$ were employed in the *ISMIS* solution.

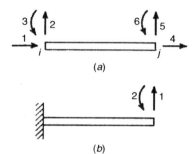

Figure 1 (a) Element and (b) system coordinate numbering.

Two versions of *ISMIS* code for solving this problem can be found on the book's website:

- Example 14_10.m: the *ISMIS* input .m-file listed below
- Example 14_10wp.m: an extended version with plotting of the two modes

The listing of *ISMIS* input .m-file <Example 14_10.m> follows.

```
runismis
disp('Example 14.10')
disp(' ')
disp('*Form Element Matrices*')
ke=framel(1,0,1,1,1)
me=framms(1,0,1,1)
disp(' ')
disp('*Remove Active Matrices*')
kaa=rmvsm(ke,5,5,2,2)
maa=rmvsm(me,5,5,2,2)
disp(' ')
disp('*Solve Eigenproblem*')
[vec,val]=eig(kaa,maa);
disp(' ')
disp('OMEGAS')
omega=diag(sqrt(val))
disp(' ')
disp('MODES (Cols.)')
vec
endismis
```

Note that special *ISMIS* commands <framel>, <framms>, and <rmvsm> (remove submatrix) were used in the solution above, together with standard MATLAB commands, including the eigensolution command <eig>. The resulting *ISMIS* output is

Example 14.10

Form Element Matrices

ke =

```
     1      0      0     -1      0      0
     0     12      6      0    -12      6
     0      6      4      0     -6      2
    -1      0      0      1      0      0
     0    -12     -6      0     12     -6
     0      6      2      0     -6      4
```

me =

```
  3.3333e-001            0            0   1.6667e-001            0            0
            0   3.7143e-001   5.2381e-002            0   1.2857e-001  -3.0952e-002
            0   5.2381e-002   9.5238e-003            0   3.0952e-002  -7.1429e-003
  1.6667e-001            0            0   3.3333e-001            0            0
            0   1.2857e-001   3.0952e-002            0   3.7143e-001  -5.2381e-002
            0  -3.0952e-002  -7.1429e-003            0  -5.2381e-002   9.5238e-003
```

Remove Active Matrices

kaa =

```
    12     -6
    -6      4
```

maa =

```
  3.7143e-001  -5.2381e-002
 -5.2381e-002   9.5238e-003
```

Solve Eigenproblem

OMEGAS

omega =

```
  3.5327e+000
  3.4807e+001
```

MODES (Cols.)

vec =

```
 -2.0195e+000   2.8145e+000
 -2.7819e+000   2.1454e+001
```

Note that the frequencies printed under "OMEGA" agree with those in Eq. 8 of Example 14.9.

- To obtain a listing of the commands included in *ISMIS*, in the MATLAB command window, type

  ```
  runismis
  help
  ```

- To obtain the description of any *ISMIS* command and its call list, in the MATLAB command window, type

  ```
  runismis
  help <Command name>
  ```

To illustrate the relationship between the number of degrees of freedom in a consistent-mass finite element model and the accuracy of the frequencies obtained using the model, *ISMIS* programs similar to that in Example 14.10 were run, and the results are tabulated in Table 14.1. Rotational degrees of freedom were retained, as illustrated in Fig. 14.16.

Table 14.1 illustrates the Rayleigh–Ritz bounds indicated previously in Table 10.1, that is, the accuracy of the frequencies deteriorates for the higher modes of a particular model, but the accuracy of each frequency is increased by increasing the number of

Table 14.1 Comparison of Consistent-Mass Finite Element Models of a Uniform Cantilever Beam with Continuum Model

	N_e						Exact
Mode	1	2	3	4	5		(Ref. [14.10])
1	3.53273	3.51772	3.51637	3.51613	3.51606	>	3.51602
2	34.8069	22.2215	22.1069	22.0602	22.0455	>	22.0345
3		75.1571	62.4659	62.1749	61.9188	>	61.6972
4		218.138	140.671	122.657	122.320	>	120.902
5			264.743	228.137	203.020	>	199.860
6			527.796	366.390	337.273	>	298.556
7				580.849	493.264	>	416.991
8				953.051	715.341	>	555.165
9					1016.20	>	713.079
10					1494.88	>	890.732

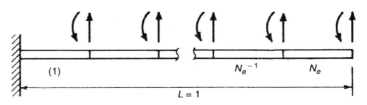

Figure 14.16 Finite element model of a uniform cantilever beam.

degrees of freedom of the model. Notice that the accuracy is quite good for a number of modes equal to the number of elements (which is half the total number of DOFs), but the frequencies for the remaining modes are poor.

14.8.2 Guyan–Irons Reduction

Next, we use the *Guyan–Irons Reduction Method* (often called *Guyan Reduction*), described in Section 14.6, to reduce out the rotations in the cantilever beam problem above. Let the active coordinates be designated by a and the coordinates to be reduced out by d, as in Section 14.6. Then the partitioned mass matrix and stiffness matrix are given by

$$\mathbf{M} = \begin{bmatrix} \mathbf{M}_{aa} & \mathbf{M}_{ad} \\ \mathbf{M}_{da} & \mathbf{M}_{dd} \end{bmatrix}, \quad \mathbf{K} = \begin{bmatrix} \mathbf{K}_{aa} & \mathbf{K}_{ad} \\ \mathbf{K}_{da} & \mathbf{K}_{dd} \end{bmatrix} \tag{14.70}$$

Using the transformation matrix $\mathbf{T}$ defined by Eq. 14.64, we get the *reduced system stiffness matrix* as given by Eq. 14.65b, repeated here:

$$\boxed{\widehat{\mathbf{K}}_{aa} = \mathbf{K}_{aa} - \mathbf{K}_{da}^{\mathrm{T}} \mathbf{K}_{dd}^{-1} \mathbf{K}_{da}} \tag{14.71}$$

Since the consistent-mass matrix has nonzero terms associated with the d-coordinates, we must use Eqs. 14.57a and 14.64 to compute the *reduced system mass matrix*, $\widehat{\mathbf{M}}_{aa}$. Thus,

$$\widehat{\mathbf{M}}_{aa} = [\mathbf{I}_{aa} \ \mathbf{T}_{da}^{\mathrm{T}}] \begin{bmatrix} \mathbf{M}_{aa} & \mathbf{M}_{ad} \\ \mathbf{M}_{da} & \mathbf{M}_{dd} \end{bmatrix} \begin{bmatrix} \mathbf{I}_{aa} \\ \mathbf{T}_{da} \end{bmatrix} \tag{14.72}$$

or

$$\boxed{\widehat{\mathbf{M}}_{aa} = \mathbf{M}_{aa} + \mathbf{M}_{ad}\mathbf{T}_{da} + (\mathbf{M}_{ad}\mathbf{T}_{da})^{\mathrm{T}} + \mathbf{T}_{da}^{\mathrm{T}}\mathbf{M}_{dd}\mathbf{T}_{da}} \tag{14.73}$$

Example 14.11 illustrates how matrix manipulations can be combined with the formulation of element matrices and assembly of system matrices to carry out Guyan Reduction using the *ISMIS* computer program. To number the system degrees of freedom in a convenient manner for using the Guyan reduction equations above, the "direct stiffness" *ISMIS* command <addmat> will be introduced in Example 14.11.

Example 14.11 Write an *ISMIS* .m-file that will reduce out the rotational degrees of freedom for the two-element beam in Fig. 1. Use $E = \rho = I = A = L = 1$ in the *ISMIS* solution.

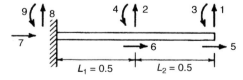

Figure 1 Two-element cantilever beam with special DOF numbering.

SOLUTION   The *ISMIS* plane frame element is used. The numbering of the displacement coordinates is in three groups: A = active DOFs, D = dependent (reduced-out)

DOFs, and C = constrained to be zero. The structure of the stiffness matrix and mass matrix is, therefore,

$$\begin{bmatrix} & 1\;2 & 3\;4 & 5...9 & \\ & AA & AD & AC & \begin{matrix}1\\2\end{matrix} \\ & DA & DD & DC & \begin{matrix}3\\4\end{matrix} \\ & CA & CD & CC & \begin{matrix}5\\\vdots\\9\end{matrix} \end{bmatrix}$$

The listing of the *ISMIS* input file <Example 14_11.m> follows.

```
runismis
%
disp('Example 14.11')
disp('Guyan Reduction')
disp('*Form Element Matrices*')
ke=framel(0.5,0,1,1,1)
me=framms(0.5,0,1,1)
disp('*Assemble Full System Stiffness Matrix*')
ks=zeros(9)
ms=zeros(9)
l1=[7 8 9 6 2 4]  % locator vectors
l2=[6 2 4 5 1 3]
ks=addmat(ks,ke,l1)  % "direct stiffness" matrix addition
ks=addmat(ks,ke,l2)
ms=addmat(ms,me,l1)
ms=addmat(ms,me,l2)
disp('*Perform Guyan Reduction on Rotations*')
kaa=rmvsm(ks,1,1,2,2)
kad=rmvsm(ks,1,3,2,2)
kda=rmvsm(ks,3,1,2,2)
kdd=rmvsm(ks,3,3,2,2)
maa=rmvsm(ms,1,1,2,2)
mad=rmvsm(ms,1,3,2,2)
mda=rmvsm(ms,3,1,2,2)
mdd=rmvsm(ms,3,3,2,2)
kddi=inv(kdd)
tda=-kddi*kda
kaah=kaa+kad*tda
maah=maa+mad*tda+(mad*tda)'+tda'*mdd*tda
disp('*Solve Translation-Only Eigenproblem*')
format long e
[vec,val2]=eig(kaah,maah)
val=diag(sqrt(val2))
endismis
```

Notice that there is one important new *ISMIS* command in the input file above. The <addmat> command implements the direct stiffness method of assembling system matrices from element matrices, as described in Section 14.4. The locator vectors provide the locator information that identifies which system row–column locations correspond to the respective ECG element coordinates.

14.8 Finite Element Solutions for Natural Frequencies and Mode Shapes

Table 14.2 Frequencies Obtained by Using Guyan Reduction to Reduce Rotations of Uniform Cantilever Beams

Mode	N_e					Exact (Ref. [14.10])
	1	2	3	4	5	
1	3.56753	3.52198	3.51699	3.51628	3.51611	3.51602
2		22.2790	22.2362	22.0946	22.0573	22.0345
3			62.6685	62.9703	62.2180	61.6972
4				123.545	124.725	120.902
5					205.277	199.860

Table 14.2 gives the frequencies of uniform beam finite element models obtained by reducing out all rotations. Note that in each case the frequency is higher than the corresponding frequency in Table 14.1, illustrating again the Rayleigh–Ritz upper-bound theorem. However, note that the two underlined values in Table 14.2 are not smaller than the values immediately to their left. That is possible, since the values in column N_e of Table 14.2 were obtained by reducing out the rotations that were present in obtaining the corresponding column of Table 14.1, not by adding a degree of freedom to the model used for obtaining column $N_e - 1$ of Table 14.2.

Although the Guyan Reduction Method does reduce the size of the eigenproblem to be solved, and often does produce accurate frequencies for a good percentage of the modes retained, two cautions should be observed: (1) the accuracy of the frequencies obtained depends on which coordinates are retained and which are reduced out, and (2) whereas **K** and **M** may be very sparse matrices, Guyan reduction destroys sparseness, and may lead to a much more expensive eigensolution even though the order of the reduced matrices $\hat{\mathbf{K}}$ and $\hat{\mathbf{M}}$ is much less than that of **K** and **M**.

However, one important use of model reduction is to reduce very large FE structural dynamics models to the much smaller size of model that agrees with the degrees of freedom used in modal testing (Chapter 18). These reduced models are called *test-analysis models* (TAMs). Guyan Reduction is only one of a number of model-reduction methods that may be used to create TAMs.

14.8.3 Lumped-Mass Models

Using the same cantilever beam as before, we now consider the accuracy of frequencies obtained by lumping the mass in the manner described in Section 14.2. The reduced stiffness matrix is again given by Eq. 14.71. The frequencies obtained are listed in Table 14.3. From the table it can be seen that for the particular lumping procedure employed (i.e., half of the mass of each element distributed to each end of that element), the lumped-mass procedure produces frequencies that converge quite slowly from below.

In a paper in which he coined the name *consistent-mass matrix*,[14.3] Archer compared frequencies of lumped-mass models and consistent-mass models of uniform free–free beams and simply supported beams, both including and excluding rotational freedoms. He concluded that "natural mode analysis and hence dynamic response analysis of beams with uniform stiffness and mass distribution is significantly improved

Table 14.3 Frequencies of a Cantilever Uniform Beam Based on Lumped-Mass Models

Mode	N_e					Exact (Ref. [14.10])
	1	2	3	4	5	
1	2.44949	3.15623	3.34568	3.41804	3.45266	3.51602
2		16.2580	18.8859	20.0904	20.7335	22.0345
3			47.0284	53.2017	55.9529	61.6972
4				92.7302	104.436	120.902
5					153.017	199.860

if the mass matrix is constructed using equations corresponding to the Rayleigh–Ritz approach (i.e., the consistent-mass approach) in place of the usual procedure of physical lumping of the structural mass at the coordinate points." In a paper published concurrently with Archer's paper, Leckie and Lindberg[14.4] presented error estimates for beam frequencies and showed that the number of elements required to achieve a desired accuracy depends on the boundary conditions if lumped-mass procedures are employed, whereas this is not the case for consistent-mass representations. Tong et al.,[14.5] established rates of convergence for mode shapes and frequencies obtained by using consistent-mass and lumped-mass finite element models. They showed that mass lumping is appropriate if the continuum model is represented by differential equations of second order, but that the consistent-mass formulation should be used for higher-order systems (e.g., beams and plates).

Up to this point, only uniform finite elements have been considered, although Fig. 14.1 indicated that a tapered beam might be modeled as an assemblage of uniform elements. It is straightforward to use the shape functions presented in Section 14.2, together with expressions for stiffness coefficients k_{ij} and mass coefficients m_{ij} (e.g., Eqs. 14.7) incorporating dimensions and physical properties that are functions of position within an element. Gallagher and Lee[14.11] compare eigensolutions for tapered beams represented by tapered elements and represented by uniform elements and showed that, for a given number of elements, much greater accuracy was achieved by using appropriately tapered elements.

The objective of the present chapter has simply been to introduce the reader to very basic concepts in finite element analysis. In Chapters 15 through 17 we present topics that relate to more advanced finite element analysis: methods for solving large eigenproblems, methods for determining dynamic response, and substructure-based methods for solving eigenproblems and dynamic response problems. In Section 1.2 a number of widely used commercial finite element codes are listed. However, instruction on the implementation of commercial FE computer codes to solve large structural dynamics problems is beyond the scope of this book.

REFERENCES

[14.1] Y. C. Fung, *An Introduction to the Theory of Aeroelasticity*, Wiley, New York, 1955.

[14.2] J. S. Przemieniecki, *Theory of Matrix Structural Analysis*, McGraw-Hill, New York, 1968.

[14.3] J. S. Archer, "Consistent Mass Matrix for Distributed Mass Systems," *Journal of the Structural Division, ASCE*, Vol. 89, 1963, pp. 161–178.

[14.4] F. A. Leckie and G. M. Lindberg, "The Effects of Lumped Parameters on Beam Frequencies," *Aeronautical Quarterly*, Vol. 14, 1963, pp. 224–240.

[14.5] P. Tong, T. H. H. Pian, and L. L. Bucciarelli, "Mode Shapes and Frequencies by Finite Element Method Using Consistent and Lumped Masses," *Computers and Structures*, Vol. 1, 1971, pp. 623–628.

[14.6] R. J. Guyan, "Reduction of Stiffness and Mass Matrices," *AIAA Journal*, Vol. 3, 1965, p. 380.

[14.7] R. G. Anderson, B. M. Irons, and O. C. Zienkiewicz, "Vibration and Stability of Plates Using Finite Elements," *International Journal of Solids Structures*, Vol. 4, 1968, pp. 1031–1055.

[14.8] C. P. Johnson et al., "Quadratic Reduction for the Eigenproblem," *International Journal of Numerical Methods in Engineering*, Vol. 15, 1980, pp. 911–923.

[14.9] R. R. Craig, Jr. and M. C. C. Bampton, "On the Iterative Solution of Semidefinite Eigenvalue Problems," *Aeronautical Journal of the Royal Aeronautical Society*, Vol. 75, 1971, pp. 287–290.

[14.10] T.-C. Chang and R. R. Craig, Jr., "Normal Modes of Uniform Beams," *Proceedings of ASCE*, Vol. 95, No. EM4, 1969, pp. 1025–1031.

[14.11] R. Gallagher and B. Lee, "Matrix Dynamic and Instability Analysis with Nonuniform Elements," *International Journal of Numerical Methods in Engineering*, Vol. 2, 1970, pp. 265–276.

[14.12] D. D. Kaña and S. Huzar, "Synthesis of Shuttle Vehicle Damping Using Substructure Test Results," *Journal of Spacecraft and Rockets*, Vol. 10, 1973, pp. 790–797.

PROBLEMS

Problem Set 14.2

14.1 Derive the four cubic shape functions in Eqs. 14.13 from Eqs. 14.10 and 14.12.

14.2 Using Eqs. 14.13 and 14.14a and b, verify the stiffness coefficient k_{12} and the mass coefficient m_{23} in Eqs. 14.15.

14.3 Stiffness coefficients can be defined in the following manner: "The stiffness coefficient k_{ij} is the *force* at degree of freedom i due to a *unit displacement* at degree of freedom j, with all other degrees of freedom constrained." Using this definition and beam-theory equations from mechanics of materials, verify that k_{12} is the transverse shear force at the left end of a beam of length L when it is deflected in the shape $\psi_2(x)$ (Fig. P14.3).

14.4 The forces $p_i(t)$ given by Eq. 14.14c are called *consistent nodal loads* and also, *fixed-end forces*. Show that the force $p_2(t)$ determined in Example 14.1 is equal to the pseudostatic fixed-end moment at the left end of the beam shown in Fig. P14.4.

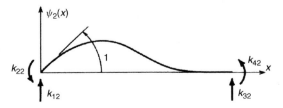

Figure P14.3

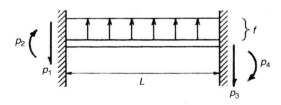

Figure P14.4

14.5 Determine the *consistent nodal loads* $p_1(t)$ through $p_4(t)$ for the triangular load distribution on the beam shown in Fig. P14.5.

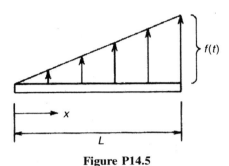

Figure P14.5

14.6 Determine the coefficients k_{11}, $k_{12} = k_{21}$, and k_{22} of the stiffness matrix **K** for the conical frustum torsion element shown in Fig. P14.6. Use Eq. 14.19a with the shape functions given in Eqs. 14.6.

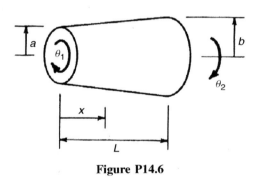

Figure P14.6

Problem Set 14.3

14.7 The **u** and $\widehat{\mathbf{u}}$ coordinates for the plane frame element, which are shown in Fig. 14.12b and listed in Eqs. 14.27, are related by the transformation equation, Eq. 14.24. By independently performing displacements $\widehat{u}_4$ and $\widehat{u}_5$ on the plane frame element in Fig. 14.12b, derive the expression represented by the u_5 row in the transformation matrix **T** in Eq. 14.28.

14.8 By using three clearly drawn sketches, determine expressions for the contributions to u_1 for the three-dimensional truss element in Fig. 14.13 from the displacements $\widehat{u}_1$, $\widehat{u}_2$, and $\widehat{u}_3$. (Your answer should correspond to the first row of Eqs. 14.29a and b.)

14.9 Using Fig. P14.9, determine the forces $\widehat{p}_1$, $\widehat{p}_2$, $\widehat{p}_3$, and $\widehat{p}_4$ which correspond to a displacement $\widehat{u}_1$ as shown. (Note that the answer corresponds to the first column of Eq. 14.36b, since $\widehat{\mathbf{p}} = \widehat{\mathbf{k}}\widehat{\mathbf{u}}$.)

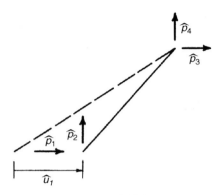

Figure P14.9

14.10 The force transformation in Eq. 14.32 was derived by using virtual work methods. It can also be derived by simple projection of force components. Using Fig. P14.10, verify the entries in Eq. 14.32, where **T** for the plane truss element is given by Eq. 14.26.

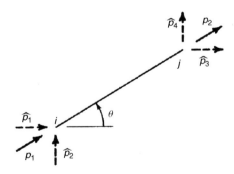

Figure P14.10

Problem Set 14.4

14.11 Each of the bars in the planar truss in Fig. P14.11 has the same cross-sectional properties, AE. (1) By using Eq. 14.36b, determine the three element stiffness matrices $\widehat{\mathbf{k}}_1$, $\widehat{\mathbf{k}}_2$, and $\widehat{\mathbf{k}}_3$. (2) Using the *Direct Stiffness Method*, assemble the 8×8 unconstrained system stiffness matrix **K**.

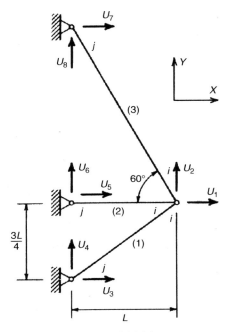

Figure P14.11

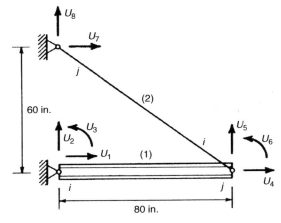

Figure P14.12

14.12 The crane boom in Fig. P14.12 consists of a uniform beam supported by a tie bar. $E = 30 \times 10^3$ ksi, $A_1 = 4.1$ in^2, $I_1 = 10.8$ in^4, and $A_2 = 0.8$ in^2. **(a)** Determine the two element stiffness matrices $\hat{\mathbf{k}}_1$ and $\hat{\mathbf{k}}_2$. **(b)** Using the *Direct Stiffness Method*, assemble the 8×8 unconstrained system stiffness matrix $\mathbf{K}$.

14.13 For the plane frame in Fig. P14.13: **(a)** Determine the three element stiffness matrices $\hat{\mathbf{k}}_1$, $\hat{\mathbf{k}}_2$, and $\hat{\mathbf{k}}_3$. **(b)** Using the *Direct Stiffness Method*, assemble the

equations of motion for the 3-DOF model which would result from neglecting axial deformation in all elements, that is, $U_5 = U_6 = 0$ and $U_4 = U_3$. The equations of motion should have the form

$$\begin{bmatrix} M_{11} & M_{12} & M_{13} \\ M_{21} & M_{22} & M_{23} \\ M_{31} & M_{32} & M_{33} \end{bmatrix} \begin{Bmatrix} \ddot{U}_1 \\ \ddot{U}_2 \\ \ddot{U}_3 \end{Bmatrix} + \begin{bmatrix} K_{11} & K_{12} & K_{13} \\ K_{21} & K_{22} & K_{23} \\ K_{31} & K_{32} & K_{33} \end{bmatrix} \begin{Bmatrix} U_1 \\ U_2 \\ U_3 \end{Bmatrix}$$

$$= \begin{Bmatrix} P_1(t) \\ P_2(t) \\ P_3(t) \end{Bmatrix}$$

Problem Set 14.6

14.14 In Example 14.7 the rotations were reduced out, yielding 2×2 stiffness and mass matrices. As noted

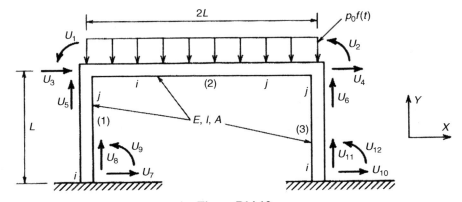

Figure P14.13

in Eq. 14.65a, the reduced mass matrix $\widehat{\mathbf{M}}_{aa}$ is just the nonzero portion $\mathbf{M}_{aa}$ of the original mass matrix. For the two-element beam in Fig. P14.14 where ρA = constant and EI = constant: (a) Determine the 4×4 consistent-mass matrix. (b) Form the transformation matrix $\mathbf{T}$ that will reduce the two rotations. (c) Determine the 2×2 reduced mass matrix $\mathbf{M}_{aa}$ by using the *Guyan Reduction Method*. (d) Use Eq. 9 of Example 14.7 to determine the natural frequencies of the 2-DOF lumped-mass model. (e) Using the mass matrix determined in part (c) and the stiffness matrix $\widehat{\mathbf{K}}_{aa}$ from Example 14.7, determine the natural frequencies of the 2-DOF Guyan-reduction model. (f) Compare your results from parts (d) and (e) with the exact frequencies of a uniform cantilever beam.

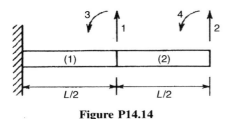

Figure P14.14

14.15 A large sign can be considered to be a uniform rigid plate welded firmly to the top of a flexible column, as depicted in Fig. P14.15. Let the total mass matrix referred to the column degrees of freedom U_1, U_2, and U_3 be

$$\mathbf{M} = \mathbf{M}_c + \widehat{\mathbf{M}}_p$$

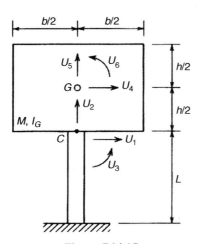

Figure P14.15

where

$$\mathbf{M}_p = \begin{bmatrix} M & 0 & 0 \\ 0 & M & 0 \\ 0 & 0 & I_G \end{bmatrix}$$

Use Eqs. 14.57, 14.60, and 14.61 to transfer the plate inertia terms from the center-of-gravity coordinates U_4, U_5, and U_6 to the end-of-column coordinates U_1, U_2, and U_3. Assume small rotation (slope) at the end of the column.

Problem Set 14.7

14.16 (a) Starting with the 3-DOF mass matrix for axial motion of the uniform bar in Fig. 1 of Example 14.8, and using the transformation matrix $\mathbf{T}$ in Eq. 3, determine $\widehat{\mathbf{M}}$ that separates the flexible coordinates from the rigid-body coordinate for the system in Example 14.8. (b) Write out the equation of motion in matrix form and solve for the modes and frequencies.

14.17 Treat the single uniform beam element in Fig. P14.17 as the "system." Let U_1 and U_2 be *e*lastic deformation coordinates, and let U_3 and U_4 be *r*igid-body coordinates. (a) Form the stiffness matrix $\mathbf{K}$ and the mass matrix $\mathbf{M}$ for the system coordinate numbering in Fig. P14.17. (b) Form the transformation matrix of Eq. 14.68 for this system. (c) Determine the mass matrix $\widehat{\mathbf{M}}$ that corresponds to the stiffness matrix $\widehat{\mathbf{K}}$ of Eq. 14.69. (d) Solve for the flexible modes of this system. Sketch these two modes.

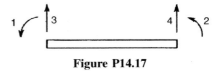

Figure P14.17

Problem Set 14.8

For problems whose number is preceded by a **C**, you are to write a computer program and use it to produce the results requested.
Note: MATLAB .m-files for some of the Example in this chapter may be found on the book's website.

C 14.18 (a) Use the *ISMIS* computer program (or another designated FE computer program) to obtain the natural frequencies and mode shapes of the five-element uniform cantilever beam in Fig. P14.18. Include both axial and transverse motion. Use $A = I = E = \rho = 1$ and $L_e = 0.20$. Use a consistent mass matrix. **(b)** Solve for the lowest three axial frequencies and the lowest three bending frequencies and compare them with exact values from Examples 13.1 and 13.3, respectively.

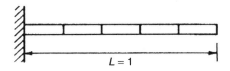

Figure P14.18

C 14.19 An early Space Shuttle model was designed to enable preliminary tests to be conducted approximating some of the aspects of the dynamics of the coupled orbiter-booster configuration.[14.12] The orbiter model, shown in Fig. P14.19, consisted of an aluminum tube ($E = 10 \times 10^6$ psi) of 1-in. outside diameter and 0.035-in. wall thickness and 60.5 in. long. Equally spaced along the tube were six rigid masses, which consisted of 0.50-in.-thick steel plates with outside diameter 6 in. and inside diameter 1 in. (to fit over the tube). Use the *ISMIS* computer program to compute the axial and transverse frequencies and mode shapes of this model if it is in the free–free condition; that is, it has no boundary constraints. Consider only in-plane motion and use the <framel> command to generate the element stiffness matrix for a typical element and the <framms> command to generate the corresponding element consistent-mass matrix. Add the three plate rigid-body inertia contributions at each of the seven system nodes.

C 14.20 Repeat Problem 14.18 using the *Guyan Reduction Method* to reduce out all rotations.

C 14.21 Repeat Problem 14.18 replacing the original consistent-mass matrix with a lumped-mass matrix.

C 14.22 The basic design of a three-story structure is shown in Fig. P14.22, where $m_1 = m_2 = 40$ kip-sec^2/ft, $m_3 = 50$ kip-sec^2/ft, $k_1 = 30 \times 10^3$ kips/ft, $k_2 = 40 \times 10^3$ kips/ft, and $k_3 = 50 \times 10^3$ kips/ft. **(a)** Determine the natural frequencies and mode shapes of this structure using the *ISMIS* computer program (or another specified FE computer program). Sketch the three mode shapes (by hand). **(b)** If the stiffness of the columns is reduced by 5%, what effect does this have on the frequencies and mode shapes?

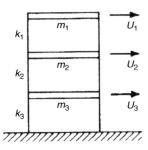

Figure P14.22

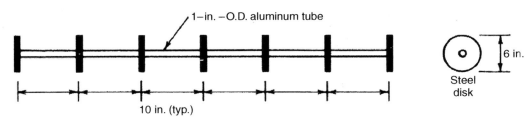

Figure P14.19

C 14.23 A rigid platform is welded to steel pipe inclined columns as shown in Fig. P14.23. $A = 2000$ mm^2, $I = 300$ cm^4, $E = 200$ GN/m^2, $m = 907$ kg, and $I_G = 302$ kg·m^2. Due to the inclination of the columns, a simple SDOF model such as the one in Problem 2.1 is not valid. Treat the structure in this problem as a planar structure. **(a)** Use the *ISMIS* computer program (or another specified FE computer program) to form a 3-DOF model of this structure using the coordinates shown in Fig. P14.23. The method of Section 14.6 can be used to relate the column displacements to the three system displacements. Neglect the mass of the columns. Solve for the modes and frequencies of this model. **(b)** Would it be valid to reduce this to an SDOF system by ignoring displacements 2 and 3? **(c)** Would it be acceptable to ignore the mass moment of inertia I_G in the 3-DOF model?

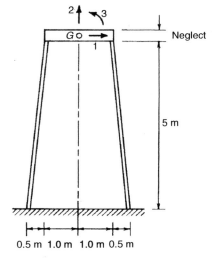

Figure P14.23

15

Numerical Evaluation of Modes and Frequencies of MDOF Systems

In Chapter 9 you learned how to solve for the modes and frequencies of a 2-DOF system by determining the roots of the characteristic (polynomial) equation, and in Chapter 10 many properties of modes and frequencies of MDOF systems were discussed. In this chapter we introduce you to the variety of procedures that are used in obtaining numerical solutions to large eigenproblems. Such problems may require the determination of modes and frequencies of systems having even hundreds of thousands of degrees of freedom. A detailed presentation of these methods, however, is beyond the scope of this book. Rather, a working knowledge of modern eigensolver approaches is desired.

Upon completion of this chapter you should:

- Be aware of the variety of numerical methods for solving the structural dynamics eigenproblem and know the principal factors to consider in choosing a method.
- Be able to apply vector iteration to determine at least a few modes and frequencies of a small system.
- Be able to solve an eigenproblem by using the MATLAB computer program, or another specified computer program, and interpret the results obtained in the computer output produced by that program.

15.1 INTRODUCTION TO METHODS FOR SOLVING ALGEBRAIC EIGENPROBLEMS

Contemporary MDOF models of structures may have from tens to hundreds of thousands (and sometimes millions) of degrees of freedom. The dynamic analysis of these structures generally involves determining some, but seldom all, of the natural frequencies (eigenvalues) and natural modes (eigenvectors) by solving the *generalized eigenvalue equation*[1]

$$(\mathbf{K} - \lambda_i \mathbf{M})\boldsymbol{\phi}_i = \mathbf{0} \tag{15.1}$$

---

[1] The *standard eigenvalue problem* has the form $\mathbf{Ax} = \lambda \mathbf{x}$, whereas the *generalized eigenvalue problem* has the form $\mathbf{Ax} = \lambda \mathbf{Bx}$, or in structural dynamics notation, $\mathbf{K}\boldsymbol{\phi} = \lambda \mathbf{M}\boldsymbol{\phi}$. In Appendix D we describe a method for converting from symmetric $\mathbf{K}$ and $\mathbf{M}$ matrices to a symmetric $\mathbf{A}$ matrix for use in the standard eigenproblem.

where $\lambda_i \equiv \omega_i^2$. The polynomial root-finding method, which was employed in Chapter 9 for determining the natural frequencies of 2-DOF systems, is only one of a number of methods for solving the algebraic eigenproblem, Eq. 15.1. Most large structural dynamics computer programs provide several far more powerful methods so that the analyst can choose the one that is most appropriate for a given problem. The purpose of this section is merely to make you aware of the variety of algebraic eigensolvers and to suggest some of the important factors that you need to consider in selecting an eigensolver. More advanced structural dynamics texts (e.g., Refs. [15.1] and [15.2]) and computational mathematics literature (e.g., Refs. [15.3] and [15.4]) provide details on eigensolvers that are particularly appropriate for structural dynamics eigenproblems.

The fundamental properties of natural frequencies and natural modes that were discussed in Chapter 10 provide the basis for a variety of eigensolution techniques. These techniques fall into three broad categories, depending on the fundamental property that is used most prominently in the solution procedure. Methods that directly employ the property stated in Eq. 15.1 are called *vector iteration* methods (or power methods). *Matrix transformation* methods constitute the second group. They employ the orthogonality properties

$$\mathbf{\Phi}^T \mathbf{K} \mathbf{\Phi} = \mathbf{\Lambda}, \qquad \mathbf{\Phi}^T \mathbf{M} \mathbf{\Phi} = \mathbf{I} \tag{15.2}$$

where $\mathbf{\Phi} = [\boldsymbol{\phi}_1 \ \boldsymbol{\phi}_2 \ \cdots \ \boldsymbol{\phi}_N]$ and $\mathbf{\Lambda} = \mathrm{diag}(\lambda_i)$, $i = 1, 2, \ldots, N$. *Polynomial root finding* is the basis of a third group of eigensolution techniques. The equation most prominently used is

$$\det(\mathbf{K} - \lambda_i \mathbf{M}) = 0 \tag{15.3}$$

or

$$\det(\mathbf{K}^{(r)} - \lambda_i^{(r)} \mathbf{M}^{(r)}) = 0 \tag{15.4}$$

where $\mathbf{K}^{(r)}$ and $\mathbf{M}^{(r)}$ were defined in Section 10.2.

In each of the three groups above there are several specific methods. For example, among the vector iteration methods are the Direct Iteration Method and the Inverse Iteration Method (the Method of Vianello and Stodola). Inverse iteration may be used with or without "spectrum shift." Matrix transformation methods include Jacobi's Method, Lanczos's Method, Givens's Method, Householder's Method, and the QR Method. Polynomial root finding is involved in determinant-search and Stürm-sequence methods.

Many practical eigensolvers, such as the Householder Method and the Subspace Iteration Method[15.1], employ techniques based on a combination of the fundamental properties above. For example, the Householder-QR-Inverse Iteration (HQRI) Method described in Ref. [15.1] combines aspects of the matrix transformation and vector iteration methods. Finally, model reduction plays a significant role in several eigensolvers: for example, subspace iteration[15.1] and AMLS (automated multilevel substructuring).[15.4]

With the foregoing methods and more from which to choose, a person wishing to use the computer to solve structural dynamics eigenproblems is obliged to consider which methods are most suited to his or her class of problems. The factors that most directly influence the choice of eigensolver are: (1) the order N of the matrices, and the bandwidths M_K and M_M of $\mathbf{K}$ and $\mathbf{M}$, respectively; and (2) the number of eigenvalues and eigenvectors that are required. Overviews by Bathé and Wilson[15.1] and Golub and Van Loan[15.5] suggest the following guidelines for choosing an eigensolver:

1. The Householder-QR-Inverse Iteration (HQRI) solution is most efficient when all eigenvalues and eigenvectors of a matrix are required and the matrix has a large bandwidth or is full.
2. The determinant search technique is very effectively used to calculate the lowest eigenvalues and corresponding eigenvectors of systems with small bandwidth. A determinant search technique is often used in the terminal phase of transformation-based methods that "nearly diagonalize" an eigenproblem.
3. The subspace iteration solution is very effective in the calculation of the lowest eigenvalues and corresponding eigenvectors of systems with large bandwidth and which are too large for the high-speed storage of the computer.
4. The Lanczos Algorithm can be an efficient technique for calculating a few extremal eigenvalues of a large, sparse eigenproblem.

A 1980 review of eigensolvers for vibration analysis is presented in Ref. [15.6], but significant research on eigensolvers for large structural dynamics problems is ongoing.[15.3,15.4]

In the next section you will be introduced to vector iteration techniques. Vector iteration can be used as a "stand-alone" eigensolver or can be combined with other procedures: for example, in the HQRI Method named above.

15.2 VECTOR ITERATION METHODS

15.2.1 Inverse Iteration

The most effective procedure for hand computation of a few eigenvalues and eigenvectors of a small (e.g., $N \leq 5$) system is, perhaps, the *Inverse Iteration Method*. Vianello (1898) used it to solve for buckling of a strut, and Stodola (1904) used it to solve for the vibration of a rotating shaft. We begin by considering Eq. 15.1, which we write in the form

$$\mathbf{K}\boldsymbol{\phi} = \lambda \mathbf{M}\boldsymbol{\phi} \quad (15.5)$$

We will assume that the fundamental (i.e., lowest) frequency is distinct and that the eigenvalues λ_i are ordered according to

$$\lambda_1 < \lambda_2 \leq \lambda_3 \leq \cdots \leq \lambda_N \quad (15.6)$$

When the left-hand side of Eq. 15.5 is equal to the right-hand side, an eigenvalue λ_i and corresponding eigenvector $\boldsymbol{\phi}_i$ have been obtained. An iterative procedure based on Eq. 15.5 can be set up by writing Eq. 15.5 in the form

$$\mathbf{K}\mathbf{v}_{s+1} = \mathbf{M}\mathbf{u}_s \quad (15.7)$$

or

$$\mathbf{v}_{s+1} = \mathbf{D}\mathbf{u}_s \quad (15.8)$$

where **D** is the *dynamical matrix* given by

$$\mathbf{D} = \mathbf{K}^{-1}\mathbf{M} \tag{15.9}$$

Then $\mathbf{u}_{s+1}$ is defined by[2]

$$\mathbf{u}_{s+1} = \lambda_{(s+1)}\mathbf{v}_{s+1} \tag{15.10}$$

where $\lambda_{(s+1)}$ is an appropriately chosen scaling factor. For example, $\lambda_{(s+1)}$ may be chosen to make the largest element of $\mathbf{u}_{s+1}$ equal to $+1$, or $\lambda_{(s+1)}$ can be chosen to make

$$\mathbf{u}_{s+1}^T \mathbf{M} \mathbf{u}_{s+1} = 1$$

Note that Eq. 15.7 can be interpreted as solving for the $(s+1)$st deflection shape produced by the inertia forces associated with the sth deflection shape. The procedure is called *inverse iteration* because it uses (symbolically) the inverse of **K** in Eq. 15.8.

We first look at Example 15.1. Then we prove that the procedure always produces convergence to the fundamental mode if $\lambda_1 < \lambda_2$, as assumed in Eq. 15.6.

Example 15.1 Use the *Inverse Iteration Method* to solve for the fundamental mode shape and natural frequency of the undamped 3-DOF system shown in Fig. 1, where $k = 1$ and $m = 1$. For this system

$$\mathbf{K} = \begin{bmatrix} 2 & -1 & 0 \\ -1 & 2 & -1 \\ 0 & -1 & 2 \end{bmatrix}, \qquad \mathbf{M} = \begin{bmatrix} 1 & 0 & 0 \\ 0 & 1 & 0 \\ 0 & 0 & 1 \end{bmatrix}$$

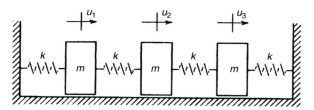

Figure 1 A 3-DOF spring–mass system.

The exact eigenvalues are $\lambda_1 = 2 - \sqrt{2} = 0.5858$, $\lambda_2 = 2.0$, $\lambda_3 = 2 + \sqrt{2} = 3.412$. Also, $\mathbf{K}^{-1}$ can be shown to be

$$\mathbf{K}^{-1} = \frac{1}{4}\begin{bmatrix} 3 & 2 & 1 \\ 2 & 4 & 2 \\ 1 & 2 & 3 \end{bmatrix} = \begin{bmatrix} 0.75 & 0.50 & 0.25 \\ 0.50 & 1.00 & 0.50 \\ 0.25 & 0.50 & 0.75 \end{bmatrix}$$

---

[2] Iteration sequence numbers of eigenvalue estimates are enclosed in parentheses (e.g., $\lambda_{(s+1)}$) to distinguish the estimates from exact eigenvalues (e.g., as listed in Eq. 15.6).

SOLUTION 1. Form the dynamical matrix

$$\mathbf{D} = \mathbf{K}^{-1}\mathbf{M} = \begin{bmatrix} 0.75 & 0.50 & 0.25 \\ 0.50 & 1.00 & 0.50 \\ 0.25 & 0.50 & 0.75 \end{bmatrix} \quad (1)$$

2. Choose a starting vector $\mathbf{u}_0$. Let

$$\mathbf{u}_0 = \begin{Bmatrix} 1 \\ 1 \\ 1 \end{Bmatrix} \quad (2)$$

3. Use Eq. 15.8 to obtain $\mathbf{v}_{s+1}$, $s = 0, 1, 2, \ldots$, and scale $\mathbf{v}_{s+1}$ to make the largest element of $\mathbf{u}_{s+1}$ equal to $+1$. The results of two iteration cycles are tabulated below.

$\mathbf{u}_0$	$\mathbf{v}_1$	$\mathbf{u}_1$	$\mathbf{v}_2$	$\mathbf{u}_2$
1.000	1.500	0.750	0.250	0.714
1.000	2.000	1.000	1.750	1.000
1.000	1.500	0.750	1.250	0.714
	$\lambda_{(1)} = 0.500$		$\lambda_{(2)} = 0.571$	

Note that the eigenvalue estimates in part (3), where scaling is based on a single element in the eigenvector, are approaching the true value, $\lambda_1 = 0.5858$, from below.

4. A better estimate of the eigenvalue can be obtained by computing the Rayleigh quotient corresponding to each shape $\mathbf{v}_{s+1}$.

$$\lambda_{(s+1)} = \frac{\mathbf{v}_{s+1}^T \mathbf{K} \mathbf{v}_{s+1}}{\mathbf{v}_{s+1}^T \mathbf{M} \mathbf{v}_{s+1}} = \frac{\mathbf{v}_{s+1}^T \mathbf{M} \mathbf{u}_s}{\mathbf{v}_{s+1}^T \mathbf{M} \mathbf{v}_{s+1}} \quad (3)$$

The denominator of Eq. 3 can also be used in the scaling equation

$$\mathbf{u}_{s+1} = \frac{\mathbf{v}_{s+1}}{(\mathbf{v}_{s+1}^T \mathbf{M} \mathbf{v}_{s+1})^{1/2}} \quad (4)$$

The following table summarizes two iteration cycles using Eqs. 3 and 4.

$\mathbf{u}_0$	$\mathbf{v}_1$	$\mathbf{u}_1$	$\mathbf{v}_2$	$\mathbf{u}_2$
1.000	1.500	0.5145	0.8575	0.5024
1.000	2.000	0.6860	1.2010	0.7037
1.000	1.500	0.5145	0.8575	0.5024
	$\lambda_{(1)} = \dfrac{5.000}{8.500} = 0.588$		$\lambda_{(2)} = \dfrac{1.706}{2.913} = 0.586$	

Although calculation of the Rayleigh quotient involves more calculation than the scaling used in step 3, you can see that even after one iteration cycle the Rayleigh quotient gives a very good upper-bound approximation to the exact value of λ_1. After two cycles of iteration, the fundamental frequency and mode shape are approximated by

$$\lambda_1 = 0.586, \qquad \boldsymbol{\phi}_1 = \begin{Bmatrix} 0.5024 \\ 0.7037 \\ 0.5024 \end{Bmatrix} \qquad \textbf{Ans. (5)}$$

As an additional step, the final estimate of eigenvector $\boldsymbol{\phi}_1$ could, of course, be rescaled to make the value of its maximum element be 1.0.

MATLAB Exercise 1: Inverse Iteration and Forward Iteration[3]   Consider the generalized eigenvalue problem studied in Example 15.1. Write a MATLAB .m-file that will (1) calculate the lowest natural frequency and associated mode shape via the *Inverse Iteration Method*, and (2) calculate the highest natural frequency and associated mode shape via the *Forward Iteration Method*. The .m-file that encodes the inverse iteration should take the following arguments as input:

- The matrices **K** and **M** for which the lowest eigenpair is sought
- An initial guess for the lowest eigenvalue
- An initial guess for the eigenvector associated with the lowest eigenvalue
- A tolerance to be used as a convergence criterion for the iteration process

The .m-file should return:

- An approximation to the lowest eigenvalue
- An approximation to the eigenvector corresponding to the lowest eigenvalue

The .m-file that encodes the Forward Iteration Method should have precisely the same structure as above, but with the modifier "lowest" changed to "highest."

Compare your numerical solution to the results that were obtained by "hand calculation" in Example 15.1. Investigate the sensitivity of the Inverse Iteration Method and the Forward Iteration Method to (a) the initial guess for the eigenvalue, and (b) the initial guess for the eigenvector.

The .m-file <mat_ex_15_1_inv.m>, which calls the inverse iteration function subroutine <inviter.m>, produces an estimate of the lowest eigenvalue of 0.5858, along with the associated eigenvector

$$\begin{Bmatrix} 0.7071 \\ 1.0000 \\ 0.7071 \end{Bmatrix}$$

The .m-file <mat_ex_15_1_for.m>, which calls the forward iteration function subroutine <poweriter.m>, produces an estimate of the largest eigenvalue of 3.4142 and the associated eigenvector

$$\begin{Bmatrix} 0.7071 \\ -1.0000 \\ 0.7071 \end{Bmatrix}$$

---

[3] In this chapter there are a number of "MATLAB Exercises." The MATLAB .m-files listed with these exercises can be found on the book's website. Variants of these exercises can be used as homework computer exercises.

Depending on the convergence criterion, the effect of varying the initial guess for the eigenvalue or the eigenvector can slow the convergence process, thus increasing the number of iterations necessary.

Reference [15.1] presents a useful computational algorithm for carrying out a computer implementation of inverse iteration including Rayleigh quotient calculations.

15.2.2 Proof of Convergence

We will now prove that the inverse iteration procedure described above produces convergence to the fundamental mode of a system. Recall (Section 10.1.9) that an arbitrary vector can be written in terms of an eigenvector expansion. Therefore, we can write $\mathbf{u}_0$ as

$$\mathbf{u}_0 = \sum_{i=1}^{N} c_i \boldsymbol{\phi}_i \tag{15.11}$$

where $\boldsymbol{\phi}_i$, $i = 1, 2, \ldots, N$, are the exact mode shapes and c_i, $i = 1, 2, \ldots, N$, are unique constants.

From Eqs. 15.1 and 15.9,

$$\mathbf{D}\boldsymbol{\phi}_i = \frac{1}{\lambda_i} \boldsymbol{\phi}_i \tag{15.12}$$

Then

$$\mathbf{D}\mathbf{u}_0 = \sum_{i=1}^{N} c_i \mathbf{D}\boldsymbol{\phi}_i = \sum_{i=1}^{N} c_i \frac{1}{\lambda_i} \boldsymbol{\phi}_i \tag{15.13}$$

As before, assume that $\lambda_1 < \lambda_2$. Each cycle of iteration using Eqs. 15.8 and 15.10 involves a solution of Eq. 15.8 and a scaling operation, Eq. 15.10. Since the scaling does not affect the convergence, we can omit the scaling step and apply Eq. 15.8 s times to get

$$\begin{aligned}\mathbf{D}^s \mathbf{u}_0 &= \sum_{i=1}^{N} c_i \left(\frac{1}{\lambda_i}\right)^s \boldsymbol{\phi}_i \\ &= \left(\frac{1}{\lambda_1}\right)^s \sum_{i=1}^{N} c_i \left(\frac{\lambda_1}{\lambda_i}\right)^s \boldsymbol{\phi}_i\end{aligned} \tag{15.14}$$

Since $\lambda_1 < \lambda_i$, $(\lambda_1/\lambda_i)^s \to 0$ for $i > 1$. Therefore,

$$\mathbf{D}^s \mathbf{u}_0 \to \left(\frac{1}{\lambda_1}\right)^s c_1 \boldsymbol{\phi}_1 \tag{15.15a}$$

and

$$\mathbf{D}^{s+1} \mathbf{u}_0 \to \left(\frac{1}{\lambda_1}\right)^{s+1} c_1 \boldsymbol{\phi}_1 \tag{15.15b}$$

so the convergence is to a vector proportional to $\boldsymbol{\phi}_1$. The convergence rate depends on the λ_1/λ_i ratios, particularly λ_1/λ_2, and on the c_i's, that is, on the makeup of the starting vector.

Equations 15.15a and b indicate that if $c_1 \neq 0$ in the starting vector $\mathbf{u}_0$, inverse iteration will always converge to the fundamental mode $\boldsymbol{\phi}_1$. This suggests that to produce convergence to $\boldsymbol{\phi}_2$ it will be necessary to remove the $\boldsymbol{\phi}_1$ component from $\mathbf{u}_0$. But since roundoff error can serve to reintroduce $\boldsymbol{\phi}_1$ into later iterates, it will be necessary to remove the $\boldsymbol{\phi}_1$ component from each $\mathbf{u}_s$ used in the iteration for $\boldsymbol{\phi}_2$. To determine $\boldsymbol{\phi}_3$ by inverse iteration, we must remove both $\boldsymbol{\phi}_1$ and $\boldsymbol{\phi}_2$ from each $\mathbf{u}_s$ used in the iteration procedure, and so on. *Gram–Schmidt orthogonalization* and associated *matrix deflation* are the names applied to procedures used for carrying out this removal of previously determined eigenvectors.

15.2.3 Gram–Schmidt Orthogonalization

Gram–Schmidt orthogonalization can be described as follows. Let

$$\widehat{\mathbf{u}}_s = \mathbf{u}_s - \alpha_1 \boldsymbol{\phi}_1 \tag{15.16}$$

and choose α_1 so that $\widehat{\mathbf{u}}_s$ is orthogonal to $\boldsymbol{\phi}_1$; that is,

$$\boldsymbol{\phi}_1^T \mathbf{M} \widehat{\mathbf{u}}_s = 0 \tag{15.17}$$

Then

$$\alpha_1 = \frac{\boldsymbol{\phi}_1^T \mathbf{M} \mathbf{u}_s}{\boldsymbol{\phi}_1^T \mathbf{M} \boldsymbol{\phi}_1}$$

or

$$\widehat{\mathbf{u}}_s = \mathbf{S}_1 \mathbf{u}_s \tag{15.18}$$

where

$$\mathbf{S}_1 = \mathbf{I} - \frac{\boldsymbol{\phi}_1 \boldsymbol{\phi}_1^T \mathbf{M}}{\boldsymbol{\phi}_1^T \mathbf{M} \boldsymbol{\phi}_1} \tag{15.19}$$

To produce convergence to $\boldsymbol{\phi}_2$, $\mathbf{u}_s$ in Eq. 15.8 is replaced by $\widehat{\mathbf{u}}_s$, which is defined by Eqs. 15.18 and 15.19. Thus,

$$\mathbf{v}_{s+1} = \mathbf{D}_2 \mathbf{u}_s \tag{15.20}$$

where

$$\mathbf{D}_2 = \mathbf{D}\mathbf{S}_1 = \mathbf{D} - \frac{\mathbf{D} \boldsymbol{\phi}_1 \boldsymbol{\phi}_1^T \mathbf{M}}{\boldsymbol{\phi}_1^T \mathbf{M} \boldsymbol{\phi}_1} \tag{15.21}$$

Combining Eqs. 15.12 and 15.21 gives

$$\mathbf{D}_2 = \mathbf{D} - \frac{1}{\lambda_1} \left(\frac{\boldsymbol{\phi}_1 \boldsymbol{\phi}_1^T \mathbf{M}}{\boldsymbol{\phi}_1^T \mathbf{M} \boldsymbol{\phi}_1} \right) \tag{15.22}$$

Procedures analogous to Eqs. 15.16 through 15.22 can be used to define $\mathbf{D}_3$, $\mathbf{D}_4$, and so on, for removing higher modes. This procedure, called *matrix deflation*[15.7], is illustrated in Example 15.2.

Example 15.2 Use a sweeping matrix to solve iteratively for the second frequency and mode of the system in Example 15.1. Use as the mode shape $\boldsymbol{\phi}_1$ the following result from Example 15.1:

$$\boldsymbol{\phi}_1 = \begin{Bmatrix} 1.000 \\ 1.400 \\ 1.000 \end{Bmatrix}$$

SOLUTION The orthogonality condition, Eq. 15.17, must be enforced. Let

$$\widehat{\mathbf{u}}_s = \begin{Bmatrix} \widehat{u}_1 \\ \widehat{u}_2 \\ \widehat{u}_3 \end{Bmatrix} \quad (1)$$

Then Eq. 15.17 becomes

$$\boldsymbol{\phi}_1^T \mathbf{M} \widehat{\mathbf{u}}_s = \begin{bmatrix} 1.000 & 1.400 & 1.000 \end{bmatrix} \begin{bmatrix} 1 & 0 & 0 \\ 0 & 1 & 0 \\ 0 & 0 & 1 \end{bmatrix} \begin{Bmatrix} \widehat{u}_1 \\ \widehat{u}_2 \\ \widehat{u}_3 \end{Bmatrix} \quad (2)$$

or

$$\widehat{u}_1 + 1.400\,\widehat{u}_2 + \widehat{u}_3 \quad (3)$$

We can treat this as a constraint on $\widehat{u}_1$ and write

$$\widehat{u}_1 = -1.400\,\widehat{u}_2 - \widehat{u}_3 \quad (4)$$

So

$$\begin{Bmatrix} \widehat{u}_1 \\ \widehat{u}_2 \\ \widehat{u}_3 \end{Bmatrix} = \begin{bmatrix} -1.400 & -1.000 \\ 1 & 0 \\ 0 & 1 \end{bmatrix} \begin{Bmatrix} \widehat{u}_2 \\ \widehat{u}_3 \end{Bmatrix} \quad (5)$$

But we can choose $\widehat{u}_2$ and $\widehat{u}_3$ arbitrarily, so we can let $\widehat{u}_2 = u_2$, $\widehat{u}_3 = u_3$ and write Eq. 5 in the form

$$\begin{Bmatrix} \widehat{u}_1 \\ \widehat{u}_2 \\ \widehat{u}_3 \end{Bmatrix} = \begin{bmatrix} 0 & -1.400 & -1.000 \\ 0 & 1 & 0 \\ 0 & 0 & 1 \end{bmatrix} \begin{Bmatrix} u_1 \\ u_2 \\ u_3 \end{Bmatrix} \quad (6)$$

where u_1 is a dummy as far as Eq. 6 is concerned. Equation 6 is in the form of Eq. 15.18. We can form

$$\mathbf{D}_2 = \mathbf{D}\mathbf{S}_1 \quad (7)$$

where $\mathbf{S}_1$ is called a *sweeping matrix* because it sweeps $\boldsymbol{\phi}_1$ out of $\hat{u}$. Using $\mathbf{D}$ from Example 15.1 and S_1 from Eq. 6, we get

$$\mathbf{D}_2 = \begin{bmatrix} 0.75 & 0.50 & 0.25 \\ 0.50 & 1.00 & 0.50 \\ 0.25 & 0.50 & 0.75 \end{bmatrix} \begin{bmatrix} 0 & -1.400 & -1.000 \\ 0 & 1 & 0 \\ 0 & 0 & 1 \end{bmatrix}$$

$$= \begin{bmatrix} 0 & -0.550 & -0.500 \\ 0 & 0.300 & 0.000 \\ 0 & 0.150 & 0.500 \end{bmatrix} \tag{8}$$

We next use the iteration equation

$$\mathbf{v}_{s+1} = \mathbf{D}_2 \mathbf{u}_s \tag{9}$$

and scale $\mathbf{u}_{s+1}$ so that its largest active element is unity. The results of four iteration cycles are tabulated below.

$\mathbf{u}_0$	$\mathbf{v}_1$	$\mathbf{u}_1$	$\mathbf{v}_2$	$\mathbf{u}_2$
1.000	(−1.500)	(−1.615)	(−0.754)	(−1.325)
1.000	0.300	0.462	0.139	0.244
1.000	0.650	1.000	0.569	1.000
	$\lambda_{(1)} = 1.539$		$\lambda_{(2)} = 1.757$	
$\mathbf{u}_2$	$\mathbf{v}_3$	$\mathbf{u}_3$	$\mathbf{v}_4$	$\mathbf{u}_4$
(−1.325)	(−0.634)	(−1.181)	−0.575	−1.106
0.244	0.073	0.136	0.041	0.079
1.000	0.537	1.000	0.520	1.000
	$\lambda_3 = 1.864$		$\lambda_{(4)} = 1.922$	

In the solution above, the numbers in parentheses are not essential to obtaining the final estimate of frequency and mode shapes.

The exact eigenvalue is $\lambda_2 = 2.0$, to which the above is converging. The eigenvector estimate, $\mathbf{u}_4$, has not yet converged to the exact (antisymmetric) eigenvector

$$\boldsymbol{\phi}_2 = \begin{Bmatrix} -1.0 \\ 0.0 \\ 1.0 \end{Bmatrix}$$

The reason for the slow convergence observed is that a symmetric starting vector $\mathbf{u}_0$ was supplied, whereas the true eigenvector is antisymmetric. The iteration procedure must therefore introduce the antisymmetry.

Orthogonalization procedures such as those described above can be used to obtain four or five modes altogether, provided that a small convergence tolerance parameter is employed in order that the calculated eigenvectors closely approximate the true eigenvectors.

15.2.4 Inverse Iteration with Spectrum Shift

An effective way to produce convergence to an eigenvalue other than the fundamental is to employ *Inverse Iteration with Spectrum Shift*. Let μ be a number in the vicinity of the frequency of interest. For example, in Fig. 15.1, μ is closer to λ_2 than to the other eigenvalues. Then, using $\mu \mathbf{M}$, modify Eq. 15.1 to give the shifted eigenproblem

$$[(\mathbf{K} - \mu\mathbf{M}) - (\lambda_i - \mu)\mathbf{M}]\boldsymbol{\phi}_i = \mathbf{0} \tag{15.23}$$

Let

$$\widehat{\mathbf{K}} = \mathbf{K} - \mu\mathbf{M}, \qquad \widehat{\lambda}_i = \lambda_i - \mu \tag{15.24}$$

Then Eq. 15.23 becomes

$$(\widehat{\mathbf{K}} - \widehat{\lambda}_i \mathbf{M})\boldsymbol{\phi}_i = \mathbf{0} \tag{15.25}$$

Figure 15.1 shows the *spectrum* (i.e., plot on a frequency axis) of λ and of $\widehat{\lambda}$. Note from the figure that if $\lambda_i < \mu$, $\widehat{\lambda}_i$ will be negative. A proof similar to that in Eqs. 15.11 through 15.15 can be used to show that inverse iteration applied to the shifted eigenproblem, Eq. 15.25, produces convergence to the λ_i closest to the shift value μ.

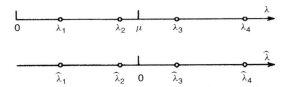

Figure 15.1 Eigenvalue spectra.

Inverse iteration with spectrum shift is a widely used procedure (e.g., in the HQRI method) for finding eigenvectors after the eigenvalues have been found by some other method. This method is also used to avoid the problem associated with the fact that $\mathbf{K}^{-1}$ (in Eq. 15.9) does not exist for a system with rigid-body modes. A small shift value can be used to remove the singularity. Other methods for applying inverse iteration to systems with rigid-body modes are described in Ref. [15.8].

MATLAB Exercise 2: Inverse Iteration with Spectrum Shift Again consider the generalized eigenvalue problem studied in Example 15.1. Write a MATLAB .m-file that performs *Inverse Iteration with Spectrum Shift* to obtain approximations of all eigenvectors of a given matrix. The .m-file should take the following arguments as inputs:

- The two matrices that define the generalized eigenproblem
- A set of initial guesses for the shifts
- A set of initial guesses for the eigenvalues
- A set of initial guesses for the eigenvectors

The .m-file should return the following:

- A list of improved approximations of the eigenvalues
- A matrix of approximated eigenvectors for the generalized eigenproblem

Investigate the sensitivity of the inverse iteration algorithm with spectrum shift to the accuracy of the guess at the set of shifts.

The .m-file <mat_ex_15_2.m> calls the function subroutine <shiftinviterv.m>, which uses shifted inverse iteration to produce the following:

$$\lambda = \begin{Bmatrix} 0.5858 \\ 2.0000 \\ 3.4142 \end{Bmatrix}, \quad \Phi = \begin{bmatrix} 0.5000 & 0.7071 & 0.5000 \\ 0.7071 & 0.0000 & -0.7071 \\ 0.5000 & -0.7071 & 0.5000 \end{bmatrix}$$

Changing the values of the spectrum shift affects the convergence rate of the iteration.

15.3 SUBSPACE ITERATION

In Section 15.2, the *Method of Inverse Iteration* is discussed in some detail. Recall that inverse iteration proves to be most effective when only a few eigenvalues and eigenvectors are desired. Modifications of the fundamental iteration to calculate multiple eigenpairs requires Gram–Schmidt orthogonalization and/or frequency shifting.

Modern applications of structural dynamics can generate governing equations that contain tens of thousands, or even millions, of degrees of freedom. As noted in Section 15.1, while the accuracy of the highest-frequency modes can be suspect in typical models, it is often the case that the lowest-frequency modes, up to some cutoff frequency, are the modes of practical interest. Early models of the space station had over 200 modes below 2 Hz that were of critical importance for design of the attitude control system. While the full, detailed model of the space station had many more eigenpairs, they were not deemed necessary for design of the attitude control system. This example makes it clear that an eigensolver that is capable of calculating a subset of eigenpairs of a large problem can be indispensable for solving large problems.

Practical and efficient eigensolver methodologies for applications in structural dynamics have been designed specifically for these reasons. The *Subspace Iteration Method* is one such algorithm that has been developed by researchers in structural mechanics.[15.1] This method can be viewed as a generalization of the Inverse Iteration Method to obtain numerical estimates of multiple eigenpairs, as opposed to a single eigenpair. The general form of the algorithm is simple to describe. The Subspace Iteration Algorithm begins by solving simultaneously for N_s static solution vectors $\mathbf{R} \equiv [\mathbf{r}_1 \cdots \mathbf{r}_{N_s}]$ for N_s user-defined linearly independent load patterns $\mathbf{F} \equiv [\mathbf{f}_1 \cdots \mathbf{f}_{N_s}]$.

$$\mathbf{KR} = \mathbf{F} \tag{15.26}$$

Strategies for selecting the user-defined load patterns are discussed shortly. It is a fundamental assumption of subspace iteration that the number of degrees of freedom embodied in the $N \times N$ stiffness matrix $\mathbf{K}$ far exceeds the number N_s of static solution vectors $\mathbf{r}_1 \cdots \mathbf{r}_{N_s}$ required. That is, $N \gg N_s$.

The collection of vectors $[\mathbf{r}_1 \cdots \mathbf{r}_{N_s}]$ is said to constitute a collection of *Ritz vectors*. They define the subspace in which approximate eigenvectors for the original eigenvalue problem are sought. The approximate eigenvectors $\mathbf{v}_1 \cdots \mathbf{v}_{N_s}$ are obtained by assuming that they can be written as linear combinations of the Ritz vectors:

$$\mathbf{v}_i = [\mathbf{r}_1 \cdots \mathbf{r}_{N_s}] \begin{Bmatrix} \psi_1 \\ \vdots \\ \psi_{N_s} \end{Bmatrix}_i = \mathbf{R}\boldsymbol{\psi}_i \qquad (15.27)$$

When the approximation in Eq. 15.27 is substituted in the original eigenvalue problem

$$[\mathbf{K} - \lambda\mathbf{M}]\boldsymbol{\phi} = \mathbf{0} \qquad (15.28)$$

an overdetermined system of equations results:

$$\mathbf{E}(\lambda)\boldsymbol{\psi} = [\mathbf{K} - \lambda\mathbf{M}]\mathbf{R}\boldsymbol{\psi}$$

This system of equations has many more rows N than unknowns N_s. A well-posed but much smaller eigenproblem is obtained by requiring that the residual $\mathbf{E}(\lambda)\boldsymbol{\psi}$ be made orthogonal to the space spanned by the Ritz vectors $[\mathbf{r}_1 \cdots \mathbf{r}_{N_s}]$. This condition is written as

$$\mathbf{R}^T[\mathbf{K} - \lambda\mathbf{M}]\mathbf{R}\boldsymbol{\psi} = \mathbf{0}$$

or

$$[\mathbf{R}^T\mathbf{K}\mathbf{R} - \lambda\mathbf{R}^T\mathbf{M}\mathbf{R}]\boldsymbol{\psi} = \mathbf{0}$$

From these equations it is apparent that we have replaced the original large eigenproblem in Eq. 15.28 with an eigenproblem of reduced dimension

$$[\widehat{\mathbf{K}} - \widehat{\lambda}_i\widehat{\mathbf{M}}]\boldsymbol{\psi}_i = \mathbf{0}, \qquad i = 1, \ldots, N_s \qquad (15.29a)$$

or

$$\widehat{\mathbf{K}}\boldsymbol{\Psi} - \widehat{\mathbf{M}}\boldsymbol{\Psi}\widehat{\boldsymbol{\Lambda}} = \mathbf{0} \qquad (15.29b)$$

where

$$\widehat{\mathbf{K}} = \mathbf{R}^T\mathbf{K}\mathbf{R}, \qquad \widehat{\mathbf{M}} = \mathbf{R}^T\mathbf{M}\mathbf{R} \qquad (15.29c)$$

and

$$\boldsymbol{\Psi} = [\boldsymbol{\psi}_1 \cdots \boldsymbol{\psi}_{N_s}], \qquad \widehat{\boldsymbol{\Lambda}} = \mathrm{diag}(\widehat{\lambda}_i) \qquad (15.29d)$$

Up to this point, the algorithm represented by Eqs. 15.27 through 15.29 simply encodes a "Ritz approximation" of the original, high-dimensional eigenvalue problem, as discussed in Section 10.2. Its main contribution is to produce an orthogonal set of vectors in the subspace spanned by the Ritz vectors $\mathbf{R}$. The Subspace Iteration Method, however, entails one additional step in a fundamental recursion that gives the algorithm its unique character. Following the solution of the reduced eigenproblem summarized in Eqs. 15.29, approximate eigenvectors of the original problem Eq. 15.28 are written as

$$\widehat{\mathbf{V}} = \mathbf{R}\boldsymbol{\Psi}$$

With these approximate eigenvectors, an improved set of (static) load vectors is calculated from the relationship

$$\widehat{\mathbf{F}} = \mathbf{MR\Psi} = \mathbf{M}\widehat{\mathbf{V}}$$

Now these load vectors can be used to reinitialize the solution procedure. The substitution of $\widehat{\mathbf{F}}$ for $\mathbf{F}$ in Eq. 15.26 enables the process to be performed recursively.

Table 15.1 summarizes the subspace iteration procedure, with an explicit description of the computational details.

Although the Subspace Iteration Algorithm is summarized completely in Table 15.1, several observations must be made to ensure that use of the algorithm produces an accurate reduced-order solution of the original eigenproblem.

Observations:

1. One of the critical issues governing the accuracy and effectiveness of the Subspace Iteration Method is the choice of the original load patterns. Practical experience has generated effective heuristics for choosing the initial load vectors.[15.1] The important criterion is that the initial load vectors must be selected such that they will excite all of the lowest-frequency modes of interest. Clearly, this selection procedure may require significant user expertise.

2. Just as the specific load vectors may be selected according to heuristic criteria, the number of Ritz vectors used in the Subspace Iteration Method is dictated by

Table 15.1 Subspace Iteration Algorithm

Step 0	Assemble $\mathbf{K}, \mathbf{M}$. Factor $\mathbf{K} = \mathbf{LU}$. Choose N_s starting (load) vectors to form $\mathbf{F}_0$. Set $j = 0$.
Step 1	Solve for the Ritz (displacement) vectors $\mathbf{R}_j$. $$\mathbf{LUR}_j = \mathbf{F}_j \quad (15.30)$$
Step 2	Form the reduced matrices. $$\widehat{\mathbf{K}}_j = \mathbf{R}_j^T \mathbf{K} \mathbf{R}_j, \quad \widehat{\mathbf{M}}_j = \mathbf{R}_j^T \mathbf{M} \mathbf{R}_j$$
Step 3	Solve the reduced eigenproblem. $$[\widehat{\mathbf{K}}_j - \lambda_m \widehat{\mathbf{M}}_j]\boldsymbol{\psi}_m = \mathbf{0}, \quad m = 1, \ldots, N_s \quad (15.31)$$
Step 4	Calculate the approximate eigenvectors. $$\widehat{\mathbf{V}}_j = \mathbf{R}_j \boldsymbol{\Psi}_j$$
Step 5	Check for convergence.
Step 6	Calculate the improved loads. $$\mathbf{F}_{j+1} = \mathbf{MR}_j \boldsymbol{\Psi}_j = \mathbf{M}\widehat{\mathbf{V}}_j \quad (15.32)$$
Step 7	Increment counter and return to step 1. $$j \leftarrow j + 1$$

user experience. There is no strict rule for selecting the number of Ritz vectors or load patterns. Reference [15.1] recommends that the dimension of the Ritz subspace satisfy

$$N_s = \min\{2N_d, N_d + 8\}$$

where the number of desired modes is N_d.

3. The reduced eigenproblem in step 3, Eq. 15.31, will be composed of fully populated $N_s \times N_s$ matrices, and all N_s eigenvalues and eigenvectors are required. Hence, completion of the Subspace Iteration Method implies that another method, suitable for small fully populated matrices, be available.
4. The final observation regarding the Subspace Iteration Method concerns the interpretation of step 6. While Eq. 15.30 can be construed as the solution for static Ritz vectors corresponding to static load shapes, Eq. 15.32 may be interpreted as producing inertial loads corresponding to the displacement vectors $[\mathbf{v}_1 \cdots \mathbf{v}_{N_s}]_j$.

MATLAB Exercise 3: Subspace Iteration   Write a MATLAB .m-file that uses the *Subspace Iteration Method* to approximate the eigenvalues and eigenvectors of a generalized eigenproblem. The .m-file should be written to be as general as possible. The .m-file should take the following arguments as input:

- The two matrices that define the generalized eigenproblem
- The N_s starting (load) vectors that form $\mathbf{F}_{(0)}$
- The tolerance that defines convergence of approximate eigenpairs

The .m-file should return the following:

- A list of approximate eigenvalues
- A list of approximate eigenvectors

Test the .m-file on the eigenproblem studied in Example 15.1. Study the sensitivity of the Subspace Iteration Method to (a) the choice of the starting load vectors, and (b) the tolerance selected for convergence.

The .m-file <mat_ex_15_3.m> calls the function subroutine <subiter.m>, which uses the Subspace Iteration Method to produce the following:

$$\lambda = \begin{Bmatrix} 0.5858 \\ 2.0000 \\ 3.4142 \end{Bmatrix}, \quad \Phi = \begin{bmatrix} 0.5000 & 0.7071 & 0.5000 \\ 0.7071 & 0.0000 & -0.7071 \\ 0.5000 & -0.7071 & 0.5000 \end{bmatrix}$$

15.4 QR METHOD FOR SYMMETRIC EIGENPROBLEMS

Thus far, the techniques discussed in this chapter have been vector iteration methods. Section 15.2 introduced the prototypical Inverse Iteration Method, and Section 15.3 summarized the more practical Subspace Iteration Method. In this section we present one of the most popular numerical solution methods for the symmetric eigenvalue problem, the *QR Method*. The QR Method, an example of a transformation method, is used to calculate eigenvalues, but it does not directly produce the corresponding eigenvectors.

15.4.1 Basic QR Algorithm

Consider the eigenvalue problem in standard form

$$(\mathbf{A} - \lambda_i \mathbf{I})\boldsymbol{\phi}_i = \mathbf{0} \tag{15.33}$$

(It is assumed that one of the techniques discussed in Section D.4.3 has been applied to the generalized eigenproblem to obtain Eq. 15.33.) The fundamental iteration comprising the QR Eigenvalue Algorithm proceeds as shown in Table 15.2. The QR factorization of the matrix $\mathbf{A}_k$ is written as $\mathbf{A}_k = \mathbf{Q}_k \mathbf{R}_k$, which is the product of an orthogonal matrix $\mathbf{Q}_k$:

$$\mathbf{Q}_k \mathbf{Q}_k^T = \mathbf{Q}_k^T \mathbf{Q}_k = \mathbf{I}$$

and an upper triangular matrix $\mathbf{R}_k$.[15.5] It is straightforward to see that the fundamental iteration in steps 1 and 2 generates a sequence of matrices:

$$\begin{aligned}\mathbf{A}_{k+1} &= \mathbf{Q}_k^T \mathbf{A}_k \mathbf{Q}_k \\ &= \underbrace{\mathbf{Q}_k^T \mathbf{Q}_{k-1}^T \cdots \mathbf{Q}_1^T}_{\overline{\mathbf{Q}}^T} \mathbf{A} \underbrace{\mathbf{Q}_1 \cdots \mathbf{Q}_{k-1} \mathbf{Q}_k}_{\overline{\mathbf{Q}}}\end{aligned} \tag{15.34}$$

With the matrix $\overline{\mathbf{Q}}$ defined in Eq. 15.34, it is possible to relate the eigenvalues of the transformed problem

$$(\overline{\mathbf{A}} - \gamma_i \mathbf{I})\boldsymbol{\psi}_i = \mathbf{0}$$

to those of the original problem Eq. 15.33.[4]

$$\begin{aligned}(\overline{\mathbf{A}} - \gamma_i \mathbf{I})\boldsymbol{\psi}_i &= (\overline{\mathbf{Q}}^T \mathbf{A} \overline{\mathbf{Q}} - \gamma_i \mathbf{I})\boldsymbol{\psi}_i \\ &= \overline{\mathbf{Q}}^T (\mathbf{A} - \gamma_i \mathbf{I}) \overline{\mathbf{Q}} \boldsymbol{\psi}_i\end{aligned}$$

Thus, if $(\gamma_i, \boldsymbol{\psi}_i)$ is an eigenpair for the transformed matrix $\overline{\mathbf{A}}$, $(\gamma_i, \overline{\mathbf{Q}}\boldsymbol{\psi}_i) = (\lambda_i, \boldsymbol{\phi}_i)$ is an eigenpair for the original problem in Eq. 15.33.

Observations:

1. The strategy of the fundamental iteration is clear, given the relation of the eigenvalues and eigenvectors of the original and transformed problems. If the eigenpairs of the transformed matrix $\overline{\mathbf{A}}$ are readily deduced or computed, the eigenpairs of the original problem can likewise be determined.

Table 15.2 QR Symmetric Eigenvalue Algorithm

Step 0	$\mathbf{A}_0 \triangleq \mathbf{A}$
Step 1	Calculate the QR decomposition of $\mathbf{A}_k$: $\mathbf{A}_k = \mathbf{Q}_k \mathbf{R}_k$
Step 2	Update the transformed matrix: $\mathbf{A}_{k+1} = \mathbf{R}_k \mathbf{Q}_k$
Step 3	Increment and return to step 1: $k \Longleftarrow k + 1$

[4]The overline ($\overline{}$) used here does not designate a complex-valued quantity.

2. There are well-documented conditions that guarantee that the fundamental iteration does indeed converge in the sense that

$$\mathbf{A}_k \longrightarrow \mathbf{\Lambda}$$

where $\mathbf{\Lambda}$ is a diagonal matrix.[15.5] If the iteration is carried out until such convergence is obtained, it is clear that the diagonal entries of the matrix $\mathbf{\Lambda}$ are approximations of the eigenvalues of the original matrix $\mathbf{A}$.

3. Standard references (e.g., Refs. [15.5] and [15.9]) document that the cost of calculating a single QR decomposition is a cubic-order calculation. That is, for an $N \times N$ matrix $\mathbf{A}$ the cost of computing the QR decomposition is $O(N^3)$ operations. In effect, this fact implies that the direct implementation of the fundamental iteration is prohibitively costly for all but the smallest eigenvalue problems.

Given the cost of a single step of the fundamental iteration summarized in Table 15.2, a practical QR Method for symmetric eigenproblems requires additional structure. Practical QR algorithms are obtained from the fundamental iteration by first choosing a sequence of orthogonal transformations that drive the transformed matrix to tridiagonal form. Subsequent orthogonal transformations need only operate on the nonzero tridiagonals. Moreover, the subsequent orthogonal transformations are selected so that they preserve the tridiagonal structure after each iteration. In this way, the cost per iteration remains small for all steps after the initial tridiagonalization. We begin our description of practical QR algorithms for the symmetric eigenvalue problem with a discussion of Householder transformations.

15.4.2 Householder Transformations

Orthogonal transformations, with additional specialized properties that enable the selective cancellation of entries in products of matrices, play a significant role in many methods for the numerical approximation of eigenvalues and eigenvectors. Jacobi Rotations, Givens Rotations, and Householder Reflections are all examples of such useful orthogonal transformations. In this section we discuss only the Householder Transformation. The interested reader is referred to Ref. [15.5] for detailed discussions of all three types of orthogonal transformations.

A *Householder Transformation* is any matrix $\mathbf{H}$ having the form

$$\mathbf{H} \triangleq \mathbf{I} - \frac{2\mathbf{u}\mathbf{u}^T}{\mathbf{u}^T\mathbf{u}}$$

for some unit vector $\mathbf{u} \in \mathbb{R}^N$, $\mathbf{u} \neq \mathbf{0}$. The $N \times N$ matrix $\mathbf{H}$ is orthogonal by inspection.

$$\begin{aligned}
\mathbf{H}\mathbf{H}^T = \mathbf{H}^T\mathbf{H} &= \left(\mathbf{I} - \frac{2\mathbf{u}\mathbf{u}^T}{\mathbf{u}^T\mathbf{u}}\right)^T \left(\mathbf{I} - \frac{2\mathbf{u}\mathbf{u}^T}{\mathbf{u}^T\mathbf{u}}\right) \\
&= \mathbf{I} - \frac{4\mathbf{u}\mathbf{u}^T}{\mathbf{u}^T\mathbf{u}} + \frac{4\mathbf{u}\mathbf{u}^T\mathbf{u}\mathbf{u}^T}{(\mathbf{u}^T\mathbf{u})^2} \\
&= \mathbf{I}
\end{aligned}$$

Just as important, it is possible to choose $\mathbf{u}$ in the definition of $\mathbf{H}$ so that for any vector $\mathbf{v} \in \mathbb{R}^N$ such that $\mathbf{v} \neq \mathbf{0}$, the product

$$\mathbf{Hv} = k\mathbf{e}_1$$

where $\mathbf{e}_1$ is the canonical basis vector $\mathbf{e}_1^T = [1 \ 0 \ 0 \ 0 \ \cdots \ 0 \ 0]$ and $k \in \mathbb{R}$. To see how this transformation is accomplished, suppose that

$$\mathbf{Hv} \triangleq \left(\mathbf{I} - \frac{2\mathbf{u}\mathbf{u}^T}{\mathbf{u}^T\mathbf{u}}\right)\mathbf{v} = k\mathbf{e}_1 \tag{15.35}$$

for some $k \in \mathbb{R}$. It must be true that

$$\mathbf{u} = b\mathbf{v} + c\mathbf{e}_1$$

for two constants $b, c \in \mathbb{R}$. In particular, it can be shown that the desired result is obtained for the choice $b = 1, c \in \mathbb{R}$.

$$\mathbf{u} = \mathbf{v} + c\mathbf{e}_1 \tag{15.36}$$

Substituting Eq. 15.36 into Eq. 15.35, we obtain

$$\mathbf{Hv} = \mathbf{v} - 2\frac{\mathbf{v}^T\mathbf{v} + c\mathbf{v}^T\mathbf{e}_1}{\mathbf{v}^T\mathbf{v} + c^2 + \mathbf{v}^T\mathbf{e}_1 2c}(\mathbf{v} + c\mathbf{e}_1)$$
$$= \left(1 - 2\frac{(\mathbf{v}^T\mathbf{v} + cv_1)}{\mathbf{v}^T\mathbf{v} + c^2 + 2cv_1}\right)\mathbf{v} - \frac{2(\mathbf{v}^T\mathbf{v} + cv_1)}{\mathbf{v}^T\mathbf{v} + c^2 + 2cv_1}c\mathbf{e}_1 \tag{15.37}$$

where we have defined $v_1 \equiv \mathbf{v}^T\mathbf{e}_1$. Recall that the goal of the construction, as stated in Eq. 15.35, is to make $\mathbf{Hv}$ proportional to $\mathbf{e}_1$. Setting the coefficient of $\mathbf{v}$ in Eq. 15.37 to zero yields

$$c^2 - \|\mathbf{v}\|^2 = 0$$

so that the choice of either

$$c = \pm\|\mathbf{v}\|$$

will generate the desired result.

15.4.3 Householder-QR Method

With the Householder transformations described in the preceding section, it is possible to apply orthogonal transformations that selectively zero out all but the tridiagonal entries of the final, transformed matrix. The process is straightforward in principle. Consider the matrix $\mathbf{A} \in \mathbb{R}^{N \times N}$ partitioned as follows:

$$\mathbf{A} \triangleq \mathbf{A}_1 = \begin{bmatrix} a_{11} & a_{12} & a_{13} & \cdots & a_{1N} \\ a_{12} & & & & \vdots \\ a_{13} & & & & \\ \vdots & & & & \\ a_{1N} & & \cdots & & a_{NN} \end{bmatrix} = \begin{bmatrix} a_{11} & \mathbf{a}_1^T \\ \mathbf{a}_1 & \overline{\mathbf{A}}_1 \end{bmatrix}$$

where $a_{11} \in \mathbb{R}$, $\mathbf{a}_1 \in \mathbb{R}^{N-1}$, and $\overline{\mathbf{A}}_1 \in \mathbb{R}^{(N-1)\times(N-1)}$. According to Section 15.4.2, it is always possible to construct a Householder transformation $\overline{\mathbf{Q}}_1^T \in \mathbb{R}^{(N-1)\times(N-1)}$ such that

$$\overline{\mathbf{Q}}_1^T \mathbf{a}_1 = k_1 \mathbf{e}_{N-1}$$

In Eq. 15.37, $\mathbf{e}_{N-1}$ is the canonical unit vector $\mathbf{e}_{N-1} = [1 \ 0 \ 0 \ \cdots \ 0]^T \in \mathbb{R}^{N-1}$ and $k_1 \in \mathbb{R}$. Construct the orthogonal transformation $\mathbf{Q}_1$ as

$$\mathbf{Q}_1^T = \begin{bmatrix} 1 & \mathbf{0} \\ \mathbf{0} & \overline{\mathbf{Q}}_1^T \end{bmatrix}$$

Upon being premultiplied by $\mathbf{Q}_1^T$ and postmultiplied by $\mathbf{Q}_1$, the matrix $\mathbf{A}$ is transformed to the form

$$\mathbf{Q}_1^T \mathbf{A} \mathbf{Q}_1 = \begin{bmatrix} a_{11} & k_1 & 0 & \cdots & 0 \\ \hline k_1 & & & & \\ 0 & & \overline{\mathbf{Q}}_1^T \overline{\mathbf{A}}_1 \overline{\mathbf{Q}}_1 & & \\ \vdots & & & & \\ 0 & & & & \end{bmatrix} = \begin{bmatrix} a_{11} & k_1 \mathbf{e}_{N-1}^T \\ k_1 \mathbf{e}_{N-1} & \mathbf{B}_1 \end{bmatrix}$$

Since $\mathbf{B}_1 \in \mathbb{R}^{(N-1)\times(N-1)}$ it too can be partitioned as

$$\mathbf{B}_1 = \begin{bmatrix} b_{11} & \mathbf{b}_1^T \\ \mathbf{b}_1 & \overline{\mathbf{B}}_1 \end{bmatrix}$$

where $b_{11} \in \mathbb{R}$, $\mathbf{b}_1 \in \mathbb{R}^{N-2}$, and $\overline{\mathbf{B}}_1 \in \mathbb{R}^{(N-2)\times(N-2)}$. As before, Section 15.4.2 guarantees that a Householder transformation $\overline{\mathbf{Q}}_2^T \in \mathbb{R}^{(N-2)\times(N-2)}$ can be constructed so that

$$\overline{\mathbf{Q}}_2^T \mathbf{b}_1 = k_2 \mathbf{e}_{N-2}$$

where $k_2 \in \mathbb{R}$ and $\mathbf{e}_{N-2}$ is the canonical unit vector $\mathbf{u}_{N-2} = [1 \ 0 \ 0 \ \cdots \ 0]^T \in \mathbb{R}^{N-2}$. Construct the orthogonal transformation $\mathbf{Q}_2$ as

$$\mathbf{Q}_2^T = \begin{bmatrix} [\mathbf{I}_{2\times 2}] & \mathbf{0} \\ \mathbf{0} & \overline{\mathbf{Q}}_2^T \end{bmatrix}$$

Then the product $\mathbf{Q}_2^T \mathbf{Q}_1^T \mathbf{A} \mathbf{Q}_1 \mathbf{Q}_2$ becomes

$$\mathbf{Q}_2^T \mathbf{Q}_1^T \mathbf{A} \mathbf{Q}_1 \mathbf{Q}_2 = \begin{bmatrix} a_{11} & k_1 & 0 & \cdots & & 0 \\ \hline k_1 & b_{11} & k_2 & 0 & \cdots & 0 \\ \hline 0 & k_2 & & & & \\ \vdots & 0 & & \overline{\mathbf{Q}}_2^T \overline{\mathbf{B}}_1 \overline{\mathbf{Q}}_2 & & \\ & \vdots & & & & \\ 0 & 0 & & & & \end{bmatrix}$$

Clearly, we can define the product

$$\mathbf{C}_1 \triangleq \overline{\mathbf{Q}}_2^T \overline{\mathbf{B}}_1 \overline{\mathbf{Q}}_2 \in \mathbb{R}^{(N-2)\times(N-2)}$$

and proceed by induction to generate a tridiagonal matrix (Table 15.3).

Table 15.3 Practical Symmetric QR Eigenvalue Algorithm

Step 0	$\mathbf{A}_0 = \mathbf{A}$
Step 1	Calculate the sequence $\{\mathbf{Q}_k\}_{k=1}^{N-1}$ of Householder transformations to tridiagonalize A_0: $$A_0 \longrightarrow A_k$$
Step 2	Solve the tridiagonal eigenproblem: $$\{\mathbf{A}_k - \gamma \mathbf{I}\}\boldsymbol{\psi} = \mathbf{0}$$ for $\{\gamma_i, \boldsymbol{\psi}_i\}_{i=1}^{N}$
Step 3	Transform calculated eigenpairs to desired eigenpairs: $$\{\gamma_i, \boldsymbol{\psi}_i\}_{i=1}^{N} \longrightarrow \{\lambda_i, \boldsymbol{\phi}_i\}_{i=1}^{N}$$

Observations:

1. Application of the specialized sequence of Householder transformations in step 1 of Table 15.3 is intended to generate and preserve the equivalent tridiagonal eigenproblem. Efficiency is achieved via the introduction and preservation of the zeros in the tridiagonal form.

2. In view of observation 1 above, practical implementations of step 1 above are nontrivial and highly structured. Details of practical implementations wherein the Householder transformations $\{\mathbf{Q}_k\}_{k=1}^{N}$ are calculated, stored, and applied only when necessary may be found in Ref. [15.5].

3. The final step of the practical QR Eigenvalue Algorithm requires that a general and efficient tridiagonal eigensolver be employed. This step can be achieved via the selective sequential application of QR factorizations that preserve the tridiagonal structure, operate only on nonzero entries, and drive the tridiagonal system to diagonal form.

MATLAB Exercise 4: QR Symmetric Eigenproblem Write a MATLAB .m-file that approximates all of the eigenvalues of a generalized eigenproblem using the *QR Method* for approximating the eigenvalues of a symmetric matrix. The .m-file should be as general as possible. The .m-file should take the following arguments as input:

- The two matrices that define the generalized eigenproblem
- The tolerance for the convergence of the approximation process

The .m-file should output one set of data:

- A list of the approximate eigenvalues for the eigenproblem

Test the .m-file on the eigenproblem studied in Example 15.1. Carry out a study of the accuracy of the approximated eigenvalues as you vary the tolerance used for the eigenvalue convergence.

The .m-file <mat_ex_15_4.m> calls the function subroutine <symqrv1.m>, which uses the Symmetric QR Method to produce the following:

$$\lambda = \begin{Bmatrix} 0.5858 \\ 2.0000 \\ 3.4142 \end{Bmatrix}$$

As the convergence tolerance is changed, the number of iterations needed to converge to a final solution also changes. Thus, the stricter the tolerance, the more iterations are necessary.

15.5 LANCZOS EIGENSOLVER

The Lanczos Method for approximating the eigenpairs of a symmetric, real matrix was introduced in 1950. In contrast to the Method of Inverse Iteration (Section 15.2) and the Subspace Iteration Method in (Section 15.3), which are both examples of vector iteration methods, the Lanczos Method falls within the class of transformation methods. Essentially, the *Lanczos Method* is a technique for transforming a given real and symmetric matrix into a tridiagonal matrix. Thus, as in the case of a practical QR algorithm, a practical Lanczos solver must employ an efficient eigensolver for a tridiagonal system during its "final phase." The transformation to tridiagonal form is achieved via a recurrence relation that generates an orthogonal basis for certain Krylov subspaces. Although the elegance of this mathematical recurrence is attractive, the recurrence relation is numerically unstable due to loss of orthogonality in the Krylov subspace basis. Initially, this was a major setback, preventing widespread use of the method. For some time, the Householder Method, discussed in Section 15.4.2, became the method of choice, owing to its superior stability properties.

Still, the Lanczos Method has several desirable features that make it well suited for the treatment of large sparse-matrix eigenvalue problems. In addition, the development of various reorthogonalization approaches to offset the loss of orthogonality associated with the fundamental recurrence relation has made the Lanczos Method a competitive algorithm for large sparse eigenproblems of the type encountered in structural dynamics.[15.2, 15.3]

15.5.1 Orthogonal Bases for Krylov Subspaces

Recall that the symmetry of a general matrix $\mathbf{A} \in \mathbb{R}^{N \times N}$ is defined by the property that

$$\mathbf{A} = \mathbf{A}^T$$

A more general notion of a symmetric matrix can be defined in terms of an *inner product*. The inner product between two vectors is defined in terms of a symmetric, positive definite matrix $\mathbf{B}$. The inner product associated with the symmetric positive definite matrix $\mathbf{B} \in \mathbb{R}^{N \times N}$ is defined for any two vectors $\mathbf{x}, \mathbf{y} \in \mathbb{R}^N$ to be the mapping

$$\langle \mathbf{x}, \mathbf{y} \rangle_{\mathbf{B}} \triangleq \mathbf{x}^T \mathbf{B} \mathbf{y}$$

The matrix $\mathbf{A}$ is said to be $\mathbf{B}$-symmetric provided that

$$\langle \mathbf{A}\mathbf{x}, \mathbf{y} \rangle_{\mathbf{B}} = \langle \mathbf{x}, \mathbf{A}\mathbf{y} \rangle_{\mathbf{B}}$$

for all choices of the vectors $\mathbf{x}, \mathbf{y} \in \mathbb{R}^N$. Two vectors $\mathbf{x}, \mathbf{y} \in \mathbb{R}^N$ are **B**-orthogonal if and only if

$$\langle \mathbf{x}, \mathbf{y} \rangle_\mathbf{B} = \mathbf{x}^T \mathbf{B} \mathbf{y} = 0$$

Of course, if **B** is chosen to be the identity matrix, these definitions are reduced to the usual notions of symmetry of a matrix and orthogonality of vectors.

The *Lanczos Method* for approximating the eigenvalues of the real, symmetric matrix **A** generates an **B**-orthonormal basis for the Krylov subspaces associated with **A**. Define

$$\mathcal{K}_j(\mathbf{r}) \stackrel{\triangle}{=} \mathrm{span}\{\mathbf{r}, \mathbf{A}\mathbf{r}, \mathbf{A}^2\mathbf{r} \cdots \mathbf{A}^{j-1}\mathbf{r}\}$$

to be the *j*th *Krylov subspace* generated by the starting vector $\mathbf{r} \in \mathbb{R}^N$. Each of the vectors $\mathbf{v}_i = \mathbf{A}^i \mathbf{r}$ for $i = 0 \cdots j - 1$ is called the *i*th *Krylov vector*. Further, suppose that the set of vectors $\{\mathbf{q}_1, \mathbf{q}_2, \ldots, \mathbf{q}_j\}$ constitute a **B**-orthonormal basis for $\mathcal{K}_j(\mathbf{r})$. That is,

$$\mathcal{K}_j(\mathbf{r}) \stackrel{\triangle}{=} \mathrm{span}\{\mathbf{r}, \mathbf{A}\mathbf{r}, \mathbf{A}^2\mathbf{r} \cdots \mathbf{A}^{j-1}\mathbf{r}\} = \mathrm{span}\{\mathbf{v}_0 \cdots \mathbf{v}_{j-1}\}$$
$$= \mathrm{span}\{\mathbf{q}_1 \cdots \mathbf{q}_j\}$$

and

$$\mathbf{q}_i^T \mathbf{B} \mathbf{q}_k = \begin{cases} 1 & i = k \\ 0 & \text{otherwise} \end{cases} \tag{15.38}$$

for $1 \leq (i, k) \leq j$.

Suppose that we are given the set of vectors $\{\mathbf{q}_1 \cdots \mathbf{q}_j\}$. The fundamental problem at hand is to find the next vector $\mathbf{q}_{j+1}$ such that

$$\mathcal{K}_{j+1}(\mathbf{r}) = \mathrm{span}\{\mathbf{v}_0 \cdots \mathbf{v}_j\}$$
$$= \mathrm{span}\{\mathbf{q}_1 \cdots \mathbf{q}_{j+1}\}$$

and the orthogonality condition Eq. 15.38 holds for $1 \leq (i, k) \leq j + 1$. The task, then, is to find a set of coefficients $\{\mu_k\}_{k=1}^{j+1}$ and the vector $\mathbf{q}_{j+1}$ such that

$$\mathbf{v}_j = \mu_{j+1} \mathbf{q}_{j+1} + \sum_{k=1}^{j} \mu_j \mathbf{q}_j \tag{15.39}$$

so that the vectors $\{\mathbf{q}_k\}_{k=1}^{j+1}$ are **B**-orthogonal. By the definition of the Krylov vectors, $\mathbf{v}_j$ is defined by the recursion equation

$$\mathbf{v}_j = \mathbf{A} \mathbf{v}_{j-1}$$

Since $\mathbf{v}_{j-1} \in \mathcal{K}_j(\mathbf{r})$, there are coefficients $\{\gamma_k\}_{k=1}^{j}$ such that

$$\mathbf{v}_{j-1} = \sum_{k=1}^{j} \gamma_k \mathbf{q}_k$$

15.5 Lanczos Eigensolver

Therefore, it is possible to write

$$\boldsymbol{v}_j = \mathbf{A}\boldsymbol{v}_{j-1} = \sum_{k=1}^{j} \gamma_k \mathbf{A}\mathbf{q}_k$$
$$= \gamma_j \mathbf{A}\mathbf{q}_j + \sum_{k=1}^{j-1} \gamma_k \mathbf{A}\mathbf{q}_k \quad (15.40)$$

But the Krylov subspaces satisfy

$$\mathbf{A}\mathcal{K}_{j-1}(\mathbf{r}) \subseteq \mathcal{K}_j(\mathbf{r})$$

so Eq. 15.40 can be written

$$\boldsymbol{v}_j = \gamma_j \mathbf{A}\mathbf{q}_j + \sum_{k=1}^{j+1} \overline{\gamma}_k \mathbf{q}_k \quad (15.41)$$

Since $\boldsymbol{v}_j \in \mathcal{K}_{j+1}(\mathbf{r}) = \text{span}\{\mathbf{q}_1, \ldots, \mathbf{q}_{j+1}\}$, we also have

$$\boldsymbol{v}_j = \sum_{k=1}^{j+1} \overline{\overline{\gamma}}_k \mathbf{q}_k$$

It is apparent that $\mathbf{q}_{j+1}$ must be selected so that

$$\mathbf{A}\mathbf{q}_j = k_{j+1}\mathbf{q}_{j+1} + a_j \mathbf{q}_j + b_j \mathbf{q}_{j-1} + c_j \mathbf{q}_{j-2} + \cdots \quad (15.42)$$

Taking the inner product of both sides of this equation with $\mathbf{q}_j$, we can show that calculation of the coefficient a_j follows easily from the **B**-orthonormality of the vectors $\{\mathbf{q}_k\}_{k=1}^{j+1}$.

$$\langle \mathbf{q}_j, \mathbf{A}\mathbf{q}_j \rangle_\mathbf{B} = k_{j+1}\langle \mathbf{q}_j, \mathbf{q}_{j+1} \rangle_\mathbf{B} + a_j \langle \mathbf{q}_j, \mathbf{q}_j \rangle_\mathbf{B} + b_j \langle \mathbf{q}_j, \mathbf{q}_{j-1} \rangle_\mathbf{B} + c_j \langle \mathbf{q}_j, \mathbf{q}_{j-2} \rangle_\mathbf{B} + \cdots$$

$$\boxed{a_j = \langle \mathbf{q}_j, \mathbf{A}\mathbf{q}_j \rangle_\mathbf{B} = q_j^\mathrm{T} \mathbf{B}\mathbf{A}\mathbf{q}_j}$$

A similar strategy can be used to obtain an expression for coefficient b_j. Taking the inner product of both sides of Eq. 15.42 with $\mathbf{q}_{j-1}$ and using the **B**-orthonormality of the vectors $\{\mathbf{q}_k\}_{k=1}^{j+1}$ yields

$$\langle \mathbf{q}_{j-1}, \mathbf{A}\mathbf{q}_j \rangle_\mathbf{B} = k_{j+1}\langle \mathbf{q}_{j-1}, \mathbf{q}_{j+1} \rangle_\mathbf{B} + a_j \langle \mathbf{q}_{j-1}, \mathbf{q}_j \rangle_\mathbf{B} + b_j \langle \mathbf{q}_{j-1}, \mathbf{q}_{j-1} \rangle_\mathbf{B}$$
$$+ c_j \langle \mathbf{q}_{j-1}, \mathbf{q}_{j-2} \rangle_\mathbf{B} + \cdots$$

$$\boxed{b_j = \langle \mathbf{q}_{j-1}, \mathbf{A}\mathbf{q}_j \rangle_\mathbf{B} = q_{j-1}^\mathrm{T} \mathbf{B}\mathbf{A}\mathbf{q}_j}$$

The vector $\mathbf{A}\mathbf{q}_{j-1}$ is derived recursively from the previous $\{\mathbf{q}_k\}_{k=1}^{j-2}$ via an expression analogous to Eq. 15.42.

$$\mathbf{A}\mathbf{q}_{j-1} = k_j \mathbf{q}_j + a_{j-1}\mathbf{q}_{j-1} + b_{j-1}\mathbf{q}_{j-2} + c_{j-1}\mathbf{q}_{j-3} + \cdots \quad (15.43)$$

Now taking the inner product of both sides with Eq. 15.43 by $\mathbf{q}_j$ and again using the **B**-orthonormality of the vectors and the **B**-symmetry of **A**, we obtain

$$\langle \mathbf{Aq}_{j-1}, \mathbf{q}_j \rangle_\mathbf{B} = k_j \langle \mathbf{q}_j, \mathbf{q}_j \rangle_\mathbf{B} + a_{j-1} \langle \mathbf{q}_{j-1}, \mathbf{q}_j \rangle_\mathbf{B} + b_{j-1} \langle \mathbf{q}_{j-2}, \mathbf{q}_j \rangle_\mathbf{B}$$
$$+ c_{j-1} \langle \mathbf{q}_{j-3}, \mathbf{q}_j \rangle_\mathbf{B} + \cdots$$

$$\boxed{k_j = b_j}$$

From the inner product of Eq. 15.42 with $\mathbf{q}_{j-2}$, the expression for c_j follows:

$$\langle \mathbf{q}_{j-2}, \mathbf{Aq}_j \rangle_\mathbf{B} = k_{j+1} \langle \mathbf{q}_{j-2}, \mathbf{q}_{j+1} \rangle_\mathbf{B} + a_j \langle \mathbf{q}_{j-2}, \mathbf{q}_j \rangle_\mathbf{B} + b_j \langle \mathbf{q}_{j-2}, \mathbf{q}_{j-1} \rangle_\mathbf{B}$$
$$+ c_j \langle \mathbf{q}_{j-2}, \mathbf{q}_{j-2} \rangle_\mathbf{B} + \cdots$$

$$\boxed{c_j = \langle \mathbf{q}_{j-2}, \mathbf{Aq}_j \rangle_\mathbf{B} = q_{j-2}^T \mathbf{BAq}_j}$$

On the other hand, as in Eq. 15.42, we can write the corresponding recursion

$$\mathbf{Aq}_{j-2} = k_{j-1} \mathbf{q}_{j-1} + a_{j-2} \mathbf{q}_{j-2} + b_{j-2} \mathbf{q}_{j-3} + c_{j-2} \mathbf{q}_{j-4} + \cdots \quad (15.44)$$

If we take the inner product of Eq. 15.44 with $\mathbf{q}_j$, we obtain

$$\langle \mathbf{Aq}_{j-2}, \mathbf{q}_j \rangle_\mathbf{B} = k_{j-1} \langle \mathbf{q}_{j-1}, \mathbf{q}_j \rangle_\mathbf{B} + a_{j-2} \langle \mathbf{q}_{j-2}, \mathbf{q}_j \rangle_\mathbf{B} + b_{j-2} \langle \mathbf{q}_{j-3}, \mathbf{q}_j \rangle_\mathbf{B}$$
$$+ c_{j-2} \langle \mathbf{q}_{j-4}, \mathbf{q}_j \rangle_\mathbf{B} + \cdots$$

By the **B**-symmetry of **A**, we conclude that

$$\boxed{c_j = \langle \mathbf{q}_{j-2}, \mathbf{Aq}_j \rangle_\mathbf{B} = \langle \mathbf{Aq}_{j-2}, \mathbf{q}_j \rangle_\mathbf{B} = 0}$$

If the process above is repeated, it is clear that the coefficients multiplying all other $\mathbf{q}_{j-3} \cdots \mathbf{q}_1$ in Eq. 15.42 are identically zero. It follows that there is a *three-term recursion* that defines the additional vector $\mathbf{q}_{j+1}$. It is given by the equation

$$b_{j+1} \mathbf{q}_{j+1} = \mathbf{Aq}_j - a_j \mathbf{q}_j - b_j \mathbf{q}_{j-1}$$

Alternatively, we can write

$$\mathbf{Aq}_j = b_{j+1} \mathbf{q}_{j+1} + a_j \mathbf{q}_j + b_j \mathbf{q}_{j-1}$$

Given the preceding discussion, an iterative technique to generate $N_L \leq N$ Lanczos vectors can be summarized as in Table 15.4.

15.5.2 Lanczos Tridiagonal Eigenproblem

The sequence of $N_L \leq N$ vectors $\{\mathbf{q}_k\}_{k=1}^{N_L}$ generated in the algorithm in Table 15.4 can be used to construct a reduced-order tridiagonal eigenproblem whose solution approximates that of the original problem in Eq. 15.33. Recall that an orthogonal basis for the Krylov subspace $\mathcal{K}_{N_L}(\mathbf{r})$ is achieved with the three-term recurrence relationship

$$\mathbf{Aq}_i = b_{i+1} \mathbf{q}_{i+1} + a_i \mathbf{q}_i + b_i \mathbf{q}_{i-1}$$

15.5 Lanczos Eigensolver

Table 15.4 Lanczos Orthogonal Basis Generation

Step 0	*Initialization*
	Choose $\mathbf{r}_0$
	Choose $\mathbf{q}_0 = \mathbf{0}$
	Choose $b_1 = (\mathbf{r}_0^T \mathbf{B} \mathbf{r}_0)^{1/2}$
	Iteration
Step 1	For $i = 1 \cdots N_L$
Step 2	$\mathbf{q}_i = \mathbf{r}_{i-1}/b_i$
Step 3	$a_i = \mathbf{q}_i^T \mathbf{B} \mathbf{A} \mathbf{q}_i$
Step 4	$\mathbf{r}_i = \mathbf{A}\mathbf{q}_i - a_i \mathbf{q}_i - b_i \mathbf{q}_{i-1}$
Step 5	$b_{i+1} = (\mathbf{r}_i^T \mathbf{B} \mathbf{r}_i)^{1/2}$
Step 6	Go to step 1.

where $\mathbf{q}_0 \triangleq \mathbf{0}$ and $i = 1 \cdots N_L$. Form the matrix

$$\mathbf{Q} = [\mathbf{q}_1 \quad \mathbf{q}_2 \quad \cdots \quad \mathbf{q}_{N_L}]$$

Then the recurrence relation can be cast in matrix form:

$$\mathbf{A}\mathbf{Q} = \mathbf{Q}\mathbf{T} + [\mathbf{0} \quad \mathbf{0} \quad \cdots \quad \mathbf{0} \quad b_{N_{\{L+1\}}} \quad \mathbf{q}_{N_{\{L+1\}}}]$$

In this equation, $\mathbf{T}$ is the tridiagonal matrix

$$\mathbf{T} = \begin{bmatrix} a_1 & b_2 & & & & \\ b_2 & a_2 & b_3 & & & \\ & b_3 & a_3 & \ddots & & \\ & & b_4 & \ddots & & \\ & & & \ddots & & b_{N_L} \\ & & & & b_{N_L} & a_{N_L} \end{bmatrix} \qquad (15.45)$$

Suppose that the eigenvectors $\mathbf{v}$ of the original problem are approximated as a linear combination of the columns of $\mathbf{Q}$. That is, the vectors $\mathbf{v}$ are written as

$$\mathbf{v} = \mathbf{Q}\boldsymbol{\psi} = [\mathbf{q}_1 \quad \cdots \quad \mathbf{q}_{N_L}] \begin{Bmatrix} \psi_1 \\ \vdots \\ \psi_{N_L} \end{Bmatrix}$$

The original eigenvalue problem then yields the overdetermined system

$$[\mathbf{A} - \mu_j \mathbf{I}]\mathbf{Q}\boldsymbol{\psi}_j = \mathbf{0} \qquad (15.46)$$

The Ritz approximation is determined by making the left-hand side of Eq. 15.46 $\mathbf{B}$-orthogonal to all of the columns of $\mathbf{Q}$. Upon premultiplying Eq. 15.46 by $\mathbf{Q}^T \mathbf{B}$, and using the $\mathbf{B}$-orthogonality of the columns of $\mathbf{Q}$, the tridiagonal eigenvalue problem is obtained:

$$[\mathbf{Q}^T \mathbf{B} \mathbf{A} \mathbf{Q} - \mu_j \mathbf{Q}^T \mathbf{B} \mathbf{Q}]\boldsymbol{\psi}_j = \mathbf{0}$$

Table 15.5 Basic Lanczos Algorithm

Step 0	Form the real symmetric matrix $\mathbf{A}$.
Step 1	Use the recurrence relation for Krylov subspaces to set up the tridiagonal matrix $$\mathbf{A}\mathbf{q}_i = b_{i+1}\mathbf{q}_{i+1} + a_i\mathbf{q}_i + b_i\mathbf{q}_{i-1}$$
Step 2	Solve the tridiagonal eigenproblem $$\{\mathbf{T} - \mu_i\mathbf{I}\}\boldsymbol{\psi}_i = 0$$
Step 3	Transform the calculated eigenpairs to the desired eigenpairs: $$\{\mu_i, \boldsymbol{\psi}_i\} \longrightarrow \{\lambda_i, \boldsymbol{\phi}_i\}$$

Table 15.6 Practical Lanczos Algorithm with Full Reorthogonalization

Step 0	Form real symmetric matrix $\mathbf{A}$.
Step 1	Use the recurrence relation for Krylov subspaces to set up the tridiagonal matrix $$\mathbf{A}\mathbf{q}_i = b_{i+1}\mathbf{q}_{i+1} + a_i\mathbf{q}_i + b_i\mathbf{q}_{i-1}$$
Step 2	At each step of the recurrence relation, reorthogonalize the Krylov basis vectors $$\mathbf{q}_i^T\mathbf{B}\mathbf{q}_k = \begin{cases} 1 & i = k \\ 0 & \text{otherwise} \end{cases}$$
Step 3	Solve the tridiagonal eigenproblem $$\{\mathbf{T} - \mu_i\mathbf{I}\}\boldsymbol{\psi}_i = 0$$
Step 4	Transform the calculated eigenpairs to the desired eigenpairs: $$\{\mu_i, \boldsymbol{\psi}_i\} \longrightarrow \{\lambda_i, \boldsymbol{\phi}_i\}$$

or

$$[\mathbf{T} - \mu_j\mathbf{I}]\boldsymbol{\psi}_j = 0 \qquad (15.47)$$

See Tables 15.5 and 15.6.

MATLAB Exercise 5: Lanczos Eigensolver Without Reorthogonalization Write a MATLAB .m-file that encodes the fundamental iteration of the *Lanczos Algorithm*. Do not include any attempt to reorthogonalize the Krylov vectors that are generated by the fundamental Lanczos iteration. The .m-file should take the following arguments as input:

- The two matrices that define the generalized eigenproblem.
- The number of eigenpairs to approximate.
- The tolerance to test for convergence of the iterative process.

- An input selector to pick the variant of the Lanczos Method. The input selector should allow different choices for the matrix **B** that defines the **B**-inner product in the fundamental Lanczos iteration.

The .m-file should return the following:

- A list of the approximate eigenvalues
- A list of the associated approximate eigenvectors

Test your .m-file on the small eigenproblem discussed in Example 15.1. Carry out a study of the accuracy of the approximated eigenvalues and eigenvectors as you vary the value of the convergence tolerance.

The .m-file <mat_ex_15_5.m> calls the function subroutine <lancgen.m>, which uses a generalized Lanczos Algorithm without reorthogonalization to produce the following results:

$$\lambda = \begin{Bmatrix} 0.5858 \\ 2.0000 \\ 3.4142 \end{Bmatrix}, \quad \Phi = \begin{bmatrix} 0.5000 & 0.7071 & 0.5000 \\ 0.7071 & 0.0000 & -0.7071 \\ 0.5000 & -0.7071 & 0.5000 \end{bmatrix}$$

Changing the tolerance changes the number of iterations it takes to reach a converged solution.

MATLAB Exercise 6: Lanczos Eigensolver with Full Reorthogonalization Write a MATLAB .m-file that encodes the *Lanczos Algorithm with Full Reorthogonalization*. The .m-file should take the following arguments as input:

- The two matrices that define the generalized eigenproblem.
- The number of eigenpairs to approximate.
- The tolerance to test for convergence of the iterative process.
- An input selector to pick the variant of the Lanczos Method. The input selector should allow different choices for the matrix **B** that defines the **B**-inner product in the fundamental Lanczos iteration.

The .m-file should return the following:

- A list of the approximate eigenvalues
- A list of the associated approximate eigenvectors

Validate your implementation of the Lanczos Algorithm with Full Reorthogonalization by considering the small generalized eigenproblem introduced in Example 15.1. Study the accuracy of the approximated eigenpairs as you vary the convergence tolerance. Compare the performance of your implementation with that of the Lanczos Algorithm without Reorthogonalization.

The .m-file <mat_ex_15_5.m> calls the function subroutine <lancgenv2.m>, which uses a generalized Lanczos algorithm with reorthogonalization to produce the following:

$$\lambda = \begin{Bmatrix} 0.5858 \\ 2.0000 \\ 3.4142 \end{Bmatrix}, \quad \Phi = \begin{bmatrix} 0.5000 & 0.7071 & 0.5000 \\ 0.7071 & 0.0000 & -0.7071 \\ 0.5000 & -0.7071 & 0.5000 \end{bmatrix}$$

Changing the tolerance changes the number of iterations it takes to reach a converged solution. The difference between the Lanczos algorithm with reorthogonalization and the Lanczos algorithm without reorthogonalization is not apparent with such a small system.

15.6 NUMERICAL CASE STUDY

Consider the planar frame structure depicted in Fig. 15.2. At each node there are three DOFs: two translations and one rotation. The frame translational DOFs (but not rotations) are restrained at both ends, that is, at nodes 1, 2, 25, and 26. The material is aluminum 6063-T83, with $E = 69 \times 10^9$ Pa and $\rho = 2700$ kg/m^3. The frame members have the following cross-sectional properties: $A = 21.2 \times 10^{-4}$ m^2 and $I = 349.0 \times 10^{-8}$ m^4.

Use the frame structure shown in Fig. 15.2 as the basis of a case study of the algorithms described in this chapter. Compute the lowest eight eigenvalues of the fixed–fixed frame structure below: (1) Using the Subspace Iteration Algorithm with a random force vector applied to the structure; (2) using the Symmetric QR Algorithm; (3) using the Lanczos Algorithm, based on an inner product that is symmetric with respect to the identity matrix, and without reorthogonalization; (4) using the Lanczos Algorithm, based on an inner product that is symmetric with respect to the identity matrix, and with full reorthogonalization; and (5) using a Shifted Inverse Iteration Algorithm. Plot the modes associated with the lowest four eigenvalues of the frame structure.

Solution The computations for this case study were performed by the MATLAB .m-file <frame3.m>. This .m-file and the associated .m-files described below are included on the book's website. The results are presented in Table 15.7 and discussed in items (1) through (5) below.[5]

(1) *Subspace Iteration Algorithm*. The lowest eight eigenvalues are listed in column (1) of Table 15.7.

(2) *Symmetric QR Algorithm*. The version of the algorithm used in this test case is encoded in the .m-file <symqrv1.m>. This algorithm and .m-file were employed in MATLAB Exercise 4. The lowest eight eigenvalues are listed in column (2) of Table 15.7.

(3) *"Basic" Lanczos Algorithm* (without any orthogonalization). This version of the algorithm uses the .m-file <lancgen.m>. The lowest eight eigenvalues produced

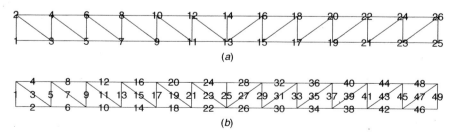

Figure 15.2 Symmetric planar frame structure: (*a*) node numbering; (*b*) element numbering.

---

[5]Only relative numerical values are shown.

Table 15.7 Eigenvalues of the Frame Structure

	(1)	(2)	(3)	(4)	(5)
1	2.960E+03	2.960E+03	2.9610E+03	2.960E+03	2.960E+03
2	1.564E+04	1.564E+04	2.9610E+03	1.564E+04	1.564E+04
3	4.372E+04	4.372E+04	2.9610E+03	4.372E+04	4.373E+04
4	8.374E+04	8.374E+04	2.9610E+03	8.374E+04	8.376E+04
5	9.718E+04	9.718E+04	2.9610E+03	9.718E+04	9.720E+04
6	1.1473E+05	1.1473E+05	2.9610E+03	1.1473E+05	1.1475E+05
7	1.1608E+05	1.1608E+05	2.9610E+03	1.1608E+05	1.1610E+05
8	1.1806E+05	1.1806E+05	2.9610E+03	1.1806E+05	1.1808E+05

by the Lanczos Algorithm, without reorthogonalization, are listed in column (3) of Table 15.7. Note that without reorthogonalization, the Lanczos Algorithm repeatedly converges to the lowest eigenvalue. This phenomenon is sometimes referred to as *ghosting*.

(4) *Lanczos Algorithm* (with full reorthogonalization). We next contrast item (3) with a numerically stable form of the Lanczos Algorithm. The lowest eight eigenvalues are evaluated using the Lanczos Algorithm with full reorthogonalization in the .m-file <lancgenv2.m>. These eight eigenvalues are listed in column (4) of Table 15.7.

(5) *Shifted Inverse Iteration*. The .m-file <shiftinviter.m> was used to evaluate the eigenvalues and mode shapes of the structure. The lowest eight eigenvalues produced by the Subspace Iteration Algorithm are listed in column (5) of Table 15.7.

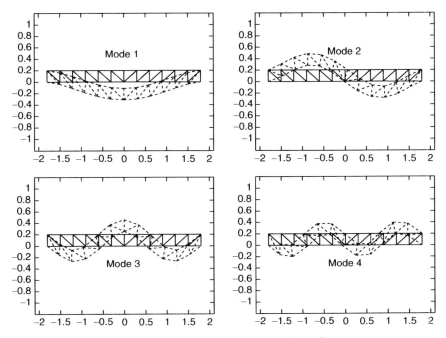

Figure 15.3 Modes 1 to 4 of the symmetric planar frame structure.

The first four mode shapes are plotted overlaid on top of the undeformed structure (Fig. 15.3). These plots were generated by the .m-file <frame3modeplot.m>.

REFERENCES

[15.1] K.-J. Bathé and E. L. Wilson, *Numerical Methods in Finite Element Analysis*, Prentice-Hall, Englewood Cliffs, NJ, 1976.

[15.2] T. J. R. Hughes, *The Finite Element Method: Linear Static and Dynamic Finite Element Analysis*, Dover, Mineola, NY, 2000.

[15.3] R. Grimes, J. Lewis, and H. Simon, "A Shifted Block Lanczos Algorithm for Solving Symmetric Generalized Eigenproblems," *SIAM Journal on Matrix Analysis and Applications*, Vol. 15, No. 1, January 1994, pp. 228–272.

[15.4] J. K. Bennighof and R. B. Lehoucq, "An Automated Multilevel Substructuring Method for Eigenspace Computation in Linear Elastodynamics," *SIAM Journal on Scientific Computing*, Vol. 25, No. 6, 2004, pp. 2084–2106.

[15.5] G. Golub and C. Van Loan, *Matrix Computations*, 3rd ed., Johns Hopkins University Press, London, 1996.

[15.6] A. Jennings, "Eigenvalue Methods for Vibration Analysis," *Shock and Vibration Digest*, Vol. 12, No. 2, 1980, pp. 3–16.

[15.7] L. Meirovitch, *Elements of Vibration Analysis*, 2nd ed., McGraw-Hill, New York, 1986.

[15.8] R. R. Craig, Jr. and M. C. C. Bampton, "On the Iterative Solution of Semidefinite Eigenvalue Problems," *Aeronautical Journal of the Royal Aeronautical Society*, Vol. 75, 1971, pp. 287–290.

[15.9] B. F. Plybon, *An Introduction to Applied Numerical Analysis*, PWS-Kent, Boston, 1992.

PROBLEMS

Problem Set 15.2

15.1 In the Inverse Iteration Method, a simple recursion was constructed using the matrix $\mathbf{D} = \mathbf{K}^{-1}\mathbf{M}$. Suppose that the structure of the recursive method stays the same but that the matrix $\tilde{\mathbf{D}} = \mathbf{M}^{-1}\mathbf{K}$ is used instead of $\mathbf{D}$. This algorithm is called the *Forward Iteration Method*. Following the same lines of reasoning as those employed in Section 15.1, show that the Forward Iteration Method converges to the dominant, or largest-magnitude, eigenvalue of the generalized eigenvalue problem.

Problem Set 15.4

15.2 In Section 15.4 the QR Method for symmetric eigenproblems was presented, and the role of Householder Transformations in realizing the QR Method was discussed in Section 15.4.2. Recall that Householder Transformations are *orthogonal transformations* that have a very useful property. They can be constructed so that premultiplication of some matrix by the transformation will zero out all entries below a selected position in a given column of the matrix. There are other orthogonal transformations, however, that can be just as efficient in some cases as the Householder Transformation. Perhaps the most common is the Givens Transformation, which has the form

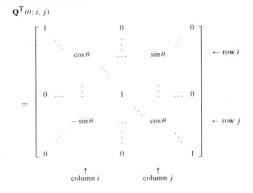

(a) Show that the transformation above is indeed *orthogonal*; that is, $\mathbf{Q}^T(\theta; i, j)\mathbf{Q}(\theta; i, j) = \mathbf{I}$. (b) Show that if we premultiply any vector $\mathbf{v}$ by $\mathbf{Q}^T(\theta; i, j)$, letting

$$\mathbf{u} = \mathbf{Q}^T(\theta; i, j)\mathbf{v}$$

the jth entry u_j of $\mathbf{u}$ can be made zero by the selection

$$\cos\theta = \frac{v_i}{\sqrt{v_i^2 + v_j^2}}, \qquad \sin\theta = \frac{-v_j}{\sqrt{v_i^2 + v_j^2}} \qquad (1)$$

(c) Graphically depict which rows of a matrix are affected when premultiplied by $\mathbf{Q}^T(\theta; i, j)$. Graphically depict which columns of a matrix are affected when postmultiplied by $\mathbf{Q}(\theta; i, j)$. (d) Discuss how a sequence of Givens Transformations can be used to calculate the QR decomposition of a matrix, in a fashion analogous to the Householder-QR Method.

For problems whose number is preceded by a **C**, you are to write a computer program and use it to produce the results requested. *Note:* MATLAB .m-files for many of the plots in this book may be found on the book's website.

C 15.3 Write a MATLAB .m-file to calculate the QR factorization of an arbitrary and, in general, fully populated matrix. Consult other references (see, e.g., Ref. 15.4) for a more numerically stable version of the formulas presented in Eq. 1 in Problem 15.2.

15.4 In our discussion of the QR Algorithm for symmetric eigenvalue problems, it was emphasized that it is prohibitively costly to calculate repeatedly the QR factorization of an arbitrary and, in general, fully populated matrix. It was shown in Section 15.4 that Householder-QR transformations could transform a symmetric matrix into a tridiagonal symmetric matrix. The Givens Transformation can be used to solve a symmetric tridiagonal eigenvalue problem efficiently. (a) Suppose that $\mathbf{A}$ is a symmetric tridiagonal matrix, and let $\mathbf{A} = \mathbf{QR}$ be its QR decomposition. Argue that $\mathbf{Q}$ has a lower bandwidth equal to 1 and that $\mathbf{R}$ has an upper bandwidth equal to 2. This argument is straightforward if you assume that a sequence of Givens Transformations has been used to calculate the QR decomposition. (b) Argue that the product $\mathbf{RQ} = \mathbf{Q}^T(\mathbf{QR})\mathbf{Q}$ is tridiagonal and symmetric. (c) Describe in pseudocode a *numerically efficient* QR Symmetric Eigenvalue Algorithm that combines two phases: an initial phase in which Householder transformations are used to drive the matrix to tridiagonal form, and a second phase in which Givens transformations are used subsequently to calculate QR decompositions on a sequence of tridiagonal matrices.

C 15.5 In MATLAB Exercise 4 in Section 15.4, the QR Symmetric Eigenvalue Algorithm was coded in an .m-file without regard to efficiency. The reader was permitted to use the MATLAB command *eig* to calculate the sequence of QR decompositions required to drive the matrix to diagonal form. The resulting method is very slow and costly. Write a MATLAB .m-file that uses the technique described in Problem 15.4 to create a *numerically efficient* QR Symmetric Eigenvalue Algorithm. First, drive the matrix to tridiagonal form by an appropriate selection of Householder transformations. Subsequently, calculate the QR decomposition of the tridiagonal, symmetric matrix using Givens transformations that operate only on the nonzero tridiagonal terms.

Problem Set 15.5

C 15.6 The Lanczos Algorithm begins by creating a tridiagonal matrix associated with a collection of Krylov subspaces. In MATLAB Exercise 5 the reader was permitted to use the MATLAB command *eig* to solve the resulting tridiagonal eigenvalue problem. Combine the results of MATLAB Exercise 5, where the .m-file encodes the Lanczos Method with Full Reorthogonalization, but solve the resulting equivalent tridiagonal eigenproblem with the Givens Transformation Method outlined in Problems 15.3 and 15.4.

16
Direct Integration Methods for Dynamic Response of MDOF Systems

In Chapter 11 *mode-superposition methods* were employed to obtain the dynamic response of linear MDOF systems by transforming the equations of motion to principal coordinates and solving the resulting set of uncoupled equations of motion. In Fig. 10.2 it was indicated that MDOF systems that cannot be handled in this manner, such as systems with nonlinearities or with coupled damping, require *direct integration* of a set of coupled equations of motion, and in Section 10.3, several procedures were described for creating a full damping matrix for use when direct integration is required, including *Rayleigh damping* and two procedures for creating a full damping matrix that corresponds to modal damping.

In this chapter we first review the topic of modeling damping in MDOF systems. Then we extend to the MDOF case the techniques for the numerical integration of linear and nonlinear systems that were applied to SDOF systems in Chapter 6. Historically, there have been two tracks taken to develop numerical integration methods that are applicable to the simulation of structural dynamics response: namely, those techniques that have been derived directly to treat a set of *second-order ordinary differential equations* representing problems in structural dynamics, and those methods that are based on an equivalent set of *first-order ordinary differential equations*. The first approach has been explored primarily by structural dynamics researchers, while the latter approach is the classical formulation explored by mathematicians and employed in various engineering disciplines.

For integrating the equations of motion for the very large finite element models that are common in structural dynamics (e.g., over 100,000 DOF), specially derived second-order numerical integration methods are in order. However, engineers who deal with the control of flexible structures may work with finite-element-based structural models (e.g., modal models) having over 100 DOFs and may have a preference for using first-order integration methods. In this chapter we discuss both first- and second-order methods.

It should be emphasized that many of the algorithms discussed in Chapter 6 can be applied with minimal or no change to the MDOF case. On the other hand, because MDOF systems generally have a broad spectrum of natural frequencies, this chapter takes a more advanced viewpoint on the entire problem of generating numerical approximations of solutions to problems in structural dynamics. We define and examine the crucial stability and accuracy properties of numerical integration algorithms and consider the implications when these algorithms are used for direct integration of the equations of motion of MDOF systems.

Upon completion of this chapter you should be able to:

- Discuss the difference between proportional and nonproportional damping, and create a damping matrix in physical coordinates that produces equivalent modal damping.
- Discuss, and in some cases derive, methods for the numerical integration of systems of second-order ordinary differential equations.
- Discuss, and in some cases derive, methods for the numerical integration of systems of first-order ordinary differential equations.
- Define the following terms or phrases: *order of accuracy, order of consistency, local truncation error, global error, 0-stability, A-stability, conditional stability, unconditional instability, absolute stability, period error, amplitude error, numerical damping.*
- Discuss the relationship between the choice of time step and the terms listed above.
- Discuss the suitability of several alternative algorithms for direct integration of the equations of motion of MDOF systems, and make a critical assessment of the benefits and drawbacks of any particular algorithm for solving a given problem.

16.1 DAMPING IN MDOF SYSTEMS

The Finite Element Method is used to create stiffness matrices and mass matrices that constitute the large MDOF math models of engineering structures. The *generalized damping matrix* was defined in Section 10.3 as

$$\bar{C} = \Phi^T C \Phi \qquad (16.1)$$

where C is the physical damping matrix, Φ is the *modal matrix* of undamped normal modes, and $\bar{C}$ is the damping matrix when the equation of motion is expressed in principal coordinates.

16.1.1 Nonproportional Damping

Damping that couples the undamped normal modes is called *nonproportional damping*. For example, automotive vehicles all have shock absorbers that provide highly localized damping. Example 16.1 illustrates how such localized damping produces a fully populated generalized damping matrix.

Example 16.1 Consider the 3-DOF spring–mass model of Example 15.1, and add localized damping between "ground" and the left-hand mass (Fig. 1, where $k = 1$ and $m = 1$). The physical damping matrix and modal matrix for this system are

$$C = \begin{bmatrix} c & 0 & 0 \\ 0 & 0 & 0 \\ 0 & 0 & 0 \end{bmatrix}, \quad \Phi = \begin{bmatrix} 1 & 1 & 1 \\ \sqrt{2} & 0 & -\sqrt{2} \\ 1 & -1 & 1 \end{bmatrix}$$

Determine the generalized damping matrix for this system, and discuss its format.

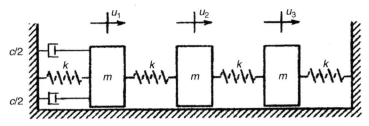

Figure 1 A 3-DOF spring–mass system with localized damping.

SOLUTION   The generalized damping matrix is given by Eq. 16.1. Then

$$\mathbf{C} = \mathbf{\Phi}^T \mathbf{C} \mathbf{\Phi}$$

$$= \begin{bmatrix} 1 & \sqrt{2} & 1 \\ 1 & 0 & -1 \\ 1 & -\sqrt{2} & 1 \end{bmatrix} \begin{bmatrix} c & 0 & 0 \\ 0 & 0 & 0 \\ 0 & 0 & 0 \end{bmatrix} \begin{bmatrix} 1 & 1 & 1 \\ \sqrt{2} & 0 & -\sqrt{2} \\ 1 & -1 & 1 \end{bmatrix} \quad (1)$$

Therefore, the generalized damping matrix is

$$\mathbf{C} = \begin{bmatrix} c & c & c \\ c & c & c \\ c & c & c \end{bmatrix} \quad \textbf{Ans. (2)}$$

The localized physical damping produces a fully populated generalized damping matrix. Thus, the localized damping couples all of the normal modes and would prohibit a mode-by-mode mode-superposition solution such as those discussed in Chapter 11. Nevertheless, the generalized damping matrix is of rank = 1.

Typically, in those cases where it is possible to define the physical damping matrix $\mathbf{C}$, the generalized damping matrix $\mathbf{C}$ is not diagonal, as is the case in Example 16.1. Several approaches have been suggested for solving for the response of system with nonproportional damping. Hurty and Rubinstein[16.1] use complex modes in a procedure analogous to the mode-superposition method of Chapter 11. Clough and Mojtahedi[16.2] compared various methods of solving for the response of systems with nonproportional damping and concluded that "the most efficient procedure is to express the response in terms of a truncated set of undamped modal coordinates and to integrate directly the resulting equations." Certainly, this approach is straightforward, uses available analytical tools (eigensolvers that handle systems with real modes, direct integration algorithms, etc.), and avoids unnecessary approximations. Various other procedures for model reduction of systems with nonproportional damping have also been proposed (e.g., use of Krylov vectors or component modes).

16.1.2   Rayleigh Damping

Section 10.3 discussed the topic, "How can damping be represented when a damping matrix $\mathbf{C}$ in physical coordinates is required?" For example, a physical damping matrix is required for solving nonlinear structural dynamics problems, where a viscous damping matrix is introduced to account for the energy dissipated by all mechanisms other than material yielding.

One procedure for defining a system damping matrix is to employ a particular form of proportional damping called *Rayleigh damping*, defined by

$$\mathbf{C} = a_0 \mathbf{M} + a_1 \mathbf{K} \qquad (16.2)$$

where, in the case of a nonlinear system, $\mathbf{K}$ could represent an initial tangent stiffness matrix, or $\mathbf{C}$ could be modified with each change in stiffness. Then, for Rayleigh damping defined by Eq. 16.2, the generalized damping matrix is a diagonal matrix given by

$$\mathbf{C} = \mathbf{\Phi}^T \mathbf{C} \mathbf{\Phi} = \text{diag}(a_0 + a_1 \omega_r^2) M_r = \text{diag}(C_r) = \text{diag}(2\zeta_r \omega_r M_r) \qquad (16.3)$$

The system damping factors are given by

$$\zeta_r = \frac{1}{2}\left(\frac{a_0}{\omega_r} + a_1 \omega_r\right) \qquad (16.4)$$

Therefore, Rayleigh damping is easy to define by choosing ζ_r for two modes and solving for a_0 and a_1. The damping in the remaining modes is then determined by Eq. 16.4. The $a_0 \mathbf{M}$ contribution to damping in Eq. 16.2 gives a contribution to ζ_r in Eq. 16.4 that is inversely proportional to ω_r. The $a_1 \mathbf{K}$ term, on the other hand, leads to a contribution to ζ_r that increases linearly with ω_r.

16.1.3 Damping Matrix C for Modal Damping

The disadvantage of Rayleigh damping is that it does not permit realistic damping to be defined for all the modes of interest. If a limited number of the lower-frequency modes are considered to be important in the response calculations, a damping matrix $\mathbf{C}$ can be formed as follows (see Section 10.3):

$$\mathbf{C} = \sum_{r=1}^{N_c} \frac{2\zeta_r \omega_r}{M_r}(\mathbf{M}\boldsymbol{\phi}_r)(\mathbf{M}\boldsymbol{\phi}_r)^T \qquad (16.5)$$

This produces a damping matrix $\mathbf{C}$ that gives modal damping ζ_r in the N_c specified modes but yields no damping in the modes $N_c + 1$ and higher.

It may, however, be desirable to provide damping in these higher-frequency modes. It is possible to modify Eq. 16.5 such that the modes $r = 1, 2, \ldots, N_c$ have specified damping ratios, and the modes $N_c + 1, N_c + 2, \ldots, N$ have damping greater than that in mode N_c. This is possible by letting

$$\mathbf{C} = a_1 \mathbf{K} + \sum_{r=1}^{N_c-1} \frac{2\widehat{\zeta}_r \omega_r}{M_r}(\mathbf{M}\boldsymbol{\phi}_r)(\mathbf{M}\boldsymbol{\phi}_r)^T \qquad (16.6)$$

where

$$a_1 = \frac{2\zeta_{N_c}}{\omega_{N_c}}, \qquad \widehat{\zeta}_r = \zeta_r - \zeta_{N_c}\frac{\omega_r}{\omega_{N_c}} \qquad (16.7)$$

Then,

$$\zeta_r = \begin{cases} \text{value specified,} & r = 1, 2, \ldots, N_c \\ \zeta_{N_c} \dfrac{\omega_r}{\omega_{N_c}}, & r = N_c + 1, N_c + 2, \ldots, N \end{cases} \quad (16.8)$$

There is further discussion of MDOF damping in Section 10.3, and Example 10.5 illustrates the use of Eq. 16.6 to create a physical damping matrix for a four-story building.

Whenever direct integration requires the use of a physical damping matrix, Rayleigh damping defined by Eq. 16.2, or modal damping defined by Eq. 16.5 or Eq. 16.6, may be employed to approximate the damping.

16.2 NUMERICAL INTEGRATION: MATHEMATICAL FRAMEWORK

Recall that the topic of Chapter 6 is the development of numerical integration methods for the SDOF standard equation of structural dynamics:

$$\begin{aligned} m\ddot{u} + c\dot{u} + ku &= p(t) \\ u(0) = u_0, \quad \dot{u}(0) &= v_0 \end{aligned} \quad (16.9)$$

Section 6.1 discusses numerical methods that are directly applicable to the second-order equation above. On the other hand, Section 6.2 presents methods for the numerical simulation of a set of first-order equations that are mathematically equivalent. The equivalent first-order equations are obtained by introducing a new dependent variable $\mathbf{x}(t) \in \mathbb{R}^2$ defined as

$$\mathbf{x(t)} \equiv \begin{Bmatrix} x_1(t) \\ x_2(t) \end{Bmatrix} \triangleq \begin{Bmatrix} u(t) \\ \dot{u}(t) \end{Bmatrix} \in \mathbb{R}^2 \quad (16.10)$$

It is easy to see that Eqs. 16.9 are equivalent to the first-order system

$$\boxed{\begin{aligned} \dot{\mathbf{u}}(t) &= \mathbf{f}(\mathbf{x}(t), t) \\ \mathbf{x}(0) &= \mathbf{x}_0 \end{aligned}} \quad (16.11)$$

where

$$\mathbf{f}(\mathbf{x}(t), t) = \mathbf{A}\mathbf{x}(t) + \mathbf{p}(t)$$

$$\mathbf{A} = \begin{bmatrix} 0 & 1 \\ -m^{-1}k & -m^{-1}c \end{bmatrix} \in \mathbb{R}^{2 \times 2} \quad (16.12)$$

$$\mathbf{p}(t) = \begin{Bmatrix} 0 \\ m^{-1} p(t) \end{Bmatrix} \in \mathbb{R}^2, \quad \mathbf{x}_0 = \begin{Bmatrix} u_0 \\ v_0 \end{Bmatrix} \in \mathbb{R}^2$$

By contrast, this chapter is concerned with the numerical simulation of MDOF equations of motion. For linear systems, these equations have the general form

$$\begin{aligned} \mathbf{M}\ddot{\mathbf{u}} + \mathbf{C}\dot{\mathbf{u}} + \mathbf{K}\mathbf{u} &= \mathbf{p}(t) \\ \mathbf{u}(0) = \mathbf{u}_0, \quad \dot{\mathbf{u}}(0) &= \mathbf{v}_0 \end{aligned} \quad (16.13)$$

One approach to defining the equivalent first-order form of the MDOF equations is to define the dependent variable $\mathbf{x}(t) \in \mathbb{R}^{2N}$ via the identity

$$\mathbf{x}(t) \triangleq \left\{ \begin{array}{c} \mathbf{u}(t) \\ \dot{\mathbf{u}}(t) \end{array} \right\} \in \mathbb{R}^{2N} \tag{16.14}$$

Then Eqs. 16.13 are equivalent to the system

$$\boxed{\begin{array}{l} \dot{\mathbf{x}}(t) = \mathbf{f}(\mathbf{x}(t), t) \\ \mathbf{x}(0) = \mathbf{x}_0 \end{array}} \tag{16.15}$$

where

$$\mathbf{f}(\mathbf{x}(t), t) = \mathbf{A}\mathbf{x}(t) + \mathbf{p}(t)$$

$$\mathbf{A} = \begin{bmatrix} \mathbf{0} & \mathbf{I} \\ -\mathbf{M}^{-1}\mathbf{K} & -\mathbf{M}^{-1}\mathbf{C} \end{bmatrix} \in \mathbb{R}^{2N \times 2N} \tag{16.16}$$

$$\mathbf{p}(t) = \left\{ \begin{array}{c} \mathbf{0} \\ \mathbf{M}^{-1}\mathbf{p}(t) \end{array} \right\} \in \mathbb{R}^{2N}, \quad \mathbf{x}_0 = \left\{ \begin{array}{c} \mathbf{u}_0 \\ \mathbf{v}_0 \end{array} \right\} \in \mathbb{R}^{2N}$$

A careful inspection of the equations of motion makes it clear that the two forms, the second- and first-order systems, have a clear formal similarity to the equations studied in Chapter 6. In fact, the first-order equations studied in Chapter 6 differ from the first-order equations in this chapter only in their dimension. In Chapter 6 the state is $\mathbf{x} \in \mathbb{R}^2$, whereas the state is $\mathbf{x} \in \mathbb{R}^{2N}$ in this chapter. *The techniques developed in Chapter 6 for the integration of first-order systems can be applied to the problems in this chapter essentially with little modification.* However, there are subtleties to the use of numerical integration methods for MDOF systems that are not encountered in the study of SDOF systems. We require a better understanding of stability and accuracy in order to carry out numerical integration of the equations governing MDOF structural dynamics problems. As noted earlier, this stems largely from the fact that N-DOF systems have N modes with N natural frequencies that generally cover a broad frequency spectrum. Also, for very large systems, efficient numerical integration requires that careful consideration be given to the form (e.g., sparseness) and dimension of the system equations.

16.2.1 Error, Consistency, and Convergence

Despite the differences in the governing equations, whether we choose to exploit directly the second-order system of ODEs or the equivalent first-order form of the ODEs, there are several critical concepts related to the performance of approximation algorithms that are common to both. These properties include notions such as *consistency, convergence, 0-stability*, and *A-stability*. In this section we define consistency and convergence of numerical methods for the approximation of solutions of ODEs. We frame the concepts so that they can be applied easily to the second- or first-order form of the systems of equations that govern structural dynamics. The following notational conventions hold in the discussions that follow:

- We seek the approximate solution defined on the mesh, or grid, defined by the nodes (i.e., discrete times) $\{t_0, t_1, \ldots, t_N\} \subset \mathbb{R}$. We denote this grid of points as

$$\boldsymbol{\tau}_N = \{t_k\}_{k=0}^N$$

- The mesh parameter, or time step, at time t_n is defined to be $h_n = t_n - t_{n-1}$ for $n = 1 \cdots N$. If h_n is constant for all n, we designate the common length of time steps to be h.
- Suppose that $z(t) \in \mathbb{R}^M$ is the solution of the governing ODE evaluated at time $t \in [t_0, t_N]$. We will compare approximate solutions to the *grid, or mesh, function* extracted from z. We define the true solution on the grid, or mesh, τ_N to be the sequence of *M-tuples* of real numbers

$$\mathcal{Z} = \{z(t_0), z(t_1), \ldots, z(t_N)\}$$

Note that we have

$$\mathcal{Z} \in \underbrace{\mathbb{R}^M \times \mathbb{R}^M \times \cdots \times \mathbb{R}^M}_{N+1 \text{ times}}$$
$$\overbrace{(\mathbb{R}^M)^{N+1}}$$

The evaluation of a grid function $\mathcal{Z}$ at node $0 \leq k \leq N$ is denoted

$$\mathcal{Z}(k) \equiv z(t_k)$$

With this notational convention in place, it is a simple matter to define a *numerical algorithm* $\mathcal{A}_N$, an *approximate solution* $\mathcal{Z}_N$, the *local truncation error* $\mathcal{R}_N$, and the *global error* $\mathcal{E}_N$ associated with the algorithm. We define the approximation to the true solution $\mathcal{Z}$ in a similar manner. A *numerical algorithm* over the grid τ_N for the approximate solution of the governing ordinary differential equation will be identified with an operator $\mathcal{A}_N$ that maps grid functions into grid functions.

$$\mathcal{A}_N : (\mathbb{R}^M)^{N+1} \longrightarrow (\mathbb{R}^M)^{N+1}$$

In other words, the operator $\mathcal{A}_N$ embodies an approximation of the differential equations, in terms of a difference equation. An *approximation* $\mathcal{Z}_N$ over the grid τ_N of the solution to the differential equation is the solution of the equation

$$\mathcal{A}_N \mathcal{Z}_N = 0 \in (\mathbb{R}^M)^{N+1}$$

while the approximation at node k (i.e., at time t_k) to the true solution $\mathcal{Z}(k)$ is given by $\mathcal{Z}_N(k)$.

Now we can define the central concepts of this section: local truncation error, global error, consistency (or accuracy) of an algorithm, and convergence of an algorithm. The *local truncation error* is the residual that is incurred when we apply the algorithm $\mathcal{A}_N$ to the true solution,

$$\mathcal{R}_N = \mathcal{A}_N \mathcal{Z}$$

so the local truncation error at the node k is

$$\mathcal{R}_N(k) = (\mathcal{A}_N \mathcal{Z})(k)$$

An algorithm is *consistent* provided the local truncation error approaches zero as the mesh parameter h goes to zero. More specifically, an algorithm $\mathcal{A}_N$ is consistent, or accurate, of order s if the local truncation error is of the order $O(h^s)$. That is, we must have

$$\mathcal{R}_N(k) = O(h^s)$$

for all $k = 1 \cdots N$. The algorithm $\mathcal{A}_N$ is *convergent* of order s provided that the global error

$$\begin{aligned} \mathcal{E}_N &= \mathcal{Z} - \mathcal{Z}_N \\ \mathcal{E}_N(k) &= \mathcal{Z}(k) - \mathcal{Z}_N(k) \end{aligned} \qquad (16.17)$$

satisfies

$$\mathcal{E}_N(k) = O(h^s) \qquad (16.18)$$

for $0 \leq k \leq N$.

Example 16.2 Recall again the vector-valued ordinary differential equations

$$\begin{aligned} \dot{\boldsymbol{x}}(t) &= \boldsymbol{f}(\boldsymbol{x}(t), t) \\ \boldsymbol{x}(0) &= \boldsymbol{x}_0 \end{aligned}$$

The Taylor Series Method of Order k, introduced in Section 6.2.1, for achieving an approximation of the differential equation above is based based on the iteration formula

$$\boldsymbol{x}_{i+1} = \boldsymbol{x}_i + h_i T_{k,h}(\boldsymbol{x}_i, t_i)$$

where $h_i = t_{i+1} - t_i$ and $i = 1 \cdots N$. In this formula, $T_{k,h}(\boldsymbol{x}_i, t_i)$ is defined as

$$T_{k,h}(\boldsymbol{x}, t) = f(\boldsymbol{x}, t) + \frac{h}{2!} \frac{d}{dt}(f(\boldsymbol{x}(t), t)) + \cdots + \frac{h^{k-1}}{k!} \frac{d^{(k-1)}}{dt^{(k-1)}}(f(\boldsymbol{x}(t), t))$$

In the problems that follow, suppose that the mesh $\boldsymbol{\tau}_N$ corresponds to a uniform mesh with $h_i = h$ for all $i = 1 \cdots N$. (a) What is the operator $\mathcal{A}_N$ that is appropriate for this example? (b) What is the local truncation error $\mathcal{R}_N(i)$ at node i for this algorithm? (c) What is the order of consistency, or approximation, for this algorithm?

SOLUTION   (a) The operator $\mathcal{A}_N$ on the mesh $\boldsymbol{\tau}_N$ is given by

$$(\mathcal{A}_N \mathcal{V}_N)(i) = \frac{\mathcal{V}_N(i+1) - \mathcal{V}_N(i)}{h} - T_{k,h}(\mathcal{V}_N(i), t_i)$$

(b) Suppose that $\mathcal{X}$ is the grid function corresponding to the true solution of the ODE. It is defined by

$$\mathcal{X}(i) = \boldsymbol{x}(t_i)$$

for $i = 1 \cdots N$. The local truncation error is just the approximation operator acting on the grid function derived from the true solution:

$$(\mathcal{A}_N \mathcal{X})(i) = \frac{1}{h} \left\{ \boldsymbol{x}(t_{i+1}) - \left[\boldsymbol{x}(t_i) + h \frac{d}{dt}(\boldsymbol{x}(t_i)) + \cdots + \frac{h^k}{k!} \frac{d^{(k)}}{dt^{(k)}}(\boldsymbol{x}(t_i)) \right] \right\}$$

For a solution $x(t)$ for $t \in [t_0 \cdots t_N]$ that is sufficiently smooth, standard error estimates for Taylor series yield

$$\frac{1}{h} O(h^{k+1}) = O(h^k)$$

(c) By definition, the order of accuracy is k.

16.2.2 Stability

In the preceding section we were able to provide reasonably simple characterizations of accuracy, consistency, and convergence. Our intuition about what constitutes convergence, for example, is evident in the mathematical definition. Moreover, the consistency or accuracy of a method is often determined via direct analysis (e.g., our example of consistency is based on the use of Taylor series).

In contrast, the notion of stability of these numerical algorithms is often less clear. Intuitively, *stability of a numerical method guarantees that small, unavoidable errors introduced into the numerical integration scheme will not be amplified by subsequent computations*. One complicating factor is that there are often several alternative and equivalent definitions of a specific notion of stability. Another layer of complexity results because there are various types of stability, with stability concepts differing in sometimes subtle ways. In this book we will encounter five different related stability definitions: *0-stability, A-stability, spectral stability, conditional stability*, and *unconditional stability*. All of these concepts of stability are encountered in structural dynamics literature and applications. We discuss each briefly.

0-Stability If perturbations are not to be amplified in subsequent computations, a fundamental way of expressing this fact is as follows: *Solutions that start close together remain close together during the time interval of interest*. This heuristic definition can be made precise. An algorithm $\mathcal{A}_N$ associated with a grid τ_N is 0-*stable* if there is an integer N_{min} and a constant C such that for any two arbitrary mesh functions $\mathcal{U}_N$ and $\mathcal{V}_N$ with $N \geq N_{min}$, we have

$$\|\mathcal{U}_N(k) - \mathcal{V}_N(k)\|_M \leq C \left\{ \|\mathcal{U}_N(0) - \mathcal{V}_N(0)\|_M + \max_{1 \leq i \leq N} \|\mathcal{A}_N \mathcal{U}_N(i) - \mathcal{A}_N \mathcal{V}_N(i)\|_M \right\}$$

Like many mathematical definitions, the characterization of 0-stability above can be intimidating at first glance. Careful inspection of the definition shows that it does capture the intuitive notion suggested above: *Solutions that start from nearby initial conditions remain close over the time interval of interest*. Suppose that the grid functions $\mathcal{U}_N$ and $\mathcal{V}_N$ are approximations to two different solutions of the same ordinary differential equation:

$$\dot{\mathbf{u}}(t) = \mathbf{f}(\mathbf{u}(t), t), \quad \mathbf{u}(0) = \mathbf{u}_0 = \mathcal{U}_N(0), \quad \forall t \in [0, T]$$
$$\dot{\mathbf{v}}(t) = \mathbf{f}(\mathbf{v}(t), t), \quad \mathbf{v}(0) = \mathbf{v}_0 = \mathcal{V}_N(0), \quad \forall t \in [0, T]$$

Since $\mathcal{U}$ and $\mathcal{V}$ are approximations to the solutions $\mathbf{u}$ and $\mathbf{v}$, respectively, they must satisfy

$$\mathcal{A}_N \mathcal{U}_N = 0, \quad \mathcal{U}_N(0) = \mathbf{u}_0$$
$$\mathcal{A}_N \mathcal{V}_N = 0, \quad \mathcal{V}_N(0) = \mathbf{v}_0$$

The inequality defining 0-stability simplifies in view of these identities. It becomes

$$\|\mathcal{U}_N(k) - \mathcal{V}_N(k)\|_{\mathbb{R}^M} \leq C\{\|\mathcal{U}_N(0) - \mathcal{V}_N(0)\|_{\mathbb{R}^M}\}$$

for $1 \leq k \leq N$. This inequality expresses the fact that approximate solutions of the governing ordinary differential equation can be made arbitrarily close for all grids τ_N with $N \geq N_{\min}$ simply by bringing the initial conditions closer together. In other words, *solutions that start close together remain close together over the time interval of interest.*

It should be noted that this definition includes, in general, some representation of the possibly nonlinear function $\mathbf{f}$ in the algorithm $\mathcal{A}$. It therefore can be difficult to ascertain 0-stability in practice. However, an extensive literature has emerged that summarizes sufficient and necessary conditions for 0-stability of many of the classes of integration schemes discussed in Chapter 6, including the Linear Multistep Methods and the family of Runge–Kutta Methods. The reader is referred to Ref. [16.3], [16.4], or [16.5] for more detailed discussions.

A-Stability The previous definition of stability dealt with the asymptotic behavior of a solution as $N \to \infty$ (or, equivalently, as $h \to 0$) on a fixed time interval. *Absolute stability* is a complementary notion of stability that seeks to understand the behavior of approximate solutions when the mesh parameter h is fixed, but $N \to \infty$. In this case the time interval is not bounded. Absolute stability is inherently different than 0-stability in another way: A *test equation* is introduced into the discussion to make the analysis tractable. Consider the linear ordinary differential equation

$$\dot{x}(t) = cx(t)$$
$$x(0) = 1 \tag{16.19}$$

where $c \in \mathbb{C}$ and $Re(c) < 0$. Since the solution is given by $x(t) = e^{ct}$, we know that the true solution satisfies $|x(t)| \to 0$ as $t \to \infty$. We can now define *absolute stability (with respect to this test problem)*. A numerical algorithm $\mathcal{A}_N$ is A-stable provided that the approximation $\{x_n\}_{n=0}^\infty$ satisfies

$$|x_n| \to 0 \text{ as } n \to \infty \tag{16.20}$$

The *region of absolute stability* is defined to be the subset of the complex plane

$$R_A = \{z \in \mathbb{C} : z = ch \text{ and Eq. 16.20 holds}\}$$

Absolute stability, and the very closely related concept of *spectral stability*, are encountered frequently in discussions of numerical methods for ordinary differential equations arising in structural dynamics. The introduction of a test problem, in lieu of the analysis of a general nonlinear function $\mathbf{f}$, makes this notion of stability much more amenable to application. In fact, in application to structural dynamics, we will investigate stability with respect to the *prototypical structural dynamics test problem*:

$$\ddot{u}(t) + 2\zeta\omega_n \dot{u}(t) + \omega_n^2 u(t) = 0$$
$$u(0) = u_0, \qquad \dot{u}(0) = v_0$$

As we have shown, this equation can be converted formally to the first-order form

$$\mathbf{x}(t) = \mathbf{A}\mathbf{x}(t) \in \mathbb{R}^{2 \times 2}$$
$$\mathbf{x}(0) = \mathbf{x}_0 \tag{16.21}$$

The study of the stability with respect to the *vector-valued* equations above is discussed in detail in Section 16.4. There we introduce the concept of spectral stability and give an example of its application in structural dynamics.

Conditional Stability and Unconditional Stability Of all of the concepts of stability, conditional stability and unconditional stability are perhaps the most straightforward. An algorithm is said to be *conditionally stable* if the stability of the numerical algorithm relies on the fact that the mesh is sufficiently fine. That is, the method is stable only if the time step satisfies $h \leq h_{\text{cr}}$. The constant h_{cr} denotes the *critical time step*, and its value depends on the algorithm parameters and on physical parameters in the simulation. If the stability of the system does not depend on some time-step constraint (i.e., the method is stable for all time-step sizes), the method is said to be *unconditionally stable*. In structural dynamics, unconditionally stable methods are generally preferable. The critical time step in a conditionally stable method may depend on quantitative properties of the system (such as the natural frequencies of the system) that are not known a priori.

16.3 INTEGRATION OF SECOND-ORDER MDOF SYSTEMS

The Constant Average Acceleration Method was presented in Chapter 6. In this chapter we discuss three related methods for integration of second-order systems: the Central Difference Method, the Newmark-β Method, and the Wilson-θ Method.[16.6,16.7] These methods are not entirely independent and we discuss the relationships among them. The first two methods are presented in a form appropriate for the direct integration of linear second-order systems, such as those whose damping matrix cannot be diagonalized by the system normal modes. The last technique is presented in a form that is appropriate for some classes of nonlinear second-order systems.

16.3.1 Central Difference Method

Perhaps the most fundamental algorithm for the approximate numerical solution of the second-order ordinary differential equations arising in applications in structural dynamics is the *Central Difference Method*. The origin of its standing among all the possible finite difference equations that can be applied to the governing equations of structural dynamics is easily understood. First, it is an exceedingly simple method to describe and hence is invaluable as an introduction to more complex algorithms. Second, the method is a *second-order* accurate algorithm. Although we discuss accuracy and stability shortly, it suffices for now to point out that vast experience in actual engineering problems has suggested that second-order accurate techniques are required in many applications. Recall the form of the second-order ordinary differential equations of structural dynamics.

$$\mathbf{M}\ddot{\mathbf{u}} + \mathbf{C}\dot{\mathbf{u}} + \mathbf{K}\mathbf{u} = \mathbf{p}(t)$$
$$\mathbf{u}(0) = \mathbf{u}_0, \qquad \dot{\mathbf{u}}(0) = \mathbf{v}_0 \tag{16.22}$$

Again, we employ the convention that $\mathbf{u}(t_n) \equiv \mathbf{u}_n$ throughout this chapter. The foundation of the *Central Difference Algorithm* is the simple finite-difference expression

$$\dot{\mathbf{u}}_n = \frac{\mathbf{u}_{n+1} - \mathbf{u}_{n-1}}{2h} \qquad (16.23a)$$

The derivative at time t_n is approximated by the slope of the line passing through the values of the function at t_{n-1} and t_{n+1}. To maintain consistency in the order of approximation, the value of the second derivative is calculated as the difference of first-order forward and backward finite differences.

$$\begin{aligned}\ddot{\mathbf{u}}_n &= \frac{\{\dot{\mathbf{u}}_{n+1/2} - \dot{\mathbf{u}}_{n-1/2}\}}{h} \\ &= \frac{\{\mathbf{u}_{n+1} - \mathbf{u}_n\}/h - \{\mathbf{u}_n - \mathbf{u}_{n-1}\}/h}{h} \\ &= \frac{\{\mathbf{u}_{n+1} - 2\mathbf{u}_n + \mathbf{u}_{n-1}\}}{h^2}\end{aligned} \qquad (16.23b)$$

When the finite-difference expressions for the first and second derivatives are substituted into the governing equation of motion evaluated at t_n, the discrete governing equation results:

$$\left(\frac{1}{h^2}\mathbf{M} + \frac{1}{2h}\mathbf{C}\right)\mathbf{u}_{n+1} + \left(\mathbf{K} - \frac{2}{h^2}\mathbf{M}\right)\mathbf{u}_n + \left(\frac{1}{h^2}\mathbf{M} - \frac{1}{2h}\mathbf{C}\right)\mathbf{u}_{n-1} = \mathbf{p}_n \qquad (16.23c)$$

Highly specialized implementations of even this simple algorithm have appeared in the literature. Table 16.1 suggests a reasonably efficient implementation that is not too tailored for highly specialized computing hardware or solution procedures.

Several observations are appropriate for the application of the Central Difference Algorithm to structural systems:

- For very large structural problems, highly efficient banded or "skyline" LU factorizations have been developed and tested. (See Ref. [16.8] for an example survey.) These algorithms would play a critical role in steps 0.5 and 0.6. Many of these techniques exploit an important feature of the LU factorization: the decomposition of a matrix having banded or skyline structure can be carried out with *nonzero fill-in* restricted to the original band or skyline. Hence, highly structured matrices yield highly structured factorizations without increasing the bandwidth or skyline.
- It is easy to show that step 0.7 in the Central Difference Algorithm is simply a Taylor series approximation *backward* in time from t_0 to t_{-1} to achieve a (second-order) consistent approximation of $\mathbf{u}_{-1}$.
- Strictly speaking, the calculation of velocity $\dot{\mathbf{u}}_n$ and acceleration $\ddot{\mathbf{u}}_n$ in step 4 are not required. These terms may be calculated at the analyst's discretion, but are not required for the propagation of the algorithm.
- The execution cost and storage requirements of this algorithm are simple to estimate if one does not employ too specialized an implementation or use hardware that is too specialized. Each step of the algorithm requires the storage of the

Table 16.1 Central Difference Method

Step 0	(0.1) Input the mass, damping and stiffness matrices $\mathbf{M}, \mathbf{C}, \mathbf{K}$.
	(0.2) Calculate the LU factorization of $\mathbf{M}$.
	(0.3) Input the initial conditions $\mathbf{u}_0$ and $\mathbf{v}_0 \equiv \dot{\mathbf{u}}_0$.
	(0.4) Set the simulation parameters including the time step h.
	(0.5) Calculate the initial acceleration from the equations of motion.
	$$\ddot{\mathbf{u}}_0 = \mathbf{M}^{-1}\{\mathbf{p}(0) - \mathbf{C}\mathbf{v}_0 - \mathbf{K}\mathbf{u}_0\}$$
	(0.6) Calculate the LU factorization of
	$$\frac{1}{h^2}\mathbf{M} + \frac{1}{2h}\mathbf{C}$$
	(0.7) Calculate the starting displacement value from the equation
	$$\mathbf{u}_{-1} = \mathbf{u}_0 - h\mathbf{v}_0 + \frac{h^2}{2}\ddot{\mathbf{u}}_0$$
Step 1	Loop for each time step, $n = 1 \ldots, t_n = t_1 \cdots$.
Step 2	Calculate the right-hand-side terms of the iteration
	$$\text{RHS}_n = \mathbf{p}_n - \left(\mathbf{K} - \frac{2}{h^2}\mathbf{M}\right)\mathbf{u}_n - \left(\frac{1}{h^2}\mathbf{M} - \frac{1}{2h}\mathbf{C}\right)\mathbf{u}_{n-1}$$
Step 3	Solve for the displacements at the next time step:
	$$\left(\frac{1}{h^2}\mathbf{M} + \frac{1}{2h}\mathbf{C}\right)\mathbf{u}_{n+1} = \text{RHS}_n$$
Step 4	Evaluate the set of velocities and accelerations, as needed.
	$$\dot{\mathbf{u}}_n = \frac{\{\mathbf{u}_{n+1} - \mathbf{u}_{n-1}\}}{2h}$$
	$$\ddot{\mathbf{u}}_n = \frac{\{\mathbf{u}_{n+1} - 2\mathbf{u}_n + \mathbf{u}_{n-1}\}}{h^2}$$
Step 5	Set $n \longrightarrow n+1$ and continue to next time step.

last two displacement vectors, which requires $O(2N)$ storage locations, where N is the number of unknowns. Each update of the displacement vector $\mathbf{u}_{n+1}$ requires the solution of a system of linear equations whose LU factorization is constant throughout the iteration, at least for linear equations. For a full storage representation of the matrix depicted in step 0.6 and in step 3, the forward and back substitution phases of the LU decomposition cost, at each time step, on the order of $O(N^2)$ floating-point operations, or *flops*, where N is the number of unknowns. (See Ref. [16.9] for a definition of the term *flop*.) By contrast, if the bandwidth is of dimension N_b and one of the LU algorithms tailored to banded matrices and banded storage methods is employed, the cost per step can be on the order of $O(NN_b)$, which is a substantial savings.

- Finally, and perhaps most important, the Central Difference Method embodied in Table 16.1 is a *conditionally stable* algorithm. The method is stable provided that the time step h is selected to be smaller than a critical step size h_{cr}, which depends on the eigenvalues of the iteration matrix.

16.3.2 Newmark-β Method

The Central Difference Method discussed in the preceding section is attractive in its simplicity in form and implementation. Recall, however, that the Central Difference Method is only conditionally stable. Some qualitative knowledge of the frequencies that are represented in the response of the system are needed so that a time step can be selected to ensure stability. For this reason, many analysts prefer methods that are *unconditionally stable*. In this section we introduce a family of methods that are defined in terms of a single parameter that can be selected so that the specific method is indeed unconditionally stable. This family of methods is known as the *Newmark-β Method*.[16.10] As in the preceding section, we will present the Newmark-β Method as it is applied to linear systems. As such, the form presented is appropriate for systems such as those that have a damping matrix **C** that is not diagonalizable by the natural modes of the system.

In Chapter 6 one of the first methods presented in the derivation of integration methods was based on Taylor series approximations of the states. The Newmark-β Method can be viewed as a generalization of this theme. The basis of the Newmark-β Method is the assumption that the system displacement vector $\mathbf{u}_{n+1}$ and its derivative $\dot{\mathbf{u}}_{n+1}$ can be written in the form

$$\begin{aligned} \dot{\mathbf{u}}_{n+1} &= \dot{\mathbf{u}}_n + \underbrace{[(1-\gamma)\ddot{\mathbf{u}}_n + \gamma\ddot{\mathbf{u}}_{n+1}]}_{\text{weighted average}} h \\ \mathbf{u}_{n+1} &= \mathbf{u}_n + \dot{\mathbf{u}}_n h + \underbrace{[(1-2\beta)\ddot{\mathbf{u}}_n + 2\beta\ddot{\mathbf{u}}_{n+1}]}_{\text{weighted average}} \frac{h^2}{2} \end{aligned} \quad (16.24)$$

Each of these equations can be viewed as a type of generalized Taylor series approximation. The weighted average is a convex combination of approximate acceleration values. A conventional Taylor series approximation would result if the terms comprising the weighted averages were simply replaced by the conventional term $\ddot{\mathbf{u}}_n$.

The Newmark-β Method is an *implicit* method, with the discrete equation of motion being enforced at the time step t_{n+1}. Thus, we require that

$$\boxed{\mathbf{M}\ddot{\mathbf{u}}_{n+1} + \mathbf{C}\dot{\mathbf{u}}_{n+1} + \mathbf{K}\mathbf{u}_{n+1} = \mathbf{p}_{n+1} \equiv \mathbf{p}(t_{n+1})} \quad (16.25)$$

When the generalized derivative expressions, Eqs. 16.24, are substituted into the discrete equation of motion, Eq. 16.25, it is possible to organize terms and write a linear equation for the new acceleration vector $\ddot{\mathbf{u}}_{n+1}$. There are many possible organizations of the calculations. It is also possible to frame the iterations in terms of $\mathbf{u}_{n+1}$, as suggested in Ref. [16.11]. Neither implementation appears to have a clear advantage in terms of efficiency. Table 16.2 summarizes a generic numerical implementation of the Newmark-β Algorithm that is not too specialized to a specific hardware configuration or software toolset.

To achieve an effective implementation of the Newmark-β Method for structural systems, the following observations should be noted:

- Perhaps the crucial difference between algorithms of the Central Difference Method and the Newmark-β Method is that the former is a *conditionally stable* algorithm, but the latter can be constructed as an *unconditionally stable* method.

Table 16.2 Newmark-β Method

Step 0	(0.1) Input the mass, damping, and stiffness matrices **M**, **C**, and **K**.
	(0.2) Calculate the LU factorization of **M**.
	(0.3) Input the initial conditions $\mathbf{u}_0$ and $\mathbf{v}_0$.
	(0.4) Set the simulation parameters including the time step h.
	(0.5) Calculate the initial acceleration from the equations of motion. $$\ddot{\mathbf{u}}_0 = \mathbf{M}^{-1}\{\mathbf{p}(0) - \mathbf{C}\mathbf{v}_0 - \mathbf{K}\mathbf{u}_0\}$$ (0.6) Calculate the LU factorization of $$[\mathbf{M} + \gamma \mathbf{C} + \beta h^2 \mathbf{K}]$$ (0.7) Calculate the starting displacement value from the equation $$\mathbf{u}_{-1} = \mathbf{u}_0 - h\mathbf{v}_0 + \frac{h^2}{2}\ddot{\mathbf{u}}_0$$
Step 1	Loop for each time step, $n = 1 \ldots, t_n = t_1 \cdots$.
Step 2	Calculate the right-hand-side terms of the iteration $$\text{RHS}_n = -\mathbf{K}\mathbf{u}_n - (\mathbf{C} + h\mathbf{K})\dot{\mathbf{u}}_n - \left[h(1-\gamma)\mathbf{C} + \frac{h^2}{2}(1-2\beta)\mathbf{K}\right]\ddot{\mathbf{u}}_n$$
Step 3	Solve for the second derivatives at the next time step: $$[\mathbf{M} + \gamma \mathbf{C} + \beta h^2 \mathbf{K}]\ddot{\mathbf{u}}_{n+1} = \text{RHS}_n$$
Step 4	Evaluate the set of displacements and velocities: $$\mathbf{u}_{n+1} = \mathbf{u}_n + \dot{\mathbf{u}}_n h + [(1-2\beta)\ddot{\mathbf{u}}_n + 2\beta \ddot{\mathbf{u}}_{n+1}]\frac{h^2}{2}$$ $$\dot{\mathbf{u}}_{n+1} = \dot{\mathbf{u}}_n + [(1-\gamma)\ddot{\mathbf{u}}_n + \gamma \ddot{\mathbf{u}}_{n+1}]h$$
Step 5	Set $n \longrightarrow n+1$ and continue to the next time step.

- Newmark originally proposed as an unconditionally stable scheme the *Constant Average Acceleration Method*, in which case $\gamma = \frac{1}{2}$ and $\beta = \frac{1}{4}$.
- As noted in the comments following the presentation of the Central Difference Method, the efficient implementation of the Newmark-β Method for large structural problems depends on specialized banded or skyline factorization methods for the matrix **M** in step 0.2 and the matrix $[\mathbf{M} + \gamma \mathbf{C} + \beta h^2 \mathbf{K}]$ in step 0.6.
- The cost per time step is quite comparable to that of the Central Difference Method. The storage required per time step, excluding the system matrices, is on the order of $O(3N)$ storage locations for the displacements, velocities, and accelerations updated at each time step t_n. Similarly, depending on the manner of storage of the system matrices, the cost in solving for the accelerations at time step t_n is either on the order of $O(N^2)$ if the matrices are kept in full-storage mode, or perhaps on the order of $O(NN_b)$ if they are maintained in banded-storage format.

16.3.3 Wilson-θ Method

The topic of finite element analysis of transient nonlinear structural behavior has been the subject of extensive research interest over the years (e.g., Refs. [16.12] and [16.13]).

16.3 Integration of Second-Order MDOF Systems

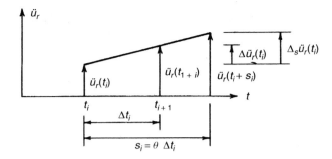

Figure 16.1 Linear acceleration assumption of the Wilson-θ method.

This brief presentation is not intended to do more than just introduce you to several approaches that can be employed to solve either nondiagonalizable linear MDOF structural dynamics problems or some nonlinear problems.

Wilson et al.[16.14] presented an incremental formulation for step-by-step integration of nonlinear equations of motion of MDOF systems. The basic assumption of the *Wilson-θ Method* is that each component $\ddot{u}_r$ of the acceleration vector $\ddot{\mathbf{u}}$ varies linearly with time over the *extended time step* $s_i = \theta \Delta t_i$, as indicated in Fig. 16.1. In Section 16.4 it is shown that θ must satisfy $\theta \geq 1.37$. Usually, the value $\theta = 1.4$ is used, although it has been shown[16.15] that the optimum value of θ is 1.420815. Based on the linear assumption above, it can be shown that

$$\boxed{\begin{aligned}\Delta_s \dot{\mathbf{u}}(t_i) &= s_i \ddot{\mathbf{u}}(t_i) + \frac{s_i}{2} \Delta_s \ddot{\mathbf{u}}(t_i) \\ \Delta_s \mathbf{u}(t_i) &= s_i \dot{\mathbf{u}}(t_i) + \frac{s_i^2}{2} \ddot{\mathbf{u}}(t_i) + \frac{s_i^2}{6} \Delta_s \ddot{\mathbf{u}}(t_i)\end{aligned}} \qquad (16.26)$$

where

$$\begin{aligned}\Delta_s \mathbf{u}(t_i) &\equiv \mathbf{u}(t_i + s_i) - \mathbf{u}(t_i) \\ \Delta_s \dot{\mathbf{u}}(t_i) &\equiv \dot{\mathbf{u}}(t_i + s_i) - \dot{\mathbf{u}}(t_i) \\ \Delta_s \ddot{\mathbf{u}}(t_i) &\equiv \ddot{\mathbf{u}}(t_i + s_i) - \ddot{\mathbf{u}}(t_i)\end{aligned} \qquad (16.27)$$

Equations 16.26 can be solved for $\Delta_s \ddot{\mathbf{u}}(t_i)$ and $\Delta_s \dot{\mathbf{u}}(t_i)$ in terms of $\Delta_s \mathbf{u}(t_i)$, giving

$$\begin{aligned}\Delta_s \ddot{\mathbf{u}}(t_i) &= \frac{6}{s_i^2} \Delta_s \mathbf{u}(t_i) - \frac{6}{s_i} \dot{\mathbf{u}}(t_i) - 3\ddot{\mathbf{u}}(t_i) \\ \Delta_s \dot{\mathbf{u}}(t_i) &= \frac{3}{s_i} \Delta_s \mathbf{u}(t_i) - 3\dot{\mathbf{u}}(t_i) - \frac{s_i}{2} \ddot{\mathbf{u}}(t_i)\end{aligned} \qquad (16.28)$$

These are substituted into the incremental equation of equilibrium

$$\mathbf{M}\,\Delta_s \ddot{\mathbf{u}}(t_i) + \mathbf{C}(t_i)\,\Delta_s \dot{\mathbf{u}}(t_i) + \mathbf{K}(t_i)\,\Delta_s \mathbf{u}(t_i) = \widehat{\mathbf{p}}(t_i + s_i) - \mathbf{p}(t_i) \qquad (16.29)$$

where a "projected load" $\widehat{\mathbf{p}}(t_i + s_i)$ given by

$$\widehat{\mathbf{p}}(t_i + s_i) = \mathbf{p}(t_i) + \theta[\mathbf{p}(t_i + \Delta t_i) - \mathbf{p}(t_i)] \qquad (16.30)$$

is needed since it is assumed that the loads are given only at t_i, $t_i + \Delta t_i$, and so on. Equations 16.28 through 16.30 can be combined to give

$$\boxed{\mathbf{K}^*(t_i) \, \Delta_s \mathbf{u}(t_i) = \mathbf{p}^*(t_i)} \tag{16.31}$$

where

$$\mathbf{K}^*(t_i) = \mathbf{K}(t_i) + \frac{3}{s_i}\mathbf{C}(t_i) + \frac{6}{s_i^2}\mathbf{M} \tag{16.32a}$$

$$\mathbf{p}^*(t_i) = \theta[\mathbf{p}(t_i + \Delta t_i) - \mathbf{p}(t_i)] + \mathbf{M}\left[\frac{6}{s_i}\dot{\mathbf{u}}(t_i) + 3\ddot{\mathbf{u}}(t_i)\right] + \mathbf{C}(t_i)\left[3\dot{\mathbf{u}}(t_i) + \frac{s_i}{2}\ddot{\mathbf{u}}(t_i)\right] \tag{16.32b}$$

Equation 16.31 is then solved for $\Delta_s \mathbf{u}(t_i)$. Once this has been obtained, Eq. 16.28a gives the acceleration increment $\Delta_s \ddot{\mathbf{u}}(t_i)$. From Fig. 16.1 it can be seen that

$$\Delta \ddot{\mathbf{u}}(t_i) = \frac{1}{\theta} \Delta_s \ddot{\mathbf{u}}(t_i) \tag{16.33}$$

With this value of $\Delta \ddot{\mathbf{u}}(t_i)$, equations similar to Eq. 16.28 give

$$\dot{\mathbf{u}}(t_i + \Delta t_i) = \dot{\mathbf{u}}(t_i) + \Delta t_i \ddot{\mathbf{u}}(t_i) + \frac{\Delta t_i}{2}\Delta \ddot{\mathbf{u}}(t_i) \tag{16.34a}$$

$$\mathbf{u}(t_i + \Delta t_i) = \mathbf{u}(t_i) + \Delta t_i \dot{\mathbf{u}}(t_i) + \frac{\Delta t_i^2}{2}\ddot{\mathbf{u}}(t_i) + \frac{\Delta t_i^2}{6}\Delta \ddot{\mathbf{u}}(t_i) \tag{16.34b}$$

To ensure dynamic equilibrium at $t_i + \Delta t_i$, the acceleration $\ddot{\mathbf{u}}(t_i + \Delta t_i)$ is obtained from the equilibrium equation

$$\ddot{\mathbf{u}}(t_i + \Delta t_i) = \mathbf{M}^{-1}[\mathbf{p}(t_i + \Delta t_i) - \mathbf{f}_D(t_i + \Delta t_i) - \mathbf{f}_S(t_i + \Delta t_i)] \tag{16.34c}$$

rather than directly from the $\Delta \ddot{\mathbf{u}}(t_i)$ of Eq. 16.33. Equations 16.34a and b give the initial conditions for the next time step.

16.4 SINGLE-STEP METHODS AND SPECTRAL STABILITY

In Section 16.3 the Central Difference Method, the Newmark-β Method, and the Wilson-θ Method were introduced. These are examples of a large group of methods available for carrying out step-by-step numerical integration of linear or nonlinear equations of motion to obtain transient response of structures. These step-by-step methods are examples of *Multivalued Single-Step Methods*. Two very important properties of these numerical integrators, stability and accuracy, are discussed next. We use the definitions, analysis, and methods introduced in Section 16.2. We first consider the use of numerical integration to solve linear SDOF problems and then discuss the implications of using these methods for direct integration of the equations of motion of MDOF systems. In this section we can now apply our knowledge regarding accuracy, consistency, convergence, and stability of numerical integration methods to assess the relative merits of these different algorithms.

16.4.1 Operator Formulation of Single-Step Integration Algorithms

For the following discussions of stability and accuracy it will be useful to formulate the step-by-step integration algorithms in operator form.[16.15] We will use the *Average Acceleration Method* of Section 6.2 as an example. The three relevant equations are the equation of motion at time $t_i + \Delta t_i$

$$m\ddot{u}_{i+1} + c\dot{u}_{i+1} + ku_{i+1} = p_{i+1} \tag{16.35}$$

and the kinematical approximations from Eqs. 6.10 and 6.11:

$$\dot{u}_{i+1} = \dot{u}_i + \frac{\Delta t_i}{2}(\ddot{u}_i + \ddot{u}_{i+1}) \tag{16.36}$$

$$u_{i+1} = u_i + \Delta t_i \dot{u}_i + \frac{\Delta t_i^2}{4}(\ddot{u}_i + \ddot{u}_{i+1}) \tag{16.37}$$

These may be written conveniently in matrix form:

$$\begin{bmatrix} k & c & m \\ 0 & 1 & \dfrac{-\Delta t_i}{2} \\ 1 & 0 & \dfrac{-\Delta t_i^2}{4} \end{bmatrix} \begin{Bmatrix} u_{i+1} \\ \dot{u}_{i+1} \\ \ddot{u}_{i+1} \end{Bmatrix} = \begin{bmatrix} 0 & 0 & 0 \\ 0 & 1 & \dfrac{\Delta t_i}{2} \\ 1 & \Delta t_i & \dfrac{\Delta t_i^2}{4} \end{bmatrix} \begin{Bmatrix} u_i \\ \dot{u}_i \\ \ddot{u}_i \end{Bmatrix} + \begin{Bmatrix} p_{i+1} \\ 0 \\ 0 \end{Bmatrix} \tag{16.38}$$

or

$$\mathbf{K}_1 \mathbf{u}_{i+1} = \mathbf{K}_0 \mathbf{u}_i + \mathbf{p}_{i+1} \tag{16.39}$$

Equation 16.38 can be solved by explicitly inverting $\mathbf{K}_1$, obtaining

$$\mathbf{L} \equiv \mathbf{K}_1^{-1} = \frac{-1}{m + c\,\Delta t_i/2 + k\,\Delta t_i^2/4} \begin{bmatrix} \dfrac{-\Delta t_i^2}{4} & \dfrac{c\,\Delta t_i}{4} & -\left(m + \dfrac{c\,\Delta t_i}{2}\right) \\ \dfrac{-\Delta t_i}{4} & -\left(m + \dfrac{k\,\Delta t_i^2}{4}\right) & \dfrac{k\,\Delta t_i}{2} \\ -1 & c & -k \end{bmatrix} \tag{16.40}$$

which is also called the *load operator*. Then the *recursion relation*

$$\boxed{\mathbf{u}_{i+1} = \mathbf{A}\mathbf{u}_i + \mathbf{L}\mathbf{p}_{i+1}} \tag{16.41}$$

can be formed, where

$$\mathbf{A} \equiv \mathbf{K}_1^{-1}\mathbf{K}_0 = \frac{1}{1 + \eta/2 + \xi/4} \begin{bmatrix} 1 + \dfrac{\eta}{2} & \Delta t_i\left(1 + \dfrac{\eta}{4}\right) & \dfrac{\Delta t_i^2}{4} \\ \dfrac{-\xi}{2\Delta t_i} & 1 - \dfrac{\xi}{4} & \dfrac{\Delta t_i}{2} \\ \dfrac{-\xi}{\Delta t_i^2} & \dfrac{-(\xi + \eta)}{\Delta t_i} & -\left(\dfrac{\eta}{2} + \dfrac{\xi}{4}\right) \end{bmatrix} \tag{16.42}$$

is the *amplification matrix*, and

$$\mathbf{F}_{i+1} \equiv \mathbf{L}p_{i+1} = \left(\frac{1}{1+\eta/2+\xi/4}\right) \begin{Bmatrix} \frac{\Delta t_i^2}{4} \\ \frac{\Delta t_i}{2} \\ 1 \end{Bmatrix} \frac{p_{i+1}}{m} \quad (16.43)$$

is the *load vector*, where

$$\xi = \frac{k\,\Delta t_i^2}{m} = \omega_n^2\,\Delta t_i^2$$
$$\eta = \frac{c\,\Delta t_i}{m} = 2\zeta\omega_n\,\Delta t_i \quad (16.44)$$

Example 16.3 Use the *Average Acceleration Method* to obtain the free vibration response of an undamped SDOF system with

$$k = m = u(0) = 1, \quad \dot{u}(0) = 0$$

Choose a reasonable time step and perform the integration from $0 \le t \le 2.0$ s.

SOLUTION

$$\omega_n = \left(\frac{k}{m}\right)^{1/2} = 1 \quad (1)$$

$$T_n = \frac{2\pi}{\omega_n} = 2\pi \quad \text{seconds} \quad (2)$$

use $\Delta t = 0.2$ s = constant, which will be shorter than the rule-of-thumb value of $T_n/10$.

$$\xi = \omega_n^2\,\Delta t^2 = 1(0.2)^2 = 0.04 \quad (3)$$
$$\eta = 2\zeta\omega_n\,\Delta t = 0 \quad (4)$$

Since this is free vibration, the recursion relation of Eq. 16.41 becomes

$$\mathbf{u}_{i+1} = \mathbf{A}\mathbf{u}_i \quad (5)$$

where $\mathbf{A}$ is given by Eq. 16.42.

$$\mathbf{A} = \frac{1}{1.01}\begin{bmatrix} 1.00 & 0.20 & 0.01 \\ -0.10 & 0.99 & 0.10 \\ -1.00 & -0.20 & -0.01 \end{bmatrix} \quad (6)$$

The equation of motion is

$$\ddot{u} + u = 0 \quad (7)$$

so the "initial condition" vector is

$$\mathbf{u}_0 \equiv \begin{Bmatrix} u(0) \\ \dot{u}(0) \\ \ddot{u}(0) \end{Bmatrix} = \begin{Bmatrix} 1 \\ 0 \\ -1 \end{Bmatrix} \tag{8}$$

Then, from Eq. 5,

$$\mathbf{u}(0.2) \equiv \mathbf{u}_1 = \mathbf{A}\mathbf{u}_0 = \frac{1}{1.01} \begin{Bmatrix} 0.99 \\ -0.20 \\ -0.99 \end{Bmatrix} = \begin{Bmatrix} 0.98020 \\ -0.19802 \\ -0.98020 \end{Bmatrix} \tag{9}$$

Continuing, we obtain

$$\mathbf{u}(0.4) \equiv \mathbf{u}_2 = \mathbf{A}\mathbf{u}_1 = \frac{1}{1.01} \begin{Bmatrix} 0.93079 \\ -0.39208 \\ -0.93079 \end{Bmatrix} = \begin{Bmatrix} 0.92158 \\ -0.38820 \\ 0.92158 \end{Bmatrix} \tag{10}$$

Table 1 summarizes the computations and compares the step-by-step solution with the exact solution, $u(t) = \cos t$.

Table 1 Average Acceleration Solution for Free Vibration

i	t_i	$\ddot{u}_i$	$\dot{u}_i$	u_i	$\cos t_i$
0	0.0	−1.00000	0.00000	1.00000	1.00000
1	0.2	−0.98020	−0.19802	0.98020	0.98007
2	0.4	−0.92158	−0.38820	0.92158	0.92106
3	0.6	−0.82646	−0.56300	0.82646	0.82534
4	0.8	−0.69861	−0.71551	0.69861	0.69671
5	1.0	−0.54309	−0.83968	0.54309	0.54030
6	1.2	−0.36606	−0.93059	0.36606	0.36236
7	1.4	−0.17454	−0.98465	0.17454	0.16997
8	1.6	0.02390	−0.99971	−0.02390	−0.02920
9	1.8	0.22139	−0.97519	−0.22139	−0.22720
10	2.0	0.41011	−0.91204	−0.41011	−0.41615

16.4.2 Stability Analysis

Let us summarize our overall procedure for analyzing the stability of the Average Acceleration Method:

1. We have introduced a test equation

$$\mathbf{M}\ddot{\mathbf{u}}(t) + \mathbf{C}\dot{\mathbf{u}}(t) + \mathbf{K}\mathbf{u}(t) = \mathbf{p}(t)$$

to assess the stability of the method. This step is analogous to our process for defining A-stability in Section 16.2.

2. We have introduced the finite-difference approximations that characterize the Average Acceleration Method.
3. We have organized the recursion as a single-step method having the form

$$\mathbf{u}_{i+1} = \mathbf{A}\mathbf{u}_i + \mathbf{L}\mathbf{p}_i$$

4. An (asymptotically) stable solution of the test problem in step 1 satisfies

$$|u(t)| \to 0 \qquad \text{as } t \to \infty$$

A numerically stable solution in step 3 of the approximation method above is required to show qualitative behavior similar to that of the true solution. We require that

$$\|\mathbf{u}_i\|_{\mathbb{R}^3} \to 0 \qquad \text{as } i \to \infty$$

This condition is identical to the requirement in the definition of A-stability discussed in Section 16.2.

As observed in Example 16.3 for free vibration, the recursion relation is

$$\mathbf{u}_{i+1} = \mathbf{A}\mathbf{u}_i \qquad (16.45)$$

so

$$\mathbf{u}_1 = \mathbf{A}\mathbf{u}_0, \qquad \mathbf{u}_2 = \mathbf{A}\mathbf{u}_1 = \mathbf{A}^2\mathbf{u}_0 \qquad (16.46)$$

and so on; that is, each integration step corresponds to raising the power of the amplification matrix by 1. The behavior of the integration operator can be characterized by its eigenvalues. Consider the eigenvalue problem

$$\mathbf{A}\boldsymbol{\phi}_r = \lambda_r \boldsymbol{\phi}_r \qquad (16.47)$$

Since $\mathbf{A}$ is a 3×3 matrix whose elements depend, in general, on ξ, η, and Δt_i (see Eq. 16.42), there will be three eigenvalues that depend on these quantities. The three eigenequations can be combined and written

$$\mathbf{A}\boldsymbol{\Phi} = \boldsymbol{\Phi}\boldsymbol{\Lambda} \qquad (16.48)$$

Then

$$\mathbf{A} = \boldsymbol{\Phi}\boldsymbol{\Lambda}\boldsymbol{\Phi}^{-1} \qquad (16.49)$$

Furthermore, further iteration cycles produce

$$\mathbf{A}^2 = \mathbf{A}\mathbf{A} = \boldsymbol{\Phi}\boldsymbol{\Lambda}^2\boldsymbol{\Phi}^{-1}$$

$$\vdots \qquad (16.50)$$

$$\mathbf{A}^s = \boldsymbol{\Phi}\boldsymbol{\Lambda}^s\boldsymbol{\Phi}^{-1}$$

From Eq. 16.50 it can be seen that if the magnitude of any of the three eigenvalues of $\mathbf{A}$ is greater than 1, there will be an amplification of the solution with each time step. This leads to the following definition of spectral stability. A Single-Step Method is *spectrally stable* if all of the eigenvalues of the iteration matrix have modulus less than, or equal to, unity. By the preceding analysis, this definition is entirely analogous to the development of the notion of *absolute stability, or A-stability*.

Integration operators of this type, which cause the solution to grow without bound regardless of the time step, are called *unconditionally (spectrally) unstable* operators. Operators that lead to bounded solutions if the time step satisfies $\Delta t_i \leq \Delta t_{cr}$ are termed *conditionally (spectrally) stable*. Operators that lead to bounded solutions regardless of the length of the time step are called *unconditionally (spectrally) stable* operators. Unconditionally stable operators are the most desirable, although conditionally stable operators may be used if the time-step restriction is observed. The average acceleration operator is unconditionally stable. The Wilson-θ Method is unconditionally stable only for $\theta \geq 1.37$. Further details on stability of operators may be found in Refs. [16.15] and [16.16].

The topic of stability of integration operators is closely related to the topic of *numerical damping* or *numerical dissipation*. The *spectral radius* of an integration operator is defined as

$$\rho = \max|\lambda_r| \qquad (16.51)$$

If $\rho > 1$, the operator is unstable, because the solution will be amplified at each time step. On the other hand, if $\rho < 1$, there will be numerical damping of the solution. Figure 16.2 shows the relationship between time step and spectral radius for several operators, including a new operator introduced in Ref. [16.17], and from this figure stability and numerical dissipation can be qualitatively inferred.

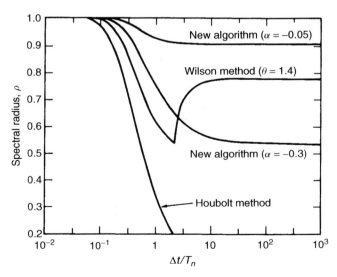

Figure 16.2 Spectral radii versus $\Delta t/T_n$. (From H. M. Hilber et al., "Improved Numerical Dissipation for Time Integration Algorithms in Structural Dynamics," *Earthquake Engineering and Structural Dynamics*, Vol. 5, 1977. Reprinted with permission from Earthquake Engineering and Structural Dynamics.)

16.4.3 Accuracy Analysis

For structural dynamics applications, the accuracy of numerical integration algorithms can be characterized by two attributes: *amplitude accuracy* and *period accuracy*. These give a qualitative understanding of the algorithm's performance that a simple order of accuracy, as discussed in Section 16.2, does not convey. Figure 16.3 shows the exact free vibration response of an undamped SDOF system and a numerical solution computed with a short time step. It can be seen that the integration operator that has been

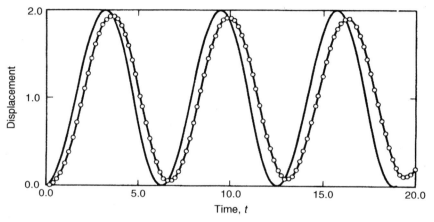

Figure 16.3 Amplitude and period error in a numerical solution. (From R. E. Nickell, "On the Stability of Approximation Operators in Problems of Structural Dynamics," *International Journal of Solids and Structures*, Vol. 7, Pergamon Press, Ltd., 1971. Reprinted with permission of Pergamon Press, Ltd.)

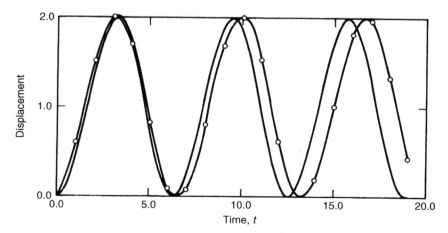

Figure 16.4 Numerical integration using the average acceleration operator (Newmark-β operator with $\beta = \frac{1}{4}$). (From R. E. Nickell, "On the Stability of Approximation Operators in Problems of Structural Dynamics," *International Journal of Solids and Structures*, Vol. 7, Pergamon Press, Ltd., 1971. Reprinted with permission of Pergamon Press, Ltd.)

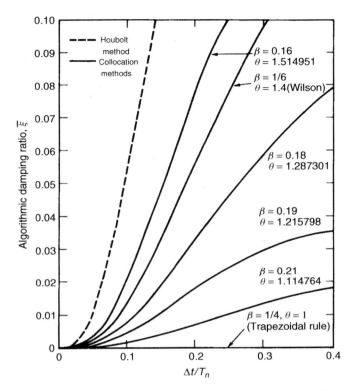

Figure 16.5 Numerical damping factors for several integration schemes. (From H. M. Hilber and T. J. R. Hughes, "Collocation, Dissipation and 'Overshoot' for Time Integration Schemes in Structural Dynamics," *Earthquake Engineering and Structural Dynamics*, Vol. 6, 1978. Reprinted with permission from Earthquake Engineering and Structural Dynamics.)

used produces an inaccurate solution, even for a short time step. The amplitude decay results from *numerical dissipation*, and there is also *period elongation*. Figure 16.4 shows a corresponding response calculation based on the Average Acceleration Algorithm. In this case there is no numerical dissipation, but there is elongation of the period of the oscillation. The period error can be reduced by decreasing the time step ratio $\Delta t/T_n$. Figures 16.5 and 16.6 present numerical damping factor and relative period error versus the time-step ratio $\Delta t/T_n$ for several integration operators. References [16.6], [16.7], [16.15], and [16.17] may be consulted for further details.

16.4.4 Direct Integration for MDOF Systems

From the preceding figures it is clear that the time step plays a key role in determining the stability and accuracy of numerical solution for the response of SDOF systems. Let us now consider the use of one of these numerical integration methods for direct integration of the equations of motion of an MDOF system. Although the equations being solved are not uncoupled equations, it is convenient to think of the response as being a superposition of the responses of the system in each of its N modes. The

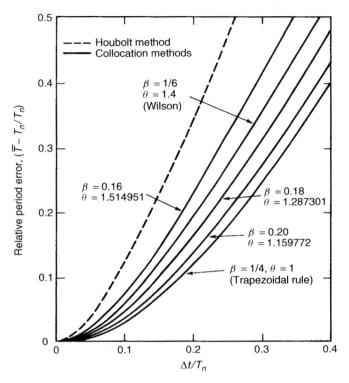

Figure 16.6 Relative period error for several integration schemes. (From H. M. Hilber and T. J. R. Hughes, "Collocation, Dissipation and 'Overshoot' for Time Integration Schemes in Structural Dynamics," *Earthquake Engineering and Structural Dynamics*, Vol. 6, 1978. Reprinted with permission from Earthquake Engineering and Structural Dynamics.)

questions to resolve are: "What numerical integration algorithm should be used?" and "What time step should be used?" These questions are obviously related.

The first consideration in selecting a method is its stability. In most cases it is desirable to use a method that is unconditionally stable.[1] This leaves a very wide selection, including the Average Acceleration Method, the Wilson-θ Method, and others. Next, it can be observed from Fig. 16.5 that the Average Acceleration Method produces no amplitude error (i.e., no numerical dissipation) regardless of the time step, and from Fig. 16.6 it can be seen to have the lowest period error of the methods represented on the figure. These characteristics make the Average Acceleration Method an excellent choice for integration of a SDOF system. However, for an MDOF system the requirements are more subtle.

Recall from the discussion following Table 16.1 that in most finite-DOF models (e.g., in finite element models), the higher frequencies and mode shapes are inaccurate, indicating poor mathematical modeling of the higher modes. Hence, in direct integration

---

[1] Explicit methods, which are not unconditionally stable, are computationally simpler than any unconditionally stable methods and are preferred when the time step is already limited by other factors, as in nonlinear analysis.

it becomes desirable to use some numerical dissipation to filter out the response of these higher modes, producing a result that is similar to that achieved by truncating the number of modes retained in a mode-superposition solution. Hence, there arises the goal of formulating a numerical integration algorithm that, in some sense, has optimal numerical dissipation properties. The Wilson-θ Method does provide for the damping of higher modes, for example, modes for which $\Delta t / T_r \geq 1.0$, but other desirable methods have also been proposed.[16.16,16.17]

The considerations of stability and accuracy discussed above apply strictly only to linear systems. The reader is urged to consult additional references on nonlinear dynamic analysis.

16.5 NUMERICAL CASE STUDY

Again, let us consider the planar frame structure in Fig. 15.2 in the numerical case study of Section 15.6. Recall the geometry of the symmetric planar frame structure. At each node there are three DOFs: two translations and one rotation. The frame translational DOFs (but not rotations) are restrained at both ends, that is, at nodes 1, 2, 25, and 26.

Use this frame structure as the basis of a case study of the numerical integration algorithms described in this chapter. Compute the transient response of the planar frame structure subject to a vertical force located at node 13. The amplitude of the vertical force is given by

$$P(t) = \begin{cases} \sin \Omega t & \text{for } t < \dfrac{t_f - t_0}{20} \\ 0 & \text{for } t > \dfrac{t_f - t_0}{20} \end{cases} \quad (16.52)$$

The initial time is $t_0 = 0$, and the final time is $t_f = 0.01$. The forcing frequency Ω is given by

$$\Omega = \frac{2\pi}{(t_f - t_0)/20}$$

Calculate the transient response of node 13 in the y direction as a function of time for the interval $[t_0, t_f]$ seconds.

Solution The computations for this case study were performed using the following MATLAB .m-files:

- frame3-int-1st.m: integrates the equations of motion in first-order form, using the full system matrices
- frame3-int-2nd-cen.m: integrates the equations of motion in second-order form, using the Central Difference Method
- frame3-int-2nd-new.m: integrates the equations of motion in second-order form, using the Newmark-β Method

The associated .m-files described above are included on the book's website. The results are presented in Figs. 16.7 and 16.8 and are discussed below.

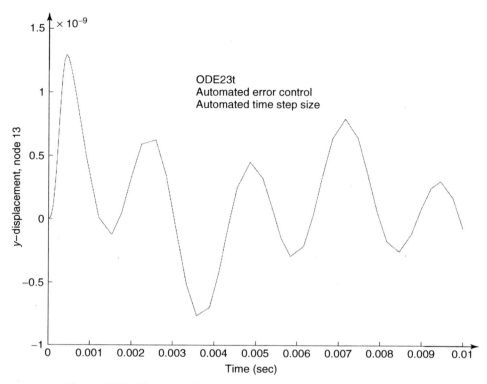

Figure 16.7 First-order integration, trapezoidal rule, MATLAB *ode23t*.

1. Figure 16.7 depicts the predicted response of node 13, in the *y* direction of the planar frame when the MATLAB built-in function *ode23t* is used in the numerical integration of the governing equations in first-order form. In this example, the MATLAB built-in function *ode23t* uses automatic rules for time-step adaptation. The theoretical details of this methodology are beyond the scope of this chapter. However, this topic has been studied extensively and reported in the numerical analysis literature. It is often available in commercial software packages for numerical analysis. For the user it can be a great asset in that there is less need to carefully assess the *spectral content* of the response. The MATLAB .m-files that have been used to generate this result include:

   - frame3-int-1st.m
   - firstODE.m

   These .m-files are included on the book's website.

2. Figure 16.8 presents a comparison of transient response results for the same system for three different numerical integration methods. The integration of the equations of motion in first-order form has been carried out in the .m-file frame-int-1st.m using the MATLAB built-in function *ode23t*. In addition, the integration of the equations of motion in second-order form has been carried out using the

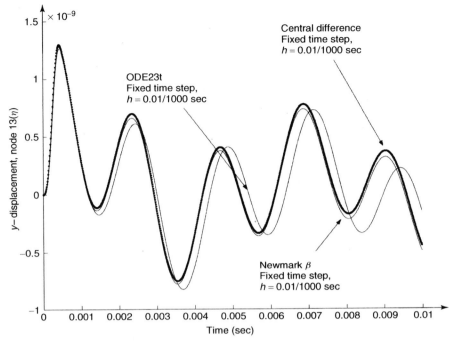

Figure 16.8 Comparison of first-order *ode23t*, central difference, and Newmark-β methods.

Central Difference Method and the Newmark-β Method ($\gamma = \frac{1}{2}$, $\beta = \frac{1}{4}$). These results have been generated using the following MATLAB .m-files:

- frame3-int-1st.m
- firstODE.m
- frame3-int-2nd-cen.m
- frame3-int-2nd-new.m

These .m-files are included on the book's website. Each method exhibits second-order truncation, and the numerical results for all of these methods are nearly identical. A uniform mesh with $h = (t_f - t_0)/1000$ has been used in each case.

REFERENCES

[16.1] W. C. Hurty and M. R. Rubinstein, *Dynamics of Structures*, Prentice-Hall, Englewood Cliffs, NJ, 1964.

[16.2] R. W. Clough and S. Mojtahedi, "Earthquake Response Analysis Considering Nonproportional Damping," *Earthquake Engineering and Structural Dynamics*, Vol. 4, 1976, pp. 489–496.

[16.3] U. M. Ascher and L. R. Petzhold, *Computer Methods for Ordinary Differential Equations and Differential-Algebraic Equations*, Society for Industrial and Applied Mathematics, Philadelphia, 1998.

[16.4] B. F. Plybon, *Applied Numerical Analysis*, PWS-Kent, Boston, 1992.

[16.5] A. Quarteroni, R. Sacco, and F. Saleri, *Numerical Mathematics*, Springer-Verlag, New York, 2000.

[16.6] K.-J. Bathé and E. L. Wilson, *Numerical Methods in Finite Element Analysis*, Prentice-Hall, Englewood Cliffs, NJ, 1976.

[16.7] T. J. R. Hughes, *The Finite Element Method: Linear Static and Dynamic Finite Element Analysis*, Dover, Mineola, NY, 2000.

[16.8] M. A. Baddourah, O. O. Storaasli, and S. Bostic, "Linear Static Structural Analysis and Vibration Analysis on High-Performance Computers," *International Journal of Computing Systems in Engineering*, Vol. 4, Nos. 4–6, December 1993, pp. 41–49.

[16.9] G. H. Golub and C. F. Van Loan, *Matrix Computations*, Johns Hopkins University Press, Baltimore, MD, 1996.

[16.10] N. M. Newmark, "A Method of Computation for Structural Dynamics," *Journal of the Engineering Mechanics Division, ASCE*, Vol. 85, 1959, pp. 67–94.

[16.11] A. K. Chopra, *Dynamics of Structures: Theory and Applications to Earthquake Engineering*, 2nd ed., Prentice Hall, Upper Saddle River, NJ, 2000.

[16.12] T. Belytschko, J. R. Osias, and P. V. Marcal, eds., *Finite Element Analysis of Transient Nonlinear Structural Behavior*, AMD Vol. 14, ASME, New York, 1975.

[16.13] K.-J. Bathé, E. Ramm, and E. L. Wilson, "Finite Element Formulation for Large Deformation Dynamic Analysis," *International Journal for Numerical Methods in Engineering*, Vol. 9, 1975, pp. 353–386.

[16.14] E. L. Wilson, I. Farhoomand, and K.-J. Bathé, "Nonlinear Dynamic Analysis of Complex Structures," *Earthquake Engineering and Structural Dynamics*, Vol. 1, 1973, pp. 241–252.

[16.15] R. E. Nickell, "On the Stability of Approximation Operators in Problems of Structural Dynamics," *International Journal of Solids and Structures*, Vol. 7, 1971, pp. 301–319.

[16.16] H. M. Hilber and T. J. R. Hughes, "Collocation, Dissipation, and 'Overshoot' for Time Integration Schemes in Structural Dynamics," *Earthquake Engineering and Structural Dynamics*, Vol. 6, 1978, pp. 99–118.

[16.17] H. M. Hilber, T. J. R. Hughes, and R. L. Taylor, "Improved Numerical Dissipation for Time Integration Algorithms in Structural Dynamics," *Earthquake Engineering and Structural Dynamics*, Vol. 5, 1977, pp. 283–292.

PROBLEMS

Problem Set 16.1

16.1 (a) Determine the Rayleigh damping coefficients a_0 and a_1 in Eq. 16.2 such that the 3-DOF system in Example 16.1 has damping factors $\zeta_1 = \zeta_2 = 0.01$. (b) What is the resulting physical damping matrix **C**? (c) What is the resulting value of ζ_3?

16.2 (a) Use Eq. 16.5 to determine a damping matrix **C** that would yield $\zeta_1 = \zeta_2 = 0.01$ and $\zeta_3 = 0.00$ for the 3-DOF system in Example 16.1. (b) Can this damping matrix be physically realized by adding discrete dashpots between the masses and/or between the masses and ground? If so, how?

16.3 (a) Use Eq. 16.5 to determine a damping matrix **C** that would yield $\zeta_1 = \zeta_2 = \zeta_3 = 0.01$ for the 3-DOF system in Example 16.1. (b) Can this damping matrix be physically realized by adding discrete dashpots between the masses and/or between the masses and ground? If so, how?

Problem Set 16.2

16.4 In Section 6.2 we showed that the *Runge–Kutta Method of Order 2* was derived by matching Taylor series terms in the discrete difference approximation of the governing ordinary differential equation. During this process, we derived the identity in Eqs. 6.33 that stipulates

$$x_{n+1} = x_n + hf_n + \frac{h^2}{2!}(f_t + ff_x)_n + \frac{h^3}{3!}(f_{tt} + 2ff_{tx} + f_{xx}f^2 + f_xf_t + f_x^2 f)_n + O(h^4)$$

for a uniform grid with time step h. See Section 6.2 to review the notation above. (a) What is the appropriate algorithm operator $\mathcal{A}_N$ for the uniform grid τ_N with constant time step h for the Runge–Kutta Method of Order 2? (b) What is the local truncation error $\mathcal{R}_N$ at node i of the uniform grid τ_N for the Runge–Kutta Method of Order 2? (c) What is the order of consistency, or accuracy, for the Runge–Kutta Method of Order 2?

16.5 The general form for the *Linear Multistep Methods* was given in Eq. 6.41 by the expression

$$x_{n+1} = \sum_{i=0}^{N} a_i x_{n-i} + \sum_{i=-1}^{M} b_i \dot{x}_{n-i}$$

One method to derive a special case of these equations has been to integrate the equations of motion to obtain

$$x_{n+1} = x_n + \int_{t_n}^{t_{n+1}} f(x(s), s) \, ds$$

and subsequently make an approximation to f by replacing it with a polynomial that interpolates f at a number of grid points. In other words, we approximate $f(x(s), s)$ by $\mathcal{P}_M(s)$ that interpolates $f(x(s), s)$ at the grid points $\{f_n, f_{n-1}, \ldots, f_{n-M}\}$. (a) If f and x are smooth enough functions, state an error bound of the form

$$|f(x(s), s) - \mathcal{P}_M(s)| \leq Ch^p$$

for all $s \in [t_{n-M}, t_{n+1}]$. (b) What is the approximation operator $\mathcal{A}_N$ over the uniform grid τ_N appropriate for the formula shown in Eq. 16.5? (c) Use the approximation error bound in Eq. 16.5 to derive is the truncation error $\mathcal{R}_N$ at node i of the uniform grid τ_N. (d) What is the approximation order of the method described in parts (a) to (c)? (e) What is the approximation order of the Adams–Bashforth method, Eq. 6.49?

Problem Set 16.3

16.6 As in Section 16.3.1, choose the canonical structural dynamics problem

$$\mathbf{M\ddot{u}} + \mathbf{C\dot{u}} + \mathbf{Ku} = \mathbf{p}(t)$$

$$\mathbf{u}(0) = \mathbf{u}_0, \qquad \mathbf{\dot{u}}(0) = \mathbf{v}_0$$

expressed in second-order form as the test problem for the study of the stability of the central difference equations. The difference equations used in deriving the *Central Difference Method* include the expression for the first-order derivative

$$\mathbf{\dot{u}}_n = \frac{\mathbf{u}_{n+1} - \mathbf{u}_{n-1}}{2h}$$

and the second-order derivative

$$\mathbf{\ddot{u}}_n = \frac{\mathbf{\dot{u}}_{n+1/2} - \mathbf{\dot{u}}_{n-1/2}}{h}$$

$$= \frac{(\mathbf{u}_{n+1} - \mathbf{u}_n)/h - (\mathbf{u}_n - \mathbf{u}_{n-1})/h}{h}$$

$$= \frac{\mathbf{u}_{n+1} - 2\mathbf{u}_n + \mathbf{u}_{n-1}}{h^2}$$

(a) What is the approximation operator $\mathcal{A}_N$ appropriate for this problem on a uniform grid τ_N with constant step size h? (b) What is the local truncation error $\mathcal{R}_N(i)$ at node i associated with the approximation operator $\mathcal{A}_N$ on the grid τ_N? (c) What is the order of approximation of the algorithm $\mathcal{A}_N$? (d) Create a single-step representation of the Central Difference Method for the second-order structural dynamics testbed equation of motion. Discuss the spectral stability of the Central Difference Method and the role of the amplification matrix for this algorithm.

Problem Set 16.4

16.7 (a) In Eq. 16.40, the load operator for the *Average Acceleration Method* is developed, and in Eq. 16.42 the amplification matrix for this method is determined. Following similar procedures, develop the load operator and the amplification matrix for the *Linear Acceleration Method*, which is obtained by setting $\theta = 1$ in Fig. 16.1. (b) For what values of the time step Δt_i is the Linear Acceleration Method (spectrally) stable?

For problems whose number is preceded by a **C**, you are to write a computer program and use it to produce the plot(s) requested. *Note:* MATLAB .m-files for many of the plots in this book may be found on the book's website.

C 16.8 Consider the undamped SDOF system of Example 16.4; that is, $k = m = u(0) = 1, \dot{u}(0) = 0$.

(a) Write a computer program to use the *Average Acceleration Method* to solve for the response of this system for $0 \leq t \leq 2T_n$ using $\Delta t_i = \frac{3}{20}T_n$. (b) Construct a table comparing your results from part (a) with the exact values of u at the corresponding times. Do your results agree with Fig. 16.4?

17
Component-Mode Synthesis

In Chapter 14 you were introduced to finite element techniques for formulating MDOF models of structures for use in structural dynamics analyses. Stiffness and mass matrices were derived for finite elements, and these were assembled to form system matrices. In Section 14.6, general techniques were introduced for reducing the order of system matrices by the use of assumed modes and constraint equations, and Guyan–Irons Reduction was illustrated in Section 14.8. In the present chapter, a class of model-reduction methods known as *component-mode synthesis* (CMS), or *substructure coupling for dynamic analysis*, is introduced. Substructuring involves dividing the structure into a number of substructures, or components (e.g., Fig. 17.1), obtaining reduced-order models of the components, and then assembling a reduced-order model of the entire structure.

CMS methods have been found to be very useful in solving very large structural dynamics problems, especially where the structure consists of several natural components, for example, body and frame components of an automobile, or the space shuttle

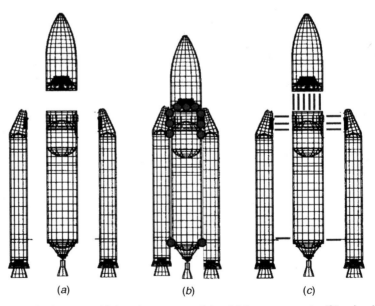

Figure 17.1 Typical space vehicle substructures: (*a*) vehicle components; (*b*) primal assembly; (*c*) dual assembly. (These figures, which are based on ESA/ESTEC models, are used with the permission of D. J. Rixen, d.j.rixen@wbmt.tudelft.nl.)

orbiter and its payloads. Recently, a *multilevel substructuring* version of component-mode synthesis has been developed for carrying out eigensolutions and frequency-response solutions based on ultralarge finite element models (e.g., finite element models having several million degrees of freedom).

In this chapter we define and illustrate many of the terms that are found in the CMS literature (fixed-interface modes, free-interface modes, constraint modes, residual flexibility, etc.); discuss general procedures for coupling substructures; and discuss several specific CMS methods, including multilevel substructuring.

Upon completion of this chapter you should be able to:

- Obtain, using a finite element model of a simple structural component, component modes of the following types: free-interface normal modes, fixed-interface normal modes, rigid-body modes, constraint modes, attachment modes, residual attachment modes, residual inertia-relief attachment modes.
- Assemble the reduced-order system matrices $\hat{K}$ and $\hat{M}$ using the generalized component coupling procedures of Section 17.5.
- Discuss the importance of including attachment modes when free-interface normal modes are used as the basis for component-mode synthesis.

17.1 INTRODUCTION TO COMPONENT-MODE SYNTHESIS

When a large, complex structural system must be analyzed for its response to dynamic excitation, some form of substructure coupling method, or component-mode synthesis (CMS) method, is usually employed. The term *component modes* is used to signify *Ritz vectors*, or *assumed modes*, that are used as *basis vectors* in describing the displacement of points within a substructure, or component. Component normal modes, or eigenvectors, are just one class of component modes. In the mid-1960s Hurty published several reports and papers on substructure coupling using fixed-interface modes (e.g., Refs. [17.1] and [17.2]). In collaboration with Hurty, Bamford created a CMS computer program that employed normal modes, rigid-body modes, constraint modes, and attachment modes.[17.3] A simplification of Hurty's method was presented by Craig and Bampton in 1968[17.4], and in the 1970s MacNeal[17.5], Rubin[17.6], Hintz[17.7], and Craig and Chang[17.8] introduced free-interface methods. A number of CMS methods are described and compared in Refs. [17.9] and [17.10] and in at least three textbooks (Refs. [17.11] to [17.13]). Damping is most often treated as modal damping imposed on the modes of the reduced-order system model. Although special CMS methods have been developed for systems with general viscous damping (e.g., Refs. [17.14] to [17.16]), these methods are not widely used, and therefore they are not discussed in this book.

Component-mode synthesis involves three basic steps: division of a structure into components, definition of sets of component modes, and coupling of the component-mode models to form a reduced-order system model. The primary uses of dynamic substructuring are (1) to couple reduced-order models of moderately complex structures (e.g., space vehicle components, as in Fig. 17.1), (2) in test verification of finite element models of components, or (3) to implement computation of the dynamics of very large finite element models (e.g., multimillion-DOF models). In this chapter we address

17.1 Introduction to Component-Mode Synthesis 533

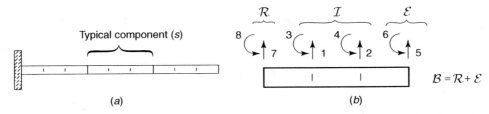

Figure 17.2 (a) Typical components and coupled system; (b) coordinate notation for typical component with redundant boundary.

primarily applications of the first type. References [17.17] to [17.19] illustrate the relationship of substructure analysis to substructure testing, and Refs. [17.20] to [17.22] are representative of the third application, multilevel substructuring.

The most general type of substructure, or component, is one that is connected to one or more adjacent components by redundant interfaces. Figure 17.2 illustrates a simple cantilever beam that is divided into three components; the middle one is a typical component with redundant interface (boundary) coordinates.

As noted in Fig. 17.2, the coordinate sets $\mathcal{I}$, $\mathcal{R}$, $\mathcal{E}$, and $\mathcal{B}$ denote interior coordinates (i.e., not shared with any adjacent component), rigid-body coordinates, excess boundary coordinates (i.e., redundant boundary coordinates), and boundary coordinates (i.e., shared with adjacent components). If a load is prescribed at an "interior" DOF, this DOF can be relabeled as a "boundary" DOF. This is done to permit static completeness of the boundary-coordinate set. The numbers of displacement coordinates in these sets are $N_i^{(s)}$, $N_r^{(s)}$, $N_e^{(s)}$, and $N_b^{(s)}$, respectively, with $N_b^{(s)} = N_r^{(s)} + N_e^{(s)}$ and $N_u^{(s)} = N_i^{(s)} + N_b^{(s)}$, where the superscript s is the label of the particular component.

The equation of motion of a typical undamped component, labeled superscript s, may be written as

$$\mathbf{M}^{(s)}\ddot{\mathbf{u}}^{(s)} + \mathbf{K}^{(s)}\mathbf{u}^{(s)} = \mathbf{f}^{(s)} + \mathbf{r}^{(s)} \tag{17.1}$$

where $\mathbf{M}^{(s)}$, $\mathbf{K}^{(s)}$, and $\mathbf{u}^{(s)}$ are the component's mass matrix, stiffness matrix, and displacement vector, respectively, in original physical coordinates. The force vector $\mathbf{f}^{(s)}$ contains the externally applied forces, and the force vector $\mathbf{r}^{(s)}$ contains the reaction forces on the component due to its connection to adjacent components at boundary degrees of freedom.

In component-mode synthesis, the component's physical displacement coordinates $\mathbf{u}$ are represented in terms of *component generalized coordinates* $\mathbf{p}$ by the Ritz coordinate transformation

$$\boxed{\mathbf{u}^{(s)} = \mathbf{\Psi}^{(s)}\mathbf{p}^{(s)}} \tag{17.2}$$

where the *component-mode matrix* $\mathbf{\Psi}^{(s)}$ is a coordinate transformation matrix of preselected *component (assumed) modes*, including modes of the following types: rigid-body modes, normal modes of free vibration (i.e., eigenvectors), constraint modes, and attachment modes. In collaboration with Hurty, Bamford defined all four of these types of modes.[17.3] Other types of assumed modes (e.g., Krylov vectors[17.23]) may also be employed as component modes.

The coordinate transformation relating component physical coordinates $\mathbf{u}^{(s)}$ to component generalized coordinates $\mathbf{p}^{(s)}$ is given by Eq. 17.2. This equation, together with the equation of motion in generalized coordinates, forms the *component modal model*. From Eqs. 17.1 and 17.2, the component equation of motion in generalized coordinates is

$$\widehat{\mathbf{M}}^{(s)}\ddot{\mathbf{p}}^{(s)} + \widehat{\mathbf{K}}^{(s)}\mathbf{p}^{(s)} = \widehat{\mathbf{f}}^{(s)} + \widehat{\mathbf{r}}^{(s)} \tag{17.3}$$

where the component mass matrix, stiffness matrix, and force vectors in component generalized coordinates are, respectively,

$$\widehat{\mathbf{M}}^{(s)} = \mathbf{\Psi}^{(s)\mathrm{T}}\mathbf{M}^{(s)}\mathbf{\Psi}^{(s)}, \quad \widehat{\mathbf{K}}^{(s)} = \mathbf{\Psi}^{(s)\mathrm{T}}\mathbf{K}^{(s)}\mathbf{\Psi}^{(s)}, \quad \widehat{\mathbf{f}}^{(s)} = \mathbf{\Psi}^{(s)\mathrm{T}}\mathbf{f}^{(s)}, \quad \widehat{\mathbf{r}}^{(s)} = \mathbf{\Psi}^{(s)\mathrm{T}}\mathbf{r}^{(s)} \tag{17.4}$$

In Sections 17.2 through 17.4, the systematic procedures used to generate FE-based component modes are described. Section 17.5 presents Lagrange-multiplier-based generalized substructure coupling procedures. This is followed, in Section 17.6, by a discussion of CMS methods that are based on fixed-interface modes, and in Section 17.7, by a discussion of CMS methods based on free-interface modes. Finally, Section 17.8 introduces briefly the concept of multilevel substructuring.

17.2 COMPONENT MODES: NORMAL, CONSTRAINT, AND RIGID-BODY MODES

The following partitioned forms of Eq. 17.1 will be useful in the derivation of component modes, first, where all boundary coordinates are kept as one set $\mathbf{u}_b$:

$$\begin{bmatrix} \mathbf{M}_{ii} & \mathbf{M}_{ib} \\ \mathbf{M}_{bi} & \mathbf{M}_{bb} \end{bmatrix} \begin{Bmatrix} \ddot{\mathbf{u}}_i \\ \ddot{\mathbf{u}}_b \end{Bmatrix} + \begin{bmatrix} \mathbf{K}_{ii} & \mathbf{K}_{ib} \\ \mathbf{K}_{bi} & \mathbf{K}_{bb} \end{bmatrix} \begin{Bmatrix} \mathbf{u}_i \\ \mathbf{u}_b \end{Bmatrix} = \begin{Bmatrix} \mathbf{0}_i \\ \mathbf{f}_b + \mathbf{r}_b \end{Bmatrix} \tag{17.5}$$

and second, where the boundary coordinates are separated into rigid-body coordinates $\mathbf{u}_r$ and excess boundary coordinates $\mathbf{u}_e$:

$$\begin{bmatrix} \mathbf{M}_{ii} & \mathbf{M}_{ie} & \mathbf{M}_{ir} \\ \mathbf{M}_{ei} & \mathbf{M}_{ee} & \mathbf{M}_{er} \\ \mathbf{M}_{ri} & \mathbf{M}_{re} & \mathbf{M}_{rr} \end{bmatrix} \begin{Bmatrix} \ddot{\mathbf{u}}_i \\ \ddot{\mathbf{u}}_e \\ \ddot{\mathbf{u}}_r \end{Bmatrix} + \begin{bmatrix} \mathbf{K}_{ii} & \mathbf{K}_{ie} & \mathbf{K}_{ir} \\ \mathbf{K}_{ei} & \mathbf{K}_{ee} & \mathbf{K}_{er} \\ \mathbf{K}_{ri} & \mathbf{K}_{re} & \mathbf{K}_{rr} \end{bmatrix} \begin{Bmatrix} \mathbf{u}_i \\ \mathbf{u}_e \\ \mathbf{u}_r \end{Bmatrix} = \begin{Bmatrix} \mathbf{0}_i \\ \mathbf{f}_e + \mathbf{r}_e \\ \mathbf{f}_r + \mathbf{r}_r \end{Bmatrix} \tag{17.6}$$

Interior forces $\mathbf{f}_i$ are omitted here, since, as noted previously, DOFs with external loads are assumed to be labeled as boundary DOFs. The coupling force $\mathbf{r}$ is discussed in Section 17.5, where coupling procedures are discussed.

The superscript s, which was used in Section 17.1 to designate a typical component, is omitted from component matrices and vectors in this section and in Sections 17.3 and 17.4, but will return later in the discussion of coupling of substructures in Sections 17.5 through 17.7.

17.2.1 Normal Modes

Component normal modes are eigenvectors and may be classified according to the interface boundary conditions specified for the component: fixed-interface normal modes, free-interface normal modes, hybrid-interface normal modes, or loaded-interface normal modes.

17.2 Component Modes: Normal, Constraint, and Rigid-Body Modes

Component *fixed-interface normal modes* are obtained by restraining all boundary DOFs and solving the following eigenproblem:

$$\left[\mathbf{K}_{ii} - \omega_j^2 \mathbf{M}_{ii}\right] \{\boldsymbol{\phi}_i\}_j = \mathbf{0}, \qquad j = 1, 2, \ldots, N_i \tag{17.7}$$

The complete set of N_i fixed-interface (flexible) normal modes from Eq. 17.7 is labeled $\boldsymbol{\Phi}_{ii}$ and assembled, according to the partitioning of Eq. 17.5, as columns of the modal matrix

$$\boxed{\underset{N_u \times N_i}{\boldsymbol{\Phi}_i} \equiv \begin{bmatrix} \boldsymbol{\Phi}_{ii} \\ \mathbf{0}_{bi} \end{bmatrix}} \tag{17.8}$$

When normalized with respect to the mass matrix $\mathbf{M}_{ii}$, the N_i fixed-interface modes satisfy

$$\boldsymbol{\Phi}_{ii}^T \mathbf{M}_{ii} \boldsymbol{\Phi}_{ii} = \mathbf{I}_{ii}, \qquad \boldsymbol{\Phi}_{ii}^T \mathbf{K}_{ii} \boldsymbol{\Phi}_{ii} = \boldsymbol{\Lambda}_{ii} = \mathrm{diag}(\omega_j^2) \tag{17.9}$$

Figure 17.3 shows the four (i.e., N_i) fixed-interface normal modes for the 8-DOF free–free beam component in Fig. 17.2*b*.

A second type of component normal mode used in CMS is the set of *free-interface normal modes*. These normal modes are defined by the equation

$$\left[\mathbf{K} - \omega_j^2 \mathbf{M}\right] \{\boldsymbol{\phi}\}_j = \mathbf{0}, \qquad j = 1, 2, \ldots (N_f = N_u - N_r) \tag{17.10}$$

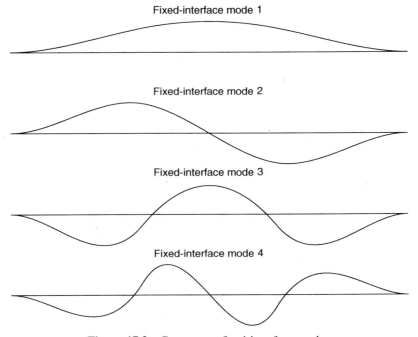

Figure 17.3 Component fixed-interface modes.

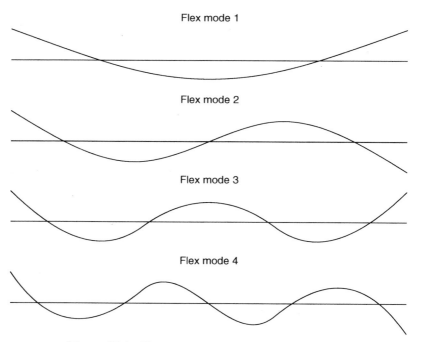

Figure 17.4 Flex modes for the free–free component.

The assembled set of N_f flexible (i.e., non-rigid-body) free-interface normal modes is

$$\underset{N_u \times N_f}{\boldsymbol{\Phi}_f} \equiv \begin{bmatrix} \boldsymbol{\Phi}_{if} \\ \boldsymbol{\Phi}_{bf} \end{bmatrix} \qquad (17.11)$$

Figure 17.4 shows the first four of the six free–free flex modes of the 8-DOF component in Fig. 17.2b. Note that for this symmetric component, these modes are called the *first symmetric mode (mode 1)*, the *first antisymmetric mode (mode 2)*, the *second symmetric mode (mode 3)*, and so on.

A third important type of component normal modes is *loaded-interface normal modes*. This includes lumped-mass loaded-interface normal modes, commonly referred to as *mass-additive normal modes*, which have been employed in modal testing of substructures.[17.17,17.18] Benfield and Hruda described CMS methods based on "consistent" mass-additive and stiffness-additive normal modes[17.24], but these methods require reduced-order models of all adjacent substructures, so they are generally of limited practical value.

17.2.2 Constraint Modes and Rigid-Body Modes

A *constraint mode* is defined as the static deformation of a structure when a unit displacement is applied to one coordinate of a specified set of "constraint" coordinates, C, while the remaining coordinates of that set are restrained, and the remaining degrees of freedom of the structure are force-free.

17.2 Component Modes: Normal, Constraint, and Rigid-Body Modes

The set of *interface constraint modes* based on unit displacement of the boundary coordinates $\mathbf{u}_b$ is a very useful CMS set, because of the ease of enforcing intercomponent compatibility when these constraint modes are employed, as explained in Section 17.6. This set, with $\mathcal{C} = \mathcal{B}$, is given by

$$\begin{bmatrix} \mathbf{K}_{ii} & \mathbf{K}_{ib} \\ \mathbf{K}_{bi} & \mathbf{K}_{bb} \end{bmatrix} \begin{bmatrix} \mathbf{\Psi}_{ib} \\ \mathbf{I}_{bb} \end{bmatrix} = \begin{bmatrix} \mathbf{0}_{ib} \\ \mathbf{R}_{bb} \end{bmatrix} \quad (17.12)$$

That is, the *interface constraint-mode matrix* $\mathbf{\Psi}_c$ is given by

$$\boxed{\underset{N_u \times N_b}{\mathbf{\Psi}_c} \equiv \begin{bmatrix} \mathbf{\Psi}_{ib} \\ \mathbf{I}_{bb} \end{bmatrix} = \begin{bmatrix} -[\mathbf{K}_{ii}^{-1} \mathbf{K}_{ib}] \\ \mathbf{I}_{bb} \end{bmatrix}} \quad (17.13)$$

From Eqs. 17.8 and 17.12 it can easily be shown that these constraint modes are stiffness-orthogonal to all of the fixed-interface normal modes; that is,

$$\mathbf{\Phi}_i^T \mathbf{K} \mathbf{\Psi}_c = \mathbf{0} \quad (17.14)$$

Figure 17.5 shows the N_b (four) constraint modes for the 8-DOF free–free beam component in Fig. 17.2b.

Rigid-body modes can be defined relative to any set of N_r coordinates that is just sufficient to restrain rigid-body motion of the component. A *rigid-body mode* is an undeformed configuration of a component obtained by setting one of a designated set $\mathcal{R}$ of displacements equal to unity, with all of the rest of this set equal to zero, and with

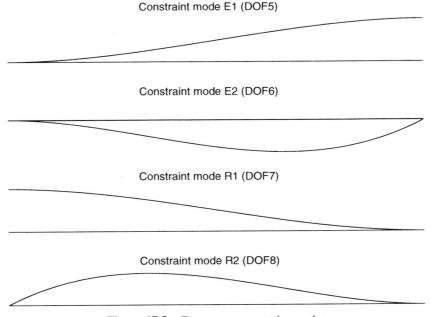

Figure 17.5 Component constraint modes.

all forces on the component equal to zero. (This requires that the stiffness matrix **K** be singular.) Although they are often considered to be zero-frequency normal modes, rigid-body modes are also a special case of (static) constraint modes.

For the purpose of substructure coupling, <u>*rigid-body modes*</u> will be defined relative to a set $\mathcal{R}$ of boundary coordinates. Then

$$\begin{bmatrix} \mathbf{K}_{ii} & \mathbf{K}_{ie} & \mathbf{K}_{ir} \\ \mathbf{K}_{ei} & \mathbf{K}_{ee} & \mathbf{K}_{er} \\ \mathbf{K}_{ri} & \mathbf{K}_{re} & \mathbf{K}_{rr} \end{bmatrix} \begin{bmatrix} \mathbf{\Psi}_{ir} \\ \mathbf{\Psi}_{er} \\ \mathbf{I}_{rr} \end{bmatrix} = \begin{bmatrix} \mathbf{0}_{ir} \\ \mathbf{0}_{er} \\ \mathbf{0}_{rr} \end{bmatrix} \qquad (17.15)$$

so the set of rigid-body modes is obtained by solving the top two row partitions of Eq. 17.15, giving

$$\mathbf{\Psi}_r \underset{N_u \times N_r}{\equiv} \begin{bmatrix} \mathbf{\Psi}_{ir} \\ \mathbf{\Psi}_{er} \\ \mathbf{I}_{rr} \end{bmatrix} = \begin{bmatrix} -\begin{bmatrix} \mathbf{G}_{ii} & \mathbf{G}_{ie} \\ \mathbf{G}_{ei} & \mathbf{G}_{ee} \end{bmatrix} \begin{bmatrix} \mathbf{K}_{ir} \\ \mathbf{K}_{er} \end{bmatrix} \\ \mathbf{I}_{rr} \end{bmatrix} \qquad (17.16)$$

where

$$\mathbf{G}_c \equiv \begin{bmatrix} \mathbf{G}_{ii} & \mathbf{G}_{ie} \\ \mathbf{G}_{ei} & \mathbf{G}_{ee} \end{bmatrix} = \begin{bmatrix} \mathbf{K}_{ii} & \mathbf{K}_{ie} \\ \mathbf{K}_{ei} & \mathbf{K}_{ee} \end{bmatrix}^{-1} \qquad (17.17)$$

is the *cantilever flexibility matrix* for the component restrained at the $\mathcal{R}$ coordinates, but nowhere else.

Redundant-interface constraint modes can then be defined for unit displacements at the redundant (excess) boundary coordinate set $\mathcal{E}$, and with the $\mathcal{R}$ coordinates fixed, by the equation

$$\begin{bmatrix} \mathbf{K}_{ii} & \mathbf{K}_{ie} & \mathbf{K}_{ir} \\ \mathbf{K}_{ei} & \mathbf{K}_{ee} & \mathbf{K}_{er} \\ \mathbf{K}_{ri} & \mathbf{K}_{re} & \mathbf{K}_{rr} \end{bmatrix} \begin{bmatrix} \mathbf{\Psi}_{ie} \\ \mathbf{I}_{ee} \\ \mathbf{0}_{re} \end{bmatrix} = \begin{bmatrix} \mathbf{0}_{ie} \\ \mathbf{R}_{ee} \\ \mathbf{R}_{re} \end{bmatrix} \qquad (17.18)$$

Therefore, the set of *redundant-interface constraint modes* is given by

$$\mathbf{\Psi}_e \underset{N_u \times N_e}{\equiv} \begin{bmatrix} \mathbf{\Psi}_{ie} \\ \mathbf{I}_{ee} \\ \mathbf{0}_{re} \end{bmatrix} = \begin{bmatrix} -[\mathbf{K}_{ii}^{-1}\mathbf{K}_{ie}] \\ \mathbf{I}_{ee} \\ \mathbf{0}_{re} \end{bmatrix} \qquad (17.19)$$

Either the set of interface constraint modes $\mathbf{\Psi}_c$ defined by Eq. 17.13, or the combined set $[\mathbf{\Psi}_r\ \mathbf{\Psi}_e]$ defined by Eqs. 17.16 and 17.19, spans the static response of the substructure to interface loading and allows for arbitrary interface displacements $\mathbf{u}_b$. Along with the prescribed interface displacement, there is accompanying displacement of the interior of the substructure, as determined by Eqs. 17.13, 17.16, and 17.19. Additional interior flexibility can be incorporated by including fixed-interface normal modes, fixed-interface Krylov vectors, or other fixed-interface assumed modes in the component mode matrix $\mathbf{\Psi}$, as illustrated in Section 17.6.[17.2,17.4,17.23]

17.3 COMPONENT MODES: ATTACHMENT AND INERTIA-RELIEF ATTACHMENT MODES

Whereas constraint modes and rigid-body modes are defined by specifying a unit displacement at one DOF, attachment modes are defined by specifying a unit force at one DOF.

17.3.1 Attachment Modes

An *attachment mode* is defined as the component displacement vector due to a single unit force applied at one of the coordinates of a given set $\mathcal{A}$. Consequently, attachment modes are just columns of the associated flexibility matrix. Attachment modes were defined by Bamford[17.3], and they get their name from their usefulness in representating the deformation of a structure to loading at the point where the attachment mode's unit force is applied (e.g., an external force, an attached mass, or an attached flexible component). In the present CMS context, we are interested in defining attachment modes to permit statically complete representation of the response of a component to forces at its interface with adjoining components.[17.7,17.8]

One difficulty encountered in using attachment modes is that many components have one to six rigid-body degrees of freedom, making it impossible to apply directly to the unrestrained component the necessary unit forces in order to compute the resulting attachment mode shapes. However, one option in this case is to select a set $\mathcal{R}$ of boundary rigid-body degrees of freedom, (mathematically) restrain the component at these DOFs, and then form *cantilever attachment modes* by applying unit loads at the redundant boundary coordinates, that is, for $\mathcal{A} = \mathcal{E}$. Then

$$\begin{bmatrix} \mathbf{K}_{ii} & \mathbf{K}_{ie} & \mathbf{K}_{ir} \\ \mathbf{K}_{ei} & \mathbf{K}_{ee} & \mathbf{K}_{er} \\ \mathbf{K}_{ri} & \mathbf{K}_{re} & \mathbf{K}_{rr} \end{bmatrix} \begin{bmatrix} \mathbf{\Psi}_{ie} \\ \mathbf{\Psi}_{ee} \\ \mathbf{0}_{re} \end{bmatrix} = \begin{bmatrix} \mathbf{0}_{ie} \\ \mathbf{I}_{ee} \\ \mathbf{R}_{re} \end{bmatrix} \qquad (17.20)$$

It can be seen that these attachment modes are just an expanded form of the columns of the right-hand partition of the flexibility matrix $\mathbf{G}_c$ of Eq. 17.17, with $\mathcal{A} = \mathcal{E}$. That is, the *cantilever attachment modes* are given by

$$\boxed{\mathbf{\Psi}_{a} \atop N_u \times N_e} \equiv \begin{bmatrix} \mathbf{\Psi}_{ie} \\ \mathbf{\Psi}_{ee} \\ \mathbf{0}_{re} \end{bmatrix} = \begin{bmatrix} \mathbf{G}_{ie} \\ \mathbf{G}_{ee} \\ \mathbf{0}_{re} \end{bmatrix} \qquad (17.21)$$

Two important topics that arise when attachment modes are to be employed to represent, in part, the flexible behavior of unrestrained components are *inertia relief* and *residual flexibility*. These two topics, and the related forms of attachment modes, are defined in Sections 17.3.2 and 17.4, respectively.

17.3.2 Inertia-Relief Attachment Modes

When free-interface normal modes are to be employed in coupling reduced-order components that have rigid-body freedom, it is essential to complement the truncated set of free-interface normal modes with a set of inertia-relief attachment modes that are based on unit forces applied at all boundary coordinates. The term *inertia relief* refers to the process of applying to the component a self-equilibrated force system $\mathbf{f}_f$, which consists of the original force vector $\mathbf{f}$ equilibrated by the rigid-body d'Alembert force vector $(-\mathbf{M}\ddot{\mathbf{u}}_r)$, where $\mathbf{u}_r$ is the rigid-body motion due to $\mathbf{f}$.

Starting with Eq. 17.1, let the displacement vector $\mathbf{u}$ be separated into its *rigid-body displacement* component and its *flexible-body displacement* component, and expressed in terms of the respective modal coordinates $\mathbf{p}_r$ and $\mathbf{p}_f$. That is, let

$$\mathbf{u} = \mathbf{\Psi}_r \mathbf{p}_r + \mathbf{\Phi}_f \mathbf{p}_f \tag{17.22}$$

where all of the N_f flexible-body normal modes are included in $\mathbf{\Phi}_f$. Then the equations

$$\mathbf{\Psi}_r^T \mathbf{M} \mathbf{\Phi}_f = \mathbf{0}, \qquad \mathbf{\Psi}_r^T \mathbf{K} \mathbf{\Phi}_f = \mathbf{0}, \qquad \widehat{\mathbf{M}}_{rr} = \mathbf{\Psi}_r^T \mathbf{M} \mathbf{\Psi}_r \tag{17.23}$$

are the appropriate orthogonality equations and the definition of the rigid-body modal mass matrix, respectively. (Note that it is not assumed that the rigid-body modes are mass-normalized, although they may be.) Since (from Eq. 17.15) $\mathbf{K}\mathbf{\Psi}_r = \mathbf{0}$, Eqs. 17.1 and 17.22 can be combined to give

$$\mathbf{M}\mathbf{\Phi}_f \ddot{\mathbf{p}}_f + \mathbf{K}\mathbf{\Phi}_f \mathbf{p}_f = \mathbf{f} - \mathbf{M}\mathbf{\Psi}_r \ddot{\mathbf{p}}_r \tag{17.24}$$

When this equation is premultiplied by $\mathbf{\Psi}_r^T$ and orthogonality is invoked, we can solve for the rigid-body modal acceleration vector, $\ddot{\mathbf{p}}_r$, and then form the *self-equilibrated force vector*

$$\mathbf{f}_f = \mathbf{f} - \mathbf{M}\ddot{\mathbf{u}}_r \equiv \mathbf{P}_r \mathbf{f} \tag{17.25}$$

where $\mathbf{P}_r$ is the *inertia-relief projection matrix*, defined by

$$\boxed{\mathop{\mathbf{P}_r}_{N_u \times N_u} = \mathbf{I} - \mathbf{M}\mathbf{\Psi}_r \widehat{\mathbf{M}}_{rr}^{-1} \mathbf{\Psi}_r^T} \tag{17.26}$$

When any force vector is premultiplied by this inertia-relief projection matrix, the resulting force system is self-equilibrated. Also, from Eq. 17.26 it can easily be verified that $\mathbf{P}_r^T$ is mass-orthogonal to the rigid-body modes; that is,

$$\mathbf{\Psi}_r^T \mathbf{M} \mathbf{P}_r^T = \mathbf{0} \tag{17.27}$$

An important type of *inertia-relief attachment mode* consists of the static-deformation shapes obtained by applying unit forces at the interface coordinates ($\mathcal{A} = \mathcal{B}$), that is, by applying the following matrix of N_b force vectors:

$$\mathbf{F}_b = \begin{bmatrix} \mathbf{0}_{ib} \\ \mathbf{I}_{bb} \end{bmatrix} \tag{17.28}$$

premultiplied by the inertia-relief projection matrix $\mathbf{P}_r$. Since the unit-force column vectors in $\mathbf{F}_b$ are self-equilibrated by the inertia-relief projection matrix, no reaction forces are required, such as there are in Eq. 17.20. Deformation of the component due to this equilibrated force system is given by

$$\widetilde{\mathbf{\Psi}}_a = \widetilde{\mathbf{G}}_c \mathbf{P}_r \mathbf{F}_b \tag{17.29}$$

17.3 Component Modes: Attachment and Inertia-Relief Attachment Modes   541

where $\widetilde{\mathbf{G}}_c$ is the *constrained flexibility matrix*, a special expanded (rank N_f) form of the cantilever flexibility matrix $\mathbf{G}_c$ in Eq. 17.17, given by[1]

$$\widetilde{\mathbf{G}}_c \equiv \begin{bmatrix} \mathbf{G}_{ii} & \mathbf{G}_{ie} & \mathbf{0}_{ir} \\ \mathbf{G}_{ei} & \mathbf{G}_{ee} & \mathbf{0}_{er} \\ \mathbf{0}_{ri} & \mathbf{0}_{re} & \mathbf{0}_{rr} \end{bmatrix} \qquad (17.30)$$

The attachment-mode set defined by Eq. 17.29 is made orthogonal to the rigid-body modes, and the resulting *inertia-relief attachment modes* are given by (see Problem 17.5)

$$\underset{N_u \times N_b}{\boldsymbol{\Psi}_a} \equiv \begin{bmatrix} \boldsymbol{\Psi}_{ib} \\ \boldsymbol{\Psi}_{bb} \end{bmatrix} = \mathbf{G}_f \mathbf{F}_b \qquad (17.31)$$

where

$$\boxed{\underset{N_u \times N_u}{\mathbf{G}_f} = \mathbf{P}_r^T \widetilde{\mathbf{G}}_c \mathbf{P}_r} \qquad (17.32)$$

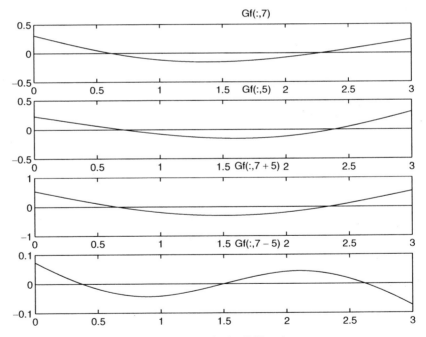

Figure 17.6 Elastic flexibility shapes.

is the *elastic flexibility matrix in inertia-relief format*. In Eq. 17.31, the $\mathbf{I}_{bb}$ matrix in $\mathbf{F}_b$ picks out the columns of the flexibility matrix $\mathbf{G}_f$ that correspond to unit forces

---

[1] This flexibility matrix is also called a *pseudoinverse* of the (singular) system stiffness matrix and is symbolized by $\widetilde{\mathbf{G}}_c = \mathbf{K}^+$.

applied at the boundary coordinates. From Eqs. 17.27 and 17.32, it can be shown that the columns of $\mathbf{G}_f$ are mass orthogonal to the rigid-body modes $\boldsymbol{\Psi}_r$. Therefore, $\mathbf{G}_f$ spans the same subspace as do the free-interface flex modes of Eq. 17.11.

The top two plots in Fig. 17.6 are the shapes that correspond to the two columns of the elastic flexibility matrix $\mathbf{G}_f$ for unit forces at the transverse DOFs at the two ends of the 8-DOF free–free beam in Fig. 17.2b, that is, at DOFs 7 and 5. It is clear that these two flexibility shapes are dominated by the contributions of free–free flex modes one and two (Fig. 17.4). This can be seen clearly by the bottom two figures, which represent symmetric and antisymmetric loadings by unit forces at the two ends of the component. However, the higher modes are also significant, especially near the ends where the forces are applied.

Example 17.1 For axial motion of the 3-DOF rod–mass system shown in Fig. 1, let the DOF sets be $\mathcal{I} = 1$, $\mathcal{A} = \mathcal{E} = 2$, and $\mathcal{R} = 3$. For a unit force applied to the left-hand mass (i.e., for $\mathbf{f}_a^T = \lfloor 0\ 1\ 0 \rfloor$): (a) Determine the attachment mode as defined by Eqs. 17.20 and 17.21. Sketch this mode. (b) Determine the self-equilibrated elastic load vector $\mathbf{Pf}_a$. (c) Determine the corresponding inertia-relief attachment mode as defined by Eqs. 17.31 and 17.32. Sketch this mode. The stiffness matrix and mass matrix for this system are, respectively,

$$\mathbf{K} = k \begin{bmatrix} 2 & -1 & -1 \\ -1 & 1 & 0 \\ -1 & 0 & 1 \end{bmatrix}, \quad \mathbf{M} = m \begin{bmatrix} 1 & 0 & 0 \\ 0 & 1 & 0 \\ 0 & 0 & 1 \end{bmatrix}$$

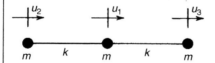

Figure 1 A 3-DOF rod–mass system undergoing axial motion.

SOLUTION (a) From Eq. 17.21, the cantilever attachment mode corresponding to a unit force at DOF 2 is

$$\boldsymbol{\psi}_a \equiv \begin{Bmatrix} \psi_{12} \\ \psi_{22} \\ 0 \end{Bmatrix} = \begin{Bmatrix} g_{12} \\ g_{22} \\ 0 \end{Bmatrix} \tag{1}$$

where the cantilever flexibility matrix for the rod fixed at DOF 3 is

$$\begin{bmatrix} g_{11} & g_{12} \\ g_{21} & g_{22} \end{bmatrix} = \begin{bmatrix} 2k & -k \\ -k & k \end{bmatrix}^{-1} = \frac{1}{k} \begin{bmatrix} 1 & 1 \\ 1 & 2 \end{bmatrix} \tag{2}$$

Therefore, the cantilever attachment mode is

$$\boldsymbol{\psi}_a = \frac{1}{k} \begin{Bmatrix} 1 \\ 2 \\ 0 \end{Bmatrix} \qquad \textbf{Ans. (a)} \tag{3}$$

which has the shape shown in Fig. 2.

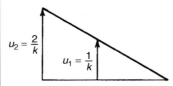

Figure 2 Cantilever attachment mode.

(b) From Eq. 17.26, the inertia-relief projection matrix is

$$\mathbf{P}_r = \mathbf{I} - \mathbf{M}\boldsymbol{\psi}_r \widehat{\mathbf{M}}_{rr}^{-1} \boldsymbol{\psi}_r^T \tag{4}$$

Equation 17.16 could be used to compute the rigid-body mode. However, for the present case, it is clear that the rigid-body mode consists of all masses displacing the same amount, so we can let

$$\boldsymbol{\psi}_r = \begin{Bmatrix} 1 \\ 1 \\ 1 \end{Bmatrix} \tag{5}$$

for which the generalized mass is

$$M_r = \boldsymbol{\psi}_r^T \mathbf{M} \boldsymbol{\psi}_r = \lfloor 1\ 1\ 1 \rfloor m \begin{bmatrix} 1 & 0 & 0 \\ 0 & 1 & 0 \\ 0 & 0 & 1 \end{bmatrix} \begin{Bmatrix} 1 \\ 1 \\ 1 \end{Bmatrix} = 3m \tag{6}$$

Therefore, the inertia-relief projection matrix is given by

$$\begin{aligned}\mathbf{P}_r &= \mathbf{I} - \mathbf{M}\boldsymbol{\psi}_r \frac{1}{M_r} \boldsymbol{\psi}_r^T \\ &= \begin{bmatrix} 1 & 0 & 0 \\ 0 & 1 & 0 \\ 0 & 0 & 1 \end{bmatrix} - m \begin{bmatrix} 1 & 0 & 0 \\ 0 & 1 & 0 \\ 0 & 0 & 1 \end{bmatrix} \begin{Bmatrix} 1 \\ 1 \\ 1 \end{Bmatrix} \frac{1}{3m} \lfloor 1\ 1\ 1 \rfloor \end{aligned} \tag{7}$$

or

$$\mathbf{P}_r = \frac{1}{3} \begin{bmatrix} 2 & -1 & -1 \\ -1 & 2 & -1 \\ -1 & -1 & 2 \end{bmatrix} \tag{8}$$

Therefore, the self-equilibrated force vector is

$$\mathbf{P}_r \mathbf{f}_a = \frac{1}{3} \begin{Bmatrix} -1 \\ 2 \\ -1 \end{Bmatrix} \qquad \text{Ans. (b)} \tag{9}$$

which is illustrated in Fig. 3.

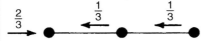

Figure 3 Equilibrated force system.

(c) From Eqs. 17.31 and 17.32 we get the following expression for the inertia-relief attachment mode corresponding to the force $\mathbf{f}_a$:

$$\boldsymbol{\psi}_a = \mathbf{G}_f \mathbf{f}_a = [\mathbf{P}_r^T \widetilde{\mathbf{G}}_c \mathbf{P}_r]\mathbf{f}_a \tag{10}$$

From Eqs. 2, 8, 9, 10, and 17.30, we get

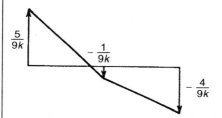

Finally, the inertia-relief attachment mode corresponding to a unit force applied at the left-hand mass is

$$\boldsymbol{\psi}_a = \frac{1}{9k} \begin{Bmatrix} -1 \\ 5 \\ -4 \end{Bmatrix} \qquad \textbf{Ans. (c)} \tag{12}$$

which has the shape shown in Fig. 4.

Figure 4 Inertia-relief attachment mode.

The reason for introducing inertia-relief attachment modes by using Eq. 17.31, rather than just using cantilever attachment modes as defined by Eq. 17.21, is the role played by Eq. 17.32 in the computation of residual flexibility, which is demonstrated in the next section.

17.4 FLEXIBILITY MATRICES AND RESIDUAL FLEXIBILITY

The complete set of component normal modes $\boldsymbol{\Phi}_n$ and the corresponding set of eigenvalues $\boldsymbol{\Lambda}_{nn}$ are identified by the subscript n, whether these are the N_i fixed-interface modes, the N_f free–free flexible (flex) modes, or some other form of component normal modes.

Let the modal mass matrix and modal stiffness matrix for modes $\boldsymbol{\Phi}_n$ be

$$\widehat{\mathbf{M}}_{nn} = \boldsymbol{\Phi}_n^T \mathbf{M} \boldsymbol{\Phi}_n = \mathrm{diag}(M_j), \qquad \widehat{\mathbf{K}}_{nn} = \boldsymbol{\Phi}_n^T \mathbf{K} \boldsymbol{\Phi}_n = \mathrm{diag}(K_j) = \mathrm{diag}(\omega_j^2 M_j) \tag{17.33}$$

respectively. Then the *elastic flexibility matrix*, $\mathbf{G}$, of the component can be expressed in the following mode-superposition format:

$$\boxed{\mathbf{G} = \boldsymbol{\Phi}_n \widehat{\mathbf{K}}_{nn}^{-1} \boldsymbol{\Phi}_n^T = \sum_{j=1}^{N_n} \boldsymbol{\phi}_j \left(\frac{1}{K_j} \boldsymbol{\phi}_j^T \right) = \mathbf{G}^T} \tag{17.34}$$

This is a very important equation. Note that each column of the jth mode's contribution to the elastic flexibility matrix has the shape of mode $\boldsymbol{\phi}_j$. That is, the jth mode's contribution to column i of $\mathbf{G}$ consists of the mode $\boldsymbol{\phi}_j$ scaled by ϕ_{ij}/K_j. For a free–free structure with N_f flexible modes, although the elastic flexibility matrix $\mathbf{G}$ of Eq. 17.34 and matrix $\mathbf{G}_f$ of Eq. 17.32 and illustrated in Fig. 17.6 are formed in different ways, they are numerically the same.

In this section we are concerned with components that have rigid-body freedom, in which case $\mathbf{G}$ is singular, with rank N_f. Regardless of whether the elastic flexibility matrix $\mathbf{G}$ is singular or not, from Eqs. 17.33b and 17.34 it can be shown that

$$\mathbf{GKG} = \mathbf{G} \tag{17.35}$$

Since model reduction is one of the major objectives in CMS, the normal mode set is usually reduced to a smaller set of *kept normal modes*, denoted by $\boldsymbol{\Phi}_k$, where $\boldsymbol{\Phi}_n = [\boldsymbol{\Phi}_k \; \boldsymbol{\Phi}_d]$.[2] The *deleted normal modes*, $\boldsymbol{\Phi}_d$, are generally all of the modes above some specified cutoff frequency. The portion of the flexibility matrix contributed by modes $\boldsymbol{\Phi}_d$ is called the *residual-flexibility matrix*. It is given by

$$\mathbf{G}_d = \boldsymbol{\Phi}_d \widehat{\mathbf{K}}_{dd}^{-1} \boldsymbol{\Phi}_d^\mathrm{T} = \mathbf{G} - \boldsymbol{\Phi}_k \widehat{\mathbf{K}}_{kk}^{-1} \boldsymbol{\Phi}_k^\mathrm{T} = \mathbf{G}_d^\mathrm{T} \tag{17.36}$$

where $\mathbf{G}$ is the total flexibility matrix. Although it is not feasible to compute or measure all of the $\boldsymbol{\Phi}_d$ modes, Eq. 17.36 is still very useful because Eq. 17.32 exists as an alternative to Eq. 17.34 for determining the elastic flexibility matrix $\mathbf{G}$.

The matrix $\mathbf{G}_d$ will always be a singular matrix because of the modes deleted in Eq. 17.36. Also, because of the mass orthogonality and stiffness orthogonality of the kept modes to the deleted modes,

$$\boldsymbol{\Phi}_k^\mathrm{T} \mathbf{M} \mathbf{G}_d = \mathbf{0} \quad \text{and} \quad \boldsymbol{\Phi}_k^\mathrm{T} \mathbf{K} \mathbf{G}_d = \mathbf{0} \tag{17.37}$$

and because of the mass orthogonality between all rigid-body modes and all flexible-body modes,

$$\boldsymbol{\Psi}_r^\mathrm{T} \mathbf{M} \mathbf{G}_d = \mathbf{0} \tag{17.38}$$

Residual-flexibility attachment modes may be defined for forces applied at the interface coordinates, that is, for $\mathcal{A} = \mathcal{B}$, by the following equation:

$$\boxed{\begin{aligned} \underset{N_u \times N_b}{\boldsymbol{\Psi}_d} &\equiv \begin{bmatrix} \boldsymbol{\Psi}_{ib} \\ \boldsymbol{\Psi}_{bb} \end{bmatrix} = \mathbf{G}_d \mathbf{F}_b \\ &= \left[\mathbf{G}_f - \boldsymbol{\Phi}_k \widehat{\mathbf{K}}_{kk}^{-1} \boldsymbol{\Phi}_k^\mathrm{T}\right] \begin{bmatrix} \mathbf{0}_{ib} \\ \mathbf{I}_{bb} \end{bmatrix} \end{aligned}} \tag{17.39}$$

where $\mathbf{G}_f$ is given by Eq. 17.32. Thus, $\boldsymbol{\Psi}_d$ contains the columns of the residual-flexibility matrix associated with the boundary DOFs.

Figure 17.7 shows the attachment mode shape for the component with a unit force at DOF7: The top figure includes all six flex modes, the middle figure is the contribution of two "kept" modes, and the bottom figure is the corresponding residual-flexibility attachment mode shape, that is, the difference between the top and middle figures. It is

---

[2] Rigid-body modes, if there are any, are not included in the normal mode sets $\boldsymbol{\Phi}_n$, $\boldsymbol{\Phi}_k$, or $\boldsymbol{\Phi}_d$.

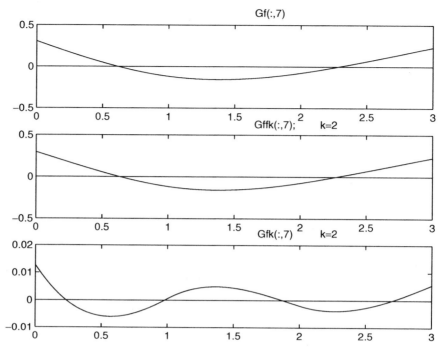

Figure 17.7 Illustration of residual flexibility.

clear that the order of magnitude of the residual flexibility is smaller than that of the flexibility of the kept modes.

Incorporation of Ψ_d into the component mode set ensures complete representation of static deflection of the component due to forces applied at interface DOFs. In this sense it is closely related to the *mode-acceleration method* for incorporating static completeness in dynamic-response computations (Section 11.4). Hintz has given an extensive discussion of the need for statically complete component mode sets in Ref. [17.7]. We return to the topic of residual flexibility in Section 17.7.

Example 17.2 This example illustrates the use of Eq. 17.39 to determine a residual inertia-relief attachment mode for the three-mass system of Example 17.1, repeated in Fig. 1. (a) Determine the normal modes of the three-mass system of Example 17.1. (b) Let $\Psi_r \equiv \phi_1$, $\Phi_k \equiv \phi_2$, and $\Phi_d \equiv \phi_3$. Determine the residual inertia-relief attachment mode associated with a unit force applied to the left-hand mass. (c) Give a physical explanation of the result you obtained in part (b).

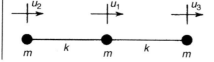

Figure 1 A 3-DOF rod–mass system undergoing axial motion.

17.4 Flexibility Matrices and Residual Flexibility

SOLUTION (a) From Example 17.1 the algebraic eigenproblem can be written as

$$\begin{bmatrix} 2k - \omega^2 m & -k & -k \\ -k & k - \omega^2 m & 0 \\ -k & 0 & k - \omega^2 m \end{bmatrix} \begin{Bmatrix} U_1 \\ U_2 \\ U_3 \end{Bmatrix} = \begin{Bmatrix} 0 \\ 0 \\ 0 \end{Bmatrix} \quad (1)$$

Then the frequencies are given by

$$|\mathbf{K} - \omega^2 \mathbf{M}| = 0$$

or

$$\begin{vmatrix} 2k - \omega^2 m & -k & -k \\ -k & k - \omega^2 m & 0 \\ -k & 0 & k - \omega^2 m \end{vmatrix} = 0 \quad (2)$$

This gives the characteristic equation

$$\omega^2 m (k - \omega^2 m)(3k - \omega^2 m) = 0 \quad (3)$$

whose roots are

$$\omega_1^2 = 0, \qquad \omega_2^2 = \frac{k}{m}, \qquad \omega_3^2 = \frac{3k}{m} \quad (4)$$

Mode shapes can be obtained from the top two equations of Eq. 1. Since the left end of the bar is DOF 2, let $U_2 = 1$. Then the mode shapes are given by the following set of equations:

$$\begin{aligned} (2k - \omega_r^2 m)U_1 - k(1) - kU_3 &= 0 \\ -kU_1 + (k - \omega_r^2 m)(1) &= 0 \end{aligned} \quad (5)$$

For mode 1, the rigid-body mode, $\omega_1^2 = 0$. So Eq. 5 becomes

$$\begin{aligned} 2kU_1 - kU_3 &= k \\ -kU_1 &= -k \end{aligned} \quad (6)$$

Therefore, as expected, the rigid-body mode is given by

$$\mathbf{U}_1^T = \lfloor 1 \ 1 \ 1 \rfloor \quad (7)$$

When normalized so that $M_1 = \boldsymbol{\phi}_1^T \mathbf{M} \boldsymbol{\phi}_1 = m$, the normalized rigid-body mode is

$$\boldsymbol{\phi}_1^T = \frac{1}{\sqrt{3}} \lfloor 1 \ 1 \ 1 \rfloor \qquad \textbf{Ans. (a)} \quad (8)$$

For mode 2, $\omega_2^2 m = k$. Therefore, Eq. 5 becomes

$$\begin{aligned} kU_1 - kU_3 &= k \\ -kU_1 &= 0 \end{aligned} \quad (9)$$

Thus, $U_1 = 0$ and $U_3 = -1$, so mode 2 is the fundamental flexible mode, which has the form

$$\mathbf{U}_2^T = \lfloor 0 \; 1 \; -1 \rfloor \tag{10}$$

When normalized so that $M_2 = \boldsymbol{\phi}_2^T \mathbf{M} \boldsymbol{\phi}_2^T = m$, the normalized mode $\boldsymbol{\phi}_2$ is given by

$$\boldsymbol{\phi}_2^T = \frac{1}{\sqrt{2}} \lfloor 0 \; 1 \; -1 \rfloor \qquad \textbf{Ans. (a)} \tag{11}$$

Finally, mode 3, normalized so that $\boldsymbol{\phi}_3^T \mathbf{M} \boldsymbol{\phi}_3 = m$, can be shown to be

$$\boldsymbol{\phi}_3^T = \frac{1}{\sqrt{6}} \lfloor -2 \; 1 \; 1 \rfloor \qquad \textbf{Ans. (a)} \tag{12}$$

(Normally, "deleted modes" are not calculated. Mode 3 is calculated and shown here to help with the explanation in part c.) Modes 2 and 3 are shown in Fig. 2.

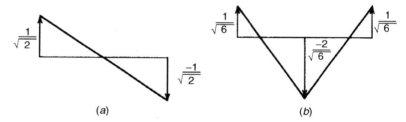

Figure 2 Flexible modes of a three-mass system: (a) mode $\boldsymbol{\phi}_2$; (b) mode $\boldsymbol{\phi}_3$.

(b) The residual inertia-relief attachment mode can be calculated from the following combination of Eqs. 17.32 and 17.39:

$$\boldsymbol{\Psi}_d = (\mathbf{P}_r^T \tilde{\mathbf{G}}_c \mathbf{P}_r) \mathbf{F}_a - (\boldsymbol{\Phi}_k \widehat{\mathbf{K}}_{kk}^{-1} \boldsymbol{\Phi}_k^T) \mathbf{F}_a \tag{13}$$

The first term on the right-hand side of Eq. 13 is the inertia-relief attachment mode that was determined in Eqs. 10 through 12 of Example 17.1. In the present problem, the kept flexible mode is mode 2, so

$$\boldsymbol{\Phi}_k = \boldsymbol{\phi}_2, \qquad \widehat{\mathbf{K}}_{kk} = \omega_2^2 M_2 = \frac{k}{m} m = k \tag{14}$$

Hence,

$$\boldsymbol{\Phi}_k \widehat{\mathbf{K}}_{kk}^{-1} \boldsymbol{\Phi}_k^T = \frac{1}{K_2} \boldsymbol{\phi}_2 \boldsymbol{\phi}_2^T = \frac{1}{k} \left(\frac{1}{2} \right) \begin{Bmatrix} 0 \\ 1 \\ -1 \end{Bmatrix} \lfloor 0 \; 1 \; -1 \rfloor = \frac{1}{2k} \begin{bmatrix} 0 & 0 & 0 \\ 0 & 1 & -1 \\ 0 & -1 & 1 \end{bmatrix} \tag{15}$$

From Example 17.1,

$$\dot{\mathbf{r}}_a = \lfloor 0 \; 1 \; 0 \rfloor^T \tag{16}$$

Finally, the residual inertia-relief attachment mode is

$$\boldsymbol{\psi}_d = \frac{1}{9k}\begin{Bmatrix} -1 \\ 5 \\ -4 \end{Bmatrix} - \frac{1}{2k}\begin{bmatrix} 0 & 0 & 0 \\ 0 & 1 & -1 \\ 0 & -1 & 1 \end{bmatrix}\begin{Bmatrix} 0 \\ 1 \\ 0 \end{Bmatrix} \quad (17)$$

or

$$\boldsymbol{\psi}_d^{\text{T}} = \frac{1}{18k}\lfloor -2 \ 1 \ 1 \rfloor \qquad \textbf{Ans. (b)} \ (18)$$

(c) As might have been expected, the residual attachment mode in the present case is exactly a multiple of $\boldsymbol{\phi}_3$, since it represents deformation of the three-mass system after removing rigid-body motion and after removing "kept" mode 2.

In determining the attachment mode, a unit force was placed at node 2 (left end). That applied force can be apportioned to the various modes as shown in Fig. 3. From Eqs. 15 and 16,

$$\boldsymbol{\Phi}_k \widehat{\mathbf{K}}_{kk}^{-1} \boldsymbol{\Phi}_k \mathbf{f}_a = \frac{1}{2k}\begin{Bmatrix} 0 \\ 1 \\ -1 \end{Bmatrix} \quad (19)$$

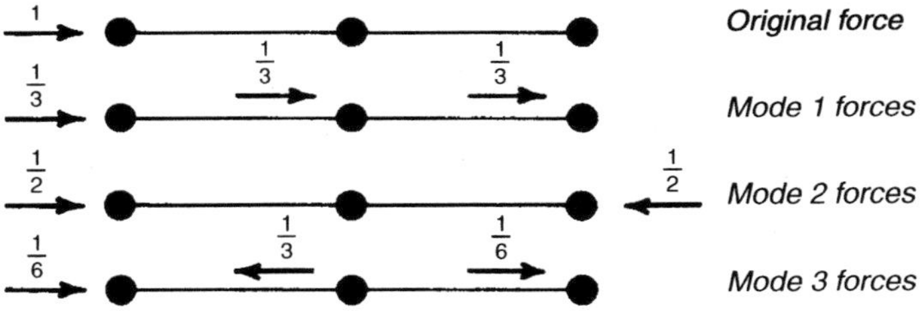

Figure 3 Modal contributions to an applied unit force.

This is exactly the deflection shape that would result from the application of the "mode 2 forces" in Fig. 3. Also, $\boldsymbol{\psi}_d$, given by Eq. 18, is exactly the deflection shape produced by the "mode 3 forces."

17.5 SUBSTRUCTURE COUPLING PROCEDURES

In this section, two versions of a *generalized substructure coupling procedure* for undamped structures are presented. Both procedures employ Lagrange multipliers to enforce component interface displacement compatibility equations (and other constraint equations, if applicable). The first coupling procedure applies to components that are assembled in the original physical-coordinate form. The second applies to coupling components after there has been a coordinate transformation to component generalized coordinates.

Let the system be composed of N_S components, labeled $\mathcal{D}^{(1)}$ through $\mathcal{D}^{(N_S)}$. We consider here the case where the interface between any two adjacent components is

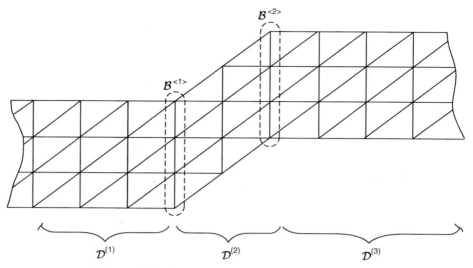

Figure 17.8 Substructures and substructure interfaces.

represented by conforming discrete grid points, as illustrated in Fig. 17.8.[3] Let these component interfaces be labeled $\mathcal{B}^{<1>}$ through $\mathcal{B}^{<N_B>}$, where angle brackets, $<\cdot>$, are used to label boundary segments.

Let $\mathbf{u}_b$ be the set of all component interface coordinates of the assembled (i.e., coupled) system. Displacement compatibility along component interfaces can be enforced through specification of the *connectivity matrices* $\mathbf{B}^{(s)}$. Then

$$\mathbf{u}_b^{(s)} = \mathbf{B}^{(s)} \mathbf{u}_b, \qquad s = 1, \ldots, N_S \qquad (17.40)$$

This implies that if $\mathcal{D}^{(p)}$ and $\mathcal{D}^{(q)}$ are neighboring substructures, the portions of $\mathbf{B}^{(p)}$ and $\mathbf{B}^{(q)}$ that correspond to their shared boundary must be identical. This compatibility information can also be expressed in the form of *interface displacement constraint equations*, written in terms of physical displacement coordinates.

$$\sum_{s=1}^{N_S} \mathbf{C}^{(s)} \mathbf{u}^{(s)} = \mathbf{C} \mathbf{u} = \mathbf{0} \qquad (17.41)$$

where the *constraint matrix* $\mathbf{C}$ and the merged *vector of component physical coordinates* have the following forms:

$$\mathbf{C}_{N_C \times N_U} = [\mathbf{C}^{(1)} \quad \mathbf{C}^{(2)} \quad \cdots \quad \mathbf{C}^{(N_S)}], \qquad \mathbf{u} = \lfloor \mathbf{u}^{(1)\mathrm{T}} \quad \mathbf{u}^{(2)\mathrm{T}} \quad \cdots \quad \mathbf{u}^{(N_S)\mathrm{T}} \rfloor^\mathrm{T} \qquad (17.42)$$

Since the interface displacement constraint equation in physical coordinates does not involve the interior coordinates $\mathbf{u}_i^{(s)}$, the substructure constraint matrices have the form

$$\mathbf{C}^{(s)} = [\mathbf{0} \quad \mathbf{C}_b^{(s)}] \qquad (17.43)$$

---

[3] References [17.25] and [17.26] generalize the procedure discussed here to nonconforming interfaces.

17.5 Substructure Coupling Procedures

Two forms of system assembly will be useful in formulating specific substructure coupling methods: System Assembly in Physical Coordinates, employed in Section 17.7, and System Assembly in Generalized Coordinates, employed in Section 17.6 and again in Section 17.7.

17.5.1 System Assembly in Physical Coordinates

The synthesis of the system equation of motion is based on Lagrange's equation of motion with Lagrange multipliers λ (see Section 8.5). The Lagrangian for the system of N_S <u>coupled</u> substructures can be written

$$\mathcal{L} = \mathcal{T} - \mathcal{V} + \lambda^T \mathbf{C}\mathbf{u} = \mathcal{T} - \mathcal{V} + \mathbf{u}^T \mathbf{C}^T \lambda \tag{17.44}$$

where $\mathcal{T}$ is the system kinetic energy and $\mathcal{V}$ is the system potential energy, given in terms of the physical coordinates by

$$\mathcal{T} = \tfrac{1}{2} \sum_{s=1}^{N_S} \dot{\mathbf{u}}^{(s)T} \mathbf{M}^{(s)} \dot{\mathbf{u}}^{(s)} \equiv \tfrac{1}{2} \dot{\mathbf{u}}^T \mathbf{M} \dot{\mathbf{u}}$$

$$\mathcal{V} = \tfrac{1}{2} \sum_{s=1}^{N_S} \mathbf{u}^{(s)T} \mathbf{K}^{(s)} \mathbf{u}^{(s)} \equiv \tfrac{1}{2} \mathbf{u}^T \mathbf{K} \mathbf{u} \tag{17.45}$$

respectively. The block-diagonal *uncoupled mass matrix* $\mathbf{M}$ and the block-diagonal *uncoupled stiffness matrix* $\mathbf{K}$ are given by

$$\mathbf{M} \equiv \mathrm{diag}(\mathbf{M}^{(s)}), \qquad \mathbf{K} \equiv \mathrm{diag}(\mathbf{K}^{(s)}) \tag{17.46}$$

Since the net work of the interconnecting boundary forces $\mathbf{r}$ within the coupled system is zero, the virtual work done on the system is given by

$$\delta \mathcal{W} = \delta \mathbf{u}^T (\mathbf{f} + \mathbf{r}) = \delta \mathbf{u}^T \mathbf{f} \tag{17.47}$$

where, corresponding to Eq. 17.42b, the merged external force vector is

$$\mathbf{f} \equiv \lfloor \mathbf{f}^{(1)T} \quad \mathbf{f}^{(2)T} \quad \cdots \quad \mathbf{f}^{(N_S)T} \rfloor^T, \qquad \text{where} \quad \mathbf{f}^{(s)} = \left\{ \begin{array}{c} \mathbf{0}_i \\ \mathbf{f}_b \end{array} \right\}^{(s)} \tag{17.48}$$

The system equations of motion can now be obtained by applying Lagrange's equation in the form

$$\frac{d}{dt} \frac{\partial \mathcal{L}}{\partial \dot{u}_j} - \frac{\partial \mathcal{L}}{\partial u_j} = f_j, \qquad j = 1, 2, \ldots, N_U \tag{17.49}$$

where u_j refers to the jth element of the merged physical displacement vector $\mathbf{u}$, and f_j refers to the corresponding jth element of the vector of externally applied forces, $\mathbf{f}$. Since they do not appear in the final expression for the virtual work, the mutually reactive interface constraint forces do not appear on the right-hand side of Eq. 17.49. Combining Eqs. 17.44, 17.45, and 17.47 with Eq. 17.49, we can write the N_U equations of motion in the following matrix form:

$$\mathbf{M}\ddot{\mathbf{u}} + \mathbf{K}\mathbf{u} = \mathbf{f} + \mathbf{C}^T \lambda \tag{17.50}$$

Comparing Eq. 17.50 with Eq. 17.1, we see that the term $\mathbf{C}^{(s)T}\boldsymbol{\lambda}$ in Eq. 17.50 corresponds to the interface constraint force vector $\mathbf{r}^{(s)}$ in Eq. 17.1. Since the substructure constraint matrices have the form given by Eq. 17.43, the actual interface constraint force acting on the boundary of component s is $\mathbf{C}_b^{(s)T}\boldsymbol{\lambda}$.

Since the N_U displacement coordinates in Eq. 17.50 are not independent but are subject to the constraint equation, Eq. 17.41, we can combine these two equations to form the $(N_U + N_C)$ set of *hybrid coupled-system equations* and write them in the following matrix form:

$$\begin{bmatrix} \mathbf{M} & \mathbf{0} \\ \mathbf{0} & \mathbf{0} \end{bmatrix} \begin{Bmatrix} \ddot{\mathbf{u}} \\ \ddot{\boldsymbol{\lambda}} \end{Bmatrix} + \begin{bmatrix} \mathbf{K} & -\mathbf{C}^T \\ -\mathbf{C} & \mathbf{0} \end{bmatrix} \begin{Bmatrix} \mathbf{u} \\ \boldsymbol{\lambda} \end{Bmatrix} = \begin{Bmatrix} \mathbf{f} \\ \mathbf{0} \end{Bmatrix} \qquad (17.51)$$

Application of Eq. 17.51 is illustrated in Section 17.7.

The following example illustrates for the three-component system in Fig. 17.8 the formation of the boundary displacement constraint equation and the physical meaning of the Lagrange multiplier vector.

Example 17.3 Treat the three components of the structure in Fig. 17.8 as *superelements*. That is, retain all original finite element DOFs, so that

$$\mathbf{u} \equiv \begin{Bmatrix} \mathbf{u}^{(1)} \\ \mathbf{u}^{(2)} \\ \hline \mathbf{u}^{(3)} \end{Bmatrix}, \quad \mathbf{u}^{(1)} \equiv \begin{Bmatrix} \mathbf{u}_i^{(1)} \\ \mathbf{u}_{b1}^{(1)} \end{Bmatrix}, \quad \mathbf{u}^{(2)} \equiv \begin{Bmatrix} \mathbf{u}_i^{(2)} \\ \mathbf{u}_{b1}^{(2)} \\ \mathbf{u}_{b2}^{(2)} \end{Bmatrix}, \quad \mathbf{u}^{(3)} \equiv \begin{Bmatrix} \mathbf{u}_i^{(3)} \\ \mathbf{u}_{b2}^{(3)} \end{Bmatrix} \qquad (1)$$

where the $\mathcal{B}^{<1>}$ and $\mathcal{B}^{<2>}$ boundary displacements of the substructures are identified by the subscripts $b1$ and $b2$, respectively (Fig. 1).

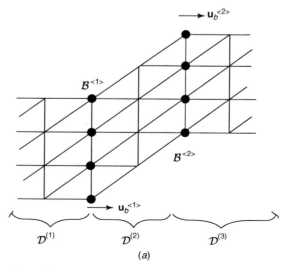

(a)

Figure 1 Interpretation of (*a*) substructure interface displacement compatibility, and (*b*) Lagrange multiplier interface constraint forces.

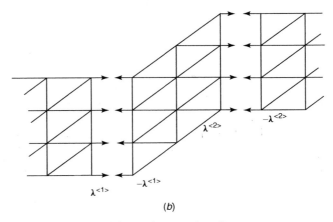

(b)

Figure 1 (*continued*)

Let the interface displacement vector and the corresponding Lagrange multiplier vector have the forms

$$\mathbf{u}_b = \begin{Bmatrix} \mathbf{u}_b^{<1>} \\ \mathbf{u}_b^{<2>} \end{Bmatrix}, \qquad \lambda = \begin{Bmatrix} \lambda^{<1>} \\ \lambda^{<2>} \end{Bmatrix} \qquad (2)$$

(a) Write the appropriate connectivity matrices as defined by Eq. 17.40. (b) Write the interface displacement constraint equation in the form prescribed by Eq. 17.41. (c) Form the $\mathbf{C}^T \lambda$ term of Eq. 17.50, and use Fig. 1b to interpret the meaning of the λ vector.

SOLUTION (a) The connectivity matrices are defined by Eq. 17.40,

$$\mathbf{u}_b^{(s)} = \mathbf{B}^{(s)} \mathbf{u}_b, \qquad s = 1, 2, 3 \qquad (3)$$

From Fig. 1a, Eq. 2a, and Eq. 3, the connectivity matrices are the *boolean* matrices (i.e., matrices of ones and zeros) in the following three equations:

$$\mathbf{u}_b^{(1)} = [\mathbf{I} \quad \mathbf{0}] \begin{Bmatrix} \mathbf{u}_b^{<1>} \\ \mathbf{u}_b^{<2>} \end{Bmatrix}, \qquad \mathbf{u}_b^{(2)} \equiv \begin{Bmatrix} \mathbf{u}_{b1}^{(2)} \\ \mathbf{u}_{b2}^{(2)} \end{Bmatrix} = \begin{bmatrix} \mathbf{I} & \mathbf{0} \\ \mathbf{0} & \mathbf{I} \end{bmatrix} \begin{Bmatrix} \mathbf{u}_b^{<1>} \\ \mathbf{u}_b^{<2>} \end{Bmatrix}$$

$$\mathbf{u}_b^{(3)} = [\mathbf{0} \quad \mathbf{I}] \begin{Bmatrix} \mathbf{u}_b^{<1>} \\ \mathbf{u}_b^{<2>} \end{Bmatrix} \qquad \textbf{Ans. (a)} \quad (4)$$

(b) The same connectivity is expressed by the following interface displacement constraint equation:

$$\begin{matrix} \text{For } \mathcal{B}^{<1>}: \\ \text{For } \mathcal{B}^{<2>}: \end{matrix} \quad \begin{bmatrix} \mathbf{0} & \mathbf{I} & \mathbf{0} & -\mathbf{I} & \mathbf{0} & \mathbf{0} & \mathbf{0} \\ \mathbf{0} & \mathbf{0} & \mathbf{0} & \mathbf{0} & \mathbf{I} & \mathbf{0} & -\mathbf{I} \end{bmatrix} \begin{Bmatrix} \mathbf{u}_i^{(1)} \\ \mathbf{u}_b^{(1)} \\ \mathbf{u}_i^{(2)} \\ \mathbf{u}_{b1}^{(2)} \\ \mathbf{u}_{b2}^{(2)} \\ \mathbf{u}_i^{(3)} \\ \mathbf{u}_b^{(3)} \end{Bmatrix} = \begin{Bmatrix} \mathbf{0} \\ \mathbf{0} \end{Bmatrix} \qquad \textbf{Ans. (b)} \quad (5)$$

Therefore, the interface displacement constraint matrix is

$$\mathbf{C} \equiv \begin{bmatrix} \mathbf{C}^{(1)} & \mathbf{C}^{(2)} & \mathbf{C}^{(3)} \end{bmatrix} = \begin{bmatrix} 0 & \mathbf{I} & 0 & -\mathbf{I} & 0 & 0 & 0 \\ 0 & 0 & 0 & 0 & \mathbf{I} & 0 & -\mathbf{I} \end{bmatrix} \qquad (6)$$

(c) Now we form $\mathbf{C}^T\boldsymbol{\lambda}$ using Eqs. 2b and 6:

$$\mathbf{C}^T\boldsymbol{\lambda} = \begin{bmatrix} 0 & 0 \\ \mathbf{I} & 0 \\ \hline 0 & 0 \\ -\mathbf{I} & 0 \\ 0 & \mathbf{I} \\ \hline 0 & 0 \\ 0 & -\mathbf{I} \end{bmatrix} \begin{Bmatrix} \boldsymbol{\lambda}^{<1>} \\ \boldsymbol{\lambda}^{<2>} \end{Bmatrix} = \begin{Bmatrix} 0 \\ \boldsymbol{\lambda}^{<1>} \\ 0 \\ -\boldsymbol{\lambda}^{<1>} \\ \boldsymbol{\lambda}^{<2>} \\ 0 \\ -\boldsymbol{\lambda}^{<2>} \end{Bmatrix} \qquad \textbf{Ans. (c) (7)}$$

Therefore, $\boldsymbol{\lambda}^{<1>}$ can be interpreted as the constraint force vector (tension positive) that enforces interface displacement compatibility at interface $\mathcal{B}^{<1>}$, and $\boldsymbol{\lambda}^{<2>}$ is the constraint force vector (tension positive) that enforces interface displacement compatibility at interface $\mathcal{B}^{<2>}$, both illustrated in Fig. 1b.

17.5.2 System Assembly in Generalized Coordinates

Let us now formulate the system assembly process using the component generalized coordinates $\mathbf{p}$. Displacement constraint equations such as Eqs. 17.41 and any other constraint equations that are to be imposed (say, N_{C_p} equations in all) can be written in terms of the $N_P = \sum_{s=1}^{N_S} N_p^{(s)}$ generalized coordinates $\mathbf{p}$ and combined to form a matrix constraint equation of the form

$$\mathbf{C}_p \mathbf{p} = \mathbf{0} \qquad (17.52)$$

For example, if there are only two substructures, Eqs. 17.2 can be combined with the boundary displacement compatibility equation

$$\mathbf{u}_b^{(1)} = \mathbf{u}_b^{(2)} \qquad (17.53)$$

to give the constraint equation

$$\begin{bmatrix} \boldsymbol{\Psi}_b^{(1)} & -\boldsymbol{\Psi}_b^{(2)} \end{bmatrix} \begin{Bmatrix} \mathbf{p}^{(1)} \\ \mathbf{p}^{(2)} \end{Bmatrix} = \mathbf{0} \qquad (17.54)$$

where $\boldsymbol{\Psi}_b^{(s)}$ refers to the boundary-coordinate partition of the component mode matrix $\boldsymbol{\Psi}^{(s)}$ of component s.

The synthesis of the system equation of motion can be based on Lagrange's equation of motion with Lagrange multipliers, where the transformation to component generalized coordinates now precedes the assembly of components.[17.8,17.12] With the vector of Lagrange multipliers now designated as $\boldsymbol{\sigma}$, the Lagrangian for the system of N_S coupled substructures can be written

$$\mathcal{L} = \mathcal{T} - \mathcal{V} + \boldsymbol{\sigma}^T \mathbf{C}_p \mathbf{p} = \mathcal{T} - \mathcal{V} + \mathbf{p}^T \mathbf{C}_p^T \boldsymbol{\sigma} \qquad (17.55)$$

17.5 Substructure Coupling Procedures

where $\mathcal{T}$ is the system kinetic energy and $\mathcal{V}$ is the system potential energy, given by

$$\mathcal{T} = \tfrac{1}{2} \sum_{s=1}^{N_S} \dot{\mathbf{p}}^{(s)\mathrm{T}} \widehat{\mathbf{M}}^{(s)} \dot{\mathbf{p}}^{(s)} \equiv \tfrac{1}{2} \dot{\mathbf{p}}^{\mathrm{T}} \mathbf{M}_p \dot{\mathbf{p}}$$
$$\mathcal{V} = \tfrac{1}{2} \sum_{s=1}^{N_S} \mathbf{p}^{(s)\mathrm{T}} \widehat{\mathbf{K}}^{(s)} \mathbf{p}^{(s)} \equiv \tfrac{1}{2} \mathbf{p}^{\mathrm{T}} \mathbf{K}_p \mathbf{p}$$
(17.56)

respectively. The block-diagonal *uncoupled mass matrix* $\mathbf{M}_p$, the block-diagonal *uncoupled stiffness matrix* $\mathbf{K}_p$, and the *vector of component-mode coordinates* $\mathbf{p}$ are given by

$$\mathbf{M}_p \equiv \operatorname{diag}(\widehat{\mathbf{M}}^{(s)}), \qquad \mathbf{K}_p \equiv \operatorname{diag}(\widehat{\mathbf{K}}^{(s)}), \qquad \mathbf{p} \equiv \lfloor \mathbf{p}^{(1)\mathrm{T}} \quad \mathbf{p}^{(2)\mathrm{T}} \quad \cdots \quad \mathbf{p}^{(N_S)\mathrm{T}} \rfloor^{\mathrm{T}}$$
(17.57)

Since the net work of the interconnecting boundary forces $\mathbf{r}$ within the coupled system is zero, the virtual work done on the system is given by

$$\delta \mathcal{W} = \delta \mathbf{u}^{\mathrm{T}} (\mathbf{f} + \mathbf{r}) = \delta \mathbf{u}^{\mathrm{T}} \mathbf{f} = \delta \mathbf{p}^{\mathrm{T}} \boldsymbol{\Psi}^{\mathrm{T}} \mathbf{f} = \delta \mathbf{p}^{\mathrm{T}} \mathbf{f}_p \tag{17.58}$$

where, corresponding to Eq. 17.57c, the transformed external force vector can be written in the following form:

$$\mathbf{f}_p \equiv \lfloor \mathbf{f}_p^{(1)\mathrm{T}} \quad \mathbf{f}_p^{(2)\mathrm{T}} \quad \cdots \quad \mathbf{f}_p^{(N_S)\mathrm{T}} \rfloor^{\mathrm{T}} \tag{17.59}$$

The system equations of motion can now be obtained by applying Lagrange's equation in the form

$$\frac{d}{dt} \frac{\partial \mathcal{L}}{\partial \dot{p}_j} - \frac{\partial \mathcal{L}}{\partial p_j} = f_{p_j}, \qquad j = 1, 2, \ldots, N_P \tag{17.60}$$

where p_j refers to the jth element of the merged displacement vector $\mathbf{p}$, and f_{p_j} refers to the corresponding jth element of the transformed vector of externally applied forces, $\mathbf{f}_p$. Since they do not appear in the final expression for the virtual work, the mutually reactive interface constraint forces do not appear on the right-hand side of Eq. 17.60. We can write the N_P equations of motion in the following matrix form:

$$\mathbf{M}_p \ddot{\mathbf{p}} + \mathbf{K}_p \mathbf{p} = \mathbf{f}_p + \mathbf{C}_p^{\mathrm{T}} \boldsymbol{\sigma} \tag{17.61}$$

The term $\mathbf{C}_p^{\mathrm{T}} \boldsymbol{\sigma}$ in Eq. 17.61 corresponds to the interface constraint force vector, expressed in generalized coordinates.

Since the N_P generalized coordinates in Eq. 17.61 are not independent but are subject to the constraint equation, Eq. 17.52, we can combine these two equations to form the $N_P + N_{C_P}$ set of coupled system equations in the following matrix form:

$$\boxed{\begin{bmatrix} \mathbf{M}_p & \mathbf{0} \\ \mathbf{0} & \mathbf{0} \end{bmatrix} \begin{Bmatrix} \ddot{\mathbf{p}} \\ \ddot{\boldsymbol{\sigma}} \end{Bmatrix} + \begin{bmatrix} \mathbf{K}_p & -\mathbf{C}_p^{\mathrm{T}} \\ -\mathbf{C}_p & \mathbf{0} \end{bmatrix} \begin{Bmatrix} \mathbf{p} \\ \boldsymbol{\sigma} \end{Bmatrix} = \begin{Bmatrix} \mathbf{f}_p \\ \mathbf{0} \end{Bmatrix}} \tag{17.62}$$

Many substructure coupling methods solve the coupled set of equations, Eq. 17.62, by introducing a linear transformation of the form[17.8, 17.12]

$$\boxed{\mathbf{p} = \mathbf{S}\mathbf{q}} \tag{17.63}$$

where $\mathbf{q}$ is the vector of independent *system generalized coordinates*. Let $\mathbf{p}$ be rearranged, if necessary, and partitioned into N_D dependent coordinates $\mathbf{p}_D$ and $N_P - N_D$ linearly independent coordinates $\mathbf{p}_I \equiv \mathbf{q}$, and let Eq. 17.52 be partitioned accordingly, giving

$$[\mathbf{C}_{DD} \quad \mathbf{C}_{DI}] \begin{Bmatrix} \mathbf{p}_D \\ \mathbf{p}_I \end{Bmatrix} = \mathbf{0} \tag{17.64}$$

where $\mathbf{C}_{DD}$ is a nonsingular square matrix. The coordinate transformation equation

$$\mathbf{p} \equiv \begin{Bmatrix} \mathbf{p}_D \\ \mathbf{p}_I \end{Bmatrix} = \begin{bmatrix} -\mathbf{C}_{DD}^{-1}\mathbf{C}_{DI} \\ \mathbf{I}_{II} \end{bmatrix} \mathbf{p}_I \equiv \mathbf{Sq} \tag{17.65}$$

defines both $\mathbf{S}$ and $\mathbf{q}$. Then the vector of independent system generalized coordinates is $\mathbf{q} \equiv \mathbf{p}_I$, and the *substructure coupling matrix* $\mathbf{S}$ is given by

$$\boxed{\mathbf{S} = \begin{bmatrix} -\mathbf{C}_{DD}^{-1}\mathbf{C}_{DI} \\ \mathbf{I}_{II} \end{bmatrix}} \tag{17.66}$$

Substitution of Eq. 17.63 into Eq. 17.61 and premultiplication of the resulting equation by $\mathbf{S}^T$ gives

$$\mathbf{M}_q \ddot{\mathbf{q}} + \mathbf{K}_q \mathbf{q} = \mathbf{f}_q + \mathbf{S}^T \mathbf{C}_p^T \boldsymbol{\sigma} \tag{17.67}$$

where

$$\boxed{\mathbf{M}_q = \mathbf{S}^T \mathbf{M}_p \mathbf{S}, \qquad \mathbf{K}_q = \mathbf{S}^T \mathbf{K}_p \mathbf{S}, \qquad \mathbf{f}_q = \mathbf{S}^T \mathbf{f}_p} \tag{17.68}$$

From Eqs. 17.64 and 17.66, it is seen that $\mathbf{C}_p \mathbf{S} = \mathbf{0}$. Therefore, the system equation of motion, Eq. 17.67, becomes simply

$$\boxed{\mathbf{M}_q \ddot{\mathbf{q}} + \mathbf{K}_q \mathbf{q} = \mathbf{f}_q} \tag{17.69}$$

Although Eq. 17.68 defines $\mathbf{M}_q$, $\mathbf{K}_q$, and $\mathbf{f}_q$ in terms of matrix multiplication operations, the system matrices and force vector can usually be assembled from the substructure matrices by the "direct stiffness" assembly procedure, as illustrated in Section 17.6.

To summarize the generalized component-mode synthesis (CMS) procedure:

1. Choose the component modes to be included in each $\boldsymbol{\Psi}^{(s)}$. This defines the corresponding component generalized coordinates $\mathbf{p}^{(s)}$.
2. Using Eqs. 17.4a, b, and c, form all $\widehat{\mathbf{M}}^{(s)}$, $\widehat{\mathbf{K}}^{(s)}$, and $\widehat{\mathbf{f}}^{(s)}$.
3. Establish which of the coordinates in $\mathbf{p}$ will be the dependent coordinates, $\mathbf{p}_D$; the remainder form $\mathbf{p}_I \equiv \mathbf{q}$.
4. Write the constraint equation (or equations) in the form of Eq. 17.54, and use Eq. 17.66 to solve for $\mathbf{S}$.
5. Determine the coupled system matrices $\mathbf{M}_q$ and $\mathbf{K}_q$ using Eqs. 17.68a and b, respectively, and use Eq. 17.68c to determine the coupled system external force vector $\mathbf{f}_q$. These are the ingredients of Eq. 17.69, the final *coupled system equation of motion*.

The two substructure assembly procedures presented above describe a single level of substructuring, which is illustrated in Sections 17.6 and 17.7. Substructuring can also be employed to treat a structure that is partitioned into several levels of substructures, as discussed briefly in Section 17.8.

17.6 COMPONENT-MODE SYNTHESIS METHODS: FIXED-INTERFACE METHODS

Most applications of component-mode synthesis employ one of two approaches, which may be called *fixed-interface-mode methods* and *free-interface-mode methods*. The former employ fixed-interface normal modes and constraint modes, as illustrated in this section and in Section 17.8. The latter employ free-interface normal modes and attachment modes, as illustrated in Section 17.7. There are also some hybrid methods. It is possible to cite here only a sample of the significant papers dealing with the use of component modes in structural dynamics.

Although there had been previous applications of component modes, Hurty's 1965 paper[17.2] provided the first comprehensive development of a finite-element-oriented CMS method based on constraint modes and fixed-interface modes. Craig and Bampton[17.4] simplified Hurty's method by treating all interface degrees of freedom together rather than requiring the interface degrees of freedom to be separated into rigid-body freedoms and redundant interface freedoms. This method has been widely adopted because of its superior accuracy, its ease of implementation, and its efficient use of computer resources.

17.6.1 Fixed-Interface Displacement Transformation

The displacement transformation of the *Craig–Bampton Method* employs a combination of fixed-interface normal modes (Eq. 17.7 and Fig. 17.3) and interface constraint modes (Eq. 17.13 and Fig. 17.5), and takes the form

$$\mathbf{u}^{(s)} \equiv \begin{Bmatrix} \mathbf{u}_i \\ \mathbf{u}_b \end{Bmatrix}^{(s)} = \begin{bmatrix} \mathbf{\Phi}_{ik} & \mathbf{\Psi}_{ib} \\ \mathbf{0} & \mathbf{I}_{bb} \end{bmatrix}^{(s)} \begin{Bmatrix} \mathbf{p}_k \\ \mathbf{p}_b \end{Bmatrix}^{(s)} \qquad (17.70)$$

where the *C-B transformation matrix* is

$$\boxed{\mathbf{\Psi}_{\text{CB}}^{(s)} = \begin{bmatrix} \mathbf{\Phi}_{ik} & \mathbf{\Psi}_{ib} \\ \mathbf{0} & \mathbf{I}_{bb} \end{bmatrix}^{(s)}} \qquad (17.71)$$

where $\mathbf{\Phi}_{ik}$ is the interior partition of the matrix of kept fixed-interface modes and $\mathbf{\Psi}_{ib}$ is the interior partition of the constraint-mode matrix.

With component fixed-interface normal modes normalized according to Eq. 17.9, the reduced component mass and stiffness matrices, Eqs. 17.4, have the special forms

$$\widehat{\mathbf{M}}_{\text{CB}}^{(s)} = \begin{bmatrix} \mathbf{I}_{kk} & \widehat{\mathbf{M}}_{kb} \\ \widehat{\mathbf{M}}_{bk} & \widehat{\mathbf{M}}_{bb} \end{bmatrix}^{(s)}, \qquad \widehat{\mathbf{K}}_{\text{CB}}^{(s)} = \begin{bmatrix} \mathbf{\Lambda}_{kk} & \mathbf{0}_{kb} \\ \mathbf{0}_{bk} & \widehat{\mathbf{K}}_{bb} \end{bmatrix}^{(s)} \qquad (17.72)$$

The zeros in the *kb* and *bk* partitions of $\widehat{\mathbf{K}}_{\text{CB}}^{(s)}$ are the result of orthogonality equation 17.14. *System Assembly in Generalized Coordinates*, based on Eqs. 17.63 through 17.69, will now be used to assemble the Craig–Bampton reduced-order coupled-system mass and stiffness matrices.[17.4]

17.6.2 Craig–Bampton Method

The bottom row of Eq. 17.70 implies that

$$\mathbf{p}_b^{(s)} \equiv \mathbf{u}_b^{(s)} \tag{17.73}$$

Therefore, in terms of component generalized coordinates, for a two-component system the interface compatibility equation, Eq. 17.53, becomes

$$\mathbf{p}_b^{(1)} = \mathbf{p}_b^{(2)} = \mathbf{u}_b \tag{17.74}$$

Equation 17.65 can be formed directly (see Problem 17.9 for application of Eqs. 17.64 and 17.66) and written in the form

$$\left\{\begin{array}{c} \mathbf{p}_{k_1}^{(1)} \\ \mathbf{p}_b^{(1)} \\ \hline \mathbf{p}_{k_2}^{(2)} \\ \mathbf{p}_b^{(2)} \end{array}\right\} = \begin{bmatrix} I & 0 & 0 \\ 0 & 0 & I \\ \hline 0 & I & 0 \\ 0 & 0 & I \end{bmatrix} \left\{\begin{array}{c} \mathbf{q}_{k_1}^{(1)} \\ \mathbf{q}_{k_2}^{(2)} \\ \mathbf{u}_b \end{array}\right\} \tag{17.75}$$

so the component coupling matrix **S** is just the "direct-stiffness assembly" matrix. For a two-component system, component mass and stiffness matrices (Eq. 17.72) are assembled to form the following <u>system</u> reduced-order mass and stiffness matrices, respectively:

$$\boxed{\widehat{\mathbf{M}}_{\text{CB}} = \begin{bmatrix} \mathbf{I}_{k_1 k_1} & \mathbf{0}_{k_1 k_2} & \widehat{\mathbf{M}}_{k_1 b}^{(1)} \\ \mathbf{0}_{k_2 k_1} & \mathbf{I}_{k_2 k_2} & \widehat{\mathbf{M}}_{k_2 b}^{(2)} \\ \widehat{\mathbf{M}}_{b k_1}^{(1)} & \widehat{\mathbf{M}}_{b k_2}^{(2)} & \widehat{\mathbf{M}}_{bb}^{(1)} + \widehat{\mathbf{M}}_{bb}^{(2)} \end{bmatrix}, \quad \widehat{\mathbf{K}}_{\text{CB}} = \begin{bmatrix} \mathbf{\Lambda}_{k_1 k_1}^{(1)} & \mathbf{0}_{k_1 k_2} & \mathbf{0}_{k_1 b} \\ \mathbf{0}_{k_2 k_1} & \mathbf{\Lambda}_{k_2 k_2}^{(2)} & \mathbf{0}_{k_2 b} \\ \mathbf{0}_{b k_1} & \mathbf{0}_{b k_2} & \widehat{\mathbf{K}}_{bb}^{(1)} + \widehat{\mathbf{K}}_{bb}^{(2)} \end{bmatrix}}$$
(17.76)

(An additional eigensolution can be performed on the *bb*-partition in the lower right-hand corner of the $\widehat{\mathbf{M}}_{\text{CB}}$ and $\widehat{\mathbf{K}}_{\text{CB}}$ matrices, reducing the corresponding terms to $\mathbf{I}_{bb}$ and $\mathbf{\Lambda}_{bb}$, respectively, and correspondingly modifying the *kb* and *bk* terms in $\widehat{\mathbf{M}}_{\text{CB}}$.)

In summary, CMS models based on the use of fixed-interface modes plus interface constraint modes are essentially reduced-order *superelements*: All physical boundary coordinates are retained in Eq. 17.63 as independent generalized coordinates, greatly facilitating component coupling. Because of the simple, straightforward procedures for formulating the component modes employed by this method; because of the straightforward way in which components are coupled to form the component-mode system model; because of the sparsity patterns of the resulting system matrices and the ease of adding adding additional modal coordinates; and because this method also produces highly accurate models with relatively few component modes[17.9]; this method has been widely used and is available in a number of commercial finite element codes (e.g., MSC/NASTRAN[17.27]). Rixen[17.28] refers to this form of component assembly as *primal assembly*, as depicted in Fig. 17.1b.

The principal drawback of the CB method is that the system submatrices related to fixed-interface modes and constraint modes are difficult to verify experimentally. As presented above, all boundary coordinates are retained in the final Craig–Bampton reduced-order system model. For two-dimensional (i.e., plate and shell) structures, and particularly for three-dimensional solids, the number of interface DOFs can become very large. Farhat and Geradin[17.25] have presented an extension of the Craig–Bampton Method that permits coupling of substructures that have incompatible interface grids and have shown how to reduce the number of interface DOFs. Craig and Hale[17.23] have presented a variant of the Craig–Bampton Method that uses fixed-interface Krylov vectors instead of fixed-interface normal modes. Constraint modes and fixed-interface normal modes also form the basis for model reduction in the AMLS multilevel substructuring method, which is discussed briefly in Section 17.8.

Fransen[17.29] has compared the mode-displacement, mode-acceleration, and modal-truncation-augmentation methods for recovering the internal loads of Craig-Bampton reduced dynamic substructure models.

17.7 COMPONENT-MODE SYNTHESIS METHODS: FREE-INTERFACE METHODS

If a reduced set of component free-interface modes is used without including a complete set of either interface constraint modes or interface attachment modes, the component-mode set is not statically complete, as indicated in Section 17.4. This is true of the "classical" CMS method of using only a set of free-interface normal modes. The accuracy of reduced-order models produced by this method is unacceptable.[17.9] However, methods that employ free-interface normal modes together with attachment modes (including residual-flexibility attachment modes and/or inertia-relief attachment modes) have been used fairly widely, especially *MacNeal's Method*[17.5] and *Rubin's Method*[17.6], and have also been used in the context of experimental verification of finite element models.[17.17–17.19]

17.7.1 Augmented Free-Interface Displacement Transformation

Let us assume that the typical component has rigid-body degrees of freedom. Then the basic displacement transformation for this class of free-interface methods employs a combination of rigid-body modes Ψ_r (from Eq. 17.16), kept free–free normal modes Φ_k (from Eq. 17.11), and residual-flexibility attachment modes Ψ_d (from Eq. 17.39). To simplify the following discussion, the rigid-body modes can be written in the following two-partition form:

$$\Psi_r_{N_u \times N_r} \equiv \begin{bmatrix} \Psi_{ir} \\ \hline \Psi_{er} \\ \mathbf{I}_{rr} \end{bmatrix} \equiv \begin{bmatrix} \Psi_{ir} \\ \Psi_{br} \end{bmatrix} \qquad (17.77)$$

These rigid-body modes are then combined with the kept free-interface flexible modes and with residual-flexibility attachment modes to give the following coordinate transformation:

$$\mathbf{u}^{(s)} \equiv \begin{Bmatrix} \mathbf{u}_i \\ \mathbf{u}_b \end{Bmatrix}^{(s)} = [\ \boldsymbol{\Psi}_r \ \ \boldsymbol{\Phi}_k \ \ \boldsymbol{\Psi}_d\ ]^{(s)} \begin{Bmatrix} \mathbf{p}_r \\ \mathbf{p}_k \\ \mathbf{p}_d \end{Bmatrix}^{(s)} \tag{17.78}$$

As noted in Section 17.1, if a load is prescribed at an "interior" DOF, this DOF is relabeled as a "boundary" DOF. From Eq. 17.78, the *augmented free-interface transformation matrix* is[4]

$$\boxed{\boldsymbol{\Psi}_{\text{RFA}}^{(s)} = [\ \boldsymbol{\Psi}_r \ \ \boldsymbol{\Phi}_k \ \ \boldsymbol{\Psi}_d\ ]^{(s)} = \begin{bmatrix} \boldsymbol{\Psi}_{ir} & \boldsymbol{\Phi}_{ik} & \boldsymbol{\Psi}_{ib} \\ \boldsymbol{\Psi}_{br} & \boldsymbol{\Phi}_{bk} & \boldsymbol{\Psi}_{bb} \end{bmatrix}^{(s)}} \tag{17.79}$$

(*Note*: As in Eq. 17.39, the columns of the residual-flexibility attachment mode matrix $\boldsymbol{\Psi}_d$ are labeled b, not d, because these modes are created by unit forces acting at the boundary degrees of freedom.)

With mass-normalized rigid-body modes $\boldsymbol{\Psi}_r$ and mass-normalized free-interface normal modes $\boldsymbol{\Phi}_k$, the component generalized mass matrix and stiffness matrix based on the free-interface transformation matrix are

$$\boxed{\widehat{\mathbf{M}}_{\text{RFA}}^{(s)} = \begin{bmatrix} \mathbf{I}_{rr} & \mathbf{0}_{rk} & \mathbf{0}_{rb} \\ \mathbf{0}_{kr} & \mathbf{I}_{kk} & \mathbf{0}_{kb} \\ \mathbf{0}_{br} & \mathbf{0}_{bk} & \widehat{\mathbf{M}}_{bb} \end{bmatrix}^{(s)}, \quad \widehat{\mathbf{K}}_{\text{RFA}}^{(s)} = \begin{bmatrix} \mathbf{0}_{rr} & \mathbf{0}_{rk} & \mathbf{0}_{rb} \\ \mathbf{0}_{kr} & \boldsymbol{\Lambda}_{kk} & \mathbf{0}_{kb} \\ \mathbf{0}_{br} & \mathbf{0}_{bk} & \widehat{\mathbf{K}}_{bb} \end{bmatrix}^{(s)}} \tag{17.80}$$

where it can be shown that $\widehat{\mathbf{K}}_{bb} = \boldsymbol{\Psi}_{bb}$. The zeros in these matrices are the result of orthogonality (e.g., Eqs. 17.37 and 17.38), and the residual flexibility term $\boldsymbol{\Psi}_{bb}$ of $\widehat{\mathbf{K}}_{\text{RFA}}^{(s)}$ is due to Eq. 17.35. The formulation above is a consistent Ritz transformation; residual-flexibility effects are included in both the stiffness and mass matrices.

17.7.2 Craig–Chang Method

A free-interface method that employs the RFA component modes in Eqs. 17.79 and 17.80 was introduced in Refs. [17.8] and [17.12]. This free-interface method has been referred to in the literature as the *Craig–Chang Method*. Due to the zeros in the third row of the RFA mass matrix and stiffness matrix in Eq. 17.80, the <u>component</u> equation of motion for $\mathbf{p}_d$ becomes

$$\widehat{\mathbf{M}}_{bb}\ddot{\mathbf{p}}_d + \widehat{\mathbf{K}}_{bb}\mathbf{p}_d = \boldsymbol{\Psi}_{bb}^T\{\mathbf{f}_b + \mathbf{r}_b\} \tag{17.81}$$

where the boundary force term comes from Eq. 17.5. Two assumptions are made: (1) that the inertia term can be neglected and a pseudostatic solution obtained for the residual-flexibility coordinates, $\mathbf{p}_d$, and (2) that the interface reaction forces alone drive

---

[4] The subscript RFA signifies rigid-body modes plus free-interface modes plus attachment modes.

17.7 Component-Mode Synthesis Methods: Free-Interface Methods

the behavior of the residual-flexibility coordinates, as is the case for free vibration. Then since $\widehat{\mathbf{K}}_{bb} = \mathbf{\Psi}_{bb} = \mathbf{\Psi}_{bb}^{T}$, Eq. 17.81 reduces to

$$\mathbf{\Psi}_{bb}\{\mathbf{p}_d - \mathbf{r}_b\} = \mathbf{0} \tag{17.82}$$

Since $\mathbf{\Psi}_{bb}$ is nonsingular,

$$\mathbf{p}_d = \mathbf{r}_b \tag{17.83}$$

When two components are coupled, the interface reaction forces acting on the two components satisfy the following force constraint equation:

$$\mathbf{r}_b^{(1)} + \mathbf{r}_b^{(2)} = \mathbf{0} \tag{17.84}$$

This can be used as an additional equation of constraint, as demonstrated in the following example.

Example 17.4 Let each of the two beam components in Fig. 1a be represented by a set of free-interface normal modes and a set of residual-flexibility attachment modes defined by forces applied at the interface (boundary) coordinates. Figure 1c illustrates the forces applied to component (1) to define its attachment modes and also illustrates one free-interface normal mode.

Let the displacement transformation for each component have the form

$$\mathbf{u}^{(s)} = [\mathbf{\Phi}_k \ \mathbf{\Psi}_d]^{(s)} \begin{Bmatrix} \mathbf{p}_k \\ \mathbf{p}_d \end{Bmatrix}^{(s)} = \begin{bmatrix} \mathbf{\Phi}_{ik} & \mathbf{\Psi}_{ib} \\ \mathbf{\Phi}_{bk} & \mathbf{\Psi}_{bb} \end{bmatrix}^{(s)} \begin{Bmatrix} \mathbf{p}_k \\ \mathbf{p}_d \end{Bmatrix}^{(s)} \tag{1}$$

Let the unassembled component generalized coordinate vector have the form

$$\mathbf{p} = \lfloor \mathbf{p}_k^{(1)\mathrm{T}} \ \mathbf{p}_d^{(1)\mathrm{T}} \mid \mathbf{p}_k^{(2)\mathrm{T}} \ \mathbf{p}_d^{(2)\mathrm{T}} \rfloor^{\mathrm{T}} \tag{2}$$

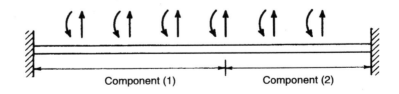

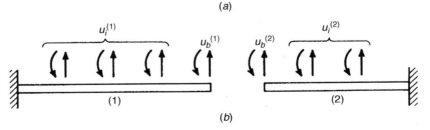

Figure 1 Component-mode synthesis: (a) coupled beam system; (b) cantilever beam components; (c) component modes.

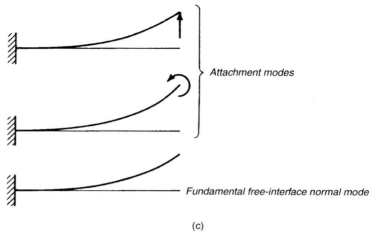

(c)

Figure 1 (*continued*)

and let the final system generalized coordinate vector have the form

$$\mathbf{q} = \lfloor \mathbf{p}_k^{(1)\text{T}} \quad \mathbf{p}_k^{(2)\text{T}} \rfloor^{\text{T}} \tag{3}$$

Determine the constraint matrix $\mathbf{C}_p$ of Eq. 17.52 that corresponds to the displacement constraint of Eq. 17.53 and the force constraint of Eq. 17.84.

SOLUTION The two equations of constraint to be enforced are, from Eq. 17.53,

$$\mathbf{u}_b^{(1)} - \mathbf{u}_b^{(2)} = \mathbf{0} \tag{4}$$

and from Eqs. 17.83 and 17.84,

$$\mathbf{p}_d^{(1)} + \mathbf{p}_d^{(2)} = \mathbf{0} \tag{5}$$

Finally, the constraint matrix has the form

$$\mathbf{C}_p = \begin{bmatrix} \mathbf{\Phi}_{bk}^{(1)} & \mathbf{\Psi}_{bb}^{(1)} & -\mathbf{\Phi}_{bk}^{(2)} & -\mathbf{\Psi}_{bb}^{(2)} \\ \mathbf{0} & \mathbf{I} & \mathbf{0} & \mathbf{I} \end{bmatrix} \quad \textbf{Ans.} \tag{6}$$

where the top row corresponds to the displacement constraint of Eq. 4 and the bottom row corresponds to the force constraint of Eq. 5.

It is left as Problem 17.11 for the transformation matrix $\mathbf{S}$ of Eqs. 17.63 through 17.66 to be determined for a similar 2-component structure.

One advantage of this method is that all $\mathbf{p}_d$ coordinates are reduced out, leaving only the kept free-interface modal coordinates in the final system model. However, it is typical of free-interface methods that the final reduced-order coupled system mass and stiffness matrices lose the sparsity exhibited by the RFA component matrices of Eqs. 17.80. That is the case for this free-interface method. One free-interface method that does preserve the sparsity of Eqs. 17.80 in the assembled reduced-order model is a method that Rixen developed recently and named the *Dual Craig–Bampton Method*.[17.28]

17.7.3 Rixen's Dual C-B Method

As noted immediately following Eq. 17.50 and illustrated in Example 17.3, the Lagrange multiplier vector $\boldsymbol{\lambda}$ contains the interface forces that enforce the N_C interface displacement constraints, Eq. 17.41. Therefore, the displacement transformation proposed by Rixen has the form

$$\mathbf{u}^{(s)} = \boldsymbol{\Psi}_r^{(s)}\mathbf{q}_r^{(s)} + \boldsymbol{\Phi}_k^{(s)}\mathbf{q}_k^{(s)} + \mathbf{G}_d^{(s)}\mathbf{C}^{(s)\mathrm{T}}\boldsymbol{\lambda} \qquad (17.85)$$

Note how this equation differs from the expression for $\mathbf{u}^{(s)}$ in Eq. 17.78, which does not require that the residual flexibility generalized coordinates $\mathbf{p}_d^{(s)}$ be identified with the constraint forces.

Starting with Eq. 17.51, repeated here,

$$\begin{bmatrix} \mathbf{M} & \mathbf{0} \\ \mathbf{0} & \mathbf{0} \end{bmatrix} \begin{Bmatrix} \ddot{\mathbf{u}} \\ \ddot{\boldsymbol{\lambda}} \end{Bmatrix} + \begin{bmatrix} \mathbf{K} & -\mathbf{C}^\mathrm{T} \\ -\mathbf{C} & \mathbf{0} \end{bmatrix} \begin{Bmatrix} \mathbf{u} \\ \boldsymbol{\lambda} \end{Bmatrix} = \begin{Bmatrix} \mathbf{f} \\ \mathbf{0} \end{Bmatrix} \qquad (17.51)$$

we use Eq. 17.85a and approximate the vector of unknowns as

$$\begin{Bmatrix} \mathbf{u} \\ \boldsymbol{\lambda} \end{Bmatrix} \equiv \begin{Bmatrix} \mathbf{u}^{(1)} \\ \vdots \\ \mathbf{u}^{(N_S)} \\ \boldsymbol{\lambda} \end{Bmatrix} = \boldsymbol{\Psi}_{\mathrm{DCB}} \begin{Bmatrix} \mathbf{q}_r^{(1)} \\ \mathbf{q}_k^{(1)} \\ \vdots \\ \mathbf{q}_r^{(N_S)} \\ \mathbf{q}_k^{(N_S)} \\ \boldsymbol{\lambda} \end{Bmatrix} \equiv \boldsymbol{\Psi}_{\mathrm{DCB}} \tilde{\mathbf{q}} \qquad (17.86)$$

where the *Dual C-B transformation matrix* has the form

$$\boldsymbol{\Psi}_{\mathrm{DCB}} = \begin{bmatrix} \boldsymbol{\Psi}_r^{(1)} & \boldsymbol{\Phi}_k^{(1)} & & & & \mathbf{G}_d^{(1)}\mathbf{C}^{(1)\mathrm{T}} \\ & & \ddots & \ddots & & \vdots \\ & & & & \boldsymbol{\Psi}_r^{(N_S)} & \boldsymbol{\Phi}_k^{(N_S)} & \mathbf{G}_d^{(N_S)}\mathbf{C}^{(N_S)\mathrm{T}} \\ & & & & & & \mathbf{I}_{cc} \end{bmatrix} \qquad (17.87)$$

Blank blocks of the transformation matrix are all zeros; dots indicate that the diagonal blocks and right-hand-column blocks follow the indicated patterns.

Combining Eqs. 17.51 and 17.86 in "Rayleigh–Ritz fashion," we get the following *reduced-order hybrid system equation of motion*[5]:

$$\tilde{\mathbf{M}}\ddot{\tilde{\mathbf{q}}} + \tilde{\mathbf{K}}\tilde{\mathbf{q}} = \tilde{\mathbf{f}} \qquad (17.88)$$

---

[5] Note that the constraint is applied directly in physical coordinates and that the transformation to component generalized coordinates includes the Lagrange multiplier vector and well as the component displacement coordinate vector. Therefore, this is not strictly a Rayleigh–Ritz transformation.

where

$$\widetilde{\mathbf{M}} = \mathbf{\Psi}_{DCB}^T \begin{bmatrix} \mathbf{M} & \mathbf{0} \\ \mathbf{0} & \mathbf{0} \end{bmatrix} \mathbf{\Psi}_{DCB}, \qquad \widetilde{\mathbf{K}} = \mathbf{\Psi}_{DCB}^T \begin{bmatrix} \mathbf{K} & -\mathbf{C}^T \\ -\mathbf{C} & \mathbf{0} \end{bmatrix} \mathbf{\Psi}_{DCB}$$
$$\widetilde{\mathbf{f}} = \mathbf{\Psi}_{DCB}^T \begin{Bmatrix} \mathbf{f} \\ \mathbf{0} \end{Bmatrix}$$
(17.89)

The reduced-order hybrid system mass matrix $\widetilde{\mathbf{M}}$ is block diagonal, and the reduced-order hybrid system stiffness matrix $\widetilde{\mathbf{K}}$ is block diagonal except for additional terms in the λ row and λ column.[17.28] Therefore, this Dual Craig–Bampton formulation preserves sparsity of the final mass and stiffness matrices. Whereas the coupling in the Craig–Bampton formulation is in the mass matrix (Eq. 17.76), the coupling terms in the Dual Craig–Bampton formulation are in the stiffness matrix.

This method does not employ a true Rayleigh–Ritz transformation of displacement coordinates, but transforms the hybrid coupled-system equations with a hybrid transformation matrix. Therefore, the eigenvalues of the reduced-order model are not guaranteed to be upper bounds and can even be negative. However, on fairly large sample problems, the method is reported to have produced excellent results.[17.28]

17.8 BRIEF INTRODUCTION TO MULTILEVEL SUBSTRUCTURING

Recently, Bennighof[6] and his graduate assistants have developed an *automated multilevel substructuring algorithm* called the *AMLS Method*.[17.20,17.21,17.30] This method uses an automated partitioning procedure that creates many levels of substructures (up to 30 or more levels) for finite element models having up to several million degrees of freedom. Then, using constraint modes and fixed-interface normal modes, in a manner similar to that discussed in Section 17.6.1, AMLS efficiently produces reduced-order models for which it calculates accurate eigensolutions and frequency-response solutions.

The development of AMLS was motivated by the need to compute frequency response, over a broad frequency range, of structures modeled with several million degrees of freedom. Direct computation of solutions of the FE model's frequency-response equations at hundreds of frequencies for models that large is prohibitive, so mode superposition has traditionally been used instead. With the modal frequency-response approach, the computational burden is shifted to solving for thousands of modes of million-DOF models. However, use of the Block Lanczos Algorithm to perform these eigensolutions requires so much data bandwidth that only vector supercomputers are capable of performing large-scale frequency-response analyses with reasonable job turnaround. With AMLS, memory requirements for substructure eigenproblems are much smaller than for the global FE eigenvalue problem. Truncation of the substructure eigensolutions defines the subspace that is used for approximating the global eigensolution needed for modal frequency-response analysis. The accuracy of the lowest-frequency approximate global eigenpairs is excellent, and the accuracy of the highest-frequency ones depends primarily on how severely the substructure eigensolutions are truncated. Consequently, AMLS enables the analyst to work with geometrically

---

[6]Professor Jeffrey K. Bennighof, <jkb@vibes.ae.utexas.edu>.

accurate FE models of complex structures generated directly from CAD representations and has made it possible for very large frequency-response problems to be solved using inexpensive computer workstation hardware rather than supercomputers.

This section, which was abstracted from Chapters 2 and 4 of Ref. 17.20, provides only a brief introduction to the concept of multilevel substructuring and indicates the significant computational benefits that have been demonstrated to date for this extension of component-mode synthesis. For further details, the reader should consult Refs. [17.20] and [17.21].

The AMLS algorithm consists of five phases:

1. Generate an FE model (e.g., use NASTRAN to generate **K** and **M**).
2. Reorder DOFs and partition **K** and **M** into substructures.
3. Transform matrices to create the reduced-order component-mode model.
4. Compute the reduced-order system eigensolution.
5. Compute the frequency-response solution.

17.8.1 Single-Level Substructuring

To point out several key differences between single- and multilevel substructuring, we begin with a rectangular plate divided into two components. In classic CMS literature, the plate in Fig. 17.9a would be said to consist of two components (or substructures) separated by a boundary. The reduced-order stiffness and mass matrices for the separate components would be generated and then assembled to form the reduced-order system model. By contrast, the AMLS implementation of multilevel substructuring starts in phase 1 with the fully assembled system stiffness and mass matrices for the plate and in phase 2 partitions them into three substructures: Subdomains 1 and 2 in Fig. 17.9a are labeled substructures 1 and 2, and the interface between subdomains 1 and 2 is also considered to be a substructure, labeled substructure 3. So the AMLS model is considered to have three substructures, whose DOFs correspond to the three displacement vectors $\mathbf{x}_1$, $\mathbf{x}_2$, and $\mathbf{x}_3$.

Figure 17.9b shows a graph of the simple relationship between the substructures. This graph is referred to as a *substructure tree diagram* and illustrates the relationships among the substructures pictured in Fig. 17.9a. Substructures 1 and 2 are referred to

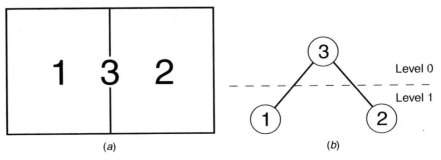

Figure 17.9 Rectangular plate partitioned into two subdomains (components): (*a*) plate divided into three substructures; (*b*) substructure tree diagram. (From [17.20].)

as the *bottom-level* substructures because there are no substructures below them on the diagram. Substructure 3 is a higher-level substructure and could also be called the *parent* of substructures 1 and 2. (In the present case, substructure 3 is also the *highest-level* substructure.) Substructure 3 is said to be at level 0; its children, substructures 1 and 2, are said to be at level 1. In typical applications of AMLS, there may be 30 or more such levels.

For the simple plate model in Fig. 17.9, the original assembled system stiffness and mass matrices would have the forms

$$\mathbf{K} = \begin{bmatrix} \mathbf{K}_{1,1} & \mathbf{0} & \mathbf{K}_{1,3} \\ \mathbf{0} & \mathbf{K}_{2,2} & \mathbf{K}_{2,3} \\ \mathbf{K}_{3,1} & \mathbf{K}_{3,2} & \mathbf{K}_{3,3} \end{bmatrix}, \quad \mathbf{M} = \begin{bmatrix} \mathbf{M}_{1,1} & \mathbf{0} & \mathbf{M}_{1,3} \\ \mathbf{0} & \mathbf{M}_{2,2} & \mathbf{M}_{2,3} \\ \mathbf{M}_{3,1} & \mathbf{M}_{3,2} & \mathbf{M}_{3,3} \end{bmatrix} \quad (17.90)$$

From the AMLS perspective, the transformation of substructure 1 can be written

$$\begin{Bmatrix} \mathbf{x}_1 \\ \mathbf{x}_2 \\ \mathbf{x}_3 \end{Bmatrix} = \begin{bmatrix} \mathbf{\Phi}_1 & \mathbf{0} & \mathbf{\Psi}_{1,3} \\ \mathbf{0} & \mathbf{I} & \mathbf{0} \\ \mathbf{0} & \mathbf{0} & \mathbf{I} \end{bmatrix} \begin{Bmatrix} \mathbf{\eta}_1 \\ \mathbf{x}_2 \\ \mathbf{x}_3 \end{Bmatrix} = \mathbf{T}^{(1)} \begin{Bmatrix} \mathbf{\eta}_1 \\ \mathbf{x}_2 \\ \mathbf{x}_3 \end{Bmatrix} \quad (17.91)$$

where $\mathbf{\Phi}_1$ is the *matrix of kept fixed-interface modes* for substructure 1, and $\mathbf{\Psi}_{1,3}$ is the *constraint-mode matrix* for constraint-mode shapes in the interior of substructure 1 due to unit displacements on the boundary DOFs that comprise substructure 3. That is, $\mathbf{\Phi}_1$ contains the N_{k_1} lowest-frequency eigenvectors of the problem,

$$\mathbf{K}_{1,1}\mathbf{\Phi}_1 = \mathbf{M}_{1,1}\mathbf{\Phi}_1\mathbf{\Lambda}_1 \quad \text{with} \quad \mathbf{\Phi}_1^T\mathbf{M}_{1,1}\mathbf{\Phi}_1 = \mathbf{I}, \quad \mathbf{\Phi}_1^T\mathbf{K}_{1,1}\mathbf{\Phi}_1 = \mathbf{\Lambda}_1 \quad (17.92)$$

and the constraint modes are defined by the equation

$$\mathbf{\Psi}_{1,3} = -\mathbf{K}_{1,1}^{-1}\mathbf{K}_{1,3} \quad (17.93)$$

The transformation of the second substructure can similarly be written

$$\begin{Bmatrix} \mathbf{\eta}_1 \\ \mathbf{x}_2 \\ \mathbf{x}_3 \end{Bmatrix} = \begin{bmatrix} \mathbf{I} & \mathbf{0} & \mathbf{0} \\ \mathbf{0} & \mathbf{\Phi}_2 & \mathbf{\Psi}_{2,3} \\ \mathbf{0} & \mathbf{0} & \mathbf{I} \end{bmatrix} \begin{Bmatrix} \mathbf{\eta}_1 \\ \mathbf{\eta}_2 \\ \mathbf{x}_3 \end{Bmatrix} = \mathbf{T}^{(2)} \begin{Bmatrix} \mathbf{\eta}_1 \\ \mathbf{\eta}_2 \\ \mathbf{x}_3 \end{Bmatrix} \quad (17.94)$$

It should be noted that AMLS performs a final eigensolution on the interface DOFs, whereas the physical boundary coordinates would be retained in the classical Craig–Bampton single-level substructure method. This third substructure transformation is

$$\begin{Bmatrix} \mathbf{\eta}_1 \\ \mathbf{\eta}_2 \\ \mathbf{x}_3 \end{Bmatrix} = \begin{bmatrix} \mathbf{I} & \mathbf{0} & \mathbf{0} \\ \mathbf{0} & \mathbf{I} & \mathbf{0} \\ \mathbf{0} & \mathbf{0} & \mathbf{\Phi}_3 \end{bmatrix} \begin{Bmatrix} \mathbf{\eta}_1 \\ \mathbf{\eta}_2 \\ \mathbf{\eta}_3 \end{Bmatrix} = \mathbf{T}^{(3)} \begin{Bmatrix} \mathbf{\eta}_1 \\ \mathbf{\eta}_2 \\ \mathbf{\eta}_3 \end{Bmatrix} \quad (17.95)$$

So the coordinate transformation of the entire model can be written

$$\begin{Bmatrix} \mathbf{x}_1 \\ \mathbf{x}_2 \\ \mathbf{x}_3 \end{Bmatrix} = \mathbf{T} \begin{Bmatrix} \mathbf{\eta}_1 \\ \mathbf{\eta}_2 \\ \mathbf{\eta}_3 \end{Bmatrix} \quad (17.96)$$

where the transformation matrix **T** has the formal structure

$$\mathbf{T} = \mathbf{T}^{(1)}\mathbf{T}^{(2)}\mathbf{T}^{(3)} = \begin{bmatrix} \Phi_1 & 0 & \Psi_{1,3}\Phi_3 \\ 0 & \Phi_2 & \Psi_{2,3}\Phi_3 \\ 0 & 0 & \Phi_3 \end{bmatrix} \quad (17.97)$$

Finally, the reduced stiffness and mass matrices are given by

$$\widehat{\mathbf{K}} = \mathbf{T}^T\mathbf{K}\mathbf{T}, \qquad \widehat{\mathbf{M}} = \mathbf{T}^T\mathbf{M}\mathbf{T} \quad (17.98)$$

17.8.2 Multilevel Substructuring

The plate in Fig. 17.10 will now be used to introduce multilevel substructuring procedures, as implemented by AMLS. This will be referred to as the S4 model. Figure 17.11 illustrates a possible substructure numbering scheme for this S4 model, and Fig. 17.12 shows the corresponding substructure tree. For convenience, we introduce notation that refers to the sets of *ancestors* (e.g., "parent," "grandparent") and *descendants* (e.g., "child," "grandchild") of a substructure. Let $\mathcal{P}_i$ be defined as the set of all substructure indices j such that j is an ancestor of substructure i. And let $\mathcal{C}_i$ be defined as the set of all substructure indices j such that j is a descendant of substructure i. Ancestors of a substructure are above it in the substructure tree; descendants are below it. From Fig. 17.11b and c it should become clear why in multilevel substructuring, boundaries

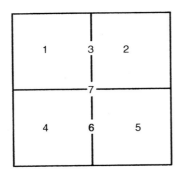

Figure 17.10 Plate partitioned into four subdomains (components). (From [17.20].)

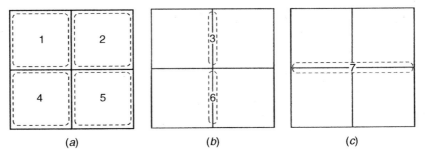

Figure 17.11 Possible substructuring for the S4 model: (a) bottom-level substructures; (b) higher-level substructures; (c) highest-level substructure. (From [17.20].)

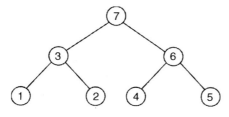

Figure 17.12 Substructure tree for the S4 model. (From [17.20].)

between substructures are also considered to be substructures. For example, although the substructure labeled 3 in Fig. 17.11b is on the boundary between substructures 1 and 2 (in Fig. 17.11a), it is the interior of the higher-level substructure whose boundary with substructure 6 is labeled substructure 7 in Fig. 17.11c.

It is important to note that in AMLS, all degrees of freedom at which external loads (including localized damping) can be applied are included in the highest-level substructure. In practice, substructures of approximately 1000 DOFs have been found to make phase 3 run efficiently.

For this problem, the stiffness and mass matrices may be partitioned as

$$\mathbf{K} = \begin{bmatrix} \mathbf{K}_{1,1} & \mathbf{0} & \mathbf{K}_{1,3} & \mathbf{0} & \mathbf{0} & \mathbf{0} & \mathbf{K}_{1,7} \\ & \mathbf{K}_{2,2} & \mathbf{K}_{2,3} & \mathbf{0} & \mathbf{0} & \mathbf{0} & \mathbf{K}_{2,7} \\ & & \mathbf{K}_{3,3} & \mathbf{0} & \mathbf{0} & \mathbf{0} & \mathbf{K}_{3,7} \\ & & & \mathbf{K}_{4,4} & \mathbf{0} & \mathbf{K}_{4,6} & \mathbf{K}_{4,7} \\ & & & & \mathbf{K}_{5,5} & \mathbf{K}_{5,6} & \mathbf{K}_{5,7} \\ & \text{symm.} & & & & \mathbf{K}_{6,6} & \mathbf{K}_{6,7} \\ & & & & & & \mathbf{K}_{7,7} \end{bmatrix}$$

(17.99)

$$\mathbf{M} = \begin{bmatrix} \mathbf{M}_{1,1} & \mathbf{0} & \mathbf{M}_{1,3} & \mathbf{0} & \mathbf{0} & \mathbf{0} & \mathbf{M}_{1,7} \\ & \mathbf{M}_{2,2} & \mathbf{M}_{2,3} & \mathbf{0} & \mathbf{0} & \mathbf{0} & \mathbf{M}_{2,7} \\ & & \mathbf{M}_{3,3} & \mathbf{0} & \mathbf{0} & \mathbf{0} & \mathbf{M}_{3,7} \\ & & & \mathbf{M}_{4,4} & \mathbf{0} & \mathbf{M}_{4,6} & \mathbf{M}_{4,7} \\ & & & & \mathbf{M}_{5,5} & \mathbf{M}_{5,6} & \mathbf{M}_{5,7} \\ & \text{symm.} & & & & \mathbf{M}_{6,6} & \mathbf{M}_{6,7} \\ & & & & & & \mathbf{M}_{7,7} \end{bmatrix}$$

Substructures are processed, starting with those on the lowest level and proceeding up the substructure tree. The interested reader can consult Ref. [17.20] for details. Here we just indicate the form of the coordinate transformation for the first substructure, the final coordinate transformation matrix, and the final reduced-order stiffness and mass matrices.

17.8 Brief Introduction to Multilevel Substructuring

The coordinate transformation for the first substructure is

$$\mathbf{T}^{(1)} = \begin{bmatrix} \mathbf{\Phi}_1 & \mathbf{0} & \mathbf{\Psi}_{1,3} & \mathbf{0} & \mathbf{0} & \mathbf{0} & \mathbf{\Psi}_{1,7} \\ \mathbf{0} & \mathbf{I} & \mathbf{0} & \mathbf{0} & \mathbf{0} & \mathbf{0} & \mathbf{0} \\ \mathbf{0} & \mathbf{0} & \mathbf{I} & \mathbf{0} & \mathbf{0} & \mathbf{0} & \mathbf{0} \\ \mathbf{0} & \mathbf{0} & \mathbf{0} & \mathbf{I} & \mathbf{0} & \mathbf{0} & \mathbf{0} \\ \mathbf{0} & \mathbf{0} & \mathbf{0} & \mathbf{0} & \mathbf{I} & \mathbf{0} & \mathbf{0} \\ \mathbf{0} & \mathbf{0} & \mathbf{0} & \mathbf{0} & \mathbf{0} & \mathbf{I} & \mathbf{0} \\ \mathbf{0} & \mathbf{0} & \mathbf{0} & \mathbf{0} & \mathbf{0} & \mathbf{0} & \mathbf{I} \end{bmatrix} \quad (17.100)$$

where, as before, $\mathbf{\Phi}_1$ refers to the matrix of kept fixed-interface normal modes of substructure 1 and $\mathbf{\Psi}_{1,j}$ refers to a matrix of constraint modes over the interior of substructure 1 due to unit displacements on the boundary of substructure 1 that is designated as substructure j. Notice that the transformation matrix for substructure 1 deviates from the identity matrix only in blocks involving substructure 1 and its ancestors 3 and 7.

The final coordinate transformation is

$$\lfloor \mathbf{x}_1^T \cdots \mathbf{x}_7^T \rfloor^T = \mathbf{T} \lfloor \boldsymbol{\eta}_1^T \cdots \boldsymbol{\eta}_7^T \rfloor^T \quad (17.101)$$

All $\boldsymbol{\eta}_i$'s except $\boldsymbol{\eta}_7$ are reduced sets of fixed-interface modal coordinates, but $\boldsymbol{\eta}_7$ must contain the full set of modal coordinates equal to the number of coordinates in $\mathbf{x}_7$ in order to preserve static completeness.

The final transformation matrix has the form

$$\mathbf{T} = \prod_{i=1}^{7} \mathbf{T}^{(i)} = \begin{bmatrix} \mathbf{T}_{1,1} & \mathbf{0} & \mathbf{T}_{1,3} & \mathbf{0} & \mathbf{0} & \mathbf{0} & \mathbf{T}_{1,7} \\ & \mathbf{T}_{2,2} & \mathbf{T}_{2,3} & \mathbf{0} & \mathbf{0} & \mathbf{0} & \mathbf{T}_{2,7} \\ & & \mathbf{T}_{3,3} & \mathbf{0} & \mathbf{0} & \mathbf{0} & \mathbf{T}_{3,7} \\ & & & \mathbf{T}_{4,4} & \mathbf{0} & \mathbf{T}_{4,6} & \mathbf{T}_{4,7} \\ & & & & \mathbf{T}_{5,5} & \mathbf{T}_{5,6} & \mathbf{T}_{5,7} \\ & \mathbf{0} & & & & \mathbf{T}_{6,6} & \mathbf{T}_{6,7} \\ & & & & & & \mathbf{T}_{7,7} \end{bmatrix} \quad (17.102)$$

A column of this transformation matrix can be thought of as representing an admissible vector in a Rayleigh–Ritz formulation. For a bottom-level substructure (i.e., 1, 2, 4, and 5), these correspond simply to substructure eigenvectors. For a higher-level substructure, a column contains row entries for the substructure in question and row entries for all of that substructure's descendants. The column for such a higher-level substructure may be called a *multilevel extended eigenvector* and is similar in concept to a constraint mode.

After the transformation at the highest level, the reduced-order system stiffness and mass matrices have the following forms:

$$\widehat{\mathbf{K}} = \begin{bmatrix} \mathbf{\Lambda}_1 & 0 & 0 & 0 & 0 & 0 & 0 \\ & \mathbf{\Lambda}_2 & 0 & 0 & 0 & 0 & 0 \\ & & \mathbf{\Lambda}_3 & 0 & 0 & 0 & 0 \\ & & & \mathbf{\Lambda}_4 & 0 & 0 & 0 \\ & & & & \mathbf{\Lambda}_5 & 0 & 0 \\ & \text{symm.} & & & & \mathbf{\Lambda}_6 & 0 \\ & & & & & & \mathbf{\Lambda}_7 \end{bmatrix}$$

(17.103)

$$\widehat{\mathbf{M}} = \begin{bmatrix} \mathbf{I} & 0 & \widehat{\mathbf{M}}_{1,3} & 0 & 0 & 0 & \widehat{\mathbf{M}}_{1,7} \\ & \mathbf{I} & \widehat{\mathbf{M}}_{2,3} & 0 & 0 & 0 & \widehat{\mathbf{M}}_{2,7} \\ & & \mathbf{I} & 0 & 0 & 0 & \widehat{\mathbf{M}}_{3,7} \\ & & & \mathbf{I} & 0 & \widehat{\mathbf{M}}_{4,6} & \widehat{\mathbf{M}}_{4,7} \\ & & & & \mathbf{I} & \widehat{\mathbf{M}}_{5,6} & \widehat{\mathbf{M}}_{5,7} \\ & \text{symm.} & & & & \mathbf{I} & \widehat{\mathbf{M}}_{6,7} \\ & & & & & & \mathbf{I} \end{bmatrix}$$

If there are rigid-body modes of the structure, these will manifest themselves as zero eigenvalues of the highest-level substructure eigenproblem, here in $\mathbf{\Lambda}_7$.

Reference [17.20] discusses the complete multilevel computing strategy, including efficiencies that result from truncating the number of substructure eigenvectors (Eq. 4.7.3) and efficiencies that result when the number of output DOFs is small compared with the total number of finite element DOFs (Eq. 4.7.4).

17.8.3 Summary of an 8.4M-DOF Case Study

Reference 17.30 presents results obtained by using AMLS to compute an eigensolution for an 8.4M-DOF full vehicle (automobile) model. The frequency range of interest was 0 to 500 Hz, and the number of output DOFs was 245. Phase 2 of AMLS divided the 8.4M-DOF model automatically into 18,373 substructures on 31 levels. The substructure size ranged up to 2190 DOFs. For the given substructure eigenproblem cutoff frequency of 3750 Hz, phase 3 computed 135,924 substructure eigenvectors and projected the system $\mathbf{K}$ and $\mathbf{M}$ onto this eigenvector subspace. Phase 4 solved the reduced eigenproblem of order 135,924, looking for more than 11,000 eigenpairs. Finally, phase 5 computed the output at the 245 output DOFs. On a four-900-MHz-processor shared-memory computer having 8 GB of physical memory and 500 GB of disk space, the complete eigensolution for this 8.4M-DOF model took just over 6 hours, of which $3\frac{1}{2}$ hours was spent in phase 3 and $1\frac{1}{2}$ hours in phase 4.

REFERENCES

[17.1] W. C. Hurty, *Dynamic Analysis of Structural Systems by Component Mode Synthesis*, Technical Report 32-530, Jet Propulsion Laboratory, Pasadena, CA, January 1964.

[17.2] W. C. Hurty, "Dynamic Analysis of Structural Systems Using Component Modes," *AIAA Journal*, Vol. 3, No. 4, 1965, pp. 678–685.

[17.3] R. M. Bamford, *A Modal Combination Program for Dynamic Analysis of Structures (Revision No. 1)*, Technical Memorandum 33-290, Jet Propulsion Laboratory, Pasadena, CA, July 1, 1967.

[17.4] R. R. Craig, Jr. and M. C. C. Bampton, "Coupling of Substructures for Dynamic Analysis," *AIAA Journal*, Vol. 6, No. 7, 1968, pp. 1313–1319.

[17.5] R. H. MacNeal, "A Hybrid Method of Component Mode Synthesis," *Journal of Computers & Structures*, Vol. 1, No. 4, December 1971, pp. 581–601.

[17.6] S. Rubin, "Improved Component-Mode Representation for Structural Dynamic Analysis," *AIAA Journal*, Vol. 13, No. 8, August 1975, pp. 995–1006.

[17.7] R. H. Hintz, "Analytical Methods in Component Modal Synthesis," *AIAA Journal*, Vol. 13, No. 8, August 1975, pp. 1007–1016.

[17.8] R. R. Craig, Jr. and C.-J. Chang, "On the Use of Attachment Modes in Substructure Coupling for Dynamic Analysis," Paper 77-045, presented at the AIAA/ASME 18th Structures, Structural Dynamics and Materials Conference, San Diego, CA, 1977.

[17.9] W. A. Benfield, C. S. Bodley, and G. Morosow, "Modal Synthesis Methods," presented at the Space Shuttle Dynamics and Aeroelasticity Working Group Symposium on Substructure Testing and Synthesis, NASA-TM-X-72318, Marshall Space Flight Center, AL, 1972.

[17.10] R. R. Craig, Jr., "Coupling of Substructures for Dynamic Analysis: An Overview," Paper AIAA-2000-1573, presented at the AIAA Dynamics Specialists Conference, Atlanta, GA, April 5–6, 2000.

[17.11] L. Meirovitch, *Computational Methods in Structural Dynamics*, Sijthoff & Noordhoff, Rockville, MD, 1980.

[17.12] R. R. Craig, Jr., *Structural Dynamics: An Introduction to Computer Methods*, Wiley, New York, 1981.

[17.13] N. M. M. Maia and J. M. M. Silva, eds., *Theoretical and Experimental Modal Analysis*, Research Studies Press, Baldock, Hertfordshire, England, and Wiley, New York, 1997.

[17.14] R. R. Craig, Jr. and Z. Ni, "Component Mode Synthesis for Model Order Reduction of Nonclassically Damped Systems," *AIAA Journal of Guidance, Control, and Dynamics*, Vol. 12, No. 4, July–August 1989, pp. 577–584.

[17.15] J. A. Morgan, C. Pierre, and G. M. Hulbert, "Calculation of Component Mode Synthesis Matrices from Measured Frequency Response Functions, Part I: Theory," *Journal of Vibration and Acoustics*, Vol. 120, No. 2, 1998, pp. 503–508.

[17.16] J. A. Morgan and R. R. Craig, Jr., "Comparison of Three Component Mode Synthesis Methods for Non-proportionally Damped Systems," Paper AIAA-2000-1653, presented at the AIAA Dynamics Specialists Conference, Atlanta, GA, April 5–6, 2000.

[17.17] J. R. Admire, M. L. Tinker, and E. W. Ivey, "Mass-Additive Modal Test Method for Verification of Constrained Structural Models," *AIAA Journal*, Vol. 31, No. 11, 1993, pp. 2148–2153.

[17.18] K. O. Chandler and M. L. Tinker, "A General Mass-Additive Method for Component Mode Synthesis," Paper AIAA-97-1381, *Proceedings of the 38th Structures, Structural Dynamics and Materials Conference*, Kissimmee, FL, April 1997, pp. 93–103.

[17.19] M. L. Tinker, "Free-Suspension Residual Flexibility Testing of Space Station Pathfinder: Comparison to Fixed-Base Results," presented at the 39th AIAA Structures, Structural Dynamics and Materials Conference, Long Beach, CA, April 1998.

[17.20] M. F. Kaplan, "Implementation of Automated Multilevel Substructuring for Frequency Response Analysis of Structures," Ph.D. dissertation, The University of Texas at Austin, Austin, TX, December 2001.

[17.21] J. K. Bennighof and R. B. Lehoucq, "An Automated Multilevel Substructuring Method for Eigenspace Computation in Linear Elastodynamics," *SIAM Journal on Scientific Computing*, Vol. 25, No. 6, 2004, pp. 2084–2106.

[17.22] C. Farhat and R. X. Roux, "A Method of Finite Element Tearing and Interconnecting and Its Parallel Solution Algorithm," *International Journal for Numerical Methods in Engineering*, Vol. 32, 1991, pp. 1205–1227.

[17.23] R. R. Craig, Jr. and A. L. Hale, "Block–Krylov Component Synthesis Method for Structural Model Reduction," *AIAA Journal of Guidance, Control, and Dynamics*, Vol. 11, No. 6, 1988, pp. 562–570.

[17.24] W. A. Benfield and R. F. Hruda, "Vibration Analysis of Structures by Component Mode Substitution," *AIAA Journal*, Vol. 9, No. 7, July 1971, pp. 1255–1261.

[17.25] C. Farhat and M. Geradin, "On a Component Mode Synthesis Method and Its Application to Incompatible Substructures," *Computers & Structures*, Vol. 51, 1994, pp. 459–473.

[17.26] C. Farhat, P.-S. Chen, and J. Mandel, "A Scalable Lagrange Multiplier Based Domain Decomposition Method for Time-Dependent Problems" *International Journal for Numerical Methods in Engineering*, Vol. 38, 1995, pp. 3831–3853.

[17.27] "Introduction to Superelements in Dynamic Analysis," Chapter 10 in *MSC/NASTRAN Reference Manual*, Version 69, MacNeal–Schwendler Corporation, Los Angeles.

[17.28] D. J. Rixen, "A Dual Craig–Bampton Method for Dynamic Substructuring," *Journal of Computational Mathematics*, Vol. 168, 2004, pp. 383–391.

[17.29] Fransen, S. H. J. A, "Data Recovery Methodologies for Reduced Dynamic Substructure Models with Internal Loads," *AIAA Journal*, Vol. 42, No. 10, 2004, pp. 2130–2142.

[17.30] M. Kim, "An Efficient Eigensolution Method and Its Implementation for Large Structural Systems," Ph.D. dissertation, The University of Texas at Austin, Austin, TX, May 2004.

PROBLEMS

The problems in this chapter employ simple axial-deformation components, and are intended to clarify the concepts and methods presented. Since several of the problems are linked together, the instructor should consider assigning all problems that are related to one another: for example, Problems 17.1, 17.2, and 17.9, or Problems 17.3, 17.8, and 17.10.

You may, if you wish use the computer (e.g., MATLAB) to solve any of the problems in this section. *ISMIS* can be used to set up the component matrices.

Problem Set 17.2

17.1 (a) Use Eq. 17.13 to determine the constraint mode for the axial-deformation component in Fig. P17.1,

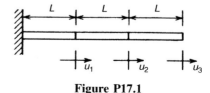

Figure P17.1

where $AE = $ constant, $\rho A = $ constant, and

$$\mathbf{u}_i \equiv \begin{Bmatrix} u_1 \\ u_2 \end{Bmatrix}, \quad \mathbf{u}_b \equiv \{u_3\}$$

(b) Sketch this constraint mode.

17.2 (a) Letting $u_3 = 0$, solve for the two fixed-interface modes of the axial-deformation component in Fig. P17.1. Normalize the modes so that $M_1 = M_2 = \rho A L$. **(b)** Sketch these two normal modes.

Problem Set 17.3

17.3 Since the axial-deformation member in Fig. P17.1 is fixed at its left end, the equation that corresponds to Eq. 17.20 to define an attachment mode with unit force at DOF 3 would be

$$\begin{bmatrix} k_{11} & k_{12} & k_{13} \\ k_{21} & k_{22} & k_{23} \\ k_{31} & k_{32} & k_{33} \end{bmatrix} \begin{Bmatrix} \psi_{1a} \\ \psi_{2a} \\ \psi_{3a} \end{Bmatrix} = \begin{Bmatrix} 0 \\ 0 \\ 1 \end{Bmatrix}$$

(a) Determine this attachment mode, $\boldsymbol{\psi}_a$. **(b)** Sketch this attachment mode.

17.4 (a) Determine the three free-interface normal modes for the bar in Fig. P17.1. Scale the three modes so that $M_1 = M_2 = M_3 = \rho A L$. **(b)** Sketch these three normal modes. **(c)** Using the modal-expansion theorem of Eqs. 10.33 and 10.34, determine the coefficients c_r in the expansion of the attachment mode determined in Problem 17.3 in terms of the free-interface modes determined in part (a). *Note*: This shows that the attachment mode is a linear combination of the three free-interface normal modes; it is not an independent vector.

17.5 (See Example 17.1.) For axial motion of the 3-DOF bar shown in Fig. P17.5, let the DOF sets be $\mathcal{A} = \mathcal{E} = \{1\}$, $\mathcal{I} = \{2\}$, and $\mathcal{R} = \{3\}$. $AE = $ constant and $\rho A = $ constant. Use a consistent mass matrix for the bar. For a unit force applied to the left end of the bar (i.e., for $\mathbf{f}_a^T = \lfloor 1\ 0\ 0 \rfloor$): **(a)** Determine the attachment mode as defined by Eqs. 17.20 and 17.21. Sketch this mode. (*Note:* The order of rows and columns of $\mathbf{K}$ are not the same here as in Eq. 17.20.) **(b)** Determine the self-equilibrated elastic load vector $\mathbf{Pf}_a$. **(c)** Determine the corresponding inertia-relief attachment mode as defined by Eqs. 17.31 and 17.32. Sketch this mode.

17.6 (See Example 17.1.) For axial motion of the 4-DOF free–free spring–mass system shown in Fig. P17.6, let the DOF sets be $\mathcal{A} = \mathcal{E} = \{1\}$, $\mathcal{I} = \{2, 3\}$, and $\mathcal{R} = \{4\}$. For a unit force applied to the left-hand mass (i.e., for $\mathbf{f}_a^T = \lfloor 1\ 0\ 0\ 0 \rfloor$): **(a)** Determine the attachment mode as defined by Eqs. 17.20 and 17.21. Sketch this mode. (*Note:* The order of rows and columns of $\mathbf{K}$ are not the same here as in Eq. 17.20.) **(b)** Determine the self-equilibrated elastic load vector $\mathbf{Pf}_a$. **(c)** Determine the corresponding inertia-relief attachment mode as defined by Eqs. 17.31 and 17.32. Sketch this mode.

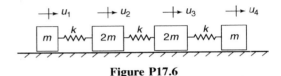

Figure P17.6

Problem Set 17.4

17.7 (See Example 17.2.) For axial motion of the 4-DOF free–free spring–mass system shown in Fig. P17.6, let the DOF sets be $\mathcal{A} = \mathcal{E} = \{1\}$, $\mathcal{I} = \{2, 3\}$, and $\mathcal{R} = \{4\}$. **(a)** Determine the four free-interface normal modes for the spring–mass system in Fig. P17.6. Normalize the modes so that $M_1 = M_2 = M_3 = M_4 = m$. **(b)** Sketch these four normal modes. **(c)** Let $\boldsymbol{\Psi}_r \equiv [\boldsymbol{\phi}_1]$, $\boldsymbol{\Phi}_k \equiv [\boldsymbol{\phi}_2]$, and $\boldsymbol{\Phi}_d \equiv [\boldsymbol{\phi}_3\ \boldsymbol{\phi}_4]$. Determine the residual inertia-relief attachment mode, corresponding to Eq. 17.39, associated with a unit force applied to the left-hand mass (i.e., for $\mathbf{f}_a^T = \lfloor 1\ 0\ 0\ 0 \rfloor$). Sketch this mode.

17.8 (a) Determine the three free-interface normal modes for the cantilever bar in Fig. P17.1. Normalize the modes so that $M_1 = M_2 = M_3 = \rho A L$. **(b)** Sketch these three normal modes. **(c)** Let $\boldsymbol{\Phi}_k \equiv [\boldsymbol{\phi}_1]$ and $\boldsymbol{\Phi}_d \equiv [\boldsymbol{\phi}_2\ \boldsymbol{\phi}_3]$. Determine the residual flexibility matrix, corresponding to Eq. 17.36. **(d)** Determine the residual flexibility attachment mode, corresponding to Eq. 17.39, for a unit force applied to the right end of the bar in Fig. P17.1 (i.e., for $\mathbf{f}_a^T = \lfloor 0\ 0\ 1 \rfloor$). Sketch this mode.

Problem Set 17.6

17.9 In Eq. 17.75 the coupling matrix $\mathbf{S}$ that relates the $\mathbf{p}$ component-mode coordinates and the $\mathbf{q}$ system

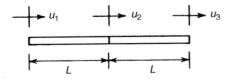

Figure P17.5

coordinates is formed "directly," where, for a two-component system, these vectors have the following forms given in Eq. 17.75:

$$\mathbf{p} \equiv \left\{ \begin{array}{c} \mathbf{p}_{k_1}^{(1)} \\ \mathbf{p}_b^{(1)} \\ \hline \mathbf{p}_{k_2}^{(2)} \\ \mathbf{p}_b^{(2)} \end{array} \right\}, \qquad \mathbf{q} \equiv \left\{ \begin{array}{c} \mathbf{q}_{k_1}^{(1)} \\ \mathbf{q}_{k_2}^{(2)} \\ \mathbf{u}_b \end{array} \right\}$$

(a) Write the displacement constraint matrix $\mathbf{C}_p$ in the format given in Eq. 17.64, letting $\mathbf{p}_b^{(1)} \equiv \mathbf{u}_b$ be the independent boundary DOF and $\mathbf{p}_b^{(2)}$ be the dependent boundary coordinate. Let $\mathbf{p}_{k_1}^{(1)} \equiv \mathbf{q}_{k_1}^{(1)}$ and $\mathbf{p}_{k_2}^{(2)} \equiv \mathbf{q}_{k_2}^{(2)}$ be independent fixed-interface modal coordinates. (b) Using Eq. 17.66, form the coupling transformation matrix $\mathbf{S}$.

17.10 For each of the two components in Fig. P17.10, use the component modes determined in Problems 17.1 and 17.2 to construct a Craig–Bampton system model like the one described in Sections 17.6.1 and 17.6.2 as follows: (a) Use Eqs. 17.72 to form $\widehat{\mathbf{K}}_{CB}^{(1)}$, $\widehat{\mathbf{K}}_{CB}^{(2)}$, $\widehat{\mathbf{M}}_{CB}^{(1)}$, and $\widehat{\mathbf{M}}_{CB}^{(2)}$, keeping only the lowest-frequency fixed-interface normal mode for each component. (b) Use Eq. 17.76a to form the system mass matrix, $\widehat{\mathbf{M}}_{CB}$. (c) Use Eq. 17.76b to form the system stiffness matrix, $\widehat{\mathbf{K}}_{CB}$. (d) Solve the 3-DOF system eigenproblem in Craig–Bampton format.

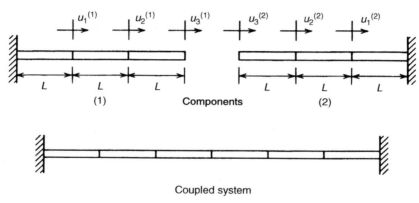

Figure P17.10

Problem Set 17.7

17.11 For each of the two components in Fig. P17.10, use the answers from Problem 17.8 to construct a Craig–Chang type of system model like the one described in Section 17.7.2. Keep one free-interface normal mode and one residual-flexibility attachment mode for each component. To correspond with the order of coordinates in Eqs. 17.63 through 17.68, let

$$\mathbf{p} \equiv \left\{ \begin{array}{c} \mathbf{p}_D \\ \mathbf{p}_I \end{array} \right\} = \left\{ \begin{array}{c} p_d^{(1)} \\ p_d^{(2)} \\ \hline p_k^{(1)} \\ p_k^{(2)} \end{array} \right\}, \qquad \mathbf{q} = \left\{ \begin{array}{c} q_k^{(1)} \\ q_k^{(2)} \end{array} \right\} \qquad (1)$$

(a) Form the connectivity matrix $\mathbf{C}_p$ as in Example 17.4, but with the $\mathbf{p}$ coordinates rearranged as in Eq. 1. The first row of the matrix should be the interface displacement constraint, and the second row should be the interface reaction-force constraint. (See Example 17.4 for an example of the constraint matrix for a similar coupled-beam problem.) (b) Following Eqs. 17.64 through 17.66, solve for the substructure coupling matrix $\mathbf{S}$ corresponding to the constraint matrix that you set up in part (a). (c) Form the component generalized mass matrices and stiffness matrices: $\widehat{\mathbf{M}}_{CC}^{(1)}$, $\widehat{\mathbf{M}}_{CC}^{(2)}$, $\widehat{\mathbf{K}}_{CC}^{(1)}$, and $\widehat{\mathbf{K}}_{CC}^{(2)}$. (d) Assemble the uncoupled mass matrix $\mathbf{M}_p$ and the uncoupled stiffness matrix $\mathbf{K}_p$, using terms from the component matrices formed in part (c), but inserting them according to the order of the $\mathbf{p}$ coordinates in Eq. 1. (e) Using the results of parts (b) and (d), and using Eqs. 17.68a and b, form the coupled-system mass matrix $\mathbf{M}_q$ and $\mathbf{K}_q$, respectively. These constitute the 2-DOF reduced-order Craig–Chang model. (f) Use the mass matrix and stiffness matrix from part (e) to solve for the two eigenvalues.

17.12 For each of the two components in Fig. P17.10, use the answers from Problem 17.8 to construct a Dual C-B system model like the one described in Section 17.7.3 as follows: **(a)** Form the connectivity matrices $\mathbf{C}^{(1)}$ and $\mathbf{C}^{(2)}$ that enforce compatibility of the component tip displacements $u_3^{(1)}$ and $u_3^{(2)}$ by writing the interface displacement constraint equation as in part (b) of Example 17.3. (*Note:* There is only one constraint equation at the one interface between the two components. Correspondingly, there is only one Lagrange multiplier for this problem.) **(b)** Using the kept free-interface normal mode and the residual flexibility matrix from part (c) of Problem 17.8 and the constraint matrices from part (a) above, form the *Dual C-B Transformation* matrix Ψ_{DCB}, as given in Eq. 17.87. (*Note:* Since both components of the coupled system have one fixed end, there will be no submatrices corresponding to rigid-body modes.) **(d)** Use Eqs. 17.89a and b to form the complete 3-DOF reduced-order hybrid system matrices $\widetilde{\mathbf{M}}$ and $\widetilde{\mathbf{K}}$, respectively. **(e)** Solve the 3-DOF Dual C-B system eigenproblem.

PART V

Advanced Topics in Structural Dynamics

18
Introduction to Experimental Modal Analysis

Several important uses of vibration testing of structures are (1) to confirm the validity of a finite element mathematical model of the structure, (2) to obtain critical design information about structural damping or potential in-service loads, (3) to monitor the operation of a machine to diagnose whether maintenance is required, and (4) to determine mass, stiffness, and/or damping modifications that would improve the performance of the structure, and so on. Vibration testing may be conducted on a full-scale structure such as the airplane undergoing ground vibration testing as pictured in Fig. 1.9a, or it may be conducted on a scale model of the structure as pictured in Fig. 1.9b.

The topic of this chapter is *experimental modal analysis* (EMA), which uses particular types of vibration tests to determine the modal properties (natural frequencies, damping factors, and mode shapes) of structures. The key components of the hardware system used in a typical modal test of a structure, or *test article*—exciter(s), input and output transducers, input and output signal conditioners, and dynamic analyzer—are discussed in Section 18.3. Figure 18.1 depicts these components in a typical ratio-calibration test configuration. From vibration test data, frequency-response functions (FRFs) are calculated, and from these, the modal properties of the structure are estimated. Experimental modal analysis is based on the *mode-superposition principle* for MDOF models of structures. Of particular relevance to the topic of EMA are the following topics from previous chapters:

- *Frequency-response functions* (FRFs) of SDOF systems (Sections 4.2 and 4.3)
- *Vibration-measuring instruments* (Section 4.5)
- *System transfer functions* (Section 5.6)
- *Frequency-domain analysis* (Chapter 7)
- *Natural frequencies and mode shapes* of MDOF models (Sections 10.1 and 10.4)
- *Mode superposition for computation of FRFs* of MDOF systems (Section 11.2)

In Section 18.1 we introduce the topic of experimental modal analysis, and in Section 18.2 we discuss frequency-response functions and their importance in EMA. A brief overview of the hardware used in modal testing is presented in Section 18.3. In Section 18.4 we discuss the computation of digital FRFs from analog input and output signals, and in Sections 18.5 and 18.6 describe processes for extracting natural frequencies, damping factors, and mode shapes from the digitized FRFs.

In this chapter we present only a very brief introduction to the topic of experimental modal analysis; more detailed coverage may be found in textbooks (e.g., Refs. [18.1] to

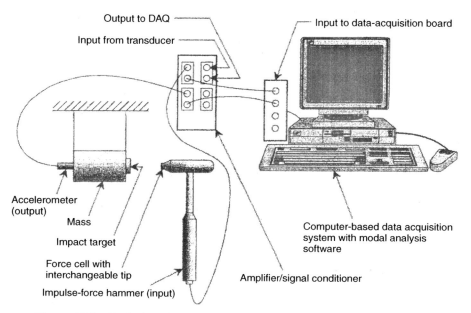

Figure 18.1 Typical modal test hardware: ratio-calibration test configuration.

[18.4]) and in the notes for various short courses (e.g., Ref. [18.5]). By kind permission of its author, Ref. [18.5] is included on the website that accompanies this book.

18.1 INTRODUCTION

In this section we emphasize the important role played in experimental modal analysis (EMA) by frequency-response functions (FRFs). Although real structures may experience various forms of damping and may exhibit nonlinear behavior, this introduction to EMA is restricted to the assumption that the structure under consideration can be modeled as a finite-DOF, linearly elastic, viscously damped system. Therefore, in the present discussion, the N-DOF mathematical model[1]

$$\mathbf{M\ddot{u}} + \mathbf{C\dot{u}} + \mathbf{Ku} = \mathbf{p}e^{i\omega t} \tag{18.1}$$

is assumed to represent, with sufficient accuracy, the dynamic behavior of a structure being tested.[2]

In Sections 11.1 and 11.2 we discuss the mode-superposition solution of Eq. 18.1 for the special case of *modal damping*. In Section 11.2 in particular we discuss the mode-superposition representation of frequency-response functions. The mode-superposition solution of Eq. 18.1 for the *steady-state response* $\mathbf{u}(t)$, based on the

---
[1]The numerical value of N, the *number of effective DOFs* of the structure, must be determined during the process of analyzing the vibration data.
[2](a) In previous discussions of FRFs, we have used an overbar ($\bar{\ }$) to denote complex quantities. That notation is dropped in this chapter. (b) It is quite common in EMA literature for $j = \sqrt{-1}$ to be used. However, we continue to use the notation i rather than j.

modes $\boldsymbol{\phi}_r$ of the undamped system, is given by

$$\mathbf{u}(t) = \mathbf{U}e^{i\omega t} = \sum_{r=1}^{N} \frac{\boldsymbol{\phi}_r \boldsymbol{\phi}_r^{\mathrm{T}} \mathbf{p}}{K_r} \frac{1}{[1-(\omega/\omega_r)^2] + i[2\zeta_r(\omega/\omega_r)]} e^{i\omega t} \quad (18.2)$$

where ω is the *forcing frequency* in rad/s, which is varied over the frequency range of interest in the modal test.

The steady-state displacement response at coordinate i due to harmonic (force) excitation of unit magnitude only at coordinate j is called the *frequency-response function* for the *response at i* due to *excitation at j*. When the FRF deals with displacement per unit force, it is called a *receptance FRF*. However, as discussed in Section 18.2, it is far more common to measure acceleration and use *accelerance* FRFs.[3] The (complex) receptance FRF has the form

$$\boxed{H_{ij}(f) \equiv H_{u_i/p_j}(f) = \sum_{r=1}^{N} \frac{\phi_{ir}\phi_{jr}}{K_r} \frac{1}{(1-r_r^2) + i(2\zeta_r r_r)}} \quad (18.3)$$

In modal testing, the forcing frequency is usually given as f hertz $= \omega/2\pi$, so the forcing frequency ratio for the rth mode is $r_r = 2\pi f/\omega_r$. Obviously, FRFs contain information about natural frequencies (in ω_r), damping factors (in ζ_r), and mode shapes (in $\boldsymbol{\phi}_r$), and these are directly related to the mass, damping, and stiffness properties of the structure. *Modal testing* is the procedure that is employed to measure FRFs, and from them to estimate these physical properties of the structure being tested.

Consider simulated modal testing of the 3-DOF cantilever beam in Fig. 18.2. The beam and its mode shapes are shown in Fig. 18.2. The test beam has *light damping*, that is, all three ζ_r's are much less than 1. In discussing simulated modal tests, we will refer to Eq. 18.3 and use the idealized receptance FRF magnitude plots (log magnitude versus log frequency) for the 3-DOF beam, shown in Fig. 18.3.[4] Each row ($i = 1, 2, 3$) is associated with the output (i.e., response) DOF i; each column ($j = 1, 2, 3$) is associated with input (i.e., excitation) DOF j. A frequency response function for which the input DOF and the output DOF are the same (i.e., $i = j$) is called a *drive-point FRF*. If the input DOF and output DOF are different (i.e., $i \neq j$), the FRF is called a *cross FRF*. For example, H_{33} is a drive-point FRF; H_{13} is a cross FRF.

Before we proceed with discussion of simple simulated modal testing of the cantilever beam depicted in Figs. 18.2, let us use Eq. 18.3 and Fig. 18.3 to make the following observations about the information that can be obtained from the FRFs:

- At every forcing frequency f, the FRF matrix is symmetric. That is, $H_{ij}(f) = H_{ji}(f)$.
- For this simple structure, the natural frequencies are widely separated.
- Since the modes are lightly damped and the natural frequencies are widely separated, when the forcing frequency is equal to one of the undamped natural frequencies (i.e., when $f = f_r$), the FRFs are dominated by that rth mode. Therefore, peaks occur in the FRFs at or very near the three natural frequencies.

---

[3] See Ref. [18.1] for alternative FRF names.
[4] Although accelerometers are the indicated output transducers in Fig. 18.3, the FRFs are shown as receptance (i.e., displacement/force) FRFs.

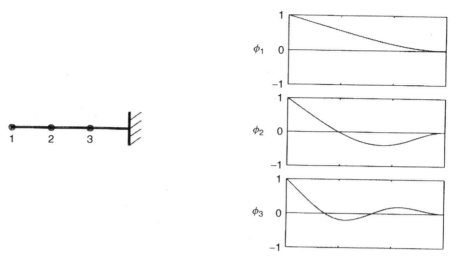

Figure 18.2 A 3-DOF beam and its mode shapes (simulated).

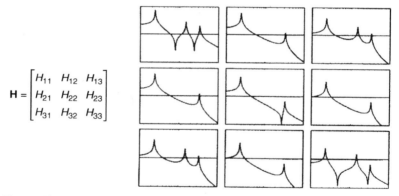

Figure 18.3 Receptance FRF magnitude plots for an idealized 3-DOF beam.

- Since as depicted in mode shape 2 in Fig. 18.2, DOF2 is at the node point of mode 2 (i.e., $\phi_{22} = 0$), there is no peak in row 2 or column 2 when $f = f_2$.

Additional characteristics of FRFs are discussed in Section 18.2.

There are three major types of modal test: *single input–multiple output* (SIMO), *multiple input–single output* (MISO), and *multiple input–multiple output* (MIMO). The ground vibration test depicted in Fig. 1.9a is a multiple input–multiple output (MIMO) modal test of an airplane, with several electrodynamic shakers providing the force inputs and accelerometers measuring the acceleration outputs at over 200 locations on the airplane. With state-of-the-art modal analysis software, data from such MIMO vibration tests can be used to identify up to 40 or more modes of the structure.

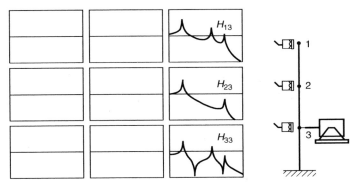

Figure 18.4 Measurement of a column of the FRF matrix.

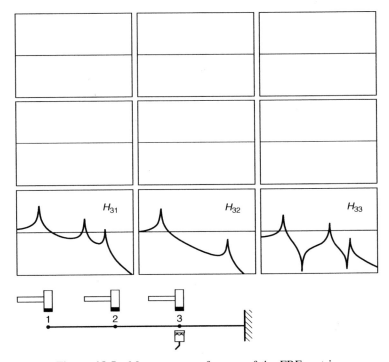

Figure 18.5 Measurement of a row of the FRF matrix.

Figure 18.4 depicts a single-input–multiple-output (SIMO) test; the single input is an electrodynamic shaker at DOF3; and there are three accelerometers, each measuring the output at one of the three output locations on the beam. Note that this produces column 3 of the 3 × 3 FRF matrix. Figure 18.5 depicts a multiple-input–single-output (MISO) test; the multiple inputs are produced by impact hammer hits at each of the three DOFs on the beam, and there is a single accelerometer measuring the output at

DOF3. Note that this produces <u>row</u> 3 of the 3 × 3 FRF matrix. The modal test hardware items depicted in Figs. 18.4 and 18.5—electrodynamic shakers, impact hammers, and accelerometers—are discussed in Section 18.3.

18.2 FREQUENCY-RESPONSE FUNCTION REPRESENTATIONS

Section 18.2.1 discusses types of FRFs and FRF display formats for SDOF systems. In Section 18.2.2 we continue the discussion of frequency-response functions (FRFs) based on MDOF models of structures with modal damping. In Section 18.2.3 the discussion of FRFs is extended to structures with general viscous damping. Other forms of damping (e.g., structural damping) are not discussed in this chapter.

18.2.1 Frequency-Response Functions for SDOF Systems

The equation of motion for an SDOF system with linear viscous damping undergoing harmonic excitation is

$$m\ddot{u} + c\dot{u} + ku = p_0 e^{i\omega t} \tag{18.4}$$

The steady-state displacement response is

$$u(t) = \frac{p_0/k}{(1-r^2) + i(2\zeta r)} e^{i\omega t} \tag{18.5}$$

and the corresponding *receptance FRF* is

$$\boxed{H_{u/p}(f) = \frac{1/k}{(1-r^2) + i(2\zeta r)}} \tag{18.6}$$

where $r = f/f_n$.

From Eq. 18.5 it can be seen that the complex displacement has the form

$$u(t) = Ue^{i\omega t} \tag{18.7}$$

So the (complex) velocity and (complex) acceleration are given by

$$v(t) = i\omega Ue^{i\omega t}, \qquad a(t) = -\omega^2 Ue^{i\omega t} \tag{18.8}$$

respectively. Therefore, the *mobility FRF* (i.e., *velocity* output per unit force input) is

$$H_{v/p}(f) = i\omega \frac{1/k}{(1-r^2) + i(2\zeta r)} \tag{18.9}$$

and the *accelerance FRF* (i.e., <u>acceleration</u> output per unit force input) is

$$H_{a/p}(f) = -\omega^2 \frac{1/k}{(1-r^2) + i(2\zeta r)} \tag{18.10}$$

Before looking at FRFs for MDOF systems, we first show various ways of plotting receptance, mobility, and accelerance FRFs for a single-DOF system with viscous damping and note some important characteristics of the various plots. Frequency-response functions have complex-number values that are functions of frequency; therefore, they can be written and plotted in magnitude-phase or real-imaginary form, as follows:

$$H(f) = |H(f)| \angle H(f) = H_{\Re}(f) + i H_{\Im}(f)$$

Figure 18.6 shows the three types of FRFs as *Bode plots*; that is, the complex frequency-response functions are plotted as log magnitude versus log frequency and linear phase angle versus log frequency. Figure 18.6a shows the SDOF receptance Bode plot; Fig. 18.6b shows the corresponding mobility Bode plot; and Fig. 18.6c shows the corresponding acceleration Bode plot.

The reader should notice carefully the characteristic differences in the three types of FRFs. Note, in particular, the differences in the slopes of the three Bode plot magnitude curves at frequencies below resonance (i.e., $r < 1$), where the stiffness term

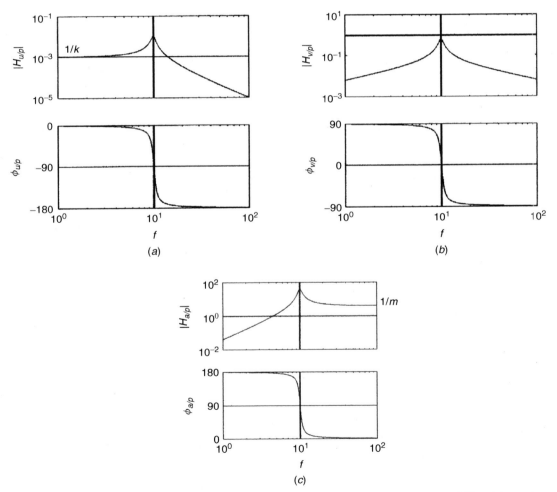

Figure 18.6 Bode plots for an SDOF system with viscous damping: (*a*) receptance; (*b*) mobility; (*c*) acceleration. $k = 1000$, $f_n = 10$, $\zeta = 0.05$.

dominates; and the differences in the slopes of the three Bode plot magnitude curves at frequencies above resonance (i.e., $r > 1$), where the mass term dominates. Also note the differences in the three phase-angle curves: The phase of the mobility is shifted up 90° relative to the receptance, and the accelerance is shifted up 90° further relative to the mobility.

It is sometimes useful to plot modal test FRFs in the form of $\Re(H)$ versus frequency and $\Im(H)$ versus frequency or as $\Im(H)$ versus $\Re(H)$. Figure 18.7a is a repeat of the accelerance FRF of Fig. 18.6c, but plotted as linear magnitude and phase versus linear frequency. Figure 18.7b is a plot of the same accelerance FRF in the format of (linear) $\Re(H)$ and (linear) $\Im(H)$ versus linear frequency. Finally, Fig. 18.7c is a Nyquist plot of the same accelerance FRF, that is, (linear) imaginary versus (linear) real, with frequency as a parameter increasing in the clockwise sense as indicated.

In experimental modal analysis it is useful to express FRFs in *partial-fraction (pole-residue) format*. Equation 18.6 can be cast in the pole-residue format of Eq. 5.61,

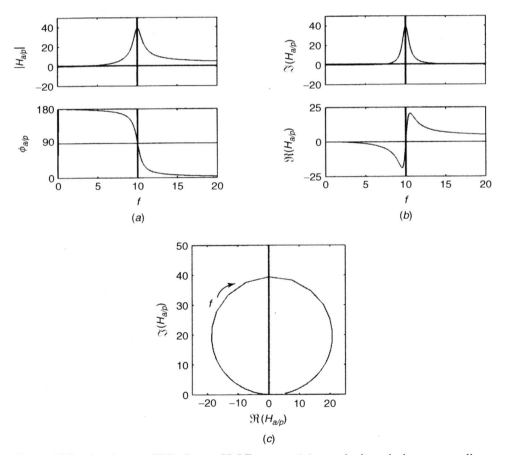

Figure 18.7 Accelerance FRFs for an SDOF system: (*a*) magnitude and phase versus linear frequency; (*b*) real and imaginary versus linear frequency; (*c*) Nyquist plot. $k = 1000$, $f_n = 10$, $\zeta = 0.05$.

as follows:

$$H_{u/p}(f) = \frac{A}{i\omega - \lambda_1} + \frac{A^*}{i\omega - \lambda_1^*} \quad (18.11)$$

where A, A^* are the (complex-conjugate) *residues*, and

$$\lambda_1, \lambda_1^* = -\zeta\omega_n \pm i\omega_n\sqrt{1-\zeta^2} = -\zeta\omega_n \pm i\omega_d \quad (18.12)$$

are the corresponding *poles*. By equating the expressions for $H(f)$ in Eqs. 18.6 and 18.11 it can be shown that the residues A, A^* are the pure imaginary quantities given by

$$A, A^* = \mp i \frac{1}{2m\omega_d} \quad (18.13)$$

18.2.2 Frequency-Response Functions of MDOF Systems Based on Real Normal Modes

Equation 18.3 gives the expression for the *receptance FRF* (i.e., <u>displacement</u> output per unit force input) for a structure with modal damping. From Eq. 18.3, the *magnitude* as a function of the forcing frequency is given by

$$|H_{ij}(f)| \equiv |H_{u_i/p_j}(f)| = \sum_{r=1}^{N} \frac{|\phi_{ir}\phi_{jr}|}{K_r} \frac{1}{\sqrt{(1-r_r^2)^2 + (2\zeta_r r_r)^2}} \quad (18.14a)$$

and the *phase (lag) angle* α_r (i.e., $\angle H_{ur} = -\alpha_r$) is given by

$$\tan\alpha_r(f) = \frac{2\zeta_r r_r}{1-r_r^2} \quad (18.14b)$$

In modal testing, the forcing frequency is usually given as f hertz $= \omega/2\pi$, so the forcing frequency ratio is $r_r = 2\pi f/\omega_r$. The receptance FRF can also be expressed in terms of its *real part*,

$$\Re(H_{ij}) = \sum_{r=1}^{N} \frac{\phi_{ir}\phi_{jr}}{K_r} \frac{1-r_r^2}{(1-r_r^2)^2 + (2\zeta_r r_r)^2} \quad (18.15a)$$

and its *imaginary part*,

$$\Im(H_{ij}) = \sum_{r=1}^{N} \frac{\phi_{ir}\phi_{jr}}{K_r} \frac{-2\zeta_r r_r}{(1-r_r^2)^2 + (2\zeta_r r_r)^2} \quad (18.15b)$$

In the same manner as derived for SDOF systems in Eqs. 18.7 through 18.10, the *mobility FRF* (i.e., <u>velocity</u> output at i per unit force input at j) for MDOF systems with modal damping is given by

$$H_{v_i p_j}(f) = i\omega \sum_{r=1}^{N} \frac{\phi_{ir}\phi_{jr}}{K_r} \frac{1}{(1-r_r^2) + i(2\zeta_r r_r)} \quad (18.16)$$

and the *accelerance FRF* (i.e., acceleration output at i per unit force input at j) is given by

$$H_{a_i p_j}(f) = -\omega^2 \sum_{r=1}^{N} \frac{\phi_{ir}\phi_{jr}}{K_r} \frac{1}{(1 - r_r^2) + i(2\zeta_r r_r)} \tag{18.17}$$

Accelerance FRFs are the FRFs used most commonly in modal testing.

Finally, as given for SDOF systems in Eq. 18.11, the partial-fraction (pole-residue) form for the receptance FRF of an underdamped MDOF system is

$$H_{ij} = \sum_{r=1}^{N} \left(\frac{A_{ijr}}{i\omega - \lambda_r} + \frac{A_{ijr}^*}{i\omega - \lambda_r^*} \right) \tag{18.18}$$

where A_{ijr}, A_{ijr}^* are the (complex conjugate) modal residues, and

$$\lambda_r, \lambda_r^* = -\zeta_r \omega_r \pm i\omega_r \sqrt{1 - \zeta_r^2} = -\zeta_r \omega_r \pm i\omega_{dr} \tag{18.19}$$

are the corresponding poles. By equating the expressions for $H_{ij}(f)$ in Eqs. 18.3 and 18.18 on a mode-by-mode basis, it can be shown that the analytical expressions for the residues of an MDOF system with modal damping are the pure imaginary quantities given by

$$A_{ijr}, A_{ijr}^* = \mp i \frac{\phi_{ir}\phi_{jr}}{2M_r \omega_{dr}} \tag{18.20}$$

where

$$M_r = \boldsymbol{\phi}_r^T \mathbf{M} \boldsymbol{\phi}_r \tag{18.21}$$

However, in modal testing practice the mass matrix is seldom known, and the modal damping factor can be estimated only roughly. Therefore, the following expression is the pole-residue equation that is most appropriate for representing the real-normal-mode form of the receptance FRF matrix in experimental modal analysis:

$$H_{ij}(f) = \sum_{r=1}^{N} \left(\frac{Q_r \phi_{ir}\phi_{jr}}{i\omega - \lambda_r} + \frac{Q_r^* \phi_{ir}\phi_{jr}}{i\omega - \lambda_r^*} \right) \tag{18.22}$$

where the values of the Q_r's are determined in the process of fitting this expression to measured FRF data.[18.4]

18.2.3 Frequency-Response Functions of MDOF Systems Based on Complex Modes

In Section 18.2.2 the discussion was limited to FRFs for systems having real normal modes, that is, to systems having modal damping. However, there are structures whose damping cannot be represented by this model and for which it is necessary to consider more general viscous damping, as discussed in Section 10.4. There it was shown that the equation of motion for a system with general viscous damping can be written in generalized state-space form (Eq. 10.96),

$$\mathbf{A}\dot{\mathbf{z}} + \mathbf{B}\mathbf{z} = \mathbf{P}e^{i\omega t} \tag{18.23}$$

18.2 Frequency-Response Function Representations

where

$$\mathbf{A} = \begin{bmatrix} \mathbf{C} & \mathbf{M} \\ \mathbf{M} & \mathbf{0} \end{bmatrix}, \qquad \mathbf{B} = \begin{bmatrix} \mathbf{K} & \mathbf{0} \\ \mathbf{0} & -\mathbf{M} \end{bmatrix}, \qquad \mathbf{z} = \begin{Bmatrix} \mathbf{u} \\ \dot{\mathbf{u}} \end{Bmatrix}, \qquad \mathbf{P} = \begin{Bmatrix} \mathbf{p} \\ \mathbf{0} \end{Bmatrix} \qquad (18.24)$$

The complex modes have the state-space form

$$\boldsymbol{\theta}_r \equiv \begin{Bmatrix} \boldsymbol{\theta}_u \\ \boldsymbol{\theta}_v \end{Bmatrix}_r = \begin{Bmatrix} \boldsymbol{\theta}_u \\ \lambda \boldsymbol{\theta}_u \end{Bmatrix}_r \qquad (18.25)$$

where λ_r is a complex scalar *eigenvalue* and $\boldsymbol{\theta}_r$ is the corresponding $2N$ state eigenvector. The subscripts u and v stand for displacement and velocity, respectively. The eigenvalue λ_r can be written in the following forms:

$$\lambda_r \equiv \mu_r + j\nu_r = -\zeta_r \omega_r + i\omega_r \sqrt{1 - \zeta_r^2} = -\zeta_r \omega_r + i\omega_{dr} \qquad (18.26)$$

Then the *natural frequency* ω_r and the *damping factor* ζ_r are given by

$$\omega_r = \sqrt{\mu_r^2 + \nu_r^2}, \qquad \zeta_r = \frac{-\mu_r}{\omega_r} \qquad (18.27)$$

It should be noted that ω_r here is called the *natural frequency*, not the undamped natural frequency.

The *modal matrix* containing the state eigenvectors has the form

$$\boldsymbol{\Theta} = [\boldsymbol{\theta}_1 \quad \boldsymbol{\theta}_2 \quad \cdots \quad \boldsymbol{\theta}_{2N}] \qquad (18.28)$$

The orthogonality equations can be combined with definitions of diagonal terms *modal* a_r and *modal* b_r in the following equations:

$$\boldsymbol{\Theta}^T \mathbf{A} \boldsymbol{\Theta} = \text{diag}(a_r), \qquad \boldsymbol{\Theta}^T \mathbf{B} \boldsymbol{\Theta} = \text{diag}(b_r) \qquad (18.29)$$

and it was shown in Section 10.4 that

$$\lambda_r = -\frac{b_r}{a_r} \qquad (18.30)$$

With harmonic excitation at forcing frequency ω, as given in Eq. 18.23, the steady-state response will have the form[5]

$$\mathbf{z} = \mathbf{Z} e^{i\omega t} \qquad (18.31)$$

where $\mathbf{Z}$ can be expressed in the following mode-superposition form:

$$\mathbf{Z} \equiv \begin{Bmatrix} \mathbf{U} \\ i\omega \mathbf{U} \end{Bmatrix} = \sum_{r=1}^{2N} \frac{\boldsymbol{\theta}_r^T \mathbf{P} \boldsymbol{\theta}_r}{a_r(i\omega - \lambda_r)} \qquad (18.32)$$

Noting the forms of $\mathbf{P}$ and $\boldsymbol{\theta}$ in Eqs. 18.24d and 18.25, respectively, we can extract the displacement partition from Eq. 18.32 and get

$$\mathbf{U} = \sum_{r=1}^{2N} \frac{\{\boldsymbol{\theta}_u\}_r^T \mathbf{p} \{\boldsymbol{\theta}_u\}_r}{a_r(i\omega - \lambda_r)} \qquad (18.33)$$

---

[5] See Section 2.7.2 of Ref. [18.1].

In analogy with Eq. 18.18, the (complex) *receptance FRF* has the form

$$H_{u_i/p_j}(f) = \sum_{r=1}^{2N} \frac{(\theta_{ui})_r (\theta_{uj})_r}{a_r(i\omega - \lambda_r)} \qquad (18.34)$$

where $(\theta_{ui})_r$ is the ith element in the displacement partition of the rth eigenvector, that is, the ith element in $\{\boldsymbol{\theta}_u\}_r$. Since the eigenvectors and eigenvalues occur in complex conjugate pairs, we can write this equation in the form

$$\boxed{H_{u_i/p_j}(f) = \sum_{r=1}^{N} \frac{(\theta_{ui})_r (\theta_{uj})_r}{a_r(i\omega - \lambda_r)} + \frac{(\theta_{ui})_r^* (\theta_{uj})_r^*}{a_r^*(i\omega - \lambda_r^*)}} \qquad (18.35)$$

18.3 VIBRATION TEST HARDWARE

Figure 18.1 shows a schematic of some of the hardware used in a typical vibration test, as configured for a ratio calibration test. Figure 1.9 shows two actual vibration test setups. In addition to the test structure and whatever supports it during the vibration test, the hardware components are[6]:

- Source of excitation force input (e.g., impact hammer, or electrodynamic shaker with signal generator and power amplifier)
- Force transducers and associated signal conditioners
- Output transducers (e.g., accelerometers) and associated signal conditioners
- Dynamic analyzer: a computer with data-acquisition (DAQ) hardware, signal analysis software, and modal analysis software; a printer; and so on.

18.3.1 Boundary Conditions

A very important consideration in vibration testing is the method used to support the structure being tested and the relationship of the boundary conditions of the test article during the vibration test to its operating environment. For example, in flight an airplane has six zero-frequency rigid-body modes. Therefore, special "low-frequency" supports, as illustrated supporting the nose landing gear of the airplane in Fig. 1.9*a* (and in Fig. 6 in the airplane GVT article on the book's website) must be used when an airplane undergoes vibration testing on the ground.[7] Smaller test articles (e.g., the aeroelastic wind tunnel model in Fig. 1.9*b*) may be supported by bungee cords, which act as low-stiffness springs, causing the natural frequencies of the "rigid-body modes" of the structure to be much lower than the lowest natural frequency of the structure's flexible modes. In some vibration tests, the test structure is supported by specially designed *flexures* that are instrumented with force cells or strain gauges to measure the reaction forces between the test article and its supporting structure.

---

[6] The reader should consult Refs. [18.1] and [18.6] for more detailed discussion of the topics in this section.
[7] See Ref. [18.7], which is included on the book's website.

Although in some modal tests the test article may be attached directly to a stiff framelike structure called a "strong-back," or to a massive reinforced concrete floor, no support can be considered to be perfectly rigid. Therefore, care must be taken in interpreting the results of any "fixed-base" modal test, since the vibration that is induced in these "rigid" support structures may affect the natural frequencies, damping factors, and mode shapes of the test article.

18.3.2 Excitation Sources

Typically, excitation is provided to a structure by *electrodynamic shakers*,[18.8] by a special force-measuring *impact hammer*,[18.9] by base excitation, or by the structure's operating environment (e.g., by traffic moving over a bridge). Figure 18.8 shows a small impact hammer with alternative tips and extra weight, and a small electrodynamic shaker with its power amplifier. In the airplane ground vibration test depicted in Fig. 1.9a, there is an electrodynamic shaker under each wingtip. Use of these impact hammers and electrodynamic shakers as excitation sources will be discussed briefly, but the topics of base excitation and operational modal testing are beyond the scope of the present chapter.

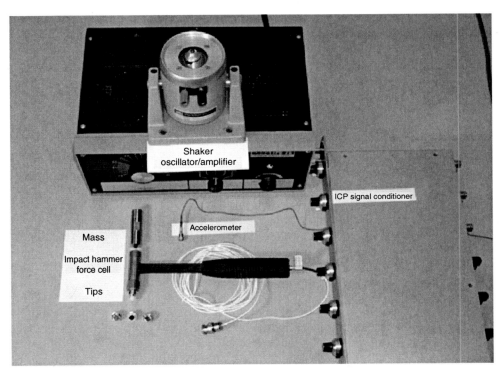

Figure 18.8 Excitation hardware: impact hammer and (small) electrodynamic shaker with power amplifier. ICP signal conditioner for force cell(s) and accelerometer(s).

Figure 18.1 illustrates a test setup for ratio calibration of a hammer–accelerometer pair, and Fig. 18.5 illustrates how an impact hammer might be used to provide an input force at a number of locations on a structure, providing data for a row of the FRF matrix. This vibration test procedure is referred to as *impulse testing*.[18.9] Note in Figs. 18.1 and 18.8 that the part of the hammer head that strikes the structure is specially fitted with a force transducer and an interchangeable tip. The force transducer senses the force and outputs an electrical signal to its associated signal conditioner during the time that the tip is in contact with the structure being tested. The signal conditioner, in turn, sends a time-dependent voltage proportional to the force to a data-acquisition channel of the dynamic analyzer.

Figure 18.9 illustrates the fact that a hard tip (e.g., a nylon or steel tip) produces a short-duration impact (solid curve in upper plot), which results in a broad spectrum (solid curve in lower plot). On the contrary, a very soft tip (e.g., a vinyl tip or soft rubber tip) produces a long-duration impact (dotted curve in upper plot), which results in a narrow spectrum of force (dotted curve in lower plot). By *spectrum* we mean the range of frequencies contained in the Fourier transform $P(f)$ of the time-dependent force signal $p(t)$. The magnitude of the peak force imparted to the structure by the hammer is roughly proportional to the mass of the hammer head times the velocity of the head at impact with the structure.

Most vibration tests that are conducted for the purpose of obtaining a modal model of the structure being tested employ electrodynamic shakers (e.g., the airplane ground vibration test in Fig. 1.9a).[18.7] A computer-generated signal is amplified and used to determine the time history of the force that is produced by the shaker. The following are types of force time histories that are commonly used: sine dwell, swept sine, sine chirp, pure random, and burst random. These are discussed briefly in Section 18.4 and in greater detail in major references on modal testing.

Electrodynamic shakers must be supported. Usually, shakers are mounted on a "firm" base, such as the built-up stands supporting the shakers in Fig. 1.9a. Sometimes a shaker is hung as a translating pendulum, with the mass of the shaker providing the reaction to the force generated by the shaker. The shaker armature is attached to the structure being tested by a "stinger" and force transducer. The stinger is a member (e.g., a thin rod) that is designed to transmit an axial force from the armature of the shaker to the force cell that is attached to the structure.

The major advantage of impact testing over vibration testing with electrodynamic shakers is that the special impact hammers, which range from small hammers weighing less than a pound to heavy sledge hammers, are relatively inexpensive and easily portable. Also, the impact hammer produces a force that is distributed continuously in frequency. However, one of the limitations of impact testing is the fact that the spectrum of frequencies that can be excited is limited to the impact hammer's "useful range," as indicated in Fig. 18.9. Of course, another important limitation of impact testing is that the force imparted to the structure by the hammer must not damage the structure. Consequently, the force content at any specific frequency within the hammer's useful frequency range is quite small. The major advantage enjoyed by electrodynamic shakers is that the direction, the level, and the spectrum of the force applied to the structure can all be controlled. Electrodynamic shakers can be operated in the sine-dwell mode

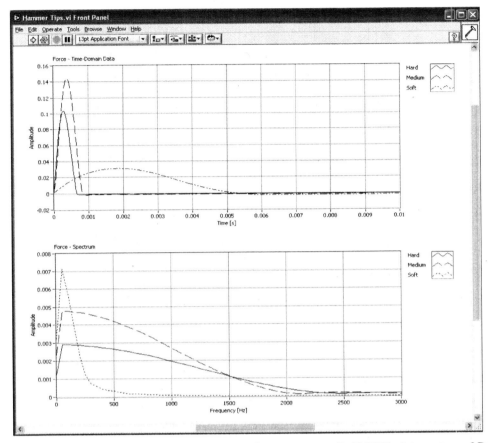

Figure 18.9 Effect of impact tip hardness on a force spectrum. (LabVIEW plot courtesy of D. Bendele. LabVIEW is a registered trademark of National Instruments Corporation.)

at a single frequency. By using several (typically, three or four) shakers, the structure can be subjected to a well-distributed force designed to excite all the vibration modes of interest.

18.3.3 Force and Motion Transducers

A *transducer* is a device that uses some principle of electromechanics to sense a quantity (e.g., force or acceleration) and to convert the sensed quantity into an electrical signal that is proportional to the sensed quantity. Section 4.5 shows how an SDOF spring–mass system can be incorporated into an accelerometer, which is a transducer that measures acceleration. A typical large-scale vibration test such as the one depicted in Fig. 1.9a might employ as many as four electrodynamic shakers with accompanying force transducers, and as many as 200 or 300 accelerometers.

The principle of piezoelectricity is employed in various designs of force transducers (also called force cells) and accelerometers, the transducers that are used to measure inputs and outputs in most vibration tests. Certain natural and human-made crystals exhibit the *piezoelectric effect*; that is, they produce an electric charge output when they are subjected to strain input. *Signal conditioning*, which is partly within the case of the transducer itself and partly external to the transducer, converts this electrical charge into an analog voltage signal (i.e., an electrical voltage that is a continuous function of time), amplifies it, and sends it to a data-acquisition input channel of a dynamic analyzer.[8]

It is important that each input transducer and each output transducer that is to be used in a vibration test be properly *calibrated* for both magnitude and phase prior to its use in a vibration test. For a vibration test that uses one accelerometer and one impact hammer to obtain FRF data, the *ratio calibration test* is a simple procedure that uses a translating pendulum mass (see Fig. 18.1) and Newton's Second Law,

$$p(t) = ma(t)$$

to calibrate the ratio[18.10]

$$\frac{A(f)}{P(f)} = \frac{\text{output acceleration}}{\text{input force}}$$

18.4 FOURIER TRANSFORMS, DIGITAL SIGNAL PROCESSING, AND ESTIMATION OF FRFs

The goal of this section is to provide the reader with a brief introduction to the following data-processing tasks that are involved in the conversion of analog force and acceleration signals into digitized frequency-response functions:

- Fourier series and Fourier transforms; time- and frequency-domain representations of signals
- Digital signal processing, including:
  - Sampled data
  - Aliasing
  - Nyquist sampling theorem
  - Leakage
  - Windows
- Noise; computation of frequency-response functions and coherence functions

In Sections 18.5 and 18.6 we discuss some computational procedures used to identify natural frequencies, damping factors, and mode shapes from collections of these FRFs.

---

[8]The reader should consult textbooks on mechanical measurements (e.g., Ref. [18.11]) for more complete discussion of transducers and their limitations, and should consult the websites of manufacturers of vibration transducers (e.g., PCB, Kistler, Bruel & Kjaer, Endevco) for specifications of available force and motion transducers.

18.4.1 Time- and Frequency-Domain Representations of Signals

The reader should review Sections 7.1 through 7.5 before reading this section. Throughout Sections 18.4 through 18.6 you will be considering the amplitude and phase of sinusoidal signals. Figure 18.10a shows two sinusoidal signals. The solid curve is a plot of the function $A \cos \omega t$; the dashed curve represents $B \cos(\omega t - \alpha) = B \cos \omega(t - \alpha/\omega)$.[9] The *amplitudes* of the two signals (i.e., functions of time) are $A = 1.0$ and $B = 0.5$, respectively. The dashed-line signal is said to *lag* the solid-line signal by a *phase lag angle* α, or, equivalently, to lag it by a *phase lag time* $\tau_\alpha = \alpha/\omega$. As shown in Fig. 18.10b, when phase angles are given for individual signals, the reference for phase is the real axis. That is, signals are phase-referenced to $\cos \omega t$. So $\sin \omega t = \cos(\omega t - \pi/2)$, as illustrated by the dashed curve in Fig. 18.10b.

We assume that general time-dependent signals consist of summations of sinusoidal signals of differing frequencies, amplitudes, and phases. We refer to the *time-domain representation* of a signal and its corresponding *frequency-domain representation*. Figure 18.11(1a) shows the time-domain representation of the function

$$f_1(t) = 3 \cos 2\pi t + 2 \cos 4\pi t$$

and its two sinusoidal components. Figures 18.11(1b,1c) show the frequency-domain representation (i.e., amplitude and phase) of that same function. Figure 18.11(2a) shows the time-domain representation of the function

$$f_2(t) = 3 \cos 2\pi t + 2 \sin 4\pi t$$

and Figs. 18.11(2b,2c) show the frequency-domain representation of this function. In both cases, the two-component signals are periodic, with period $T_1 = T_2 = 1$ s. Although they consist of components having the same amplitude and the same frequency, $f_1(t)$ and $f_2(t)$ are significantly different because of the difference in the phase angles of the second component.

We turn now to the topic of *Fourier integral transforms*, which was introduced in Section 7.3 as the procedure for representing nonperiodic signals in the frequency

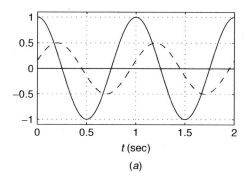
(a)

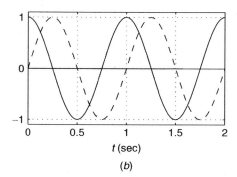
(b)

Figure 18.10 Magnitude and phase of sinusoidal signals.

---

[9] In modal testing work, the frequency of a sinusoidal signal is almost always given as f hertz, but the frequency $\omega = 2\pi f$ rad/s will often be used in mathematical expressions, as here.

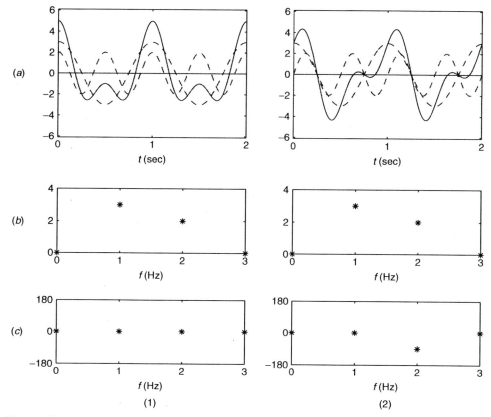

Figure 18.11 Two-component periodic signals: (*a*) time-domain representations; (*b*) frequency-domain representations—magnitude; (*c*) frequency-domain representations—phase. (1) Signal $f_1(t)$ and its two components; (2) signal $f_2(t)$ and its two components.

domain. Equations 7.23 and 7.24 give the following symmetric form for the *Fourier transform pair*, that is, the *direct Fourier transform* and the *inverse Fourier transform*, written in terms of the frequency $f = \omega/2\pi$:[10]

$$P(f) \equiv \mathcal{F}[p(t)] = \int_{-\infty}^{\infty} p(t) e^{-i(2\pi f t)} \, dt \qquad (18.36)$$

$$p(t) \equiv \mathcal{F}^{-1}[P(f)] = \int_{-\infty}^{\infty} P(f) e^{i(2\pi f t)} \, df \qquad (18.37)$$

Recall that the Fourier transform of a rectangular pulse of duration $2T$ was determined in Examples 7.6 and 7.7. Example 7.7 is essentially repeated here.

---

[10] $P(f)$ is usually a complex quantity. However, as in previous sections of this chapter, the overline notation (¯) has been dropped.

Example 18.1 Let $p(t)$ be the rectangular pulse defined by the even function

$$p(t) = \begin{cases} 0, & t < -T \\ p_0, & -T \leq t \leq T \\ 0, & t > T \end{cases}$$

(a) Determine the Fourier transform of this rectangular pulse. Express the transform as a function of the frequency variable f. (b) Plot the Fourier transform.

SOLUTION   (a) From Eq. 18.36,

$$P(f) = \int_{-\infty}^{\infty} p(t) e^{-i(2\pi f t)} \, dt = \int_{-T}^{T} p_0 e^{-i(2\pi f t)} \, dt \tag{1}$$

Therefore,

$$P(f) = \frac{p_0}{-i 2\pi f} (e^{-i 2\pi f T} - e^{i 2\pi f T}) \tag{2}$$

which can also be written in terms of the sinc function, $\operatorname{sinc} \theta = (\sin \theta)/\theta$.

$$P(f) = 2 p_0 T \frac{\sin 2\pi f T}{2\pi f T} \qquad \textbf{Ans. (a)} \tag{3}$$

(b) The Fourier transform $P(f)$ is therefore a real function. In Fig. 1b its amplitude, $|P(f)|$, is plotted versus the frequency variable f. This can be compared with the plot in Fig. 1 of Example 7.6.

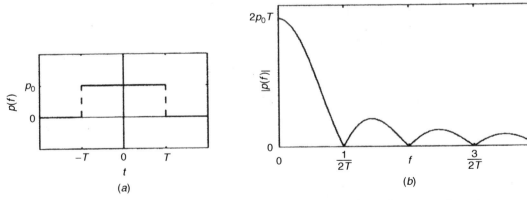

Figure 1   (*a*) Time- and (*b*) frequency-domain representations of a rectangular pulse.

18.4.2   Digital Signal Processing

In this subsection we consider the computational steps that are performed in the *dynamic analyzer* (Fig. 18.1) from the input of analog force and motion signals up to the computation of digital FRFs, and in Section 18.4.3 we discuss the estimation of digital FRFs. These processes are shown schematically in Fig. 18.12. In Sections 18.5 and 18.6 we discuss how the parameters of a modal model are extracted from the digital FRFs.

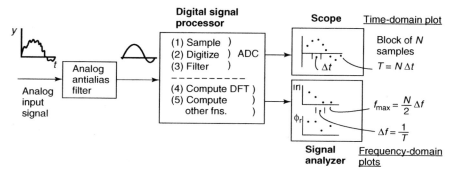

Figure 18.12 Stages of digital signal processing.

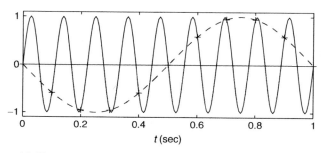

Figure 18.13 Example of aliasing. $f = 9$ Hz, $f_s = 10$ S/s, $f_a = 1$ Hz.

The key digital signal-processing (DSP) assumptions are:

- The signal is periodic in time, with period T_0.
- The signal is *sampled* at N times that are Δt apart, where Δt = sampling interval, $f_s \equiv 1/\Delta t$ = sampling frequency in samples per second (S/s), $T_0 = N\,\Delta t$ = sampling period, and N = block size.
- The sampling interval Δt is small enough and the block size N is large enough to adequately represent the signal being processed.

But how fast does the signal have to be sampled? This leads to the concept of aliasing in the time domain, or simply, *aliasing*. Figure 18.13 shows a sinusoidal signal with true frequency $f = 9$ Hz. The signal is sampled at a sampling frequency of $f_s = 10$ S/s. Note that it is possible to fit an apparent signal of $f_a = 1$ Hz through the 10 sampled points, which are indicated by plus signs. The apparent signal is called an *alias*. The cause of aliasing is that the sampling frequency was not fast enough. It can be shown that[18.1,18.11] a signal must be sampled at a rate that is <u>greater than twice the</u> frequency f_{max} of the highest-frequency component in the signal being sampled. That is, $f_s > 2 f_{max}$. This is called the *Nyquist sampling theorem*, and the frequency

$$f_{Nyq} = \frac{f_s}{2} \qquad (18.38)$$

is called the *Nyquist frequency*.

18.4 Fourier Transforms, Digital Signal Processing, and Estimation of FRFs

Obviously, if an analog input signal consists of components at many frequencies, some above one-half the sampling frequency of the *analog-to-digital converter* (ADC) and some below, the sampled signal will not accurately represent the original analog signal. Components of the signal from above $f_s/2$ will be aliased to lower frequencies and added to the lower-frequency content of the original signal. Thus, the resulting sampled signal will not accurately represent the original analog signal. So what is the "cure" for aliasing? Note in Fig. 18.12 that the first stage is labeled *analog antialias filter*. This is a low-pass analog filter that filters the original analog signal so that only frequency content below a specified *cutoff frequency* is allowed to pass through to be sampled by the ADC. Since antialias filters do not have a sharp cutoff frequency, most analyzers only use the sampled data below $c_d f_{\text{Nyq}}$, where $c_d = 0.5$ to 0.625.[18.1]

The following parameters are involved in digital signal processing:

$f_{d\max}$ = maximum displayed frequency = maximum usable frequency

N_d = number of lines displayed

c_d = frequency-domain display factor

N_f = number of frequency lines calculated

N = frame size = number of time samples

Δf = frequency resolution

f_s = sampling frequency

Δt = sampling interval

T_0 = total sample duration = $1/\Delta f$ always!

The first three parameters are primary quantities; the remaining six are computed from those three. Example 18.2 illustrates the relationships among these parameters.

Example 18.2 Let $c_d = \frac{5}{8}$ be the display parameter set by the analyzer's manufacturer, and let

$$N_d = 320 \text{ lines}, \qquad f_{d\max} = 2560 \text{ Hz}$$

be the two user-selected parameters. Calculate the corresponding values of the following parameters: N_f, N, Δf, f_s, Δt, and T_0.

SOLUTION

$$N_f = \frac{N_d}{c_d} = 512 \text{ frequency lines (512 real, 512 imaginary)}$$

$$N = 2N_f = 1024 \text{ time samples}$$

$$\Delta f = \frac{f_{d\max}}{N_d} = 8 \text{ Hz}$$

$$f_s = N\Delta f = 8192 \text{ S/s}$$

$$\Delta t = \frac{1}{f_s} = 1.221 \times 10^{-4} \text{ s/S}$$

$$T_0 = \frac{1}{\Delta f} = 0.125 \text{ s}$$

Note that the frequency resolution and the acquisition time for one block of data are always the reciprocal of each other; that is,

$$\Delta f = \frac{1}{T_0} \qquad (18.39)$$

Next we consider the problem of *leakage*, another important consideration in digital signal processing. Whereas aliasing occurs when the sampling interval Δt is not small enough, leakage is a direct consequence of the fact that only a finite number of sampled points of the time history can be saved (e.g., $N = 1024$), coupled with the assumed periodicity of the signal. Figure 18.14(1a) shows a sampled 4-Hz sinusoidal signal that is *periodic in the window* $T_0 = N \Delta t$, with $\Delta t = \frac{1}{32}$ s, $N = 32$ samples, and $T_0 = 1$ s. Figure 18.14(1b) shows the corresponding magnitude of the Fourier transform of the sampled signal in Fig. 18.14(1a), and Fig. 18.14(1c) is the phase plot of this Fourier transform. Since the 4-Hz sinusoidal signal is a cosine wave, the phase angle is zero. There is a line in the Fourier transform at every $\Delta f = 1/T_0 = 1$ Hz. Therefore, the Fourier transform of the signal sampled consists of a single line at the true frequency of that signal, 4 Hz. The Fourier transforms in Figs. 18.14(1b,c) and (2b,c) were computed using the fft function in MATLAB. Note that the magnitude of the 4-Hz sinusoidal signal, which is 1 in the time-domain representation in Fig. 18.14(1a), is 16 in the frequency-domain representation in Fig. 18.14(1b). The reason for this is that the algorithm used in MATLAB to compute the FFT multiplies the transform by one-half of the block size, in the present case by $N/2 = 16$.

Figure 18.14(2a) shows a 4.5-Hz cosine signal, also sampled at intervals of $\Delta t = \frac{1}{32}$ s. This signal is <u>not</u> *periodic in the window* $T = 1$ s. Therefore, its Fourier transform cannot consist of a single line at the true frequency of that signal, since there is no line in the Fourier transform at that frequency. Both amplitude and phase of the Fourier transform of this signal have "leaked" into several frequency lines on each side of 4.5 Hz, the true frequency. Therefore, this phenomenon is called *leakage*. Significant errors can occur in the frequency, damping, and mode shape results obtained from FRFs based on sampled data that have been affected by leakage.

The "cure" for the problem of leakage is not as straightforward as is the cure for aliasing. There are two primary approaches that can be employed in the case of leakage. The best approach is to tailor the time history of the input force signal to cause both input and output to be either fully contained within the *analysis time record* $N \Delta t$, or to actually be periodic in this time window (e.g., the *burst random* input time history that is used in modal testing with electrodynamic shakers). A second approach is to apply a *time-domain window* to the input signal and a time-domain window to the output signal (e.g., the *force window* that is applied to the input signal, and the *exponential window* that is applied to the output signal in an impact test). Further discussion of windowing is beyond the scope of this book, but information on leakage and windowing may be found in the major references on modal testing and in the Q & A section in several issues of *Sound & Vibration*.[18.12]

18.4.3 Estimation of Frequency-Response Functions

Although the mathematical representation of a single-input-single-output frequency response function $H(f)$, as given by Eq. 7.27, is

$$U(f) = H(f)P(f) \qquad (18.40)$$

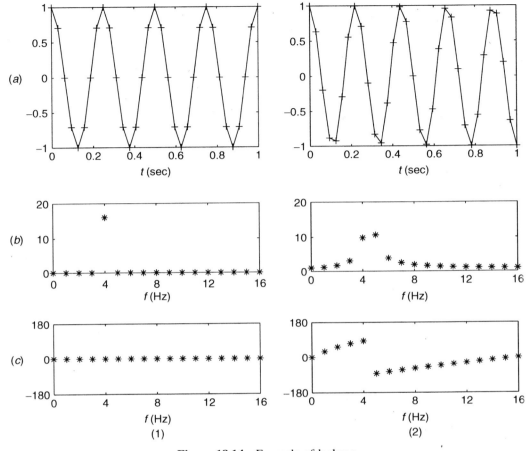

Figure 18.14 Example of leakage.

we need to consider procedures for estimating FRFs from measured signals. Figure 18.15 shows a model that is frequently used to assess how best to calculate FRFs from "noisy" input and output signals.[11]

The *auto-power spectrum* of a time-domain signal $a(t)$ is

$$G_{AA} = A(f)A^*(f) \qquad (18.41)$$

and the *cross-power spectrum* of time-domain signals $a(t)$ and $b(t)$ is

$$G_{AB} = A(f)B^*(f) \qquad (18.42)$$

where $A(f)$ and $B(f)$ are the Fourier transforms of signals $a(t)$ and $b(t)$, respectively, and $(\cdot)^*$ denotes complex conjugate. Note that the auto-power spectrum is a real function of frequency, but the cross-power spectrum is a complex function that carries both magnitude and phase information.

---
[11]SISO and MIMO estimation procedures are discussed in Refs. [18.1] and [18.2], and Ref. [18.13] gives illustrations.

If $H(f)$ represents a linear system and $P(f)$ and $U(f)$ are, respectively, the transforms of the true input and output signals, we can form

$$U(f)P^*(f) = H(f)P(f)P^*(f) \quad \text{and} \quad U(f)U^*(f) = H(f)P(f)U^*(f)$$

by postmultiplying Eq. 18.40 by $P^*(f)$ and $U^*(f)$, respectively. Therefore, $H(f)$ can be calculated from either

$$H_0(f) = \frac{G_{UP}}{G_{PP}} \quad \text{or} \quad H_0(f) = \frac{G_{UU}}{G_{PU}} \tag{18.43}$$

However, we do not have true input and output signals, but must use the measured signals, which contain added noise.

To minimize the effect of the noise, we obtain the following *averaged auto-power spectrum* from different segments $(\cdot)_n$ of the time-domain signal $a(t)$:

$$\tilde{G}_{AA} = \frac{1}{N_{avg}} \sum_{n=1}^{N_{avg}} A_n(f) A_n^*(f) \tag{18.44}$$

and the following *averaged cross-power spectrum* of time-domain signals $a(t)$ and $b(t)$:

$$\tilde{G}_{AB} = \frac{1}{N_{avg}} \sum_{n=1}^{N_{avg}} A_n(f) B_n^*(f) \tag{18.45}$$

If the input noise $m(t)$ and output noise $n(t)$ are not correlated with each other or with the true input signal or true output signal, then with sufficient averaging, the respective auto- and cross-spectra become

$$\tilde{G}_{\hat{U}\hat{P}} = \frac{1}{N_{avg}} \sum_{n=1}^{N_{avg}} (U+n)_n (P+m)_n^* \longrightarrow G_{UP}$$

$$\tilde{G}_{\hat{U}\hat{U}} = \frac{1}{N_{avg}} \sum_{n=1}^{N_{avg}} (U+n)_n (U+n)_n^* \longrightarrow G_{UU} + G_{NN} \tag{18.46}$$

$$\tilde{G}_{\hat{P}\hat{P}} = \frac{1}{N_{avg}} \sum_{n=1}^{N_{avg}} (P+m)_n (P+m)_n^* \longrightarrow G_{PP} + G_{MM}$$

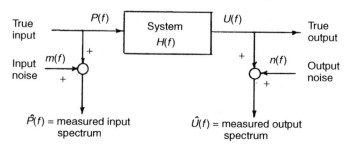

Figure 18.15 Measurement system with noise sources.

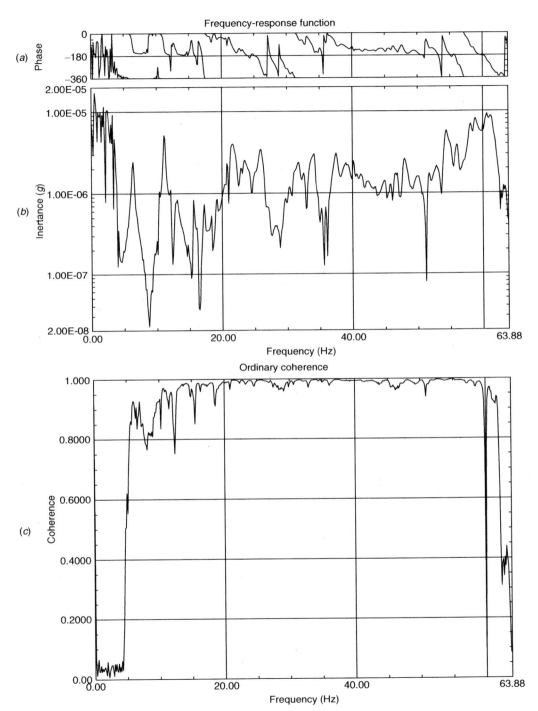

Figure 18.16 H_1 FRF Bode plot: (*a*) phase; (*b*) magnitude; (*c*) associated coherence plot. (Courtesy of Boeing—Huntington Beach.)

Finally, using averaged spectra of measured input and output signals, and using the two equations for $H(f)$ in Eq. 18.43, we define the following three *FRF estimators*:

$$H_1 \triangleq \frac{\tilde{G}_{\hat{U}\hat{P}}}{\tilde{G}_{\hat{P}\hat{P}}} = H_0 \left(\frac{1}{1 + G_{MM}/G_{PP}} \right) \tag{18.47}$$

$$H_2 \triangleq \frac{\tilde{G}_{\hat{U}\hat{U}}}{\tilde{G}_{\hat{P}\hat{U}}} = H_0 \left(1 + \frac{G_{NN}}{G_{UU}} \right) \tag{18.48}$$

and

$$H_v = \sqrt{H_1 H_2} \tag{18.49}$$

From Eqs 18.47 and 18.48, we can see that:

- $H_1(f) \leq H_0(f)$; that is, at every frequency f, $H_1(f)$ is a lower bound to the true FRF.
- $H_1(f)$ in affected only by input noise $m(t)$. Therefore, H_1 is poorest near resonances, where the input is small.
- $H_2(f) \geq H_0(f)$; that is, at every frequency f, $H_2(f)$ is an upper bound to the true FRF.
- $H_2(f)$ in affected only by output noise $n(t)$. Therefore, $H_2(f)$ is poorest near antiresonances, where the output is small.
- $H_v(f)$ is an average of $H_1(f)$ and $H_2(f)$.

A measure of the amount of the output signal that is due to the input signal is the *ordinary coherence*, called γ^2, which is given by

$$\gamma^2(f) = \frac{H_1(f)}{H_2(f)} \tag{18.50}$$

It can be shown that $\gamma^2 \leq 1.0$, with $\gamma^2 = 1.0$ when there is no noise on the input or output. Most modal analyzers provide for calculating and displaying the coherence along with the associated FRF, as illustrated in Fig. 18.16.

Reference works on modal testing should be consulted for corresponding discussions of the effects of noise on the calculation of FRFs and coherence from tests that involve multiple inputs.

18.5 MODAL PARAMETER ESTIMATION

The first phase of experimental modal analysis is the acquisition of analog time-domain records of inputs and responses (Section 18.3). The second phase is digital signal processing, involving sampling, filtering, calculation of Fourier transforms, calculation of frequency response functions, and so on (Section 18.4). The third phase consists of estimating and validating the *modal parameters* of the test article, including the complex-valued *poles* ($\lambda_r = \sigma_r + i\omega_{dr}$), the *modal vectors* ($\boldsymbol{\theta}_r$), the *modal scaling* (modal mass M_r or modal a_r), and the *modal participation vectors*. This phase is called *modal parameter estimation*. Reference [18.5], which can be found on the book's website, presents a good overview of the entire modal analysis process.

Modern techniques of experimental modal analysis were made possible by the development of the fast Fourier transform (FFT) computer algorithm in 1965 and by the introduction of minicomputers in the late 1960s and early 1970s. Today, a number of modal parameter estimation methods are available, and the analyst can choose which one is best for the application at hand. Modal parameter estimation procedures can be identified according to several categories, including the following:

- *Modal parameter versus direct parameter*. Direct-parameter methods are used to compute an MCK model (Eq. 18.1) or a state-space model (Eq. 18.23) from which modal parameters are then calculated.
- *Frequency domain versus time domain*. Frequency-domain methods use FRFs or auto- and cross-spectra; time-domain methods use impulse-response functions.
- *SDOF versus MDOF*. SDOF methods identify one mode at a time on one FRF at a time; MDOF methods identify several modes on one or more FRFs at a time.
- *Single input versus multiple input*. Multiple-input (polyreference) testing is required when a high-order model is required or when the test structure has closely spaced frequency modes.

In the present section we discuss briefly methods for estimating natural frequencies, damping factors, and modal participation vectors; in Section 18.6 we consider methods for estimating mode shapes and for validating modal models. The objective of the coverage in these two sections is just to give the reader an overview of the processes involved in modal parameter estimation. For a more complete description of the various parameter estimation algorithms that are currently available, the reader should consult a modal analysis reference work (e.g., Refs. [18.1] to [18.4]), or a technical paper or comprehensive dissertation on modal parameter estimation (e.g., Refs. [18.14] to [18.19]).

18.5.1 Basic Equations

Frequency-response functions that have been calculated from measured input and vibration response data by the procedures discussed in Section 18.4.3 provide the primary data that are used in experimental modal analysis to calculate estimates of the modal parameters. Expressions for FRFs for MDOF systems were developed in Sections 18.2.2 and 18.2.3. Based on Eqs. 18.22 and 18.35, we can write the following expression for the receptance FRF matrix[12]:

$$[H(i\omega)] \equiv [H_{u/p}(i\omega)] = \sum_{r=1}^{N} \frac{Q_r \boldsymbol{\theta}_r \boldsymbol{\theta}_r^T}{i\omega - \lambda_r} + \frac{Q_r^* \boldsymbol{\theta}_r^* \boldsymbol{\theta}_r^{*T}}{i\omega - \lambda_r^*} \quad (18.51)$$

---

[12]Since it is customary to develop modal parameter estimation algorithms in terms of receptance rather than accelerance, that practice is followed here.

where, for generality, the complex-mode form has been selected. In practice, in a modal test there will be N_i inputs, N_o outputs, and N active modes. Then, for parameter identification purposes, Eq. 18.51 is written in the form

$$[H(i\omega)]_{N_o \times N_i} = [V]_{N_o \times 2N} \left\lceil \frac{1}{i\omega - \lambda_r} \right\rfloor_{2N \times 2N} [L]_{2N \times N_i} \qquad (18.52)$$

where

$$[V] = [\boldsymbol{\theta}_1 \cdots \boldsymbol{\theta}_N \quad \boldsymbol{\theta}_1^* \cdots \boldsymbol{\theta}_N^*] = \text{modal vector matrix}$$

$$[L] = \lceil Q \rfloor [V]^T = \text{modal participation vector matrix}$$

Figure 18.17 shows a simulated mobility plot for a 4-DOF system with widely spaced natural frequencies. The solid curve includes the contributions of all four modes of this system; the dashed curves are the mobility FRFs for (a) the first mode only, (b) modes 2 and 3 only, and (c) mode 4 only. Typically, modal parameter estimation is concentrated on a limited-frequency band containing N_λ observable modes, represented by modes 2 and 3 in this figure. All modes having natural frequencies below this band are approximated by a single term, called the *lower residual* term or *residual inertia* term. All of the modes above the band of interest are represented by a single term called the *upper residual* or *residual flexibility*. Then Eq. 18.52 takes the form

$$[H(i\omega)]_{N_o \times N_i} = [V]_{N_o \times 2N_\lambda} \left\lceil \frac{1}{i\omega - \lambda_r} \right\rfloor [L]_{2N_\lambda \times N_i} - \frac{1}{\omega^2}[LR]_{N_o \times N_i} + [UR]_{N_o \times N_i}$$

$$(18.53)$$

where

$$[V] = [\boldsymbol{\theta}_L \cdots \boldsymbol{\theta}_U \quad \boldsymbol{\theta}_L^* \cdots \boldsymbol{\theta}_U^*] = \text{modal vector matrix } (2N_\lambda \text{ vectors})$$

$$[L] = \lceil Q \rfloor [V]^T = \text{modal participation vector matrix}$$

$$[LR] = \text{matrix of lower residual terms}$$

$$[UR] = \text{matrix of upper residual terms}$$

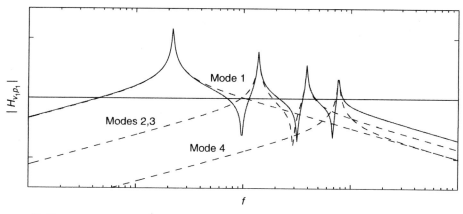

Figure 18.17 Mobility FRF of lightly damped 4-DOF system. Solid curve, all four modes; dashed curves, mode 1, modes 2 and 3, and mode 4.

The residual terms in Eq. 18.53, especially the residual flexibility term, make it possible to match the measured FRFs more closely over the frequency range of interest, particularly near antiresonances.

Time-domain modal parameter estimation is based on impulse response functions. Impulse-response functions are seldom measured directly but are calculated from the corresponding frequency-response functions by use of the inverse FFT algorithm. The impulse-response function matrix corresponding to Eq. 18.53 is

$$[h(t)]_{N_o \times N_i} = [V]_{N_o \times 2N_\lambda} \lceil e^{\lambda_r t} \rfloor_{2N_\lambda \times 2N_\lambda} [L]_{2N_\lambda \times N_i} \tag{18.54}$$

The residual terms in Eq. 18.53 are not accounted for in this expression for the impulse response function matrix. In Sections 18.5.2 and 18.5.3 we consider SDOF parameter-estimation methods, and in Section 18.5.4 we return to Eqs. 18.52 through 18.54, which are the basis for MIMO frequency- and time-domain parameter-estimation methods.

18.5.2 SDOF Parameter-Estimation Methods

In Section 4.6 we discussed the use of frequency response data to determine the natural frequency and damping factor of a lightly damped SDOF system. The procedure applied there was to use the receptance (output displacement/input force) FRF magnitude and phase plots for the lightly damped SDOF system; to use the forcing frequency corresponding to 90° phase lag as the value of the undamped natural frequency f_n; and to use Eq. 4.68 and the frequencies at the *half-power points* to estimate the damping factor ζ. This procedure is referred to as the *half-power method*. Since the forcing frequency at which the magnitude of the FRF reaches its peak is very near the forcing frequency corresponding to 90° phase lag, this frequency is often used as the undamped natural frequency f_n. Also, the procedure can be applied to SDOF acceleration FRFs as well as to receptance FRFs, with proper account taken of the difference in phase angles between receptance and acceleration. You should reread Section 4.6 before continuing to read the present section.

A method that is closely related to the half-power method is the *circle-fit method*. When applied to FRFs for MDOF systems, it is called the *method of Kennedy and Pancu*.[18.20] We begin by examining a Nyquist plot of the mobility FRF for a lightly damped SDOF system with viscous damping; Fig. 18.18 corresponds to the Bode plot in Fig. 18.6b. In Ref. [18.1] it is shown that the receptance FRF for an SDOF system with structural damping (given by Eq. 4.81) plots as a circle, and that the mobility FRF for an SDOF system with viscous damping (given by Eq. 18.9) also plots as a circle. For small values of damping factor (e.g., $\zeta < 0.1$), the acceleration FRF for an SDOF system with viscous damping (given by Eq. 18.10) also plots very nearly as a circle.

The frequency resolution for the data plotted in Fig. 18.18 is $\Delta f = 0.1$ Hz, with plus signs marking the data points from 9.5 to 10.5 Hz, proceeding clockwise from top to bottom. From Eq. 18.10, the undamped natural frequency is the forcing frequency at the point where the circle crosses the positive real axis. This is also the point on the circle that is farthest from the origin, and it is the point that corresponds to the greatest arc length traversed per Δf. The two 45° points on the circle are actually the half-power points that were used in Section 4.6 in the discussion of the half-power method.

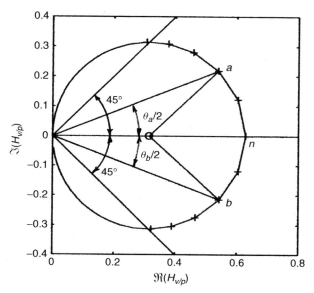

Figure 18.18 Simulated mobility FRF of a lightly damped SDOF system. $k = 1000$, $f_n = 10.0$ Hz, and $\zeta = 0.05$.

Based on the fact that the mobility FRF plot is a circle with center on the positive real axis, we can use the trigonometric properties of a circle with FRF data of two points a and b at equal frequency spacings just below and just above resonance, respectively, to estimate the viscous damping factor ζ. From Fig. 18.18, note that the angles $\theta_a/2$ and $\theta_b/2$ are related to the FRF values $H(f_a)$ and $H(f_b)$ as follows:

$$\tan\frac{\theta}{2} = \frac{|\Im(H)|}{|\Re(H)|} \tag{18.55}$$

From Eq. 18.9,

$$\tan\frac{\theta_a}{2} = \frac{1 - r_a^2}{2\zeta r_a}, \quad \tan\frac{\theta_b}{2} = \frac{r_b^2 - 1}{2\zeta r_b} \tag{18.56}$$

Then the damping factor is given by

$$2\zeta = \frac{r_b^2 - r_a^2}{r_a \tan(\theta_a/2) + r_b \tan(\theta_b/2)} \tag{18.57}$$

Since points a and b are taken at equal frequency spacings on either side of the resonance point $r_n = 1$, $r_a + r_b = 2$, and Eq. 18.57 can be simplified as

$$\boxed{\zeta = \frac{f_b - f_a}{f_a \tan(\theta_a/2) + f_b \tan(\theta_b/2)}} \tag{18.58}$$

For this SDOF system, Eq. 18.56 can be used to calculate the two tangent values. Example 18.3 shows how to apply Eq. 18.58 to calculate the damping factor for the SDOF system whose mobility FRF is shown in Fig. 18.18.

Example 18.3 Use Eq. 18.58 to estimate the damping factor for the SDOF system whose mobility FRF is shown in Fig. 18.18. The data from this figure are:

$$f_a = 9.8 \text{ Hz}, \quad H_{v/p}(9.8 \text{ Hz}) = 0.540 + i\, 0.218$$

$$f_n = 10.0 \text{ Hz}, \quad H_{a/p}(10.0 \text{ Hz}) = 0.628 + i\, 0.0$$

$$f_b = 10.2 \text{ Hz}, \quad H_{a/p}(10.2 \text{ Hz}) = 0.543 - i\, 0.215$$

SOLUTION   Equation 18.58 can be used to estimate the viscous damping factor.

$$\zeta = \frac{f_b - f_a}{f_a \tan(\theta_a/2) + f_b \tan(\theta_b/2)} \tag{1}$$

$$= \frac{10.2 - 9.8}{9.8(0.218/0.540) + 10.2(0.215/0.543)} = 0.050 \qquad \textbf{Ans.} \tag{2}$$

The damping value that was used in creating the FRF plot in Fig. 18.18 was $\zeta = 0.050$.

If points a and b are taken at $\theta_a = \theta_b = 45°$, Eq. 18.58 reduces to Eq. 4.68, the equation used to calculate the damping factor by the half-power method. It must be stressed that Eq. 18.58 for estimating the value of the damping factor is strictly valid only for SDOF systems with light damping. However, it can also be used to estimate damping factors from FRFs of an MDOF system with widely spaced frequencies or even, in some cases, moderately spaced natural frequencies, as illustrated in the following section.

18.5.3 Local MDOF Parameter-Estimation Method

In some cases it is possible to adapt an SDOF modal parameter estimation method to MDOF systems. The word *local* in the section title above refers to parameter estimates that are calculated for one mode at a time on one FRF at a time. Figure 18.19 is the driving-point mobility FRF for a 4-DOF system with two closely spaced natural frequencies, f_2 and f_3. Note in the Nyquist plot of Fig. 18.19c that the mode 2 response has not yet become negligible when the mode 3 response starts to become significant. Because their natural frequencies are widely separated from the natural frequencies of these two closely spaced center modes, the response of mode 1 and the response of mode 3 are nearly circular and are nearly centered on the positive real axis. Mode 1 is the large circle in Fig. 18.19b. The damping factor for this mode could be estimated by using the procedure of Example 18.3.

In Ref. [18.20], Kennedy and Pancu extended to lightly damped MDOF systems the idea of using Nyquist plots to estimate natural frequencies and mode shapes, even for modes as closely spaced as modes 2 and 3 in Fig. 18.19. Their basic assumption was that the response of an MDOF system very near a resonance frequency is due predominantly to that single mode. On a driving-point FRF, each point on the Nyquist plot where the spacing of equal-Δf data points is a local maximum is considered to be the response at one of the system's natural frequencies, f_r, and an arc of a circle is

610 Introduction to Experimental Modal Analysis

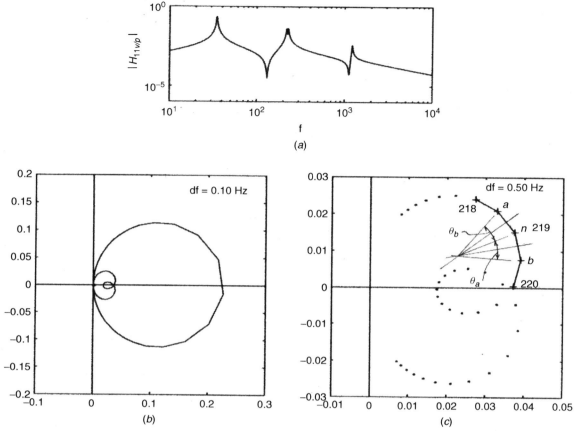

Figure 18.19 Simulated mobility FRFs for a lightly damped 4-DOF system. $f_1 = 35.0$ Hz, $f_2 = 218.4$ Hz, $f_3 = 227.4$ Hz, $f_4 = 1220$ Hz; $\zeta_r = 0.01$, $r = 1, 4$; for 11 to 999 Hz, $\Delta f = 0.50$ Hz.

fitted to the data points in its vicinity.[13] For viscous-damped systems, Eq. 18.58 is then used to estimate that mode's damping factor, ζ_r. Example 18.4 illustrates this process.

Example 18.4 Use Eq. 18.58 to calculate the damping factor for mode 2 of the 4-DOF system whose driving-point mobility FRF $H_{v_1 p_1}$ is shown in Fig. 18.19c. The following data have been estimated from this figure:

$f_a = 218.5$ Hz, $\theta_a = 25.0°$; $f_n = 219.0$ Hz; $f_b = 219.5$ Hz, $\theta_b = 27.0°$

SOLUTION   First, point n is considered to be the data point where the Δf spacing has a local maximum, indicating mode 2. Then points a, b, and n are considered to

---

[13]If three points are used to fit a circle, the coordinates of the center of the circle and the radius of the circle are determined directly by three simultaneous equations. If more than three neighboring FRF points are used, a least-squares solution is used to obtain the best fit to the coordinates of the center of the circle and its radius.

be points that lie on a circle, and straight lines are drawn between points *a* and *n* and between points *n* and *b*. Perpendicular lines drawn from the centers of secant lines *an* and *bn*, respectively, intersect at the center *O* of the (assumed) circle. Lines drawn from points *a*, *b*, and *n* to this center make it possible to measure angles θ_a and θ_b. Equation 18.58 can now be used to estimate the viscous-damping factor ζ_2.

$$\zeta_2 = \frac{f_b - f_a}{f_a \tan(\theta_a/2) + f_b \tan(\theta_b/2)} \qquad (1)$$

$$= \frac{219.5 - 218.5}{218.5 \tan(12.5°) + 219.5 \tan(13.5°)} = 0.010 \qquad (2)$$

Therefore, the estimated natural frequency and damping factor for mode 2 are

$$f_2 = 219.0 \text{ Hz}, \qquad \zeta_2 = 0.010 \qquad \textbf{Ans.} \quad (3)$$

The damping values used in creating the FRF plot in Fig. 18.19 were $\zeta_r = 0.010$.

The results obtained in Example 18.4 would be considered to be "excellent." Although natural frequencies of MDOF systems can often be estimated to within a percent or so, estimated damping factors based on actual measured FRF data are considered to be "good" if independent estimates of damping agree to within 30%.

The procedure illustrated in Example 18.4 is not readily applicable to cross FRFs because sign changes in mode shapes can lead to large angular sweeps of FRF plots per Δf of frequency change at frequencies between the true natural frequencies, as is the case in Fig. 1*b* of Example 11.3. It is also not applicable to driving-point FRFs when two natural frequencies are so close, or the damping levels are so high, that it is impossible to separate the two modes, as was the case in Fig. 1*a* of Example 11.3.

Although, in Refs. [18.1] and [18.2] and other references, variants of the Kennedy–Pancu circle-fit method, like the procedure illustrated above, have been devised for determining the damping factors of MDOF systems with light damping and "moderately closely spaced" frequencies, these procedures generally require time-consuming operator interaction. Since more accurate MDOF parameter-estimation algorithms that require minimal operator interaction are available in commercial modal analysis software, modal parameter estimation based on multi-input multi-output FRFs of MDOF systems is best carried out with a true MIMO parameter-estimation algorithm.

18.5.4 Global MIMO Modal-Parameter-Estimation Methods

A *global* modal-parameter-estimation algorithm is one that processes simultaneously either the FRFs (frequency-domain methods) or the impulse-response functions (time-domain methods) obtained by measuring N_o outputs due to N_i inputs. These methods generally involve a two-stage process. In the first stage, a matrix polynomial is formulated from Eq. 18.52 or 18.54. From it, the poles and modal participation factors are estimated. In a second stage, the modal vectors are estimated. Reference [18.2] provides a synopsis of these methods and cites original sources.

Among the time-domain methods that are available are the following: the Polyreference Least-Squares Complex Exponential Method, the Eigensystem Realization Algorithm (ERA), and the Time-Domain Direct Parameter Identification (TDPI) Algorithm.

Global frequency-domain methods include: the Orthogonal Polynomial (OP) Method, the Frequency-Domain Direct Parameter Identification (FDPI) Algorithm, and the Complex-Mode Indicator Function (CMIF) Method. Finally, recent research has produced the Polyreference Least-Squares Complex Frequency-Domain Method (Poly-MAX) and has shown excellent results in industrial applications.[18.19,18.20] It is beyond the scope of this book to go into details on these global system-identification methods.

18.6 MODE SHAPE ESTIMATION AND MODEL VERIFICATION

We turn now to the use of FRF plots for estimating mode shapes of MDOF systems. Example 11.2 provides FRF plots for a 2-DOF system with widely separated natural frequencies; Example 11.3 provides FRF plots for a related 2-DOF system with closely spaced natural frequencies. Real structural systems such as airplanes and automobiles, have many closely spaced frequencies (e.g., 10 or more natural frequencies occurring within a span of 1 Hz). However, to simplify the discussion of FRFs for MDOF systems, we employ a 4-DOF system with widely separated natural frequencies and without any rigid-body modes.

Figure 18.20a shows a 4-DOF beam similar to the 3-DOF beam in Fig. 18.2. Figure 18.20b shows plots of the imaginary parts of four acceleration (i.e., acceleration/force) FRFs that might be obtained by placing an accelerometer at DOF1 and tapping the beam with a force-measuring hammer at each of the four DOF locations. From Eq. 18.17, the FRFs in Fig. 18.20b are given by the imaginary parts of the acceleration FRFs:

$$H_{a_i p_j} = -\omega^2 \sum_{r=1}^{4} \frac{\phi_{ir}\phi_{jr}}{K_r} \frac{1}{(1 - r_r^2) + i(2\zeta_r r_r)} \tag{18.59}$$

with output DOF $i = 1$.

18.6.1 Mode Shapes of Lightly Damped MDOF Systems: Quadrature Peak-Picking Method

When the natural frequencies are widely separated, as in Fig. 18.20, and the forcing frequency is equal to one of the undamped natural frequencies, $f = f_r$, the FRFs are dominated by the corresponding rth mode. Then, from Eq. 18.59, the complex acceleration FRF is approximately

$$H_{a_i p_j}(f = f_r) \sim i \frac{\phi_{jr}}{2K_r \zeta_r} \phi_{ir} = i \frac{\phi_{ir}}{2K_r \zeta_r} \phi_{jr} \tag{18.60}$$

Thus, mode shape r can be obtained by examining the <u>imaginary part</u> of the FRFs for any output measurement point i for which $\phi_{ir} \neq 0$, while the structure is subjected to harmonic excitation at forcing frequency $f = f_r$ at each DOF j, that is, from a row of the FRF matrix. Mode shape r can also be obtained by examining the <u>imaginary part</u> of the FRFs at each output measurement point i while the structure is subjected to harmonic excitation at forcing frequency $f = f_r$ at any DOF j for which $\phi_{jr} \neq 0$, that is, from a column of the FRF matrix.

18.6 Mode Shape Estimation and Model Verification

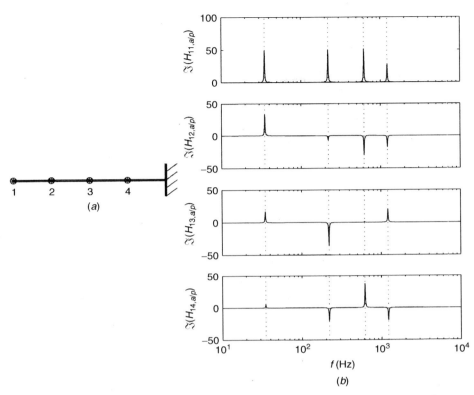

Figure 18.20 (a) 4-DOF beam; (b) simulated row 1 of $\Im(H_{a_i/p_j})$.

Example 18.5 Use Eq. 18.60 to estimate the fundamental mode shape of the 4-DOF system whose acceleration FRFs are shown in Fig. 18.20.

SOLUTION The first peaks in each of the four plots of $\Im(H)_{a_1 p_j}$ occur at the frequency $f_1 = 35$ Hz and have the following values:

$$\boldsymbol{\phi}_1 \sim \lfloor 49.8 \ \ 32.8 \ \ 16.9 \ \ 4.84 \rfloor^T \tag{1}$$

When scaled to make the tip deflection 1.00, the fundamental mode is estimated to be

$$\boldsymbol{\phi}_1 = \lfloor 1.00 \ \ 0.66 \ \ 0.34 \ \ 0.10 \rfloor^T \qquad \textbf{Ans.} \ (2)$$

This does have the same shape as mode 1 in Fig. 18.2.

The mode-shape estimation method illustrated in Example 18.4 is called *quadrature peak picking*, because the values used to estimate the mode shapes are the peak values of the respective quadrature responses (i.e., plots of the imaginary parts of frequency-response functions). Unlike magnitude plots, the quadrature responses carry the sign as well as the amplitude of the mode shape at each DOF.

Note that for this structure with widely separated frequencies, Eq. 18.60 indicates that a mode shape can be obtained from any row or column that does not correspond to a measurement point where that mode is zero. For example, mode shape 3 can be estimated from the FRF values at $f = f_3$ in any row or any column of the imaginary-part plots. Specifically, Fig. 18.20 shows how the peaks of the imaginary part of the FRFs in row 1 can be used to determine the shape of mode 1. It can be seen that the shapes of modes 2 and 3 that can be estimated from Fig. 18.20 agree with the shapes of modes 2 and 3 of the similar 3-DOF beam in Fig. 18.2.

This method of using peaks of the imaginary part of FRFs to determine mode shapes is applicable only to lightly damped simple structures, such as simple beams, that have widely separated natural frequencies. Methods for determining mode shapes from the FRFs of real structures are listed in Section 18.5.4.[14]

18.6.2 Model Verification

A number of computational tools have been developed to assist the analyst in the task of determining a valid modal model (natural frequencies, damping factors, and mode shapes). The following tools are used in determining the proper model order and in separating valid poles from "computational" poles:

- Mode indicator function (MIF)
- Complex-mode indicator function (CMIF)
- Pole stabilization diagram
- Others

Once the estimates of the modal parameters and mode shapes have been determined, the following tools are used to validate the model:

- Comparison of synthesized FRFs with the measured FRFs
- Modal assurance criterion (MAC) between analysis modes and test modes
- Orthogonality
- Sine-dwell testing
- Others

It is beyond the scope of this chapter to discuss these analysis and model verification tools in detail. The subject is discussed in Ref. [18.8] and in the set of tutorial notes (Ref. [18.5]), both of which are included on the website that accompanies this book. References [18.21] to [18.27] list several sources for information about commercial experimental modal analysis software and related software. Many original papers on experimental modal analysis have appeared in proceedings of annual meetings of the International Modal Analysis Conference (IMAC).[18.28]

---

[14] Instructional demos of modal parameter estimation and mode shape animation may be accessed from the Downloads tab of the ME'scopeVES pull-down menu at the website of Vibrant Technology, Inc., http://www.vibetech.com.

REFERENCES

[18.1] D. J. Ewins, *Modal Testing: Theory, Practice and Application*, 2nd ed., Research Studies Press, Baldock, Hertfordshire, England, 2000.

[18.2] W. Heylen, S. Lammens, and P. Sas, *Modal Analysis Theory and Testing*, Katholieke Universiteit Leuven, Leuven, Belgium, 1999.

[18.3] N. M. M. Maia and J. M. M. Silva, *Theoretical and Experimental Modal Analysis*, Wiley, New York, 1997.

[18.4] R. J. Allemang, *Vibrations: Experimental Modal Analysis*, Report UC-SDRL-CN-20-263-663/664, University of Cincinnati, Cincinnati, OH, February 1995.

[18.5] P. Avitabile, *TUTORIAL NOTES: Structural Dynamics and Experimental Modal Analysis*, Modal Analysis and Controls Laboratory, University of Massachusetts, Lowell, MA, 2000 (included on CD as P._Avitabile_TUTORIAL_NOTES.pdf).

[18.6] K. G. McConnell, *Vibration Testing: Theory and Practice*, Wiley, New York, 1995.

[18.7] C. R. Pickrel, "Airplane Ground Vibration Testing: Nominal Modal Model Correlation," *Sound and Vibration*, Vol. 36, No. 11, November 2002, pp. 18–23.

[18.8] G. F. Lang, "Electrodynamic Shaker Fundamentals," *Sound and Vibration*, Vol. 31, No. 4, April 1997, pp. 14–23.

[18.9] W. G. Halvorsen and D. L. Brown, "Impulse Technique for Structural Frequency Response Testing," *Sound & Vibration*, November 1977, pp. 8–21.

[18.10] D. Corelli, "Ratio Calibration: The Right Choice for Modal Analysis," *Sound & Vibration*, Vol. 18, No. 1, January 1984, pp. 27–30.

[18.11] T. G. Beckwith, R. D. Marangoni, and J. H. Lienhard, *Mechanical Measurements*, 5th ed., Addison-Wesley, Reading MA, 1993.

[18.12] D. Formenti, "Q & A: Structural Analysis," *Sound & Vibration* (in 7 issues), May–October 1983.

[18.13] G. T. Rocklin, J. Crowley, and H. Vold, "A Comparison of H_1, H_2, and H_v Frequency Response Functions," *Proceedings of International Modal Analysis Conference III*, 1985, pp. 272–278.

[18.14] R. J. Allemang, D. L. Brown, and W. Fladung, "Modal Parameter Estimation: A Unified Matrix Polynomial Approach," *Proceedings of International Modal Analysis Conference XII*, 1994, pp. 501–514.

[18.15] F. Lembregts, "Frequency Domain Identification Techniques for Experimental Multiple Input Modal Analysis," Ph.D. dissertation, Katholieke Universiteit Leuven, Leuven, Belgium, 1988.

[18.16] P. Verboven, "Frequency-Domain System Identification for Modal Analysis," Ph.D. dissertation, Vrije Universiteit Brussels, Brussels, Belgium, 2002.

[18.17] F. Deblauwe, D. L. Brown, and R. J. Allemang, "Polyreference Time Domain Technique" *Proceedings of International Modal Analysis Conference V*, 1987, pp. 832–845.

[18.18] P. Guillaume et al., "A Poly-reference Implementation of the Least-Squares Complex Frequency-Domain Estimator," *Proceedings of International Modal Analysis Conference XXI*, Kissimmee, FL, February 2003.

[18.19] B. Peeters et al., "Automotive and Aerospace Applications of the PolyMAX Modal Parameter Estimation Method," *Proceedings of International Modal Analysis Conference XXII*, Dearborn, MI, January 2004.

[18.20] C. C. Kennedy and C. D. P. Pancu, "Use of Vectors in Vibration Measurement and Analysis," *Journal of the Aeronautical Sciences*, Vol. 14, 1947, pp. 603–625.

[18.21] LMS International, http://www.lmsintl.com.

[18.22] SD Tools, http://www.sdtools.com.
[18.23] Vibrant Technology, Inc., http://www/vibetech.com.
[18.24] Axiom EduTech, http://www.vibratools.com.
[18.25] National Instruments, http://www.ni.com.
[18.26] Dynamic Design Solutions, http://www.ferntools.com.
[18.27] $m + p$ international, inc., http://www.mpihome.com.
[18.28] Society for Experimental Mechanics, http://www.sem.org.

PROBLEMS

Ideally, this chapter will be used as resource material in conjunction with laboratory exercises that make use of signal-processing hardware/software and modal-analysis software.

19

Introduction to Active Structures

For many years, the field of structural dynamics has been concerned with developing modeling methods, quantitative analyses, and numerical techniques to study structural dynamic systems that are linear, or even nonlinear, in character. Classical continuum mechanics, elasticity theory, and structural mechanics have provided the theoretical foundation for this work. More recently, there has been an interest in the application of control theory to problems in structural dynamics. Within the past decade or so, an ever-growing number of structural dynamics problems involve systems that are controlled by *active materials*, sometimes defined as those materials that exhibit non-negligible deformation or material property change with the application of thermal fields, electrical fields, or magnetic fields. These materials include electrorheological fluids, magnetorheological fluids, shape memory alloys, magnetic shape memory alloys, electrostrictive materials, magnetostrictive materials, and piezoelectric materials. These materials can be used as components in the design of structural systems that contain conventional structural components. Such systems are often called *smart structures, intelligent structures,* or *active structures*, in that they can be designed to perform tasks that conventional structures cannot carry out.

A large quantity of scientific literature has accumulated that studies the specifics of these materials and structures. In this chapter we study a single active material that is encountered in many active structural systems: piezoelectric material.

Upon completion of this chapter you should be able to:

- Describe the underlying physics that couples crystalline shape change and electrical fields in piezoelectric materials.
- Discuss the direct piezoelectric effect and the converse piezoelectric effect.
- Discuss the general form of the constitutive equations of linear piezoelectricity, and contrast their form with conventional constitutive laws of linear elasticity.
- Use Newton's Laws to derive the equations of motion of structural dynamic systems with piezoelectric components.
- Use a modified version of the Extended Hamilton's Principle to derive the equations of motion of structural dynamic systems that contain piezoelectric components.

19.1 INTRODUCTION TO PIEZOELECTRIC MATERIALS

19.1.1 History and Applications

Piezoelectric materials produce electric charge (voltage) if their crystalline structure is deformed by an external force. This effect is called the *direct effect* of piezoelectricity.

On the other hand, piezoelectric materials change their crystalline structure with the application of an external electric field. This phenomenon is known as the *converse effect* of piezoelectricity. It is a well-known and fascinating fact that Pierre and Jacques Curie discovered the direct effect of piezoelectricity in 1880 while conducting experiments that involved a number of naturally occurring materials, including quartz, topaz, and Rochelle salts. The converse piezoelectric effect was first predicted by Lippmann in 1881 via methods of analytical thermodynamics. Subsequently, the Curies confirmed the validity of the converse piezoelectric effect.

For nearly three decades, subsequent study of piezoelectric effects in different crystals resulted in a mathematically rich theory. Still, applications of this amazing class of materials was extremely limited. With the advent of World War I, however, Langevin designed a prototypical sonar for generating a high-frequency chirp signal underwater.[19.1] This research effort initiated an investigation of the use of piezoelectrics in ultrasonics that continues to this day. It is important to note that this initial task of generating a signal underwater is an example of the application of the converse effect of piezoelectricity. Over time, the use of the converse effect of piezoelectricity has become critical in applications that require oscillators with a relatively large amplitude and very slow rate of decay. If a piezoelectric crystal is subject to a sinusoidal electric current that is in resonance with a natural structural frequency of the crystal, the crystal will exhibit the desired properties: large-amplitude response with a low decay rate (low rate of internal dissipation of energy). Timing circuits that rely on such resonators appear in devices as widely varying as quartz watches and global positioning systems.[19.2]

Applications of piezoelectric materials in the sensing and measurement industry are today also widespread. Commercial accelerometers for vibration measurement, as well as pressure or force sensors, are sold by a number of vendors around the world (see Section 18.3). These sensors have become so advanced that the task of conditioning and amplifying the raw sensor signal have, to some degree, been rendered routine. The technical aspects of the signal conditioning required are so mature that they are often implemented in hardware or software provided by the vendor. Still, at the foundation of these sensor packages, the direct effect of piezoelectricity is used for generation of the measurement signal.

19.1.2 Qualitative Response Properties

An understanding of the physical foundations of piezoelectricity in piezoceramics begins by considering microscopic length scales of these materials. Eventually, we progress to the discussion of the macroscopic properties of the materials.

A ceramic is a crystalline solid, the crystal structure being composed of individual building blocks, the *unit cells*, regular arrays of positively and negatively charged ions. In some cases the centers of the positive and negative charges coincide. An example of this situation is depicted in Fig. 19.1, showing barium titanate, $BaTiO_3$, a common piezoelectric material. The unit cell is a face-centered cubic cell whose corners are occupied by barium ions, whose faces are occupied by oxygen ions, and whose center is occupied by a titanium ion. This configuration, in which the centers of charge coincide, is known as the *paraelectric phase* of the crystal. Figure 19.1 depicts the structure of the unit cell of $BaTiO_3$ when the temperature is raised above the *Curie temperature*, T_c. On the other hand, if the temperature is below the Curie temperature, the energetically

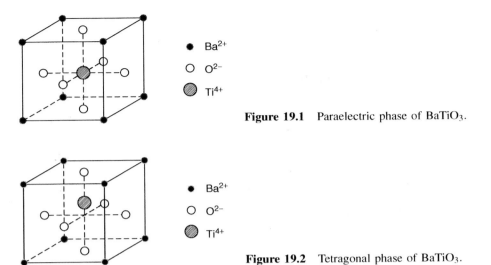

Figure 19.1 Paraelectric phase of BaTiO$_3$.

Figure 19.2 Tetragonal phase of BaTiO$_3$.

preferred unit cell has the tetragonal structure depicted in Fig. 19.2. The tetragonal structure is characterized by the fact that one cell parameter, the side length c, is elongated. The centers of the positively charged ions and negatively charged ions do not coincide in this asymmetric unit cell. The unit cell is said to be *polarized* or is said to exhibit a dipole moment. The *polarization vector P* is a measure of the strength of the dipole moment.

The origin of the direct and converse piezoelectric effects can be understood via consideration of the case where the centers of the positive charge and negative charge do not coincide in the unit cell. Let us first study the origin of the *converse piezoelectric effect*. Suppose that a voltage difference is applied across the faces of the unit cell that are perpendicular to the polarization vector. If the corresponding electrical field is oriented in the direction of the polarization vector, the applied voltage will tend to move the centers of charge apart. The result will be that the unit cell is distorted, or deformed, to lengthen in this direction. Alternatively, if the electric field is oriented in the direction opposite the polarization vector, the centers of charge will be shifted closer to one another. This movement will result in a contraction of the unit cell along the polarization direction. The *direct piezoelectric effect* can be understood in an analogous fashion. Physical deformation of the unit cell induces a shift in the centers of charge. Because the current is equal to the time derivative of the electric charge, the deformation induces a current.

Unfortunately, these simple considerations of a single unit cell cannot be extended to most commercially available piezoelectric materials, which are usually fabricated as a polycrystalline material. That is, a sample of the piezoelectric material is usually composed of many grains, each having its own differently oriented crystal lattice. Furthermore, the polarization vectors of the unit cells within the grains are not oriented in a unique direction. Rather, to minimize intergranular stresses, multiple domains, each domain with identically oriented polarization vectors, are formed within each grain. It is important to note that the polarization vectors may vary in direction from one domain to another.

As a sample of a typical polycrystalline piezoelectric material is lowered below the Curie temperature, the various domains that evolve will have polarization vectors that vary randomly from one domain to another. The net result is that a polycrystalline piezoelectric below the Curie temperature does not have an effective macroscopic polarization for the entire solid. This state of the polycrystalline piezoceramic is called the thermally depoled, or virgin, state.

It was a major breakthrough in the synthesis of piezoelectric materials when it was learned that an effective overall polarization of certain polycrystalline materials could be induced with the application of an external electrical field. A *ferroelectric material* is defined to be any material that is comprised of polarized crystallites for which the orientation of the dipole moment can be reoriented with the application of an external electrical field. If an electric field greater than the *coercive field strength* E_c is applied to a ferroelectric material, the orientation of the dipole moment will switch such that the polarization vector is (nearly) aligned with the external field. It is a remarkable and critical fact that once the applied external electrical field is turned off, the orientation of the dipoles in the various domains remain in the switched state. Jaffe [19.3] compares this process to that of magnetization. A polycrystalline ferroelectric ceramic will consequently exhibit an effective macroscopic polarity, one that can be described by equations of piezoelectricity that are consistent with the framework and assumptions of conventional continuum mechanics. A polycrystalline ferroelectric ceramic is often called simply a *piezoceramic*.

19.2 CONSTITUTIVE LAWS OF LINEAR PIEZOELECTRICITY

From the qualitative discussion in Section 19.1, it is not difficult to piece together the form of constitutive laws for linear piezoelectricity, at least in the one-dimensional case. The study of the constitutive laws of linear piezoelectricity are considerably more complex for general three-dimensional continua.

19.2.1 Constitutive Laws in One Dimension

Consider the piezoceramic depicted in Fig. 19.3. This figure depicts schematically the positive sign conventions for the electric field $\mathbb{E}$, the polarization P, and the stress σ.

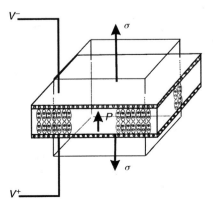

Figure 19.3 Piezoceramic specimen with a collinear electric field and polarization vector.

Since the electrical field is defined to be the negative of the gradient of the potential (voltage), the arrow depicting the orientation of the electrical field points from higher voltage potential to lower. The electrical field is oriented in the same direction as the polarization vector. As noted earlier, the centers of charge in the dipole moments within the piezoceramic will separate. A positive strain is induced along the direction of the polarization vector. In the case that the specimen is free to expand, unrestrained by constraints, the stress is $\sigma = 0$. If, on the other hand, the electric field is applied in alignment with the polarization but the specimen is constrained such that there is no displacement, a compressive stress will result. If it is assumed that the stress σ, strain ϵ, and electric field $\mathbb{E}$ satisfy a linear relationship, then

$$c_1 \sigma + c_2 \epsilon + c_3 \mathbb{E} = 0 \qquad (19.1)$$

for three constants c_1, c_2, and c_3. One of the conventional forms for this relationship is written

$$\sigma = C\epsilon - e\mathbb{E} \qquad (19.2)$$

In this equation, C is the *Young's modulus* of linear elasticity and e is referred to as a *piezoelectric coupling coefficient*. Some simple thought experiments make it clear that Eq. 19.2 encodes the converse piezoelectric effect in a linear constitutive law. If the specimen is free, application of the electric field $\mathbb{E}$ yields a strain. If the electric field is absent, the stress and strain obey the classical Hooke's law.

The direct piezoelectric effect is represented in the linear constitutive law

$$\mathbb{D} = e\epsilon + \gamma \mathbb{E} \qquad (19.3)$$

In this equation, $\mathbb{D}$ is the *dielectric displacement* and is measured in coulombs/meter^2. Recall that the current $i(t)$ is defined to be the time derivative of the charge $q(t)$.

$$i(t) = \frac{dq(t)}{dt} \qquad (19.4)$$

In the present example, the current $i(t)$ is related to the dielectric displacement $\mathbb{D}$ via the relationship

$$i(t) = \frac{d(A\mathbb{D}(t))}{dt} \qquad (19.5)$$

where A is the area of the electrodes. Thus, a simple thought experiment shows that Eq. 19.5, with Eq. 19.3, embodies the direct piezoelectric effect in a linear constitutive law. In the absence of an electric field (i.e., with $\mathbb{E} = 0$) deformation of the specimen results in a current.

19.2.2 Constitutive Laws in Three Dimensions

For the simple one-dimensional case, we have indicated that the linear constitutive laws for piezoelectric materials can be written

$$\begin{aligned} \sigma &= C\epsilon - e\mathbb{E} \\ \mathbb{D} &= e\epsilon + \gamma \mathbb{E} \end{aligned} \qquad (19.6)$$

The linear piezoelectric constitutive equations for a general three-dimensional continuum can be obtained from many sources (e.g., Ref. [19.3] or [19.4]). Using the summation convention, one form of the *constitutive laws for a three-dimensional piezoelectric specimen* can be written

$$\sigma_{ij} = c^E_{ijkl}\epsilon_{kl} - e_{kij}\mathbb{E}_k$$
$$\mathbb{D}_k = e_{kij}\epsilon_{ij} + \gamma^\epsilon_{kl}\mathbb{E}_l \tag{19.7}$$

It is important to note that these are tensor equations. There are also many other alternative forms for the constitutive relations of piezoelectricity, depending on which field quantities are chosen as dependent variables and which as independent variables. The derivation of these relations from thermodynamic considerations can be found in Ikeda's book.[19.5] One of the most common alternative forms of the linear equations of piezoelectricity expresses the strains in terms of the stress and electric field in Eqs. 19.7.

$$\epsilon_{ij} = s^E_{ijkl}\sigma_{kl} + d_{kij}\mathbb{E}_k$$
$$\mathbb{D}_k = d_{kij}\sigma_{ij} + \gamma^T_{kl}\mathbb{E}_l \tag{19.8}$$

In many applications of the constitutive relations above, it has become conventional to reindex these equations and depict them in matrix form. The following notation is commonly employed:

Tensor indices ij or kl	Matrix indices p or q
11	1
22	2
33	3
23 or 32	4
31 or 13	5
12 or 21	6

Thus, the tensor relationship, expressed in the summation convention, is given by Eq. 19.7, whereas the matrix form of the *same* constitutive laws is written in the form

$$\sigma_p = c^E_{pq}\epsilon_q - e_{pk}\mathbb{E}_k$$
$$\mathbb{D}_k = e_{kq}\epsilon_q + \gamma^\epsilon_{kl}\mathbb{E}_l \tag{19.9}$$

Note carefully that the symbol σ_p denotes the pth entry of a column vector of length 6 whose *value* is the appropriate stress tensor component. For example, σ_4 is the fourth entry of a column vector whose entry is the stress tensor component σ_{23}.

The advantage of expressions such as Eqs. 19.9 is that they are simple to depict in matrix form, as in Eq. 19.10. Crystallographic considerations for piezoelectric materials are considerably more complicated than those for simple linearly elastic materials. In this book, we consider one particular crystal symmetry class. Polarized ferroelectric ceramic effectively exhibits the symmetry of a hexagonal crystal of the class 6*mm*.

19.2 Constitutive Laws of Linear Piezoelectricity

For the hexagonal crystal of class 6*mm*, the strain-stress relationship has the following form[19.4]:

$$\left\{ \begin{Bmatrix} \epsilon_{11} \\ \epsilon_{22} \\ \epsilon_{33} \\ 2\epsilon_{23} \\ 2\epsilon_{31} \\ 2\epsilon_{12} \end{Bmatrix} \\ \begin{Bmatrix} \mathbb{D}_1 \\ \mathbb{D}_2 \\ \mathbb{D}_3 \end{Bmatrix} \right\} = \begin{bmatrix} S_{11} & S_{12} & S_{13} & 0 & 0 & 0 & 0 & 0 & d_{31} \\ S_{12} & S_{11} & S_{13} & 0 & 0 & 0 & 0 & 0 & d_{31} \\ S_{13} & S_{13} & S_{33} & 0 & 0 & 0 & 0 & 0 & d_{33} \\ 0 & 0 & 0 & S_{44} & 0 & 0 & 0 & d_{15} & 0 \\ 0 & 0 & 0 & 0 & S_{44} & 0 & d_{15} & 0 & 0 \\ 0 & 0 & 0 & 0 & 0 & S_{66} & 0 & 0 & 0 \\ 0 & 0 & 0 & 0 & d_{15} & 0 & \gamma^T_{11} & 0 & 0 \\ 0 & 0 & 0 & d_{15} & 0 & 0 & 0 & \gamma^T_{22} & 0 \\ d_{31} & d_{31} & d_{33} & 0 & 0 & 0 & 0 & 0 & \gamma^T_{33} \end{bmatrix} \left\{ \begin{Bmatrix} \sigma_{11} \\ \sigma_{22} \\ \sigma_{33} \\ \tau_{23} \\ \tau_{31} \\ \tau_{12} \end{Bmatrix} \\ \begin{Bmatrix} \mathbb{E}_1 \\ \mathbb{E}_2 \\ \mathbb{E}_3 \end{Bmatrix} \right\} \quad (19.10)$$

Commercial piezoelectric specimens typically are comprised of polarized polycrystalline ferroelectric ceramics. The coefficient S_{66} in the constitutive matrix is not independent, $S_{66} = 2(S_{11} - S_{12})$. The structure of all possible coupling matrices, tabulated for all crystal classes that exhibit piezoelectric effects, may be found in a number of sources, including Refs. [19.3] to [19.5].

19.2.3 Reduction of the Three-Dimensional Case to a One-Dimensional Case

Two specific physical examples of the general linear constitutive laws described in Section 19.2.2 occur frequently in applications. More specifically, these special cases arise in the development of simple models of piezoelectric sensors and actuators. We review them in this section to provide a quick reference for the analysis of more complicated active structural systems. Consider Fig. 19.4, which depicts the same physical specimen shown in Fig. 19.3, except that a coordinate system has been defined so that the polarization (or poling) direction coincides with the 3-axis. This choice is made to

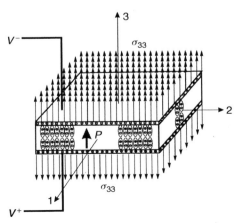

Figure 19.4 Piezoceramic specimen: poling along the 3-axis, stress along the 3-axis.

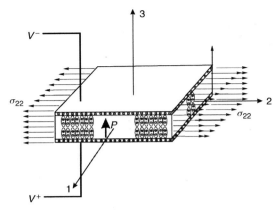

Figure 19.5 Cross-polarized piezoceramic specimen: poling direction along the 3-axis, stress along the 2-axis.

coincide with the setup of the constitutive laws discussed in Section 19.2.1. Let us also make the following three assumptions in this section:

(A1) The only nonnegligible stress is the normal stress in the 3-axis direction.
(A2) The material is a polarized ferroelectric ceramic, so it therefore exhibits the symmetry of a hexagonal crystal of class 6*mm*.
(A3) The polarization vector is directed along the positive 3-axis.

In this case, the general constitutive law having the form

$$\sigma_{ij} = C^E_{ijkl}\epsilon_{kl} - e_{kij}\mathbb{E}_k$$
$$\mathbb{D}_k = e_{kij}\epsilon_{ij} + \gamma_{kl}\mathbb{E}_l \tag{19.11}$$

takes the particular form

$$\sigma_{33} = C^E_{3333}\epsilon_{33} - e_{333}\mathbb{E}_3$$
$$\mathbb{D}_3 = e_{333}\epsilon_{33} + \gamma_{33}\mathbb{E}_3 \tag{19.12}$$

In other words, we have simply identified the crystallographic indices of the coefficients appearing in Eqs. 19.11. (It is important to note that the tensor constant e_{333} is often simply denoted e_{33} in matrix form.)

Figure 19.5 depicts a second uniaxial loading case, but now the nonnegligible stress is orthogonal to the polarization vector. In this case the general constitutive law reduces to the particular form

$$\sigma_{22} = C^E_{2222}\epsilon_{22} - e_{322}\mathbb{E}_3$$
$$\mathbb{D}_3 = e_{322}\epsilon_{22} + \gamma_{33}\mathbb{E}_3 \tag{19.13}$$

19.3 APPLICATION OF NEWTON'S LAWS TO PIEZOSTRUCTURAL SYSTEMS

In Chapter 12 it was pointed out that many alternatives exist for deriving the equations of motion for mechanical and structural systems. Examples of these alternatives

19.3 Application of Newton's Laws to Piezostructural Systems

include those based on Newton's Laws, Lagrange's Equations, the Principle of Virtual Displacements, and Hamilton's Principle. In this section we demonstrate that Newton's Laws are directly applicable to piezostructural systems. The problems are made more complex, however, in that the constitutive laws now involve additional fields. For linearly elastic structures, the constitutive laws relate the stress and strain fields. The constitutive laws for linear piezoelectricity relate stress, strain, electrical fields, and electric displacement fields, for example. The implications of this added complexity are best illustrated by an example.

Example: Axial Deformation of a Cross-Polarized Piezoelectric Rod This example considers one of the simplest types of piezoelectric structure. It illustrates the method by which more complex structural systems are "actuated" using piezoelectric components. The geometry of the structural system is depicted in Fig. 19.6. When we apply Newton's Second Law to the free-body diagram extracted from the specimen, as shown in Fig. 19.7, we obtain the expression

$$p(y,t)\Delta y + \underbrace{\sigma_2(y+\Delta y, t)}_{\sigma_2(y,t)+\frac{\partial \sigma_2}{\partial y}(y,t)\Delta y} \underbrace{A(y+\Delta y)}_{A(y)+\frac{\partial A}{\partial y}(y)\Delta y} - \sigma_2(y,t)A(y) = \rho A \Delta y \frac{\partial^2 v}{\partial t^2}(y,t)$$

We next expand the Taylor series expression for $A\sigma_2$ and retain only the first-order term. Then

$$\rho A \frac{\partial^2 v}{\partial t^2}(y,t) - \frac{\partial}{\partial y}(A\sigma_2(y,t)) = p(y,t) \tag{19.14}$$

The constitutive law for the ferroelectric piezoelectric ceramic is precisely one of the special "one-dimensional" cases studied in the Section 19.2.3: namely, the one illustrated in Fig. 19.5.

$$\sigma_2(y,t) = E(y)\epsilon_2(y,t) - e_{32}\chi_e(y)\mathbb{E}_3 \tag{19.15}$$

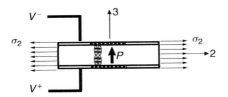

Figure 19.6 Axial piezoelectric rod with lateral electrodes.

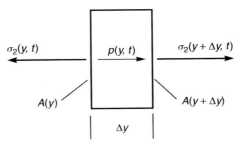

Figure 19.7 Free-body diagram of a piezoelectric rod with lateral electrodes.

The extent of the electroded region is indicated by the characteristic function $\chi_e(y)$, which takes a value of 1 over the electroded region and is equal to zero otherwise. The dependence of the electric field on space and time is quite simple in this example. It is assumed that the electric field exists only between the electrodes. This is a reasonable approximation, as the electrical field attenuates rapidly outside the electroded region. It is assumed that the voltage varies linearly across the specimen. The voltage takes the value of V^- at $z = h/2$ and the value of V^+ at $z = -h/2$. The linear expression for the voltage is just

$$V(z,t) = \frac{V^-(t) - V^+(t)}{h} z + \frac{1}{2}[V^-(t) - V^+(t)] \tag{19.16}$$

Calculation of the electrical field follows from the definition

$$\mathbb{E} = -\nabla V \tag{19.17}$$

From the gradient of the voltage, it is easy to see that the $\mathbb{E}_3$ component of the electrical field $\mathbb{E}$ is determined by the equation

$$\mathbb{E}_3 \hat{k} = -\frac{V^-(t) - V^+(t)}{h} \hat{k} \tag{19.18}$$

This expression is simplified by noting that the electric field depends only on the voltage difference; that is,

$$\mathbb{E}_3 = \frac{V^+(t) - V^-(t)}{h} = \frac{\Delta V(t)}{h} \tag{19.19}$$

The component $\mathbb{E}_3$ of the electric field can now be substituted into the constitutive law. We obtain

$$\sigma_2(y,t) = E(y)\epsilon_2(y,t) - e_{32}\chi_e(y)\frac{\Delta V(t)}{h} \tag{19.20}$$

The appropriate strain-displacement equation, comparable to Eq. 12.1, is

$$\epsilon_2(y,t) = \frac{\partial v}{\partial y}(y,t) \tag{19.21}$$

This can be substituted into the constitutive law, which can then be substituted for the stress σ_2 in the equation of motion. We obtain the following final form of the governing partial differential equation:

$$\boxed{\rho A \frac{\partial^2 v}{\partial t^2}(y,t) - \frac{\partial}{\partial y}\left(AE \frac{\partial v}{\partial y}(y,t)\right) = -\frac{e_{32} \Delta V(t)}{h} \frac{d}{dy}(\chi_e(y)) + p(y,t)} \tag{19.22}$$

Several observations can be made regarding this example:

- If the dielectric constant e_{32} is zero, the equation simplifies to the conventional equation of motion for a rod undergoing axial loading, Eq. 12.8.
- The forcing term due to the presence of the piezoceramic is *linear* in the voltage difference $\Delta V(t)$.

- The equations derived in this example are an example of the *strong form* of the governing differential equations. The forcing term due to the presence of the piezoceramic includes a spatial derivative that does not exist in a "classical sense." That is, the expression $d\chi_e(y)/dy$ cannot be calculated because $\chi_e(y)$ is not differentiable at the edges of the patch in a classical sense. This term must be understood as a "generalized function," or the equation must be interpreted in the *weak sense*.
- A finite element model could be created, as illustrated in Section 19.5.2.

19.4 APPLICATION OF EXTENDED HAMILTON'S PRINCIPLE TO PIEZOELECTRICITY

Recall from Section 12.4 that numerous problems of analytical dynamics and structural dynamics can be formulated effectively by employing the *Extended Hamilton's Principle*,

$$\delta \int_{t_0}^{t_1} \mathcal{L}\, dt + \int_{t_0}^{t_1} \delta \mathcal{W}_{\text{nc}}\, dt = 0 \qquad (19.23)$$

In this energy principle, $\mathcal{L} = \mathcal{T} - V$ is the Lagrangian and $\delta\mathcal{W}_{\text{nc}}$ is the virtual work of the nonconservative forces. The principle asserts that the variation of the integral of the Lagrangian added to the integral of the virtual work of the nonconservative forces must be equal to zero for all variations, consistent with the kinematic constraints. It is interesting and useful in applications that an analogous result holds for linear piezoelectric continua. Consider a piezoelectric body that is subject to surface stresses τ on portions of the surface denoted S_τ and has prescribed surface charges q over portions of the surface denoted S_q. The *electrical enthalpy (density)* for the piezoelectric continuum is defined to be

$$\boxed{\mathcal{H} \equiv \mathcal{U} - \mathbb{E}_i \mathbb{D}_i} \qquad (19.24)$$

where the summation convention is employed in the term $\mathbb{E}_i \mathbb{D}_i$. In this equation, $\mathcal{U}$ is the internal energy density due to stored mechanical and electrical energy. The components of the dielectric displacement are given by $\mathbb{D}_i$, while $\mathbb{E}_i$ denotes the components of the electrical field vector. Define the modified *Lagrangian for a linear piezoelectric continuum* to be

$$\mathcal{L}_H = \int_\Omega \left\{\tfrac{1}{2}\rho(v \cdot v) - \mathcal{H}\right\} d\Omega = \mathcal{T} - \int_\Omega \{\mathcal{U} - \mathbb{E}_i \mathbb{D}_i\}\, d\Omega \qquad (19.25)$$

where Ω is the volume of the piezoelectric continuum. The *Extended Hamilton's Principle for Linear Piezoelectricity* then holds that

$$\boxed{\delta \int_{t_0}^{t_1} \mathcal{L}_H\, dt + \int_{t_0}^{t_1} \delta \mathcal{W}_{\text{nc}}\, dt = 0} \qquad (19.26)$$

for all variations consistent with the kinematic constraints.

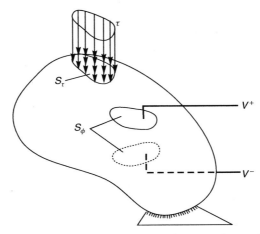

Figure 19.8 Piezoceramic continuum subject to surface traction and imposed charge distribution.

It should be noted that the virtual work of the external loads $\delta \mathcal{W}_{nc}$ includes the electrical "loads" in addition to the mechanical loads in this case. For example, the virtual work done by the electrical loads and the mechanical loads, as depicted in Fig. 19.8, is given by

$$\delta \mathcal{W}_{nc} = \int_{S_\tau} \tau_k \delta u_k \, dS + \int_{S_\phi} (-q) \delta\phi \, dS$$

In this equation τ is the vector of surface tractions acting over the surface region S_τ, u is the vector of displacements throughout the piezoelectric region as evaluated at the surface region S_τ, q is the surface charge over the surface regions S_ϕ, and ϕ is the electrical potential over the surface regions S_ϕ.

For computational and analytical purposes, we require an explicit expression for the electrical enthalpy density. The most common, and elementary, assumption is that the electrical enthalpy density is a quadratic function of the mechanical strains and the electrical field. From either Ref. [19.4] or [19.6], we find that

$$\mathcal{H} = \tfrac{1}{2} C_{ijkl} \epsilon_{ij} \epsilon_{kl} - e_{ijk} \mathbb{E}_i \epsilon_{jk} - \tfrac{1}{2} \gamma_{ij} \mathbb{E}_i \mathbb{E}_j \tag{19.27}$$

On the other hand, the linear constitutive laws in Eqs. 19.7 enable us to write

$$\mathcal{H} = \mathcal{U} - \mathbb{E}_i \mathbb{D}_i = \mathcal{U} - \mathbb{E}_i (e_{ikl} \epsilon_{kl} + \gamma_{ij} \mathbb{E}_j) = \mathcal{U} - e_{ikl} \mathbb{E}_i \epsilon_{kl} - \gamma_{ij} \mathbb{E}_i \mathbb{E}_j \tag{19.28}$$

Comparing Eqs. 19.27 and 19.28, we obtain the following concise expression for the internal energy density function:

$$\mathcal{U} = \tfrac{1}{2} C_{ijkl} \epsilon_{ij} \epsilon_{kl} + \tfrac{1}{2} \gamma_{ij} \mathbb{E}_i \mathbb{E}_j \tag{19.29}$$

Example: Extended Hamilton's Principle for a Piezoelectric Rod We now study the structural system shown in Fig. 19.6, using the Extended Hamilton's Principle for Piezoelectricity. The kinetic energy can easily be expressed in terms of the displacement v

19.4 Application of Extended Hamilton's Principle to Piezoelectricity

along the length of the rod. It has the familiar form

$$T = \tfrac{1}{2} \int_0^L \rho(y) A(y) \dot{v}^2(y,t)\, dy \tag{19.30}$$

The internal energy $\mathcal{U}$ is given by

$$\mathcal{U} = \tfrac{1}{2} \int_0^L A(y) E(y) \epsilon_2^2(y,t)\, dy + \tfrac{1}{2} \int_0^L A \gamma_{33} \mathbb{E}_3^2\, dy \tag{19.31}$$

The only way in which this example differs from the conventional application of the extended Hamilton's principle is that we must now calculate the contribution of the electric field and dielectric displacement in the sum $\mathbb{E}_i \mathbb{D}_i$ that appears in the modified Lagrangian.

$$\int_\Omega \mathbb{E}_i \mathbb{D}_i\, d\Omega = \int_\Omega \mathbb{E}_3 \mathbb{D}_3\, d\Omega = \int_\Omega \mathbb{E}_3 (e_{32} \epsilon_2 + \gamma_{33} \mathbb{E}_3)\, d\Omega$$

$$\int_\Omega \mathbb{E}_3 \mathbb{D}_3\, d\Omega = \int_0^L \frac{A\,\Delta V(t)}{h} \chi_e(y) \left[e_{32} \frac{\partial v}{\partial y}(y,t) + \gamma_{33} \frac{\Delta V(t)}{h} \chi_e(y) \right] dy$$

Calculation of the variation of the modified Lagrangian $\mathcal{L}_H$ entails the evaluation of

$$\delta \int_{t_1}^{t_2} \mathcal{L}_H\, dt = \delta \int_{t_1}^{t_2} \left\{ \frac{1}{2} \int_0^L \rho A \dot{v}^2\, dy - \frac{1}{2} \int_0^L A \gamma_{33} \mathbb{E}_3^2\, dy - \int_0^L AE \left(\frac{\partial v}{\partial y} \right)^2 dy \right.$$
$$\left. + \int_0^L \frac{A\,\Delta V(t)}{h} \chi_e(y) \left[e_{32} \frac{\partial v}{\partial y}(y,t) + \gamma_{33} \frac{\Delta V(t)}{h} \chi_e(y) \right] dy \right\} dt \tag{19.32}$$

This calculation is made simple by recalling the following two identities that were employed in the application of the conventional Extended Hamilton's Principle. The boundary terms in the first equation vanish since the variations δv vanish at t_1 and t_2. The boundary terms in the second equation provide the boundary conditions on the displacement field for the elastic rod, as in the conventional case.

$$\delta \left[\frac{1}{2} \int_{t_1}^{t_2} \int_0^L \rho A \dot{v}^2\, dy \right] = \int_0^L \left\{ \rho A \dot{v}\, \delta v \big|_{t_1}^{t_2} - \int_{t_1}^{t_2} \rho A \ddot{v}\, \delta v\, dt \right\} dy$$

$$\delta \left[\frac{1}{2} \int_{t_1}^{t_2} \int_0^L AE \left(\frac{\partial v}{\partial y} \right)^2 dy\, dt \right] = \int_{t_1}^{t_2} \left\{ AE \frac{\partial v}{\partial y} \delta v \Big|_0^L - \int_0^L \frac{\partial}{\partial y} \left(AE \frac{\partial v}{\partial y} \right) \delta v\, dy \right\} dt$$

The variation of the term containing the contribution of the electrical field $\mathbb{E}$ and dielectric displacement $\mathbb{D}$ can be carried out in the usual manner using integration by parts:

$$\delta \int_{t_1}^{t_2} \int_0^L \frac{A\,\Delta V(t)}{h} \chi_e(y) \left[e_{32} \frac{\partial v}{\partial y}(y,t) + \gamma_{33} \frac{\Delta V(t)}{h} \chi_e(y) \right] dy\, dt$$

$$= \int_{t_1}^{t_2} \int_0^L \frac{A\,\Delta V(t)}{h} \chi_e(y) e_{32} \frac{\partial (\delta v)}{\partial y}(y,t)\, dy\, dt \tag{19.33}$$

$$= \int_{t_1}^{t_2} \left\{ \frac{A\,\Delta V(t)}{h} \chi_e(y) e_{32}\, \delta v \Big|_0^L - \int_0^L e_{32} \frac{A\,\Delta V(t)}{h} \frac{d(\chi_e(y))}{dy} \delta v\, dy \right\} dt$$

The first term on the right of Eq. 19.33 is zero because it is assumed that the electroded region does not extend to the edges of the bar at $y = 0$ and $y = L$. At these locations the characteristic function $\chi_e(y)$ is equal to zero. When these identities are collected and substituted into the variational statement, we obtain

$$0 = \int_{t_1}^{t_2} \int_0^L \left\{ -\rho A \ddot{v} + \frac{\partial}{\partial y}\left(AE \frac{\partial v}{\partial y}\right) - e_{32} \frac{A \, \Delta V(t)}{h} \frac{d(\chi_e(y))}{dy} + p \right\} \delta v \, dy \, dt \\ - \int_{t_1}^{t_2} \left\{ AE \frac{\partial v}{\partial y} \delta v \bigg|_0^L \right\} dt \quad (19.34)$$

From this variational equation, we see that the governing equation is precisely the same as the equation obtained via Newton's Method in Section 19.3. The boundary conditions are encoded in the last term in the form of a variational boundary condition. Any appropriate set of displacement or force boundary conditions can be imposed via the variational boundary condition. Of course, if there are force-type boundary conditions, these must be accounted for in the virtual work of nonconservative forces.

19.5 ACTIVE TRUSS MODELS

As shown in Fig. 19.9, an *active strut* may replace, at least geometrically, a typical element in the truss or frame. The dynamics of the *active truss* can then, in some instances, be modified by feedback control of the dynamics of the active strut. In this section we examine how to model the dynamics of such an active strut.

19.5.1 Modeling a Piezoceramic Stack

A number of standard *solid-state actuators* have been designed over the years that employ ferroelectric ceramics to carry out the transduction of electrical energy into

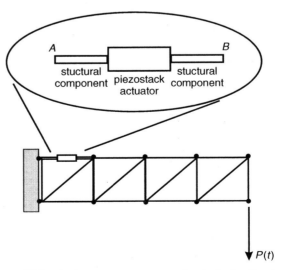

Figure 19.9 Active strut geometry and a typical active truss.

mechanical energy. These devices vary widely in shape, size, maximum stroke, maximum load, and operating bandwidth. A quick survey of commercial products reveals a broad array of commercially available actuators.

Roughly speaking, piezoelectric ceramics have a very broad operating bandwidth and exhibit precise positioning response and repeatability. It is not uncommon that piezoceramics operate from $O(10\,\text{Hz})$ to $O(1\,\text{MHz})$. They can achieve precision that is measured in *nanometers*. These actuators may be well suited for applications involving the positioning of mirrors for precise applications in optics, but their *stroke* is severely limited. Consequently, the same actuators may not be well suited for applications involving structural systems whose length scale might be measured in meters. It is a common task in designing piezoceramic actuators to "amplify the stroke" of the underlying piezoceramic crystal. One common actuator design that addresses the limitation in achievable stroke is the stack actuator depicted schematically in Fig. 19.10. This illustration is merely symbolic in nature, to facilitate a discussion of the equations of motion of the piezostack. A commercially available stack actuator is depicted in Fig. 19.11.

As shown in Fig. 19.10, there are several interesting features of the stack actuator design. The stack is created by bonding thin layers of piezoelectric material together such that their polarization directions alternate along the major axis. Electrodes separate the layers nearly entirely. Small gaps must separate the electrodes, of course; otherwise, a short circuit would result. When a voltage difference is applied across the electrodes, a complicated electrical field and mechanical response result. For example, the electrical field and mechanical response in the vicinity of the small electrode gaps and "corners" in the specimen exhibit large gradients. However, away from the edges, at the core or center of the device, the electrical field and mechanical deformation can be approximated as a one-dimensional phenomenon. For example, from Eq. 19.12 the constitutive law in layer 1 of the stack can be written as

$$\sigma_{33} = E\epsilon_{33} - e_{33}\mathbb{E}_3$$

$$\mathbb{D}_3 = e_{33}\epsilon_{33} + \gamma_{33}\mathbb{E}_3$$

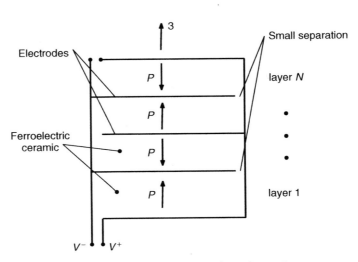

Figure 19.10 Longitudinal cross section of a stack actuator.

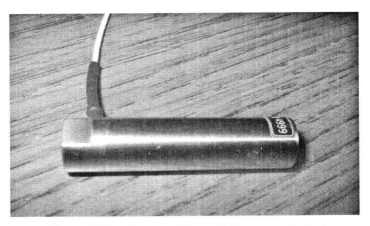

Figure 19.11 Commercially available stack actuator.

In contrast, the constitutive law in layer 2, to be consistent with the global coordinate system and polarization vector depicted, can be written as

$$\sigma_{33} = E\epsilon_{33} + e_{33}\mathbb{E}_3$$

$$\mathbb{D}_3 = -e_{33}\epsilon_{33} - \gamma_{33}\mathbb{E}_3$$

The signs in the constitutive law can be checked by simple thought experiments. For example, suppose that the strain is zero and a positive electrical field (in the global coordinate system) is applied. Is the stress that results compression, or is it tension? We leave this as an exercise for the reader. Extract a thin slice from the first layer and draw a free-body diagram where the independent variable is z ($\equiv 3$). It is easy to establish that the equation of motion for the layer can be written as

$$\rho A \ddot{w}^{(1)}(z) = \frac{\partial}{\partial z}\left(AE\frac{\partial w^{(1)}}{\partial z}(z)\right) - Ae\frac{\partial}{\partial z}(\mathbb{E}_3^{(1)}) + p(z,t) \qquad (19.35)$$

A superscript has been introduced to distinguish this equation as applying to the first layer only. But the voltage in the first layer is given by

$$\phi^{(1)}(z) = \frac{V^- - V^+}{h}z + V^+ \qquad (19.36)$$

so that the electric field is simply

$$\mathbb{E}_3^{(1)}(z) = -\frac{\partial}{\partial z}(\phi^{(1)}(z)) = \frac{\Delta V}{h} \qquad \forall z \in (0, h) \qquad (19.37)$$

The final governing equation for the first layer is

$$\rho A \ddot{w}^{(1)}(z,t) = \frac{\partial}{\partial z}\left(AE\frac{\partial w^{(1)}}{\partial z}(z,t)\right) - \frac{Ae}{h}\Delta V \frac{d}{dz}(\chi_1(z)) + p(z,t) \qquad (19.38)$$

19.5 Active Truss Models

If we carry out the same analysis for the second layer, we find that the governing equation is analogous.

$$\rho A \ddot{w}^{(2)}(z) = \frac{\partial}{\partial z}\left(AE\frac{\partial w^{(2)}}{\partial z}(z)\right) + Ae\frac{\partial}{\partial z}(\mathbb{E}_3^{(2)}) + p(z,t) \qquad (19.39)$$

Of course, the superscripts have been changed to refer to the second layer. Otherwise, there is a single sign difference for this layer in comparison to the first layer. But the voltage in the second layer is given by

$$\phi^{(2)}(z) = -\frac{V^- - V^+}{h}z - V^+ + 2V^- \qquad (19.40)$$

In the second layer, the electric field is given by

$$\mathbb{E}_3^{(2)} = -\frac{\Delta V}{h} \qquad (19.41)$$

Substituting the electric field in the equation of motion, we obtain

$$\rho A \ddot{w}^{(2)}(z,t) = \frac{\partial}{\partial z}\left(AE\frac{\partial w^{(2)}}{\partial z}(z,t)\right) - \frac{Ae}{h}\Delta V \frac{d}{dz}(\chi_2(z)) + p(z,t) \qquad (19.42)$$

It is clear from the study of these two layers that the same procedure will yield equations having <u>precisely the same form for all layers</u>. The equation for each layer is subject to initial conditions and boundary conditions. In particular, continuity must be enforced between layers. The following set of boundary conditions must be imposed at the respective interfaces between layers:

$$\begin{aligned} w^{(1)}(h,t) &= w^{(2)}(h,t) \\ w^{(2)}(2h,t) &= w^{(3)}(2h,t) \\ &\vdots \end{aligned} \qquad (19.43)$$

This collection of partial differential equations is cumbersome to use in applications. A compact representation the stack actuator is desired. Fortunately, it is relatively simple to collect this family of governing equations and derive a single "effective" governing equation for the entire stack. First, note that the electric field, voltage, and piezoelectric coupling coefficient e_{33} are rapidly oscillating over the length of the stack. This fact is depicted schematically in Fig. 19.12. It is obvious from these figures that the electric field $\mathbb{E}_3(z)$ and the piezoelectric coupling coefficient $e_{33}(z)$ can be expressed, respectively, as the sums

$$\begin{aligned} \mathbb{E}_3(z) &= \sum_{i=1}^{N}(-1)^{i+1}\chi_i(z)\frac{\Delta V}{h} \\ e_{33}(z) &= \sum_{i=1}^{N}(-1)^{i+1}\chi_i(z)e \end{aligned} \qquad (19.44)$$

In these equations, $\chi_i(z)$ is the characteristic function of the ith layer. As shown in the figure, the characteristic function $\chi_i(z)$ is equal to 1 if z is in layer i, and the function

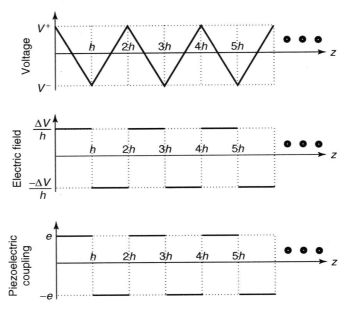

Figure 19.12 Properties along a longitudinal cross section of stack actuator.

takes the value zero otherwise. For any $z \in [0, L]$, the governing equation can therefore be written

$$\rho A \ddot{w}(z,t) = \frac{\partial}{\partial z}\left(AE \frac{\partial w}{\partial z}(z,t)\right) \\ - \frac{\partial}{\partial z}\left(A \sum_{i=1}^{N}(-1)^{i+1}\chi_i(z)e \cdot \sum_{j=1}^{N}(-1)^{j+1}\chi_j(z)\frac{\Delta V}{h}\right) + p(z,t) \quad (19.45)$$

But this equation simplifies because the collection of characteristic functions $\{\chi_i(z)\}_{i=1}^{N}$ satisfies

$$\chi_i(z) \cdot \chi_j(z) = \begin{cases} 0 & \text{if } i \neq j \\ \chi_i(z) & \text{if } i = j \end{cases} \quad (19.46)$$

and, of course,

$$\sum_{i=1}^{N} \chi_i(z) = \chi_{[0,L]}(z) \quad (19.47)$$

The final *effective governing equation* is

$$\boxed{\rho A \ddot{w}(z,t) = \frac{\partial}{\partial z}\left(AE \frac{\partial w}{\partial z}(z,t)\right) - \frac{Ae}{h}\Delta V \frac{d}{dz}(\chi_{[0,L]}(z)) + p(z,t)} \quad (19.48)$$

19.5.2 Finite Element Modeling of an Active Strut

In Section 19.5.1 the physics of the stack actuator was reviewed and the partial differential equation that governs its axial deformation was derived. The stack actuator is often used to create an *active strut* that can be used in conjunction with conventional structural members to create an active truss or frame, as shown in Fig. 19.9.

The equations that describe the behavior of a purely axial active strut of the type depicted in Fig. 19.9 are relatively straightforward to derive using the principles outlined in this chapter. To cast the equations in finite element form, we also make use of the discussion of analytical mechanics principles in Chapter 8 and finite element methods introduced in Chapter 14. An actual strut would probably be discretized into a number of axial-deformation elements. Here, we simply provide an overview of the steps in the derivation of a FE model, particularly the steps that involve piezoelectric terms.

The Extended Hamilton's Principle for Piezoelectricity states that of all the kinematically feasible motions of the system, the actual motion will be that which renders the modified action integral stationary,

$$\delta \int_{t_0}^{t_1} \left\{ \mathcal{T} - \int_\Omega \mathcal{H} \, d\Omega \right\} dt = 0$$

for all variations consistent with the constraints. Under the assumption that the equations of linear piezoelectricity are applicable and that only the axial deformations are nonnegligible, the kinetic energy and strain energy for the active strut element can be written

$$\mathcal{T}_e = \frac{1}{2} \sum_i \sum_j m_{ij} \dot{u}_i \dot{u}_j$$

$$\mathcal{V}_e = \frac{1}{2} \sum_i \sum_j k_{ij} u_i u_j \tag{19.49}$$

As discussed in Chapter 14, the classical finite element approximation of the displacement $u(x, t)$ is used[1]:

$$u(x, t) = \sum_{i=1}^{N_{\text{dof}}} N_i(x) u_i(t) \tag{19.50}$$

so that the element mass and stiffness matrices are given by the familiar expressions

$$m_{ij} = \int_0^L \rho A N_i(x) N_j(x) \, dx, \qquad k_{ij} = \int_0^L A E N'_i(x) N'_j(x) \, dx \tag{19.51}$$

[For a single axial-deformation element, the finite element shape functions $N_i(x)$ would be the linear functions $\psi_1(x)$ and $\psi_2(x)$ given in Eqs. 14.6. The integral indicated over L could, in actual fact, consist of integration over a typical element length L_e and assembly of "system" matrices for a stack of axial-deformation elements that form the strut.]

---

[1] The axial element is taken here to be oriented in the x-axis ($\equiv$ 1-axis) direction.

It is an underlying assumption in the modeling of an active strut that the piezoelectric effect is evident only in the stack, that is, in the active element. That is, we assume that the expansions

$$\mathbb{E}_1(x) = \sum_{i=1}^{N_{\text{layers}}} (-1)^{i+1} \chi_i(x) \frac{\Delta V}{h}$$

$$e_{11}(x) = \sum_{i=1}^{N_{\text{layers}}} (-1)^{i+1} \chi_i(x) e$$
(19.52)

hold for the electric field $\mathbb{E}_1$ and piezoelectric coupling coefficient e_{11}, respectively. The piezoelectric contribution to the modified Hamiltonian is, consequently,

$$\mathbb{D} \cdot \mathbb{E} = (e_{11}\epsilon_{11} + \gamma_{11}\mathbb{E}_1) \cdot \mathbb{E}_1$$

$$= \left\{ \left(\sum_{i=1}^{N_{\text{layers}}} (-1)^{i+1} \chi_i(x) e \right) \left(\sum_{j=1}^{N_{\text{dof}}} N_j'(x) u_j(t) \right) \right.$$

$$\left. + \gamma_{11} \sum_{i=1}^{N_{\text{layers}}} (-1)^{i+1} \chi_i(x) \frac{\Delta V}{h} \right\} \left\{ \sum_{k=1}^{N_{\text{layers}}} (-1)^{k+1} \chi_i(x) \frac{\Delta V}{h} \right\}$$
(19.53)

where $e \equiv e_{11}$ here and in the remainder of this finite element model derivation.

When the kinetic energy, the strain energy, and the piezoelectric contribution are substituted into the piezoelectric Hamiltonian, the variation with respect to the displacements yields

$$0 = \delta \int_{t_0}^{t_1} [\mathcal{T} - (\mathcal{U} - \mathbb{D} \cdot \mathbb{E})] \, dt + \int_{t_0}^{t_1} \delta \mathcal{W}_{\text{nc}} \, dt$$

$$= \int_{t_0}^{t_1} \left\{ \sum_{j=1}^{N_{\text{dof}}} [-m_{ij} \ddot{u}_j(t) - k_{ij} u_j(t)] - \frac{Ae \, \Delta V}{h} \int_0^L N_i(x) \frac{d}{dx}(\chi_s(x)) dx \right.$$

$$\left. + \int_0^L p(x,t) N_i(x) \, dx \right\} \delta u_i$$
(19.54)

Since this equality must hold for every virtual displacement δu_i, $i = 1 \cdots N_{\text{dof}}$, consistent with the kinematic constraints, the equations of motion for the active truss are simply

$$\sum_{j=1}^{N_{\text{dof}}} [m_{ij} \ddot{u}_j(t) + k_{ij} u_j(t)] = \frac{Ae \, \Delta V}{h} \int_0^L N_i(x) \frac{d}{dx}(\chi_s(x)) dx + \int_0^L p(x,t) N_i(x) \, dx$$

for $i = 1 \cdots N_{\text{dof}}$. Of course, this equation can be written more succinctly in matrix form

$$\boxed{\mathbf{M}\ddot{\mathbf{U}}(t) + \mathbf{K}\mathbf{U}(t) = \mathbf{B} \, \Delta V(t) + \mathbf{P}(t)}$$
(19.55)

where **M**, **K** are the conventional mass and stiffness matrices. Vector **B** is the so-called *control influence vector* and **P**(t) is the vector of nodal loads, which are given by

$$B_i = -\frac{Ae}{h}\int_0^L N_i(x)\frac{d}{dx}(\chi_s(x))dx, \qquad P_i = \int_0^L p(x,t)N_i(x)\,dx \qquad (19.56)$$

respectively. Again, integration over L could actually consist of forming separate element matrices and assembling system matrices to form a matrix equation of motion of the form given in Eq. 19.55.

19.6 ACTIVE BEAM MODELS

In addition to the piezoelectric stack actuator, piezoelectric materials can be used directly to induce bending in beam structures. One of the most common configurations employs symmetrically located piezoelectric patches that are bonded directly to a host beam structure. Consider the piezostructural system in Fig. 19.13. This figure depicts a uniform, linearly elastic, Bernoulli–Euler beam having Young's modulus E_H, cross-sectional moment of inertia I_H, and length L, respectively. Two symmetrically opposed piezoelectric patches are located on opposite faces of the beam. They are uniform in shape with Young's modulus, cross-sectional moment of inertia, and width equal to E_p, I_p, and W, respectively. The orientation of the polarization vectors, voltage across the two patches, and electrical field orientation are depicted in Fig. 19.14. A key assumption is that the beam and the bonded cross sections deform together in a state of pure bending. The *strain-displacement equation* that corresponds to this assumption is

$$\epsilon_{22} = -z\frac{\partial^2 w}{\partial y^2} \qquad (19.57)$$

for $z \in [-h_2, h_2]$, that is, over the entire depth of the host beam and the bottom and top patches, where patches exist.

A review of the constitutive laws for a piezoelectric continuum shows that the top and bottom patches have constitutive laws that are given by

$$\sigma_{22}^{(T)} = E_p\epsilon_{22} + e_{32}\mathbb{E}_3^{(T)}, \qquad \sigma_{22}^{(B)} = E_p\epsilon_{22} + e_{32}\mathbb{E}_3^{(B)} \qquad (19.58)$$

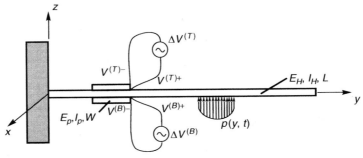

Figure 19.13 Linearly elastic beam with symmetrically bonded piezoelectric patches.

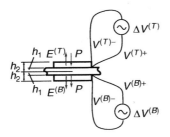

Figure 19.14 Detail of a beam with symmetrically bonded piezoelectric patches.

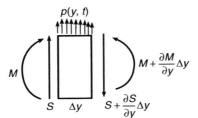

Figure 19.15 Free-body diagram of a section of a beam with stress resultants.

respectively, while the center layer, or *host beam*, has the simpler constitutive law

$$\sigma_{22}^{(H)} = E_H \epsilon_{22} \tag{19.59}$$

In the usual way, we take a free-body diagram of length Δy from the beam at any station y along its length. We can apply the methodology from Section 12.2 to rederive the equations of motion for the beam in terms of the stress resultants. The free-body diagram with associated stress resultants is depicted in Fig. 19.15.[2] For convenience, we repeat the derivation of the equation of motion for transverse deflection of the beam using Newton's Method. Summing forces in the z direction, we obtain the equation of motion for displacement transverse to the beam, expressed in terms of the distributed lateral load $p(y, t)$ and the internal transverse shear resultant $S(y, t)$.

$$S - \left(S + \frac{\partial S}{\partial y} \Delta y\right) + p(y, t) \, \Delta y = \rho A \, \Delta y \, \ddot{w} \tag{19.60}$$

This equation simplifies to

$$\rho A \ddot{w} = \frac{\partial S}{\partial y} + p(y, t) \tag{19.61}$$

If we neglect rotatory inertia and sum moments about the left edge of the FBD depicted in Fig. 19.15, taking counterclockwise as positive, the equation for rotation is derived in terms of the bending moment $M(y, t)$, the transverse shear $S(y, t)$, and the distributed

---

[2] The notation and sign conventions in the present derivation may differ from those in Chapter 12 (e.g., y instead of x).

load $p(y, t)$.

$$-M + \underbrace{p(y,t)\Delta y \frac{\Delta y}{2}}_{\text{higher order}} - \left(S + \underbrace{\frac{\partial S}{\partial y}\Delta y}_{\text{higher order}}\right)\Delta y + \left(M + \frac{\partial M}{\partial y}\Delta y\right) = 0 \quad (19.62)$$

When we eliminate the higher-order terms, the internal shear can be related to the bending moment:

$$S = \frac{\partial M}{\partial y} \quad (19.63)$$

We obtain the final equation of motion for transverse deflection of the beam in terms of the stress resultants when we combine Eqs. 19.61 and 19.63.

$$\rho A \ddot{w} - \frac{\partial^2 M}{\partial y^2} = p(y,t) \quad (19.64)$$

In Section 12.2 we proceeded to incorporate the moment–curvature equation that combined the definition of the bending-moment resultant, the constitutive law (stress–strain equation), and the strain–displacement equation. Here, we follow a similar process by substituting the stress expressions into the definition for the bending-moment stress resultant M. In this case, the normal stress varies with the transverse position coordinate z as follows:

$$\sigma_{22} = \begin{cases} \sigma_{22}^{(T)}, & h_1 \leq z < h_2 \\ \sigma_{22}^{(H)}, & -h_1 \leq z < h_1 \\ \sigma_{22}^{(B)}, & -h_2 \leq z < -h_1 \end{cases} \quad (19.65)$$

where the thickness of the core beam is $2h_1$ and the thickness of each piezoelectric patch is h_2. The bending moment M can therefore be expressed as

$$M = -\iint z\sigma_{22}\,dA = \underbrace{-\iint_{A_B} z\sigma_{22}^{(B)}\,dA}_{\mathcal{I}_B} \underbrace{-\iint_{A_H} z\sigma_{22}^{(H)}\,dA}_{\mathcal{I}_H} \underbrace{-\iint_{A_T} z\sigma_{22}^{(T)}\,dA}_{\mathcal{I}_T}$$
$$(19.66)$$

Each of the integrals can be evaluated using the constitutive law for the stress in the appropriate layer over which the integral is expressed (Eqs. 19.58 and 19.59), combined with the strain–displacement equation (Eq. 19.57). Equations 19.57 and 19.59 can be combined with the definition of $\mathcal{I}_H$, the integral over the host layer, to give

$$\mathcal{I}_H \equiv \iint_{A_H} z\sigma_{22}^{(H)}\,dA = \iint_{A_H} z(E_H \epsilon_{22}^{(H)})\,dA = -E_H I_H \frac{\partial^2 w}{\partial y^2} \quad (19.67)$$

Since the constitutive laws governing the top and bottom layers include a voltage term, expressions are required for the distribution of the voltage over each of these layers. Assuming a linear variation in the z direction of the voltage over the top layer, we can derive

$$V^{(T)}(z) = c_1 z + c_2, \qquad V^{(T)+} = c_1 h_1 + c_2, \qquad V^{(T)-} = c_1 h_2 + c_2 \qquad (19.68)$$

These equations can be combined to yield the formula

$$V^{(T)}(z) = \frac{\Delta V^{(T)}}{h_1 - h_2} z + \text{constant} = -\frac{\Delta V^{(T)}}{\Delta h} z + \text{constant} \qquad (19.69)$$

Similarly, an assumption of a linear variation of voltage over the bottom layer yields the final expression for the voltage over that layer:

$$V^{(B)}(z) = \frac{\Delta V^{(B)}}{h_2 - h_1} z + \text{constant} = \frac{\Delta V^{(B)}}{\Delta h} z + \text{constant} \qquad (19.70)$$

The associated electrical field supported over the dielectric is equal to the negative gradient of the voltage expressions. Therefore,

$$\begin{aligned}\mathbb{E}_3^{(B)} &= -\frac{d}{dz}[V^{(B)}(z)] = -\frac{\Delta V^{(B)}}{h_2 - h_1} = -\frac{\Delta V^{(B)}}{\Delta h} \\ \mathbb{E}_3^{(T)} &= -\frac{\Delta V^{(T)}}{h_1 - h_2} = \frac{\Delta V^{(T)}}{\Delta h}\end{aligned} \qquad (19.71)$$

Assuming that the middle layer is perfectly conducting, we have $V^{(T)+} = V^{(B)+}$. The voltage relationship becomes

$$\Delta V^{(T)+} = \Delta V^{(B)+} \equiv \Delta V(t)$$

when we choose the drive voltage to satisfy $V^{(B)-}(t) = V^{(T)-}(t)$ for all $t > 0$.

We are now prepared to evaluate the stress resultants over the bottom and top layers. The integral $\mathcal{I}_B$ can be expanded into the expression

$$\begin{aligned}\mathcal{I}_B &\equiv \iint_{A_B} z \sigma_{22}^{(B)} \, dA = \iint_{A_B} z \left(E_p \epsilon_{22} + e_{32} \mathbb{E}_3^{(B)} \right) dA \\ &= E_p \iint_{A_B} z^2 \, dA \, \epsilon_{22} + e_{32} \iint_{A_B} z \, dA \left(-\frac{\Delta V^{(B)}}{\Delta h} \right) \\ &= -E_p I_B \frac{\partial^2 w}{\partial y^2} + e_{32} Q_B \left(-\frac{\Delta V^{(B)}}{\Delta h} \right)\end{aligned}$$

So

$$\mathcal{I}_B = -E_p I_B \frac{\partial^2 w}{\partial y^2} - \frac{e_{32} Q_B}{\Delta h} \Delta V^{(B)} \qquad (19.72)$$

where[3]

$$I_B \equiv \iint_{A_B} z^2 \, dA, \qquad Q_B \equiv \iint_{A_B} z \, dA$$

---

[3] Note that z is measured from the middle plane of the beam, not from the middle plane of the patch.

Similarly, the integral over the top layer becomes

$$\mathcal{I}_T \equiv \iint_{A_T} z\sigma_{22}^{(T)}\, dA = -E_p I_T \frac{\partial^2 w}{\partial y^2} + \frac{e_{32} Q_T}{\Delta h}\Delta V^{(T)} \tag{19.73}$$

Substituting the integrals $\mathcal{I}_T$ and $\mathcal{I}_B$ into the equation of motion yields the final equation for the piezoelectrically actuated beam:

$$\boxed{\rho A \frac{\partial^2 w}{\partial t^2} = -\frac{\partial^2}{\partial y^2}\left(EI \frac{\partial^2 w}{\partial y^2}\right) + B\frac{d^2}{dy^2}[\chi_p(y)]\Delta V(t)} \tag{19.74}$$

In this equation, it is important to note that the bending stiffness is a function of the variable y. For the bare beam, $EI(y) = E_H I_H$, but the bending stiffness is greater over that part of the beam covered by the piezoceramic patches, where it is given by

$$EI(y) = E_p I_B \chi_p(y) + E_H I_H + E_p I_T \chi_p(y)$$

The control influence operator is scaled by the constant B. This term is written as

$$B \triangleq \frac{e_{32}}{h_2 - h_1}(Q_T - Q_B)$$

But Q_T and Q_B are straightforward to calculate for many simple cross sections. For a rectangular patch having width W and thickness $h_2 - h_1$, these terms are simply

$$Q_T = \iint_{A_T} z\, dA = \underbrace{\tfrac{1}{2}(h_1 + h_2)}_{\text{location of centroid}}\underbrace{\{W(h_2 - h_1)\}}_{\text{patch c--s area}} = -Q_B$$

Thus, for the special case that the piezoceramic patch is a uniform rectangular area having width W and thickness $h_2 - h_1$, the scaling factor B for the control influence operator reduces to the equation

$$B = e_{32} W(h_1 + h_2)$$

This result may be compared to the corresponding result in Ref. [19.7]. (It is important in comparisons to note that the sign on B is somewhat arbitrary; switching the polarity from the power supply to the electrodes on the piezoceramic will necessarily induce a change in sign of the control influence term.)

19.7 ACTIVE COMPOSITE LAMINATES

In the past few sections we have considered the response of one-dimensional piezostructural systems, that is, axial deformation of struts and bending of beams. At least as common in applications are those piezostructural systems that are based on platelike structures consisting of *active composite laminates*. It is interesting that for the most part, the theoretical development of active structural composites follows that for conventional structural composites. The assumptions that underlie our treatment of active composite laminates are, for the most part, the same as those that form the foundation of classical laminated plate theory.[19.8]

642 Introduction to Active Structures

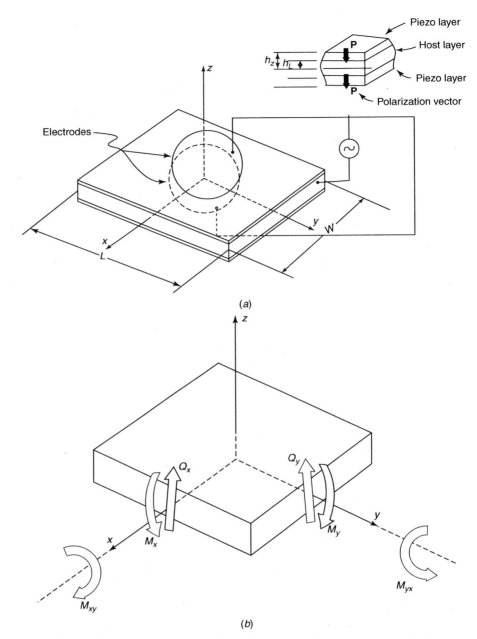

Figure 19.16 (a) Schematic of laminated plate; (b) plate stress resultants.

As depicted in Fig. 19.16, we will consider the transverse displacement $w(x, y; t)$ of a laminated plate whose middle surface is designated as the $x-y$ plane. In this book we make some rather simple restrictions on the alignment of the principal structural axes of the host structure and the axes of piezoelectricity of the piezoceramic. This simplifies the presentation, and we subsequently comment on the extension to more general geometries.

- The poling axis of the piezoelectric is aligned with the global 3-axis.
- The principal axes of the structural plies are aligned with the principal axes of the piezoelectric layers.
- Fibers normal to the midplane of the composite layers remain normal to the midplane during the deformation.
- The thickness of the laminate is negligible in the 3-direction, so that we can approximate the stress σ_{33} as $\sigma_{33} = 0$.
- Different plies are bonded together perfectly, so that no slip occurs at ply interfaces.
- Interlaminar stresses are neglected.
- The thickness and weight of the electrode are neglected.
- The electric field and electric displacement are nonnegligible only in the global 3-direction.
- The materials comprising the laminate satisfy the constitutive equations of linear piezoelectricity.

19.7.1 Symmetric Bimorph, Uniform Layers

Figure 19.17 depicts one of the most common bending actuators that is commercially available today, the *symmetric bimorph*. The actuator in this figure is composed of three layers: two piezoelectric layers bonded to a center, linearly elastic layer. Note carefully the orientation of the polarization vectors and electric fields for this geometry. It is clear that (for the voltage polarity, electric field orientation, and polarization shown) the bottom layer seeks to expand laterally, whereas the top layer seeks to contract. The net effect is that the structure will bend upward. If the polarity of the voltage and electric field are reversed, the composite will tend to bend downward.

Classical Kirchhoff–Love plate theory dictates that the strain–displacement equations can be expressed as

$$\epsilon_x = -z\frac{\partial^2 w}{\partial x^2}, \qquad \epsilon_y = -z\frac{\partial^2 w}{\partial y^2}, \qquad \epsilon_{xy} = -z\frac{\partial^2 w}{\partial x \partial y} \qquad (19.75)$$

where $w(x, y; t)$ is the displacement of the middle surface of the plate in the z direction. The constitutive relationships can be derived in two steps. First, we can suppose that there is no electric field and ask what the stress–strain relationship would be in this case. In the absence of an electric field, each layer is assumed to be in a state of plane stress. That is, we assume that the normal stress in the z direction is negligible throughout the laminate; that is, $\sigma_z(x, y; t) \equiv 0$. From conventional linear elasticity theory and the assumptions stated in this section, the constitutive law for the ferroelectric piezoceramic

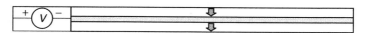

Figure 19.17 Voltage applied to a bimorph structure.

layers on the outer faces of the laminate can be summarized as

$$\left\{\begin{array}{c} \sigma_x \\ \sigma_y \\ \tau_{xy} \\ \mathbb{D}_3 \end{array}\right\} = C \left\{\begin{array}{c} \epsilon_x \\ \epsilon_y \\ 2\epsilon_{xy} \\ \mathbb{E}_3 \end{array}\right\} \qquad (19.76)$$

in which the constitutive matrix for the ferroelectric piezoceramic system is

$$C = \begin{bmatrix} \dfrac{E_p}{1-\nu_p^2} & \dfrac{\nu_p E_p}{1-\nu_p^2} & 0 & -e_{31} \\ \dfrac{\nu_p E_p}{1-\nu_p^2} & \dfrac{E_p}{1-\nu_p^2} & 0 & -e_{31} \\ 0 & 0 & \dfrac{E_p}{2(1+\nu_p)} & -e_{36} \\ e_{31} & e_{31} & e_{36} & \gamma_{33}^\epsilon \end{bmatrix} \qquad (19.77)$$

The constitutive relationship for the linearly elastic layer in the center has the same form, except that the subscript p is replaced by s in this case, and the last row and column are filled with zeros. Since the voltage is assumed to vary linearly in the z direction over the electroded region of each piezoelectric patch, the electric field strength is a constant over each such area. The uniform electric field strength $\mathbb{E}_3$ can be derived to be $\Delta V(t)/h_p$ over each piezoelectric layer. Recall that the direction of the electric field is derived from the fact that it is equal to the *negative* gradient of the voltage potential. The stresses acting on a cross section parallel to the x–z plane or to the y–z plane of the laminated plate are then easy to derive, although the expressions are tedious to write down. We have

$$\sigma_x(z) = \begin{cases} -\dfrac{E_s z}{1-\nu_s^2}\left(\dfrac{\partial^2 w}{\partial x^2} + \nu_s \dfrac{\partial^2 w}{\partial y^2}\right), & 0 < |z| < \dfrac{h_s}{2} \\ -\dfrac{E_p z}{1-\nu_p^2}\left(\dfrac{\partial^2 w}{\partial x^2} + \nu_p \dfrac{\partial^2 w}{\partial y^2}\right) - e_{31}\dfrac{\Delta V(t)}{h_p}, & \dfrac{h_s}{2} < |z| < \dfrac{h_s}{2} + h_p \end{cases}$$

$$\sigma_y(z) = \begin{cases} -\dfrac{E_s z}{1-\nu_s^2}\left(\dfrac{\partial^2 w}{\partial y^2} + \nu_s \dfrac{\partial^2 w}{\partial x^2}\right), & 0 < |z| < \dfrac{h_s}{2} \\ -\dfrac{E_p z}{1-\nu_p^2}\left(\dfrac{\partial^2 w}{\partial y^2} + \nu_p \dfrac{\partial^2 w}{\partial x^2}\right) - e_{31}\dfrac{\Delta V(t)}{h_p}, & \dfrac{h_s}{2} < |z| < \dfrac{h_s}{2} + h_p \end{cases}$$

$$\tau_{xy}(z) = \begin{cases} -\dfrac{E_s z}{1+\nu_s}\dfrac{\partial^2 w}{\partial x \partial y}, & 0 < |z| < \dfrac{h_s}{2} \\ -\dfrac{E_p z}{1+\nu_p}\dfrac{\partial^2 w}{\partial x \partial y} - e_{36}\dfrac{\Delta V(t)}{h_p}, & \dfrac{h_s}{2} < |z| < \dfrac{h_s}{2} + h_p \end{cases}$$

(19.78)

From Fig. 19.16b we can think of the moment M_x as a combination of the contributions from the outer layers and the midlayer, respectively. Consistent with Kirchhoff–Love plate theory, the bending moments can be calculated as follows:

$$
\begin{aligned}
M_y &= 2\int_0^{h_s/2} \sigma_x(x,y,z;t)z\,dz + 2\int_{h_s/2}^{(h_s/2)+h_p} \sigma_x(x,y,z;t)z\,dz \\
&= -\frac{h_s^3}{12}\frac{E_s}{1-v_s^2}\left(\frac{\partial^2 w}{\partial x^2} + v_s\frac{\partial^2 w}{\partial y^2}\right) \\
&\quad - h_p\left(\frac{h_s^2}{2} + h_s h_p + \frac{2}{3}h_p^2\right)\frac{E_p}{1-v_p^2}\left(\frac{\partial^2 w}{\partial x^2} + v_p\frac{\partial^2 w}{\partial y^2}\right) \\
&\quad - e_{31}v(t)\chi(x,y)(h_s + h_p)
\end{aligned}
\quad (19.79)
$$

Similarly, we can write

$$
\begin{aligned}
M_x &= -\frac{h_s^3}{12}\frac{E_s}{1-v_s^2}\left(\frac{\partial^2 w}{\partial y^2} + v_s\frac{\partial^2 w}{\partial x^2}\right) \\
&\quad - h_p\left(\frac{h_s^2}{2} + h_s h_p + \frac{2}{3}h_p^2\right)\frac{E_p}{1-v_p^2}\left(\frac{\partial^2 w}{\partial y^2} + v_p\frac{\partial^2 w}{\partial x^2}\right) \\
&\quad - e_{31}v(t)\chi(x,y)(h_s + h_p)
\end{aligned}
\quad (19.80)
$$

and

$$
M_{xy} = -\left[\frac{h_s^3}{12}\frac{E_s}{1+v_s} + h_p\left(h_s^2 2 + h_s h_p + \frac{2}{3}h_p^2\right)\frac{E_p}{1+v_p}\right]\frac{\partial^2 w}{\partial x\,\partial y} \quad (19.81)
$$
$$
- e_{36}v(t)\chi(x,y)(h_s + h_p)
$$

The moments M_x, M_y, and M_{xy} can be combined with the force equilibrium equations for the composite plate to obtain a governing equation in terms of the displacement alone. Recall that by summing the shear forces acting on a differential element of the plate, it is possible to obtain the equation of motion

$$
\rho h \frac{\partial^2 w}{\partial t^2} = \frac{\partial Q_x}{\partial x} + \frac{\partial Q_y}{\partial y} \quad (19.82)
$$

Also, from the equilibrium equations of conventional Kirchhoff–Love thin plate theory, the shear force resultants and moment resultants are related by the equations

$$
Q_x = \frac{\partial M_x}{\partial x} + \frac{\partial M_{xy}}{\partial y}, \qquad Q_y = \frac{\partial M_{xy}}{\partial x} + \frac{\partial M_y}{\partial y} \quad (19.83)
$$

By combining these two sets of equilibrium equations, it is possible to derive the following equation of motion for thin plates that is expressed in terms of the moment resultants alone:

$$
\frac{\partial^2 M_x}{\partial x^2} + 2\frac{\partial^2 M_{xy}}{\partial x\,\partial y} + \frac{\partial^2 M_y}{\partial y^2} = \rho h \frac{\partial^2 w}{\partial t^2} \quad (19.84)
$$

We now have all the ingredients required to derive the equations of motion for transverse displacement of the plate in terms of the displacement alone. To simplify the notation, let us introduce the following constants:

$$D_{11} = \frac{E_s h_s^3}{12(1-\nu_s^2)} + \frac{E_p}{1-\nu_p^2} h_p \left(\frac{h_s^2}{2} + h_s h_p + \frac{2}{3} h_p^2 \right) \tag{19.85}$$

$$D_{12} = \frac{\nu_s E_s h_s^3}{12(1-\nu_s^2)} + \frac{\nu_p E_p}{1-\nu_p^2} h_p \left(\frac{h_s^2}{2} + h_s h_p + \frac{2}{3} h_p^2 \right) \tag{19.86}$$

$$D_{66} = \frac{E_s h_s^3}{12(1+\nu_s)} + \frac{E_p}{1+\nu_p} h_p \left(\frac{h_s^2}{2} + h_s h_p + \frac{2}{3} h_p^2 \right) \tag{19.87}$$

It is not difficult to show that

$$D_{11} = D_{12} + D_{66} \tag{19.88}$$

It is a tedious and algebraically complicated exercise, but one that is not too difficult in principle, to derive the following equation of motion governing the transverse displacement w of the plate:

$$\begin{aligned}\rho h \frac{\partial^2 w}{\partial t^2} + D_{11} \left(\frac{\partial^4 w}{\partial x^4} + 2 \frac{\partial^4 w}{\partial x^2 \partial y^2} + \frac{\partial^4 w}{\partial y^4} \right) \\ = -(h_s + h_p) v(t) \left[e_{31} \frac{\partial^2 \chi(x,y)}{\partial x^2} + e_{31} \frac{\partial^2 \chi(x,y)}{\partial y^2} + 2 e_{36} \frac{\partial^2 \chi(x,y)}{\partial x \partial y} \right]\end{aligned} \tag{19.89}$$

In this equation, $v(t) \equiv \Delta V(t)$, and the variable $\chi(x,y)$ denotes the characteristic function of the symmetrically placed electrodes on the outer layers of the composite. Sometimes we call D_{11} the effective stiffness. Comparing this equation with the Eq. 13.59 in Section 13.6, Eq. 19.89 is more complicated because of the piezoelectric effect. However, the left-hand side keeps the same form.

REFERENCES

[19.1] http://scienceworld.wolfram.com/biography/Langevin.html.

[19.2] G. Gautschi and B. Mandel, *Sensors and Tools in Engine and Vehicle Engineering*, ATZ/MTZ-Sonderausgabe, Kistler, Amherst, NY, http://www.cimwareukandusa.com/DigiFactory.html, 1998.

[19.3] C. Jaffe, *Piezoelectric Ceramics*, Academic Press, London, 1971.

[19.4] H. F. Tiersten, *Linear Piezoelectric Plate Vibrations: Elements of the Linear Theory of Piezoelectricity and the Vibrations of Piezoelectric Plates*, Plenum Press, New York, 1969.

[19.5] T. Ikeda, *Fundamentals of Piezoelectricity*, Oxford University Press, Oxford, 1990.

[19.6] Y. Y. Yu, *Vibrations of Elastic Plates*, Springer-Verlag, New York, 1995.

[19.7] H. T. Banks, R. C. Smith, and Y. Wang, *Smart Material Structures Modeling, Estimation and Control*, Masson, Paris, 1996.

[19.8] L. P. Kollar and G. P. Springer, *Mechanics of Composite Structures*, Cambridge University Press, Cambridge, 2003.

PROBLEMS

Problem Set 19.1

19.1 A vibration absorber is a common device used to eliminate unwanted disturbances in a host structure. A conventional vibration absorber is depicted in Fig. P19.1a, and a variant that includes a piezoceramic layer is depicted in Fig. P19.1b. There are two primary components of this system: the host structure and the proof mass actuator. In this figure, displacement of the host structure is denoted by y_s and displacement of the proof mass actuator is denoted by y_a. In a conventional vibration absorber, the system components, including the mass of the actuator m_a, the spring constant k_a, and damping constant c_a, are chosen to reject input disturbances from the base motion. (a) Derive the time-domain equations governing the system response of the conventional vibration absorber. (b) Derive the equations governing the system response of piezoceramic vibration absorber when the switch is closed. (c) Derive the equations governing the system response of the piezoceramic vibration absorber when the switch is open. (d) Discuss the ramifications of having an active layer in the piezoceramic vibration absorber in comparison to the conventional vibration absorber.

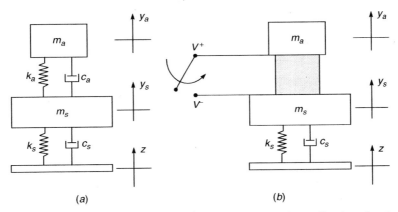

Figure P19.1 (a) Conventional vibration absorber; (b) piezo vibration absorber.

19.2 (a) Consider the conventional vibration absorber in Fig. P19.1a. For the special <u>undamped</u> case, that is, $c_a = c_s = 0$, derive the transfer function from the input disturbance to the structural response. (b) Consider the piezoceramic vibration absorber in Fig. P19.1b. For the special <u>undamped</u> case, that is, when $c_a = 0$ and the internal energy dissipation of the piezoceramic layer is zero, derive the transfer function from the input disturbance to the structural response.

Problem Set 19.3

19.3 Recall the axial rod with lateral electrodes that was first considered in Section 19.3. It is shown again in Fig. P19.3 for convenience. In Section 19.3, the partial differential equation of motion governing this distributed-parameter system was derived. Use the *Assumed-Modes Method* to derive the general form of a finite-dimensional (i.e., MDOF) model of this piezostructural system, a model that would be appropriate for use in simulations of the response of this system. Assume that the input voltage is a known, prescribed function of time.

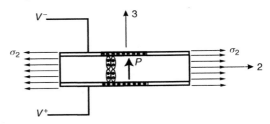

Figure P19.3 Axial piezoelectric rod with lateral electrodes.

Problem Set 19.6

19.4 Recall the Bernoulli–Euler beam with symmetrically configured piezoceramic patches considered first in Section 19.6. It is shown again in Fig. P19.4. In Section 19.6 the partial differential equations governing this infinite dimensional system were derived. Use the *Assumed-Modes Method* to derive the general form of a finite-dimensional, MDOF model of the piezostructural system that is appropriate for simulation. Assume that the input voltage is a known, prescribed function of time.

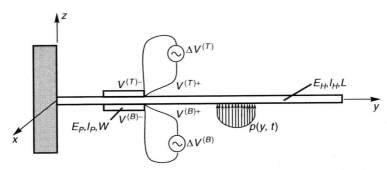

Figure P19.4 Linearly elastic beam with symmetrically bonded piezoelectric patches

19.5 Consider the active structural composite system depicted in Fig. P19.5. This composite structure is composed of $2N+1$ thin layers, and the overall composite can be viewed as a thin plate. The center layer is linearly elastic and the adjacent layers are all comprised of piezoelectric material. The thickness of the middle layer is denoted h_s. The thickness of the ith piezoelectric layer is $h_{p,i}$, where $i = -N \cdots N$. The composite is symmetric, which means that the structural and material properties of the ith and $-i$th layers are identical. The elastic modulus, Poisson's ratio, and the density of the ith layer are given by $E_{p,i}$, $\nu_{p,i}$ and $\rho_{p,i}$, respectively. The constants E_s, ν_s, and ρ_s are the elastic modulus, Poisson's ratio, and density of the center layer, respectively. The function $\chi(x, y)$ is the characteristic function of the electroded region for each layer. Derive the equations governing the dynamics of the multilayer composite when there is a prescribed input voltage $\Delta V(t)$.

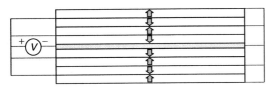

Figure P19.5 Uniformly electroded multilayer composite.

19.6 Another common configuration for an active composite in applications is the simple two-layer structure depicted in Fig. P19.6. Suppose that we can accurately approximate the deformation of the composite by assuming that it is in a state of pure bending. Note that the composite is not symmetric about its midplane. Consequently, the neutral plane is not identical to the center plane of the composite. (**a**) Find the location of the neutral plane under the assumption that the system is in a

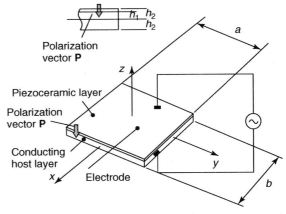

Figure P19.6 Two-layer active composite structure.

state of pure bending. **(b)** Find the equations governing the dynamics of the two-layer composite structure.

19.7 Consider the active structural composite system depicted in Fig. P19.7. This composite structure is composed of three thin layers, is a symmetric composite, but the layers are non-uniform. The thickness of the middle layer is h_s, while the thicknesses of the piezoceramic layers are h_p. The constants E_s, ν_s, and ρ_s are the elastic modulus, Poisson's ratio, and the density of the center layer. The corresponding constants for the piezoceramic layers are E_p, ν_p, and ρ_p. Derive the equations governing the dynamics of the 3-layer, symmetric, but non-uniform layer when there is a prescribed input voltage equal to $\Delta V(t)$.

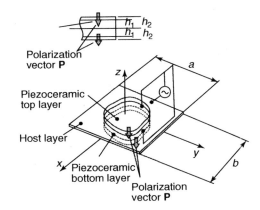

Figure P19.7 Three-layer active composite structure.

20
Introduction to Earthquake Response of Structures

Although the structural dynamics analysis techniques presented so far in this book have applicability to a wide range of structural dynamics problems, one problem that merits special consideration is the analysis and design of structures to withstand earthquake excitation. Many of the ingredients of the analysis of earthquake response of structures that were covered in previous chapters are thus drawn together in the present chapter. However, the treatment here is intended only as a brief introduction to this important topic. Readers who require more thorough coverage are encouraged to consult a reference on earthquake engineering[20.1–20.3] or a structural dynamics text with more extensive coverage of earthquake analysis and design of structures.[20.4,20.5] Upon completion of this chapter you should be able to:

- Define the three types of response spectra—$\mathcal{S}_a$, $\mathcal{S}_v$, and $\mathcal{S}_d$—in this chapter and describe how they are obtained.
- Use shock spectra to estimate the maximum displacement and maximum spring force for a lumped-mass SDOF system.
- Use shock spectra to estimate the maximum displacement and maximum base shear of an SDOF assumed-modes model of a column subjected to base acceleration.
- Use shock spectra to estimate the maximum displacement and maximum base shear of a lumped-mass MDOF model of a structure subjected to base acceleration.

20.1 INTRODUCTION

Although thousands of earthquakes occur each year, and although they are widely distributed over the Earth's surface, comparatively few of these cause significant damage to property or loss of life. Those that are of interest to the structural engineer are those that can cause structural damage; these are called *strong-motion earthquakes*. The majority of earthquakes, both large and small, occur in two zones, or belts. One of these is the Circum-Pacific Belt, which extends around the Pacific Ocean. The other is the Alpide belt, which extends from the Himalayan mountain range through Iran and Turkey to the Mediterranean Sea. Figure 20.1 illustrates the concentration of seismic activity in these two zones.[20.6] In the continental United States the most active seismic zone is along the California coast and is associated with the San Andreas fault. The Richter scale

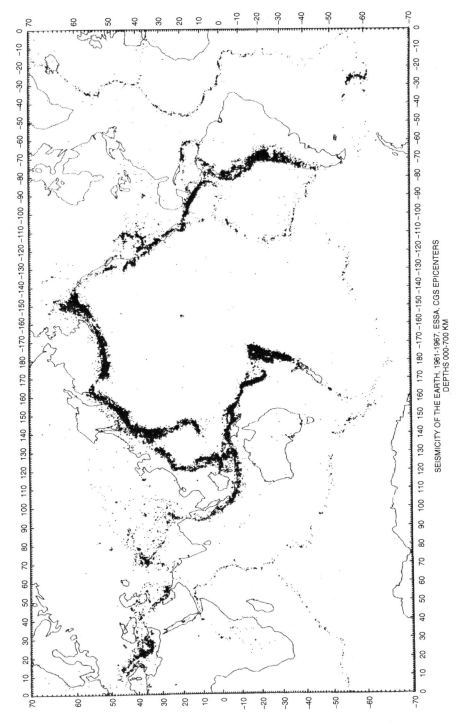

Figure 20.1 Earthquakes recorded during 1961 through 1967. (From M. Barazangi and J. Dorman, "Seismicity of the Earth 1961–1967" (Ref. [20.6]). Seismological Society of America, 1969.)

provides a convenient means of classifying earthquakes according to size. Earthquakes of magnitude 5.0 or greater on the Richter scale generate ground motions that are severe enough to be damaging to structures. For example, the San Francisco earthquake of 1906 registered 8.2 on the Richter scale (Ref. [20.1], p. 77). Figure 20.2 shows a typical record of acceleration recorded during a strong-motion earthquake and velocity and displacement records obtained by integration of the acceleration record.[20.7] Two fundamental difficulties encountered in earthquake response analysis are the random nature of the excitation[20.4,20.8,20.9] and the nonlinear nature of the response.[20.10,20.11] It is beyond the scope of this book to treat either of these topics in detail. Rather, earthquake engineers have found that a deterministic approach based on response spectra provides valuable insight into the response of structures to earthquake excitation, so this is the technique that is presented here.[20.8],[20.12]–[20.15] Not only is a consideration of earthquake response important for the design of building structures, but the design of mechanical and electrical equipment to be housed in these buildings must take into account dynamic response to earthquake excitation as transmitted to the equipment through the structure of the building. The response of a structure to earthquake excitation is a base motion problem (see Sections 4.4 and 5.5). In Section 20.2 we treat the response of an SDOF system to earthquake-type base motion, and in Section 20.3 extend the analysis to MDOF systems.

20.2 RESPONSE OF A SDOF SYSTEM TO EARTHQUAKE EXCITATION: RESPONSE SPECTRA

The earthquake response problem is essentially a base motion problem similar to those introduced in Sections 4.4 and 5.5. Figure 20.3 shows an SDOF system subjected to ground motion. This is the simplest structural model that captures the base excitation. Since only linear response is treated in this chapter, Fig. 20.3 represents a linear structure with spring constant k. It is assumed that the base is rigid and has a translational motion $z(t)$. The equations of motion for absolute motion $u(t)$ and relative motion $w(t)$ were found in Example 2.2 to be

$$m\ddot{u} + c\dot{u} + ku = c\dot{z} + kz \tag{20.1}$$

and

$$m\ddot{w} + c\dot{w} + kw = -m\ddot{z} \tag{20.2}$$

In Section 5.5 expressions for the maximum relative displacement w_{max} and maximum absolute acceleration (of an undamped system) were obtained. It is customary to ignore the minus sign in Eq. 20.2 [in effect reversing the sense of $z(t)$] and to neglect the small difference between ω_n and ω_d. Thus, the Duhamel integral solution of Eq. 20.2 becomes (see also Eq. 5.42)

$$w(t, \omega_n, \zeta) = \frac{1}{\omega_n} W(t) \tag{20.3}$$

where

$$W(t) = \int_0^t \ddot{z}(\tau) e^{-\zeta \omega_n (t-\tau)} \sin \omega_n (t - \tau) \, d\tau \tag{20.4}$$

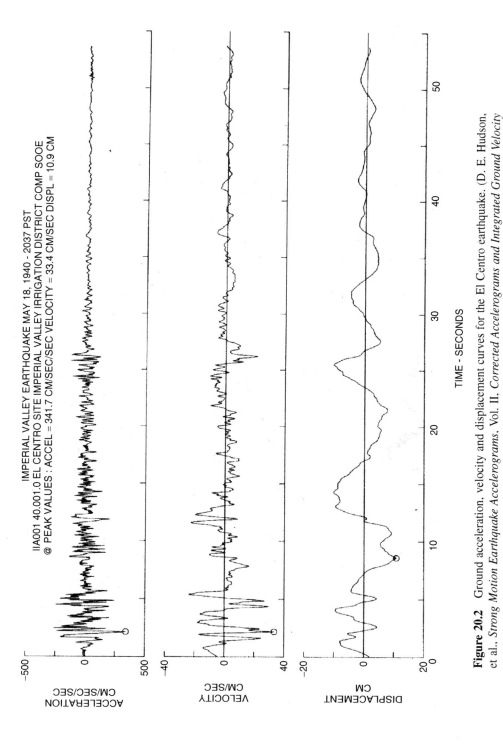

Figure 20.2 Ground acceleration, velocity and displacement curves for the El Centro earthquake. (D. E. Hudson, et al., *Strong Motion Earthquake Accelerograms.* Vol. II. *Corrected Accelerograms and Integrated Ground Velocity and Displacement Curves. Part A.* Earthquake Engineering Research Laboratory, California Institute of Technology, Pasadena, CA. 1971.)

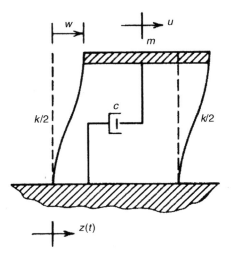

Figure 20.3 SDOF system subjected to base motion.

The maximum value of relative displacement occurs at time t_m. This is customarily given the symbol S_d and is called the *spectral displacement*.

$$S_d(T, \zeta) = w_{\max} = \frac{1}{\omega_n} W(t_m) \qquad (20.5)$$

A plot of S_d versus the natural period of the system, $T = 2\pi/\omega_n$, is called the *displacement response spectrum*. The integral in Eq. 20.4 is seen to have the dimensions of velocity. The maximum value of this integral is called the *spectral pseudovelocity*, S_v. A plot of S_v versus period T is called the *pseudovelocity response spectrum*.

$$S_v(T, \zeta) = W(t_m) = \omega_n S_d \qquad (20.6)$$

In Eq. 5.46 it was shown that for an undamped system the maximum absolute acceleration is given by $\ddot{u}_{\max} = \omega_n^2 w_{\max}$. Although this relationship does not hold for a damped system, the maximum absolute acceleration of a lightly damped system can be approximated by the *spectral pseudoacceleration*, S_a, where

$$S_a(T, \zeta) = \omega_n^2 S_d = \omega_n S_v \qquad (20.7)$$

A plot of S_a versus period T is called the *pseudoacceleration response spectrum*. The usefulness of S_a stems from the fact that the maximum spring force is given by kS_d, and thus

$$(f_s)_{\max} = kS_d = \frac{k}{\omega_n^2} S_a = mS_a \qquad (20.8)$$

that is, the maximum spring force is obtained by multiplying the spectral pseudoacceleration, S_a, by the mass. Figure 20.4 shows a pseudovelocity response spectrum plotted to linear scales.[20.16] This gives the maximum pseudovelocity as a function of the natural period of the structure for several values of damping. The sharp peaks and valleys in Fig. 20.4 are the result of local resonances and antiresonances of the ground motion. For design purposes these irregularities can be smoothed out and a

20.2 Response of a SDOF System to Earthquake Excitation: Response Spectra **655**

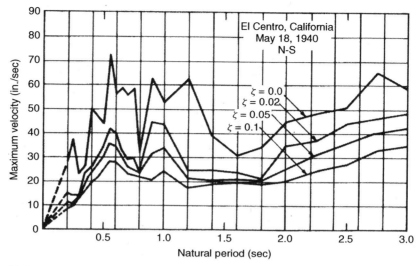

Figure 20.4 Pseudovelocity response spectrum for the N-S component of the El Centro earthquake of May 18, 1940. (From G. S. Housner, "Strong Ground Motion," in *Earthquake Engineering*, R. L. Wiegel, ed., Prentice-Hall, Englewood Cliffs, NJ, 1970.)

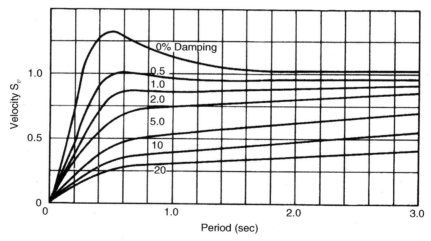

Figure 20.5 Average velocity response spectrum, El Centro earthquake of May 18, 1940. (From G. S. Housner, "Design Spectrum," in *Earthquake Engineering*, R. L. Wiegel, ed., Prentice-Hall, Englewood Cliffs, NJ, 1970.)

number of different response spectra averaged after normalizing them to a standard intensity. Figure 20.5 gives such an average velocity response spectrum.[20.17] One of the advantages of using pseudovelocity and pseudoacceleration is a result of the simple relationships given by Eq. 20.7. These relationships make it possible to plot all three response spectra simultaneously on a tripartite log-log graph, as shown in Fig. 20.6. Earthquake response spectra such as those given in Figs. 20.4 through 20.6 make it

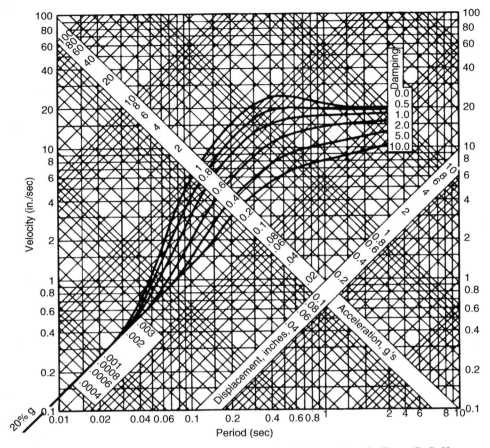

Figure 20.6 Tripartite plot of design spectrum scaled to 20% g at $T = 0$. (From G. S. Housner, "Design Spectrum," in *Earthquake Engineering*, R. L. Wiegel, ed., Prentice-Hall, Englewood Cliffs, NJ, 1970.)

possible to calculate the maximum response of any SDOF system to the earthquake from which they were derived.[20.17] Example 20.1 illustrates such calculations for a lumped-mass SDOF system.

Example 20.1 An SDOF frame structure such as the one shown in Fig. 20.3 has a period $T = 1.0$ sec and a damping factor of 5%. The weight is $W = mg = 1500$ lb. Using Fig. 20.6, determine (a) the maximum relative displacement, w_{max}, and (b) the maximum base shear force, f_s.

SOLUTION For $T = 1.0$ s and $\zeta = 0.05$, Fig. 20.6 gives the following response values:

$$S_d = 1.7 \text{ in.}, \quad S_v = 10.5 \text{ in./sec}, \quad S_a = 0.17g = 66 \text{ in./sec}^2$$

(a) From Eq. 20.5,
$$w_{max} = S_d = 1.7 \text{ in.} \quad \text{Ans.(a)} \quad (1)$$

(b) From Eq. 20.8,
$$(f_s)_{max} = mS_a = \frac{W}{g}S_a = 1500(0.17) = 255 \text{ lb} \quad (2)$$

$$(f_s)_{max} = 255 \text{ lb} \quad \text{Ans.(b)} \quad (2)$$

In Example 2.9 a generalized-parameter model was obtained for transverse vibration of a Bernoulli–Euler beam. That model can easily be extended to beamlike structures that can be modeled by an SDOF generalized-parameter model. Figure 20.7 shows such a system. Following Example 2.9, let the relative displacement $w(x, t)$ be represented by the following assumed-modes form:

$$w(x, t) = \psi(x)q_w(t) \quad (20.9)$$

If $\psi(x)$ is normalized so that $\psi(L) = 1$, $q_w(t)$ represents the relative motion, $w(L, t)$, at the top of the column. From Eq. 2.32, the virtual work is given by

$$\delta W^* = \delta W_{nc} - \delta V + \delta W_{inertia} \quad (20.10)$$

If there are no distributed external loads, as is the case in Fig. 20.7, $\delta W_{nc} = 0$. The strain energy is a function of the relative displacement, so from Eq. 2.34b,

$$\delta V = \int_0^L (EI \, w'') \, \delta w'' \, dx \quad (20.11)$$

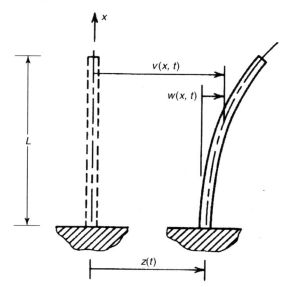

Figure 20.7 Earthquake excitation of an SDOF generalized-coordinate cantilever column.

The inertial force depends on the absolute acceleration. Thus,

$$\delta W_{\text{inertia}} = \int_0^L -\rho A (\ddot{w} + \ddot{z}) \, \delta w \, dx \tag{20.12}$$

Combining Eqs. 20.9 through 20.12, we obtain

$$(m\ddot{q}_w + kq_w + \mu \ddot{z}) \, \delta q_w = 0 \tag{20.13}$$

Since $\delta q_w \neq 0$, the equation of motion for the SDOF generalized-parameter system is

$$m\ddot{q}_w + kq_w = -\mu \ddot{z} \tag{20.14}$$

where, as before,

$$m = \int_0^L \rho A \psi^2 \, dx, \qquad k = \int_0^L EI \, (\psi'')^2 \, dx \tag{20.15}$$

The base acceleration causes an effective force

$$p_{\text{eff}} = -\mu \ddot{z} \tag{20.16}$$

where μ is an *earthquake participation factor* given by

$$\mu = \int_0^L \rho A \psi \, dx \tag{20.17}$$

If damping proportional to relative velocity is included in the generalized-parameter model, Eq. 20.14 becomes

$$m\ddot{q}_w + c\dot{q}_w + kq_w = -\mu \ddot{z} \tag{20.18}$$

(Note that the right-hand side of Eqs. 20.14 and 20.18 involve μ rather than m, which appears in Eq. 20.2.)

For a lumped-mass SDOF system, the equation of relative motion is Eq. 20.2 and the maximum relative displacement is given by Eq. 20.5. By comparing Eq. 20.18 to Eq. 20.2, it can be seen that for a distributed mass system,

$$(q_w)_{\max} = \frac{\mu}{m} S_d = \frac{\mu}{m\omega_n} S_v \tag{20.19}$$

where Eq. 20.6 has also been used.

It is of interest to determine the base shear and overturning moment due to earthquake-related base motion. The base shear is determined here. From Eqs. 12.23 and 12.24 the shear could be expressed in terms of the third derivative of $w(x, t)$. However, since the assumed-modes solution in Eq. 20.9 is only approximate, the third derivative would not provide an acceptable approximation of the shear. In Eq. 20.8 it was shown that the maximum spring force can be obtained for a lumped-mass SDOF system by multiplying the mass by the spectral pseudoacceleration, which is ω_n^2 times the spectral displacement. For the generalized-parameter model of Fig. 20.7, we can define an *effective acceleration* $\ddot{w}_e$ by the equation

$$\ddot{w}_e = \omega_n^2 w(x, t) \tag{20.20}$$

20.2 Response of a SDOF System to Earthquake Excitation: Response Spectra

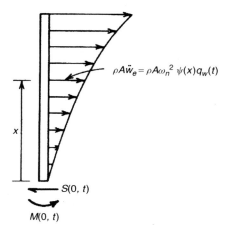

Figure 20.8 Effective inertial loads on a cantilever column.

Combining Eqs. 20.9 and 20.20, we get

$$\ddot{w}_e = \omega_n^2 \psi(x) q_w(t) \tag{20.21}$$

From Fig. 20.8 the base shear $S(0, t)$ is given by

$$S(0, t) = \omega_n^2 q_w(t) \int_0^L \rho A \psi(x)\, dx = \mu \omega_n^2 q_w(t) \tag{20.22}$$

From Eqs. 20.19, 20.22, and 20.7, the maximum base shear is given by

$$S_{\max}(0, t) = \mu \omega_n^2 \frac{\mu}{m} S_d = \frac{\mu^2}{m} S_a \tag{20.23}$$

The base overturning moment can be determined from Fig. 20.8 in a similar fashion.

Example 20.2 A uniform cantilever column similar to those shown in Figs. 20.7 and 20.8 has a weight W, a period of 1.0 sec, and a damping factor of 5%. Assume a deflection shape function

$$\psi(x) = \left(\frac{x}{L}\right)^2$$

and use the response spectrum of Fig. 20.6 to determine (a) the maximum displacement at the top of the column, and (b) the maximum base shear as a fraction of the total weight of the column.

SOLUTION (a) Since the period and damping factor are the same as those of Example 20.1,

$$S_d = 1.7 \text{ in.}, \qquad S_v = 10.5 \text{ in./sec}, \qquad S_a = 0.17g$$

From Eqs. 20.9 and 20.19,

$$w_{\max}(L, t) \equiv (q_w)_{\max} = \frac{\mu}{m} S_d \tag{1}$$

where from Eqs. 20.15a and 20.17,

$$m = \int_0^L \rho A \psi^2 \, dx = \frac{W}{gL} \int_0^L \psi^2 \, dx = \frac{W}{g}\left(\frac{1}{5}\right) \tag{2}$$

$$\mu = \int_0^L \rho A \psi \, dx = \frac{W}{gL} \int_0^L \psi \, dx = \frac{W}{g}\left(\frac{1}{3}\right) \tag{3}$$

Therefore,

$$w_{\max_t}(L, t) \equiv (q_w)_{\max} = \frac{1/3}{1/5}(1.7) \tag{4}$$

or

$$w_{\max_t}(L, t) = 2.8 \text{ in.} \qquad \textbf{Ans. (a)} \tag{5}$$

(b) From Eq. 20.23,

$$S_{\max}(0, t) = \frac{\mu^2}{m} S_a = \frac{(1/3)^2 (W/g)^2}{(1/5)(W/g)}(0.17g) \tag{6}$$

so

$$S_{\max}(0, t) = \frac{5}{9}(0.17W) = 0.094W \tag{7}$$

or

$$S_{\max}(0, t) = 9.4\% \, W \qquad \textbf{Ans. (b)} \tag{8}$$

20.3 RESPONSE OF MDOF SYSTEMS TO EARTHQUAKE EXCITATION

The methods described in Chapters 11 and 16 for determining the response of MDOF systems to dynamic excitation can, of course, be applied to earthquake problems. That is, for specified ground motion time histories, the response of MDOF systems can be determined by using mode superposition or direct integration of the equations of motion. However, some of the simplifications made possible for SDOF systems by the introduction of response spectra can be extended to MDOF systems. In this section, relatively simple MDOF systems such as the lumped-mass multistory building model of Fig. 20.9 are treated. In Section 20.4 some of the limitations of this simple model are discussed briefly. Because response spectra are available for SDOF systems, the mode-superposition method of Chapter 11 will be applied.

For the building model of Fig. 20.9, it is assumed that there is no rotation of the base and that the base moves as a rigid body with displacement $z(t)$. The equations of motion can easily be shown to have the form

$$\mathbf{M\ddot{v}} + \mathbf{C\dot{w}} + \mathbf{Kw} = \mathbf{0} \tag{20.24}$$

if viscous damping proportional to the relative displacement is assumed. In vector form,

$$\mathbf{v}(t) = \mathbf{w}(t) + \mathbf{1}z(t) \tag{20.25}$$

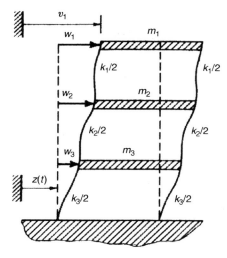

Figure 20.9 Multistory building subjected to base motion.

where **1** is the column vector of 1's. Hence, Eq. 20.24 becomes

$$\mathbf{M\ddot{w}} + \mathbf{C\dot{w}} + \mathbf{Kw} = \mathbf{p}_{\text{eff}}(t) \tag{20.26}$$

where

$$\mathbf{p}_{\text{eff}}(t) = -\mathbf{M1}\ddot{z}(t) \tag{20.27}$$

The use of mode superposition to solve for the response of MDOF systems was treated extensively in Chapter 11. In particular, Eqs. 11.8 through 11.10 apply to MDOF systems with general viscous damping, but Eqs. 11.11 and 11.12 restrict the damping to a form for which the equations of motion in principal coordinates are uncoupled. That is, a mode-superposition solution of the form

$$\mathbf{w}(t) = \mathbf{\Phi}\boldsymbol{\eta}(t) = \sum_{r=1}^{N} \boldsymbol{\phi}_r \eta_r(t) \tag{20.28}$$

leads to the uncoupled equations of motion

$$M_r \ddot{\eta}_r + 2M_r \omega_r \zeta_r \dot{\eta}_r + \omega_r^2 M_r \eta_r = f_r(t), \qquad r = 1, 2, \ldots, N \tag{20.29}$$

where

$$f_r(t) = \boldsymbol{\phi}_r^T \mathbf{M1}\ddot{z}(t) \tag{20.30}$$

(As before, the negative sign in Eq. 20.27 is neglected in Eq. 20.30.) By analogy with Eqs. 20.16 and 20.17, we can define a *modal earthquake participation factor* μ_r such that

$$\mu_r = \boldsymbol{\phi}_r^T \mathbf{M1} = \mathbf{1}^T \mathbf{M} \boldsymbol{\phi}_r \tag{20.31}$$

$$f_r(t) = \mu_r \ddot{z}(t) \tag{20.32}$$

By analogy with Eqs. 20.2 through 20.4,

$$\eta_r(t) = \frac{\mu_r}{M_r \omega_r} W_r(t) \tag{20.33}$$

where

$$W_r(t) = \int_0^T \ddot{z}(\tau) e^{-\zeta_r \omega_r (t-\tau)} \sin \omega_r (t - \tau) \, d\tau \tag{20.34}$$

Finally, from Eqs. 20.28 and 20.33,

$$\mathbf{w}(t) = \sum_{r=1}^N \boldsymbol{\phi}_r \frac{\mu_r W_r(t)}{M_r \omega_r} \tag{20.35}$$

Although Eq. 20.35 gives a very straightforward means of determining the time history of the relative motion of a MDOF system subjected to earthquake excitation, it involves the integration of the various modal response equations of the form of Eq. 20.34. Even though the response of a many-DOF system can be represented acceptably by the response of the first few modes, Eq. 20.35 represents a tedious procedure which only produces the response to one particular base motion acceleration history, $z(t)$.

The maximum relative displacements due to the rth mode can be expressed in terms of response spectra by

$$\max_t |\mathbf{w}_r(t)| = |\boldsymbol{\phi}_r| \frac{|\mu_r|}{M_r \omega_r} S_v(T_r, \zeta_r) \tag{20.36}$$

where the absolute value notation $|\mathbf{w}_r(t)|$ and $|\boldsymbol{\phi}_r|$ implies the magnitude of each DOF. The difficulty arises from the fact that each mode in Eq. 20.35 may reach its maximum, given by Eq. 20.36, at a different time, and hence

$$\max_t |\mathbf{w}(t)| \neq \sum_{r=1}^N \max_t |\mathbf{w}_r(t)| \tag{20.37}$$

A very satisfactory estimate of the maximum of any response quantity can frequently be obtained by the root-mean-square (rms) method, that is,

$$\max_t |Q(t)| \approx \left[\sum_{r=1}^N \left(\max_t |Q_r(t)| \right)^2 \right]^{1/2} \tag{20.38}$$

where $Q(t)$ is any response quantity. Hence, the maximum relative displacement response can be approximated by the root-mean-square (rms) response

$$\max_t |\mathbf{w}(t)| \approx \left\{ \sum_{r=1}^N \left[\frac{|\boldsymbol{\phi}_r||\mu_r|}{M_r \omega_r} S_v(T_r, \zeta_r) \right]^2 \right\}^{1/2} \tag{20.39}$$

The rms value may overestimate or underestimate the true value of maximum response. This estimate is generally good for systems with well-separated frequencies, but may be quite poor if the system has closely spaced frequencies. Methods for estimating the maximum response of such systems for which Eq. 20.38 does not hold are discussed in Ref. [20.2].

To calculate the modal forces resulting from earthquake excitation of a MDOF system such as the building in Fig. 20.9, an effective modal acceleration can be defined which, by analogy with Eq. 20.20, is given by

$$(\ddot{\mathbf{w}}_r)_e = \omega_r^2 \mathbf{w}_r \tag{20.40}$$

The inertia force due to the effective acceleration is given by $(\mathbf{m}\ddot{\mathbf{w}}_r)_e$. For a structure such as the one in Fig. 20.9, the contribution of the rth mode to the base shear is, by analogy with Eq. 20.22,

$$\begin{aligned} S_r(t) &= \sum_{i=1}^{N} m_i (\ddot{w}_{ir})_e = \sum_{i=1}^{N} \omega_r^2 m_i w_{ir} \\ &= \mathbf{1}^T \mathbf{M} \boldsymbol{\phi}_r \omega_r^2 \eta_r(t) \\ &= \mathbf{1}^T \mathbf{M} \boldsymbol{\phi}_r \frac{\mu_r \omega_r}{M_r} W_r(t) \\ &= \frac{\mu_r^2 \omega_r}{M_r} W_r(t) \end{aligned} \tag{20.41}$$

The maximum base shear due to the rth mode can thus be expressed as

$$\max_t |S_r(t)| = \frac{\mu_r^2}{M_r} S_a(T_r, \zeta_r) \tag{20.42}$$

As with the displacement maximum, the base shear maximum can be approximated by the root-mean-square value

$$\max_t |S(t)| \approx \left\{ \sum_{r=1}^{N} \left[\frac{\mu_r^2}{M_r} S_a(T_r, \zeta_r) \right]^2 \right\}^{1/2} \tag{20.43}$$

Example 20.3 For the four-story building of Example 11.4, determine the following: (a) S_d, S_v, and S_a for each mode, if each mode has 5% damping and if Fig. 20.6 gives the response spectra; (b) a root-mean-square estimate of the maximum displacement of the top mass; and (c) a root-mean-square estimate of the maximum base shear.

SOLUTION (a) From the ω_r values given in Example 11.4, the periods T_r are calculated. The corresponding points on the 5% damping curve of Fig. 20.6 are determined and the values of S_d, S_v, and S_a are read from the curves. These values are tabulated below.

r	T_r (sec)	S_d (in.)	S_v (in./sec)	S_a (g's)
1	0.47	0.60	8.0	0.28
2	0.21	0.14	4.0	0.30
3	0.15	0.07	2.7	0.28
4	0.11	0.03	1.6	0.24

(b) From Eq. 20.39,

$$\max_t |w_1(t)| \approx \left\{\sum_{r=1}^{N}\left[\left(\frac{|\phi_{1r}||\mu_r|}{M_r\omega_r}\right)S_v(T_r,\zeta_r)\right]^2\right\}^{1/2} \quad (1)$$

From Eq. 20.32,

$$\mu_r = \boldsymbol{\phi}_r^T\mathbf{M1} = \mathbf{1}^T\mathbf{M}\boldsymbol{\phi}_r = \sum_{i=1}^{4}m_i\phi_{ir} \quad (2)$$

Therefore, using values from Example 11.4, we get

$$\mu_1 = 4.2565, \quad \mu_2 = -1.5919, \quad \mu_3 = -1.3425, \quad \mu_4 = -0.6526$$

Then

$$\max_t |w_1(t)| \approx \left[\left(\frac{1(4.2565)(8.0)}{2.873(13.294)}\right)^2 + \left(\frac{1(1.5919)(4.0)}{2.177(29.660)}\right)^2 \right.$$
$$\left. + \left(\frac{0.9015(1.3425)(2.7)}{4.367(41.079)}\right)^2 + \left(\frac{0.1544(0.6526)(1.6)}{3.642(55.882)}\right)^2\right]^{1/2} \quad (3)$$

$$\max_t |w_1(t)| \approx 0.897 \text{ in.} \qquad\qquad\qquad\qquad\qquad \textbf{Ans. (b)} \quad (4)$$

Note that the maximum response due to mode 1 alone is 0.8916 in. Hence, mode 1 contributes practically all of the response due to this earthquake excitation.

(c) The maximum base shear can be approximated by using Eq. 20.43:

$$\max_t |S(t)| \approx \left\{\sum_{r=1}^{4}\left[\frac{\mu_r^2}{M_r}S_a(T_r,\zeta_r)\right]^2\right\}^{1/2} \quad (5)$$

$$\max_t |S(t)| \approx \left[\left(\frac{(4.2565)^2(0.28)(386)}{2.873}\right)^2 + \left(\frac{(1.5919)^2(0.30)(386)}{2.177}\right)^2 \right.$$
$$\left. + \left(\frac{(1.3425)^2(0.28)(386)}{4.367}\right)^2 + \left(\frac{(0.6526)^2(0.24)(386)}{3.642}\right)^2\right]^{1/2} \quad (6)$$

$$\max_t |S(t)| \approx 696 \text{ kips} \qquad\qquad\qquad\qquad\qquad \textbf{Ans. (c)} \quad (7)$$

Again, most of the base shear is contributed by the first mode.

20.4 FURTHER CONSIDERATIONS

The treatment of response of structures to earthquake excitation presented in this chapter is, of necessity, brief and limited in scope. As has been indicated, however, the tools necessary for linear analysis have been presented in detail and some of the computational tools necessary for nonlinear analysis have also been presented. The references at the end of the chapter provide resources for the study of more difficult problems in design of earthquake-resistant structures: for example, deterministic analysis of linear response, nondeterministic analysis of both linear and nonlinear response, multiple-support excitation, soil–structure interaction, and base rotation.

Whereas much attention has been given to the analysis and design of building structures to resist earthquakes, design of the equipment housed within the buildings to withstand dynamic excitation has received far less attention. Equipment such as pumps, compressors, piping systems, and power generators must be functional in the aftermath of a disaster such as a major earthquake. Reference [20.18] treats one class of problems associated with equipment design, the design of lightly damped relatively lightweight equipment.

Some of the aspects of design of earthquake-resistant structures are also present in other structural design problems. For example, the design of spacecraft and space payloads for transient environments shares some of the design aspects of earthquake problems: namely, transient-type base excitation; stochastic-type excitation, which is relatively poorly defined for early flights of a particular booster-payload system; and so on. Reference [20.19], for example, presents a generalized modal shock spectra method for spacecraft loads analysis. Reference [20.20] presents a survey of the problem of design of space payloads for transient environments.

REFERENCES

[20.1] R. L. Wiegel, ed., *Earthquake Engineering*, Prentice-Hall, Englewood Cliffs, NJ, 1970.

[20.2] N. M. Newmark and E. Rosenbleuth, *Fundamentals of Earthquake Engineering*, Prentice-Hall, Englewood Cliffs, NJ, 1971.

[20.3] S. Okamoto, *Introduction to Earthquake Engineering*, Wiley, New York, 1973.

[20.4] R. W. Clough and J. Penzien, *Dynamics of Structures*, 2nd ed., McGraw-Hill, New York, 1993.

[20.5] A. K. Chopra, *Dynamics of Structures: Theory and Applications to Earthquake Engineering*, 2nd ed., Prentice Hall, Upper Saddle River, NJ, 2001.

[20.6] M. Barazangi and J. Dorman, "World Seismicity Maps Compiled from ESSA Coast and Geodetic Survey, Epicenter Data, 1961–1967," *Bulletin of the Seismological Society of America*, Vol. 59, No. 1, 1969, pp. 369–380.

[20.7] D. E. Hudson et al., *Strong Motion Earthquake Accelerograms*, Vol. II, *Corrected Accelerograms and Integrated Ground Velocity and Displacement Curves, Part A*, Report EERL 71-50, Earthquake Engineering Research Laboratory, California Institute of Technology, Pasadena, CA, 1971.

[20.8] J. M. Biggs, *Introduction to Structural Dynamics*, McGraw-Hill, New York, 1964.

[20.9] A. H-S. Ang, "Probability Concepts in Earthquake Engineering," *Applied Mechanics in Earthquake Engineering*, AMD Vol. 8, ASME, New York, 1974, pp. 225–259.

[20.10] M. A. Sosen, "Hysteresis in Structural Elements," *Applied Mechanics in Earthquake Engineering*, AMD Vol. 8, ASME, New York, 1974, pp. 63–98.

[20.11] W. D. Iwan, "Application of Nonlinear Analysis Techniques," *Applied Mechanics in Earthquake Engineering*, AMD Vol. 8, ASME, New York, 1974, pp. 135–161.

[20.12] M. A. Biot, "Analytical and Experimental Methods in Engineering Seismology," *Transactions of ASCE*, Vol. 108, 1943, pp. 365–408.

[20.13] G. W. Housner, "Characteristics of Strong-Motion Earthquakes," *Bulletin of the Seismological Society of America*, Vol. 37, 1947, pp. 19–29.

[20.14] D. E. Hudson, "Response Spectrum Techniques in Engineering Seismology," *Proceedings of the World Conference on Earthquake Engineering*, Berkeley, CA, 1956.

[20.15] N. M. Newmark, "Current Trends in the Seismic Analysis and Design of High-Rise Structures," Chapter 16 in *Earthquake Engineering*, R. L. Wiegel, ed., Prentice-Hall, Englewood Cliffs, NJ, 1970.

[20.16] G. W. Housner, "Strong Ground Motion," Chapter 4 in *Earthquake Engineering*, R. L. Wiegel, ed., Prentice-Hall, Englewood Cliffs, NJ, 1970.

[20.17] G. W. Housner, "Design Spectrum," Chapter 5 in *Earthquake Engineering*, R. L. Wiegel, ed., Prentice-Hall, Englewood Cliffs, NJ, 1970.

[20.18] J. L. Sackman and J. M. Kelly, *Rational Design Methods for Light Equipment in Structures Subjected to Ground Motion*, Report DCB/EERC-78/19, Earthquake Engineering Research Laboratory, University of California, Berkeley, CA, 1978.

[20.19] M. Trubert and M. Salama, *A Generalized Modal Shock Spectra Method for Spacecraft Loads Analysis*, Publication 79-2, Jet Propulsion Laboratory, Pasadena, CA, 1979.

[20.20] B. K. Wada, "Design of Space Payloads for Transient Environments," *Survival of Mechanical Systems in Transient Environments*, AMD Vol. 36, ASME, New York, 1979.

PROBLEMS

Problem Set 20.2

20.1 Using the values from the $\zeta = 0$ curve of Fig. 20.5, sketch a curve of S_a versus T_n for $0.2 \leq T_n \leq 2.0$ s.

20.2 An SDOF frame structure such as the one shown in Fig. 20.3 has a period of 0.8 sec and a damping factor of 2%. The mass is $m = 10$ lb-sec^2/in. (a) Using Fig. 20.5, determine the value of the spectral velocity S_v that applies to this structure. (b) Determine the corresponding values of S_a and S_d. (c) Determine the maximum displacement of the mass. (d) Determine the maximum spring force.

20.3 Using Fig. 20.8, determine an expression for the maximum overturning moment $M(0, t)$.

20.4 Repeat Example 20.2 if the period of the column is 0.8 s, the damping factor of 2%, and $\psi(x)$ is chosen to have the form

$$\psi(x) = 1 - \cos\frac{\pi x}{2L}$$

Problem Set 20.3

20.5 For the three-story building of Problem 11.8, whose natural frequencies and mode shapes are given in that problem statement: (a) Determine the values of S_d, S_v, and S_a for each mode, if each mode has 2% damping, and if Fig. 20.6 gives the response spectra for the earthquake excitation under consideration. (b) Determine a root-mean-square estimate of the maximum displacement of the top mass. (c) Determine a root-mean-square estimate of the maximum base shear.

20.6 Reduce the stiffnesses of the building in Fig. P11.8 to 50% of their present value, increase the intensity of the response given in Fig. 20.6 by a factor of 2, and repeat Problem 20.5.

A

Units

Problems in structural dynamics are based on Newton's Second Law:

$$\text{force} = (\text{mass})(\text{acceleration}) \tag{A.1}$$

and these quantities, plus others directly related to them, must be expressed in a consistent system of units. At the present time, engineering practice is in the process of a conversion from English engineering units (United States Customary System) to the International System of Units (SI).[A.1] In English engineering units dimensional homogeneity is obtained in Eq. A.1 when the force is given in lb_f, the mass in slugs, and the acceleration in ft/sec^2. This is the English ft-lb_f-sec system of units. The slug is a derived unit, and from Eq. A1, 1 slug = 1 lb_f-sec^2/ft.

The units of force and mass frequently lead to confusion because of the use of the term *weight* as a quantity to mean either force or mass. On the one hand, when one speaks of an object's "weight," it is usually the mass, that is, the quantity of matter, that is referred to. On the other hand, in scientific and technological usage the term *weight of a body* has usually meant the force that if applied to the body would give it an acceleration equal to the local acceleration of free fall. This is the "weight" that would be measured by a spring scale. If the "mass" of a body is given in pounds (lb_m), it must be divided by the acceleration of gravity $g \approx 32.2$ ft/sec^2 to obtain the mass as referred to in Newton's Second Law, that is, a force of 1 lb_f is exerted on a mass of 1 lb_m by the gravitational pull of the Earth.

In structural dynamics and vibrations the in.-lb_f-sec system is frequently used. In this system $g = (32.2 \text{ ft/sec}^2)(12 \text{ in./ft}) = 386 \text{ in./sec}^2$ is used.

The International System of Units (SI) is a modern version of the metric system.[A.1] It is a *coherent system*[1] with seven *base units*, for which names, symbols, and precise definitions have been established. Many *derived units* are defined in terms of the base units (Table A.1). Symbols have been assigned to each, and in some cases, they have been given names. In addition, there are *supplementary units*.

The base units are regarded as dimensionally independent. The ones of interest in structural dynamics are the meter (m), kilogram (kg),[2] and second (s). One great advantage of SI is that there is one and only one unit for each physical quantity: the meter for length, the kilogram for mass, the second for time, and so on.

---

[1] A coherent system of units is one in which there are no numerical factors that must enter into an equation employing numerical values; for example, Eq. A.1 is valid whether it is written in symbols or in numerical values—there is an implied multiplicative factor of 1 on the right-hand side.

[2] In SI units the kilogram is a unit of mass. The kilogram-force (from which the suffix "force" is often omitted) should not be used.

Table A.1 Some Derived SI Units and Supplementary SI Units

Quantity	Unit (name)	Symbol	Formula
Derived units			
Frequency	hertz	Hz	s^{-1}
Force	newton	N	$kg \cdot m/s^2$
Pressure, stress	pascal	Pa	N/m^2
Energy, work	joule	J	$N \cdot m$
Supplementary unit			
Phase angle	radian	rad	

The unit of length, the *meter*, "is the length equal to 1 650 763.73 wavelengths in vacuum of the radiation corresponding to the transition between levels $2p_{10}$ and $5d_5$ of the krypton-86 atom. The *kilogram* is the unit of mass; it is equal to the mass of the international prototype of the kilogram."[A.1] This prototype kilogram is preserved by the International Bureau of Weights and Measures near Paris, France. The *second* is the unit of time. It "is the duration of 9 192 631 770 periods of the radiation corresponding

Table A.2 Prefixes for Multiples and Submultiples of SI Units

Multiple	Prefix	Symbol	Submultiple	Prefix	Symbol
10^{12}	tera	T	10^{-1}	deci	d
10^{9}	giga	G	10^{-2}	centi	c
10^{6}	mega	M	10^{-3}	milli	m
10^{3}	kilo	k	10^{-6}	micro	μ
10^{2}	hecto	h	10^{-9}	nano	n
10	deca	dc	10^{-12}	pico	p
			10^{-15}	femto	f
			10^{-18}	atto	p

Table A.3 Examples of Conversions[a] from English Units to SI Units

To convert from:	To:	Multiply by:
foot (ft)	meter (m)	3.048 000 E−01
horsepower (550 ft-lb_f/sec)	watt (W)	7.456 999 E+02
inch (in.)	meter (m)	2.540 000 E−02
kip (1000 lb_f)	newton (N)	4.448 222 E+03
pound-force (lb_f)	newton (N)	4.448 222 E+00
pound-force-inch (lb_f-in.)	newton meter (N·m)	1.129 848 E−01
pound-force/inch (lb_f/in.)	newton per meter (N/m)	1.751 268 E+02
pound-force/inch² (psi)	pascal (Pa)	6.894 757 E+03
pound-mass (lb_m)	kilogram (kg)	4.535 924 E−01
pound-mass/inch³ (lb_m/in³)	kilogram per meter³ (kg/m³)	2.767 990 E+04
slug	kilogram (kg)	1.459 390 E+01

[a] Conversions should be handled with careful regard to the implied correspondence between the accuracy of the data and the number of digits.

to the transition between the two hyperfine levels of the ground state of the cesium-133 atom."[A.1]

Reference A.1 gives extensive tables of conversion factors and a number of rules to be followed in writing numbers and their units in correct SI form, Table A.2 gives the prefixes for the multiples and submultiples of SI units, and Table A.3 provides conversion factors for conversion from English units to SI units.

A number of recommendations have been made in order to standardize the writing of numbers, symbols, and so on, in SI.[A.1] The following recommendations apply to situations encountered frequently.

A.1 SI PREFIXES

   a. When expressing a quantity by a numerical value and a unit, prefixes should preferably be chosen so that the numerical value lies between 0.1 and 1000.
   b. Normally, the prefix should be attached to a unit in the numerator. One exception to this is when the kilogram is one of the units.
   c. No space is used between the prefix and the unit symbol.

A.2 UNIT SYMBOLS

   a. Unit symbols should be printed in roman (upright) type regardless of the type style used in the surrounding text.
   b. Unit symbols are not followed by a period except when used at the end of a sentence.
   c. In the complete expression for a quantity, a space should be left between the numerical value and the unit symbol.

A.3 UNITS FORMED BY MULTIPLICATION AND DIVISION

With unit names:

   a. *Product.* Use a space (newton meter) or a hyphen (newton-meter).
   b. *Quotient.* Use the word per and not the solidus (/).
   c. *Powers.* Use the modifier squared or cubed placed after the unit name (meter per second squared).

With unit symbols:

   a. *Product.* Use a centered dot (N·m).
   b. *Quotient.* Use one of the following forms:

$$\text{m/s} \quad \text{or} \quad \text{m} \cdot \text{s}^{-1} \quad \text{or} \quad \frac{\text{m}}{\text{s}} \tag{A.2}$$

A.4 NUMBERS

Outside the United States the comma is sometimes used as a decimal marker. To avoid this potential source of confusion, recommended international practice calls for separating the digits into groups of three, counting from the decimal point toward the left and the right, and using a small space to separate the groups.

REFERENCE

[A.1] An American National Standard, ASTM/IEEE, *Standard Metric Practice*, ASTM E 380-76, IEEE Std. 268-1976, ANSI Z210.1-1976, Institute of Electrical and Electronics Engineers, New York, 1976.

B

Complex Numbers

In this textbook complex numbers are encountered primarily in Section 4.3 in solving for the frequency response of an SDOF system, in Section 10.4 in solving the state-space eigenproblem for an MDOF system with arbitrary viscous damping, and in Chapter 18 in the representation of frequency response functions (FRFs) for experimental modal analysis studies. In all cases, a few ideas about representation of complex numbers and a few simple algebraic operations with complex numbers is all that is required. Those topics are summarized in this appendix.

B.1 RECTANGULAR AND POLAR REPRESENTATIONS OF COMPLEX NUMBERS

Complex numbers can be represented as two-dimensional vectors in a plane, called the *complex plane* or the *Argand plane*. Figure B.1 illustrates the two ways to represent the complex number $\overline{C}$.[1] The *rectangular representation* has the form

$$\overline{C} \equiv C_\Re + iC_\Im \tag{B.1}$$

$C_\Re$ is called the *real part* and $C_\Im$ the *imaginary part* of the complex number $\overline{C}$. The symbol i is a unit vector that indicates that the imaginary component is 90° counterclockwise from the real component.[2] $C_\Re$ is also called the *coincident* (co) *part* and $C_\Im$ the *quadrature* (quad) *part* of the complex number $\overline{C}$, and the rectangular representation is then referred to as the *co-quad representation*.

The *polar representation* has the form

$$\overline{C} \equiv C e^{i\theta} \tag{B.2}$$

where from Fig. B.1, C is the *magnitude* of $\overline{C}$, given by

$$C = \sqrt{C_\Re^2 + C_\Im^2} \tag{B.3}$$

and θ is the *phase angle* of $\overline{C}$, given by

$$\theta \equiv \angle \overline{C} = \tan^{-1} \frac{C_\Im}{C_\Re} \tag{B.4}$$

---

[1] The overline notation is used here for clarity but is not essential if it is otherwise clear that a symbol represents a complex number.
[2] Some authors use the symbol j instead of i.

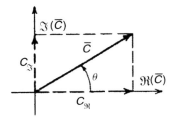

Figure B.1 Complex number representations.

From Fig. B.1, the rectangular and polar representations are also related by

$$C_\Re = C \cos\theta, \qquad C_\Im = C \sin\theta \tag{B.5}$$

B.2 ALGEBRA OF COMPLEX NUMBERS

Two complex numbers are represented by

$$\overline{A} \equiv A_\Re + iA_\Im = Ae^{i\theta_A}, \qquad \overline{B} \equiv B_\Re + iB_\Im = Be^{i\theta_B} \tag{B.6}$$

Addition (subtraction) of the two complex numbers is most easily carried out using the rectangular form:

$$\overline{C} = \overline{A} + \overline{B} = (A_\Re + B_\Re) + i(A_\Im + B_\Im) \tag{B.7}$$

that is,

$$C_\Re = A_\Re + B_\Re, \qquad C_\Im = A_\Im + B_\Im \tag{B.8}$$

Multiplication and division of complex numbers is carried out most easily using the polar form. First, let $\overline{C}$ be the product of two complex numbers

$$\overline{C} = \overline{A}\,\overline{B} \tag{B.9}$$

Then the magnitude and phase angle of the product are

$$C = AB, \qquad \angle \overline{C} = \angle \overline{A} + \angle \overline{B} \tag{B.10}$$

Similarly, for the quotient,

$$\overline{C} = \frac{\overline{A}}{\overline{B}} \tag{B.11}$$

$$C = \frac{A}{B}, \qquad \angle \overline{C} = \angle \overline{A} - \angle \overline{B} \tag{B.12}$$

A common example of this operation that appears frequently in this book has the form

$$\overline{z} = \frac{1}{x + iy} \tag{B.13}$$

To express this complex number in rectangular form, multiply both numerator and denominator by the complex conjugate of the denominator, noting that $i \cdot i = -1$. This gives

$$\bar{z} = \frac{1}{x+iy} \frac{x-iy}{x-iy} = \frac{x-iy}{x^2+y^2} = \frac{x}{x^2+y^2} - i\frac{y}{x^2+y^2} \qquad (B.14)$$

Powers and roots of complex numbers can be handled by extensions of the multiplication operation. That is,

$$\begin{aligned}\overline{C} = (\overline{A})^n &\longrightarrow C = A^n, \qquad \angle \overline{C} = n(\angle \overline{A}) \\ \overline{C} = (\overline{A})^{1/n} &\longrightarrow C = A^{1/n}, \qquad \angle \overline{C} = \frac{\angle \overline{A}}{n}\end{aligned} \qquad (B.15)$$

For example,

$$\sqrt{i} = 1 \cdot e^{i\pi/4} = \frac{\sqrt{2}}{2}(1+i) \qquad (B.16)$$

C

Elements of Laplace Transforms

The Laplace transform is a mathematical tool that is used in the study of the response of linear systems with constant coefficients, such as electrical circuits and vibrating structures.[C.1–C.4] The method is particularly useful for solving initial-value problems, where the system is described by ordinary differential equations and the response is to be determined for all time following the time at which stated initial conditions are imposed on the system.

The basic idea behind the Laplace transformation procedure is to transform ordinary differential equations, together with prescribed initial conditions, into algebraic expressions that can be solved more easily for the transformed system response. An inverse Laplace transform procedure is then required to produce the system response as a function of time.

Since most structural systems are modeled by hundreds, thousands, or even millions of degrees of freedom, the Laplace transformation procedure is of relatively limited usefulness for solving response problems in the field of structural dynamics. However, Laplace transform symbolism does play an important role in experimental structural dynamics, as indicated in Chapter 18. Therefore, this appendix provides a brief introduction to the aspects of Laplace transform theory that are most relevant to Section 5.6 and Chapter 18. The reader should consult a reference work on the Laplace transform (e.g., Ref. [C.1]) for a more comprehensive treatment of the subject.

C.1 DEFINITION OF THE ONE-SIDED LAPLACE TRANSFORM

We consider the (real) function $f(t)$ defined for all time $t > 0$. The Laplace transform of $f(t)$ is written symbolically as $F(s) \equiv \mathcal{L}[f(t)]$, where s is a complex variable referred to as the *Laplace variable*. The *one-sided Laplace transform* of $f(t)$ is defined by the definite integral

$$F(s) \equiv \mathcal{L}[f(t)] = \int_0^\infty e^{-st} f(t)\, dt \qquad \text{(C.1)}$$

where e^{-st} is called the *kernel* of the transformation. Because this is a definite integral with t as the variable of integration, the resulting transform is a function of the Laplace variable s, a complex variable that can be written

$$s = \Re(s) + i\Im(s) = \sigma + i\omega \qquad \text{(C.2)}$$

The Laplace domain can therefore be represented by the complex plane with coordinates (σ, ω), and it can be said that the function $f(t)$ in the *time domain* is transformed into the function $F(s)$ in the *Laplace domain*.

C.2 TRANSFORMATION OF ORDINARY DIFFERENTIAL EQUATIONS

We wish to use the Laplace transformation procedure to solve the prototype SDOF equation of motion (Eq. 3.1),

$$m\ddot{u} + c\dot{u} + ku = p(t) \tag{C.3}$$

with given *excitation* $p(t)$ and subject to specified *initial displacement* $u(0)$ and *initial velocity* $\dot{u}(0)$, respectively. Therefore, we need expressions for the Laplace transforms of the derivatives du/dt and d^2u/dt^2. By using the Laplace transform definition, Eq. C.1, and performing simple integration by parts, we obtain the following Laplace transform of the first time derivative:

$$\mathcal{L}\left[\frac{du(t)}{dt}\right] = \int_0^\infty e^{-st}\frac{du(t)}{dt}\,dt = e^{-st}u(t)\Big|_0^\infty + s\int_0^\infty e^{-st}u(t)\,dt \tag{C.4}$$
$$= sU(s) - u(0)$$

Similarly, it is straightforward to show that the Laplace transform of the second time derivative is

$$\mathcal{L}\left[\frac{d^2u(t)}{dt^2}\right] = \int_0^\infty e^{-st}\frac{d^2u(t)}{dt^2}\,dt = s^2 U(s) - su(0) - \dot{u}(0) \tag{C.5}$$

The Laplace transform of the excitation $p(t)$ is simply

$$P(s) \equiv \mathcal{L}[p(t)] = \int_0^\infty e^{-st} p(t)\,dt \tag{C.6}$$

Transforming both sides of the time-domain equation of motion and rearranging, we obtain the following *equation of motion in the Laplace domain*:

$$(ms^2 + cs + k)U(s) = P(s) + m\dot{u}(0) + (ms + c)u(0) \tag{C.7}$$

which can also be written in the form (see Eq. 5.54)

$$Z(s)U(s) = P(s) + m\dot{u}(0) + (ms + c)u(0) \tag{C.8}$$

where $Z(s)$ is called the *system impedance function*. For the SDOF vibrating system,

$$Z(s) = ms^2 + cs + k = m(s^2 + 2\zeta\omega_n s + \omega_n^2) \tag{C.9}$$

Equation C.7 is an algebraic equation, which can, therefore, be easily solved for the *transformed response* $U(s)$.

$$\boxed{U(s) = H(s)[P(s) + m\dot{u}(0) + (ms + c)u(0)]} \tag{C.10}$$

where $H(s)$ is the *system transfer function*, given by

$$H(s) = \frac{1}{m(s^2 + 2\zeta\omega_n s + \omega_n^2)} \tag{C.11}$$

C.3 INVERSE LAPLACE TRANSFORM

The Laplace transformation process has been employed to convert a differential equation for $u(t)$ in the *time domain* (Eq. C.3) into an algebraic expression for $U(s)$ in the *Laplace domain* (Eq. C.10). To obtain the time-dependent response $u(t)$, it is therefore necessary to perform an *inverse Laplace transformation* of $U(s)$, an operation that is represented symbolically by

$$u(t) = \mathcal{L}^{-1}[U(s)] \tag{C.12}$$

Mathematically, the inverse Laplace transform operation $\mathcal{L}^{-1}[\cdot]$ involves a line integral in the complex s-domain.[C.1] Fortunately, for structural dynamics applications the form of the right-hand side of Eq. C.10 is such that a much easier process may be used to carry out the inversion indicated in Eq. C.12. By the method of partial fractions, the transformed response $U(s)$ can be decomposed into a sum of terms, each with an inverse that can be obtained from a table of Laplace transform pairs such as Table C.1 in Section C.6.

C.4 COMPLEX SHIFTING THEOREM

Let function $\widehat{f}(t)$ have the form

$$\widehat{f}(t) = f(t)e^{at} \tag{C.13}$$

where a is a real or complex number. The Laplace transform of $\widehat{f}(t)$ is given by

$$\begin{aligned} \widehat{F}(s) &= \int_0^\infty [f(t)e^{at}]e^{-st}\,dt \\ &= \int_0^\infty f(t)e^{-(s-a)t}\,dt = F(s-a) \end{aligned} \tag{C.14}$$

where $F(s)$ is the Laplace transform of the function $f(t)$. Therefore, we have the following Laplace transform pairs:

$$\begin{array}{c} f(t) \Longleftrightarrow F(s) \\ f(t)e^{at} \Longleftrightarrow F(s-a) \end{array} \tag{C.15}$$

Equation C.15 states that the effect of multiplying a function $f(t)$ by e^{at} in the time domain is to shift the Laplace transform $F(s)$ of $f(t)$ by the amount a in the complex s domain, and vice versa. This is called the *complex shifting theorem*.

The method of partial fractions is discussed in the next section. First, however, let us take note of two key Laplace transform pairs that will be needed in that discussion and that follow from the complex shifting theorem. First, consider

$$\mathcal{L}[e^{\lambda t}] = \int_0^\infty e^{\lambda t}e^{-st}\,dt = \int_0^\infty e^{(\lambda-s)t}\,dt = \left.\frac{e^{(\lambda-s)t}}{\lambda-s}\right|_0^\infty = \frac{1}{s-\lambda} \tag{C.16}$$

Therefore, we have the following important Laplace transform pair:

$$\boxed{e^{\lambda t} \Longleftrightarrow \frac{1}{s - \lambda}} \qquad \text{(C.17)}$$

Next, it takes only a straightforward integration to show that

$$\mathcal{L}[t^{p-1}] = \frac{(p-1)!}{s^p} \qquad \text{(C.18)}$$

Consequently, using the complex shifting theorem, Eq C.15b, we obtain the following Laplace transform pair:

$$\boxed{\frac{t^{p-1}}{(p-1)!} e^{\lambda t} \Longleftrightarrow \frac{1}{(s - \lambda)^p}} \qquad \text{(C.19)}$$

(Note that for $p = 1$, Eq. C.19 reduces to Eq. C.17.)

C.5 METHOD OF PARTIAL FRACTIONS

The *method of partial fractions* can be used to convert the transformed response $U(s)$ from the form

$$U(s) = \frac{A(s)}{B(s)} \qquad \text{(C.20)}$$

into the equivalent partial-fraction expansion, which is a sum of simpler terms. For structural dynamics applications it is sufficient to assume that both $A(s)$ and $B(s)$ are polynomials and that the numerator polynomial $A(s)$ is of lower degree than that of the denominator polynomial $B(s)$.

The *poles* of $U(s)$ are the roots of the polynomial equation

$$B(s) = 0 \qquad \text{(C.21)}$$

In structural dynamics these poles will generally be either from linear factors of the type $s - \lambda_k$ or from quadratic factors of the type $(s + \sigma_k)^2 + (\omega_k)^2$. In either case, the poles may be distinct or they may be repeated. Therefore, the denominator polynomial $B(s)$ can be decomposed into factors of the following four types:

1. For each linear factor of $B(s)$ of the form $s - \lambda_k$, there corresponds a partial fraction of the form

$$\frac{c_k}{s - \lambda_k}, \qquad c_k = \text{constant}$$

If any of the poles appear as nonrepeated complex conjugate pairs, a partial-fraction expansion of type 3 can be applied to these poles.

2. For each repeated linear factor of the form $(s - \lambda_k)^m$, there corresponds a partial-fraction expansion of the form

$$\frac{c_{k1}}{(s - \lambda_k)^m} + \frac{c_{k2}}{(s - \lambda_k)^{m-1}} + \cdots + \frac{c_{km}}{s - \lambda_k}, \qquad c_{k1}, c_{k2}, \ldots, c_{km} = \text{constants}$$

If any of the poles appear as repeated complex conjugate pairs, a partial-fraction expansion of type 4 can be applied to these poles.

3. For each quadratic factor of $B(s)$ of the form $[(s - \sigma_k)^2 + (\omega_k)^2]$, there corresponds a partial-fraction expansion of the form

$$\frac{f_k s + g_k}{(s - \sigma_k)^2 + (\omega_k)^2}, \quad f_k, g_k = \text{constants}$$

4. For each repeated quadratic factor of $B(s)$ of the form $[(s - \sigma_k)^2 + (\omega_k)^2]^n$, there corresponds a partial fraction that combines the features of types 2 and 3.

Here we discuss the method of partial fractions for three cases: case 1: where $U(s)$ has only *simple poles*; case 2: where $U(s)$ has one or more *multiple poles*; and case 3: where $U(s)$ has one or more sets of *complex-conjugate poles* resulting from quadratic factors of $B(s)$. Example Problem 5.9 illustrates the fourth case listed above.

C.5.1 System with Simple Poles

Let $B(s)$ be a polynomial of degree n in s, and let $s = \lambda_k$, $k = 1, \ldots, n$ denote its n roots. If all of these roots are distinct, $B(s)$ can be written in the form

$$B(s) = (s - \lambda_1)(s - \lambda_2) \cdots (s - \lambda_n) = \prod_{k=1}^{n}(s - \lambda_k) \tag{C.22}$$

where $\prod$ is the product symbol. Then the *partial-fraction expansion* of Eq. C.20 has the form

$$\boxed{U(s) = \frac{c_1}{s - \lambda_1} + \frac{c_2}{s - \lambda_2} + \cdots + \frac{c_n}{s - \lambda_n} = \sum_{k=1}^{n} \frac{c_k}{s - \lambda_k}} \tag{C.23}$$

where the coefficients c_k are given by the formula

$$\boxed{c_k = [(s - \lambda_k)U(s)]|_{s=\lambda_k}} \tag{C.24}$$

From Eq. C.23, together with the Laplace-transform pair given in Eq. C.17, we get the following *time-domain response for a system with n simple poles*:

$$\boxed{u(t) = \mathcal{L}^{-1}[U(s)] = \sum_{k=1}^{n} c_k e^{\lambda_k t}} \tag{C.25}$$

An example of the application of partial-fraction expansion with simple poles is given in Example 5.8.

C.5.2 System with a Higher-Order Pole

Next, consider the case of a system for which the denominator polynomial $B(s)$ has one repeated root of order m; that is, $B(s)$ can be written in the form

$$B(s) = (s - \lambda_1)^m (s - \lambda_2)(s - \lambda_3) \cdots (s - \lambda_n) \tag{C.26}$$

where the repeated root is designated as λ_1. Then $U(s)$ is said to have one *pole of order m*, in addition to several simple poles such as the ones considered above. In this case, the partial-fraction expansion of $U(s)$ takes the form

$$U(s) = \frac{A(s)}{B(s)} = \frac{c_{11}}{(s-\lambda_1)^m} + \frac{c_{12}}{(s-\lambda_1)^{m-1}} + \cdots + \frac{c_{1m}}{s-\lambda_1} \\ + \frac{c_2}{s-\lambda_2} + \frac{c_3}{s-\lambda_3} + \cdots + \frac{c_n}{s-\lambda_n}$$ (C.27)

From Eq. C.27, together with the Laplace-transform pairs given in Eqs. C.17 and C.19, we get the following *time-domain response for a system with one pole of order m and $n-1$ simple poles*:

$$u(t) = \mathcal{L}^{-1}[U(s)] = \left[c_{11} \frac{t^{m-1}}{(m-1)!} + c_{12} \frac{t^{m-2}}{(m-2)!} + \cdots + c_{1m} \right] e^{\lambda_1 t} + \sum_{k=2}^{n} c_k e^{\lambda_k t}$$ (C.28)

where the coefficients c_k of the simple poles are given by Eq. C.24, and the coefficients c_{1p} associated with the multiple root λ_1 are given by the formula

$$c_{1p} = \frac{1}{(p-1)!} \frac{d^{p-1}}{ds^{p-1}} [(s-\lambda_1)^m U(s)] \bigg|_{s=\lambda_1}, \quad p = 1, 2, \ldots, m$$ (C.29)

C.5.3 System with Quadratic Factors in the Denominator Polynomial

In structural dynamics applications, many of the poles λ_k occur in complex-conjugate pairs. Note that the quadratic factor $[(s+\sigma_k)^2 + (\omega_k)^2]$ leads to complex conjugate poles

$$[(s-\sigma_k)^2 + (\omega_k)^2] = (s - \sigma_k + i\omega_k)(s - \sigma_k - i\omega_k)$$ (C.30)

For each quadratic term in the denominator of this form, there will be a partial fraction of the form

$$\frac{f_k s + g_k}{(s-\sigma_k)^2 + (\omega_k)^2}, \quad f_k, g_k = \text{constants}$$ (C.31)

This form is useful because the inverse transforms can be obtained directly from transform pairs 12 and 13 in Table C.1.

For each quadratic factor, therefore, two coefficients need to be determined. In Example C.1 we repeat Example 5.8, keeping the quadratic factor in the denominator.[1]

Example C.1 A undamped spring–mass oscillator is subjected to a step input of magnitude p_0 starting at $t = 0$. The initial conditions are $u(0) = \dot{u}(0) = 0$. Use partial-fraction expansion and table look-up to determine an expression for the total response, $u(t)$, of this system. Retain the quadratic term in the denominator.

---

[1] Reference [C.3] discusses the case where there are several sets of complex-conjugate poles.

SOLUTION The differential equation is
$$m\ddot{u} + ku = p_0 \quad \text{for } t > 0 \tag{1}$$

From transform pair 2 of Table C.1, the transformed step input is
$$P(s) = \frac{p_0}{s} \tag{2}$$

Combining Eq. 2 with Eqs. C.10 and C.11, we obtain the transformed response
$$U(s) = H(s)P(s) = \frac{p_0}{ms(s^2 + \omega_n^2)} = \frac{p_0 \omega_n^2}{k} \frac{1}{s(s^2 + \omega_n^2)} \tag{3}$$

The inverse Laplace transform of $U(s)$ has the form
$$u(t) = \frac{p_0}{k} \mathcal{L}^{-1}\left[\frac{\omega_n^2}{s(s^2 + \omega_n^2)}\right] \tag{4}$$

Since there is no one $F(s)$ in Table C.1 from which to obtain the inverse transform needed in Eq. 4, let us write this function in the partial-fraction form
$$F(s) = \frac{\omega_n^2}{s(s^2 + \omega_n^2)} = \frac{c_1}{s - \lambda_1} + \frac{f_2 s + g_2}{s^2 + \omega_n^2} \tag{5}$$

where the three poles are the roots of the *characteristic equation*
$$s(s^2 + \omega_n^2) = (s - \lambda_1)(s - \lambda_2)(s - \lambda_3) = 0 \tag{6}$$

There is one simple pole λ_1 and one complex-conjugate pair $\lambda_2, \lambda_3 = \lambda_2^*$, which are
$$\lambda_1 = 0, \quad \lambda_2 = +i\omega_n, \quad \lambda_2^* = -i\omega_n \tag{7}$$

Equation C.24, together with Eqs. 5 and 7a, can be used to compute the coefficient c_1.
$$c_1 = [(s - \lambda_1)F(s)]|_{s=\lambda_1} = \frac{\omega_n^2}{\omega_n^2} = 1 \tag{8}$$

To determine values for the remaining coefficients, f_2 and g_2, we can combine Eqs. 5, 7a, and 8 in the following manner:
$$\frac{\omega_n^2}{s(s^2 + \omega_n^2)} - \frac{1}{s} = \frac{-s}{s^2 + \omega_n^2} = \frac{f_2 s + g_2}{s^2 + \omega_n^2} \tag{9}$$

Equating coefficients of like powers of s in the last two numerators, we get
$$f_2 = -1, \quad g_2 = 0 \tag{10}$$

Therefore, the transformed response is
$$U(s) = \frac{p_0}{k}\left(\frac{1}{s} - \frac{s}{s^2 + \omega_n^2}\right) \tag{11}$$

Finally, using transform pairs 2 and 13, we can immediately write the response of the undamped SDOF system to a step input as
$$u(t) = \frac{p_0}{k}(1 - \cos\omega_n t) \quad \textbf{Ans.} \tag{12}$$

C.6 LAPLACE TRANSFORM PAIRS

See Table C.1.

Table C.1 Laplace Transform Pairs

	$F(s)$	$f(t)$
1	1	$\delta(t) =$ Dirac delta function
2	$\dfrac{1}{s}$	$u(t) =$ unit step function
3	$\dfrac{1}{s^2}$	t
4	$\dfrac{1}{s^n}$ $(n = 1, 2, \ldots)$	$\dfrac{t^{n-1}}{(n-1)!}$
5	$\dfrac{1}{s-a}$	e^{at}
6	$\dfrac{a}{s^2 + a^2}$	$\sin at$
7	$\dfrac{s}{s^2 + a^2}$	$\cos at$
8	$\dfrac{a}{s^2 - a^2}$	$\sinh at$
9	$\dfrac{s}{s^2 - a^2}$	$\cosh at$
10	$\dfrac{1}{(s^2 + a^2)^2}$	$\dfrac{1}{2a^3}(\sin at - at\cos at)$
11	$\dfrac{s}{(s^2 + a^2)^2}$	$\dfrac{1}{2a}(t\sin at)$
12	$\dfrac{a}{(s-b)^2 + a^2}$	$e^{bt}\sin at$
13	$\dfrac{s-b}{(s-b)^2 + a^2}$	$e^{bt}\cos at$
14	$\dfrac{1}{s^2 + 2\zeta\omega s + \omega^2}$	$\dfrac{1}{\omega_d}e^{-\zeta\omega t}\sin\omega_d t,$
15	$\dfrac{s + 2\zeta\omega}{s^2 + 2\zeta\omega s + \omega^2}$	$e^{-\zeta\omega t}\left[\cos\omega_d t + \dfrac{\zeta}{(1-\zeta^2)^{1/2}}\sin\omega_d t\right]$
		where $\omega_d = \omega(1-\zeta^2)^{1/2}$

REFERENCES

[C.1] J. L. Schiff, *The Laplace Transform: Theory and Applications*, Springer-Verlag, New York, 1999.

[C.2] R. A. Gabel and R. A. Roberts, *Signals and Systems*, 3rd ed., Wiley, New York, 1987.

[C.3] J. G. Reid, *Linear System Fundamentals: Continuous and Discrete, Classic and Modern*, McGraw-Hill, New York, 1983.

[C.4] L. Meirovitch, *Fundamentals of Vibrations*, McGraw-Hill, New York, 2001.

D

Fundamentals of Linear Algebra

This book requires many fundamental operations and representations from linear algebra, which can be a topic of considerable complexity. In this appendix we will cover the bare minimum of topics from linear algebra such that the text is self-explanatory and self-contained. Only topics relevant to the discussion of structural dynamics in this volume are presented. The reader is naturally referred to the collection of excellent texts on the topic for a complete discussion. As is evident by the number of references to them throughout this volume, Refs. [D.1] to [D.4] are an excellent starting points for students who wish to know more about this topic.

D.1 VECTOR SPACES AND LINEAR OPERATORS

A *vector space* is a set $\mathcal{V}$ with two operations defined on all the elements $v \in \mathcal{V}$. These operations are usually referred to as *vector addition* and *scalar multiplication*. Vector addition combines two elements of $\mathcal{V}$ to generate a third, usually different element of $\mathcal{V}$. Symbolically, we can represent this relationship as follows:

$$+: \mathcal{V} \times \mathcal{V} \to \mathcal{V}$$

$$+: (\mathbf{u}, \mathbf{v}) \mapsto \mathbf{w}$$

Of course, in practice, vector addition is written simply as $\mathbf{w} = \mathbf{u} + \mathbf{v}$. Scalar multiplication combines a constant from a *scalar field*, which we denote $\mathbb{F}$, with an element of $\mathcal{V}$, thereby generating a new element of $\mathcal{V}$. The symbolic representation of this operation is just

$$\bullet: \mathbb{F} \times \mathcal{V} \to \mathcal{V} \qquad (D.1)$$

$$\bullet: (\alpha, \mathbf{v}) \mapsto \mathbf{w} \qquad (D.2)$$

The usual representation of this operation is $\mathbf{w} = \alpha \bullet \mathbf{v}$, or more simply, $\mathbf{w} = \alpha \mathbf{v}$. The scalar field $\mathbb{F}$ is either the set of real numbers $\mathbb{R}$ or the set of complex numbers $\mathbb{C}$. Thus, we speak of a *real vector space* or a *complex vector space*, respectively. If some explanation really does not depend on a specific choice of the scalar field, we use the generic symbol $\mathbb{F}$ in the discussion.

The rigorous definition of a vector space can be quite complicated from an axiomatic viewpoint. An axiomatic definition of a vector space can be found in many sources (see, e.g., Ref. [D.5]). From a practical viewpoint, we need to understand one fundamental fact about vector spaces in this book. *Vector spaces must be closed with respect to the operations of vector addition and scalar multiplication.* This idea is made clear by the definitive example of a vector space that pervades this book: the vector space $\mathbb{R}^N$.

Example: The Vector Space $\mathbb{R}^N$ Let $\mathcal{V}$ be the collection of all N-tuples of real numbers. By convention, organize each such N-tuple in the form of a column vector.

$$\mathbf{u} = \begin{Bmatrix} u_1 \\ u_2 \\ \vdots \\ u_N \end{Bmatrix}, \quad \mathbf{v} = \begin{Bmatrix} v_1 \\ v_2 \\ \vdots \\ v_N \end{Bmatrix} \tag{D.3}$$

For each such N-tuple $\mathbf{u}$, $\mathbf{v}$, we have that $\mathbf{u}, \mathbf{v} \in \mathbb{R}^N$. For any scalar $\alpha \in \mathbb{F} \equiv \mathbb{R}$, the scalar field, we define multiplication of a vector by a scalar *componentwise*. That is, we define

$$\alpha \mathbf{u} \equiv \begin{Bmatrix} \alpha u_1 \\ \alpha u_2 \\ \vdots \\ \alpha u_N \end{Bmatrix} \tag{D.4}$$

It is obvious that $\mathcal{V}$ is closed with respect to the multiplication of vectors by scalars in this definition. Vector addition is also defined componentwise:

$$\mathbf{u} + \mathbf{v} = \begin{Bmatrix} u_1 \\ u_2 \\ \vdots \\ u_N \end{Bmatrix} + \begin{Bmatrix} v_1 \\ v_2 \\ \vdots \\ v_N \end{Bmatrix} = \begin{Bmatrix} u_1 + v_1 \\ u_2 + v_2 \\ \vdots \\ u_N + v_N \end{Bmatrix} \tag{D.5}$$

Again, it is obvious that vector addition is closed with respect to this definition.

With analogous definitions of the addition of N-tuples of complex numbers in $\mathbb{C}^N$ and the multiplication of an N-tuple of complex numbers by a complex number $\alpha \in \mathbb{C}$, the set $\mathbb{C}^N$ of complex numbers is a vector space. The set of N-tuples of real numbers $\mathbb{R}^N$ and N-tuples of complex numbers $\mathbb{C}^N$, with the conventional definitions of vector addition and multiplication by a scalar outlined above, are the vector spaces encountered in this book. Numerous other examples can be found in Ref. [D.5]. We now introduce several definitions that are commonly encountered in this book that are determined by the vector space structure of either $\mathbb{R}^N$ or $\mathbb{C}^N$.

Definition 1 *Let $\mathcal{V}$ be a vector space and let $\{v_1, v_2, \ldots, v_N\}$ be a collection of vectors contained in $\mathcal{V}$. This set of vectors is* linearly independent *if and only if the identity*

$$\alpha_1 v_1 + \alpha_2 v_2 + \cdots + \alpha_N v_N = \mathbf{0}$$

holds only for the choice $\alpha_i = 0$, where $i = 1 \cdots N$.

This definition is not very intuitive at first sight. Suppose, to the contrary of the conditions in the definition, that *there is* a collection of constants such that

$$\alpha_1 v_1 + \alpha_2 v_2 + \cdots + \alpha_N v_N = \mathbf{0}$$

but that there is at least one index i such that $\alpha_i \neq 0$. This fact implies that v_i can be expressed as a linear combination of the remaining $N - 1$ vectors:

$$v_i = \sum_{k \neq i} -\frac{\alpha_k}{\alpha_i} v_k$$

We say that the set $\{v_1, v_2, \ldots, v_N\}$ is *linearly dependent* in this case. A subset $\mathcal{U}$ of a vector space $\mathcal{V}$ is a *vector subspace* of $\mathcal{V}$ if and only if $\mathcal{U}$ is a vector space with respect to the operations of vector addition and scalar multiplication defined on $\mathcal{V}$. The subspace $\mathcal{U}$ is said to inherit these operations from $\mathcal{V}$.

Definition 2 *Let $\{v_1, v_2, \ldots\}$ be a collection of vectors in the vector space $\mathcal{V}$. The span of the set of vectors is the collection of all finite linear combinations of these vectors. That is, it is the set of all $u \in \mathcal{V}$ that can be expressed as*

$$u = \sum_{k=1}^{s} c_k v_k$$

for constants $c_k \in \mathbb{F}$ and some integer $s \in \mathbb{N}$.

Definition 3 *The* dimension *of a vector space $\mathcal{V}$ is equal to the size of the largest linearly independent set of vectors that can be extracted from $\mathcal{V}$.*

It is easy to see that the span of a set of vectors extracted from a vector space $\mathcal{V}$ is always a subspace of $\mathcal{V}$. If the set of vectors $\{v_1, v_2, \ldots, v_s\}$ extracted from $\mathcal{V}$ is also linearly independent, the *dimension* of the subspace is equal to the integer s.

We end this section with a discussion of the most important set of operators in this book that act on elements of vector spaces. These are the linear operators.

Definition 4 *Suppose that T maps elements of a vector space $\mathcal{U}$ into the vector space $\mathcal{V}$. This fact is denoted*

$$T : \mathcal{U} \longrightarrow \mathcal{V}$$

The operator T is a linear mapping *if and only if we have*

$$T(\alpha u + \beta v) = \alpha T(u) + \beta T(v)$$

for any two scalars $\alpha, \beta \in \mathbb{F}$ and for any two vectors $u, v \in \mathcal{V}$.

The *range* of the linear transformation T is the collection of vectors $v \in \mathcal{V}$ such that there exists a $u \in \mathcal{U}$ so that

$$v = T(u)$$

The *rank* of a linear operator is the dimension of its range. The *domain* of the linear transformation T is the collection of all vectors $u \in \mathcal{U}$ such that $T(u)$ is well defined. These two sets are denoted

$$\text{range}(T) = \{v \in \mathcal{V} : \exists u \in \mathcal{U} \quad \text{s.t.} \quad v = T(u)\} \tag{D.6}$$

$$\text{domain}(T) = \{u \in \mathcal{U} : T(u) \in \mathcal{V}\} \tag{D.7}$$

It is not difficult to show that these two sets, range(T) and domain(T), are in fact subspaces of $\mathcal{V}$ and $\mathcal{U}$, respectively. The *nullspace* of a linear operator T is the collection of all vectors in the domain(T) that are mapped into the zero vector $\mathbf{0} \in$ range(T).

$$\text{nullspace}(T) = \{u \in \mathcal{U} : T(u) = \mathbf{0} \in \mathcal{V}\} \tag{D.8}$$

It is not difficult to prove that nullspace(T) is a subspace of domain(T).

Again, there are many examples of linear operators on a vector space. The most important example of a linear operator T mapping a vector space $\mathcal{U}$ into a vector space $\mathcal{V}$ in this book is that of a matrix in $\mathbb{R}^{m \times n}$ acting on the vector space $\mathbb{R}^n$ and yielding an element in $\mathbb{R}^m$. An interesting type of converse result should also be noted. *Any* linear operator acting between finite-dimensional vector spaces can be represented in terms of a matrix!

D.2 MATRIX COMPUTATIONS

A matrix is an array of real or complex numbers arranged in rows and columns.

$$A = \begin{bmatrix} a_{11} & a_{12} & \cdots & a_{1t} \\ a_{21} & a_{22} & & \\ \vdots & & & \vdots \\ a_{s1} & a_{s2} & \cdots & a_{st} \end{bmatrix} \tag{D.9}$$

In this example the matrix A has s rows and t columns. The matrix A is said to have dimension $s \times t$ in this case. This fact can be expressed more succinctly as

$$A \in \mathbb{R}^{s \times t} \quad \text{or} \quad A \in \mathbb{C}^{s \times t}$$

depending on whether A is comprised of real or complex elements, respectively. It is clear from Eq. D.9 that an arbitrary element located at the position in the ith row and the jth column is denoted a_{ij}. A matrix A is *square* if $s = t$; that is, it has the same number of columns as rows. A matrix A is *symmetric* if $a_{ij} = a_{ji}$ for all indices i, j. A matrix is said to be *diagonal* if $a_{ij} = 0$ for all indices such that $i \neq j$. If A is an $s \times t$ matrix, the transpose $A^T = B$ matrix is defined by switching the rows and columns of A. That is, we have that $B = A^T$ provided that

$$b_{ji} = a_{ij}$$

for $1 \leq i \leq s$ and $1 \leq j \leq t$. Thus, another characterization of a symmetric matrix is possible. A matrix A is symmetric if and only if $A = A^T$. A matrix is *lower triangular* if whenever $j > i$, we have $a_{ij} = 0$. A matrix is similarly *upper triangular* if whenever $i > j$, we have $a_{ij} = 0$. A *row vector* is a matrix B that has dimension $1 \times t$ for some integer t. If A has dimension $s \times 1$, it is called a *column vector*. These two types of matrices are depicted as

$$B = \lfloor b_1 \ b_2 \ \cdots \ b_t \rfloor$$

or
$$A = \begin{Bmatrix} a_1 \\ a_2 \\ \vdots \\ a_s \end{Bmatrix}$$

respectively. It follows immediately that the transpose of a row vector is a column vector, and vice versa. A matrix is *positive semidefinite* if the quadratic form $v^T A v \geq 0$ for all choices of v. A matrix A is *positive definite* if the quadratic form $v^T A v \geq 0$ for all $v \in \mathcal{V}$, and equality holds only in the case that $v = 0$.

Matrices may be added if and only if they have exactly the same dimensions. Thus, addition of two matrices $A, B \in \mathbb{F}^{s \times t}$ is defined componentwise. That is, we have

$$C \equiv A + B \in \mathbb{F}^{t \times s}$$

if and only if

$$c_{ij} = a_{ij} + b_{ij}$$

for $1 \leq i \leq s$ and $1 \leq j \leq t$. Any matrix, having any dimension, can be multiplied by a scalar to yield a matrix having the same dimension. For example, if $A \in \mathbb{F}^{s \times t}$, $C \equiv \alpha A$ if and only if

$$c_{ij} = \alpha a_{ij}$$

for $1 \leq i \leq s$ and $1 \leq j \leq t$. As in the definition of addition of matrices, the multiplication of matrices by a scalar α is defined componentwise.

Matrix multiplication is, perhaps, a less intuitively defined operation than that of other simple algebraic operations. However, it is simple to define mathematically. A matrix $C = A \cdot B = AB$ is well defined provided that $A \in \mathbb{F}^{s \times t}$ and $B \in \mathbb{F}^{t \times w}$ for three integers s, t, w. We write $C = AB$ if and only if

$$c_{ij} = \sum_{l=1}^{t} a_{il} b_{lj}$$

for $1 \leq i \leq s$ and $1 \leq j \leq w$. Some important observations should be made at this point. The product AB is well defined only if the number of columns of A is equal to the number of rows of B. The product $C = AB$ has as many rows as A and as many columns as B. Note also that matrix multiplication is definitely *not commutative*!

$$AB \neq BA \tag{D.10}$$

Even if one side of Eq. D.10 is well defined, the other side of the equation may not be. For this reason, the fact that matrix multiplication is not commutative, we distinguish between the notions of *left (or pre-) multiplication* and *right (or post-) multiplication* of matrices. In the expression

$$C = AB$$

we say that A *left-multiplies* B or that B *right-multiplies* A.

The *inverse*, A^{-1}, of a square matrix A appears often throughout this book. It is defined in terms of the multiplicative *identity matrix*, denoted here by the special symbol I. The identity matrix is simply a square diagonal matrix whose diagonal entries are all equal to 1. The identity matrix is aptly named: We have $AI = A$ and $IA = A$ for all suitably dimensioned matrices. In general, the inverse matrix A^{-1} does not exist for every matrix A. First, suppose that A is a nonsquare matrix. Any matrix B such that

$$BA = I$$

is said to be a *left inverse* of A. Similarly, any matrix B such that

$$AB = I$$

is called a *right inverse* of A. Neither the right inverse nor the left inverse need be unique for a general nonsquare matrix. A square matrix A is invertible, or *nonsingular*, provided that there is a matrix B such that

$$BA = AB = I$$

Such a matrix B is the inverse of A:

$$B = A^{-1}$$

If it happens that the matrix A^{-1} exists, we say that the square matrix A is invertible, or *nonsingular*. Alternatively, we could simply define an invertible matrix as a matrix for which the left inverse is equal to the right inverse. A matrix is *orthogonal* if and only if $A^{-1} = A^T$. In this case, we can also write that $AA^T = A^TA = I$.

There are many alternative tests to determine whether a square matrix A is invertible, or nonsingular. The rank of a linear operator has been defined as the dimension of its range. The range of a matrix is equal to the span of its columns. For a matrix A, the rank is therefore equal to the number of its linearly independent columns. A matrix $A \in \mathbb{R}^{m \times m}$ is invertible if and only if it has *full rank*, or in other words, the rank of A is equal to m. (Sometimes, authors define the row rank and column rank to be the number of linearly independent rows, or columns, respectively. For finite-dimensional matrices, however, these numbers are the same.) Historically, the *determinant* has been used to determine whether a matrix is invertible. The determinant of an upper or lower triangular matrix is equal to the product of its diagonal entries. The determinant of a general matrix is defined inductively, and it is more complicated to describe. In any event, the determinant has lost favor for the determination of the invertibility or rank of a general matrix A. The preferred method to determine the rank of a matrix is to count the number of nonzero *singular values*. The Singular Value Decomposition is discussed in Section D.3.1.

A primary use of the inverse A^{-1} is that it enables the solution of systems of linear equations. Consider the system of linear equations

$$Ax = b$$

where $A \in \mathbb{R}^{m \times m}$, $x \in \mathbb{R}^{m \times 1}$, and $b \in \mathbb{R}^{m \times 1}$. If the matrix A is invertible, this equation has a unique solution achieved by premultiplying both sides of the equation by A^{-1}:

$$x = A^{-1}b \tag{D.11}$$

It is important to note that the result in Eq. D.11 is often meant to be interpreted *symbolically*, and not literally in a computational sense, in numerical analysis literature. Generally speaking, the explicit calculation of the matrix A^{-1} and its subsequent application to the right-hand side of a system of linear equations may not be very computationally efficient. In some of the sections that follow, we present decompositions or factorizations of the matrix A that enable the efficient solution of systems of linear equations without the explicit calculation of the matrix A^{-1}.

Manipulation of matrices for computation is often facilitated in this book by *conformal partitioning*. In one of the simplest examples of this technique, a matrix $B \in \mathbb{F}^{s \times t}$ is partitioned into t column vectors each having length s:

$$B = \begin{bmatrix} b_1 & b_2 & \cdots & b_t \end{bmatrix}$$

In this equation each b_k is a column vector in $\mathbb{F}^s$ for $1 \leq k \leq t$. It is also evident that it is easy to partition a transposed matrix A^T into row vectors. Suppose that $A \in \mathbb{F}^{s \times r}$, so that $A^T \in \mathbb{F}^{r \times s}$. We can write

$$A^T = \begin{bmatrix} a_1^T \\ a_2^T \\ \vdots \\ a_r^T \end{bmatrix}$$

We can now interpret matrix multiplication in terms of these partitions. The product $C = A^T B$ can be expanded in the form

$$A^T B = \begin{bmatrix} a_1^T \\ a_2^T \\ \vdots \\ a_r^T \end{bmatrix} \begin{bmatrix} b_1 & b_2 & \cdots & b_t \end{bmatrix} = \begin{bmatrix} a_1^T b_1 & a_1^T b_2 & \cdots & a_1^T b_t \\ a_2^T b_1 & a_2^T b_2 & & a_2^T b_t \\ \vdots & & & \\ a_r^T b_1 & a_r^T b_2 & \cdots & a_r^T b_t \end{bmatrix}$$

We expect that the multiplication of $A^T \in \mathbb{F}^{r \times s}$ and $B \in \mathbb{F}^{s \times t}$ should yield a matrix in $\mathbb{F}^{r \times t}$. This is exactly what occurs. Each entry in the product $A^T B$ is an inner product of the form $a_i^T b_j$ where a_i^T is a row vector of length s and b_j is a column vector of length s for all the indices i, j satisfying $1 \leq i \leq r$ and $1 \leq j \leq t$. In other words, *matrix multiplication can be viewed as a matrix formed by the dot product of the rows of the first matrix with the columns of the second matrix.*

D.3 COMMON MATRIX DECOMPOSITIONS

In many cases, matrix computations depend on the existence and use of certain matrix decompositions, or factorizations. The matrix decomposition algorithms start with a rather general matrix, and through the use of fundamental or canonical operations, transform the general matrix to one (or a few) having a specialized structure. The original matrix may be transformed into a product of matrices that are orthogonal, triangular, or diagonal, for example. In this section we do not discuss the detailed algorithms used to achieve these decompositions. The reader is referred to any of the excellent texts on this subject. Rather, the goal of this section is to provide the reader with an understanding of the structure and utility of the most common decompositions encountered in the book.

D.3.1 Singular Value Decomposition

One of the most useful factorizations of a general matrix $A \in \mathbb{R}^{m \times n}$ is the *singular value decomposition*. This factorization has many uses in matrix computations, including (1) the construction of an orthogonal basis for the range of a matrix, (2) the construction of an orthogonal basis for the nullspace of a matrix, (3) the determination of the dimension of the range and nullspace of a matrix, (4) the determination of the rank of a matrix, (5) the determination of the condition number of a matrix, and (6) the solution of linear systems of equations. It is an integral part of a number of system identification algorithms in experimental modal analysis. The diversity of applications of this factorization provides a clear indication of its utility in practical computations.

Definition 5 *Every matrix $A \in \mathbb{R}^{m \times n}$ has a unique representation, called its* Singular Value Decomposition, *that is, the product*

$$A = U \Sigma V^T$$

where $U \in \mathbb{R}^{m \times m}$ is an orthogonal matrix and $V \in \mathbb{R}^{n \times n}$ is an orthogonal matrix. The matrix of singular values, Σ, is partitioned into the form

$$\Sigma = \begin{bmatrix} \sigma \\ 0 \end{bmatrix}$$

$$\sigma = \begin{bmatrix} \sigma_1 & & & \\ & \ddots & & \\ & & \sigma_d & \\ & & & 0 \\ & & & & \ddots \\ & & & & & 0 \end{bmatrix} \in \mathbb{R}^{n \times n}$$

and the diagonal entries $\sigma_1 \geq \sigma_2 \geq \cdots \geq \sigma_d > 0$ are known as the singular values. Moreover, the number of nonzero singular values induces a partitioning of the orthogonal matrices U and V:

$$U = \begin{bmatrix} U_1 & U_2 \end{bmatrix} \quad \text{where} \quad U_1 \in \mathbb{R}^{m \times d}$$
$$V = \begin{bmatrix} V_1 & V_2 \end{bmatrix} \quad \text{where} \quad V_2 \in \mathbb{R}^{n \times (n-d)}$$

The conformal partitioning is important in that it provides important orthogonal bases for certain subspaces associated with the matrix A. The columns of the matrix U_1 form an orthogonal basis for the range of the matrix A. Also, the columns of V_2

form an orthonormal basis for the nullspace of the matrix A. A comparison of the relative costs of these fundamental decompositions can, again, be given in terms of the number of *floating-point operations*, or *flops*, required to complete the factorization. A singular value decomposition is a cubic order complexity operation in terms of the dimensions of the matrix A. Although there are many alternative implementations and selections of which matrices need to be computed explicitly, the computational cost of the algorithm can be as high as $O(4m^2n + 8mn^2 + 9n^3)$ *flops*.[D.1] The discussions that follow will show that this is indeed an expensive decomposition to compute.

D.3.2 QR Decomposition

The Singular Value Decomposition is a costly algorithm. But it is robust in the sense that it performs very well even for very poorly conditioned matrices. Another useful decomposition that can provide some of the same information about a general matrix A is the QR Decomposition. It is somewhat less expensive to compute.

Definition 6 *The* QR Decomposition *of a matrix* $A \in \mathbb{R}^{m \times n}$ *expresses the matrix* A *as the product*

$$A = QR$$

where Q is an $m \times m$ orthonormal matrix and R is an $m \times n$ upper triangular matrix.

The similarity in the information provided by the QR Decomposition is evident if the matrix Q is partitioned carefully. Suppose that A has full column rank and that $m \geq n$. Then we can partition the matrix Q in the form

$$Q = [Q_1 \ Q_2]$$

where

$$Q_1 \in \mathbb{R}^{m \times n}$$
$$Q_2 \in \mathbb{R}^{m \times (m-n)}$$

We can deduce that

$$\text{range}(A) = \text{span}(Q_1)$$
$$\text{nullspace}(A) = \text{span}(Q_2)$$

The QR Decomposition can require on the order of $O(2n^2(m - n/3))$ *flops*.[D.1] Again, it must be emphasized that there are numerous variants of implementation, each of which has a different asymptotic cost in flops. In comparison to the Singular Value Decomposition, the QR Decomposition (1) generates an orthogonal basis for the nullspace of the matrix A, (2) generates an orthogonal basis for the range of the matrix A, and (3) can be used to solve linear systems of equations. It does not provide as much detailed information as the Singular Value Decomposition regarding the condition number or singularity of the matrix A.

D.3.3 LU Decomposition

Although the Singular Value Decomposition and the QR Decomposition provide detailed information about the structure of a given matrix A, they can be expensive to compute. The SVD can be on the order of four to five times as expensive to compute as the QR Decomposition, and the QR Decomposition is itself costly. If we seek to solve a system of *exactly determined linear equations* having the form

$$Ax = b \qquad (D.12)$$

where $A \in \mathbb{R}^{m \times m}$, $x, b \in \mathbb{R}^m$, there are simpler and less costly alternatives. One of the most popular methods used to solve this problem is the LU Decomposition of the matrix A.

Definition 7 *The* LU Decomposition *of a square matrix $A \in \mathbb{R}^{m \times m}$ expresses the matrix A as the product*

$$A = LU \qquad (D.13)$$

where L is a unit *lower triangular matrix and U is an upper triangular matrix.*

There can exist many LU Factorizations of a matrix A, or none at all. If however, the LU Decomposition exists and A is invertible, the LU factorization is unique. Under these circumstances, the solution of the system in Eqs. D.12 is carried out in two phases. The linear system is written in the form

$$L\underbrace{Ux}_{y} = b$$

The solution of the intermediate linear system of equations

$$Ly = b \qquad (D.14)$$

$$y = L^{-1}b \qquad (D.15)$$

is first obtained. This first phase is called *forward elimination*. Subsequently, the solution of the linear system of equations

$$Ux = y \qquad (D.16)$$

$$x = U^{-1}y \qquad (D.17)$$

is carried out. This process is called *backward substitution*. Considering this process in retrospect, we see that we have calculated

$$x = U^{-1}L^{-1}b \qquad (D.18)$$

$$x = A^{-1}b \qquad (D.19)$$

It is important to note that we have never explicitly formed and applied the inverse A in the solution of the system Eqs. D.12. Instead, we have (1) calculated the LU factorization of the matrix A, (2) solved the system of equations in Eq. D.14 that are expressed in terms of the lower triangular matrix L, and (3) solved the system of equations in Eq. D.16 that are expressed in terms of the upper triangular matrix U. The LU factorization of a square matrix $A \in \mathbb{R}^{m \times m}$ has a computational cost on the order of $O(2m^3/3)$ *flops*. The solution of a system of equations that have a coefficient matrix that is either lower or upper triangular can be carried out in the order of $O(m^2)$ *flops*.

Consider some simple examples to show that this technique is superior in most cases to the explicit formation and application of the inverse matrix A^{-1}. How can we calculate A^{-1}? An *expression* for the inverse A^{-1} can be obtained from either the SVD, QR Decomposition or LU Decomposition. We have

$$A^{-1} = V\sigma^{-1}U^T \qquad (D.20)$$

$$A^{-1} = R^{-1}Q^T \qquad (D.21)$$

$$A^{-1} = U^{-1}L^{-1} \qquad (D.22)$$

So a typical method for constructing A is to create one of the factorizations above, say the LU Decomposition. Then we solve the matrix equation

$$LU = I \qquad (D.23)$$

$$I = U^{-1}L^{-1} \qquad (D.24)$$

where I is the identity matrix of appropriate order. So the total cost of forming and applying the matrix inverse A^{-1} is comprised of (1) creating the appropriate factorization, (2) solving the matrix equation in Eq. D.23, and finally, (3) applying A^{-1} to b to obtain x. The first of these steps has the same computational cost as the method in Eqs. D.14 and D.16. The second step costs on the order of $O(m^2)$ *flops for each column in the matrix I*. Step (2) is therefore also on the order of $O(m^3)$ operations in cost. Finally, the matrix multiplication of A^{-1} times b costs on the order of $O(m^2)$ flops. As a result, when an LU Factorization is used to compute A^{-1}, the cost is over twice that of the process outlined in Eqs. D.14 through D.23.

D.3.4 Cholesky Decomposition

The existence of the LU Decomposition is not guaranteed for a general matrix A. However, the matrices that are generated in typical structural dynamics models are highly structured. A consistent mass matrix M is guaranteed to be symmetric and positive definite. A stiffness matrix K is guaranteed to be symmetric and positive semi-definite. If the matrix K is associated with a structure that has no rigid-body modes, the matrix is symmetric and positive definite. The LU factorization can be used to derive a closely related decomposition that is exceptionally well suited to these matrices. This decomposition is the Cholesky Factorization.

Definition 8 *Suppose that A is a symmetric, positive definite matrix. The* Cholesky Factorization *of A always exists and is the unique representation having the form*

$$A = LL^T \qquad (D.25)$$

where L is a lower triangular matrix with positive diagonal elements.

The reader should carefully note that the matrix L in Eq. D.13 in the LU Decomposition is a *unit* lower triangular matrix, whereas the matrix L in Eq. D.25 is a lower triangular matrix with a positive diagonal. The same symbol L is used in both definitions since the meaning will be clear from the context.

D.4 EIGENVALUE PROBLEM

In Chapter 15 we discuss a number of the practical methods for solving algebraic eigenvalue problems that arise in structural dynamics. In this section we review some of the most fundamental properties of the standard eigenvalue problem, the generalized eigenvalue problem, and the techniques for mapping the generalized form into the standard form.

D.4.1 Standard Eigenvalue Problem

The standard eigenvalue problem associated with a real matrix $A \in \mathbb{R}^{m \times m}$ seeks to find all pairs consisting of a scalar and a *nontrivial* vector $(\lambda, v) \in \mathbb{C} \times (\mathbb{R}^m - \mathbf{0})$ that satisfy the equation

$$Av = \lambda v \tag{D.26}$$

Any scalar $\lambda \in \mathbb{C}$ satisfying Eq. D.26 is referred to as an *eigenvalue* of A, and any such *nontrivial* vector v is an *eigenvector* of A. Sometimes we refer to the couple (λ, v) as an *eigenpair* of the matrix A. The essential properties of the algebraic eigenvalue problem are deduced by rewriting Eq. D.26 in the form

$$(A - \lambda I)v = \mathbf{0} \tag{D.27}$$

Equation D.27 always has a trivial solution, that is, $v = \mathbf{0}$. By definition, an eigenvector is a *nontrivial solution* $v \neq 0$ of Eq. D.27. If, however, the matrix $(A - \lambda I)$ is nonsingular, the solution is unique. We already know one solution of Eq. D.27, the trivial solution $v = 0$. Eigenvalues must be precisely those λ that render the matrix $(A - \lambda I)$ singular. Consequently, eigenvalues are determined to be the roots of the characteristic polynomial associated with A:

$$p(\lambda) = \det(A - \lambda I) = 0 \tag{D.28}$$

If $A \in \mathbb{R}^{m \times m}$, all the coefficients of the characteristic polynomial are real. It is an mth order polynomial with m roots in $\mathbb{C}$ by the fundamental theorem of algebra. Moreover, since the coefficients of the polynomial $p(\lambda)$ are all real, any complex roots must appear in complex conjugate pairs.

Thus far, the only qualification on the matrix A in our discussion is that it must have entries that are real numbers. The matrix need not be symmetric. However, as we have noted repeatedly, we will be particularly interested in cases in which the matrix A is real and symmetric. There is a powerful result available that provides much more information about eigenpairs for these matrices.

Theorem 1 *Suppose that the real matrix $A \in \mathbb{R}^{m \times m}$ is symmetric. Then the m eigenvalues are real. There exist m mutually perpendicular eigenvectors associated with the* matrix A. The m eigenvectors are also A-orthogonal in the sense that

$$v_i^T A v_j = \begin{cases} 0 & \text{if } i \neq j \\ \lambda_i & \text{otherwise} \end{cases}$$

One implication of this theorem is that the eigenvectors of a real symmetric matrix can be used as a basis for $\mathbb{R}^m$. The proof of this theorem follows the proof in Section 10.1.9.

D.4.2 Generalized Eigenvalue Problem

Most mathematics literature that studies eigenvalue problems treats the standard eigenvalue problem. However, most eigenproblems that arise in structural dynamics appear naturally in the form of a generalized eigenvalue problem. We can easily see how this occurs. The general form of the equations of motion for the transient, undamped response of a linear structural dynamics model with m degrees of freedom can be written

$$M\ddot{u}(t) + K u(t) = 0 \qquad \text{(D.29)}$$

where M is an $m \times m$ mass matrix, K is an $m \times m$ stiffness matrix, and u is an m-vector of degrees of freedom. If we substitute $u(t) = U e^{\gamma t}$ into Eq. D.29 and set $-\lambda = \gamma^2$, we obtain

$$KU = \lambda M U \qquad \text{(D.30)}$$

$$(KU - \lambda M)U = 0 \qquad \text{(D.31)}$$

The *generalized eigenvalue problem* then seeks all pairs (λ, U) consisting of a scalar and a nontrivial vector $U \neq 0$ that satisfy Eq. D.30. Because the matrices K, M are real and symmetric in problems arising in structural dynamics, again, much more can be said about the structure of the solutions to the generalized eigenvalue problem summarized above.

Theorem 2 *Suppose that M, K are real, symmetric matrices and that M is nonsingular. Then there are m eigenvalues of the generalized eigenvalue problem and they are real. There exist m eigenvectors of the generalized eigenvalue problem that are M-orthogonal and K-orthogonal in the sense that*

$$U_i^T M U_j = \begin{cases} 0 & \text{if } i \neq j \\ m_i & \text{otherwise} \end{cases}$$

and

$$U_i^T K U_j = \begin{cases} 0 & \text{if } i \neq j \\ k_i & \text{otherwise} \end{cases}$$

For each $i = 1 \cdots m$, we have $\lambda_i = k_i/m_i$.

D.4.3 Transformation of a Generalized Eigenproblem to Standard Eigenproblem Form

Some numerical methods for the solution of eigenvalue problems are designed to treat the generalized eigenvalue problem directly. The standard eigenvalue problem then can be treated as a special case by setting $M = I$ in the eigenvalue solver. The Lanczos Method described in Chapter 15 is one of the numerical methods that can treat the generalized eigenvalue problem directly. However, the vast majority of research on the solution of eigenproblems has studied the standard eigenvalue problem. Just as important, the most stable, robust, and rapid numerical methods for the solution of standard algebraic eigenvalue problems are designed for matrices A that are symmetric. In this section we discuss two methods for converting generalized eigenvalue problems that arise in structural dynamics into the form of a standard eigenvalue problem.

Consider first the case when we have a *consistent* $m \times m$ mass matrix M appearing in the generalized eigenvalue problem. Since M is guaranteed to be real, symmetric, and positive definite, it has a unique Cholesky Factorization:

$$M = LL^T \tag{D.32}$$

where L is an $m \times m$ lower triangular matrix with a positive diagonal. When the Cholesky decomposition is substituted into the generalized eigenvalue problem, we can write

$$(K - \lambda LL^T)U = 0 \tag{D.33}$$

Now, introduce a new set of variables,

$$v = L^T U \tag{D.34}$$

into Eq. D.33 and premultiply by L^{-1}. The result is an eigenvalue problem in standard form:

$$(L^{-1}KL^{-T} - \lambda I)v = 0 \tag{D.35}$$

$$(A - \lambda I)v = 0 \tag{D.36}$$

where the matrix $A = L^{-1}KL^{-T}$ is easily verified to be a symmetric matrix.

The transformation generated in Eq. D.35 is advantageous in that it is well defined in all but the most unusual structural dynamics problems. It is reasonable to expect that the mass matrix M is real, symmetric, and positive definite, whether it is a consistent-mass matrix or a lumped-mass matrix (except for a special lumped-mass matrix such as the one in Fig. 10.1). Another approach exists for transforming the generalized eigenvalue problem in Eq. D.30 into the standard form. This technique is applicable only if the structure under consideration does not have rigid-body modes. In this case, in which the structural system does not have rigid-body modes, the matrix K is real, symmetric, and positive definite. It has a unique Cholesky Factorization of the form

$$K = LL^T \tag{D.37}$$

As in the last transformation, we introduce a new set of unknowns

$$v = L^T U \tag{D.38}$$

Substitute the new set of variables v into Eq. D.30, and premultiply by L^{-1}. A new standard eigenproblem is derived:

$$(I - \lambda L^{-1} M L^{-T})v = 0 \tag{D.39}$$

$$(A - \mu I)v = 0 \tag{D.40}$$

where $\mu = 1/\lambda$ and $A = L^{-1} M L^{-T}$. Again, it is straightforward to show that the matrix A is symmetric and positive definite. The identity $\mu = 1/\lambda$ emphasizes the fact that this method is appropriate only when the eigenvalues of the structural dynamics problem are bounded away from zero.

REFERENCES

[D.1] G. H. Golub and C. F. Van Loan, *Matrix Computations*, 3rd ed., Johns Hopkins University Press, Baltimore, MD, 1996.

[D.2] A. Quarteroni, R. Sacco, and F. Saleri, *Numerical Mathematics*, Springer-Verlag, New York, 2000.

[D.3] G. Strang, *Linear Algebra and Its Applications*, 3rd ed., Harcourt Brace Jovanovich, San Diego, CA, 1988.

[D.4] H. Anton, *Elementary Linear Algebra*, 8th ed., Wiley, New York, 2000.

[D.5] J. T. Oden and L. Demkowicz, *Applied Functional Analysis*, Prentice Hall, Englewood Cliffs, NJ, 1996.

E

Introduction to the Use of MATLAB

The purpose of this appendix is to cover the topics that will enable you to use the problem-solving language MATLAB in solving many of the homework exercises in this book.[E.1] Over the past decade there has been a dramatic increase in the use of object-oriented higher-level problem-solving languages. Problem-solving languages differ in several respects from conventional programming languages such as FORTRAN, BASIC, or C. Typically, the higher-level problem-solving languages provide the ability to manipulate complex mathematical objects, such as vectors and matrices, as intrinsic entities. Addition, subtraction, multiplication, etc. of objects like matrices are carried out via single commands, for example. The advantage of this approach is that complex computational sequences can be broken down into their logical constituents, and they can be programmed very efficiently. Just a few lines of MATLAB commands can provide the functionality of hundreds of lines of a conventional program language. (Of course, the problem-solving languages invoke lower-level programs written in conventional languages!) One disadvantage of this approach is that the higher-level problem-solving languages usually incur a nontrivial computational overhead. This computational cost can be significant, even prohibitive, in some applications. Effectively, the computational overhead limits the size of problems that can be addressed via the problem-solving languages. However, the problem-solving languages are easier to learn, have highly intuitive interfaces, and are interpreted at the command line. This last property means that it is not required to process a program through the compiler and linkage editor associated with most conventional languages.

In this appendix we discuss only the command-line interface to MATLAB. This is the most basic interface to MATLAB. There are two modes of operation when working at the command-line level in MATLAB. One mode is *interactive*, where each command input is interpreted and executed immediately after the user inputs <Enter>. The second mode of operation makes use of *.m-files*, which are user programs consisting of MATLAB commands. The former mode of operation is simpler and provides quick answers: for example, a simple plot of a short mathematical expression. The latter mode is far more powerful and can yield complex solutions to diverse problems.

The reader is advised that highly intuitive graphical graphical user interfaces (GUIs) to MATLAB have been developed. The program MATLAB also includes the ability to create problem-solving simulations via block diagrams via the associated program **Simulink**. The program MATLAB also enables direct interfacing to data acquisition hardware. The reader should see the MATLAB home page for a detailed discussion of these products and options.[E.1]

E.1 INTRODUCTORY SYNTAX, COMMANDS, AND MATRIX ALGEBRA

Variables of all types can be entered in the command line, as illustrated in Fig. E.1. The command input line in MATLAB is preceded by the symbol >>. No express *variable type casting* is needed. That is, we need not specify whether a variable is an integer, real variable, or logical variable before we set the variable. Note that integer and real variables can be defined at the command line. No previous commands have been entered to specify the type of each variable. A semicolon after a command line entry *suppresses* all output from that command. Otherwise, the output of the command is listed after the command-line input. The MATLAB command *who* provides a list of all the variables that are currently included in the workspace (Figs. E.2 and E.3). In other words, it lists the names of all the variables you have defined up to this point. Note that the output of all the variable assignments in Fig. E.1 has been suppressed with the use of a semicolon, ;, terminating each line. Figure E.4 illustrates the case when an input line is not terminated with a ;, and the command output appears on the screen. The reader is warned that lengthy output (particularly associated with matrices) can result if the command input line is not terminated with a semicolon. The *clear variable* command will remove *variable* from the current list of variables in the workspace. The variant *clear all* eliminates all variables from the current workspace. Figure E.3 shows the operation of the *clear* command when used to eliminate a single variable, and when used to eliminate all variables, in the workspace.

Matrices in MATLAB are treated as logical entities and manipulated via simple commands. Matrices can be entered by explicitly listing their elements within brackets, $A = [a_1 \quad a_2 \quad \cdots]$. Rows of a matrix are terminated by the semicolon, ;. The entries within a row are separated by blank spaces or by commas, and every row must have the same number of entries (see matrix A in Fig. E.4 and matrix B in Fig. E.5a).

```
%
%   set the parameters
%
alpha=0;
beta=0;
t0=0.0;
tf=1.e-2;
h=1.e-5;
Nsteps=(tf-t0)/h;
P=1.0;
Forcetime=(tf-t0)/20;
Omega=2*pi/Forcetime;
Node=13;
DOF=2;
Nmodes=70;
```

Figure E.1 Explicit variable entry without *type casting*. Code fragment from the *.m-file* frame3-int-2nd-cen.m.

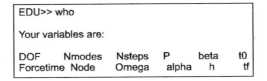

Figure E.2 Output of the *who* command.

E.1 Introductory Syntax, Commands, and Matrix Algebra

```
EDU>> clear DOF
EDU>> who

Your variables are:

Forcetime  Node    Omega   alpha   h    tf
Nmodes     Nsteps  P       beta    t0

EDU>> clear all
EDU>> who
EDU>>
```

Figure E.3 Output of the *clear* command.

```
EDU>> A=[ 1 2 3;4 5 6]

A =

   1   2   3
   4   5   6

EDU>>
```

Figure E.4 Explicit matrix entry.

```
EDU>> B=[4 5 6 ; 7 8 9]        EDU>> C=A+B

B =                            C =

   4   5   6                      5    7    9
   7   8   9                     11   13   15

         (a)                           (b)

EDU>> D=4*A                    EDU>> E=C-A

D =                            E =

   4    8   12                    4   5   6
  16   20   24                    7   8   9

         (c)                           (d)
```

Figure E.5 Fundamental matrix algebra: (*a*) definition of matrix B; (*b*) matrix addition; (*c*) multiplication of a scalar times a matrix; (*d*) subtraction of matrices.

The addition, multiplication, and subtraction of matrices is carried out in the same way that we usually apply these operations to scalars (Fig. E.5*b*, *c*, *d*). Simple syntax makes it easy to access individual rows and columns of matrices as objects themselves. For example, $E(:, 1)$ extracts the first column of the matrix E, and $E(1, :)$ extracts the first row of the matrix E. Similarly, a submatrix can be extracted easily. The syntax $E(1:2, 2:3)$ extracts the rows having addresses starting with 1 and ending with 2, and columns starting with addresses 2 and ending with 3 of the matrix E. The command *size* is a particularly useful matrix utility command. It returns the numbers of rows and columns in a matrix. These commands are illustrated in Fig. E.6.

In addition to these simple mathematical operations on matrices, there are a number of advanced Marx operations that are nearly as simple to apply. These operations

```
EDU>> F=E(:,1)
                                    EDU>> G=E(1,:)
F =
                                    G =
  4
  7                                   4   5   6
       (a)                                 (b)

EDU>> H=E(1:2,2:3)
                                    EDU>> size(H)
H =
                                    ans =
  5   6
  8   9                               2   2
       (c)                                 (d)
```

Figure E.6 Accessing submatrices by ranges of indices: (*a*) extract a column from E; (*b*) extract a row from E; (*c*) extract a 2×2 submatrix from E; (*d*) determine the *size* of a 2×2 matrix.

Table E.1 Advanced Array Operations Available in Core MATLAB

Operation	Description
lu	Calculate the LU decomposition of a matrix
svd	Calculate the singular value decomposition of a matrix
eig	Calculate the eigenvalues and eigenvectors of a matrix
qr	Calculate the QR decomposition of a matrix
zeros	Create a matrix of zeros of prescribed dimension
eye	Create an identity matrix of any prescribed dimension

include the operations listed in Table E.1. The syntax of the commands in this table is surprisingly easy to learn and use and is readily available from the MATLAB *help* utility or a MATLAB *User Guide*. For example, the command $A = eye(10, 10)$ creates a 10×10 identity matrix, $B = zeros(6, 1)$ defines a 6×1 column vector of zeros, and $[U, S, V] = svd(A)$ returns the orthogonal matrix U, the diagonal matrix S, and the orthogonal matrix V that make up the Singular Value Decomposition of matrix A. The user is referred to Section D.3.1 for a discussion of the Singular Value Decomposition.

Examples of the operations listed in Table E.1 are given throughout this appendix. For example, the use of the *eye* command to create a 3×3 identity matrix is shown in Fig. E.14. Use of the *eig* command to use the built-in capability of MATLAB to calculate eigenvalues and eigenvectors is shown in the code fragment in Fig. E.14. Use of the *zeros* command to create a vector of zeros is illustrated in the .m-file shown in Fig. E.16.

E.2 INPUT AND OUTPUT OPERATIONS

In addition to the direct input of the elements of a matrix, as described above, the program MATLAB has a number of essential built-in functions to archive the results of computations by saving them to files. When using these capabilities, it is essential to

```
EDU>> diary
EDU>> alpha=0          alpha=0

alpha =                alpha =

    0                      0

EDU>> beta=0           beta=0

beta =                 beta =

    0                      0

EDU>> t0=0.0           t0=0.0

t0 =                   t0 =

    0                      0

EDU>> tf=1.e-2         tf=1.e-2

tf =                   tf =

    0.0100                 0.0100

EDU>> diary off        diary off
     (a)                   (b)
```

Figure E.7 Syntax and output of the *diary* command: (*a*) command line input/output using the *diary*; (*b*) output file *diary* saved to working directory.

```
EDU>> clear all
EDU>> %
EDU>> %    run a program that generates a time signal:
EDU>> %
EDU>> frame3_int_2nd_cen;
EDU>> who

Your variables are:

A      De       M        Omega    dyn      n        t        yn
Ainv   E        Minv     P        h        p        t0       ynm1
C      Forcetime Nmodes  alpha    i        p0       tf       ynp1
D      K        Node     beta     ii       pn       tn
DOF    Ke       Nsteps   ddyn     m        pt       y

EDU>> tcen=t;
EDU>> ycen=y;
EDU>> save signals tcen ycen
EDU>> %
EDU>> % run a program that generates another time signal
EDU>> %
EDU>> frame3_int_2nd_new;
EDU>> tnew=t;
EDU>> ynew=y;
EDU>> save signals tnew ynew -append
EDU>> clear all
EDU>> who
EDU>> load signals
EDU>> who

Your variables are:

tcen  tnew  ycen  ynew
```

Figure E.8 Syntax and use of the *save* and *load* commands.

be able to manipulate and examine the current working directory. There are a number of standard commands that operate exactly as they do in the DOS or UNIX operating systems to accomplish this task. For example, the DOS command *dir* will provide a list of the files in the current working directory. The UNIX command *pwd* will display the working directory path name. The DOS or UNIX command *cd* will change the current working directory when invoked from the MATLAB command line.

These tools enable manipulation of the working directory. Saving results to the working directory is possible using a variety of MATLAB commands. One of the simplest operations makes use of the *diary*, illustrated in Fig. E.7a, b. Once the diary is turned on, the *diary* simply saves all commands and their associated output to an ascii file. This file can be consulted later to recall operations or results or to create a report. The simplest syntax of the *diary* command invokes the command *diary*. All subsequent commands and output are written to the ascii file named *diary* in the current working directory. When the command *diary off* is invoked, no further commands or output are written to the diary. An alternative form of the *diary* command is useful for distinguishing between different sections or sessions of computational work. The *diary filename* command directs all subsequent input commands and associated output to the ascii file *filename*. Again, the command *diary off* terminates output of the file *filename*. In this way, output can be directed to several different diaries.

Perhaps the most common output command is the *save* command, illustrated in Fig. E.8. A general form of its syntax is given by *save filename variable1 variable2* .... This commands saves the contents of the variables *variable1, variable2, ...* in the file *filename.mat*. Note that the default extension for the output file is *.mat*. All *.mat* files are stored in MATLAB-specific format but unlike a simple text file, cannot be opened for reading. Files with the extension *.mat* provide fast and efficient input–output storage format for MATLAB results. It is possible to add more information to a given file continuously. We enter the command *save filename variable3-append* to append the variable *variable3* to the *.mat* file *filename*.

Often, we seek to save MATLAB computations in formats that are readable by other programs or can be inspected by the user as a text file. In this case, another useful variant of the *save* command is available. By entering the command *save filename*

```
EDU>> frame3_int_2nd_cen;
EDU>> who

Your variables are:

A        De       M         Omega     dyn       n         t         yn
Ainv     E        Minv      P         h         p         t0        ynm1
C        Forcetime Nmodes   alpha     i         p0        tf        ynp1
D        K        Node      beta      ii        pn        tn
DOF      Ke       Nsteps    ddyn      m         pt        y

EDU>> tcen=t;
EDU>> ycen=y;
EDU>> sigarray(:,1)=tcen;
EDU>> sigarray(:,2)=ycen;
EDU>> save signals.txt sigarray -ascii
```

Figure E.9 Saving ascii-readable variables in a file using the *save* command.

```
0.0000000e+000  0.0000000e+000
1.0000000e-005  4.5463097e-013
2.0000000e-005  1.8111654e-012
3.0000000e-005  4.5022749e-012
4.0000000e-005  8.9390187e-012
5.0000000e-005  1.5504173e-011
6.0000000e-005  2.4546007e-011
7.0000000e-005  3.6372604e-011
8.0000000e-005  5.1246819e-011
9.0000000e-005  6.9381944e-011
1.0000000e-004  9.0938167e-011
1.1000000e-004  1.1601986e-010
1.2000000e-004  1.4467375e-010
1.3000000e-004  1.7688801e-010
1.4000000e-004  2.1259227e-010
1.5000000e-004  2.5165853e-010
1.6000000e-004  2.9390303e-010
1.7000000e-004  3.3908898e-010
1.8000000e-004  3.8693009e-010
```

Figure E.10 The ascii file generated by the code in Fig. E.9.

```
EDU>> clear all
EDU>> load kframe3v3_femlabsym;
EDU>> load mframe3v3_femlabsym;
EDU>> who

Your variables are:

De  Ke

EDU>> size(De)

ans =

    70    70

EDU>> size(Ke)

ans =

    70    70
```

Figure E.11 Importing large matrices archived using the *load* command.

variable1 variable2-ascii, the output variables *variable1, variable2* are saved in the file *filename* in ascii-readable format (Figs. E.9 and E.10).

Once a set of variables has been archived in either *ascii* format or a *.mat* file, they can be input using the *load* command (Fig. E.11).

E.3 CONDITIONALS

As in any programming language, the use of *conditional expressions* is essential to writing complex computational algorithms. These constructs allow us to choose alternative actions based on a logical expression or to continue processing similar actions based on a logical expression. Perhaps the most common conditional statement in MATLAB is the *if* statement and its associated structure. The *if* statement, illustrated in Fig. E.12,

```
if DOF == 1                    % force in the x direction
    ii=(Node-2);
    p(ii,1)=1;
elseif DOF == 2                % force in the y direction
    ii=22+(Node-2);
    p(ii,1)=1;
elseif DOF ==3                 % moment about a node
    ii=44+(Node-2);
    p(ii,1)=1;
else
    ierror=1;                  % signal an error
end
```

Figure E.12 The *if* statement structure in a code fragment extracted from frame3-int-2nd-cen.m.

has the following general structure:

$$
\begin{aligned}
&\text{if} \quad \text{logical expression} \\
&\qquad \text{executable expressions} \\
&\qquad \vdots \\
&\text{elseif} \quad \text{logical expression} \\
&\qquad \text{executable expressions} \\
&\qquad \vdots \\
&\text{else} \\
&\qquad \text{executable expressions} \\
&\qquad \vdots \\
&\text{end}
\end{aligned}
\tag{E.1}
$$

Another of the most widely used conditional statements is the *for* loop. The general structure of the *for* loop is given as follows:

$$
\begin{aligned}
&\text{for} \quad \text{logical expression} \\
&\qquad \text{executable expressions} \\
&\qquad \vdots \\
&\text{end}
\end{aligned}
\tag{E.2}
$$

An example of the use of the *for* loop in conjunction with the *if* conditional is illustrated in the code fragment in Fig. E.13, extracted from *autoframe.m*. Finally, we

```
for i=1:26
    if (i==1)
        x(i)=-1.8;
        y(i)=0;
    else
        if (mod(i,2)==0)
            x(i)=x(i-1);
            y(i)=0.2;
        else
            x(i)=x(i-1)+0.3;
            y(i)=0;
        end
    end
end
```

Figure E.13 The *if* statement structure in a code fragment extracted from autoframe.m.

also have frequent occasions in the text to make use of the *while* loop. The general structure of the *while* loop is given as follows:

$$\begin{aligned} &\text{while} \quad \text{logical expression} \\ &\qquad\qquad \text{executable expressions} \\ &\qquad\qquad \vdots \\ &\text{end} \end{aligned} \qquad (E.3)$$

E.4 WRITING FUNCTIONS AND .*m-files*

Programming in MATLAB can be very easy, but at the same time it can enable development of complex algorithms in a relatively few lines of code. Programming and the invocation of user-written algorithms from the command line is usually achieved via the use of .*m-files*. The use of .*m-files* is easy to describe. MATLAB .*m-files* are typically of two types: .*m-file procedures* or *macros*, and .*m-file functions*. .*M-file procedures* or *macros* are simply collections of MATLAB commands that are organized in a file. The file should be named *filename.m*. It should be located in the current working directory or in the path that is searched for .*m-files* by the MATLAB interpreter. The collection of commands is executed simply by invoking the name of the .*m-file*, *filename*, at the command line.

If a user has written an .*m-file filename.m* and invokes it using >>*filename* at the command line, the first location that MATLAB searches for *filename.m* is in the current working directory. However, MATLAB does not stop there. It searches all the directories in the current *PATH*. There are several commands available to the user to modify which directories are searched by the MATLAB interpreter when it encounters an unrecognized .*m-file*. See the *help* on the topic of *Adding Directories to the* MATLAB *Search Path*.

It should be noted that comments are used extensively in well-written .*m-files* so that they can be reused by many users, as illustrated in Fig. E.14. Any text preceded on the left by a % symbol is interpreted as a comment and ignored by MATLAB. This enables a programmer to comment out entire lines or to place comments and annotations to the right of executable lines. The command *disp('here is some text')* outputs the text message as the .*m-file* is being executed. This command can be useful for allowing the user to see the progress of execution of an .*m-file*, and it can make the format of the output more structured. See Fig. E.15, for example. During the execution of an .*m-file*, variables can be input from the keyboard into an .*m-file* by use of the *input* command. The command *nsteps = input('Enter the number of time steps')* will prompt the user for a number and wait for keyboard input. This number will be assigned to the variable *nsteps* within the .*m-file* and can then be used to control the execution of the .*m-file*.

Once an .*m-file procedure* is executed, all the variables defined in the .*m-file procedure* remain active in the workspace. This fact is made clear in Fig. E.15. Before the .*m-file procedure* is invoked, the *clear all* command eliminates all variables from the workspace. After the .*m-file procedure* is invoked, the MATLAB command *who* shows that all the variables defined within the .*m-file procedure* remains in the workspace after its execution.

The second type of MATLAB .*m-file* is illustrated by the .*m-file function*. The .*m-file function* has a slightly different declaration syntax. An example is shown in Fig. E.16.

```
%
%%%%%%%%%%%%%%%%%%%%%%%%%%%%%%%%%%%%%%%%%%%%%%%%
%  Fundamentals of Structural Dynamics
%     Roy Craig and Andrew Kurdila
%     September 4, 2004
%
%
%  MATLAB Example 15.3  (was 16.3)
%  Changed to Chapter 15 on 3/7/05.
%  Calculation of eigenvalues / eigenvectors by
%  subspace iteration method
%
%%%%%%%%%%%%%%%%%%%%%%%%%%%%%%%%%%%%%%%%%%%%%%%%
%
%  Create the stiffness and mass matrix as defined in Example 16.1
%
clear all;
k=[2,-1,0;-1,2,-1;0,-1,2];
m=eye(3,3);
disp('======================================================================')
disp('   Define the stiffness and mass matrices                      ')
disp('======================================================================')
disp('The stiffness matrix is defined in the example to be')
k
disp('The mass matrix is defined in the example to be ')
m
%
%%%%%%%%%%%%%%%%%%%%%%%%%%%%%%%%%%%%%%%%%%%%%%%%
%  Check the eigenvalues by the matlab built in function
%%%%%%%%%%%%%%%%%%%%%%%%%%%%%%%%%%%%%%%%%%%%%%%%
%
[v,d]=eig(k,m);
disp('======================================================================')
disp('   Eigenvalues and Eigenvectors Calculated by Matlab built in function   ')
disp('======================================================================')
disp('  Eigenvalues are ....')
diag(d)
disp('  Mass normalized eigenvectors are given by ...')
v
%
%%%%%%%%%%%%%%%%%%%%%%%%%%%%%%%%%%%%%%%%%%%%%%%%
%  use Subspace Iteration to calculate approximation to 1 mode
%%%%%%%%%%%%%%%%%%%%%%%%%%%%%%%%%%%%%%%%%%%%%%%%
%
[vv,dd]=subiter(k,m,3,1.e-7)
%
%%%%%%%%%%%%%%%%%%%%%%%%%%%%%%%%%%%%%%%%%%%%%%%%
%  Test the iteration
%%%%%%%%%%%%%%%%%%%%%%%%%%%%%%%%%%%%%%%%%%%%%%%%
%
disp('The approximation to the  eigenvalue is ')
dd
disp('The approximation to the eigevector is ...')
vv
%
%%%%%%%%%%%%%%%%%%%%%%%%%%%%%%%%%%%%%%%%%%%%%%%%
%  mass normalize, as used in Matlab Function
%%%%%%%%%%%%%%%%%%%%%%%%%%%%%%%%%%%%%%%%%%%%%%%%
dum=vv'*m*vv;
vv=vv/sqrt(dum);
end
disp(' Mass normalized eigenvectors are given by...')
vv
%
```

Figure E.14 Example *.m-file* procedure, mat-ex-15-3.m.

```
EDU>> clear all
EDU>> mat_ex_15_3
===============================================================
 Define the stiffness and mass matrices
===============================================================
The stiffness matrix is defined in the example to be

k =

     2    -1     0
    -1     2    -1
     0    -1     2

The mass matrix is defined in the example to be

m =

     1     0     0
     0     1     0
     0     0     1

===============================================================
 Eigenvalues and Eigenvectors Calculated by Matlab built in function
===============================================================
Eigenvalues are ....

ans =

    0.5858
    2.0000
    3.4142

Mass normalized eigenvectors are given by ...

v =

    0.5000   -0.7071   -0.5000
    0.7071    0.0000    0.7071
    0.5000    0.7071   -0.5000


error =

   14.0000


error =

   8.9687e-014


vv =

   -0.1464   -0.3536    0.8536
    0.2071   -0.0000    1.2071
   -0.1464    0.3536    0.8536


dd =

    3.4142
    2.0000
    0.5858

The approximation to the eigenvalue is

dd =

    3.4142
    2.0000
    0.5858

The approximation to the eigevector is ...

vv =

   -0.1464   -0.3536    0.8536
    0.2071   -0.0000    1.2071
   -0.1464    0.3536    0.8536

Mass normalized eigenvectors are given by...

vv =

   -0.5000 + 0.0000i  -0.7071 - 0.0000i   0.5000 + 0.0000i
    0.7071 + 0.0000i  -0.0000 - 0.0000i   0.7071 + 0.0000i
   -0.5000 + 0.0000i   0.7071 - 0.0000i   0.5000 - 0.0000i

EDU>> who

Your variables are:

ans  d  dd  dum  k  m  v  vv

EDU>>
```

Figure E.15 Example .m-file procedure, mat-ex-15-3.m.

```
function f=firstODE(t,x)
global M C K Minv P Omega Node DOF Forcetime
[m2,n2]=size(x);
m=m2/2;
x1=x(1:m,1);
%size(x1)
x2=x(m+1:m2,1);
%size(x2)
p=zeros(m,1);
%
%   create the force
%
if (Node < 3) | (Node > 24)
    disp('The Frame Example Contains Only Free Nodes Numbered From 3 to 24')
    exit
end
if DOF == 1                 % force in the x direction
    ii=(Node-2);
    p(ii,1)=1;
elseif DOF == 2             % force in the y direction
    ii=22+(Node-2);
    p(ii,1)=1;
elseif DOF ==3              % moment about a node
    ii=44+(Node-2);
    p(ii,1)=1;
end
if t > Forcetime
    p=zeros(m,1);
end
%
%   create the forces
%
%f=zeros(m2,1);
f(1:m,1)=x2;
% size(Minv)
% size(K)
% size(x1)
% size(C)
% size(x2)
% size(p)
pt=P*sin(Omega*t)*p;
f(m+1:m2,1)=-Minv*K*x1-Minv*C*x2+Minv*pt;
```

Figure E.16 Example *.m-file* function, firstODE.m.

E.5 GRAPHICAL OUTPUT

There is an extensive well-designed interface to graphics capabilities in MATLAB. This simple survey makes no attempt to capture the breadth of the graphics capabilities in MATLAB. Instead, we focus on a few simple two-dimensional plotting capabilities that have been used throughout the text, in the homework, and in the examples. The most common feature of the MATLAB graphics package that has been used in this book is the *plot* command. The format of this command varies greatly. The most common format that we employ in this book is the following: *PLOT(X1, Y1, S1, X2, Y2, S2, X3, Y3, S3, ...)*. The input to the *plot* command can be grouped in sets of three inputs. The triple *X1, Y1, S1* defines an array of abscissa values in *X1*, an array or ordinate values in *Y1*, and a symbol or representation code in *S1*. The symbol code *S1* can be quite varied. The symbol code is constructed from entries extracted from the three columns.

Table E.2 does not exhaust the possible choices. The reader is referred to the online *help* in MATLAB for additional variants. For example, the command

$$plot(X1, Y1,'r.', Y2, Y2,'b*', X3, Y3,'k-.')$$

will plot the set of $X1, Y1$ pairs as red dots, plot the set of $X2, Y2$ pairs as blue stars, and pass a black dash-dotted line through the set of $X3, Y3$ pairs.

Figure E.17 displays a typical command line to generate a plot of computational results. The MATLAB *.m-file procedures frame3-int-2nd-cen.m* and *frame3-int-2nd-new.m* are invoked to generate approximate transient response of node 13 in the symmetric frame studied in the examples in Chapters 15 and 16. The variables *tcen,ycen,tnew*, and *ynew* are archived in a *.mat* file with the *save* command. Subsequently, they are

```
EDU>> frame3_int_2nd_cen;
EDU>> tcen=t;
EDU>> ycen=y;
EDU>> save simsignals tcen ycen
EDU>> dir sim*.mat

simsignals.mat

EDU>> frame3_int_2nd_new;

EDU>> who

Your variables are:

A        DOF      K      Node    beta    h      p       t0     yn
Ainv     De       Ke     Nsteps  ddyn    i      p0      tf     ynm1
Beta     E        M      Omega   ddynp1  ii     pnp1    tn     ynp1
C        Forcetime Minv  P       dyn     m      pt      tnp1
D        Gamma    Nmodes alpha   dynp1   n      t       y

EDU>> tnew=t;
EDU>> ynew=y;
EDU>> save simsignals tnew ynew -append
EDU>> clear all
EDU>> who
EDU>> dir simsignals.*

simsignals.mat

EDU>> load simsignals
EDU>> who

Your variables are:

tcen  tnew  ycen  ynew

EDU>> plot(tnew,ynew,'.b',tcen,ycen,'.-k')
EDU>> xlabel('time(sec)')
EDU>> ylabel('y-displacement, node 13 (m)')
EDU>> title('Comparison of Central Difference and Newmark \beta')
```

Figure E.17 Example command line input to calculate and plot Central Difference and Newmark-β simulation comparison.

Table E.2 Colors, Markers, and Line Styles for the Plot Command

	Color		Marker		Line style
b	blue	.	point	-	solid line
g	green	o	circle	:	dotted line
r	red	x	x-mark	-.	dash-dotted line
c	cyan	+	plus	–	dashed line
m	magenta	*	star		
y	yellow	s	square		
k	black	d	diamond		

input via the *load* command into a workspace that has been emptied out using the *clear all* command. The *plot* command represents the output of the Newmark-β Integration Method (stored in the variables *tnew* and *ynew*) with a blue dot. The output of the Central Difference Method (stored in the variables *tcen* and *ycen*) are depicted with a black, dash-dotted linestyle. The *xlabel* command annotates the abscissa and the *ylabel* command annotates the ordinate. The entire plot is given a label with the *title* command. The output of the code fragment is depicted in Fig. E.18. Procedures for sizing plots and procedures for grouping multiple plots together in an array can be found in several

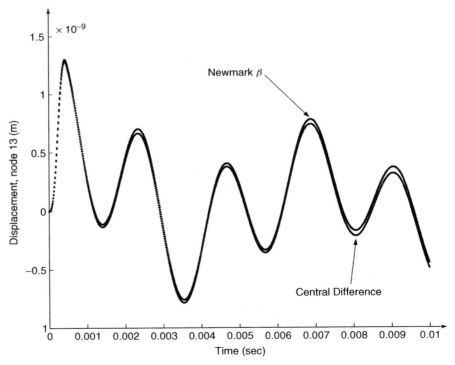

Figure E.18 Output generated by command-line input in Fig. E.17.

example *.m-files* that were used to create figures in this book and that are provided on the book's website.

E.6 ARCHITECTURE OF ODE INTEGRATION

One of the most useful capabilities of MATLAB that is used in this book is the systematic and very general ability to integrate systems of ordinary differential equations that are cast in *canonical first-order form*. The canonical first-order form that is used by the MATLAB software assumes that the system of ordinary differential equations is in the form

$$\dot{y}(t) = f(t, y(t)) \quad \in \mathbb{R}^n$$
$$y(0) = y_0 \quad \in \mathbb{R}^n$$
(E.4)

A number of algorithms that are a standard part of MATLAB can be used to integrate these equations forward in time. These algorithms include:

1. ODE45: solve nonstiff differential equations, medium-order method.
2. ODE23: solve nonstiff differential equations, low-order method
3. ODE113: solve nonstiff differential equations, variable-order method
4. ODE15S: solve stiff differential equations and DAEs, variable-order method
5. ODE23S: solve stiff differential equations, low-order method
6. ODE23T: solve moderately stiff ODEs and DAEs, trapezoidal rule
7. ODE23TB: solve stiff differential equations, low-order method

The reader is referred to the MATLAB *help* facility to obtain a detailed discussion of all of the variants for integrating these equations. One of the best design features of the MATLAB architecture for the integration of ordinary differential equations is that any of these integration methods can be invoked with essentially the same syntax. The simplest form of this syntax is exemplified by the following MATLAB statement:

$$[T, Y] = ODE45(ODEFUN, TSPAN, Y0)$$

In this equation, the MATLAB function *ODE45* can be replaced by any of the integration methods listed above. The array of discrete times at which the solution of the equations has been approximated is returned in the array T. The approximate solution values at this set of discrete times is returned in the array Y. The range of times over which the solution is sought is input in the variable *TSPAN*. This array can have two formats. We can input either

$$TSPAN = [t_0 \ t_f] \quad \text{or} \quad TSPAN = [t_0 \ t_1 \ t_2 \ \ldots \ t_f]$$

In the first variant, just the beginning time t_0 and the final time t_f for the interval of integration are specified. In the second variant, all of the discrete times at which a solution is sought are specified. The variable *Y0* is a column vector in $\mathbb{R}^n$ that specifies the initial values $y(0)$ appearing in Eq. E.4. A specific system of equations in the canonical form written in Eq. E.4 is encoded in the function *ODEFUN*. The MATLAB *.m-file function ODEFUN* must have the following syntax:

$$F = ODEFUN(T, Y)$$

```
%
%   frame3_int_1st.m
%       This m-file carries out the direct integration of the equations of
%   motion
%           [M]\ddot{u} + [C]\dot{u} + [K] \dot{u} = F(t)
%
%   [M]     The mass matrix, sparse format, mframe3_femlab.mat
%   [K]     The stiffness matrix, sparse format, kframe_femlab.mat
%   [C]     The damping matrix.  Chosen to be the proportional matrix
%
%           [C]=\alpha [M] + \beta [K]
%
%
%
close all
clear all
%
global M C K Minv P Omega Node DOF Forcetime
%
load kframe3v3_femlabsym
load mframe3v3_femlabsym
%
%   set the parameters
%
alpha=0;
beta=0;
t0=0.0;
tf=1.e-2;
tspan=[t0 tf];
P=1.0;
Forcetime=(tf-t0)/20;
Omega=2*pi/Forcetime;
Node=13;
DOF=2;
%
%   Factor the mass matrix
%
K=full(Ke);
M=full(De);
[m,n]=size(K);
Minv=inv(M);
%
%   create the proportional damping matrix
%
C=alpha*K+beta*M;
%
%   integrate the equations of motion
%
y0=zeros(2*m,1);
options = odeset('AbsTol',1.e-7,'RelTol',1.e-3);
%
%       Fixed Time Steps
%
 tspan=(0:1000)*(tf/1000);
%
%       Automated Time Steps
%       tspan=[t0,tf];
[t,y]=ode23t(@firstODE,tspan,y0,options);
%
%   plot the requested DOF
%
%   create the force
%
if (Node < 3 )| (Node > 24)
    disp('The Frame Example Contains Only Free Nodes Numbered From 3 to 24')
    exit
end
if DOF == 1             % force in the x direction
    ii=(Node-2);
elseif DOF == 2         % force in the y direction
    ii=22+(Node-2);
elseif DOF ==3          % moment about a node
    ii=44+(Node-2);
end
plot(t,(y(:,ii))')
```

Figure E.19 Example *.m-file procedure frame3-int-1st.m* to integrate ordinary differential equations.

The *.m-file procedure* in Fig. E.19 illustrates the use of this syntax. The function $ODEFUN(T, Y)$ takes as input a single time T and column vector $Y \in \mathbb{R}^n$. The function $ODEFUN(T, Y)$ must return a column vector $F \in \mathbb{R}^n$ of the value of the derivatives $f(t, y(t))$ appearing in Eq. E.4.

There are several features of this *.m-file procedure* that should be noted. First, the procedure invokes the MATLAB general integration procedure $ODE23T$. The general integration procedure $ODE23T$ calls the function *firstODE.m* that appears in Fig. E.16. Finally, the calling sequence to the function $ODEFUN(T, Y) \equiv firstODE.m$ is very restrictive; only the time T and current value of the state Y are passed into the function *firstODE.m*. Other important variables are defined as *global* variables in the *.m-file procedure frame3-int-1st.m*. These are then available in the *.m-file function firstODE.m* as *global* variables. Some of the *global* variables that are passed in this way include M, C, K, $MINV$, for example.

REFERENCE

[E.1] The Mathworks, http://www.mathworks.com.

Index

Acceleration, 1, 8
  angular, 23
  frequency-response function, 103
  relative to inertial reference frame, 22
Accelerometer, 103–104, 581, 584
Active structures, 617–646. *See also* Smart structures; Piezoelectricity
Admissible function(s), 43, 228
Admissible vector, 569
Algebraic eigenproblem, numerical solution of, 469–498
  case study, of planar frame structure, 496–498
  combined matrix transformation and vector iteration method
    Householder-QR-Inverse Iteration (HQRI) Method, 470–471
  matrix transformation methods, 470, 483–496
    Automated Multilevel Substructuring (AMLS), 564–570
    Givens Rotations, 470, 485
    Householder Method, 470, 485–486, 489
    Householder-QR Method, 486–489
    Jacobi Rotations, 470, 485
    Lanczos Eigensolver, 470–471, 489–497. *See also* Lanczos Eigensolver Method
    QR Method, for symmetric eigenproblems, 470, 483–489, 496
  polynomial root-finding methods, 470
  vector iteration methods, 470–483
    Direct Iteration Method, 470, 474
    Forward Iteration Method, 474
    Inverse Iteration Method, 470–475
    Inverse Iteration with Spectrum Shift, 479–480, 497
    Method of Vianello and Stodola, *see* Algebraic eigenproblem ... , Inverse Iteration Method
    Subspace Iteration Method, 470, 480–483, 496
Algebraic eigenvalue problem, 249, 251, 283, 326. *See also* Eigenvalue problem
  generalized, 469, 694
  generalized state-space form, 308
  standard, 469, 693–694
  transformation of generalized form to standard form, 695–696
Amplitude
  modal, of steady-state vibration of viscous-damped MDOF systems, 331
  of complex response, 93
  of free vibration of underdamped SDOF systems, 63
  of steady-state response of undamped SDOF system, 82
  of steady-state response of underdamped SDOF system, 91, 94
Analysis, 1. *See also* Analytical model
Analytical mechanics, 22, 218, 379. *See also* Work-energy methods
Analytical model, 2–7
Antiresonance frequency, 272, 607
Apollo Saturn V, 4–5
Argand plane, *see* Complex plane
Assumed mode(s), 43, 532. *See also* Component modes; Shape function(s)
  fixed-interface, 538
Assumed-Modes Method, 41–50, 228–238
  examples of use of, 229–236
    axial deformation, 229–231
    bending of Bernoulli-Euler beam, 233–236
  for model reduction of MDOF system, 300–301
  global, 417–418
  procedure for applying, 231–233
Assumed-modes model, 228–238.
  modes and frequencies based on, 258–264
Automated Multilevel Substructuring (AMLS) Method, 564–570
Average Acceleration Method, 155–159, 178–181
Avitabile, Peter, 16
Axial deformation
  assumptions, 368
  of a straight, slender member, 367–370
  of a uniform rod, 44–46

Base excitation, *see also* Earthquake excitation
  of viscous-damped MDOF system, 214–216
  of viscous-damped SDOF system, 98–101
    absolute motion, 98–101
    relative motion, 98–99
Basis vectors, 451
Bathé, K-J. and Wilson, E. L., 471
Beat phenomenon, 254–258
Bernoulli-Euler beam theory, 373–379, 385, 401
  assumptions of, 374–375
Bode plot, 90, 341
Boundary conditions
  for axial-deformation of a finite element, 420
  for finite element (FE) model, 445–446

Boundary conditions *(Continued)*
  for torsion element, 425
  for transverse deflection of a beam finite element, 421
  for vibration test articles, 590. *See also* Vibration test
  generalized, for Bernoulli-Euler beam, 402
  generalized, for Timoshenko beam, 384
  geometric, 42, 44, 228, 380, 382, 384, 400, 407
  natural, 228–229, 382, 384, 400, 407
  of bar undergoing axial deformation, 369–370, 389–390
  of beam undergoing transverse vibration, 376–377, 393–395
  of 1-dimensional continuous system, 404
  of torsion rod, 372–373, 380, 382
  prescribed, *see* Boundary conditions, geometric
  system displacements, 445
  system forces, 445
  redundant, 533
Boundary force(s), 560

Central Difference Method, 511–514
Characteristic equation, 57–58, 62, 66, 140, 249, 251, 283, 308, 390, 394, 396, 547
  roots of, 390, 394, 396, 547
Characteristic polynomial, 286, 288
  roots of, 286, 288
Circle-fit method, 107
Compatibility, interface displacement, 554
Complex frequency response
  of SDOF system with structural damping, 111–112
  of viscous-damped MDOF system in physical coordinates, 331–332
  of viscous-damped SDOF system to harmonic excitation, 93–99
  of viscous-damped SDOF system to periodic excitation, 193–195
  plot, of viscous-damped MDOF system in physical coordinates, 332. *See also* Nyquist plot
Complex frequency-response function, 332. *See also* Transfer function
Complex modes, 250, 307, 309–310, 502
  interpretation of, 310–316
  orthogonality of, 309–310
Complex numbers, 671–673
  algebra of, 94, 672–673
  polar form, 60–61, 94, 671
  rectangular form, 60–61, 94, 671
Complex plane, 60–61. *See also* Rotating vector; $\bar{s}$-plane
Component modal model, 534
Component modes, 532–543, 558
  attachment mode(s), 533, 539
    cantilever, 539, 542–543
  inertia-relief, 539–541
  residual-flexibility, 545, 559–560
  residual inertia-relief, 548–549
  constraint mode(s), 533, 536–538
    interface, 537, 566, 569
    redundant-interface, 538
  Krylov vectors, 533
  normal modes, 533–538
    antisymmetric, 536, 542
    flex, 536, 545
    fixed-interface, 534–535, 566, 569
    free-interface, 534–536, 560
    hybrid-interface, 534
    loaded-interface, 534, 536
    mass-additive, 536
    symmetric, 536, 542
  orthogonality of, 537, 541, 560
  rigid-body, 533, 536–538, 543, 547, 559–560
Component-mode synthesis (CMS), 531–570. *See also* Component modes; Substructure coupling
  coupling procedures, 549–570
  fixed-interface methods, 557–559
    Craig-Bampton Method, 557–559
    fixed-interface transformation, 557–558
    Hurty Method, 557
  free-interface methods, 557, 559–564
    Craig-Chang Method, 560–562
    Dual Craig-Bampton Method, 562–564
    MacNeal's Method, 559
    Rubin's Method, 559
  generalized coupling procedure, 536
  Lagrange-multiplier based, 534
  multilevel, 532, 564–570. *See also* Multilevel substructuring
Component-mode system model, 558
Consistent-mass matrix, 429, 461. *See also* Lumped-mass matrix; Matrix(ces), consistent-mass
Consistent nodal loads, 463. it *See also* Fixed-end forces
Constitutive equation, 368. *See also* Force-deformation behavior; Piezoelectricity, piezoelectric materials
Constraint, 34–35
  matrix, 550
Constraint equations, 220, 238–240. *See also* Lagrange multipliers
  for interface reaction forces, 561
  in matrix form, 447–448, 562
  interface displacement, 550, 552–553
  interface, in component modal coordinates, 554
  used to reduce DOFs, 447–451
Continuous system(s), 4, 42, 365–413
  free vibration of, 388–409
  mathematical models of, 367–385

INDEX 717

Convergence
    of inverse iteration to fundamental eigenvalue, 475–476
Convolution integral, 125–126. *See also* Duhamel integral
Coordinates, component
    boundary, 533–534
    component-mode, 555
    component physical, 550
    excess boundary, 533
    flexible-body, 540
    generalized, 533
    interface displacement, 553
    interior, 533
    redundant boundary, 533
    residual-flexibility, 560
    rigid-body, 533, 540
    transformation from modal coordinates to physical coordinates, 534
Coordinates, system
    dependent generalized, 556
    independent generalized, 556
Coupling
    inertia, 218
    stiffness, 214
Craig–Bampton Method, 557–559

Damping
    coefficient of viscous, 27
    Coulomb, 70–72
    equivalent viscous, 107–110
    modal, 327, 580
    numerical, 521
    of linear viscous dashpot, 27
    proportional, 303, 313–314
    Rayleigh, 303
    structural, 111–112
    substructure, 532
    uncoupled, in MDOF systems, 302–307
Damping factor, 56–57, 62
    experimental determination of, 74–77, 608–609
    from complex eigenvalue, 310, 589
    Kennedy-Pancu Method for experimental determination of, 609–610
    modal, 303–304, 327
Damping matrix for modal damping, 304–307
    augmented, 305–307
d'Alembert force, 22. *See also* Inertia force
Deflection diagram, 38. *See also* Deformation diagram
Deformation diagram, 38, 40
Degree(s) of freedom (DOF), 4
    active system, 445
    constrained system, 445
    effective, in modal test, 580
Design spectra, plot of, for earthquake analysis, 656
Difference equation, 506

Differential equation, eigenvalue
    for axial free vibration, 389
Differential equation(s) of motion, *see also* Equation(s) of motion
    partial, for axial deformation of a linearly elastic bar, 369
    partial, for axial deformation of a piezoelectric rod, 626
    partial, for axial deformation of a piezoelectric stack, 634
    partial, for torsional deformation of a linearly elastic rod, 372, 382
    partial, for transverse motion of a composite piezoelectric laminated plate, 646
    partial, for transverse motion of a piezoelectrically actuated beam, 641
    partial, for transverse vibration of a beam with constant axial force, 379
    partial, for transverse vibration of a Bernoulli-Euler beam, 376
    partial, for transverse vibration of a Timoshenko beam, 384, 400
Digital signal processing, 594, 597–600
    aliasing, 594, 598
    assumptions, 598
    calculation of coherence functions, 594
    calculation of frequency-response functions, 594
    calculation of modal parameters, 594
    coherence, 603–604
    leakage, 594, 600–601
    noisy signals, 594, 601–604
    Nyquist sampling theorem, 594, 598
    sampled data, 594, 598
    stages of, 598
    time-domain windows, 600
Direct Stiffness Method, 438–444
    applied to component-mode synthesis, 556
    implemented in *ISMIS*, 460
Discrete-parameter models, 4. *See also* Lumped-parameter models
Displacement coordinate(s), 34. *See also* Coordinates, component; Generalized coordinates; Virtual displacement
    active system, 445, 447
    component generalized, 533
    elastic deformation, 451–453
    notation, for 3-dimensional frame element, 429
    planar truss element, referred to element axes (ECE), 430–432
    planar truss element, referred to global axes (ECG), 430–432
    plane frame element, referred to element axes (ECE), 434
    plane frame element, referred to global axes (ECG), 434
    rigid-body, 451–453

Displacement coordinate(s) *(Continued)*
  system, 438
    constrained, 445
    dependent, 447
    3-dimensional truss element, 435
Displacement method, 430. *See also* Force method
Displacement transformation matrix, 436, 438
  for Dual C-B Method, 563
  for planar truss element, 432–434
  for plane frame element, 434–435
  for 3-dimensional truss element, 435–436
  multilevel substructuring, 569
  used to reduce DOFs, 447
  used to transform element force vector and element mass and stiffness matrices, 436
Displacement vector, 214
Distortion, of accelerometer signal
  amplitude, 103–104
  phase, 103–104
Distributed-parameter systems, *see* Continuous system(s)
Divergence, 68–70
Duhamel integral
  for earthquake response, 652
  for undamped SDOF system, 125–128
  for underdamped SDOF system, 127
Dynamic load factor, 118
Dynamic response, *see* Response of . . .
Dynamics
  of particles, 21–23
  of rigid bodies in plane motion, 23–24

Earthquake excitation, 14, 96
Earthquake participation factor, 658
  modal, 661
Earthquake response of structures, 650–665
  based on MDOF models, 660–665
  based on SDOF models, 652–660
Eigenfunction(s), 390, 394, 396–397. *See also* Mode shape(s)
Eigenpair(s), 309, 326, 480
Eigensolvers
  Automated Multilevel Substructuring (AMLS) Method, 564–570
    AMLS algorithm, 565
    Block Lanczos Algorithm, 564
Eigenvalue problem, *see also* Algebraic eigenvalue problem; Differential equation, eigenvalue; Natural frequencies and mode shapes
  generalized, 469
  ordinary differential equation (ODE)
    for axial vibration of a uniform bar, 389
    for transverse vibration of a Bernoulli-Euler beam, 392, 402
    for transverse vibration of a thin, flat plate, 407
    for transverse vibration of a Timoshenko beam, 401

reduced-order, 300
semidefinite, 268
standard, 469
state, for undamped system with rigid-body modes, 317–320
state, for viscous-damped system with rigid-body modes, 320–321
tridiagonal, 493
Eigenvalue(s), 249, 283. *See also* Natural frequency(ies)
  complex, 308–310
  of uniform bar undergoing axial vibration, 389
  state-space, interpretation of, 310
Eigenvalue separation property, 301–302
Eigenvector expansion, 475
Eigenvector(s), 250, 283
  complex, 309. *See also* Complex modes
  generalized, 316–321
  Lanczos, 493
  state, 319
  state-space, interpretation of, 310–317
Elastic–plastic behavior, 177
El Centro earthquake, 653
Energy
  kinetic, 7
    of bar undergoing axial deformation, 230
    of beam, including rotatory inertia, 383
    of beam undergoing transverse deflection, 234
    of continuous system, 219–220, 230
    of finite elements and system, 441
    of particles, 22, 221
    of rigid bodies in plane motion, 24
    of system of substructures, 551, 554
    of torsion rod, 381, 425
    used to transform element mass matrix, 436
  potential, 7, 43, 219–221, 380
    of system of substructures, 551, 554
  strain, 25, 219
    of bar undergoing axial deformation, 25, 43, 229
    of beam, including bending strain energy and shear strain energy, 383
    of beam undergoing transverse deflection, 26–27, 43, 233
    of finite elements and system, 440
    of linear spring, 25
    of torsion rod, 25–26, 380–381, 425
    used to transform element stiffness matrix, 436
Energy methods, 379. *See also* Analytical mechanics; Work-energy methods
Equation(s) of motion, *see also* Differential equation(s) of motion
  for base excitation, 98
  for component residual-flexibility, 560
  in complex form, 93, 98
  in generalized state-space form, 307–308, 588
  in Laplace domain, 138

in principal coordinates, 270, 302, 327
linearized, 32
modal, 327, 343
  of undamped MDOF systems with rigid-body modes, 353
nonlinear, 32, 227
of component in generalized coordinates, 534
of component in physical coordinates, 533
of hybrid coupled-system, 552
  reduced-order, 563
of lumped-mass model, in partitioned-matrix form, 449
of MDOF system, 226–227
of MDOF system, in matrix form, 211, 231, 233, 236, 238, 510
of prototype SDOF system, 56
of system of components
  in generalized coordinate form, 555
  in hybrid form, 552
of thin, flat plate, 406
of 2-DOF system, in matrix form, 248–249
of undamped SDOF system, 9, 29, 41, 45–46, 58
of undamped SDOF system with harmonic excitation, 82
of undamped 2-DOF system with harmonic excitation, 270
of viscous-damped SDOF system, 62, 87, 93, 148, 159
  with ideal step input, 117
reduced-order system, 447
uncoupled, 270, 304, 327
Equilibrium
dynamic, 22, 24. *See also* d'Alembert force
static, 29
Equivalent viscous damping, 107–110
Euler's formula, 58, 60, 63
Expansion theorem
in terms of eigenfunctions of continuous systems, 404–405
in terms of eigenvectors of MDOF systems, 295–296, 475
Experimental determination of SDOF system parameters
damping factor
  by half-amplitude method, 75–77
  by half-power method, 106–107
  by logarithmic decrement method, 74–75
  by use of frequency-response data, 106–107
time constant, 77
undamped natural frequency
  by free-vibration method, 73–74
  by static-displacement method, 72–73
  by use of frequency-response data, 105–106
Experimental modal analysis, 12, 579–614. *See also* Vibration testing
Extended Hamilton's Principle, 219, 223. *See also* Hamilton's Principle

applied to flexure of a beam including shear deformation and rotatory inertia, 382–385
applied to piezoelectric structures, 627–630
applied to torsion of a rod with circular cross section, 379–382

Fast Fourier transform (FFT)
algorithm, 202
MATLAB computation of, 203–204
Finite-difference expressions for derivatives, 511
Finite element computer program(s), 6, 462
Finite Element Method (FEM)
element lumped-mass matrix, for beam element, 429–430
element stiffness and mass matrices and element force vector
  for axial deformation of a uniform bar element, 419–421
  for planar bending of a uniform Bernoulli-Euler beam element, 421–424
  for 3-dimensional motion of a uniform frame element, 426–429
  for torsion of a uniform rod element, 425–426
Finite element models
consistent-mass, 453
ISMIS-generated, 455–462
lumped-mass, 453
used in multilevel substructuring, 532
Finite element solutions for natural frequencies and mode shapes, 453–462
Fixed-end forces, 463
Flexibility, *see also* Matrix(ces), flexibility
interior, 538
residual, 539, 541
Floating-point operations (flops), 512
Flutter, 68
Force-deformation behavior
of elastic elements, 21–27
  axial-deformation member, 25
  cantilever beam, 26
  linear spring, 24–25
  simply supported beam, 26–27
  torsion rod, 25–26, 33
of viscous-damping element, 21, 27
Forced vibration of SDOF system, 10–12. *See also* Response of ...
Force method, 430. *See also* Displacement method
Force transmissibility, 96–98
Force vector. *See also* Load vector
effective, 216
elastic, 354
element, referred to global coordinates, 443
element, referred to local element coordinates, 443
interface constraint, 555
modal, 270
rigid-body inertia, 354

Force vector *(Continued)*
  self-equilibrated, 540, 543
  system, 443
Force vector polygon, 87–88, 95
Forcing frequency, 82, 88, 104, 581
Fourier integral
  to represent nonperiodic excitation, 195–197
Fourier series, 594
  complex, 189–195
  real, 184–189
Fourier transform pair(s), 196–199, 596
  table of, 197
Fourier transform(s), 594
  direct, 596
  discrete, 200–202
  fast, *see* Fast Fourier transform
  inverse, 196, 201, 596
  of force and motion experimental data, 592, 595–597
  of rectangular pulse, 197, 596–597
Free-body diagram, 8, 22, 24, 28, 30, 32, 33, 69, 213–217, 368, 370–371, 374–375, 378
Free vibration of continuous systems, 388–409
  axial vibration, 389–391
  torsional vibration, 391–392
  transverse vibration of Bernoulli-Euler beams, 392–397
  transverse vibration of thin, flat plates, 405–409
Free vibration of MDOF systems, 281–321
Free vibration of SDOF systems, 9–10, 56–80
  undamped systems, 9–10, 58–61
  viscous-damped systems, 61–66
    critically damped ($\zeta = 1$), 62, 64–65
    overdamped ($\zeta > 1$), 62, 65–66
    underdamped ($\zeta < 1$), 62–64, 74–77
Free vibration of 2-DOF systems, 250, 253–254
Frequency(ies), *see also* Forcing frequency; Natural frequency(ies); Period
  fundamental, Rayleigh quotient used to approximate upper bound to, 405
  repeated, 408
Frequency domain, 195, 199. *See also* Fourier integral; Time domain
Frequency-domain representation of data, 595–602
Frequency ratio, 83
  modal, 331
Frequency response
  calculated by Automated MultiLevel Substructuring (AMLS) Method, 564–570
  calculated by mode superposition, 268–272
  of undamped 2-DOF system, 268–272
Frequency-response function(s) (FRFs)
  acceleration, 103, 581, 584–586
    of MDOF systems, based on real normal modes, 588
  complex, 94–96, 98–99, 142, 198–199
    of MDOF system, imaginary part, 335, 587
    of MDOF system, in physical coordinates, 331
    of MDOF system, in principal coordinates, 331
    of MDOF system, real part, 335, 587
    relationship to unit impulse-response function, 199
  cross, 581
  drive-point, 581, 609–610
  effect of signal noise on, 601–604
  estimation, from digitized time-domain data, 600–604
  H_1, H_2, and H_v FRF estimates, 604
  magnitude of, 587
  mobility, 584, 587, 610
    of MDOF systems, based on real normal modes, 587–588
    of MDOF systems, based on complex modes, 588–590
    of MDOF systems, based on real normal modes, 587–588
    of SDOF systems with structural damping, 112
    of SDOF systems with viscous damping, 88–90, 198–199
    of undamped SDOF system, 83
    of undamped 2-DOF system, 271
  partial-fraction format, 586
  phase angle, 587
  pole-residue format, 586–588
  receptance, 581, 584, 587
    of damped MDOF systems, based on complex modes, 590, 605–606
    of MDOF systems, based on real normal modes, 587
  single-input-single-output (SISO), 600
Frequency-response function (FRF) matrix
  column of, 583
  row of, 583
Frequency-response plots. *See also* Frequency-response functions; Bode plot; Nyquist plot
  acceleration, Bode plot for SDOF system, 585–586
  circle, 608
  complex, of viscous-damped 2-DOF systems
    imaginary part, plotted versus frequency, 337, 341, 586
    real part, plotted versus frequency, 336, 340, 586
  mobility, 606, 608
    Bode plot for SDOF system, 585
  Nyquist, of MDOF system, 609–610
  Nyquist, of 2-DOF system with closely spaced frequencies, 339–340
  Nyquist, of 2-DOF system with widely separated frequencies, 335–336
  of absolute-motion response to base excitatiion, 98

INDEX **721**

of relative-motion response to base excitatiion, 99
of transmissibility, 98
receptance, Bode plot for SDOF system, 585
receptance, of simulated 3-DOF beam, 582
used to determine parameters of viscous-damped SDOF systems, 105–107
Friction
  coefficient of kinetic, 71
    Coulomb, 71–72
    sliding, 71
Fundamental frequency, 254, 453
Fundamental period, 254

Generalized coordinates, 36, 43, 220–222, 228–229, 232, 234
Generalized displacement vector, *see* Generalized coordinates
Generalized-parameter model, 42. *See also* Assumed-Modes Method
Generalized parameters
  consistent mass coefficient, 231, 233–234, 236
    for Bernoulli-Euler beam, 422, 424
    for torsion element, 426
  external force(s), 36–37, 49–50, 220, 222, 231–234
    for Bernoulli-Euler beam, 422
    for torsion element, 426
  geometric stiffness coefficient, 48–50, 236
  mass coefficient, 47, 50
  stiffness coefficient, 47–48, 50, 230–231, 233–235
    for Bernoulli-Euler beam, 422
    for torsion element, 426
  viscous damping coefficient, 47, 50, 237
  viscous damping matrix, 297
Generalized state-space form, 308
Givens transformation, 498–499
Golub, G. H. and Van Loan, C. F., 471
Gram–Schmidt orthogonalization procedure, 294, 476, 480
Ground vibration test, 12–13
Guyan–Irons reduction, 431, 459–461
Guyan Reduction Method, 451, 461. *See also* Guyan-Irons reduction

Half-amplitude method, 75–77
Half-power method, 106–107
Hamilton's Principle, 218–219, 223, 238. *See also* Extended Hamilton's Principle
  applied to a torsion rod, 379–382
Hammer, impact, 583
Harmonic excitation
  of undamped SDOF systems, 82–86
  of undamped 2-DOF system, 269–271
  of viscous-damped SDOF systems, 87–101, 126

Householder transformation, 485–486, 498

Impedance function for viscous-damped SDOF system, 138
Impulse-response function
  for undamped SDOF system, 123–125
  for underdamped SDOF system, 124, 126
  relationship to frequency-response function, 199
Inertia force, 2, 22–24
Inertial reference frame, 22
Inertia relief, 539–541
Initial conditions, 9, 250, 326–328
  in modal coordinates, 327–328
Inner product, 489–492
Integration of first-order ODEs, 159–171
  explicit multistep methods, 166–168
    Adams-Bashforth method, 167–168
  for MDOF systems, 500, 504–505
  implicit multistep methods, 169–170
    Adams-Moulton method, 170
  linear multistep methods, 165–170
  MATLAB example of RK2 and Taylor series integration, 163–164
  Runge-Kutta methods, 161–165
  state-space form, 159
  Taylor series methods, 160–164
Integration of second-order ODEs
  Average Acceleration Method, 155–159
  Average Acceleration Aethod, to integrate nonlinear equations, 178–181
  based on interpolation of excitation function, 148–155
    MATLAB example, 154–155
    piecewise-constant interpolation, 149, 154–155
    piecewise-linear interpolation, 149–155
  for MDOF systems, 500, 510–516
Interfaces, substructure, 550
International Modal Analysis Conference (IMAC), 614
ISMIS, 15, 453
  commands, 456, 458–460. *See also* MATLAB commands
  plane frame element, 459–460

Jordan form, 316, 319

Kennedy-Pancu method, 609–610. *See also* Circle-fit method
Kinematics of deformation, 21, 39. *See also* Deformation diagram; Strain-displacement behavior
Kirchoff–Love theory for thin, flat plates, 406
  assumptions of, 406
Krylov
  subspace(s), 489–492, 494
  basis vector(s), 490, 494, 533

Lagrange's Equations, 5, 220–223
    applied to constrained systems, 238–240
    applied to continuous models, 228–238. *See also*
        Assumed-Modes Method
    applied to lumped-parameter models, 223–227
    applied to substructure coupling in generalized
        coordinates, 555
    applied to substructure coupling in physical
        coordinates, 551
Lagrange multipliers, 238–240, 549, 551, 554
Lagrange multiplier vector, 552
Lagrangian function, 219, 554
Lagrangian mechanics, 22. *See also* Analytical
    mechanics
Lanczos Eigensolver Method, 489–496
    with full reorthogonalization, 494–495
    without reorthogonalization, 494–495
Lanczos vectors, 492–493
Laplace domain, 139
Laplace transform(s), 674–681
    complex shifting theorem, 676–677
    inverse, 676
    method of partial fractions, 677–680
        for system with higher-order poles, 678–679
        for system with quadratic poles in the
            denominator, 679–680
        for system with simple poles, 678
    one-sided Laplace transform, 137, 674–675
    pairs, table of, 681
    used to solve linear differential equations,
        138–142, 675
Linear algebra, 682–696
    definition of matrix operations, 685–688
    eigenvalue problem, *see* Algebraic eigenvalue
        problem
    matrix decompositions
        Cholesky decomposition, 692–693
        LU decomposition, 691–692
        QR decomposition, 690
        Singular value decomposition, 689–690
    vector spaces and linear operators, 682–685
        definitions, 683–685
Load operator, for Average Acceleration Method,
    517
Load vector, 211, 214
    for Bernoulli-Euler beam element, 423
Loading, 1
    dynamic, 1
    element, 443–444
    prescribed, 1
    random, 1
    system, 443–444
Locator information, 442–443, 446
Locator matrix, 439–441
Locator vector(s), 439, 460
Logarithmic decrement, 74–75. *See also*
    Half-amplitude method

LU factorization, 511–512, 514
Lumped-mass matrix, *see also* Consistent-mass
    matrix
    for beam element, 429–430, 449
    for finite element models, 448–451
    in partitioned-matrix form, 449
Lumped-mass models, 461–462
    frequencies of, 462
Lumped-parameter models, 24–41. *See also*
    Discrete-parameter model
    application of Lagrange's Equations to, 223–227
    application of Newton's Laws to, 27–34,
        212–218
    application of the Principle of Virtual
        Displacements to, 39–41
    damping element of, 27
    elastic elements of, 24–27
    modes and frequencies based on, 264–266

Magnification factor(s), 83, 85, 88–90
Mass-spring-dashpot SDOF system, 28–31
Mathematical model(s), 2–9
    N-DOF, used to represent experimental modal
        model, 580
    of MDOF systems, 211–247
    of SDOF systems, 28–34
MATLAB, 15–16, 154, 163, 175
    array operations, 700
    commands, 456. *See also* ISMIS, commands
    conditional expressions, 703
    eigensolution command, 456, 470
    graphical output, 708–711
    integration of ordinary differential equations,
        711–713
    introduction to the use of, 697–713
        input and output operations, 700–703
        syntax, commands, and matrix algebra,
            698–703
    matrix decomposition operations, 700
    writing functions and .m-files, 705–708
Matrix(ces)
    amplification, of Average Acceleration Method,
        518
    assembly of system matrices, 438–444. *See also*
        Direct Stiffness Method
    augmented free-interface transformation, 560
    banded, 511
    boolean, 553
    B-orthogonal, 490–491
    B-symmetric, 489, 492
    cantilever flexibility, 538
    component mass, 557
    component-mode, 533, 554. *See also* Component
        modes
    component stiffness, 557
    connectivity, of substructures, 550, 553

consistent mass, 231, 454, 461. *See also* Consistent-mass matrix
   for Bernoulli-Euler beam element, 423
constraint-mode, 557
Craig-Bampton transformation, 557. *See also* Component-mode synthesis
damping
   augmented modal, 503
   generalized, 501
   modal, 503
   nonproportional, 501–502
   proportional, 502
   Rayleigh, 502–503
displacement transformation, *see* Displacement transformation matrix
dynamical, 284, 472
eigenvalue, 294
element mass, 419, 436–438. *See also* Transformation, of matrices . . .
   expanded to system DOFs, 441
   for Bernoulli-Euler beam, 423
   for torsion element, 426
   for uniform axial-deformation element, 421
   for uniform 3-dimensional frame element, 428
element stiffness, 419, 436–438. *See also* Transformation, of matrices . . .
   expanded to system DOFs, 440
   for Bernoulli-Euler beam, 423
   for torsion element, 426
   for uniform axial-deformation element, 421
   for uniform 3-dimensional frame element, 427
factorization, 511. *See also* LU factorization
fixed-interface mode, 557
fixed-interface transformation, 557–558
flexibility, 349, 355, 544–545
   cantilever component, 538, 542
   constrained component, 541
   elastic, in inertia-relief form, 541
   elastic, in mode-superposition format, 544–545
   pseudoinverse of singular component stiffness matrix, 541
   residual, 545
inertia-relief, 354
inertia-relief attachment mode, 540, 544
inertia-relief projection, 540, 543
interface constraint, 550
interface constraint-mode matrix, 537
locator matrix and locator vector, 439
lumped-mass, for beam element, 429–430, 454
mass, 211, 214
   modal, 544
   reduced-order hybrid system, 564
   reduced-order system, 567, 570
modal, 294, 501
   complex, 589
partitioned multilevel-substructure system, 568

positive definite, 282, 489
positive semidefinite, 282
reduced system mass, 451, 568
reduced system stiffness, 449, 568
single-level substructure coupling, 567
singular, 268, 282
skyline, 511
stiffness, 211, 214, 230
   for Bernoulli-Euler beam element, 423
   modal, 544
   reduced-order hybrid system, 564
   reduced-order system, 567, 570
sweeping, 477–478
system mass, 441
   reduced-order, 449, 459, 558
   uncoupled, 551, 555
system stiffness, 440
   reduced-order, 449, 459, 558
   uncoupled, 551, 555
transformation, *see* Displacement transformation matrix
tridiagonal, 493–494
viscous damping, 211
Matrix deflation, 476–477
Mechanics of deformable solids, 5. *See also* Stress-strain behavior; Strain-displacement behavior
Mesh, 505–507
Modal matrices; modal vectors
   eigenvalue matrix, 294
   generalized damping matrix, 303
   modal coordinates, *see* Principal coordinates
   modal damping matrix, 327
   modal force vector, 270, 297, 327
   modal mass matrix, 270, 294, 297, 302, 327
   modal matrix, 294, 326
   modal participation vector matrix, 606
   modal stiffness matrix, 270, 294, 297, 302, 327
   modal vector matrix, complex modes, 606
   principal coordinate vector, 270
Modal parameter estimation, 604–613
   categories of, 605
   basic equations for, 605–607
Modal parameters
   modal damping factor, 327
   modal mass, 284, 326, 402
   modal stiffness, 284, 326, 403
Modal response, 342, 344
Modal static displacement, 348
Modal testing, 581
   hardware, 580
   multiple input–multiple output (MIMO), 582
   multiple input–single output (MISO), 582
   simulated, 581–584
   single input–multiple output (SIMO), 582
Mode-acceleration method, 349–359, 546

724 INDEX

Mode-acceleration solution, 343
 for internal stresses, 351
 for internal stresses in MDOF systems with rigid-body modes, 355
 for response of an undamped MDOF system, 349–351
 for response of an undamped MDOF system with rigid-body modes, 354–359
Mode-displacement method, 342–349
Mode-displacement solution, 342
 for internal stresses, 351, 353
 for response of an undamped MDOF system, 343–344
 for steady-state response of an undamped MDOF system, 344–347
 for stresses in systems with rigid-body modes, 353
 for undamped MDOF systems with rigid-body modes, 353
Modeling, 1–7. See also Model(s)
Model reduction, 298–301, 447–451, 531. See also Assumed-Modes Method; Guyan-Irons reduction; Rayleigh Method; Rayleigh–Ritz Method
Model(s), 2–12. See also Analytical model; Discrete-parameter models; Lumped-parameter models; Mathematical model(s)
 beam-rod, 4–5
 component-mode, 532, 534
 component-mode system, 558
 continuous, see Model(s), distributed-parameter
 distributed-parameter, 4, 211
 finite element, see Finite Element Method (FEM)
 lumped-mass, 4
 modal, based on vibration test, 592
 prototype MDOF, 211
 prototype SDOF, 7–12, 29
 reduced-order system, 532
Model verification, 612, 614
Mode number, 391
Modes
 complex conjugate pairs, 590
 complex, in state-space form, 589
Mode shape estimation, 612
 quadrature peak-picking method, 612–613
Mode shape(s), 249–250, 283. See also Eigenvector(s); Mode shape estimation
 distinct-frequency case, 284–289
 generalized state rigid-body, 318
 linearly independent, 290, 293. See also Gram–Schmidt procedure
 normalized, 283–284
 of MDOF systems with distinct roots, 284–289
 of MDOF systems with repeated roots, 289–294
 of simulated 3-DOF cantilever beam, 582
 of uniform bar undergoing axial vibration, 390–391
 of uniform beams undergoing transverse vibration, 394, 397
 repeated-frequency case, 289–294
 rigid-body, 290–293, 317–321
 symmetric and antisymmetric, 295, 395
Mode superposition for undamped systems with rigid-body modes, 353–359
Mode-superposition method, 268–272, 296–297, 302–303, 500. See also Mode-acceleration method; Mode-displacement method
 employing complex modes of the damped structure, 589–590
 employing modes of the undamped structure, 326–327
 for dynamic response of MDOF systems, 325–359
 for free vibration of undamped MDOF systems, 328–330
 for frequency-response analysis of MDOF systems with modal damping, 330–341
 for FRFs based on experimental data, 580–581
Mode-superposition principle, 579
Modulus of elasticity, 368, 374
 shear, 371
Moment-curvature equation, 375
Moment of inertia, 375
 polar, 371, 381. See also Torsion, constant
Multilevel substructuring, 533, 563–565, 567–570
 AMLS case study, 570
 AMLS Method, 564–565, 567–570
Multiple-degree-of-freedom (MDOF) systems, 211–240, 281–321, 325–359, 415–571

Natural frequencies and mode shapes, see also Algebraic eigenproblems, numerical solution of
 based on assumed-modes models, 258–264
 based on finite element models, see Finite element solutions …
 based on lumped-mass models, 264–268
 of continuous systems, 402
 of structures with arbitrary viscous damping, 307–321. See also Complex modes
 of 2-DOF systems, 249–268
 of 2-DOF systems with rigid-body modes, 266–268
 of undamped systems with rigid-body modes, 317–320
 of uniform beams undergoing transverse vibration, 393–397
 of viscous-damped systems with rigid-body modes, 320–321
Natural frequency(ies)
 closely spaced, 254, 257
 damped circular, of SDOF system, 63
 damped modal, of MDOF system, 342

of structure with arbitrary viscous damping, 310, 589
of undamped MDOF systems, 249, 283
of uniform beams in transverse vibration, 394, 396
undamped circular, of MDOF system, 249
undamped circular, of SDOF system, 9, 56, 59, 72–74
Natural mode(s), 283. *See also* Mode shape(s); Natural frequencies and mode shapes
of continuous systems, properties of, 401–405
Neutral axis, 374
Newmark-β Method, 513–514, 525
Newtonian mechanics, 21, 218. *See also* Vectorial mechanics
Newton's Laws, 5, 21–23
applied to continuous systems, 367–379
axial deformation, 367–370
torsion of rods with circular cross section, 371–373
transverse vibration of Bernoulli-Euler beams, 373–379
applied to lumped-parameter models, 27–34
applied to undamped SDOF systems, 8
Nodal loads, *see* Consistent nodal loads; Fixed-end forces
Node lines, of flat plate, 408
Node point, 391
Nonlinear equation of motion, 32
Nonlinear MDOF problems, 514–515
Nonlinear SDOF systems, 171–181
first-order formulation of, 174–175
geometric nonlinearity, 171–173, 175
material nonlinearity, 173–174. *See also* Elastic–plastic behavior
MATLAB example, 175
second-order formulation of, 176–181
Nonperiodic excitation, response of SDOF systems to, 117–146
Normalization, of modes, 402–403
Normal-mode method, *see* Mode-superposition method
Normal modes, of continuous systems, 402
Numerical dissipation, resulting from numerical integration, 523
Numerical evaluation of response of SDOF systems, *see* Integration of first-order ODEs; Integration of second-order ODEs
Numerical integration for dynamic response of MDOF systems, 500–530
case study of, for plane frame structure, 525–527
criteria for evaluating algorithms, 523–525
first-order equations for, 504–505
mathematical framework for, 504–510
error, consistency, and convergence, 505–508
stability, 508–510
second-order methods for, 504, 510–516

accuracy analysis of, 522–523
amplitude accuracy of, 522
Average Acceleration Method, 514, 517–520, 522–524
Central Difference Method, 510–514, 525, 527
implicit, 513
Newmark-β Method, *see* Newmark-β Method
numerical dissipation due to, 521, 523
period elongation due to, 522–523
stability analysis of, 519–521
Taylor series approximation, 507, 511, 513
Wilson-θ Method, *see* Wilson-θ Method
stability of, *see* Stability of numerical integration algorithms
Nyquist FRF plot
for MDOF system with closely spaced frequencies, 339–340
for MDOF system with viscous damping, 332, 335–336, 339–340
for SDOF system with structural damping, 112
for SDOF system with viscous damping, 96, 586

Operator formulation of single-step numerical integration algorithms, 517–517–518
Orthogonal basis vectors, 489
Orthogonality, 270, 289, 326
of complex modes, 309–310
of component modes, 537, 541, 560
with respect to the mass distribution, 404
with respect to the mass matrix, 289, 293, 326
with respect to the stiffness distribution, 404
with respect to the stiffness matrix, 289, 326
of modes of continuous systems, 403–404
Orthogonal transformations, 485
Orthonormal modes, 404
Orthonormal vectors, 294
Overshoot, 118

Parameter estimation, 198
frequency-domain algorithms for, 200, 607
time-domain algorithms for, 200, 607
Parameter-estimation methods
global MDOF, 611–612
local MDOF, 609–611
SDOF, 607–609
Partial-fraction format, 142
Pendulum system, 31–32
inverted simple, 68–70
Period
of viscous-damped SDOF system, 63, 75
of undamped SDOF system, 10, 59
Periodic excitation, 184–195. *See also* Fourier series
steady-state response to, 187–189
Periodic function, 184–185
Periodic motion, 85

Phase angle
    modal, of steady-state vibration of viscous-damped MDOF systems, 331
    of free vibration of underdamped SDOF systems, 63
    of steady-state response of underdamped SDOF systems, 88-91
Pickrel, Charles, 16
Piezoelectricity, 617–646
    constitutive laws of linear, 620–624
    piezoelectric effect, 594, 617–618
        converse, 617–619
        direct, 617, 619, 621
    piezoelectric materials, 617
        barium titanate, 618–619
        constitutive law in one dimension, 620–621
        constitutive laws in three dimensions, 621–622
        constitutive laws reduced from three dimensions to one dimension, 623–624
        dielectric displacement in, 621
        electrical enthalpy in, 627
        piezoceramic, 618, 620, 625–626
    piezoelectric structures
        active beam models, 637–641
        active composite laminates, 641–646
        active strut, 630
        active truss models, 630–634
        application of Extended Hamilton's Principle to, 627–630. *See also* Virtual work
        application of Newton's Laws to, 624–627
        axial deformation of a piezoelectric rod, 625–627
        equations of motion for, *see* Differential equation(s) of motion
        finite element modeling of active strut, 635–637
        Lagrangian for, 627
        piezoceramic stack, 630–632
        sensors and actuators, 623, 631–632
        symmetric bimorph, 643–646
Plane motion, 23–24
Pole-residue format, 142
Poles, 140–142
Principal coordinates, 270, 294, 302, 325–327
Principle of Virtual Displacements, 21–22, 37
    applied to continuous models, 41–50, 218, 220. *See also* Assumed-Modes Method
    applied to lumped-parameter models, 34–41
Proportional damping, 303, 313. *See also* Rayleigh damping
Prototype SDOF system, 9, 29, 56
Pseudostatic response, 343, 349, 354, 560
Ramp response of undamped SDOF systems, 11–12, 121–123
Ratio calibration test, 580, 594
Rayleigh damping, 303
Rayleigh Method, 298–299

Rayleigh quotient, 298–299, 405, 473
Rayleigh–Ritz bounds, 455, 461
    for consistent-mass FE models, 458
Rayleigh–Ritz Method, 299–302, 569
Real modes, 250
Recurrence relation, 492, 494
Recursion relation
    for Average Acceleration Operator, 517
    3-term, for generating Lanczos vectors, 492
Reference frame
    element, 430
    global, 430
Relative motion, 30–31
Residual flexibility, 606–607
Residual inertia, 606
Resonance, 86
Response of MDOF systems
    by the mode-superposition method, *see* Mode-acceleration method; Mode-displacement method
Response of undamped SDOF systems to harmonic excitation, 82–86
    steady-state response, 82–83
    total response, 84–85
    with excitation at resonance frequency, 86
Response of undamped SDOF systems to nonperiodic excitation
    ideal step input, 117–119, 140–141
    impulse loading, 123–124
    ramp loading, 121-123
    rectangular pulse loading, 119–121
Response of viscous-damped SDOF systems to harmonic excitation, 87–107
    steady-state response, 87–93. *See also* Complex frequency response; Steady-state response
    total response, 90–92
Response of viscous-damped SDOF systems to nonperiodic excitation
    Duhamel integral method for total response, 127
    Laplace transform solution for, 138–139
    unit impulse response, 124
Response spectra, 128–136
    for earthquake response analysis, 652–660
Response spectrum
    displacement, 654
    pseudoacceleration, 654
    pseudovelocity, 654
Ride quality, *see* Base excitation, of viscous-damped SDOF system
Rigid-body modes, 266–268, 288–293, 570
    separated from elastic modes by coordinate transformation, 451–453
Rigidity, flexural, 406
Ritz approximation, 493
Ritz basis vectors, 451, 480–481, 532
Ritz transformation, 533, 560, 564

Roots of the characteristic equation, 62, 64, 66, 283, 286, 288, 291. *See also* Poles
Rotating vector, 60–61, 63, 87, 93, 95
Rotatory inertia, 374, 383, 385
Rotatory inertia correction term, 401

Sampling, effects of, 201
Seismic transducer, 102
Shape function(s), 43–44, 46, 228, 232, 235
    for axial-deformation of a finite element, 420
    for transverse deflection of a beam finite element, 421–424
Shear correction term, 401
Shear deformation, 383, 385
Shear force, transverse, in beam, 374
Shear modulus of elasticity, 25
Signal conditioning, 594
Signals
    periodic, 596
    sinusoidal, 595
Simple harmonic motion, 10, 59–60
    amplitude of, 10, 59, 61
    phase angle of, 59, 61
Single-degree-of-freedom (SDOF) systems, 7–12, 19–205
    mathematical models of, 21–55
Slenderness ratio, effective, 401
Smart structures, *see* Active structures; Piezoelectricity
Spectral radius, 521
Spectrum, 479
    auto-power, of time-domain signal, 601–604
    cross-power, of time-domain signal, 601–604
    of force generated by impact hammer, 592–593
$\bar{s}$-plane, 66–67
Spring, 8, 24–27
Spring constant
    of lumped-parameter models, 24–27
Spring-mass oscillator, 8–11
Stability of motion, 66–70
    asymptotically stable, 67–68
    represented in the $\bar{s}$-plane, 67
    stable, 67–68
    unstable, 67–70. *See also* Divergence; Flutter
Stability of numerical integration algorithms, 508–510, 519–521. *See also* Numerical integration for dynamic response of MDOF systems
    absolute, 509
    A-stability, 509, 519–521
    conditional, 510, 512–513
    0-stability, 508–509
    spectral, 516, 520–521
    unconditional, 510, 513, 524
Starting transient, 91–92
Starting vector, 478

State-space form, 159. *See also* Generalized state-space form; Integration of first-order ODEs
State vector, 307
Static completeness, 546
Static condensation, 448–451
    transformation matrix, 449
Static displacement, 82–83
Steady-state response
    determined by complex frequency-response method, 93–96, 193–195
    experimental FRFs used to represent, 580
    in complex mode-superposition form, 589
    of undamped MDOF system, 344–347
    of undamped SDOF system, 82–83
    of undamped 2-DOF system, 271
    of viscous-damped MDOF system, 332
    of viscous-damped SDOF system, 87–92, 584
        amplitude, 87, 91
        magnification factor, 88–90
        phase (lag) angle, 87, 91
    to periodic excitation, 193–195
Step response
    of undamped SDOF system, 118
    of underdamped SDOF system, 118
    of viscous-damped SDOF system, 117–119
Stiffness coefficient, *see* Generalized parameters; Spring constant
Strain-displacement equation, 5, 368, 375
Stresses
    dynamic, by mode superposition, 351–352
    of undamped MDOF systems with rigid-body modes, by mode-acceleration method, 355
    of undamped MDOF systems with rigid-body modes, by mode-displacement method, 353
Stress-strain behavior, 5
Structural damping, 111–112
Substructure coupling, *see also* Component-mode synthesis; System assembly
    multilevel, 567–570
    single-level, 565–567
Substructure tree diagram, 565, 567–568
Superelements, 558
System assembly, 551–564
    in generalized coordinates, 551, 554–564
    in physical coordinates, 551–554
    procedures
        dual assembly, 531
        primal assembly, 531, 559

Tapered-beam finite element model, 462
Taylor series, *see* Integration of first-order ODEs
Test-analysis model (TAM), 461
Testing, *see* Vibration testing
Time constant, 77
Time domain, 195, 199

Time-domain representation of data, 595–600. *See also* Frequency-domain representation of data
Timoshenko beam theory, 382–385
Torsion
  assumptions, 371
  constant, 25, 381
  of rod with circular cross section, 371–373
  of shaft-disk system, 33–34
  of uniform circular rod, 25–26
Total response
  of undamped SDOF system, 84–85
  of underdamped SDOF system, 90
Transducer(s), 101, 594
  force and motion, used in vibration test, 593–594
  piezoelectric sensors and actuators, 623
Transfer function, 136, 139, 142. *See also* Complex frequency-response function; Laplace transform
Transformation, linear, from component modal coordinates to component physical coordinates, 555–556
Transformation matrix
  multilevel substructuring, 569
  single-level substructuring, 567
Transformation, of matrices and vectors from ECE reference to ECG reference, *see also* Displacement transformation matrix
  element force vector, 436
  element mass matrix, 436–437
    planar truss, 437–438
  element stiffness matrix, 436–437
    planar truss, 437
Transmissibility, *see* Force transmissibility
Transverse displacement, of beam, 374
Truncation, effects of, 201
Truncation error, local, 506–507

Unit impulse response, *see* Impulse-response function
Units used in structural dynamics, 667-670

Vectorial mechanics, 21. *See also* Newtonian mechanics
Vector-response plot, *see* Nyquist FRF plot
Vectors
  basis, 451
  linearly independent, 290, 295
Vibration
  forced, of undamped SDOF system, 10–12
  forced, of undamped 2-DOF systems, 268–275
  free, of undamped SDOF system, 9–10
  free, of undamped 2-DOF systems, 248–268
Vibration absorber, 272–273
Vibration isolation, 96–101. *See also* Base excitation . . . ; Force transmissibility
Vibration-measuring instruments, 101–104. *See also* Accelerometer; Transducer(s) Vibrometer
Vibration properties of MDOF systems, 281–321
Vibration test hardware, 590–594
  dynamic analyzer, 597
  excitation sources for, 591–593
    base excitation, 591
    electrodynamic shaker, 591–592, 600
    impact hammer, 591–592
  force and motion transducers, 593–594
  signal conditioning, 591, 598–599
Vibration testing, 1, 12. *See also* Experimental determination of SDOF system parameters; Experimental modal analysis; Ground vibration test
  boundary conditions used in, 590–591
    flexures, 590
    consideration of rigid-body modes in, 590
    fixed-base, 591
  impulse testing, 592
  of full-scale structures, 12–13
  of reduced-scale physical models, 12–13
Vibrations, 1
Vibrometer, 102–103
Virtual displacement
  of a continuous system, 42, 218–219, 379
  of a rigid body, 34–35
Virtual work, 36–37
  of conservative forces, 43
  of electrical loads and mechanical loads, 628
  of nonconservative forces, 43, 219–220, 380
  of substructure interface constraint forces, 551
  used to form transformation matrices, 436
Viscous damping
  coefficient of, 27
  element, 27

Website, vii, xiv, 15, 78, 113, 143, 154, 182, 205, 207, 257, 274, 276, 312, 322, 324, 409, 453, 456, 466, 474, 496, 499, 525–527, 580, 590, 605, 614, 711
Wilson-θ Method, 514–516, 521, 524–525
Work-energy methods, 22. *See also* Energy methods